Earth at Night, City Lights

The Americas

These images of Earth at night from NASA's Suomi-NPP "Marble" series use a collection of satellite-based observations, stitched together in a seamless mosaic of our planet. This view is based on instrumentation that observes light emanating from the ground. Notice how strongly major cities show up in the image.

Africa, Europe, and the Middle East

Defense Meteorological Satellite Program (DMSP) NASA/GSFC

So Many Options for Your Human Geography Class!

Students today want options when it comes to their textbooks. *Human Geography* gives students the flexibility they desire, offering a wide range of formats for the book and a large array of media and online learning resources. Find a version of the book that works best for YOU!

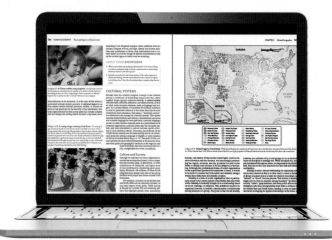

Whether it's on a laptop, tablet, smartphone, or other wired mobile devices, *Human Geography* lets students access media and other tools for learning human geography.

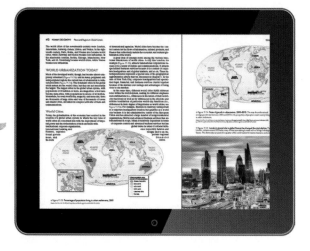

Human Geography Plus MasteringGeography with eText ISBN 0-321-98896-5/978-0-321-98896-6

Available at no additional charge with MasteringGeography, the Pearson eText version of *Human Geography,* 7th Edition, gives students access to the text whenever and wherever they are online.

Features of Pearson eText:
- Now available on smartphones and tablets.
- Seamlessly integrated videos and other rich media.
- Fully accessible (screen-reader ready).
- Configurable reading settings, including resizable type and night reading mode.
- Instructor and student note-taking, highlighting, bookmarking, and search.

Human Geography CourseSmart eTextbook
ISBN 0-321-99429-9/978-0-321-99429-5

CourseSmart eTextbooks are an alternative to purchasing the print textbook, where students can subscribe to the same content online and save up the 40% off the suggested list price of the print text.

Human Geography Books à La Carte
ISBN 0-321-98713-6/978-0-321-98713-6

Books à la Carte features the same exact content as *Human Geography* in a convenient, three-hole-punched, binder-ready, loose-leaf version. Books à la Carte offers a great value for students—this format costs 35% less than a new textbook package.

Pearson Custom Library: You Create Your Perfect Text http://www.pearsoncustomlibrary.com

Human Geography is available on the Pearson Custom Library, allowing instructors to create the perfect text for their courses. Select the chapters you need, in the sequence you want. Delete chapters you don't use: students pay only for the materials chosen.

MasteringGeography Student Study Area
No matter the format, with each new copy of the text, students will receive full access to the Study Area in **MasteringGeography**™, providing a wealth of Videos, **MapMaster** Interactive Maps, Animations, *In the News* readings, Flashcards, Practice Quizzes, and much more.

A distinctly modern review of HUMAN GEOGRAPHY

▲

***Human Geography: Places and Regions in Global Context,* Seventh Edition** fosters awareness of current issues and developing trends from a geographic perspective, providing a solid foundation in human geography.

A Critical Exploration of HUMAN GEOGRAPHY

NEW! Chapter 6: Language, Communication, and Belief focuses on how both language and religion reflect and influence societies, as well as how they spread around the world, and how they permeate politics and social life.

6

LANGUAGE, COMMUNICATION, AND BELIEF

As an infant—as young, perhaps, as 8 months old—you had to determine the internal structure of a system that possesses tens of thousands of individual elements. Each of the elements is derived from the same collection of materials and combined into larger units. Those units can be put together into an infinite set of combinations, although only a limited set of those joined units are correct within the context of the system. How does an infant proceed? Fortunately, we tend to learn this system effortlessly: The system is language, and it is composed of words, sounds, and sentences.

But now imagine that you're a deaf child, 6 or 7 years old. You have reached this age not fully understanding what it means to be deaf. Imagine how much more difficult the mission of acquiring language will be for you. Of course, there will not necessarily be sounds involved in forming your language, but there must be something else to take the place of sound that will allow you to communicate the words and the sentences you wish to convey.

Imagine further in this already challenging scenario that it's 1970 and you live in Managua, Nicaragua, and there are no teachers at your school who know sign language. What is perhaps even more remarkable than the capacity of the hearing infant's ability to comprehend and eventually use language is the capacity of a group of deaf children, assembled in a collective but without the aid of a sign language instructor, to develop their own language and be able communicate with each other.[1]

These children developed the Nicaraguan Sign Language. It is a unique example of how language emerges and becomes populated with a structure, words, and sentences. The deaf children created the language, not with the help of their teachers or their parents or any other adults but through their interactions with each other. Independently, they constructed a natural sign language that contains the kinds of grammatical regularities that are key to all languages. And, since the

▲ Speech and hearing impaired students use sign language to answer their teacher's question in a classroom at the Xa Dan school in Hanoi

[1]Adapted from J. R. Saffran, A. Senghas, and J. S. Trueswell, 2001, *The Acquisition of Language by Children*, Proceedings of the National Academy of Sciences, 98, 23: npn.

LEARNING OUTCOMES

■ Describe how language both reflects and influences the way different groups understand and interpret the world.

■ Compare and contrast different forms of communication, including standard language, slang, dialects, social media, and nonverbal modes of expression.

■ Interpret how different geographies impact the spread or preservation of language and how different groups use language to give or change a place's meaning.

■ Describe the global distribution of the world's religions—how they developed in specific regions and how they proliferated around the world.

■ Recognize the difference between religions and religious movements around the world, and analyze the impacts of both on political and social life.

■ Interpret the importance of space to religion in pilgrimages and sacred spaces in every culture.

186   187

5.3 Spatial Inequality   The Global Gender Gap

In 1990, the United Nations published the first of its annual Human Development Reports. The report analyzes how economic growth and human development are inextricably tied and provides statistics about changes in both over time as well as suggestions for how to improve them (Figure 5.D). Since 1990, the report has taken the position that women are at a structural disadvantage compared to men and in its 1997 report, stated baldly, "No society treats its women as well as its men." While the differences between women's and men's pay in the developed world is a common topic of discussion and concern (where in the United States for every $1 men earn, women earn 77 cents), in the developing world, women experience deep deprivation, exploitation, and harm. The following are ten examples of gender inequality globally.[3]

1. Women everywhere experience a gender wage gap whether in the developed or the developing world.
2. Women in many parts of the world experience limited mobility from not being allowed to drive on public roads to refusing to go out by themselves at night for fear of attack or rape.
3. One in every three women around the world is likely to be beaten, coerced into sex, or otherwise abused sometime in her lifetime.
4. In some countries, a male child is more valuable than a female child and parents who don't want a girl may either abort the fetus or kill the child after birth.
5. In some countries, women are legally prohibited from owning land.
6. According to the United Nations, women do two-thirds of the world's work, receive ten percent of the world's income and own one percent of the means of economic production.
7. Women have more limited access to health care than men while one women dies in childbirth every minute of every day.
8. Forced marriages and the lack of legal access to divorce limits many women's life chances.
9. Despite making up half the global population, women hold only 15.6 percent of elected seats in national parliaments or congresses.
10. Women make up more than two-thirds of the world's illiterate adults.

1. What is the gender gap?
2. In what ways would a narrowing of the gender gap improve the lives of women around the globe?

[3]Adapted from Molly Edmonds, 2014, "Examples of Gender Inequality around the World", http://www.discovery.com/tv-shows/curiosity/topics/examples-gender-inequality-around-world.htm (accessed June 29, 2014).

◄ Figure 5.D The geography of the global gender gap Shown in this graphic are (a) key indicators as well as (b) a map of the gender inequality index globally.

NEW! Spatial Inequality features highlight the growing imbalances and inequalities in today's global society relative to the chapter's major themes.

6.2 Spatial Inequality   Geographies of Literacy

At a very basic level, **literacy** is the ability to read and write. Being able to read and write allows us to determine more readily the course of our lives as we push beyond simply comprehending language and reproducing it to transforming who we are and what we are able to do in the world. UNESCO defines literacy as the "ability to identify, understand, interpret, create, communicate and compute, using printed and written materials associated with varying contexts. Literacy involves a continuum of learning in enabling individuals to achieve their goals, to develop their knowledge and potential, and to participate fully in their community and wider society."[2]

In the United States, one of the wealthiest countries on Earth, there are 44 million people—many of whom are incarcerated—who are **functionally illiterate**. Functional illiteracy means that an individual's reading and writing skills are inadequate to manage daily living or hold down a job that requires reading skills beyond a basic level. **Figure 6.C** is a variation

◄ Figure 6.C School-to-prison pipeline Illiteracy among U.S. youth is the result of a combination of initial conditions but the result is often prison. One of the largest illiterate populations in the country is the prison population.

○ Fourth graders reading below level
○ Fourth graders reading at or above level

◄ Figure 6.D Race, literacy and prison Fourth-grade reading level is one of the measures private prison firms use to determine how large to build their prisons. Illiteracy and crime are highly related.

FROM SCHOOL TO PRISON
STUDENTS OF COLOR FACE HARSHER DISCIPLINE AND ARE MORE LIKELY TO BE PUSHED OUT OF SCHOOL THAN WHITES.

40% OF STUDENTS EXPELLED FROM U.S. SCHOOLS EACH YEAR ARE BLACK.

70% OF STUDENTS INVOLVED IN "IN-SCHOOL" ARRESTS OR REFERRED TO LAW ENFORCEMENT ARE BLACK LATINO.

3.5X BLACK STUDENTS ARE THREE AND A HALF TIMES MORE LIKELY TO BE SUSPENDED THAN WHITES.

2X BLACK AND LATINO STUDENTS ARE TWICE AS LIKELY TO NOT GRADUATE HIGH SCHOOL AS WHITES.

68% OF ALL MALES IN STATE AND FEDERAL PRISON DO NOT HAVE A HIGH SCHOOL DIPLOMA.

FROM FOSTER CARE TO PRISON
YOUTH OF COLOR ARE MORE LIKELY THAN WHITES TO BE PLACED IN THE FOSTER CARE SYSTEM.

50% OF CHILDREN IN THE FOSTER CARE SYSTEM ARE BLACK OR LATINO.

30% OF FOSTER CARE YOUTH ENTERING THE JUVENILE JUSTICE SYSTEM ARE PLACEMENT-RELATED BEHAVIORAL CASES.

25% OF YOUNG PEOPLE LEAVING FOSTER CARE WILL BE INCARCERATED WITHIN A FEW YEARS AFTER TURNING 18.

50% OF YOUNG PEOPLE LEAVING FOSTER CARE WILL BE UNEMPLOYED WITHIN A FEW YEARS AFTER TURNING 18.

70% OF INMATES IN CALIFORNIA STATE PRISON ARE FORMER FOSTER CARE YOUTH.

THE COLOR OF MASS INCARCERATION AND LITERACY LEVELS

[2]"The Plurality of Literacy and Its Implications for Policies and Programs," UNESCO Education Sector Position Paper, 2004, p. 13.

Structured Learning Path

The Seventh Edition of *Human Geography: Places and Regions in Global Context* provides an active structured learning path to help guide students toward mastery of key human geography concepts.

Learning Outcomes in each chapter opener guide students through the main learning goals for the chapter.

LEARNING OUTCOMES

- *Explain* why populations change, where those changes occur, and what the implications of population change are for the future of different places around the globe.

- *Identify* the two most important factors in population dynamics, birth and death, and how they shape population characteristics.

- *Analyze* how geography is a powerful force in the incidence of health and disease.

- *Demonstrate* how the movement of populations is affected by both push and pull factors, and explain how these factors are key to understanding new settlement patterns.

APPLY YOUR KNOWLEDGE

1. What can we learn by studying cultural traits? How does looking at cultural complexes help us better understand the relationship between humans and the spaces in which they live?

2. Identify two traits that are characteristic of the cultural group to which you belong. Are the traits related to the country or region in which you live? Describe the relationship or explain why there is none.

UPDATED! Apply Your Knowledge questions are integrated throughout the chapter sections, giving students a chance to stop and practice/apply their understanding. The first of these paired questions is now a lower-level knowledge-based reading question, while the second is a higher-level application question.

1. Look around you both at home and in stores. What souvenirs do you find? What do they remind you of? What geographies—of landscapes, emotions, peoples, and travels—do these material objects recall for you or for their collectors?

2. How else do we connect fact and fiction in our daily lives? Think of an example of something significant and influential that is nevertheless not really "real." How does this connection between fact and fiction influence you?

NEW! Active learning questions are now included in all boxed features so that students can check their understanding as they read.

LEARNING OUTCOMES REVISITED

- Describe why populations change, where those changes occur, and what the implications of population change are for the future of different places around the globe.

  Population geographers bring to demography a special perspective—the spatial perspective—that emphasizes description and explanation of the "where" of population distribution, patterns, and processes. The distribution of population is a result of many factors, such as employment opportunities, culture, water supply, climate, and other physical environment characteristics. Geographers explore these patterns of distribution and density, as well as population composition in order to comprehend the complex geography of populations. Understanding the reasons for and implications of variation in patterns and composition provides geographers with insight into population change and the potential

- Demonstrate how the movement of population is affected by both push and pull factors and explain how these factors are key to understanding new settlement patterns.

  In general terms, migrants make their decisions to move based on push factors and pull factors. Remember that push factors are events and conditions that impel an individual to move from a location. Pull factors are forces of attraction that influence migrants to move to a particular location. Mobility is the capacity to move from one place to another, either permanently or temporarily. Migration, in contrast, is an actual long-distance move to a new location. Permanent and temporary changes of residence can occur for a variety of reasons. Striving for economic betterment or escaping from adverse political conditions, such as war or oppression, are the most frequent causes. Push

Learning Outcomes Revisited found at the end of each chapter summarizes chapter content correlated to the Learning Outcomes stated in each chapter opener.

NEW! The end-of-chapter **Data Analysis** activities feature takes students beyond traditional review material. Students further their understanding as they manipulate media, collect data, and use interactive mapping.

DATA ANALYSIS

In this chapter we have looked at a central component of human-environmental interactions: the geography of food and agriculture, from the global to the household and individual level. In looking at this basic aspect of life—producing and consuming food—the issue of space, economy, and politics play a huge role as seen in the debates over the Green Revolution, the Biorevolution, food sovereignty, anti-GMO resistance movements and the concept of "food deserts." To look closer at how and where we produce food, watch the story of Ron Finley, a "guerilla gardener" in South Central Los Angeles and answer these questions.]

LA Guerilla Gardeners

http://goo.gl/RgNYYO

1. What does Ron Finley say about fast food *versus* drive-by-shootings in his communities?

2. Why is "food the problem, and food is the solution"?

3. Where in the city does Finley plant his gardens?

4. What is Los Angeles Green Grounds and how do they work?

5. How is gardening like art? How does Finley talk about soil?

6. How does guerilla gardening change a community? How are children a vital component of this process?

7. What does Finley say about flipping the script and making gardening "gangster"?

8. Do an Internet search on "guerrilla gardeners." What other cities have guerrilla gardener groups? Does your city? Would you consider starting a guerilla garden?

Cutting-Edge Cartography & Visual Program

The superior cartography of *Human Geography: Places and Regions in Global Context* comprises scores of rich, diverse, and fully updated maps that help professors better teach their students the important spatial elements inherent to human geography.

Current data

Up-to-date information gives readers access to the most current demographic statistics and data.

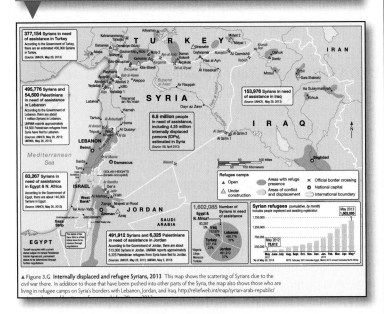

▲ Figure 3.G Internally displaced and refugee Syrians, 2013 This map shows the scattering of Syrians due to the civil war there. In addition to those that have been pushed into other parts of the Syria, the map also shows who are living in refugee camps on Syria's borders with Lebanon, Jordan, and Iraq. http://reliefweb.int/map/syrian-arab-republic/

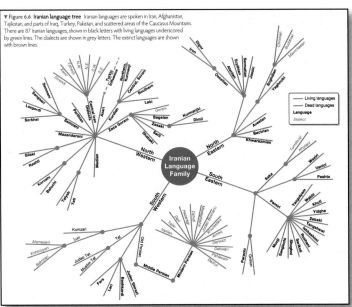

▼ Figure 6.6 Iranian language tree Iranian languages are spoken in Iran, Afghanistan, Tajikistan, and parts of Iraq, Turkey, Pakistan, and scattered areas of the Caucasus Mountains. There are 87 Iranian languages, shown in black letters with living languages underscored by green lines. The dialects are shown in grey letters. The extinct languages are shown with brown lines.

Mental maps & diagrams

These graphics depict people's perceptions of concepts and geography, highlighting the ways in which everyday phenomena and data can be mapped.

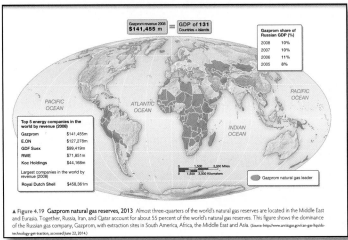

▲ Figure 4.19 Gazprom natural gas reserves, 2013 Almost three-quarters of the world's natural gas reserves are located in the Middle East and Eurasia. Together, Russia, Iran, and Qatar account for about 55 percent of the world's natural gas reserves. This figure shows the dominance of the Russian gas company, Gazprom, with extraction sites in South America, Africa, the Middle East and Asia. (Source: http://www.arcticgas.gov/can-gas-liquids-technology-get-traction, accessed June 22, 2014.)

Compound figures

The book features many compound figures that combine maps with photographs and/or illustrations. These figures capture student interest by integrating spatial, real-world, and conceptual information.

▲ Figure 2.2 The Premodern World (a) This map shows the distribution of the Greek city-states and Carthaginian colonies and the spread of the Roman empire from 218 B.C.E. to 117 C.E. (Source: Adapted from R. King et al., The Mediterranean. London: Arnold, 1977, pp. 59 and 64.)

(b) Many of the roads that were built by Romans became established as major routes throughout Europe. This photo shows the cardos maximus (main road) at Tharros in Sardinia, Italy.

(c) The Romans were great engineers. This photograph shows the Roman aqueduct in Segovia, Spain.

Engaging, Relevant Applications

Provocative applications increase student interest, fostering awareness of current issues and developing trends that impact the world and their lives.

UPDATED!
Geography Matters explore contemporary real-world applications of key chapter concepts and themes. Authored by expert contributors, the *Geography Matters* features demonstrate to students that the focus of human geography is on real-world problems.

UPDATED! Visualizing Geography incorporate edgy, modern applications and visualizations of geographic data. These interesting and challenging visualizations are unique, and set apart Knox & Marston's visual program.

Window on the World take a key concept and explore its application in a particular location. This feature helps students to appreciate the relevance of geographic concepts to world events, and brings some far-flung places closer to their comprehension.

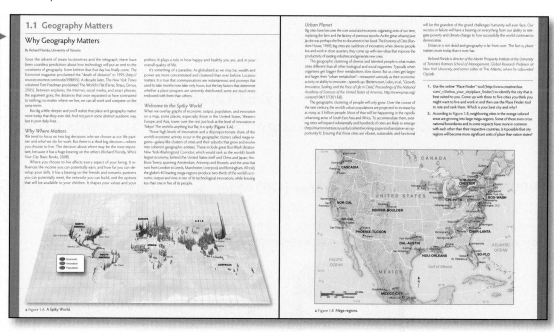

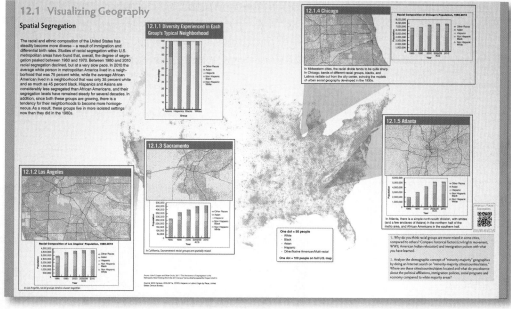

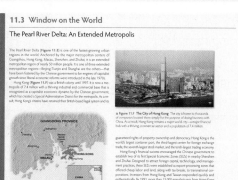

Continuous Learning Before, During & After Class with MasteringGeography

MasteringGeography™ delivers engaging, dynamic learning opportunities—focusing on course objectives and responsive to each student's progress—that are proven to help students absorb geography course material and understand challenging geographic processes and concepts.

BEFORE CLASS

Mobile Media & Reading Assignments Ensure Students Come to Class Prepared

NEW! mobile-ready Quick Response (QR) codes integrated throughout the chapters give students instant access to online data sets, readings, and media.

Interactive Album of Map Projections

http://goo.gl/tTWBvc

Renewable Energy Institute International

http://goo.gl/mZd0Mn

International Network for Urban Agriculture

http://goo.gl/pOjbe2

NEW! Dynamic Study Modules personalize each student's learning experience. Created to allow students to acquire knowledge on their own and be better prepared for class discussions and assessments, this mobile app is available for iOS and Android devices.

Pearson eText in MasteringGeography gives students access to the text whenever and wherever they can access the internet.

Features of Pearson eText:

- Now available on smartphones and tablets.
- Seamlessly integrated videos and other rich media.
- Fully accessible (screen-reader ready).
- Configurable reading settings, including resizable type and night reading mode.
- Instructor and student note-taking, highlighting, bookmarking, and search functionality.

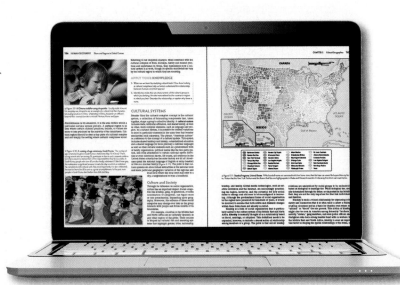

Reading Questions ensure that students complete the assigned reading before class and stay on track with reading assignments. Reading Questions are 100% mobile ready and can be completed by students on mobile devices.

DURING CLASS
Learning Catalytics & Engaging Media

> " *My students are so busy and engaged answering Learning Catalytics questions during lecture that they don't have time for Facebook.* "

DECLAN DE PAOR,
OLD DOMINION UNIVERSITY

What has professors and students excited? **Learning Catalytics**, a "bring your own device" student engagement, assessment, and classroom intelligence system, allows students to use their smartphone, tablet, or laptop to respond to questions in class. With Learning Catalytics, you can:

- Assess students in real-time using open-ended question formats to uncover student misconceptions and adjust lecture accordingly.

- Automatically create groups for peer instruction based on student response patterns, to optimize discussion productivity.

Enrich Lecture with Dynamic Media
Teachers can incorporate dynamic media into lecture, such as Videos, MapMaster Interactive Maps, and Geoscience Animations.

AFTER CLASS
Easy to Assign, Customizable, Media-Rich, & Automatically Graded Assignments

NEW! Geography Videos from such sources as the BBC and the *Financial Times* are now included in addition to the videos from Television for the Environment's *Life and Earth Report* series in MasteringGeography.

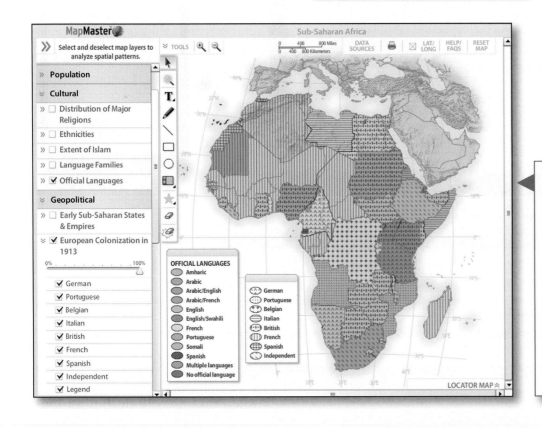

MapMaster Interactive Map Activities are inspired by GIS, allowing students to layer various thematic maps to analyze spatial patterns and data at regional and global scales. This tool include zoom and annotation functionality, with hundreds of map layers leveraging recent data from sources such as NOAA, NASA, USGS, United Nations, and the CIA.

NEW! GeoTutors. These highly visual & data-rich coaching items with hints and specific wrong answer feedback help students master the toughest topics in geography.

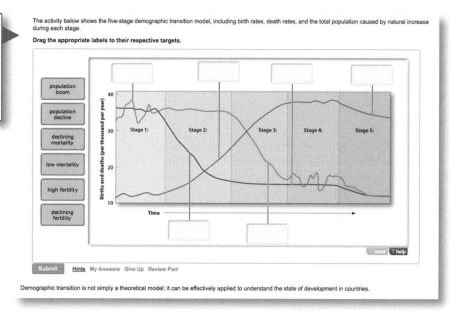

UPDATED! Encounter (Google Earth) activities provide rich, interactive explorations of human geography concepts, allowing students to visualize spatial data and tour distant places on the virtual globe.

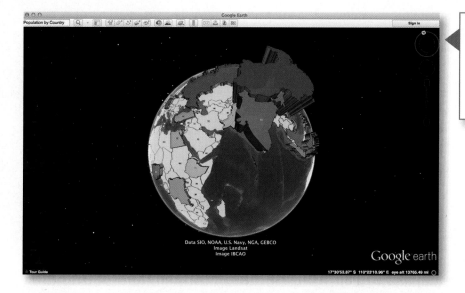

Map Projections interactive tutorial media help reinforce and remediate students on the basic yet challenging fundamental map projection concepts.

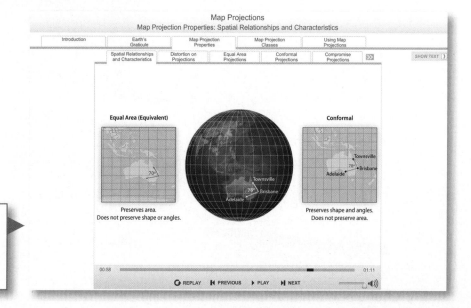

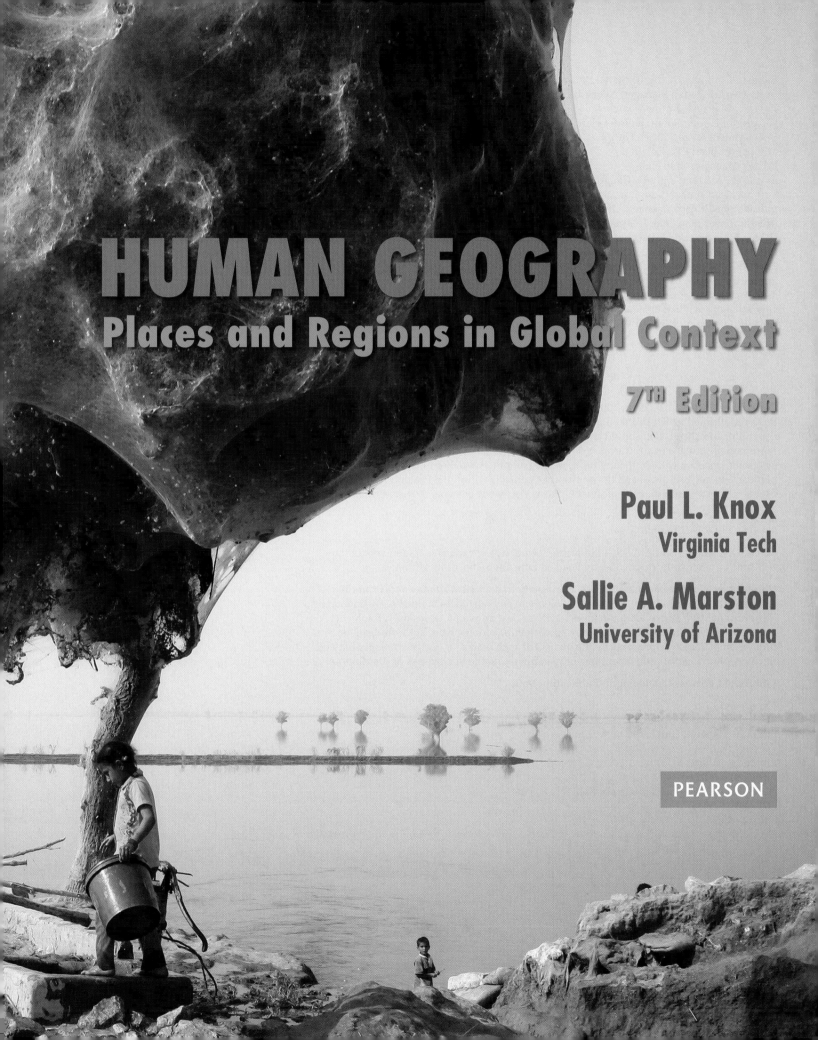

HUMAN GEOGRAPHY
Places and Regions in Global Context

7TH Edition

Paul L. Knox
Virginia Tech

Sallie A. Marston
University of Arizona

PEARSON

Senior Geography Editor: Christian Botting
Executive Marketing Manager: Neena Bali
Program Manager: Anton Yakovlev
Director of Development: Jennifer Hart
Development Editor: Karen Gulliver
Media Producer: Ziki Dekel
Project Manager: Sean Hale
Editorial Assistant: Amy De Genaro
Full Service/Composition: Lumina Datamatics, Inc.
Full Service Project Manager: Lindsay Bethoney
Illustrations: Kevin Lear, International Mapping
Design Manager: Derek Bacchus
Interior and Cover Design: Tamara Newnam
Photo & Text Permissions Manager: Rachel Youdelman
Photo Researcher: Lauren McFalls, Lumina Datamatics, Inc.
Text Permissions Researcher: Mark Schaefer
Operations Specialist: © Maura Zaldivar-Garcia
Cover Photo Credit: Russell Watkins/Department for International Development/Reuters

Library of Congress Cataloging-in-Publication Data

Knox, Paul L.
  Human geography : places and regions in global context / Paul L. Knox, Virginia Tech,
Sallie A. Marston, University of Arizona. — Seventh edition.
    pages cm
  Includes index.
  ISBN 978-0-321-98424-1
  1. Human geography. I. Marston, Sallie A. II. Title.
  GF41.K56 2015
  304.2—dc23
                                                        2014012298

1 2 3 4 5 6 7 8 9 10—CRK—15 14 13 12 11

www.pearsonhighered.com

ISBN-10: 0-321-98424-2; ISBN-13: 978-0-321-98424-1 (Student Edition)
ISBN-10: 0-321-98761-6; ISBN-13: 978-0321-98761-7 (Instructor's Review Copy)

Brief Contents

Contents

1 Geography Matters 3

2 The Changing Global Context 31

3 Geographies of Population and Migration 65

4 People and Nature 107

7 Interpreting Places and Landscapes   229

Behavior, Knowledge, and Human Environments   230

8 Geographies of Economic Development   259

Patterns of Economic Development   260

Technological Change and Economic Development   260 • The Unevenness of Economic Development   260 • Measuring Levels of Economic Development   261 • Resources and Development   262

9 Geographies of Food and Agriculture **299**

12 City Spaces: Urban Structure   429

About Our Sustainability Initiatives

Pearson recognizes the environmental challenges facing this planet, as well as acknowledges our responsibility in making a difference. This book is carefully crafted to minimize environmental impact. The binding, cover, and paper come from facilities that minimize waste, energy consumption, and the use of harmful chemicals. Pearson closes the loop by recycling every out-of-date text returned to our warehouse.

Along with developing and exploring digital solutions to our market's needs, Pearson has a strong commitment to achieving carbon-neutrality. As of 2009, Pearson became the first carbon- and climate-neutral publishing company. Since then, Pearson remains strongly committed to measuring, reducing, and offsetting our carbon footprint.

The future holds great promise for reducing our impact on Earth's environment, and Pearson is proud to be leading the way. We strive to publish the best books with the most up-to-date and accurate content, and to do so in ways that minimize our impact on Earth. To learn more about our initiatives, please visit **www.pearson.com/responsibility**.

Pearson - Social Impact

http://goo.gl/2VZuMM

Preface

A nation, like a person, has a mind—a mind that must be kept informed and alert, that must know itself, that understands the hopes and needs of its neighbors—all the other nations that live within the narrowing circle of the world.

Franklin Roosevelt, Third Inaugural Address, Monday, January 20, 1941

Most people have an understanding of what their own lives are like and know a good deal about their own neighborhood and perhaps even something of the larger city and state in which they live. Yet, even as the countries and regions of the world become more interconnected, most of us still know very little about the lives of people on the other side of our country, or in other societies, or about the ways the lives of those people connect to our own. To change the world, to make it a better place for all people, we need to understand not just our little corner of it, but the whole of it—the broad sweep of human geography that constitutes the larger world of which our small corners are just a part.

This book provides an introduction to human geography that will help young men and women to understand critically the world in which they live. To study human geography, to put it simply, is to study the dynamic and complex relationships between peoples and the worlds they inhabit. Our book gives students the basic geographical tools and concepts needed to understand the complexity of places and regions, and to appreciate the interconnections between their own lives and those of people in different parts of the world—to make the world a better place.

NEW TO THE SEVENTH EDITION

The seventh edition of *Places and Regions in Global Context* represents a thorough revision. Every part of the book was examined carefully with the goal of keeping topics and data current while also improving the clarity of the text and the graphics. We have also sought to enhance the utility of the book for both instructors and students.

- A new chapter (Chapter 6) on *Language, Communication, and Belief* gives a greater focus on how both language and religion reflect and influence societies, as well as how they spread around the world, and how they permeate politics and social life.

- *Spatial Inequality* features highlight the growing imbalances and inequalities in today's global society relative to the chapter's major themes.

- The reimagined *Geography Matters* features, authored by expert contributors from the community, present contemporary research and hot topics in geography subfields.

- *Data Analysis* activities at the end of each chapter give students the chance to put their understanding of key themes in the chapter into practice. Students further their understanding as they manipulate media, collect data, and use interactive mapping.

- Newly redesigned *Visualizing Geography* features consistently incorporate edgy, modern applications and visualizations of current spatial data.

- Active learning assessments are now included in all boxed features so that students can check their understanding.

- The seventh edition also incorporates a comprehensive updating of all of the data, maps, photographs, and illustrative examples.

- We have added or expanded upon quite a few topics, including climate change and issues of sustainability; rising sea levels; spatial inequality; gender and economic development; place, space and scale; landscape and art; cultural heritage; urban regeneration; and urban environmental problems; conflict zones; changing demographic issues; food, health and place; gender and sexuality; and what the near future is likely to be. These changes are designed to ensure that we offer the most up-to-date coverage in the field of human geography.

- A renewed focus on fundamentals gives students access not only to the new ideas, concepts, and theories that address the changes mentioned earlier in this text, but also to the fundamentals of human geography: the principles, concepts, theoretical frameworks, and basic knowledge.

- Over 150 new Geography Videos from the BBC and the *Financial Times* are now included in MasteringGeography. Students can access the videos on their own in the Study Area, and teachers can assign the videos with assessment activities.

- Learning Catalytics™ is a "bring your own device" student engagement, assessment, and classroom intelligence system.

  With Learning Catalytics™ you can:

  - Assess students in real time, using open-ended tasks to probe student understanding.

  - Understand immediately where students are and adjust your lecture accordingly.

  - Improve your students' critical-thinking skills.

  - Access rich analytics to understand student performance.

  - Add your own questions to make Learning Catalytics™ fit your course exactly.

- Dynamic Study Modules personalize each student's learning experience. Created to allow students to acquire knowledge on their own and be better prepared for class discussions and assessments, this mobile app is available for iOS and Android devices.

OBJECTIVE & APPROACH

The objective of the book is to introduce the study of human geography by providing not only a body of knowledge about the creation of places and regions, but also an understanding of the interdependence of places and regions in a globalizing world. The approach is aimed at establishing an intellectual foundation that will enable a lifelong and life-sustaining geographical imagination: an essential tool for today's students in order to confront tomorrow's global, national, regional, and local challenges.

The book takes a fresh approach to human geography, reflecting the major changes that have recently been impressed on global, regional, and local landscapes. These changes include the globalization of industry and the related rapid rise of China and India as economic powerhouses, the upwelling of ethnic regionalisms on the heels of decolonization and the formation of new states, the movement of peoples around the world in search of better lives, the physical restructuring of cities, the transformation of traditional agricultural practices throughout much of the world, global environmental change and the movement for sustainability, the eruptions of war and the struggles for peace, and the emerging trend toward transnational political and economic organizations. The approach used in *Places and Regions in Global Context* provides access not only to the new ideas, concepts, and theories that address these changes, but also to the fundamentals of human geography: the principles, concepts, theoretical frameworks, and basic knowledge that are necessary to more specialized studies.

The most distinctive feature of this approach is that it emphasizes the interdependence of both places and processes in different parts of the globe. In overall terms, this approach is designed to provide an understanding of relationships between global processes and the local places in which they unfold. It follows that one of the chief organizing principles is how globalization frames the social and cultural construction of particular places and regions. This approach has several advantages. For example:

- It captures aspects of human geography that are among the most compelling in the contemporary world—the geographical bases of cultural diversity and their impacts on everyday life.

- It encompasses the salient aspects of new emphases in academic human geography—the new emphasis on sustainability and its role in the social construction of spaces and places.

- It makes for an easier marriage between topical and regional material by emphasizing how processes link them—technological innovation and the varying ways technology is adopted and modified by people in particular places.

- It facilitates meaningful comparisons between places in different parts of the world—how the core-generated industrialization of agriculture shapes gender relations in households both in the core and the periphery.

In short, the textbook is designed to focus on geographical processes and to provide an understanding of the interdependence among places and regions without losing sight of their individuality and uniqueness.

Several important themes are woven into each chapter, integrating them into the overall approach:

- the relationship between global processes and their local manifestations;

- the interdependence of people and places, especially the interactive relationships between core regions and peripheral regions;

- the continuing transformation of the political economy of the world system, and of nations, regions, cities, and localities;

- the social and cultural differences that are embedded in human geographies (especially the differences that relate to race, ethnicity, gender, age, and class).

CHAPTER ORGANIZATION

The organization of the book is innovative in several ways. First, the chapters are organized so that the conceptual framework—why geography matters in a globalizing world—is laid out in Chapters 1 and 2 and then deployed in thematic chapters (Chapters 3 through 12). Second, the conceptual framework of the book requires the inclusion of two introductory chapters rather than the usual one. The first describes the basics of a geographic perspective; the second explains the value of the globalization approach.

Third, the distinctive chapter ordering within the book follows the logic of moving from less complex to more complex systems of human social and economic organization, always highlighting the interaction between people and the world around them. The first thematic chapter (Chapter 3) focuses on human population. Its early placement in the book reflects the central importance of people in understanding geography. Chapter 4 deals with the relationship between people and the environment as it is mediated by technology. This chapter explores human–environment relations and establishes a central theme: that all human geographical issues are about how people negotiate their environment—whether the natural or the built environment.

The chapter on nature, society, and technology is followed by Chapter 5 on cultural geography. The intention in positioning the cultural chapter here is to signal that culture is the primary medium through which people operate and understand their place in the world. Chapter 6, new to this edition, gives a focus on how both language and religion reflect and influence societies, as well as how they spread around the world, and how they permeate politics and social life. In Chapter 7, the impact of cultural processes on the landscape is explored, together with the ways in which landscape shapes cultural processes.

In Chapter 8, the book begins the move toward more complex concepts and systems of human organization by concentrating on economic development. The focus of Chapter 9 is agriculture. The placement of agriculture after economic development reflects the overall emphasis on globalization. This chapter shows how processes of globalization and economic development have led to the industrialization of agriculture at the expense of more traditional agricultural systems and practices.

The final three thematic chapters cover political geography (Chapter 10), urbanization (Chapter 11), and city structure (Chapter 12). Devoting two chapters to urban geography, rather than a more conventional single chapter, is an important indication of how globalization increasingly leads to urbanization of the world's people and places.

Features

The hallmark feature of our book is the global framework that promotes a strong connection between topical and regional material by emphasizing how their processes are linked (e.g., technological innovation and the varying ways technology is adopted and modified by people and places). This makes for a contemporary approach to human geography, reflecting many trends in the discipline, such as the globalization of industry, the upwelling of ethnic regionalisms on the heels of decolonization and new state formation, and the trend toward transnational political and economic organizations. The global framework also facilitates meaningful comparisons between people and places in different parts of the world, such as how the core-generated industrialization of agriculture shapes gender relations in households, both in the core and in the periphery. It allows us to present the relevant aspects of different emphases in academic human geography (e.g., geopolitics and its role in the social construction of spaces and places). At the same time, a focus on fundamentals gives students access not only to the new ideas, concepts, and theories that address the changes mentioned earlier in this text but also to the fundamentals of human geography: the principles, concepts, theoretical frameworks, and basic knowledge. The book's thematic structure weaves several important themes into every chapter: the interdependence of people and places, especially the interactive relationships between the core and the periphery; social and cultural differences that are embedded in human geographies, especially the differences that relate to race, ethnicity, gender, age, and class; the relationship between global processes and their local manifestations; and the continuing transformation of the political economy of the world system and nations, regions, cities, and localities.

To signal the freshness of the approach, the book features a superior cartographic program, consisting of rich, diverse, and fully updated maps that help professors better teach their students the important spatial elements inherent to human geography. The cartography program features numerous compound figures that combine maps with photographs and line drawings.

The pedagogy of the book employs four different boxed features—"Geography Matters," "Visualizing Geography," "Spatial Inequality," and "Window on the World."

Geography Matters features, authored by expert guest contributors, show how geographers are using their unique perspectives and contemporary geographic tools to solve real-world problems, emphasizing how geography and the geographic method matter in the world today.

Visualizing Geography boxes treat key concepts of the chapter through modern applications and visualizations of data, helping students "see" the ways geography shapes their daily lives.

Spatial Inequality features highlight the growing imbalances and inequalities in today's global society relative to the chapter's major themes.

Window on the World boxes take a key concept and explore its application in a particular location. This feature allows students to appreciate the relevance of geographic concepts to world events and brings far-flung places closer to their comprehension.

These features are explored through each chapter's *learning path*, which guides the reader through each chapter's themes, starting with the *Learning Outcomes* at the start of each chapter, and continuing with each *Apply Your Knowledge* question pairing to ensure a student's understanding of key elements of each section. Every chapter includes *Future Geographies*, which details what global and local elements may take shape in the years ahead. We conclude each chapter with *Learning Outcomes Revisited*, designed to encapsulate the imperative elements of the chapter's narrative.

CONCLUSION

The idea for this book evolved from conversations between the authors and colleagues about how to teach human geography in colleges and universities. Our intent was to find a way not only to capture the exciting changes that are rewriting the world's landscapes and reorganizing the spatial relationships between people but also to demonstrate convincingly why the study of geography matters. Our aim was to show why a geographical imagination is important, how it can lead to an understanding of the world and its constituent places and regions, and how it has practical relevance in many spheres of life.

ACKNOWLEDGMENTS

We are indebted to many people for their assistance, advice, and constructive criticism in the course of preparing this book. Among those who provided comments on drafts of the various editions of this book are the following professors:

David Aagesen *(SUNY: Geneseo)*
Christopher A. Airriess *(Ball State University)*
Stuart Aitken *(University of California at San Diego)*
Matthew Anderson *(Montana State University, Billings)*
Kevin Archer *(University of South Florida)*
Sarah Bednarz *(Texas A&M University)*

Brian J. L. Berry *(University of Texas at Dallas)*
Brian W. Blouet *(College of William and Mary)*
George O. Brown, Jr. *(Boston College)*
Michael P. Brown *(University of Washington)*
Henry W. Bullamore *(Frostburg State University)*
Edmunds V. Bunske *(University of Delaware)*
Craig Campbell *(Youngstown State University)*
Dylan Clark *(University of Colorado)*
David B. Cole *(University of Northern Colorado)*
Mario Cora *(University of Phoenix)*
Jerry Crampton *(George Mason University)*
Christine Dando *(University of Nebraska, Omaha)*
Fiona M. Davidson *(University of Arkansas)*
Ronald Davidson *(California State University, Northridge)*
Jeff DeGrave *(University of Wisconsin at Eau Claire)*
Daniel Dempsey *(College of the Redwoods)*
Benjamin Dixon *(SUNY College at Oneonta)*
Vernon Domingo *(Bridgewater State College)*
Patricia Ehrkamp *(Miami University)*
Nancy Ettlinger *(The Ohio State University)*
Emily Fekete *(University of Kansas)*
Paul B. Frederic *(University of Maine)*
Kurtis G. Fuelhart *(Shippensburg University)*
Gary Fuller *(University of Hawaii at Manoa)*
Wilbert Gesler *(University of North Carolina)*
Melissa Gilbert *(Temple University)*
Jeffrey Allman Gritzner *(University of Montana)*
David Gwynn *(Michigan State University)*
Joshua Hagen *(Marshall University)*
Stephen Healy *(Worcester State University)*
Douglas Heffington *(Middle Tennessee State University)*
Andrew Herod *(University of Georgia)*
Nik Heynen *(University of Georgia)*
Peter Hugill *(Texas A&M University)*
David Icenogle *(Auburn University)*
Mary Jacob *(Mount Holyoke College)*
Wendy Jepson *(Texas A&M University)*
Douglas L. Johnson *(Clark University)*
Jin-Kyu Jung *(University of North Dakota)*
Colleen E. Keen *(Minnesota State University)*
Paul Kelley *(University of Nebraska, Lincoln)*
Thomas Klak *(Miami University)*
James Kus *(California State University, Fresno)*
David Lanegran *(Macalester College)*
James Lindberg *(University of Iowa)*
Max Lu *(Kansas State University)*
John C. Lowe *(George Washington University)*
Donald Lyons *(University of North Texas)*
Brian McCabe *(University of New Mexico at Valencia)*
James McCarthy *(Penn State University)*
Neusa H. McWilliams *(University of Toledo)*
Katie Meehan *(University of Oregon)*
John Milbauer *(Northeastern State University)*
Byron Miller *(University of Cincinnati)*
Roger Miller *(University of Minnesota)*
Andrew Milson *(University of Texas at Arlington)*
Don Mitchell *(Syracuse University)*
Wendy Mitteager *(SUNY College at Oneonta)*

Woodrow W. Nichols, Jr. *(North Carolina Central University)*
Richard Pillsbury *(Georgia State University)*
James Proctor *(University of California at Santa Barbara)*
Mark Purcell *(University of Washington)*
Jeffrey Richetto *(University of Alabama)*
Andrew Schoolmaster *(University of North Texas)*
David Schul *(The Ohio State University, Marion)*
Alex Standish *(Rutgers University)*
Debra Straussfogel *(University of New Hampshire)*
Amy Trauger *(University of Georgia)*
Johnathan Walker *(James Madison University)*
Gerald R. Webster *(University of Alabama)*
Lisa Westwood *(Ruidoso Branch Community College)*
Joseph S. Wood *(George Mason University)*
Wilbur Zelinsky *(Penn State University)*
Sandra Zupan *(University of Kentucky)*

Special thanks go to our project manager, Sean Hale, and Karen Gulliver, our smart and assiduous development editor, Jay McElroy for his dynamic work on *Visualizing Geography*, as well as to Jennifer McCormack for her swift and insightful research assistance and her contributions to the active learning assessments. We thank as well the rest of our Pearson and wider publishing team, including Christian Botting, Anton Yakovlev, Amy De Genaro, and Lindsay Bethoney at Lumina Datamatics. For photo research we thank Lauren McFalls of Lumina Datamatics, and for the graphics program we thank Kevin Lear of International Mapping.

We are very grateful for the *Geography Matters* expert contributing authors who expanded each chapter's themes in exciting ways for the print and online materials of this seventh edition: Paul Adams (University of Texas, Austin), John Agnew (University of California, Los Angeles), Brian Blouet (College of William and Mary), Tim Creswell (Northeastern University), Elizabeth Currid-Halkett (University of Southern California, Price), Vincent Del Casino (University of Arizona), Dydia DeLyser (California State University, Fullerton), Richard Florida (University of Toronto), Jeffrey Garmany (King's College, London), Nik Heynen (University of Georgia), Ron Johnston (University of Bristol, UK), Tracey Osborne (University of Arizona), Mitch Rose (National University of Wales, Aberystwyth), Phil Steinberg (Durham University, UK), Peter Taylor (Northumbria University, UK).

In addition to his work on the *Geography Matters*, we also would like to acknowledge the terrific help from Paul Adams in helping craft Chapter 6: *Language, Communication, and Belief*.

Finally, a number of colleagues gave generously of their time and expertise in guiding our thoughts, making valuable suggestions, and providing materials: Alejandro A. Alonso (University of Southern California), Martin Cadwallader (University of Wisconsin), John Paul Jones, III (University of Arizona), Cindi Katz (City University of New York), Diana Liverman (University of Arizona), Ian Shaw (University of Glasgow), Harriet Hawkins (Royal Holloway, University of London).

Paul L. Knox
Sallie A. Marston

About the Authors

Paul L. Knox

Paul Knox received his PhD in Geography from the University of Sheffield, England. After teaching in the United Kingdom for several years, he moved to the United States to take up a position as professor of urban affairs and planning at Virginia Tech. His teaching centers on urban and regional development, with an emphasis on comparative study. He has received the university's award for teaching excellence. He has written several books on aspects of economic geography, social geography, and urbanization; serves on the editorial board of several scientific journals; and is co-editor of a series of books on world cities. In 2008, Professor Knox received the Distinguished Scholarship Award from the Association of American Geographers. He is currently a University Distinguished Professor in the College of Architecture and Urban Studies at Virginia Tech.

Sallie A. Marston

Sallie Marston received her PhD in Geography from the University of Colorado, Boulder. She is a full professor in the School of Geography and Development at the University of Arizona. Her undergraduate teaching focuses on political and cultural geography through innovative forms of pedagogy. She is the recipient of the College of Social and Behavioral Sciences' Outstanding Teaching Award as well as the Graduate College's Graduate Mentor Award. She is the co-editor of five books and author or co-author of over 75 journal articles and book chapters and received the Association of American Geographers Lifetime Achievement Award. She directs a service- learning course at the University of Arizona that places student-interns in school and community gardens as a way of supporting innovative teaching and learning initiatives. She serves on the editorial board of several scientific journals.

Digital & Print Resources

The seventh edition provides a complete human geography program for teachers and students.

For Students & Teachers

MasteringGeography™ with Pearson eText

The Mastering platform is the most widely used and effective online homework, tutorial, and assessment system for the sciences. It delivers self-paced tutorials that provide individualized coaching, focuses on course objectives, and is responsive to each student's progress. The Mastering system helps teachers maximize class time with customizable, easy-to-assign, and automatically graded assessments that motivate students to learn outside of class and arrive prepared for lecture. MasteringGeography offers:

- Assignable activities that include GIS-inspired Map-Master™ interactive maps, Encounter Human Geography Google Earth Explorations, Videos, Geoscience Animations, Map Projection Tutorials, GeoTutor coaching activities on the toughest topics in the geosciences, Dynamic Study Modules that provide each student with a customized learning experience, end-of-chapter questions and exercises, reading quizzes, Test Bank questions, and more.

- A student Study Area with GIS-inspired MapMaster™ interactive maps, Videos, Geoscience Animations, Web links, glossary flashcards, "In the News" RSS feeds, chapter quizzes, an optional Pearson eText, and more.

Pearson eText gives students access to the text whenever and wherever they can access the Internet.

Features of Pearson eText:

- Now available on smartphones and tablets.
- Seamlessly integrated videos and other rich media.
- Fully accessible (screen-reader ready).
- Configurable reading settings, including resizable type and night reading mode.
- Instructor and student note-taking, highlighting, bookmarking, and search.

Teaching College Geography: A Practical Guide for Graduate Students and Early Career Faculty (0136054471) This two-part resource provides a starting point for becoming an effective geography teacher from the very first day of class. Part One addresses "nuts-and-bolts" teaching issues. Part Two explores being an effective teacher in the field, supporting critical thinking with GIS and mapping technologies, engaging learners in large geography classes, and promoting awareness of international perspectives and geographic issues.

Aspiring Academics: A Resource Book for Graduate Students and Early Career Faculty (0136048919) Drawing on several years of research, this set of essays is designed to help graduate students and early career faculty start their careers in geography and related social and environmental sciences. *Aspiring Academics* stresses the interdependence of teaching, research, and service—and the importance of achieving a healthy balance of professional and personal life—while doing faculty work. Each chapter provides accessible, forward-looking advice on topics that often cause the most stress in the first years of a college or university appointment.

Practicing Geography: Careers for Enhancing Society and the Environment (0321811151) This book examines career opportunities for geographers and geospatial professionals in business, government, nonprofit, and educational sectors. A diverse group of academic and industry professionals share insights on career planning, networking, transitioning between employment sectors, and balancing work and home life. The book illustrates the value of geographic expertise and technologies through engaging profiles and case studies of geographers at work.

Learning Catalytics™ Learning Catalytics™ is a "bring your own device" student engagement, assessment, and classroom intelligence system. With Learning Catalytics™, you can:

- Assess students in real time, using open-ended tasks to probe student understanding.
- Understand immediately where students are and adjust your lecture accordingly.
- Improve your students' critical thinking skills.
- Access rich analytics to understand student performance.
- Add your own questions to make Learning Catalytics™ fit your course exactly.
- Manage student interactions with intelligent grouping and timing. Learning Catalytics™ has grown out of 20 years of cutting-edge research, innovation, and implementation of interactive teaching and peer instruction. Available integrated with MasteringGeography.

FOR STUDENTS

Goode's World Atlas 23rd Edition (0133864642) *Goode's World Atlas* has been the world's premiere educational atlas since 1923, and for good reason. It features over 250 pages of maps, from definitive physical and political maps to important thematic maps that illustrate the spatial aspects of many important topics. The 23rd edition includes over 160 pages of digitally produced reference maps, as well as new thematic maps on global climate change, sea level rise, CO_2 emissions, polar ice fluctuations, deforestation, extreme weather events, infectious diseases, water resources, energy production, and more.

Television for the Environment *Earth Report* **Geography Videos on DVD** (0321662989) This three-DVD set is designed to help students visualize how human decisions and behavior have affected the environment and how individuals are taking steps toward recovery. With topics ranging from the poor land management promoting the devastation of river systems in Central America to the struggles for electricity in China and Africa, these 13 videos from Television for the Environment's global *Earth Report* series recognize the efforts of individuals around the world to unite and protect the planet.

Dire Predictions: Understanding Global Warming 2nd Edition **by Michael Mann, Lee R. Kump** (0133909778) Periodic reports from the Intergovernmental Panel on Climate Change (IPCC) evaluate the risk of climate change brought on by humans. But the sheer volume of scientific data remains inscrutable to the general public, particularly to those who may still question the validity of climate change. In just over 200 pages, this practical text presents and expands upon the essential findings of the IPCC's 5th Assessment Report in a visually stunning and undeniably powerful way to the lay reader. Scientific findings that provide validity to the implications of climate change are presented in clear-cut graphic elements, striking images, and understandable analogies.

Pearson's Encounter Series Pearson's Encounter Series provides rich, interactive explorations of geoscience concepts through Google Earth™ activities, covering topics in regional, human, and physical geography. For those who do not use MasteringGeography, explorations are available in print workbooks and in online quizzes at **www.mygeoscienceplace.com**. Each exploration consists of a worksheet, online quizzes whose results can be emailed to instructors, along with a corresponding Google Earth™ KMZ file.

- *Encounter Human Geography* by Jess C. Porter (0321682203)

- *Encounter World Regional Geography* by Jess C. Porter (0321681754)

- *Encounter Physical Geography* by Jess C. Porter and Stephen O'Connell (0321672526)

FOR TEACHERS

TestGen® Computerized Test Bank (Download Only) (0321987551) TestGen® is a computerized test generator that lets instructors view and edit *Test Bank* questions, transfer questions to tests, and print the test in a variety of customized formats. This *Test Bank* includes over 2,000 multiple-choice, true/false, and short-answer/essay questions. Questions are correlated to the revised US National Geography Standards and Bloom's Taxonomy to help instructors better map the assessments against both broad and specific teaching and learning objectives. The *Test Bank* is also available in Microsoft Word®, and is importable into Blackboard. **www.pearsonhighered.com/irc**

Instructor's Resource DVD (0321987624) The *Instructor Resource Center on DVD* provides high-quality electronic versions of photos and illustrations from the book, as well as customizable PowerPoint™ lecture presentations, Classroom Response System questions in PowerPoint and the *Instructor Resource Manual* and *Test Bank* in Microsoft Word® and TestGen formats. The DVD includes all of the illustrations and photos from the text in presentation-ready JPEG files. For easy reference and identification, all resources are organized by chapter. **www.pearsonhighered.com/irc**

Instructor Resource Manual (Download Only) (0321987632) For download only, the *Instructor Resource Manual* is intended as a resource for both new and experienced instructors. It includes a variety of lecture outlines, additional source materials, teaching tips, advice about how to integrate visual supplements (including MasteringGeography resources), and various other ideas for the classroom. **www.pearsonhighered.com/irc**.

Instructor's Resource Materials (Download Only) (0321987578) This Instructor Resource content is also available online via the Instructor Resources section of MasteringGeography and at **www.pearsonhighered.com/irc**.

LEARNING OUTCOMES

- *Explain* how the study of geography has become essential for understanding a world that is more complex, interdependent, and changing faster than ever before.

- *Identify* four examples of how places influence inhabitants' lives.

- *State* the differences among major map projections and describe their relative strengths and weaknesses.

- *Explain* how geographers use geographic information systems (GIS) to merge and analyze data.

- *Summarize* the five concepts that are key to spatial analysis and describe how

they help geographers to analyze relationships between peoples and places.

- *Describe* the importance of distance in shaping human activity.

- *Summarize* the three concepts that are key to regional analysis and explain how they help geographers analyze relationships between peoples and places.

▲ Flower market in Karnataka, India.

GEOGRAPHY MATTERS

In Buenos Aires, Argentina, rioting teenagers ransacked and robbed working-class neighborhood grocery stores in 2012, leaving 22 dead and more than 200 injured. It was one of more than 50 riots worldwide between 2007 and 2014 where food was the principal issue.[1] The problem of food shortages and rising food prices in Argentina and in many other places is a reflection of the increasing geographic interdependence of the world. The situation is partly the result of increasing food consumption in other parts of the world, especially in booming China and India, where many have stopped growing their own food and have the cash to buy a lot more of it. Increasing meat consumption has helped drive up demand for feed grain, and this in turn has driven up the price of grain everywhere. Speculators in international commodity markets have joined the fray, further accelerating price rises. Another key linkage concerns energy prices: High oil prices push up fertilizer prices, while the cost of transporting food from farm to market adds to food costs. The popularity of biofuels as an alternative to hydrocarbons is straining food supplies, especially in the United States, where generous federal subsidies for ethanol have lured farmers away from growing crops for food. Compounding all this is climate change. Harvests in many countries have been seriously disrupted by more frequent extreme weather events. In 2013–2014, there were prolonged droughts in Argentina, California, eastern Brazil, Texas, parts of the Mediterranean, and the Sahel region of Africa; catastrophic floods in Canada and Central Europe; severe hailstorms in Germany; and a record-breaking typhoon that killed more than 6,000 in the Philippines.

Human geography is about recognizing and understanding the interdependence among places and regions, without losing sight of the uniqueness of each specific place. **Places** are specific geographic settings with distinctive physical, social, and cultural attributes. **Regions** are territories that encompass

[1]The World Bank Group, Poverty Reduction and Economic Management Network, *Food Price Watch*, 17, Washington, D.C., May 2014.

many places, all or most of which share attributes different from the attributes of places elsewhere. Maps are also important tools for introducing geographers' ideas about the way that places and regions are made and altered.

WHY GEOGRAPHY MATTERS

The importance of geography as a subject of study is becoming more widely recognized as people everywhere struggle to understand a world that is increasingly characterized by instant global communications, rapidly changing international relationships, unexpected local changes, and growing evidence of environmental degradation. Many more schools now require courses in geography than just a decade ago, and the College Board has added the subject to its Advanced Placement program. Meanwhile, many employers are coming to realize the value of employees with expertise in geographical analysis and an understanding of the uniqueness, influence, and interdependence of places. Through an appreciation of the diversity and variety of the world's peoples and places, geography provides real opportunities not only to contribute to local, national, and global development but also to understand and promote multicultural, international, and feminist perspectives in the world.

Most people want to understand the intrinsic nature of the world in which we live. Geography enables us to understand where we are both literally and figuratively. Geography provides knowledge of Earth's physical and human systems and of the interdependency of living things and physical environments. That knowledge, in turn, provides a basis for people to cooperate in the best interests of our planet. Geography also captures the imagination: It stimulates curiosity about the world and the world's diverse inhabitants and places. By obtaining a better understanding of the world, people can overcome closed-mindedness, prejudice, and discrimination.

APPLY YOUR KNOWLEDGE

1. Why do you think studying geography is critical in today's world?

2. List three reasons why a corporate employer would feel it is important for prospective employees to have some knowledge of geography.

WHY PLACES MATTER

An appreciation of the diversity and variety of peoples and places is a theme that runs through all of *human geography,* the study of the spatial organization of human activity and of people's relationships with their environments. This theme is inherently interesting to nearly all of us. *National Geographic* magazine has become a venerable institution by bringing us monthly updates of the seemingly endless variety of landscapes and communities around the world. More than 5 million households, representing about 19 million regular readers, subscribe to this magazine for its intriguing descriptions and striking photographs. Millions more read it occasionally in offices, lobbies, waiting rooms, or online.

Yet many Americans often seem content to confine their interest in geography to the pages of glossy magazines, to television documentaries, or to one-week packaged vacations. It has become part of the conventional wisdom—both in the United States and around the world—that many Americans have little real appreciation or understanding of people and places beyond their own daily routines. This is perhaps putting it too mildly. Surveys have revealed widespread ignorance among a high proportion of Americans, not only of the fundamentals of the world's geography but also of the diversity and variety within the United States itself. In surveys of young adults in Canada, France, Germany, Great Britain, Italy, Japan, Mexico, Sweden, and the United States, Americans come in next to last in terms of geographic literacy. Neither wars nor natural disasters appear to have compelled the majority of young Americans to absorb knowledge about international places in the news.

So although most people in the United States are fascinated by different places, relatively few have a systematic knowledge of them. Fewer still understand how different places came to be the way they are or why places matter in the broader scheme of things. This lack of understanding is unfortunate because geographic knowledge can take us far beyond a simple glimpse of the inherently interesting variety of peoples and places.

The Influence of Places

Places are dynamic, with changing properties and fluid boundaries that are the product of the interplay of a wide variety of environmental and human factors. This dynamism and complexity is what makes places so fascinating for readers of *National Geographic*. It is also what makes places so important in shaping people's lives and in influencing the pace and direction of change. Places provide the settings for people's daily lives and their social relations (patterns of interaction among family members, at work, in social life, in leisure activities, and in political activity). It is in these settings that people learn who and what they are, how they are expected to think and behave, and what life is likely to hold for them.

Places exert a strong influence, for better or worse, on people's physical well-being, opportunities, and lifestyle choices. Living in a small town dominated by petrochemical industries, for example, means a higher probability of being exposed to air and water pollution, having a limited range of job opportunities, and having a relatively narrow range of lifestyle options because of a lack of amenities such as theatres, specialized stores and restaurants, and recreational facilities (**Figure 1.1**). Living in a central neighborhood of a large metropolitan area, on the other hand, usually means having a wider range of job opportunities and a greater choice of lifestyle options because of the variety of amenities accessible within a short distance (**Figure 1.2**). But it also means, among other things, living with a relatively high exposure to crime.

◄ **Figure 1.1** **Quality of life** Heavy industry adjacent to a housing development.

▼ **Figure 1.2** **Central city neighborhood** Shopping on Newbury Street in Boston.

The Meaning of Places

Places also contribute to people's collective memory and become powerful emotional and cultural symbols. Consider the evocative power for most Americans of places like Times Square in New York; the Mall in Washington, D.C.; Hollywood Boulevard in Los Angeles; and Graceland in Memphis. And for many people, ordinary places have special meaning: a childhood neighborhood, a college campus, a baseball stadium, or a family vacation spot. This layering of meanings reflects the way that places are *socially constructed*—given different meanings by different groups for different purposes. Places exist and are constructed by their inhabitants from a subjective point of view.

The meanings given to a place may be so strong that they become a central part of the identity of the people experiencing them. Your **identity** is the sense that you make of yourself through your subjective feelings based on your everyday experiences and social relations. Your own neighborhood, for example, is probably heavily laden with personal meaning and sentiment for you. But your neighborhood may well be viewed very differently, perhaps unsympathetically, by outsiders. This distinction is useful in considering the importance of understanding spaces and places from the viewpoint of the insider—the person who normally lives in and uses a particular place—as well as from the viewpoint of outsiders (including geographers).

Finally, places are the sites of innovation and change, of resistance and conflict (**Figure 1.3**). The unique characteristics of specific places can provide the preconditions for new agricultural practices (such as the development of seed agriculture and the use of plow and draft animals that sparked the first agricultural revolution in the Middle East in prehistoric times—see Chapter 9); new modes of economic organization (such as the high-tech revolution that began in Silicon Valley

◄ **Figure 1.3** Tahrir Square, Cairo, Egypt
The site of major anti-government demonstrations in 2011 that led to the fall of Egypt's President Mubarak; and of celebrations in 2014 (shown here) of the inauguration of President el-Sissi.

in the late twentieth century); new cultural practices (e.g., the punk movement that began in disadvantaged British housing projects); and new lifestyles (e.g., the hippie lifestyle that began in San Francisco in the late 1960s). It is in specific locales that important events happen, and it is from them that significant changes spread.

Nevertheless, the influence of places is by no means limited to the occasional innovative change. Because of their distinctive characteristics, places always modify and sometimes resist the imprint of even the broadest economic, cultural, and political trends. Consider, for example, the way that a global cultural trend—rock 'n' roll—was modified in Jamaica to produce reggae. And how in Iran and North Korea rock 'n' roll has been resisted by the authorities, with the result that it has acquired an altogether different kind of value and meaning for the citizens of those countries. Similarly, Indian communities in London developed Bhangra—a "world beat" composite of traditional Punjabi music, Bollywood (Hindi) movie scores, and Western disco. Cross-fertilization with local music cultures in New York and Los Angeles has produced Bhangra rap.

To consider a different illustration, think of the ways some communities have declared themselves "nuclear-free" zones: places where nuclear weapons and nuclear reactors are unwelcome or even banned by local laws. By establishing such zones, individual communities are seeking to challenge trends toward using nuclear energy and maintaining nuclear arms. They are, to borrow a phrase, "thinking globally and acting locally." Similarly, some communities have established "GM-free" zones, taking a stance against genetically modified crops and food. In adopting such strategies, they hope to influence thinking in other communities so that eventually their challenge could result in a reversal of established trends.

In summary, places are settings for social interaction that, among other things,

- structure the daily routines of people's economic and social life;

- provide both opportunities and constraints in terms of people's long-term social well-being;

- provide a context in which everyday, commonsense knowledge and experience are gathered;

- provide a setting for processes of socialization; and

- provide an arena for contesting social norms.

APPLY YOUR KNOWLEDGE

1. How does place affect identity?

2. Explain how and why a particular place has mattered to you. How might others' experience or perception of that same place differ from yours? How does your place influence your health or job prospects?

STUDYING HUMAN GEOGRAPHY

The study of geography involves the study of Earth as created by natural forces and modified by human action. This, of course, covers an enormous amount of subject matter. There are two main branches of geography: physical and human. **Physical geography** deals with Earth's natural processes and their outcomes. It is concerned, for example, with climate, weather patterns, landforms, soil formation, and plant and animal ecology. **Human geography** deals with the spatial organization of human activities and with people's relationships to their environments. This involves looking at natural physical environments insofar as they influence, and are influenced by, human activity. To that end, the study of human geography must cover a wide variety of phenomena. These include, for example, agricultural production and food security, population change, the ecology of human diseases, resource management, environmental pollution, regional planning, and the symbolism of places and landscapes.

Regional geography combines elements of both physical and human geography. Regional geography is concerned with the way that unique combinations of environmental and human factors produce territories with distinctive landscapes and cultural attributes. The concept of region is used by geographers to apply to larger-sized territories that encompass many neighboring places, all or most of which have similar attributes distinct from the attributes of other places.

Geographical Relationships

What is distinctive about the study of human geography is not so much the phenomena that are studied as the way they are approached. The contribution of human geography is to reveal *how and why geographical relationships are important* in relation to a wide spectrum of natural, social, economic, political, and cultural phenomena. Thus, for example, human geographers are interested not only in patterns of agricultural production but also in the geographical relationships and interdependencies that are both causes and effects of such patterns. To put it in concrete terms, geographers are interested not only in what specialized agricultural subregions (e.g., the dairy farming area of Jutland, Denmark) are like but also in the role of subregions such as Jutland in national and international agro-food systems (their interdependence with producers, distributors, and consumers in other places and regions—see Chapter 8).

Geography is to a great extent an applied discipline as well as a means of understanding the world. Geographers employed in business, industry, and government are able to use geographic theories and techniques to understand and solve a wide variety of specific problems. A great deal of the research undertaken by geography professors has an applied focus.

Once data have been obtained through some form of observation, the next important step is to portray and describe them through *visualization* or *representation.* This can involve a variety of tools, including written descriptions, charts, diagrams, tables, mathematical formulas, and maps (see Box 1.2: "Visualizing Geography: Maps"). Visualization and representation are important activities because they allow large amounts of information to be explored, summarized, and presented to others. They are nearly always a first step in the analysis of geographical relationships, and they are important in conveying the findings and conclusions of geographic research.

At the heart of geographic research, as with other kinds of research, is the *analysis* of data. The objective of analysis, whether of quantitative or qualitative data, is to discover patterns and establish relationships so that hypotheses can be established and models can be built. Models, in this sense, are abstractions of reality that help explain the real world. They require tools that allow us to generalize about things. Once again, we find that geographers are like other social scientists in that they utilize a wide range of analytical tools, including conceptual and linguistic devices, maps, charts, and mathematical equations.

In many ways, therefore, the tools and methods of human geographers are parallel to those used in other sciences, especially the social sciences. In addition, geographers increasingly use some of the tools and methods of the humanities—interpretive analysis and inductive reasoning, for example—together with ethnographic research (the systematic recording of human cultures) and textual analysis. One of the most distinctive tools in the geographer's kit bag is geographic information systems (GIS).

THE BASIC TOOLS AND METHODS OF HUMAN GEOGRAPHERS

In general terms, the basic tools employed in geography are similar to those in other disciplines. Like other social scientists, human geographers usually begin with observation. Information must be collected and data recorded. This can involve many different methods and tools. Fieldwork (surveying, asking questions, using scientific instruments to measure and record things), laboratory experiments, and archival searches all are used by human geographers to gather information about geographical relationships. Geographers also use **remote sensing**, the collection of information about parts of Earth's surface by means of aerial photography or satellite imagery designed to record data on visible, infrared, and microwave sensor systems (**Figure 1.4**). For example, agricultural productivity can be monitored by remotely sensed images of crops, and energy efficiency can be monitored by remotely sensed levels of heat loss from buildings.

▲ **Figure 1.4 Remotely sensed images** Remotely sensed images can provide new ways of seeing the world, as well as unique sources of data on all sorts of environmental conditions. Such images can help explain problems and processes. Aerial photographs, for example, can be helpful in explaining what would otherwise require expensive surveys and detailed cartography. They are especially useful in working with multidisciplinary teams. This example shows the lower Connecticut River near the town of Old Lyme. The photograph was taken during the Connecticut River Marsh Restoration Project.

1.1 Geography Matters

Why Geography Matters

By Richard Florida, University of Toronto

Since the advent of steam locomotives and the telegraph, there have been countless predictions about how technology will put an end to the constraints of geography. Some believe that that day has finally come. *The Economist* magazine proclaimed the "death of distance" in 1995 (http://www.economist.com/node/598895). A decade later, *The New York Times* columnist Tom Friedman proclaimed The World Is Flat (Farrar, Straus, Giroux, 2005). Between airplanes, the Internet, social media, and smart phones, the argument goes, the distances that once separated us have contracted to nothing; no matter where we live, we can all work and compete on the same terms.

But dig a little deeper and you'll realize that place and geography matter more today than they ever did. And not just in some abstract academic way, but in your daily lives.

Why Where Matters

We tend to focus on two big decisions: *who* we choose as our life partner and *what* we do for work. But there is a third big decision—where you choose to live. The decision about *where* may be the most important, because it has a huge bearing on the others (Richard Florida, *Who's Your City*. Basic Books, 2008).

Where you choose to live affects every aspect of your being. It influences the income you can potentially earn, and how far you can develop your skills. It has a bearing on the friends and romantic partners you can potentially meet, the networks you can build, and the options that will be available to your children. It shapes your values and your politics. It plays a role in how happy and healthy you are, and in your overall quality of life.

It's something of a paradox. As globalized as we may be, wealth and power are more concentrated and clustered than ever before. Location matters. It is true that communications are instantaneous and journeys that used to take months now take only hours, but the key factors that determine whether a place prospers are unevenly distributed; some are much more endowed with them than others.

Welcome to the Spiky World

When we overlay graphs of economic output, population, and innovation on a map, some places, especially those in the United States, Western Europe, and Asia, tower over the rest. Just look at the level of innovation in Tokyo! The world is anything but flat; it is spiky (**Figure 1.A**).

Those high levels of innovation and a disproportionate share of the world's economic activity occur in the geographic clusters called mega-regions—galaxy-like clusters of cities and their suburbs that grow and evolve into coherent geographic entities. These include great Bos-Wash (Boston-New York-Washington) Corridor, which would rank as the world's fourth largest economy, behind the United States itself and China and Japan; Am-Bruss-Twerp spanning Amsterdam, Antwerp and Brussels; and the area that runs from London to Leeds, Manchester, Liverpool, and Birmingham. All told, the globe's 40 leading mega-regions produce two-thirds of the world's economic output and nine in ten of its technological innovations, while housing less than one in five of its people.

▲ Figure 1.A A Spiky World.

Urban Planet

Big cities have become the core social and economic organizing units of our time, replacing the farm and the factory of previous epochs. As the great urbanist Jane Jacobs was perhaps the first to document in her book *The Economy of Cities* (Random House, 1969), big cities are cauldrons of innovation; when diverse people live and work in close quarters, they come up with new ideas that improve the productivity of existing industries and generate new ones.

This geographic clustering of diverse and talented people is what makes cities different than all other biological and social organisms. Typically when organisms get bigger their metabolisms slow down. But as cities get larger and larger their "urban metabolism"—measured variously as their economic activity or ability to innovate—speeds up. (Bettencourt, Lobo, et al., "*Growth, Innovation, Scaling, and the Pace of Life in Cities,*" *Proceedings of the National Academy of Sciences of the United States of America*, http://www.pnas.org/content/104/17/7301.full).

The geographic clustering of people will only grow. Over the course of the next century, the world's urban populations are projected to increase by as many as 5 billion people. Most of that will be happening in the rapidly urbanizing areas of South East Asia and Africa. To accommodate them, existing cities will expand substantially and hundreds of cities are likely to emerge (http://marroninstitute.nyu.edu/content/working-papers/urbanization-as-opportunity1). Ensuring that those cities are vibrant, sustainable, and functional will be the grandest of the grand challenges humanity will ever face. Our success or failure will have a bearing on everything from our ability to mitigate poverty and climate change to how successfully the world continues to democratize.

Distance is not dead and geography is far from over. The fact is, place matters more today than it ever has.

Richard Florida is director of the Martin Prosperity Institute at the University of Toronto's Rotman School of Management, Global Research Professor at New York University, and senior editor at The Atlantic, *where he cofounded* CityLab.

1. Use the online "Place Finder" tool (http://www.creativeclass.com/_v3/whos_your_city/place_finder/) to identify the city that is best suited to you. Come up with three to five cities you think you might want to live and work in and then use the Place Finder tool to rate and rank them. Which is your best city and why?

2. According to **Figure 1.B**, neighboring cities in the orange colored areas are growing into large mega-regions. Some of these even cross national boundaries and in some respects have more in common with each other than their respective countries. Is it possible that city regions will become more significant units of place than nation states?

▲ Figure 1.B Mega-regions.

1.2 Visualizing Geography

Maps

Maps not only describe data but also serve as important sources of data and tools for analysis. Because of their central importance to geographers, they can also be objects of study in their own right.

They express particular interpretations of the world, and affect how we understand the world– and how we see ourselves in relation to others. All maps are "social products," and generally reflect the power of the people who draw them up. The design of maps- what they include, what they omit, and how their content is portrayed- inevitably reflects the experiences, priorities, interpretations, and intentions of their authors.

1.1 Map scales

A **map scale** is the ratio between linear distance on a map and linear distance on Earth's surface, usually expressed in terms of a *representative fraction* (in the example of Figure 1.1.1A, 1/25,000) or ratio (1:25,000). *Small-scale* maps are maps based on small representative fractions (for example, 1/10,000,000). *Large-scale* maps are maps based on larger representative fractions (e.g., 1/10,000).

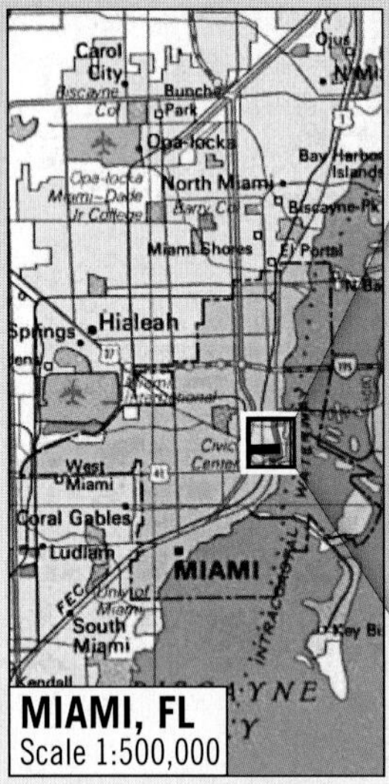

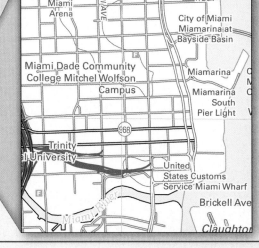

Figure 1.1.1A
Representative fraction:1:500,000 or 1/500,000
Relatively small scale map of Miami area shows less detail.

MIAMI, FL
Scale 1:500,000

Figure 1.1.1B
Representative fraction:1:24,000 or 1/24,000
Relatively large scale map of the same area shows a higher level of detail.

1.2 Thematic maps

Maps that are designed to represent the spatial dimensions of particular conditions, processes, or events are called **thematic maps**. These can be based on any one of a number of devices that allow cartographers or map makers to portray spatial variations or spatial relationships. One of these is the **isoline**, a line (similar to a contour) that connects places of equal data value (for example precipitation, as in **Figure 1.2.1**). Maps based on isolines are known as **isopleth maps**.

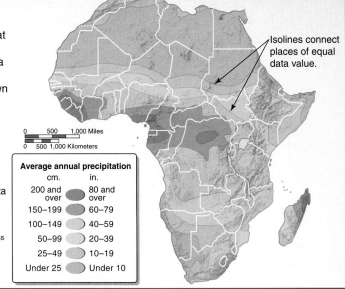

Isolines connect places of equal data value.

Figure 1.2.1
Isoline maps. Isoline maps portray spatial information by connecting points of equal data value. Contours on topographic maps are isolines. This map shows average annual precipitation for the continent of Africa.
(*Source:* Reprinted with permission of Pearson, from D. Hess and D. Tasa, McKnight's Physical Geography: A Landscape Appreciation, 10th edition, 2011, p. 36.)

Average annual precipitation

cm.	in.
200 and over	80 and over
150–199	60–79
100–149	40–59
50–99	20–39
25–49	10–19
Under 25	Under 10

0 500 1,000 Miles
0 500 1,000 Kilometers

1.3 Map projections

A **map projection** is a systematic rendering on a flat surface of the geographic coordinates of features found on Earth's surface. Because Earth's surface is curved, it is impossible to represent on a flat plane without some distortion. Cartographers have devised different techniques for projecting latitude and longitude onto a flat surface, and the resulting representations each have advantages and disadvantages. None of them can represent distance correctly in all directions, though many can represent compass bearings, or area without distortion.

Figure 1.3.1 Mollweide projection
Relative sizes are true, but shapes are distorted.

From globe to flat map. Conversion of the globe to a flat map projection requires a decision about which properties to preserve and the amount of distortion that is acceptable.

Globe

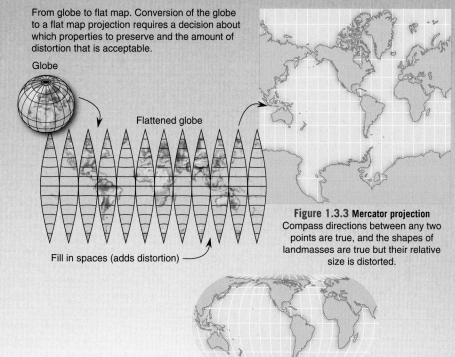

Flattened globe

Fill in spaces (adds distortion)

Figure 1.3.3 Mercator projection
Compass directions between any two points are true, and the shapes of landmasses are true but their relative size is distorted.

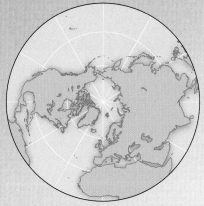

Figure 1.3.2 Azimuthal Equidistant projection
Distances measured from the center of the map are true, but direction, area, and shape are increasingly distorted as the distance from the center point increases.

Figure 1.3.4 The Robinson projection
Distance, direction, area, and shape are all distorted in an attempt to balance the properties of the map. It is designed purely for appearance and is best used for thematic and reference maps at the world scale.
(*Source*: After E. F. Bergman, Human Geography: Cultures, Connections, and Landscapes, © 1995 by Pearson, p. 12.)

One particular kind of map projection that is sometimes used in small-scale thematic maps is the **cartogram**. In this kind of projection, space is transformed according to statistical factors, with the largest mapping units representing the greatest statistical values.

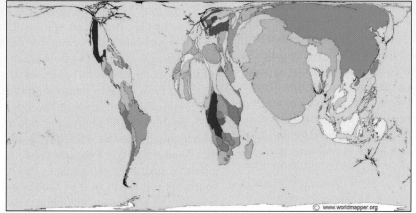

Animation: Map Projections
http://goo.gl/7SISq1

Interactive Album of Map Projections
http://goo.gl/tTWBvc

Figure 1.3.5 A cartogram of the world
The relative size of countries is based not on area but on the proportion of people with extremely low incomes. The deliberate distortion of the shapes of the continents dramatically emphasizes spatial variations.
Source: © Copyright SASI Group (University of Sheffield) and Mark Newman (University of Michigan).

1. How do market segmentation programs merge geographic data with consumer preferences?

2. Can you think of other areas of everyday life that could employ a geographic segmentation program for analysis, e.g., voting, environmental sustainability, food access, etc.?

Geographic Information Systems (GIS)

The combination of high-performance computing and computerized record keeping has led to an unprecedented increase in the volume and popularity of geographic data. **Geographic information systems (GIS)** have rapidly grown to become one of the most important methods of geographic analysis, particularly in the military and commercial worlds. The software in GIS incorporates programs to store and access spatial data, to manipulate those data, and to draw maps.

The primary requirement for data to be used in GIS is that the locations for the variables—the characteristics under consideration—are known. Location may be annotated by *x, y,* and *z* coordinates of longitude, latitude, and elevation, or by such systems as ZIP codes or highway mile markers. Any variable that can be located spatially can be fed into a GIS. Data capture—putting the information into the system—is the most time-consuming component of GIS work. Different sources of data, using different systems of measurement, scales, and systems of representation, must be integrated with one another; changes must be tracked and updated. Many GIS operations in the United States, Europe, Japan, and Australia have begun to contract out such work to firms in countries where labor is cheaper. India has emerged as a major data-conversion center for GIS.

The most important aspect of GIS, from an analytical point of view, is that they allow data from several different sources, on different topics, and at different scales, to be merged. This allows analysts to emphasize the spatial relationships among the objects being mapped. A geographic information system makes it possible to link, or integrate, information that is difficult to associate through any other means.

Applications GIS technology can render visible many aspects of geography that were previously unseen. GIS can, for example, produce incredibly detailed maps based on millions of pieces of information—maps that could never have been drawn by human hands. One example of such a map is the satellite image reconstruction of the vegetation cover of the United States shown in **Figure 1.5**. At the other extreme of spatial scale, GIS can put places under the microscope, creating detailed new insights using huge databases and effortlessly browsable media.

Many advances in GIS have come from military applications. GIS allows infantry commanders to calculate line of sight from tanks and defensive emplacements, allows drones and cruise missiles to fly below enemy radar, and provides a comprehensive basis for military intelligence. Beyond the military, GIS technology allows an enormous range of problems to be addressed. For instance, it can be used to decide how to manage farmland, to monitor the spread of infectious diseases, to monitor tree cover in metropolitan areas, to assess changes in ecosystems, to analyze the impact of proposed changes in the boundaries of legislative districts, to identify the location of potential business customers, to identify the location of potential criminals, and to provide a basis for urban and regional planning.

Some of the most influential applications of GIS have resulted from geodemographic research. **Geodemographic research** uses census data and commercial data (such as sales data and property records) about the populations of small districts in creating profiles of those populations for market

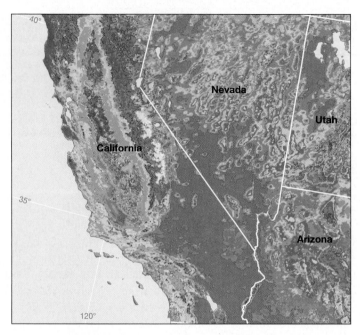

▲ Figure 1.5 **Map of land cover** This extract is from a map of land cover in the United States that was compiled from several data sets using GIS technology. These included 1-kilometer resolution, Advanced Very High Resolution Radiometer (AVHRR) satellite imagery, and digital data sets on elevation, climate, water bodies, and political boundaries. Each of the 159 colors on the U.S. map represents a specific vegetation region. The purples and blues represent various subregions of western coniferous forests, the yellows are grasslands, and the reds are shrublands. The gray-brown region is the barren area of the Mojave Desert.

(*Source:* United States Geological Survey, Map of Seasonal Land Cover Regions, 1993; see T. Loveland, J. W. Merchant, J. F. Brown, D. O. Ohlen, B. C. Reed, P. Olson, and J. Hutchinson, "Seasonal Land Cover Regions of the United States," *Annals, Association of American Geographers, 85,* 1995, 339–355.)

research. Market research consultants, using statistical methods similar to those used by urban social geographers, are able to identify the geography of different types of households according to their distinctive consumption patterns and preferences as well as their socioeconomic and demographic attributes and their typical residential settings. The digital media used by GIS make such applications very flexible.

As maps have become more commonplace, more people and more businesses have become more spatially aware. Nevertheless, some critics have argued that GIS has been exploited by those who already possess power and control to increase the level of surveillance of the population. The fear is that GIS may be helping to create a world in which people are not treated and judged by who they are and what they do, but more by where they live. People's credit ratings, ability to buy insurance, and ability to secure a mortgage, for example, are all routinely judged in part by GIS-based analyses that take into account the attributes and characteristics of their neighbors.

APPLY YOUR KNOWLEDGE

1. What are some advantages and disadvantages of the increasing use of GIS in society?

2. Choose the geographic representations and tools that you would use to illustrate alternative views of a local issue that is currently the subject of public discussion in your community. Explain your choice.

SPATIAL ANALYSIS

The study of many geographic phenomena can be approached in terms of their arrangement as points, lines, areas, or surfaces on a map. This is known as **spatial analysis**. *Location, distance, space, accessibility,* and *spatial interaction* are five concepts that are key to spatial analysis. Although these concepts may be familiar from everyday language, they require some elaboration.

Location

Location is often nominal; that is, it is expressed solely in terms of the names given to regions and places. We speak, for example, of Washington, D.C., or of Georgetown, a location within Washington, D.C. Location can also be used as an absolute concept, whereby locations are fixed mathematically through coordinates of latitude and longitude (**Figure 1.6**).

Latitude refers to the angular distance of a point on Earth's surface, measured in degrees, minutes, and seconds north or south from the equator. The equator is assigned a value of 0°. Lines of latitude around the globe run parallel to the equator, which is why they are sometimes referred to as *parallels.* **Longitude** refers to the angular distance of a point on Earth's surface, measured in degrees, minutes, and seconds east or west from the *prime meridian.* The prime meridian is the line that passes through both poles and through Greenwich, England, which is assigned a value of 0° (see Box 1.3: "Window on the World: Greenwich, England"). Lines of longitude, called *meridians,* run from the North Pole (latitude 90° north) to the South Pole (latitude 90° south). Georgetown's coordinates are precisely 38°55'N, 77°00'E.

Thanks to the **Global Positioning System (GPS)**, it is very easy to determine the latitude and longitude of any given point. The Global Positioning System consists of 31 satellites that orbit Earth on precisely predictable paths, broadcasting highly accurate time and locational information. The GPS is owned by the U.S. government, but the information transmitted by the satellites is freely available to everyone around the world. Nevertheless, other systems are in development: the European Union's Galileo positioning system, India's Indian Regional Navigational Satellite System, and China's Compass Navigation System. Basic GPS receivers can relay latitude, longitude, and height to within 100 meters day or night, in all weather conditions, in any part of the world where there is an unobstructed line of sight to four or more GPS satellites. The most precise GPS receivers, costing thousands of dollars, are accurate to within a centimeter. The GPS has dramatically increased the accuracy and efficiency of collecting spatial data. In combination with GIS and remote sensing, GPS has revolutionized mapmaking—think, for example, of Google Maps—and spatial analysis. The introduction of tracking technology in cell phones and other devices, and the consequent proliferation of apps for everything from navigation to shopping, has meanwhile had a transformational effect on many aspects of life among more affluent populations.

Site and Situation   Location can also be *relative,* fixed in terms of site or situation. **Site** refers to the physical attributes of a location: its terrain, its soil, vegetation, and water sources, for example. **Situation** refers to the location of a place relative to other places and human activities: its accessibility to routeways, for example, or its nearness to population centers. The location of telecommunications activities in Denver, Colorado, provides a good example of the significance of the geographic concepts of site and situation. Because of its site and situation, Denver is a major center for cable television and associated specialized support companies. Denver's site, 1.6 kilometers (1 mile) above sea level, is important because it gives commercial transmitters and receivers a better "view" of communications satellites. Its situation, on the 105th meridian and equidistant between the telecommunications satellites

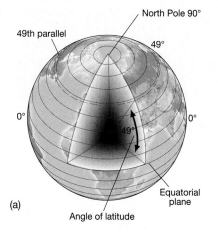
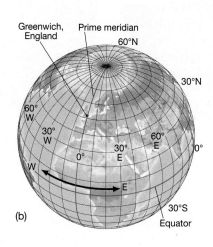
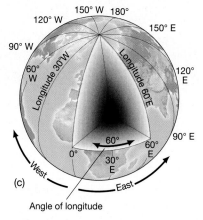

▲ **Figure 1.6 Latitude and longitude** Lines of latitude and longitude provide a grid that covers Earth, allowing any point on Earth's surface to be accurately referenced. Latitude is measured in angular distance (i.e., degrees and minutes) north or south of the equator, as shown in (a). Longitude is measured in the same way, but east and west from the prime meridian, a line around Earth's surface that passes through both poles (North and South) and the Royal Observatory in Greenwich, just to the east of central London, in England. Locations are always stated with latitudinal measurements first. The location of Paris, France, for example, is 48°51'N and 2°20'E, as shown in (b).

Source: (a) and (c), adapted from R. W. Christopherson, *Geosystems: An Introduction to Physical Geography,* 2nd ed., © 1994, pp. 13 and 15. (b), adapted from E. F. Bergman, *Human Geography: Cultures, Connections, and Landscapes,* © 1995, Figs. 1–10 and 1–13.

1.3 Window on the World

Greenwich, England

Greenwich is the historic reference point for the standardized measurement of both space and time. Historically, Greenwich represented a prime defensive site with excellent river access. It is situated on the most dramatic outer curve of the River Thames, where the river's current has scoured a deep channel. Archaeological digs have found remains of prehistoric, Roman, Saxon, and Danish settlements. The making of the district, though, was royal patronage, and the result was a collection of buildings in a distinctive setting that have been added to UNESCO's World Heritage List.

Greenwich was first developed as a royal residence in the fifteenth century. The Tudor monarchs Henry VIII, Mary I, and Elizabeth I were all born there. It was Henry VIII who intensified the naval presence along this stretch of the Thames, founding two new shipyards as his navy grew. In the seventeenth century, the hillside behind the royal palaces was landscaped into Greenwich Park by King Charles II. On the brow of the hill the king sited his Royal Observatory (**Figure 1.C**), designed by Christopher Wren with the assistance of scientist Robert Hooke. With the increasing importance of British maritime trade and naval power in the seventeenth century, the role of astronomy in navigation became critically important. Charles II founded the Royal Observatory to solve the problem of finding longitude at sea. As a result of the observatory's preeminence in the field, together with British imperial power and influence, Greenwich Meridian and Greenwich Mean Time were adopted as global standards for the measurement of space and time in 1884 (**Figure 1.D**).

▲ Figure 1.C Royal Observatory, Greenwich

▲ Figure 1.D The Greenwich Meridian

that are in geostationary orbit over the Pacific and Atlantic oceans, allows it to send cable programming directly not just to the whole of the Americas but also to Europe, the Middle East, India, Japan, and Australia—to every continent, in fact, except Antarctica. This precludes "double-hop" transmission (in which a signal goes up to a satellite, then down, then up and down again), which increases costs and decreases picture quality. Before the location of telecommunications facilities in Denver, places east or west of the 105th meridian had to double-hop some of their transmissions because satellite dishes could not have a clear "view" of both the Pacific and Atlantic telecommunications satellites. (**Figure 1.7**).

Mental Maps Finally, location also has a *cognitive* dimension, in that people have cognitive images of places and regions, compiled from their own knowledge, experiences, and impressions. **Cognitive images** (sometimes referred to as *mental maps*) are psychological representations of locations that spring from people's individual ideas and impressions of these locations. These representations can be based on direct

▲ **Figure 1.E Maritime Greenwich** Queen's House and the old Royal Naval College.

Meanwhile, King Charles had decided to build a new palace by the river. Building began in 1696 and continued throughout much of the eighteenth century, during which the complex was changed from a palace to a royal hospital for seamen. The result was Britain's most outstanding group of classical buildings (**Figure 1.E**). The hospital complex became a symbol not only of intellectual and artistic refinement but also of Britain's growing naval power. At the height of British sea power, in 1873, the complex was converted to house the Royal Naval College. This brought the navy's officer class, its families and its suppliers to Greenwich, shaping the character of the district for more than a century before the hospital complex was converted again, this time for use by the University of Greenwich and the Trinity Laban Conservatoire of Music and Dance.

1. What is the historical significance of Greenwich? Why do you think it represents the global standard for measuring time and space?

2. Do you think there is a legacy on how we view the world by centering Greenwich, London, England? What if we had used Tokyo, Japan, Baghdad, or Iraq as the center? What if we flipped the globe from the perspective of New Zealand and had the South Pole on top? How do orientations of the "center" affect our perspectives on world power and influence?

experiences, on written or visual representations of actual locations, on hearsay, on imagination, or on a combination of these sources. Location in these cognitive images is fluid, depending on a given individual's changing information and perceptions of the principal landmarks in their environment.

Some things may not be located in a person's cognitive image at all. **Figure 1.8** shows a cognitive image of Washington, D.C. Georgetown is shown on this mental map, even though it is some distance from the residence of the person who sketched her image of the city. Less well-known and less distinctive places do not appear in this particular image.

Distance

Distance is also useful as an *absolute* physical measure, whose units we may count in kilometers or miles. Distance can also be a *relative* measure, expressed in terms of time, effort, or cost. It can take more or less time, for example, to travel 10 kilometers from point A to point B than it does to travel 10 kilometers from point A to point C. Similarly, it can cost more or less. Geographers have to recognize that distance can sometimes be in the eye of the beholder. It can seem longer or shorter, more or less pleasant, to travel from A to B as compared to traveling

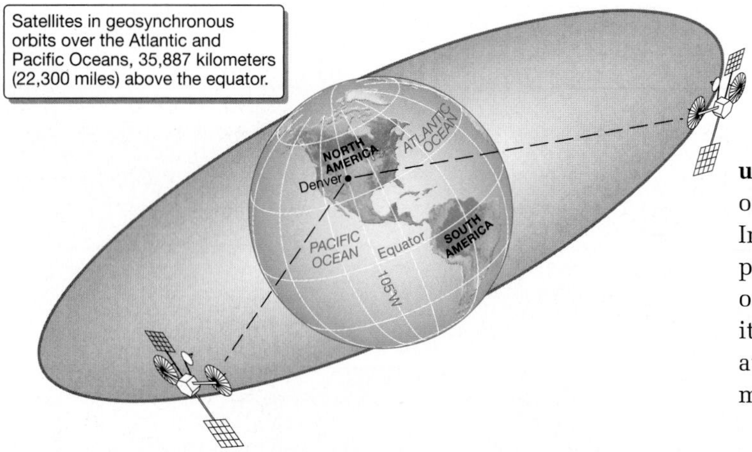

Satellites in geosynchronous orbits over the Atlantic and Pacific Oceans, 35,887 kilometers (22,300 miles) above the equator.

◀ **Figure 1.7 The importance of site and situation** Denver's site (its altitude) and situation (mid-continent) make it an ideal location to receive signals from widely spaced satellites.

utility of particular locations. The utility of a specific place or location is its usefulness to a particular person or group. In practice, utility is thought of in different ways by different people in different situations. The emphasis may be on cost, on profitability, on prestige, on security, or on ease of mobility, for example, or more likely on some combination of these attributes. However place utility is determined, people in most circumstances tend to *seek to maximize the net utility of*

from A to C. This is **cognitive distance**, the distance that people perceive as existing in a given situation. Cognitive distance is based on people's personal judgments about the degree of spatial separation between points.

Distance is a fundamental factor in determining real-world relationships and this is a central theme in geography. It was once described as the "first law" of geography: "Everything is related to everything else, but near things are more related than distant things." Waldo Tobler, the geographer who put it this way, is one of many who have investigated the friction of distance, the deterrent or inhibiting effect of distance on human activity. The **friction of distance** is a reflection of the time and cost of overcoming distance.

What these geographers have established is that these effects are not uniform—that is, they are not directly proportional to distance itself. This is true whether distance is measured in absolute terms (i.e., kilometers) or in relative terms (i.e., time- or cost-based measures). The deterrent effects of extra distance tend to lessen as greater distances are involved. Thus, for example, while there is a big deterrent effect in having to travel 2 kilometers instead of 1 to get to a grocery store, the deterrent effect of the same extra distance (1 kilometer) after already traveling 10 kilometers is relatively small.

Distance-Decay and the Utility of Places

This sort of relationship creates what geographers call a distance-decay function. A **distance-decay function** describes the rate at which a particular activity or phenomenon diminishes with increasing distance. Typically, the farther people have to travel, the less likely they are to do so. Distance-decay functions reflect people's behavioral response to opportunities and constraints in time and space. As such, they reflect the

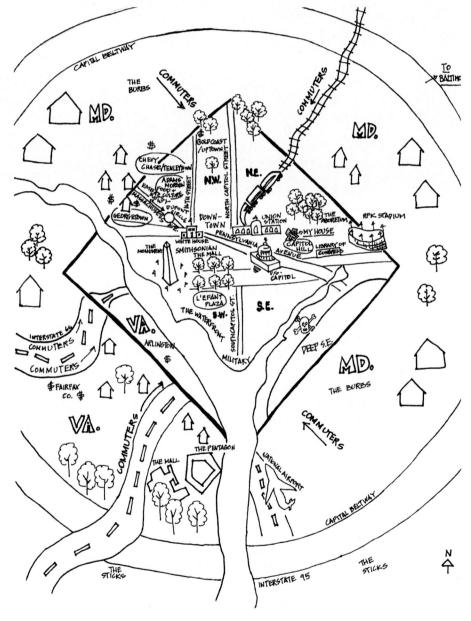

▲ **Figure 1.8 One person's cognitive image of Washington, D.C.** This sketch was drawn by Rasheda DuPree, an urban affairs major at Virginia Tech, as part of a class exercise in recalling locations within students' hometowns. Rasheda has included many of Washington's most prominent landmarks and some of its distinctive districts, including Georgetown. In contrast, there are no recorded locations in the city's southeastern quarter (marked by a skull and crossbones in her sketch) or in the eastern outskirts (marked as "the burbs").

location. Because we seek to maximize the net utility of location, a great deal of human activity is influenced by what University of Washington geographer Richard Morrill once called the "nearness principle." According to this principle—a more explicit version of Tobler's first law—people will seek to

- maximize the overall utility of places at minimum effort;

- maximize connections between places at minimum cost; and

- locate related activities as close together as possible.

The result is that patterns of behavior between people and places come to take on fairly predictable, organized patterns.

APPLY YOUR KNOWLEDGE

1. What is Tobler's law of geography and how does it relate to distance?

2. Provide three examples of the inhibiting effect that distance has on human activity.

Space

Like distance, space can be measured in absolute, relative, and cognitive terms. Human geographers talk about space in various ways. **Absolute space** is a mathematical space described through points, lines, areas, planes, and configurations whose relationships can be fixed precisely through mathematical reasoning. Several ways of analyzing space mathematically are of use to geographers. The conventional way is to view space as a container, defined by rectangular coordinates and measured in absolute units of distance (e.g., kilometers or miles). But some dimensions of space and aspects of spatial organization do not lend themselves to description simply in terms of distance. **Topological space** is measured not in terms of conventional measures of distance, but by the nature and degree of connectivity between locations. The connectivity of people and places is important: whether they are linked, how they are linked, and so on. These attributes determine the flows of people and things (goods, information) and the centrality of places. The map of the Metro system in Milan, Italy (**Figure 1.9**) is a topological map, showing how specific points are joined within a particular network. As most Milanese know, the Metro system gives Duomo a very high degree of connectivity because trains on both the M1 (red) and M3 (yellow) lines stop there. Duomo is therefore relatively central within the "space of flows" of passenger traffic. Missori—nearby in absolute terms and on the M3 line—is much less central, however, and much less the focus of passenger flows.

Socioeconomic space can be described in terms of sites and situations, routes, regions, and distribution patterns. In these terms, spatial relationships are fixed through measures of time, cost, profit, and production, as well as through physical distance. For human geographers, some of the most important aspects of socioeconomic space have to do mapping, analyzing, and understanding the inequalities among places and regions (see Box 1.4: "Spatial Inequality"). Experiential or **cultural space** is the space of people with common ties, described through the places, territories, and settings whose attributes carry special meaning for particular groups. Finally, **cognitive space** is defined and measured in terms of people's values, feelings, beliefs, and perceptions about locations, districts, and regions. Cognitive space can be described, therefore, in terms of behavioral space—landmarks, paths, environments, and spatial layouts.

Accessibility

Because it is a fundamental influence on the utility of locations, distance is an important influence on people's behavior. **Accessibility** is generally defined by geographers in terms of relative location: the opportunity for contact or interaction from a given point or location in relation to other locations. It implies proximity, or nearness, to something. Distance is one aspect of accessibility, but it is by no means the only important aspect.

Connectivity is also an important aspect of accessibility because contact and interaction are dependent on channels of communication and transportation: streets, highways, telephone lines, and wave bands, for example. Effective accessibility is a function not only of distance but also of the configuration of networks of communication and transportation. Commercial airline networks provide many striking examples of this. Cities that operate as airline hubs are much more accessible than cities that are served by fewer flights and fewer airlines. Charlotte, North Carolina, for example (a U.S. Airways hub), is more accessible from Albany, New York, than from Richmond, Virginia, even though Richmond is 400 kilometers (248 miles) closer to Albany than Charlotte. To get to Richmond from Albany, airline passengers must travel to Charlotte or another hub and change—a journey that takes longer and often costs more.

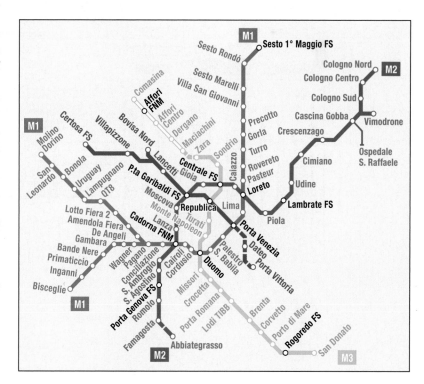

▲ **Figure 1.9 Topological space** The map of the Metro system in Milan, Italy, is a topological map, showing how specific points are joined within a particular network. The most important aspects of networks of any kind, from the geographer's viewpoint, are their connectivity attributes. These attributes determine the flows of people and things (goods, information) and the centrality of places.

1.4 Spatial Inequality

Assessing Spatial Inequality

Identifying the spatial component of socioeconomic inequalities is crucial not only to the study of human geography but also to the improvement of conditions within society as a whole. Whatever the basic causes of inequality, it is clearly a problem that affects locational, as well as occupational, social, and demographic groups, and if we do not expect to discriminate against people on the bases of race, religion, color, or social class, neither should we discriminate against people on the basis of location.

There are many potential ways of assessing spatial inequality, and each chapter of this book will examine a different aspect. It is never simple or straightforward. How well-off we are may, in a capitalist society, be largely dependent on our monetary income and assets, but our real income and wealth depends not only on what money we receive but what we can do with it (what we can buy, and at what price), as well as on those goods that we do not purchase directly, such as fresh air, or a congenial community. Accordingly, geographers and others have attempted to compile composite measures of social well-being, such as the Social Progress Index depicted at the international level in **Figure 1.F**. Note that geographers almost invariably look at the conditions of populations as a whole, as in this example. But we must always remember that in most cases there is likely to be considerable diversity within countries, regions, and cities. Investigating patterns of spatial inequality at different scales of analysis is a necessary step in understanding how they came about, how they are changing, and what their effects may be.

1. What is spatial inequality? Can you think of an example in your everyday life?

2. Pull up the "Social Progress Index" on your computer and scroll over the interactive map. What do you notice about different regions of the world? How is the index rank tabulated?

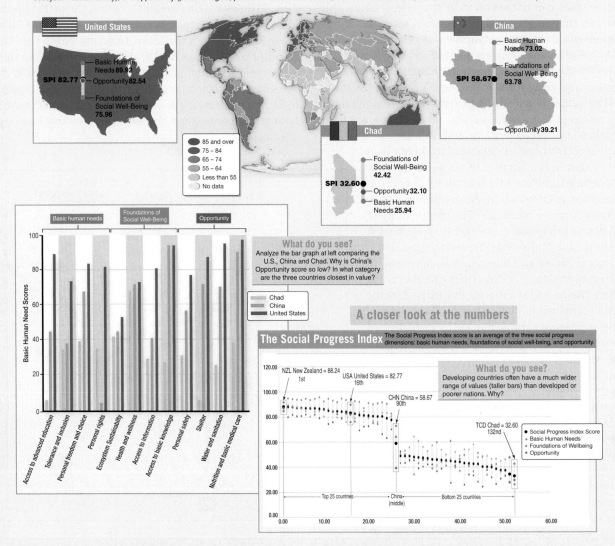

▲ **Figure 1.F Global inequality in 2014** The Social Progress Index is a measure that combines indicators of basic human needs (nutrition and basic medical care, water and sanitation, shelter, and personal safety), the foundations of social well-being (access to basic knowledge, access to information and communications, health and wellness, and ecosystem sustainability), and opportunity (personal rights, personal freedom and choice, tolerance and inclusion, and access to advanced education). *Source:* http://www.socialprogressimperative.org/data/spi

Accessibility is often a function of economic, cultural, and social factors. In other words, relative concepts and measures of distance are often as important as absolute distance in determining accessibility. A nearby facility, such as a health care clinic, is accessible to us only if we can actually afford the cost of getting there, in other words, only if it seems close according to our own standards of distance, if we can afford to use the facility, if we feel that it is socially and culturally acceptable for us to use it, and so on. To take another example, a day care center may be located just a few blocks from a single-parent family, but the center is not truly accessible if it opens after the parent has to be at work, if the cost of care is too high, or if the parent feels that the staff, children, or other parents at the center are from an incompatible social or cultural group.

Spatial Interaction

Interdependence between places and regions can be sustained only through movement and flows. Geographers use the term **spatial interaction** as shorthand for all kinds of movement and flows involving human activity. Freight shipments, commuting, shopping trips, telecommunications, electronic cash transfers, migration, and vacation travel are all examples of spatial interaction. The fundamental principles of spatial interaction can be reduced to four basic concepts: complementarity, transferability, intervening opportunities, and diffusion.

Complementarity For any kind of spatial interaction to occur between two places, there must be a demand in one place and a supply that matches, or complements, it in the other. This complementarity can be the result of several factors. One important factor is the variation in physical environments and resource endowments from place to place. For example, a heavy flow of vacation travel from Swedish cities to Mediterranean resorts is largely a function of climatic complementarity. To take another example, the flow of crude oil from Saudi Arabia (with vast oil reserves) to Japan (with none) is a function of complementarity in natural resource endowments.

A second factor contributing to complementarity is the international division of labor that derives from the evolution of the world's economic systems. The more developed countries of the world have sought to establish overseas suppliers for their food, raw materials, and exotic produce, allowing these more developed countries to specialize in more profitable manufacturing and knowledge-based industries (see Chapter 2). Through a combination of colonialism, imperialism, and sheer economic dominance on the part of the more developed countries, less powerful countries have found themselves with economies that directly complement the needs of the more developed countries. Among the many flows resulting from this complementarity are shipments of sugar from Barbados to the United Kingdom, bananas from Costa Rica and Honduras to the United States, palm oil from Cameroon to France, automobiles from France to Algeria, school textbooks from the United Kingdom to Kenya, and investment capital from the United States to less developed countries.

A third contributory factor to complementarity is specialization and **economies of scale**. Places, regions, and countries can derive economic advantages from the efficiencies created through specialization, which allows for larger-scale operations. Economies of scale are cost advantages to manufacturers in high-volume production; the average cost of production falls with increasing output. Among other things, fixed costs (e.g., the cost of renting or buying factory space, which is the same—fixed—whatever the level of output from the factory) can be spread over higher levels of output so that the average cost per unit of production falls. Economic specialization results in complementarities, which in turn contribute to patterns of spatial interaction. For example, Israeli farmers specialize in high-value fruit and vegetable crops for export to the European Union, which in return exports grains and root crops to Israel.

Transferability Another precondition for interdependence between places is *transferability*, which depends on the frictional (or deterrent) effects of distance. Transferability is a function of two things: the costs of moving a particular item, measured in real money and/or time, and the ability of the item to bear these costs. If, for example, the costs of moving a product from one place to another make it too expensive to sell successfully at its destination, then that product does not have transferability between those places.

Transferability also varies over time, with successive innovations in transport and communications technologies and successive waves of **infrastructure** development (canals, railways, harbor installations, roads, bridges, etc.). New technologies and new or extended infrastructures alter the geography of transport costs and the transferability of particular things between particular places. *As a result, the spatial organization of many different activities is continually changing and readjusting.* The consequent tendency toward a shrinking world gives rise to **time-space convergence**, the rate at which places move closer together in travel or communication time or costs. Time-space convergence results from a decrease in the friction of distance as space-adjusting technologies have, in general, brought places closer together over time (**Figure 1.10**). Overland travel between New York and Boston, for example, has been reduced from 3.5 days (in 1800) to 5 hours (in the 2000s) as the railroad displaced stagecoaches and was in turn displaced by interstate automobile travel. Other important space-adjusting innovations include air travel and air cargo; telegraphic, telephonic, and satellite communications systems; national postal services; package delivery services; and modems, fiber-optic networks, and electronic-mail software.

What is most significant about the latest developments in transport and communication is that they are not only global in scope but also are able to penetrate to local scales. As this penetration occurs, some places that are distant in kilometers are becoming closer together, while some that are close in terms of absolute space are becoming more distant in terms of their ability to reach one another electronically. Much depends on the mode of communication—the extent to which people in different places are "plugged in" to new technologies. The shrinking of space has important implications for people's everyday conceptions of space and distance and for their level of knowledge about other places.

Intervening Opportunity While complementarity and transferability are preconditions for spatial interaction, intervening opportunities are more important in determining the *volume* and *pattern* of movements and flows. Intervening

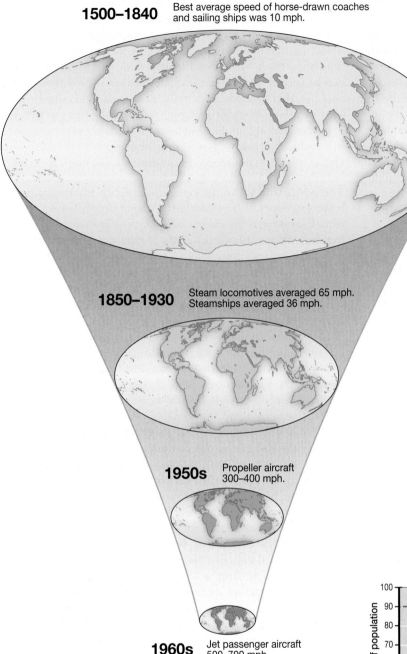

1500–1840   Best average speed of horse-drawn coaches and sailing ships was 10 mph.

1850–1930   Steam locomotives averaged 65 mph. Steamships averaged 36 mph.

1950s   Propeller aircraft 300–400 mph.

1960s   Jet passenger aircraft 500–700 mph.

▲ **Figure 1.10   Time-space convergence** The effects of changing transportation technologies in "shrinking" the world in terms of travel time.

aggregate number of hotel rooms and vacation apartments. We can therefore state the principle of intervening opportunity as follows: Spatial interaction between an origin and a destination will be proportional to the number of opportunities at that destination and inversely proportional to the number of opportunities at alternative destinations.

Spatial Diffusion   Disease outbreaks, technological innovations, political movements, and new musical fads all originate in specific places and subsequently spread to other places and regions. The way that things spread through space and over time—**spatial diffusion**—is one of the most important aspects of spatial interaction and is crucial to an understanding of geographic change.

Diffusion seldom occurs in an apparently random way, jumping unpredictably all over the map. Rather, it occurs as a function of statistical probability, which is often based on fundamental geographic principles of distance and movement. The diffusion of a contagious disease, for example, is a function of the probability of physical contact, modified by variations in individual resistance to the disease. The result is typically a "wave" of diffusion that describes an S-curve, with a slow buildup, rapid spread, and final leveling off (**Figure 1.11**).

It is possible to recognize several spatial tendencies in patterns of diffusion. In *expansion diffusion* (also called *contagion diffusion*), a phenomenon spreads because of the proximity of carriers, or agents of change, who are fixed in their location. An example would

opportunities are alternative origins and/or destinations. Such opportunities are not necessarily situated directly between two points or even along a route between them. Thus, to take one of our previous examples, for Swedish families considering a Mediterranean vacation in Greece, resorts in Spain, southern France, and Italy are all likely to be intervening opportunities because they can probably be reached more quickly and cheaply than resorts in Greece.

The size and relative importance of alternative destinations are important aspects of the concept of intervening opportunity. For our Swedish families, Spanish resorts probably offer the greatest intervening opportunity because they contain the largest

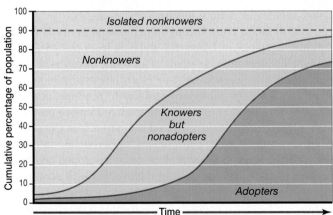

▲ **Figure 1.11   Spatial diffusion** The spatial diffusion of many phenomena tends to follow an S-curve of slow buildup, rapid spread, and leveling off. In the case of the diffusion of an innovation, for example, it usually takes a while for enough potential adopters to find out about the innovation, and even longer for a critical mass of them to adopt it. After that the innovation spreads quite rapidly, until most of the potential adopters have been exposed. (Adapted from D. J. Walmsley and G. J. Lewis, *Human Geography: Behavioral Approaches.* London: Longman, 1984, Fig. 5.3, p. 52.)

be the diffusion of an agricultural innovation, such as the use of hybrid seed stock, among members of a local farming community.

With *hierarchical diffusion* (also known as cascade diffusion), a phenomenon can be diffused from one location to another without necessarily spreading to people or places in between. An example would be the spread of a fashion trend from large metropolitan areas to successively smaller cities, towns, and rural settlements.

Many patterns and processes of diffusion reflect both expansion and hierarchical diffusion, as different aspects of human interaction come into play in different geographic settings. The diffusion of outbreaks of communicable diseases, for example, usually involves a combination of expansion and hierarchical diffusion.

APPLY YOUR KNOWLEDGE

1. What is a spatial interaction?

2. Referring to spatial analysis concepts, discuss a national or international environmental issue. Provide examples of how complementarity, transferability, intervening opportunities, and diffusion each relate to the issue you have chosen.

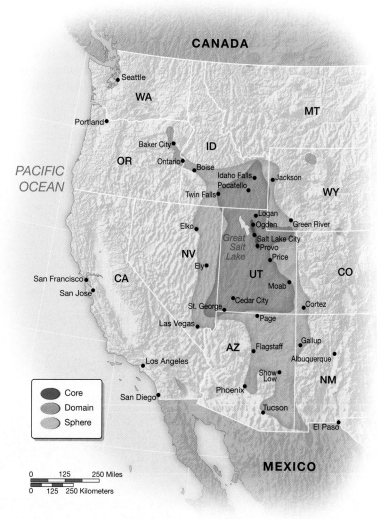

REGIONAL ANALYSIS

Not all geographic phenomena are most effectively understood through spatial analysis. Geographers also seek to understand the complex relationships between peoples and places in terms of the similarities and differences among and between them and the identities and qualities associated with them. Here the key concepts are regionalization, landscape, and sense of place.

Regionalization

The geographer's equivalent of *scientific classification* is **regionalization**, with individual places or areal units being the objects of classification. Geographers are especially interested in **functional regions** (sometimes referred to as nodal regions)—regions that, while they may exhibit some variability in certain attributes, share an overall coherence in structure and economic, political, and social organization. The coherence and distinctive characteristics of a region are often stronger in some places than in others. This point is illustrated by geographer Donald Meinig's *core-domain-sphere* model, which he set out in his classic essay on the Mormon region of the United States (**Figure 1.12**). In the core of a region the distinctive attributes are very clear; in the domain they are dominant but not to the point of exclusivity; in the sphere they are present but not dominant.

People's own conceptions of place, region, and identity may generate strong feelings of regionalism and sectionalism that feed back into the processes of place-making and regional differentiation. **Regionalism** is a term used to describe situations in which different religious or ethnic groups with distinctive identities coexist within the same state boundaries, often concentrated within a particular region and sharing strong feelings of collective identity. If such feelings develop into an extreme devotion to regional interests and customs, the condition is known as **sectionalism**. Regionalism often involves ethnic groups whose aims include autonomy from a national state and the development of their own political power (see Chapter 10). In certain cases, enclaves of ethnic minorities are claimed by the government of a country other than the one in which they reside. **Irredentism** is the assertion by the government of a country that a minority living outside its formal borders belongs to it historically

◄ Figure 1.12 **The Mormon culture region** Cultural attributes often gradually shade from one region to another, rather than having a single, clear-cut boundary. Geographer Donald Meinig's work on the Mormon culture region of the United States identified a "core" region, which exhibits all the attributes of Mormon culture; a "domain," where not all these attributes may be present (or may be less intense); and a "sphere," where some attributes of Mormon culture are present but often as a minority. (Adapted from D. Meinig, "The Mormon Culture Region: Strategies and Patterns in the Geography of the American West," *Annals, Association of American Geographers*, 55, 1965, pp. 191–220.)

and culturally. In some circumstances, as with Russia's claims on Russian-speaking enclaves in eastern Ukraine in 2014, irredentism can lead to armed conflict.

Landscape

Geographers think of landscape as a comprehensive product of human action such that every landscape is a complex repository of society. It is a collection of evidence about our character and experience, our struggles and triumphs as humans. To better understand the meaning of landscape, geographers have developed different categories of landscape types based on the elements contained within them.

Ordinary landscapes (or vernacular landscapes, as they are sometimes called) are the everyday landscapes that people create in the course of their lives together. From crowded city centers to leafy suburbs and quiet rural villages, landscapes are lived in and changed and influence and change the perceptions, values, and behaviors of the people who live and work in them.

Symbolic landscapes, by contrast, represent particular values or aspirations that builders and financiers want to impart to a larger public. For example, the neoclassical architecture of the buildings of the federal government in Washington, D.C., along with the streets, parks, and monuments of the capital, constitute a symbolic landscape intended to communicate a sense of power but also of democracy in its imitation of the Greek city-state.

Some landscapes become powerfully symbolic of national identity. Nation-building depends heavily on stories of golden ages, enduring traditions, heroic deeds, shared hardships, and dramatic destinies, all located in traditional (or promised) homelands with hallowed sites and scenery. Landscapes thus become a way of picturing a nation. With the creation of modern Italy during the *Risorgimento* ("revival through unification"—1815–1861), for example, the classical Tuscan landscape (**Figure 1.13**) became emblematic of Italy itself and has attracted landscape painters, romantic poets, and novelists ever since. Similarly, the West of Ireland (**Figure 1.14**) came to symbolize the whole of Ireland to Irish nationalists in the early twentieth century—partly because it was seen as the region least affected by British colonization, but also because its bare and rugged landscape seemed to contrast so strikingly with the more bucolic rural landscapes (**Figure 1.15**) by which England was popularly imagined.

Geographers now recognize that there are many layers of meaning embedded in the landscape. Landscapes reflect people's dreams and ideas as well as their material lives. The messages embedded in landscapes can be read as signs about values, beliefs, and practices, though not every reader will take the same message from a particular landscape (just as people may differ in their interpretation of a passage from a book). In short, landscapes both produce and communicate meaning, and one of our tasks as geographers is to interpret those meanings.

Sense of Place

Everyday routines experienced in familiar settings allow people to derive a pool of shared meanings. Often this carries over into people's attitudes and feelings about themselves and their locality. When this happens, the result is a

▼ **Figure 1.13 The power of place** In some countries, particular landscapes have become powerfully symbolic of national identity. In Italy it is the classical landscape of Tuscany, with its scattered farms and villas, elegant cypress trees, silvery-green olive trees, and rolling fields with a rich mixture of cereals, vegetables, fruit trees, and vines.

▲ **Figure 1.14 The national landscape of Ireland** The rugged landscapes of the West of Ireland have come to symbolize the whole country to many people, both within Ireland and beyond. This photograph shows part of Ballinacregga, on Inishmore.

▼ **Figure 1.15 Welford on Avon, England** The well-ordered and picturesque landscape of the southern parts of rural England have long been taken to be emblematic of England as a whole and of the values and ideals of its people—even though urban and industrial development, together with modern agricultural practices, have brought about significant changes to both landscapes and society.

self-conscious sense of place. A **sense of place** refers to the feelings evoked among people as a result of the experiences and memories they associate with a place and to the symbolism they attach to that place. It can also refer to the character of a place as seen by outsiders: its distinctive physical characteristics and/or its inhabitants.

For *insiders*, this sense of place develops through shared dress codes, speech patterns, public comportment, and so on. A crucial concept here is that of the **lifeworld**, the taken-for-granted pattern and context for everyday living through which people conduct their day-to-day lives without conscious attention. People become familiar with one another's vocabulary, speech patterns, dress codes, gestures, and humor as a result of routine encounters and shared experiences in bars and pubs, cafés and restaurants, shops and street markets, and parks. This is known as **intersubjectivity**: shared meanings that are derived from everyday practice. Elements of daily rhythms (such as mid-morning grocery shopping with a stop for coffee, the *aperitivo* en route from work to home, and the after-dinner stroll) are all critical to the intersubjectivity that is the basis for a sense of place within a community (**Figure 1.16**). The same is true of elements of weekly rhythms, such as street markets and farmers' markets, and of seasonal rhythms, such as festivals.

These rhythms, in turn, depend on certain kinds of spaces and places: not only streets, squares, and public open spaces but also "third places" (after home, first, and workplace, second): the sidewalk cafés, pubs, post offices, drugstores, corner stores, and family-run *trattoria* that are the loci of routine activities and sociocultural transactions. Third places accommodate "characters," "regulars," and newcomers, as well as routine patrons and, like public spaces, facilitate casual encounters as well as settings for sustained conversations (**Figure 1.17**). The

nature and frequency of routine encounters and shared experiences depend a great deal on the attributes of these spaces and places.

A sense of place also develops through familiarity with the history and symbolism of particular elements of the physical environment—a mountain or lake, the birthplace of someone notable, the site of some particularly well-known event, or the expression of community identity through art. Sometimes it is deliberately fostered by the construction of symbolic structures such as monuments and statues. Often it is a natural outcome of people's familiarity with one another and their surroundings.

Because of this consequent sense of place, insiders feel at home and "in place."

For *outsiders,* a sense of place can be evoked only if local landmarks, ways of life, and so on are distinctive enough to evoke a significant common meaning for people who have no direct experience of them. Central London, for example, is a setting that carries a strong sense of place to outsiders who have a sense of familiarity with the riverside panoramas, busy streets, and distinctive monuments and historic buildings.

▲ **Figure 1.16 Intersubjectivity** Routine encounters such as this, in Chiavenna, Italy, help to develop a sense of community and a sense of place among residents.

APPLY YOUR KNOWLEDGE

1. What is the difference between an "ordinary" landscape and a "symbolic" landscape?

2. What are the most distinctive characteristics of the region in which you live? How would you describe its landscapes? What, from your perspective, gives your community a sense of place?

◄ **Figure 1.17** A 'third place' The Five Bells pub, in Spitalfields, London.

DEVELOPING A GEOGRAPHICAL IMAGINATION

A **geographical imagination** allows us to understand changing patterns, processes, and relationships among people, places, and regions. Developing this capacity is increasingly important as the pace of change around the world increases to unprecedented levels.

It is often useful to think of places and regions as representing the cumulative legacy of successive periods of change. Following this approach, we can look for superimposed layers of development, or evidence of the imprint of different phases of local development (see **Box 1.3**, "Windows on the World: South Beach, Miami Beach"). We can show how some patterns and relationships last, while others are modified or obliterated. We can show how different places bear the imprint of different kinds of change, perhaps in different sequences and with different outcomes. To do so, we must be able to identify the kinds of changes that are most significant.

We can prepare our geographical imagination to deal with an important aspect of spatial change by making a distinction between the *general* and the *unique.* This distinction helps account for geographical diversity and variety because it provides a way of understanding how and why one kind of change can result in a variety of spatial outcomes. Because the *general effects* of a particular change always involve some degree of modification as they are played out in different environments, *unique outcomes* result.

Although we can usually identify some general outcomes of major episodes of change, there are almost always some unique outcomes, too. Let us take two related examples. The Industrial Revolution of nineteenth-century Europe provides a good example of a major period of change. A few of the general spatial outcomes were increased urbanization, regional specialization in production, and increased interregional and international trade. At one level, places could be said to have become increasingly alike: generic coalfield regions, industrial towns, ports, downtowns, worker housing, and suburbs.

It is clear, however, that these general outcomes were mediated by the different physical, economic, cultural, and social attributes of different places. Beneath the dramatic overall changes in the geography of Europe, new layers of diversity and variety also existed. Industrial towns developed their own distinctive character as a result of their manufacturing specialties, their politics, the personalities and objectives of their leaders, and the reactions and responses of their residents. Downtowns were differentiated from one another as the general forces of commerce and land economics played out across different physical sites and within different patterns of land ownership. Various local socioeconomic and political factors gave rise to different expressions of urban design. Meanwhile, some places came to be distinctive because they were almost entirely bypassed by this period of change, their characteristics making them unsuited to the new economic and spatial order (**Figure 1.18**).

The second example of general and unique outcomes of change is the introduction of the railroad, one of the specific changes involved in the Industrial Revolution. In general terms, the railroad contributed to time-space convergence, to the reorganization of industry into larger market areas, to an increase in interregional and international trade, and to the interconnectedness of urban systems. Other unique outcomes, however, have also contributed to distinctive regional geographies. In Britain, the railroad was introduced to an environment that was partially industrialized and densely settled. The increased efficiencies provided by the railroad helped to turn Britain's economy into a highly integrated and intensively urbanized national economy. In Spain, however, the railroad was introduced to an environment that was less urbanized and industrialized and less able to afford the costs of railroad construction. The result was that the relatively few Spanish towns connected by the railroads gained a massive comparative advantage. As a result, Spain's space-economy was much less integrated than Britain's, with an urban system dominated by just a few towns and cities.

▲ **Figure 1.18 Hersbrück, Germany** Hersbrück was once a prosperous regional center on an overland trade route—the "Golden Road"—between Nuremberg and Prague. After 1806, when Napoleon redrew the political map of Europe, the reorganization of the European economy, together with the onset of the Industrial Revolution, left Hersbrück somewhat isolated and economically disadvantaged. Hersbrück was never drawn into the industrial development of Germany and is not well connected to the transportation infrastructure of canals, railways, or major highways.

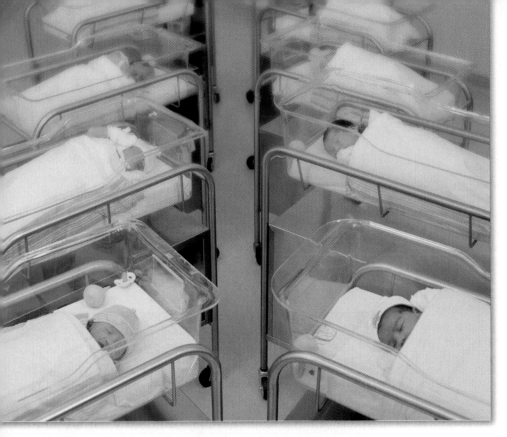

◀ **Figure 1.19 The demographic future** We can predict a great deal about many aspects of human geography in the next few decades because we know about the geography of population growth.

APPLY YOUR KNOWLEDGE

1. What is a "general" change and a "unique" outcome? Think of an example that has not been discussed.

2. Describe the region in which you live, emphasizing the imprint of different periods of development. Which features of the region can be said to be the result of general spatial effects, and which are unique?

FUTURE GEOGRAPHIES

Places and regions are constantly in a state of change. This raises the question of what we may expect to see in the future. An understanding of geographic processes and principles, together with a knowledge of past and present spatial patterns, allows us to make informed judgments about future geographies: an important dimension of applying our geographical imagination.

Whereas much of the world had remained virtually unchanged for decades, even centuries, the Industrial Revolution and long-distance, high-speed transportation and communications brought a rapid series of rearrangements to the countryside and to towns and cities in many parts of the world. Today, with a globalized economy and global telecommunications and transportation networks, places have become much more interdependent, and still more of the world is exposed to increasingly urgent imperatives to change.

Will social networking sites bring about new patterns of human interaction? Will we be able to cope with the environmental stresses of increasing industrialization and rapid population growth? Will the United States retain its position as the world's most powerful and influential nation? What kind of problems will the future bring for local, regional, and international development? What new technologies are likely to have the most impact in reshaping human geographies? Will globalization undermine regional cultures? These are just a few of the many questions that spring from the key themes in human geography.

As we begin to look to the future, we can appreciate that some dimensions of human geography are more certain than others. We can only guess, of course, at some aspects of the future. Two of the most speculative realms are those of politics and technology, which are both likely to spring surprises at any time. On the other hand, in some ways the future is already here, embedded in the world's institutional structures and in the dynamics of its populations (**Figure 1.19**). We know, for example, a good deal about the demographic trends of the next quarter century, given present populations, birth and death rates, and so on. We also know a good deal about the distribution of environmental resources and constraints, about the characteristics of local and regional economies, and about the legal and political frameworks within which geographic change will probably take place. The tools and concepts of human geography allow us to understand change in terms of local place-making processes that are subject to certain broader principles of spatial organization as well as the overall framework of the global economy. In subsequent chapters, as we look more closely at specific aspects of human geography, we shall be able to see how geographic process and principles can suggest future patterns and pathways of change.

APPLY YOUR KNOWLEDGE

1. How has your space changed already through time? Think about how you use different technologies to communicate, travel, and map out your everyday routine. What is different about today from five years ago?

2. Make a list of likely future changes to the geography of the region in which you live.

◼ CONCLUSION

Human geography is the systematic study of the location of peoples and human activities across Earth's surface and of their relationships to one another. An understanding of human geography is important both from an intellectual point of view (i.e., understanding the world around us) and a practical point of view (e.g., contributing to environmental quality, human rights, social justice, business efficiency, political analysis, and government policymaking).

Human geography reveals how and why geographical relationships matter in terms of cause and effect in relation to economic, social, cultural, and political phenomena. Human geographers strive to recognize these wider processes and broad geographical patterns without losing sight of the uniqueness of specific places.

Geography matters because it is in specific places that people learn who and what they are and how they should think and behave. Places are also a strong influence, for better or worse, on people's physical well-being, their opportunities, and their lifestyle choices. Places also contribute to peoples' collective memory and become powerful emotional and cultural symbols. Places are the sites of innovation and change, of resistance and conflict.

To investigate specific places we must be able to frame our studies of them within the compass of the entire globe. This is important for two reasons. First, the world consists of a complex mosaic of places and regions that are interrelated and interdependent in many ways. Second, place-making forces—especially economic, cultural, and political forces that influence the distribution of human activities and the character of places—are increasingly operating at global and international scales. In Chapter 2, we describe the changing global context that has shaped places and regions around the world.

LEARNING OUTCOMES REVISITED

- **Explain how the study of geography has become essential for understanding a world that is more complex, interdependent, and changing faster than ever before.**

  Geography matters because it enables us to understand where we are both literally and figuratively. Geography provides an understanding of the interdependency of people and places and an appreciation of how and why certain places are distinctive or unique. From this knowledge, we can begin to understand the implications of future spatial patterns.

- **Identify four examples of how places influence inhabitants' lives.**

  Specific places provide the settings for people's daily lives. Places are settings for social interaction that, among other things, structure the daily routines of people's economic and social lives; provide both opportunities for—and constraints on—people's long-term social well-being; establish a context in which everyday commonsense knowledge and experience are gathered; provide a setting for processes of socialization; and provide an arena for contesting social norms.

- **State the differences among major map projections and describe their relative strengths and weaknesses.**

  The choice of map projection depends largely on the purpose of the map. Equidistant projections allow distance to be represented as accurately as possible but in only one direction (usually north-south). Conformal projections render compass directions accurately but tend to exaggerate the size of northern continents. Equivalent projections portray areas on Earth's surface in their true proportions but result in world maps on which many locations appear squashed and have unsatisfactory outlines.

- **Explain how geographers use geographic information systems (GIS) to merge and analyze data.**

  New technologies combine high-performance computing, global positioning systems (GPS), and computerized record keeping. The most important aspect of these technologies, from an analytical point of view, is that they allow data from several different sources, on different topics and at different scales, to be merged.

- **Summarize the five concepts that are key to spatial analysis and describe how they help geographers to analyze relationships between peoples and places.**

  The study of many geographic phenomena can be approached in terms of their arrangement as points, lines, areas, or surfaces on a map. This is known as spatial analysis. Location, distance, space, accessibility, and spatial interaction are five concepts that are key to spatial analysis. Each of these concepts is multifaceted and can be applied to different spatial scales. Together, they provide a powerful set of tools for describing and analyzing places and regions.

- **Describe the importance of distance in shaping human activity.**

  The first law of geography is that "Everything is related to everything else, but near things are more related than are distant things." Human activity is influenced by the "nearness principle," according to which people tend to seek to maximize the overall utility of places at minimum effort, to maximize connections between places at minimum cost, and to locate related activities as close together as possible.

- **Summarize the three concepts that are key to regional analysis and explain how they help geographers analyze relationships between peoples and places.**

  The key concepts of regional analysis are regionalization, landscape, and sense of place. Regionalization is the geographer's equivalent of scientific classification; landscapes embody many layers of meaning and reflect the influence of past processes of change, while sense of place derives from everyday routines experienced in familiar settings. Geographers also seek to understand the complex relationships between peoples and places in terms of the similarities and differences among and between them and the identities and qualities associated with them.

KEY TERMS

absolute space (p. *17*)
accessibility (p. *17*)
cognitive distance (p. *16*)
cognitive image (p. *14*)
cognitive space (p. *17*)
cultural space (p. *17*)
distance-decay function (p. *16*)
economies of scale (p. *19*)
friction of distance (p. *16*)
functional region (p. *21*)
geodemographic research (p. *12*)
geographical imagination (p. *25*)
geographic information systems (GIS) (p. *12*)
Global Positioning System (GPS) (p. *13*)

human geography (p. *6*)
identity (p. *5*)
infrastructure (p. *19*)
intersubjectivity (p. *23*)
irredentism (p. *21*)
latitude (p. *13*)
lifeworld (p. *23*)
longitude (p. *13*)
map projections (p. *11*)
map scale (p. *10*)
ordinary landscapes (p. *22*)
physical geography (p. *6*)
place (p. *3*)
region (p. *3*)
regional geography (p. *6*)

regionalism (p. *21*)
regionalization (p. *21*)
remote sensing (p. *7*)
sectionalism (p. *21*)
sense of place (p. *23*)
site (p. *13*)
situation (p. *13*)
socioeconomic space (p. *17*)
spatial analysis (p. *13*)
spatial diffusion (p. *20*)
spatial interaction (p. *19*)
symbolic landscapes (p. *22*)
time-space convergence (p. *19*)
topological space (p. *17*)
utility (p. *16*)

REVIEW & DISCUSSION

1. Consider the term *symbolic landscapes.* Identify and list five examples of symbolic landscapes in your town (see **Figures 1.13–1.15**). For each of the landscapes, interpret the values, beliefs, and aspirations that it embodies. If necessary, find a picture of the specific landscape to aid with your observations.

2. Decide on a location that all members of your group are familiar with, such as your downtown, city hall, or student union. As individuals, take 10 minutes to make a cognitive sketch of the agreed-upon location (see **Figure 1.8** for an example of a cognitive sketch). Come back together as a group and compare your drawings. How did the cognitive sketches in your group differ? How were they similar? List five things that are most striking about your sketches.

3. Consider the features that give your specific location a sense of place by creating a list of at least ten things. Make sure your list is specific; take into consideration thinks like vocabulary, speech patterns, clothing, and jokes that are trending, as well as physical geography. Once this is complete, consider how an "outsider" would interact with and understand your place.

UNPLUGGED

1. Consider spatial interaction from the point of view of your own life. Take an inventory of food you consume in a day, noting, where possible, the location where each item of food was produced, bought, and consumed. Was the store where you bought the food part of a regional, national, or international chain? Where was the food processed? How far did it travel from point of production to the grocery store to your plate? How would you go about representing the journey your food traveled on a thematic map?

2. Describe, as exactly and concisely as possible, the site (see p. 13) of your campus. Then describe its situation (see p. 13). Think of three reasons why the campus is sited and situated where it is. Would there be a better location in your community or further afield? If so, why?

3. Choose a local landscape, one with which you are familiar, and write a short essay (500 words, or two double-spaced typed pages) about how the landscape has evolved over time. Note especially any evidence that physical environmental conditions have shaped any of the human elements in the landscape, as well any evidence of people having modified the physical landscape.

4. **Figures 1.13–1.15** show examples of landscapes that have acquired a strong symbolic value because of the buildings, events, people, or histories with which they are associated. Find five photographs of landscapes that have strong symbolic value to a large number of citizens of your own region or country, and state in 25 words or less why each setting has acquired such value.

DATA ANALYSIS

Prezi.com
http://goo.gl/tCtGxt

In this chapter, the concepts of place, identity and the dynamics of time and space were introduced. Seeing the world from a "geographic" perspective is a view relevant to everyone, and allows us to analyze how place impacts our daily lives. American urban planner Kevin Lynch once commented, "Every citizen has long-standing associations with some part of his city and this image is soaked in memories and meanings." And geographer Doreen Massey said, "Really, thinking spatially means looking beyond ourselves, a recognition of others." Interrogate those ideas by creating a conceptual map of your identity from the spaces you have occupied—the places you were born, grew up, and now study and work. To start, use a conceptual mapping program like Prezi (http://prezi.com). Chart your personal story alongside your geography and the relationships you have formed from those places. Pull media like photos, maps, videos, and music into your conceptual map and answer the following questions:

1. Where were you born? What languages and/or accents are spoken there? Were you born in a hospital?

2. What neighborhood did you live in when you started school? Who were your friends? Where did you spend your summers? Did you travel? How far did you go and what forms of transportation did you use to get there?

3. Where did you attend high school? Did you play sports, work, participate in drama and art scenes? Where did you "hang out"? Did you have a skate board route, a bike path, a swimming pool, or a favorite club or music scene that was part of your routine? How did any of these places and activities contribute to your identity?

4. Who are the influential people in your life? Where do they live? How did you meet them? What did they teach/mentor/share with you that impacted you as a person?

5. Where are you now? Are you attending school close to your city of origin or in a completely new country? What places are important to your education? Where do you relax? Do you have important relationships that are long distance?

6. Finally, where do you want to go? What are the countries and/or cities you want to visit or live? Where do you want to work and build a family/life? Why are these places desirable to you?

Mastering Geography™

Looking for additional review and test prep materials? Visit the Study Area in MasteringGeography™ to enhance your geographic literacy, spatial reasoning skills, and understanding of this chapter's content by accessing a variety of resources, including **MapMaster** interactive maps, Videos, *In the News* RSS feeds, flashcards, web links, self-study quizzes, and an eText version of *Human Geography*.

LEARNING OUTCOMES

- *Summarize* the distinctive stages of the evolution of the modern world-system.

- *Analyze* how and why the new technologies of the Industrial Revolution helped bring about the emergence of a global economic system.

- *Examine* the changing patterns of interdependence among different world regions.

- *Compare* the three tiers that constitute the modern world-system.

- *Explain* how the growth and internal development of the world's core regions could take place only with the foodstuffs, raw materials, and markets provided by the colonization of the periphery.

- *Identify* an example of each of the four key issues caused by globalization—environmental, health, core-periphery disparity, and security issues.

2

THE CHANGING GLOBAL CONTEXT

▲ Yuyuan Garden in Shanghai, a sixteenth-century Ming Dynasty garden, adjacent to the city's modern business district.

The story is told of a little Korean girl who arrives in Los Angeles, sees a McDonald's restaurant, tugs at her mother's sleeve, and says, "Look, Mother, they have McDonald's in this country, too."

It has become a cliché about the twenty-first century that everywhere will come to look like everywhere else, with the same McDonalds, Pizza Hut, and Kentucky Fried Chickens, the same television programming with Hollywood movies and TV series, and the same malls selling the same Nike shoes, Philips electronics, and GAP clothing. Another cliché is that instantaneous global telecommunications, satellite television, and the Internet will soon overthrow all but the last vestiges of geographical differentiation in human affairs. Large corporations, according to this view, will no longer have strong ties to their home country, scattering their activities around the world in search of low-cost, low-tax locations. Employees will work as effectively from home, car, or beach as they could in the offices that need no longer exist. Events halfway across the world will be seen, heard, and felt with the same immediacy as events across town. National differences and regional cultures will dissolve, the cliché has it, as a global marketplace brings a uniform dispersion of people, tastes, and ideas. Such developments are, in fact, highly unlikely. Even in the Information Age, geography will still matter and may well become more important than ever. Places and regions will undoubtedly change as a result of the new global context of the Information Age. But geography will still matter because of several factors: transport costs, different resource endowments, fundamental principles of spatial organization, people's territorial impulses, the resilience of local cultures, and the legacy of the past.

In this chapter, we take a long-term, big-picture, perspective on changing human geographies, emphasizing the continuing interdependence among places and regions. We show how geographical divisions of labor have evolved with the growth of a worldwide system of trade and politics and with the changing opportunities provided by successive technology systems. As a result of this evolution, the world is now structured around a series of core regions, semiperipheral regions, and peripheral regions; and globalization seems to be intensifying many of the differences among places and regions, rather than diminishing them.

THE PREMODERN WORLD

The essential foundation for human geography is an ability to understand places and regions as components of a constantly changing global system. In this sense, all geography is historical geography. Built into every place and each region is the legacy of major changes in world geography. The world is an evolving, competitive, political-economic system that has developed through successive stages of geographic expansion and integration. This evolution has affected the roles of individual places in different ways. It has also affected the nature of the interdependence among places. This explains why places and regions have come to be distinctive and how this distinctiveness has formed the basis of geographic variability. To understand the sequence of major changes in human geography, we need to begin with the hearth areas of the first agricultural revolution.

Hearth Areas

The first agricultural revolution involved a transition from hunter-gatherer groups to agricultural-based **minisystems** that were both more extensive and more stable. A minisystem is a society with a *reciprocal* social economy. That is, each individual specializes in particular tasks (e.g., tending animals, cooking, or making pottery) and freely gives any excess product to others. The recipients reciprocate in turn by giving up the surplus product of their own specialization. Such societies are found only in subsistence-based economies. Because they do not have (or need) an extensive physical infrastructure, minisystems are limited in geographic scale.

The transition to minisystems began in the Proto-Neolithic (or early Stone Age) period, between 9000 and 7000 B.C.E., and was based on a series of technological preconditions: the use of fire to process food, the use of grindstones to mill grains, and the development of improved tools to prepare and store food. The key breakthrough was the evolution and diffusion of a system of slash-and-burn agriculture (also known as "swidden" cultivation—see Chapter 9). **Slash-and-burn** is a system of cultivation in which plants are harvested close to the ground, the stubble left to dry for a period, and then ignited. The burned stubble provides fertilizer for the soil. Another important breakthrough was the domestication of cattle and sheep, a technique that had become established in a few regions by Neolithic times.

These agricultural breakthroughs could take place only in certain geographic settings: where natural food supplies were plentiful; where the terrain was diversified (thus offering a variety of habitats and species); where soils were rich and relatively easy to till; and where there was no need for large-scale irrigation or drainage. Archaeological evidence suggests that the breakthroughs took place independently in several agricultural hearth areas and that agricultural practices diffused slowly outward from each (**Figure 2.1**). **Hearth areas** are geographic settings where new practices have developed and from which they have spread. The main agricultural hearth areas were situated in four broad regions:

- **In the Middle East:** in the so-called Fertile Crescent around the foothills of the Zagros Mountains (parts of present-day Iran and Iraq), along the floodplains of the Tigris and Euphrates rivers, around the Dead Sea Valley (Jordan and Israel), and on the Anatolian Plateau (Turkey).

- **In South Asia:** along the floodplains of the Ganga (Ganges), Brahmaputra, Indus, and Irrawaddy rivers (Assam, Bangladesh, Burma, and northern India).

- **In China:** along the floodplain of the Huang He (Yellow) River.

- **In the Americas:** in Mesoamerica (the middle belt of the Americas that extends north to the North American Southwest and south to the Isthmus of Panama) around Tamaulipas and the Tehuacán Valley (Mexico), in Arizona and New Mexico, and along the western slopes of the Andes in South America.

The transition to food-producing minisystems had several important implications for the long-term evolution of the world's geographies:

1. It allowed much higher population densities and encouraged the proliferation of settled villages.

2. It brought about a change in social organization, from loose communal systems to systems that were more highly organized on the basis of kinship. Kin groups provided a natural way of assigning rights over land and resources and of organizing patterns of land use.

3. It allowed some specialization in nonagricultural crafts, such as pottery, woven textiles, jewelry, and weaponry.

4. Specialization led to a fourth development: the beginnings of barter and trade between communities, sometimes over substantial distances.

Most minisystems vanished a long time ago, although some remnants have survived to provide material for Discovery Channel and National Geographic specials. Examples of these residual and fast-disappearing minisystems are the bushmen of the Kalahari, the hill tribes of Papua New Guinea, and the tribes of the Amazon rain forest. They contribute powerfully to regional differentiation and sense of place in a few enclaves around the world, but their most important contribution to contemporary human geographies is that they provide a stark counterpoint to the landscapes and practices of the rest of the contemporary world.

APPLY YOUR KNOWLEDGE

1. What are some of the features of the hearth areas that allowed minisystems to thrive?

2. List and describe three examples of traditional crafts that were originally developed in the agricultural hearth areas of Arizona and New Mexico.

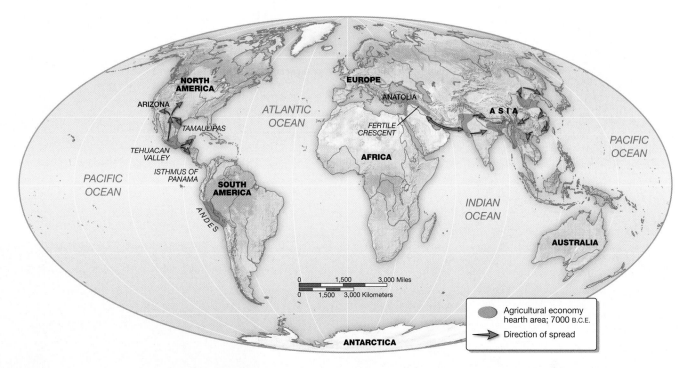

▲ **Figure 2.1 Hearth areas** The first agricultural revolution took place independently in several hearth areas where hunter-gatherer communities experimented with locally available plants and animals, eventually leading to their domestication. From these hearth areas, improved strains of crops, domesticated animals, and new farming techniques diffused slowly outward. Farming supported larger populations than hunting and gathering could, and the extra labor developed other specializations, such as making pottery and jewelry. The agricultural hearth area in Central America gave rise to the Aztec and Mayan world-empires; the hearth areas of the Andes in South America developed into the Inca world-empire.

The Growth of Early Empires

The higher population densities, changes in social organization, craft production, and trade brought about by the first agricultural revolution provided the preconditions for the emergence of several "world-empires." A **world-empire** is a group of minisystems that have been absorbed into a common political system. In world-empires, wealth flows from producer classes to an elite class in the form of taxes or tribute. This redistribution of wealth is most often achieved through military coercion, religious persuasion, or a combination of the two. The best-known world-empires were the largest and longest lasting of the ancient civilizations—Egypt, Greece, China, Byzantium, and Rome (**Figure 2.2a**). These world-empires brought important new elements to the evolution of the world's geographies. One was the emergence of *colonization*; the other was *urbanization* (see Chapter 10).

Colonization **Colonization**—the physical settlement in a new territory of people from a colonizing state—was in part an indirect consequence of the operation of the **law of diminishing returns**. This law refers to the tendency for productivity to decline after a certain point with the continued addition of capital and/or labor to a given resource base. World-empires could support growing populations only if overall levels of productivity could be increased. While some productivity gains could be achieved through better agricultural practices, harder work, and improvements in farm technology, a fixed resource base meant that as populations grew, overall levels of productivity fell. For each additional person working the land, the gain in production per worker was less.

The usual response of empire builders to these diminishing returns was to enlarge the resource base by colonizing nearby land. This colonization had immediate spatial consequences in terms of establishing dominant/subordinate spatial relationships between world-empires and colonies. Colonization played a role in establishing hierarchies of settlements and creating improved transportation networks as well. The military underpinnings of colonization also meant that new towns and cities came to be carefully sited for strategic and defensive reasons.

The legacy of these important changes is still apparent in today's landscapes. The clearest examples are in Europe, where the Roman world-empire colonized an extensive territory that was controlled through a highly developed system of towns and connecting roads. Most of today's important Western European cities had their origins as Roman settlements. In quite a few, it is possible to trace the original street layouts. In some, it is possible to glimpse remnants of Roman defensive city walls, paved streets, aqueducts, viaducts, arenas, sewage systems, baths, and public buildings (**Figure 2.2b and 2.2c**). In the modern European countryside, we can still read the legacy of the Roman world-empire in arrow-straight roads built by Roman engineers and maintained by successive generations.

▲ **Figure 2.2 The Premodern World** (a) This map shows the distribution of the Greek city-states and Carthaginian colonies and the spread of the Roman empire from 218 B.C.E. to 117 C.E. (Source: Adapted from R. King et al., The Mediterranean. London: Arnold, 1977, pp. 59 and 64.)

(b) Many of the roads that were built by Romans became established as major routes throughout Europe. This photo shows the cardos maximus (main road) at Tharros in Sardinia, Italy.

(c) The Romans were great engineers. This photograph shows the Roman aqueduct in Segovia, Spain.

Some colonial world-empires were exceptional in that they were based on a particularly strong central state, with totalitarian rulers who were able to organize large-scale communal land-improvement schemes. These world-empires were found in China, India, the Middle East, Central America, and the Andean region of South America. Most of them relied heavily on slave labor. Their development of large-scale land-improvement schemes (particularly irrigation and drainage schemes) as the basis for agricultural productivity has led some scholars to characterize them as *hydraulic societies*. Today, their legacy can be seen in the landscapes of terraced fields in places like Sikkim, India, and East Java, Indonesia (**Figure 2.3**).

Early Geographers These early world-empires were also significant in developing a base of geographic knowledge. Greek scholars, for example, developed the idea that places embody fundamental relationships between people and the

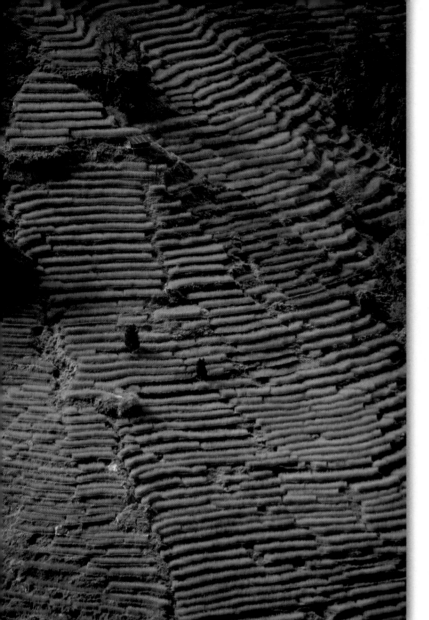

▲ **Figure 2.3** Terraced rice fields in Sikkim, India.

natural environment and that the study of geography provides the best way of addressing the *interdependencies* between places and between people and nature. The Greeks were also among the first to appreciate the practical importance and utility of geographic knowledge, not least in politics, business, and trade. The word *geography* is in fact derived from the Greek language, the literal translation meaning "earth-writing" or "earth-describing." As Greek civilization developed, descriptive geographical writing came to be an essential tool for recording information about sea and land routes and for preparing colonists and merchants for the challenges and opportunities of faraway places.

Urbanization   Towns and cities became essential as centers of administration for early world-empires. Towns served as military garrisons and as theological centers for the ruling classes, who used a combination of military and theological authority to hold their empires together. While these early world-empires were successful, they gave rise not only to monumental capital cities but also to a whole series of smaller settlements, which acted as intermediate centers in the flow of tribute and taxes from colonized territories.

The most successful world-empires, such as the Greek and Roman, established quite extensive urban systems. In general, the settlements in these urban systems were not very large—typically ranging from a few thousand inhabitants to about 20,000. The seats of empire grew quite large, however. The Mesopotamian city of Ur, in present-day Iraq, for example, has been estimated to have reached a population of around 200,000 by 2100 B.C.E., and Thebes, the capital of Egypt, is thought to have had more than 200,000 inhabitants in 1600 B.C.E. Athens and Corinth, the largest cities of ancient Greece, had populations between 50,000 and 100,000 by 400 B.C.E. Rome at the height of the Roman Empire (around C.E. 200) may have had as many as a million inhabitants. The most impressive thing about these cities, though, was not so much their size as their degree of sophistication: elaborately laid out, with paved streets, piped water, sewage systems, massive monuments, grand public buildings, and impressive city walls.

The Geography of the Premodern World

Figure 2.4 shows the generalized framework of human geographies in the Old World as they existed around 1400 C.E. The following characteristics of this period are important:

1. Harsher environments in continental interiors were still characterized by isolated, subsistence-level, hunting-and-gathering minisystems.

2. The dry belt of steppes and desert margins stretching across the Old World from the western Sahara to Mongolia was a continuous zone of pastoral minisystems (minisystems based on herding animals, usually moving with the animals from one grazing area to another).

3. The principal areas of sedentary agricultural production (with permanently settled farmers) extended in a discontinuous arc from Morocco to China, with two main outliers (not shown on the map in Figure 2.4), in the central Andes and in Mesoamerica.

The dominant centers of global civilization were China, northern India (both of them hydraulic society variants of world-empires), and the Ottoman Empire of the eastern Mediterranean. They were all linked by the Silk Road, a series of overland trade routes between China and Mediterranean Europe (**Figure 2.5**). From Roman times until Portuguese navigators found their way around Africa and established seaborne trade routes, the Silk Road provided the main East-West trade route between Europe and China. This shifting trail of caravan tracks facilitated the exchange of silk, spices, and porcelain from the East and gold, precious stones, and Venetian glass from the West. The ancient cities of Samarkand, Bukhara, and Khiva stood along the Silk Road, places of glory and wealth that astonished Western travelers, such as Marco Polo in the thirteenth century. These cities were East-West meeting places for philosophy, knowledge, and religion, and in their prime they were known for producing scholars in mathematics,

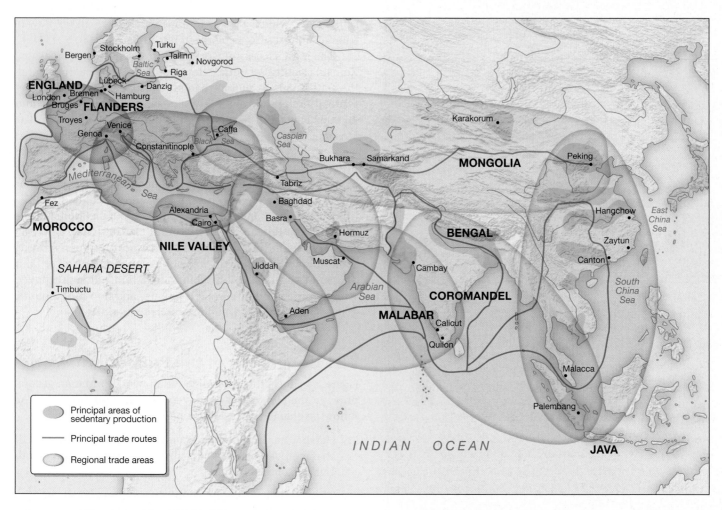

▲ **Figure 2.4** **The precapitalist Old World, circa 1400 C.E.** Principal areas of sedentary agricultural production are shaded. Some long-distance trade took place from one region to another, but for the most part it was limited to a series of overlapping regional circuits of trade. (Source: Adapted from R. Peet, Global Capitalism: Theories of Societal Development. New York: Routledge, 1991; J. Abu-Lughod, Before European Hegemony: The World-System a.d. 1200–1350. New York: Oxford University Press, 1989; and E. R. Wolf, Europe and the People Without History. Berkeley: University of California Press, 1983.)

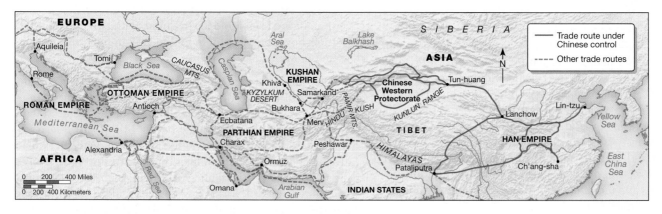

▲ **Figure 2.5** **The Silk Road** This map shows the trade routes of the Silk Road as they existed between 112 B.C.E. and 100 C.E.

music, architecture, and astronomy, such as Al Khoresm (780–847), Al Biruni (973–1048), and Ibn Sind (980–1037). The cities' prosperity was marked by impressive feats of Islamic architecture.

By 1400 C.E., other important world-empires had developed in Southeast Asia, in Muslim city-states of coastal North Africa, in the grasslands of West Africa, around the gold and copper mines of East Africa, and in the feudal kingdoms and merchant towns of Europe. Over time, all of these more developed realms were interconnected through trade, which meant that several emerging centers of **capitalism** came into existence. Capitalism is a form of economic and social organization

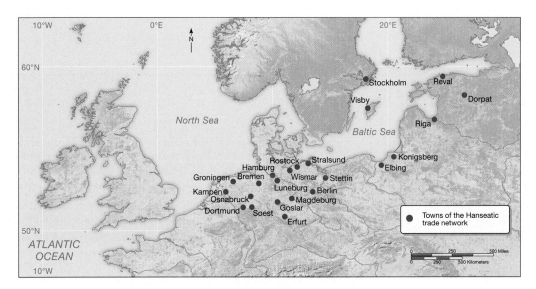

◄ **Figure 2.6 Principal towns of the Hanseatic League** These towns formed a trading network of city-states that flourished between the thirteenth and seventeenth centuries.

characterized by the profit motive and the control of the means of production, distribution, and exchange of goods by private ownership.

Port cities were particularly important to capitalism. Among the leading centers were the city-state of Venice; the Hanseatic League of independent city-states in northwestern Europe (**Figure 2.6**); and Cairo, Calicut, Canton, and Malacca in North Africa and Asia. Traders in these port cities began to organize the production of agricultural specialties, textiles, and craft products in their respective hinterlands. The **hinterland** of a town or city is its sphere of economic influence—the area from which it collects products to be exported and through which it distributes imports. By the fifteenth century, several established regions of budding capitalism existed: northern Italy, Flanders, southern England, the Baltic Sea region, the Nile Valley, Malabar, Coromandel, Bengal, northern Java, and southeast coastal China.

The Expansion of Geographic Knowledge Between roughly 500 and 1400 C.E., geographic knowledge was preserved and expanded by Chinese and Islamic scholars. Chinese maps of the world from the same period were more accurate than those of European cartographers because the Chinese were able to draw on information brought back by imperial China's admirals who successfully navigated much of the Pacific and Indian oceans. Chinese geographers recognized, for example, that Africa was a southward-pointing triangle, whereas on European and Arabic maps of the time it was always represented as pointing eastward.

With the rise of Islamic power in the Middle East and the Mediterranean in the seventh and eighth centuries C.E., centers of scholarship emerged in places such as Baghdad, Damascus, Cairo, and Granada, Spain. Here, surviving Greek and Roman texts were translated into Arabic by scholars such as Al-Battani, Al-Farghani, and Al-Khwarizmi. These Islamic scholars were also able to draw on Chinese geographical writing and cartography brought back by traders along the Silk Road. The requirement that the Islamic

religious faithful undertake at least one pilgrimage to Mecca created a demand for travel guidebooks. It also brought scholars from all over the Arab world into contact with one another, stimulating considerable debate over different philosophical views of the world and of people's relationship with nature.

APPLY YOUR KNOWLEDGE

1. Explain how the "law of diminishing returns" contributed to the colonization of new territories by the Roman empire. How was such colonization linked to urbanization?

2. Do you see remnants of those colonial relationships of capitalism still operating in the world, for example, between different countries?

AN INTERDEPENDENT WORLD GEOGRAPHY

When exploration beyond European shores began to be seen as an important way of opening up new opportunities for trade and economic expansion, a modern world-system emerged. A **world-system** is an *interdependent* system of countries linked by political and economic competition. The term *world-system*, which was coined by historian Immanuel Wallerstein, is hyphenated to emphasize the interdependence of places and regions around the world.

By the sixteenth century, new techniques of shipbuilding and navigation had begun to bind more and more places and regions together through trade and political competition. As a result, more and more peoples around the world were exposed to one another's technologies and ideas. Their different resources, social structures, and cultural systems resulted in quite different pathways of development, however. Some societies were incorporated into the new, European-based international economic system faster than others;

2.1 Geography Matters

The Expansion and Disintegration of States

Brian W. Blouet, College of William and Mary in Virginia

During the eighteenth and nineteenth centuries the size of various world states tended to enlarge. The reasons for this territorial expansion were various: for protection, the enlargement of national markets and, particularly in the nineteenth century, the development of new technologies, including the railroad and the telegraph, made it practical to integrate and administer much larger states than before. In the United States, Canada, and the Russian and German land empires, this expansion allowed information, people, and goods to move more easily over longer distances than previously.

The break-up of the Austro-Hungarian, Russian, German, and Ottoman empires following World War I reversed this enlargement trend. Many new, smaller states appeared in the wake of war, aided by the idea of self-determination. Once introduced, self-determination—the nationality principle—would not go back in the box. Subsequently, many of the new post World War I states were further divided, such as in Yugoslavia, Czechoslovakia, and the USSR.

In today's age of electronic communication, it is easy to share a separatist message and promote a sense of regional or national identity. When these nationalist sentiments are on the rise, the practical, economic facts are often ignored. The start-up costs for a new state are high—a constitution has to be drawn up and agreed upon, a parliamentary body must be created, the machinery of government developed, a central bank established, and a new currency minted.

Breaking Up is Hard to Do: Scotland and the United Kingdom

Many of these factors were evident in the 2014 campaign led by the Scottish Nationalist Party for the Scots to exit the UK—independence was rejected in a referendum (**Figure 2.A**).

▲ **Figure 2.A A rally for Scottish independence.** While 44.7% of the votes in the Scottish Independence Referendum said "Yes" to an independent Scotland, supporters of independence were not able to get the simple majority of votes needed to secede from the United Kingdom.

The reasons for failure of Scotland's bid for independence were several. What currency would a newly independent Scotland use? Would Scotland become a member of the EU? Brussels—the *de facto* seat of the European Parliament—was unequivocal that Scotland would be required to apply for EU membership. Even though the region had been part of the EU, it could not automatically retain membership. If Scotland's secession from the UK was instantly accepted by the EU, other regions might be encouraged to break up countries—places such as Catalonia, Galacia, and the Basque country in Spain. Belgium might split into a northern Flemish region and a southern Walloon area, and those northern parts of Italy that might want to dump the slower-growing south would have a precedent.

The leader of the Scottish Nationalist Party, Alex Salmond, assumed that Scotland would automatically be an EU member and that Scotland's currency would be the pound sterling in a currency union. But as the British treasury in London explained, if Scotland continued to use the pound sterling, then London would still oversee the Scottish budget—and Scotland would not be fully independent. London was not going to let Scotland spend pounds without accountability as happened with Portugal, Ireland, and Greece—countries that spent Euros they did not own, and had to be rescued by the European central bank.

England and Scotland had come together in 1603 when James VI of Scotland (a Stuart) legitimately succeeded to the English throne as James I of England. Today, Queen Elizabeth II is queen of England *and* queen of Scotland, where the monarch has an official residence at Holyrood Palace, in Edinburgh. Even if Scotland exited the UK, Elizabeth II would still remain queen of Scotland until the Scots voted for a republic.

Politically, England, Wales, and Scotland came together in the 1707 Act of Union. Scotland retained its legal system, the Church of Scotland, and administration of its educational institutions. The Scottish parliament was dissolved, and Scotland elected MPs to the Westminster Parliament in London. Scottish merchants, particularly those in Glasgow, promoted the 1707 Act of Union at a time when many Scots were trying to put the Stuarts back on the throne—merchants wanted to trade with the American colonies, own plantations in the West Indies, trade slaves and import goods into Glasgow, where families became wealthy merchants, bankers, and entrepreneurs.

Once Scotland joined England and Wales in an expanding British empire, many Scots secured imperial roles as governors, military commanders, teachers, clergymen, and merchants.

At home, Scotland industrialized into textiles, coal, iron, steel, and shipbuilding. The Firth of Clyde became a world-renowned shipbuilding region. But today, the great nineteenth century industries are now gone or downsized, causing resentment with the process of globalization. Scotland is now a service economy with major financial institutions, because when Scots made money in the nineteenth century they invested with insurance companies still based in Edinburgh. And these major financial institutions indicated they might move south of the border if Scotland voted itself out of the UK and the EU.

The costs of separating from the UK would have been considerable for everyone. Scotland receives far more from the British treasury than it contributes in taxes. Taxes on Scots would likely have been raised, especially since Scotland's share of North Sea oil is a depreciating asset. The Scots would have had to enlarge the machinery of government, and appoint ambassadors and consuls to represent the country overseas. Meanwhile the UK would have lost prestige and the value of the pound would have dropped, as it did briefly when polls indicated the possibility of an independent Scotland. Britain's nuclear submarines, based in western Scotland, would have had to be moved to a new site constructed at great cost. Nevertheless, despite these and other practical issues, the impulse in Scotland to separate from the UK may continue to draw supporters to the Scottish Nationalist Party.

The lesson to be drawn from Scotland's independence movement is that geography matters. National boundaries, once fixed, are difficult to change. Meanwhile, national and cultural identities evolve, technologies change, and emigration and immigration alter the composition of populations. As a result, human geographies are complex and dynamic.

---

1. What is an example of successful "contraction" of states, and why did it succeed?

2. Why would the EU oppose the fragmentation of its member states?

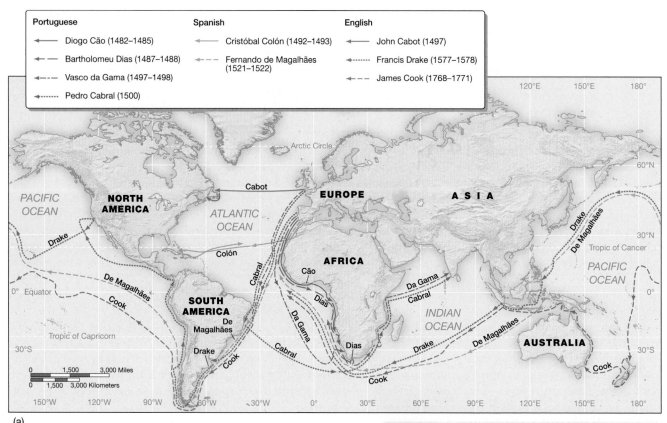

(a)

▲ **Figure 2.7** (a) Principal voyages of exploration. (b) The European Age of Discovery

(b)

some resisted incorporation; and some sought alternative systems of economic and political organization. Australia and New Zealand, for example, were discovered by Europeans only in the late eighteenth century and were barely penetrated, if at all, by the European world-system. Regions not yet absorbed into the world-system are called **external arenas**.

Trade and Merchant Capitalism   With the emergence of this modern world-system at the beginning of the sixteenth century, a whole new geography began to emerge. Although several regions of budding capitalist production existed, and although imperial China could boast of sophisticated achievements in science, technology, and navigation, it was European merchant capitalism that reshaped the world. Several factors motivated European overseas expansion. A relatively high-density population and a limited amount of cultivable land meant that it was a continuous struggle to provide enough food. Meanwhile, the desire for overseas expansion was intensified both by competition among a large number of small monarchies and by inheritance laws that produced large numbers of impoverished aristocrats with little or no land of their own. Many of these landless nobles were eager to set out for adventure and profit.

Added to these motivating factors were the enabling factors of innovations in shipbuilding, navigation, and gunnery. In the mid-1400s, for example, the Portuguese developed a cannon-armed ship—the caravel—that could sail anywhere, defend itself against pirates, pose a threat to those who

were initially unwilling to trade, and carry enough goods to be profitable. Naval power enabled the Portuguese and the Spanish to enrich their economies with gold and silver plundered from the Americas. The quadrant (1450) and the astrolabe (1480) enabled accurate navigation and mapping of ocean currents, prevailing winds, and trade routes. Europeans embarked on a succession of voyages of discovery (**Figure 2.7**), seeking out new products and new markets. The European voyages of discovery can be traced to Portugal's Prince Henry the Navigator (1394–1460), who set up a school of navigation and financed numerous expeditions with the objective of circumnavigating Africa to establish a profitable sea route for spices from India. The knowledge of winds, ocean currents, natural harbors, and watering places amassed by Henry's captains was an essential foundation

▼ **Figure 2.8** Triangular Trade
The Triangular Trade system that operated until the early nineteenth century, when the British took action to outlaw slavery throughout the British Empire.

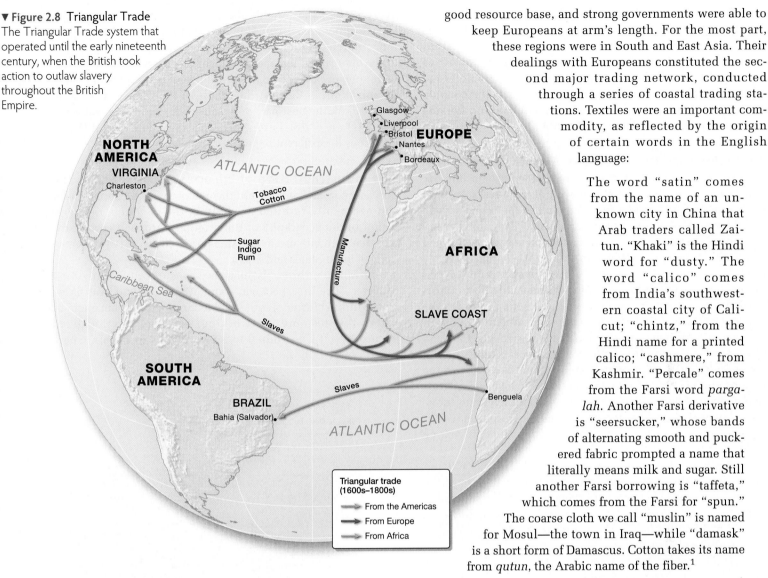

good resource base, and strong governments were able to keep Europeans at arm's length. For the most part, these regions were in South and East Asia. Their dealings with Europeans constituted the second major trading network, conducted through a series of coastal trading stations. Textiles were an important commodity, as reflected by the origin of certain words in the English language:

> The word "satin" comes from the name of an unknown city in China that Arab traders called Zaitun. "Khaki" is the Hindi word for "dusty." The word "calico" comes from India's southwestern coastal city of Calicut; "chintz," from the Hindi name for a printed calico; "cashmere," from Kashmir. "Percale" comes from the Farsi word *pargalah.* Another Farsi derivative is "seersucker," whose bands of alternating smooth and puckered fabric prompted a name that literally means milk and sugar. Still another Farsi borrowing is "taffeta," which comes from the Farsi for "spun." The coarse cloth we call "muslin" is named for Mosul—the town in Iraq—while "damask" is a short form of Damascus. Cotton takes its name from *qutun,* the Arabic name of the fiber.[1]

Within Europe, meanwhile, innovations in business and finance (e.g., banking, loan systems, credit transfers, commercial insurance, and courier services) helped increase savings, investment, and commercial activity. European merchants and manufacturers also became adept at **import substitution**— copying and making goods previously available only by trading. The result was the emergence of Western Europe as the core region of a world-system that had penetrated and incorporated significant portions of the rest of the world.

For Europe, this overseas expansion stimulated still further improvements in technology. These included new developments in nautical mapmaking, naval artillery, shipbuilding, and sailing. The whole experience of overseas expansion also provided a great practical school for entrepreneurship and investment. In this way, the self-propelling growth of merchant capitalism was intensified and consolidated.

For the periphery, European overseas expansion meant dependency (as it has ever since for many of the world's

for the subsequent voyages of Cristóbal Colón (Columbus), da Gama, de Magalhães (Magellan), and others. The end of the European Age of Discovery was marked by Captain James Cook's voyages to the Pacific.

Equipped with better maps and navigation techniques, Europeans sent adventurers in search of gold and silver and also to commandeer land, decide on its use, and exploit coerced labor to produce high-value crops (such as sugar, cocoa, cotton, and indigo) on **plantations**, large landholdings that usually specialize in the production of one particular crop for market. From the sixteenth to the early nineteenth centuries, trade was dominated by two systems. One was the Triangular Trade system among Europe, West Africa, and the Caribbean and the eastern seaboard of North America (**Figure 2.8**). West African states sold their slaves to merchants who transported them to plantation owners in North America and Brazil. Sugar, rum, and indigo was shipped from Caribbean plantations to Europe and tobacco and cotton were shipped from Virginia and the Carolinas to Europe. The triangle was completed as European goods were exported to West Africa.

Those regions whose populations were resistant to European disease and which also had high population densities, a

[1]B. Wallach, *Understanding the Cultural Landscape.* New York: Guilford Press, 2005, p. 148.

peripheral regions). At worst, territory was forcibly occupied and labor systematically exploited. At best, local traders were displaced by Europeans, who imposed their own terms of economic exchange. Europeans soon destroyed most of the Muslim shipping trade in the Indian Ocean, for example, and went on to capture a large share of the oceangoing trade within Asia, selling Japanese copper to China and India, Persian carpets to India, and Indian cotton textiles to Japan.

Technology and Its Limitations

As revolutionary as these changes were, however, they were constrained by a technology that rested on wind and water power, on wooden ships and structures, and on wood for fuel. Grain mills, for example, were built of wood and powered by water or wind. They could generate only modest amounts of power and only at sites determined by physical geography, not human choice. Within the relatively small European landmass, wood grown for structural use and for fuel competed for acreage with food and textile fiber crops.

More important, however, were the inherent limits in the size and strength of timber (which imposed structural limits on the size of buildings), the diameter of waterwheels, the span of bridges, and so on. In particular, it imposed limits on the size and design of ships, which in turn imposed limits on the volume and velocity of world trade. The expense and relative inefficiency of horse- or ox-drawn wagons for overland transportation also meant that for a long time the European world-system could penetrate into continental interiors only along major rivers.

After 300 years of evolution, roughly between 1450 and 1750, the world-system had incorporated only parts of the world. The principal spheres of European influence were Mediterranean North Africa, Portuguese and Spanish colonies in Central and South America, Indian ports and trading colonies, the East Indies, African and Chinese ports, the Greater Caribbean, and British and French territories in North America. The rest of the world functioned more or less as before, with slow-changing geographies based on modified minisystems and world-empires that were only partially and intermittently penetrated by market trading.

APPLY YOUR KNOWLEDGE

1. What are some of the social and technological reasons that Europeans sought to colonize other parts of the world?

2. Search the Internet to find a map of Brazil in the 1600s as well as a current map of Brazil. How would you interpret the historical map in terms of the world-system of the time? How has it changed over time?

Core and Periphery in the New World-System

With the new production and transportation technologies of the Industrial Revolution (from the late 1700s), capitalism began to grow into a global system that reached into virtually every part of the inhabited world and into virtually every aspect of people's lives. It is important to recognize that the Industrial Revolution was really an extended transition to new forms of organization and new technologies and that its effects were uneven, reflecting the influence of principles of spatial organization. In Europe, the cradle of the Industrial Revolution, it took the best part of a century for industrialization to work its way across European landscapes, with very different outcomes for different regions.

The Industrialization of Europe

The Industrial Revolution began in England toward the end of the eighteenth century and eventually resulted not only in the complete reorganization of the geography of the original European core of the world-system but also in an extension of the world-system core to the United States and Japan.

In Europe, three distinctive waves of industrialization occurred. The first, between about 1790 and 1850, was based on the initial cluster of industrial technologies (steam engines, cotton textiles, and ironworking) and was very localized (**Figure 2.9**). It was limited to a few regions in Britain where

▲ **Figure 2.9 The spread of industrialization in Europe** European industrialization began with the emergence of small industrial regions in several parts of Britain, drawing on local mineral resources, water power, and nascent industrial technologies. As new technologies emerged, industrialization spread to other regions with the right locational attributes: access to raw materials and energy sources, good communications, and large labor markets.

industrial entrepreneurs and workforces had first exploited key innovations and the availability of key resources (coal, iron ore, water). Although these regions shared the common impetus of certain key innovations, each of them retained its own technological traditions and industrial style. From the start, then, industrialization was a regional-scale phenomenon.

The second wave of industrialization, between about 1850 and 1870, involved the diffusion of industrialization to most of the rest of Britain and to parts of northwest Europe, particularly the coalfields of northern France, Belgium, and Germany. This second wave also brought a certain amount of change to the geographies of first-wave industrial regions as new technologies (steel, machine tools, railroads, steamships) brought new opportunities, new locational requirements, new business structures, and new forms of societal organization. Railroads and steamships made more places accessible, bringing their resources and their markets into the sphere of industrialization. These new activities brought some significant changes to the logic of industrial location.

The importance of railway networks, for example, attracted industry away from smaller towns on the canal systems toward larger towns with good rail connections. The importance of steamships for coastal and international trade attracted industry to larger ports. At the same time, the importance of steel produced concentrations of heavy industry in places that had nearby supplies of coal, iron ore, and limestone. By the outbreak of the First World War in 1914, industrialization had spread to parts of southern and eastern Europe where there were coalfields, concentrations of population, and good transport connections. The scale of industry increased as improved technologies and transportation made larger markets accessible to firms. Local, family firms became small companies that were regional in scope. Small companies grew to become powerful firms serving national markets. Specialized business, legal, and financial services emerged within larger cities. The growth of new occupations transformed the structure of social classes, and this transformation in turn was reflected in the politics and landscapes of industrial regions.

APPLY YOUR KNOWLEDGE

1. What are both the positive and negative effects that industrialization has had on the globe?

2. Can you think of other examples of how technology changed regional and global markets?

Colonialism and Imperialism

Human geographies were recast again, this time with a more interdependent dynamic. New production technologies, based on more efficient energy sources, helped raise levels of productivity and create new and better products that stimulated demand, increased profits, and created a pool of capital for further investment. New transportation technologies triggered successive phases of geographic expansion, allowing for internal development

as well as for external colonization and **imperialism** (the deliberate exercise of military power and economic influence by powerful states to advance and secure their national interests—see Chapter 8).

Since the seventeenth century, the world-system has been consolidated, with stronger economic ties between countries. It has also been extended, with all the world's countries eventually becoming involved to some extent in the interdependence of the capitalist system. Although there have been some instances of resistance and adaptation, the overall result is that a highly structured relationship between places and regions has emerged. This relationship is organized around three tiers: *core, semiperipheral,* and *peripheral* regions. These broad geographic divisions have evolved—and are still evolving—through a combination of processes of private economic competition and competition among states.

The **core regions** of the world-system at any given time are those that dominate trade, control the most advanced technologies, and have high levels of productivity within diversified economies. As a result, they enjoy relatively high per capita incomes. The first core regions of the world-system were the trading hubs of Holland and England, joined soon afterward by France. By the end of the nineteenth century, the core of the world-system had extended to include the United States and Japan; and today it includes Scandinavia and most of Western Europe (**Figure 2.10**). But the continuing success of core regions depends on their dominance and exploitation of other regions. This dominance in turn depends on the participation of these other regions within the world-system. Initially, such participation was achieved by military enforcement, then by European colonialism.

Colonialism involves the establishment and maintenance of political and legal domination by a state over a separate and alien society. This domination usually involves some colonization (i.e., the physical settlement of people from the colonizing state) and always results in economic exploitation by the colonizing state. After World War II, the sheer economic and political influence of the core regions was sufficient to maintain their dominance without political and legal control, and colonialism was gradually phased out.

Regions that have remained economically and politically unsuccessful throughout this process of incorporation into the world-system are peripheral. **Peripheral regions** are characterized by dependent and disadvantageous trading relationships, by primitive or obsolescent technologies, and by undeveloped or narrowly specialized economies with low levels of productivity.

Transitional between core regions and peripheral regions are semiperipheral regions. **Semiperipheral regions** are able to exploit peripheral regions but are themselves exploited and dominated by core regions. They consist mostly of countries that were once peripheral. The existence of this semiperipheral category underlines the fact that neither peripheral status nor core status is necessarily permanent. The United States and Japan both achieved core status after having been peripheral; Spain and Portugal, part of the original core in the sixteenth

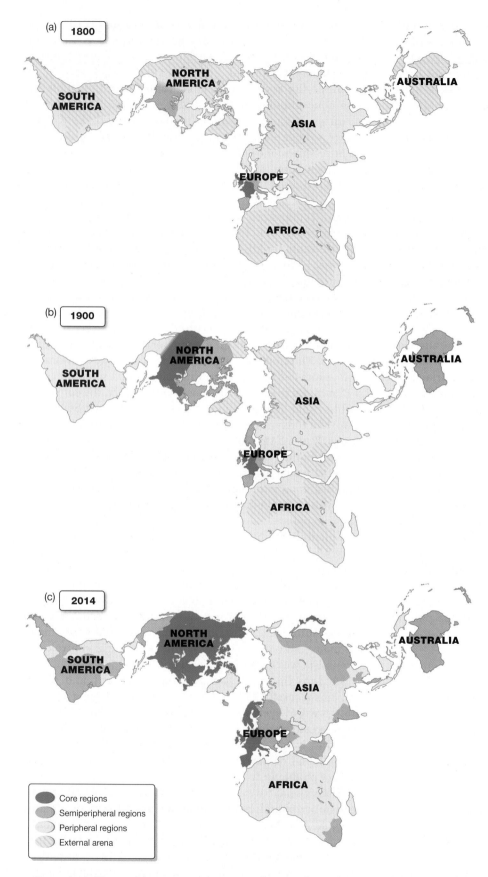

▲ Figure 2.10 The world-system core, semiperiphery, and periphery in (a) 1800, (b) 1900, and (c) 2014.

century, became semiperipheral in the nineteenth century but are now once more part of the core. Quite a few countries, including Brazil, India, Mexico, South Korea, and Taiwan, have become semiperipheral after first having been incorporated into the periphery of the world-system and then developing a successful manufacturing sector.

An important determinant of these changes in status is the effectiveness of states in ensuring the international competitiveness of their domestic producers. They can do this in several ways:

- Manipulating markets (e.g., protecting domestic manufacturers by charging taxes on imports)

- Regulating their economies (e.g., enacting laws that help establish stable labor markets)

- Creating physical and social infrastructures (spending public funds on road systems, ports, educational systems, etc.).

Because some states are more successful than others in pursuing these strategies, the hierarchy of the three geographical tiers is not rigid. Rather, it is fluid, providing a continually changing framework for geographical transformation within individual places and regions.

Leadership Cycles The colonization and imperialism that accompanied the expansion of the world-system was closely tied to the evolution of world leadership cycles. **Leadership cycles** are periods of international power established by individual states through economic, political, and military competition. In the long term, success in the world-system depends on economic strength and competitiveness, which brings political influence and pays for military strength. With a combination of economic, political, and military power, individual states can dominate the world-system, setting the terms for many economic and cultural practices and imposing their particular ideology by virtue of their preeminence. The modern world-system has so far experienced several distinct leadership cycles. In simplified terms, they have involved dominance by Portugal (for most of the sixteenth century), the Netherlands (for the first three-quarters of the seventeenth century), Great Britain (from the early eighteenth century through the early twentieth century), and the United States (from the 1950s)—see Chapter 11.

This kind of dominance is known as hegemony. **Hegemony** refers to domination over the world economy, exercised—through a combination of economic, military, financial, and cultural means—by one national state in a particular historical epoch. Over the long run, the costs of maintaining this kind of power and influence tend to weaken the hegemony. This phase of the cycle, when the dominant nation is weakened, is known as imperial overstretch. It is followed by another period of competitive struggle, which brings the possibility of a new dominant world power.

APPLY YOUR KNOWLEDGE

1. What are core and peripheral regions? How do they relate to leadership cycles?

2. Look in recent issues of national newspapers or news magazines and select a current international issue involving former colonial territories. Explain how the issue may relate to the legacy of colonialism. (*Hint:* You might consider oil in the Middle East or copper mining in the Democratic Republic of the Congo.)

Organizing the Periphery

The growth and internal development of the core regions simply could not have taken place without the foodstuffs, raw materials, and markets provided by the colonization of the periphery and the incorporation of more and more territory into the sphere of industrial capitalism. Early in the nineteenth century, the industrial core nations embarked on the penetration of the world's inland midcontinental grassland zones to exploit them for grain or livestock production. This led to the settlement, through the emigration of European peoples, of the temperate prairies and pampas of the Americas, the veld in southern Africa, the Murray-Darling Plain in Australia, and the Canterbury Plain in New Zealand. At the same time, as the demand for tropical plantation products (e.g., sugar, cotton, coffee, cocoa, and rubber) increased, most of the tropical world came under the political and economic control—direct or indirect—of one or another of the industrial core nations. In the second half of the nineteenth century, and especially after 1870, there was a vast increase in the number of colonies and the number of people under colonial rule.

The International Division of Labor   The fundamental logic behind all this colonization was economic: the need for an extended arena for trade, an arena that could supply foodstuffs and raw materials in return for the industrial goods of the core. The outcome was an international division of labor, driven by the needs of the core and imposed through its economic and military strength. This **division of labor** involved the specialization of different people, regions, and countries in certain kinds of economic activities. In particular, colonies began to specialize in the production of commodities meeting certain criteria:

- where an established demand existed in the industrial core (e.g., for foodstuffs and industrial raw materials);

- where colonies held a **comparative advantage** in specializations that did not duplicate or compete with the domestic suppliers within core countries (e.g., tropical agricultural products like cocoa and bananas simply could not be grown in core countries).

The result was that colonial economies were founded on narrow specializations that were oriented to and dependent upon the needs of core countries. Examples of these specializations were many: bananas in Central America; cotton in India; coffee in Brazil, Java, and Kenya; copper in Chile; cocoa in Ghana; jute in East Pakistan (now Bangladesh); palm oil in West Africa; rubber in Malaya (now Malaysia) and Sumatra; sugar in the Caribbean islands; tea in Ceylon (now Sri Lanka); tin in Bolivia; and bauxite in Guyana and Surinam. Most of these specializations persist today. For example, 45 of the 55 countries in sub-Saharan Africa still depend on just three products—tea, cocoa, and coffee—for more than half of their export earnings.

This new global economic geography took some time to establish, and the details of its pattern and timing were heavily influenced by technological innovations. The incorporation of the temperate grasslands into the commercial orbit of the core countries, for example, involved changes in regional landscapes resulting from critical innovations—such as barbed wire, the railroad, and refrigeration.

The single most important innovation stimulating the international division of labor, however, was the development of metal-hulled, oceangoing steamships (**Figure 2.11**). This development was cumulative, with improvements in engines, boilers, transmission systems, fuel systems, and construction

▼ **Figure 2.11 Isambard Brunel** Born in 1806, Brunel was designer of the Great Western, the first steamship purpose-built for crossing the Atlantic, and the largest passenger ship in the world from 1837 to 1839.

materials adding up to produce dramatic improvements in carrying capacity, speed, range, and reliability. The Suez Canal (opened in 1869) and the Panama Canal (opened in 1914) were also critical, providing shorter and less hazardous routes between core countries and colonial ports of call. By the eve of World War I, the world economy was effectively integrated by a system of regularly scheduled steamship trading routes. This integration, in turn, was supported by the second most important innovation stimulating the international division of labor: a network of telegraph communications that enabled businesses to monitor and coordinate supply and demand across vast distances on an hourly basis.

The international division of labor brought about a substantial increase in trade and a huge surge in the overall size of the capitalist world economy. The peripheral regions of the world contributed a great deal to this growth. By 1913, Africa and Asia provided more *exports* to the world economy than either North America or the British Isles. Asia alone was *importing* almost as much, by value, as North America. The industrializing countries of the core bought increasing amounts of foodstuffs and raw materials from the periphery, financed by profits from the export of machinery and manufactured goods. Britain, the hegemonic power of the period, drew on a trading empire that was truly global (**Figure 2.12**).

Patterns of international trade and interdependence became increasingly complex. Britain used its capital to invest not only in peripheral regions but also in profitable industries in other core countries, especially the United States. At the same time, these other core countries were able to export cheap manufactured items to Britain. Britain financed the purchase

of these goods, together with imports of food from its dominion states (Canada, South Africa, Australia, and New Zealand) and colonies, through the export of its own manufactured goods to peripheral countries. India and China, with large domestic markets, were especially important. A widening circle of exchange and dependence developed, with constantly shifting patterns of trade and investment.

APPLY YOUR KNOWLEDGE

1. What is the international division of labor and what was technology's role in establishing this system?

2. Look at some of the clothes and products that you possess and see where they were made. List the materials that go into making your clothing. Can you speculate about where the cotton was grown? Who drove it to the factory? Who produced the cotton? Who sewed the clothing together? How do these questions relate to the principles of division of labor and comparative advantage?

Imperialism: Imposing New Geographies on the World The incorporation of the periphery was by no means entirely motivated by this basic logic of free trade and investment. Although Britain was the dominant power in the late nineteenth century, several other European countries (notably Germany, France, and the Netherlands), together with the United States—and later Japan—were competing for global influence. This competition developed into a scramble for territorial and commercial domination. The core countries engaged in preemptive geographic expansionism to protect their established interests and to limit the opportunities of others. They also wanted to secure as much of the world as possible—through a combination of military oversight, administrative control, and economic regulation—to ensure stable and profitable environments for their traders and investors.

This combination of circumstances defined a new era of imperialism in the final quarter of the nineteenth century. Africa, more than any other peripheral region, was given an entirely new geography. It was carved up into a patchwork of European colonies and protectorates in just 34 years, between 1880 and 1914 (**Figure 2.13**), with little regard for either physical geography or the preexisting minisystems and world-empires. Whereas European interest had previously focused on coastal trading stations and garrison ports, it now extended to the entire continent.

Within just a few years, the whole of Africa became incorporated into the

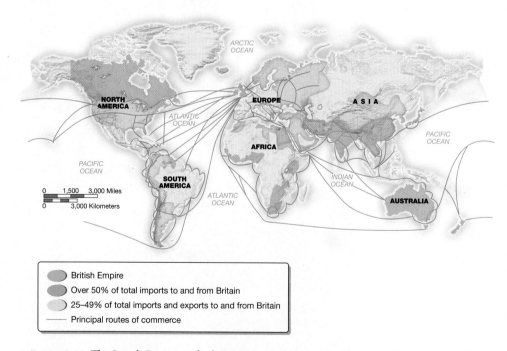

▲ **Figure 2.12 The British Empire in the late 1800s** Protected by the all-powerful Royal Navy, the British merchant navy established a web of commerce that collected food for British industrial workers and raw materials for its industries, much of it from colonies and dependencies appropriated by imperial might and developed by British capital. Its trading empire was so successful that Britain became the hub of trade for other states. (Source: Adapted from P. Hugill, World Trade Since 1431. Baltimore: Johns Hopkins University Press, 1993, p. 136.)

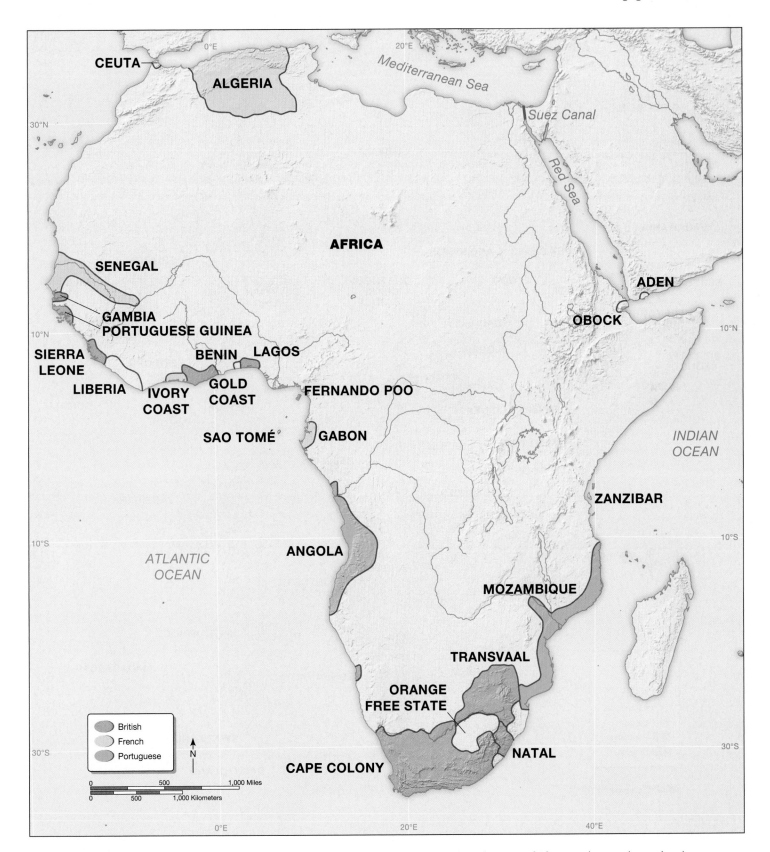

▲ **Figure 2.13a The colonization of Africa, 1880** In 1880 European colonization was limited to a few parts of Africa, mainly around coastal trading centers.

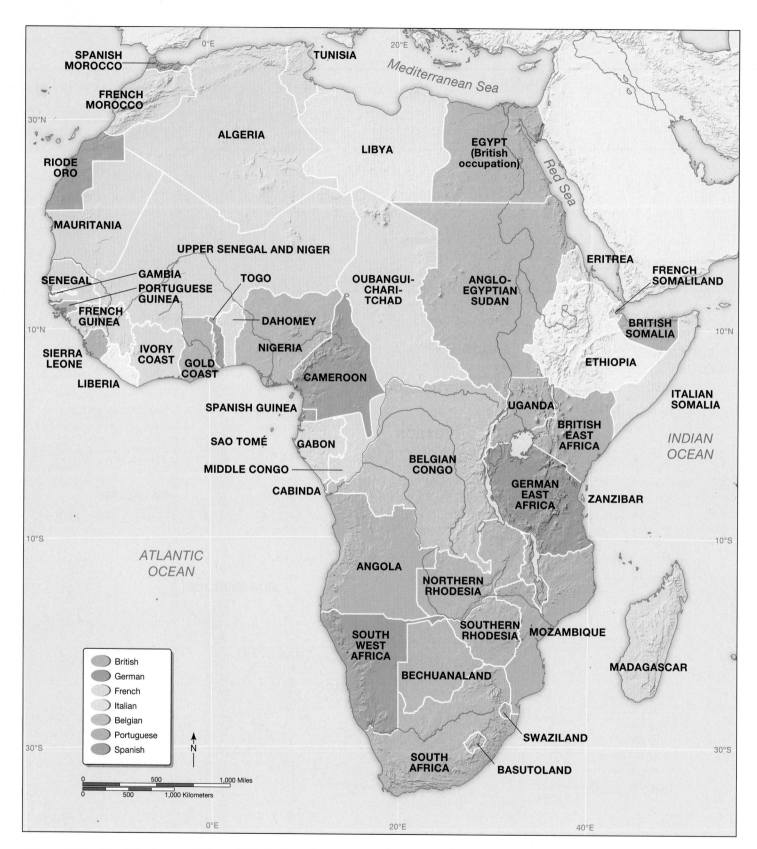

▲ **Figure 2.13b** **The Colonization of Africa, 1914** By 1914 almost the entire continent had been colonized by European powers.

Source: After Christopher, A. J., Colonial Africa. Beckenham: Croom Helm, 1984, Figures 2.1 and 2.2.

modern world-system, with a geography that consisted of a hierarchy of three kinds of spaces:

- One consisted of regions and localities organized by European colonial administrators and European investors to produce commodities for the world market.

- A second consisted of zones of production for local markets where peasant farmers produced food for consumption by laborers engaged in commercial mining and agriculture.

- The third consisted of widespread regions of subsistence agriculture whose connection with the world-system was as a source of labor for the commercial regions.

Meanwhile, the major powers also jostled and squabbled over small Pacific islands that had suddenly become valuable as strategic coaling stations for their navies and merchant fleets. Resistance from indigenous peoples was quickly brushed aside by imperial navies with ironclad steamers, high-explosive guns, and troops with rifles and cannons. European weaponry was so superior that Otto von Bismarck, the founder and first chancellor (1871–1890) of the German empire, referred to these conflicts as "sporting wars." Between 1870 and 1900, European countries added almost 22 million square kilometers (8.5 million square miles) and 150 million people to their spheres of control—20 percent of the Earth's land surface and 10 percent of its population.

The imprint of imperialism and colonization on the geographies of the newly incorporated peripheries of the world-system was immediate and profound. The periphery was rendered almost entirely dependent on European and North American capital, shipping, managerial expertise, financial services, and news and communications. Consequently, it also became dependent on European cultural products: language, education, science, religion, architecture, and planning. All of these influences were etched into the landscapes of the periphery in a variety of ways as new places were created, old places were remade, and regions were reorganized.

Geography and Imperialism The discipline of geography played an important role in providing a "scientific" rationale for the domination of peripheral countries by Europeans and North Americans. Prominent geographers argued that civilization and successful economic development are largely the result of "invigorating" temperate climates with marked seasonal variations and varied weather but without prolonged extremes of heat, humidity, or cold. Conversely, tropical climates, they asserted, limit people's vitality. This kind of reasoning reflects an underlying ethnocentrism and environmental determinism. **Ethnocentrism** is the attitude that one's own race and culture are superior to those of others. **Environmental determinism** is a doctrine holding that human activities are shaped and constrained by the environment. Most of the geographic writing in the nineteenth and early twentieth centuries was strongly influenced by this assumption that the physical attributes of geographical settings are the root not only of people's physical differences but also of differences in people's economic vitality, cultural activities, and social structures.

The Struggle for Independence

The imperial world order began to disintegrate shortly after World War II, however. The United States emerged as the new hegemonic power, the dominant state within the world-system core. The Soviet Union and China, opting for alternative, socialist paths of development for themselves and their satellite countries, were seen as a Second World, withdrawn from the capitalist world economy. Their pursuit of alternative political economies was based on radically different values.

By the 1950s, many of the old European colonies began to seek political independence. Some of the early independence struggles were very bloody because the colonial powers were initially reluctant to withdraw from colonies where strategic resources or large numbers of European settlers were involved. In Kenya, in the early 1950s, for example, a militant nationalist movement known as the Mau Mau launched a campaign of terrorism, sabotage, and assassination against British colonists. They killed more than 2,000 white settlers between 1952 and 1956; in return, 11,000 Mau Mau rebels were killed by the colonial army and 20,000 put into detention camps by the colonial administration. By the early 1960s, however, the process of decolonization had become relatively smooth. (In 1962, Jomo Kenyatta, who had been jailed as a Mau Mau leader in 1953, became prime minister of a newly independent Kenya.) The periphery of the world-system now consisted of politically independent states, some of which adopted a policy of nonalignment; that is, they were not formally aligned with or against either the United States or the Soviet Union.

Neocolonialism and Transnational Corporations As newly independent peripheral states struggled from the 1960s onward to be free of their economic dependence through industrialization, modernization, and trade, the capitalist world-system became increasingly integrated and interdependent. The old imperial patterns of international trade broke down and were replaced by more complex patterns. Nevertheless, the newly independent states were still influenced by many of the old colonial links and legacies that remained intact. The result was a neocolonial pattern of international development.

Neocolonialism refers to economic and political strategies by which powerful states in core economies indirectly maintain or extend their influence over other areas or people. Instead of formal, direct rule (colonialism), controls are exerted through such strategies as international financial regulations, commercial relations, and covert intelligence operations. Through neocolonialism, the human geographies of peripheral countries continued to be heavily shaped by the linguistic, cultural, political, and institutional influence of the former colonial powers as well as by their investment and trading activities.

At about the same time, a new form of imperialism was emerging. This was the *commercial imperialism* of giant corporations. These corporations had grown within the core countries through the elimination of smaller firms by mergers and takeovers. By the 1960s, quite a few of them had become so big that they were *transnational* in scope, having established overseas subsidiaries, taken over foreign competitors, or simply bought into profitable foreign businesses.

These **transnational corporations** have investments and activities that span international boundaries, with subsidiary companies, factories, offices, or facilities in several countries. Examples of transnational corporations include Airbus, BP, Halliburton, News Corporation, Siemens, and the Virgin Group. By 2007, over 79,000 transnational corporations were operating, 90 percent of which were headquartered in the core states. These corporations control about 790,000 foreign affiliates and account for the equivalent of 11 percent of world **Gross Domestic Product (GDP)** and one-third of world exports.

Transnational corporations have been portrayed as imperialist by some geographers because of their ability and willingness to exercise their considerable power in ways that adversely affect peripheral states. They have certainly been central to a major new phase of geographical restructuring that has been taking place for the last 35 years or so. This phase has been distinctive because an unprecedented amount of economic, political, social, and cultural activity has spilled beyond the geographic and institutional boundaries of states. It is a phase of *globalization,* a much fuller integration of the economies of the worldwide system of states and a much greater interdependence of individual places and regions from every part of the world-system.

APPLY YOUR KNOWLEDGE

1. How did the doctrine of "environmental determinism" support European rationale for conquest in other parts of the world?

2. Provide an example of how neocolonialism reinforces the power and influence of core countries. Please be specific with your example. (*Hint:* You might want to consider milk production in Jamaica or oil extraction in Nigeria.) What is the role of transnational corporations in neocolonialism?

CONTEMPORARY GLOBALIZATION

Globalization is the increasing interconnectedness of different parts of the world through common processes of economic, environmental, political, and cultural change. As we have seen, globalization has been underway since the inception of the modern world-system in the sixteenth century. In the nineteenth century, when the competitive system of states fostered the emergence of international agencies and institutions, global networks of communication, a standardized system of global time, international law, and internationally shared notions of citizenship and human rights, the basic framework of modern globalization came into being. Global connections today, though, differ in at least four important ways from those in the past:

1. They function at much greater *speed* than ever before.

2. Globalization operates on a much larger *scale,* leaving few people unaffected and wielding its influence in even the most remote places.

3. The *scope* of global connections is much broader and has *multiple dimensions:* economic, technological, political, legal, social, and cultural, among others.

4. The interactions and interdependencies among numerous global actors have created a new level of *complexity* for the relationships between places and regions.

Over the past 35 years, telecommunication technologies, corporate strategies, and institutional frameworks have combined to create a dynamic new geographical framework. Emerging information technologies have helped create a complex and frenetic international financial system, while transnational corporations are now able to transfer their production activities from one part of the world to another in response to changing market conditions and changing transportation and communications technologies (see Chapter 8). Now products, markets, and organizations are both spread and linked across the globe. Governments, in their attempts to adjust to this situation, have sought new ways of dealing with the consequences of globalization, including unprecedented international political and economic alliances such as NAFTA and the European Union (see Chapter 10).

The economic basis of contemporary globalization depends on myriad commodity chains that crisscross global space. **Commodity chains** are networks of labor and production processes that originate in the extraction or production of raw materials and whose end result is the delivery and consumption of a finished commodity (see Box 2.2, "Visualizing Geography: Commodity Chains"). These networks often span countries and continents, linking into vast global assembly lines the production and supply of raw materials, the processing of raw materials, the production of components, the assembly of finished products, and the distribution of finished products. As we shall see in Chapter 8, these global assembly lines are increasingly important in shaping places and regions.

Globalization also has important cultural dimensions (see Chapter 5). One is quite simply the diffusion around the world of all sorts of cultural forms, practices, and artifacts that had previously been confined to specific places or regions. Examples include "ethnic" and regional cuisine, "world" music, Caribbean carnivals, and "charismatic" Christian sects. Another dimension of cultural globalization derives from consumer culture: everything that is sold in international markets, from sneakers, replica soccer shirts, and automobiles to movies and rock concert tours. This had led some observers to believe that globalization is producing a new set of universally shared images, practices, and values—literally, a global culture.

All this adds up to an intensified global connectedness and the beginnings of the world as an interdependent system. Or, to be more precise, this is how it adds up for the 900 million or so of the world's people who are directly tied to global systems of production and consumption and who have access to global networks of communication and knowledge. All of us in this globalizing world are in the middle of a major reorganization of the world economy and a radical change in our relationships to other people and places.

At first glance it might seem that globalization will render geography obsolete—especially in the more developed parts of the world. High-tech communications and the global marketing of standardized products seem as if they might soon wash away the distinctiveness of people and places, permanently diminishing the importance of differences between places. Far from it. The new mobility of money, labor, products, and ideas actually increases the significance of place in some very important ways:

- The more universal the diffusion of material culture and lifestyles, the more valuable regional and ethnic identities become.

- The faster the information highway takes people into cyberspace, the more they feel the need for a subjective setting—a specific place or community—they can call their own.

- The greater the reach of transnational corporations, the more easily they are able to respond to place-to-place variations in labor markets and consumer markets. As a result, economic geography has to be reorganized more frequently and more radically.

- The greater the integration of transnational governments and institutions, the more sensitive people have become to local cleavages of race, ethnicity, and religion.

For some places and regions, globalization is a central reality; for others, it is still a marginal influence. While some places and regions have become more closely interconnected and interdependent as a result of globalization, others have been bypassed or excluded. In short, there is no one experience of globalization. The reality is that globalization is variously embraced, resisted, subverted, and exploited as it makes contact with specific cultures and settings. In the process, places are modified or reconstructed rather than destroyed or homogenized.

Key Issues in a Globalizing World

The integrated global system has increased awareness of a set of common problems—climate change, pollution, disease, crime, poverty, and inequality—that many see as a consequence of globalization. The globalization of the contemporary world—its causes and effects on specific aspects of human geographies at different spatial scales—is a recurring theme through the rest of this book. Here, we note in broad outline the principal issues associated with contemporary globalization.

Environmental Issues   The sheer scale and capacity of the world economy means that humans are now capable of altering the environment at the global scale. The "footprint" of humankind extends to more than four-fifths of Earth's surface (**Figure 2.14**). Many of the important issues facing modern society are the consequences—intended and unintended—of human modifications of our physical environment.

Humans have altered the balance of nature in ways that have brought economic prosperity to some areas and created environmental dilemmas and crises in others. For example,

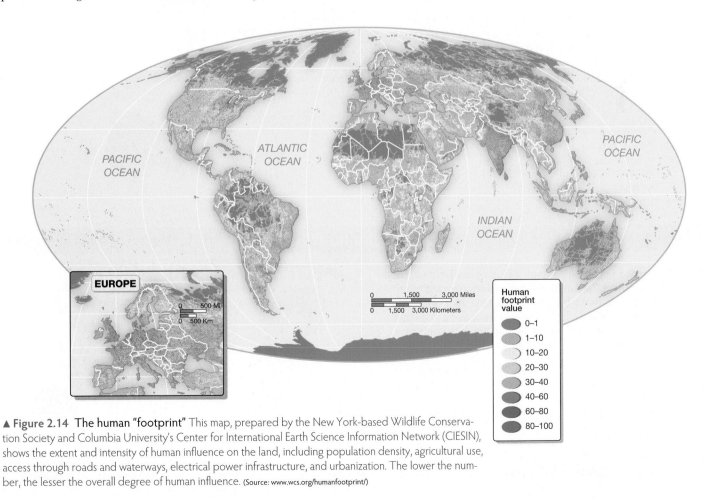

▲ **Figure 2.14 The human "footprint"** This map, prepared by the New York–based Wildlife Conservation Society and Columbia University's Center for International Earth Science Information Network (CIESIN), shows the extent and intensity of human influence on the land, including population density, agricultural use, access through roads and waterways, electrical power infrastructure, and urbanization. The lower the number, the lesser the overall degree of human influence. (Source: www.wcs.org/humanfootprint/)

2.2 Visualizing Geography

Commodity Chains

Global commodity chains link the progression of a commodity from design through procurement of raw materials and production to import or export to the point of sale, distribution for sale, marketing, and advertising.

Almost every mass-marketed manufactured product involves a complex commodity chain. Here we look at the manufacture of cell phones.

2.1 Suppliers and Manufacturers

Advances in telecommunications, management techniques, transportation, finance, and other services to industry have made possible the segmentation of corporate production lines and services. Manufacturing companies now design a product in one country, have it produced by contractors in various countries continents apart, sell the product with its brand name by telephone or Internet almost anywhere in the world, and have other contractors deliver it. These services—design, sales, financing, and delivery—can be undertaken without the various actors ever meeting face to face. Advances in technology and management permit the reproduction and standardization of services and products on a global basis.

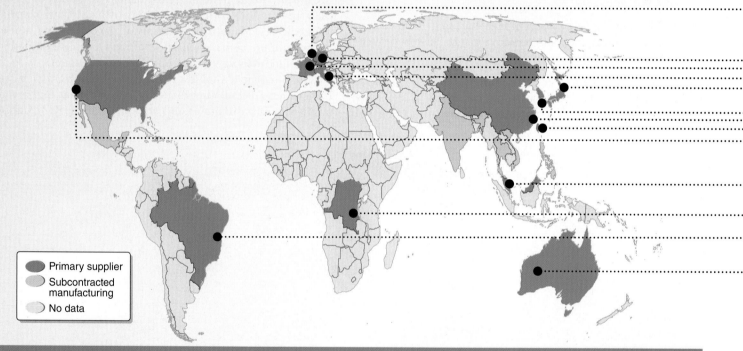

- Primary supplier
- Subcontracted manufacturing
- No data

2.2 Electronic Waste

Much of the world's electronic waste, including cell phones, finds its way back to China. Tons of discarded cell phones and other electronic waste also make their way into China via Hong Kong and Vietnam. Informal e-waste treatment methods such as open burning and acid stripping of electronic components leads to severe health and environmental hazards.

More than **70%** of mobile devices today can be reused but only **14-17%** are recycled annually.

Proper recycling of 1 million cell phones can recover:

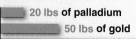

- 20 lbs **of palladium**
- 50 lbs **of gold**

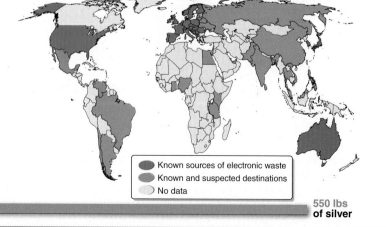

- Known sources of electronic waste
- Known and suspected destinations
- No data

550 lbs **of silver**

2.3 The iPhone Commodity Chain

COMMODITIES
- Design
- Raw materials
- Production
- Marketing/Advertising

AUSTRALIA
- Rare-earth minerals used in electronic components

BRAZIL
- Rare-earth minerals used in electronic components

CHINA
- Yttrium, lanthanum, neodymium, and many other rare-earth minerals used in electronic components, color screen, glass polishing, and vibration unit
  Currently, more than **90%** of rare earth minerals are mined in China. Cleaner, more efficient processes are being developed in the U.S. and other countries.
- Main chassis
  Final assembly
  China's **Foxconn** is the world's largest electronics contract manufacturer. It's estimated that Foxconn makes 40% of consumer electronics in the world. Foxconn makes electronics for many leading brands such as **Apple, Hewlett-Packard, Dell, Nintendo, Motorola, Amazon, Nokia, Sony, and Samsung**.

DEMOCRATIC REPUBLIC OF CONGO (DRC)
- Tin, tantalum, tungsten and gold used in electronic components
  The DRC is the **fifth largest supplier of tin ore**, and according to a U.S. Geological Survey, about 10 percent of tungsten--the mineral used to make cell phones vibrate--is imported to the United States. Armed rebel groups connected with violent crimes profit from trade of these minerals.

FRANCE AND ITALY
- Gyroscope (allows vertical or horizontal display)

GERMANY
- Accelerometer (detects direction and acceleration)

JAPAN
- iSight camera
- Retina display
- Fingerprint sensor
- Flash memory

NETHERLANDS
- M7 motion coprocessor (interprets data from accelerometer, gyroscope, and compass)

SINGAPORE
- Plastic chassis for iPhone 5c

SOUTH KOREA
- Microprocessors

TAIWAN
- RF modules (radio frequency modules used to transmit/receive radio signals)

UNITED STATES
- Yttrium, lanthanum, neodymium, and many other rare-earth minerals used in electronic components, color screen, glass polishing, and vibration unit
- **Product design**
  The majority of cell phones are designed in the U.S. while nearly all manufacturing takes place overseas, mainly in China and other Asian countries.
- Gorilla glass display
- Audio components
- Transmit modules
- RF switches (radio frequency device used for signal routing)
- Touchscreen controller
- FaceTime camera chip
- Chiat/Day advertising agency

1.6 billion smartphones of all brands are projected to be sold by 2016. The largest areas of growth will be China and India.

The average user in the U.S. upgrades their cell phone every 21.7 months, which amounts to **130 million** devices being discarded each year **in the U.S. alone**. Users in other countries keep their devices much longer on average:
Germany: 45.7 months
Brazil: 80.8 months
India: 93.6 months

References:
http://www.cnet.com/news/digging-for-rare-earths-the-mines-where-iphones-are-born/
Forbes.com
Apple.com
FinancesOnline.com
http://www.statisticbrain.com/iphone-5-sales-statistics/
e-cycle.com
EPA Website: http://www.step-initiative.org/news.php?id=0000000163
http://www.e-stewards.org/the-e-waste-crisis/
Huffington Post
New York Times
Alliance of American Manufacturing
United Nations University

Public Radio International

http://goo.gl/AQYx0k

1. Describe the three types of commodity chains and think of an example commodity produced by each process.

2. Take a closer look at a commodity you use every day—your favorite coffee, your tablet or laptop, your lipstick or tennis shoes--and trace the chain of production from company home, design, raw materials, production and market. How traveled is your favorite commodity?

clearing land for settlement, mining, and agriculture provides livelihoods and homes for some but also transforms human populations, wildlife, and vegetation. The inevitable by-products—garbage, air and water pollution, hazardous wastes, and so forth—place enormous demands on the capacity of physical systems to absorb and accommodate them.

Climate change as a result of human activity—in particular, our burning of fossil fuels, agriculture, and deforestation that cause emissions of carbon dioxide (CO_2) and other "greenhouse" gases—also has profound implications for environmental quality. Without concerted action to reduce greenhouse gas emissions, the global average surface temperature is likely to rise by a further 1.8–4.0°C this century. Since preindustrial times, even the lower end of this range would experience a temperature increase above 2°C, the threshold beyond which irreversible and possibly catastrophic changes become far more likely. Projected global warming this century is likely to trigger serious consequences for humanity and other life forms. These consequences may include a rise in sea levels of between 18 and 59 centimeters (which will endanger coastal areas and small islands) and a greater frequency and severity of extreme weather events.

In addition to the specter of global warming, we are facing serious global environmental degradation through deforestation, desertification, acid rain, loss of genetic diversity, smog, soil erosion, groundwater depletion, and the pollution of rivers, lakes, and oceans. The fate of Lake Baikal, in Russia (**Figure 2.15**), provides a distressing example. It is a place of incredible beauty—"the Pearl of Siberia"—that has long been emblematic of the pristine wilderness of the region. The lake holds 20 percent of the world's freshwater and is home to 2,500 species, many of them found nowhere else, such as the world's only exclusively freshwater seal.

The lake has warmed 1.21°C (2.18°F) since 1946 due to climate change, almost three times faster than global air temperatures. The lake's purity and unique ecosystem have also been compromised by environmental mismanagement. When thousands of the lake's freshwater seals began dying in 1997, the lake's fragile ecology came under international scrutiny, and in 1998 the lake was designated a World Heritage Site by UNESCO, the UN cultural agency. In 2007, the Russian government declared the Baikal region a Special Economic Zone, to encourage tourism. Nevertheless, it remains to be seen whether Russia can solve its environmental problems at a time when its economy is still in transition.

Environmental issues such as these point to the importance of sustainability. **Sustainability** is about the interdependence of the economy, the environment, and social well-being. This is often couched in terms of the "three Es" of sustainable development, referring to the environment, the economy, and equity in society (**Figure 2.16**). The oft-quoted definition of sustainable development from the Brundtland Report, which examined the issues on the international scale, is "development that meets the needs of the present without compromising the ability of future generations to meet their own needs."[2]

[2]World Commission on Environment and Development, *Our Common Future* (Brundtland Report), Oxford, UK: Oxford University Press, 1987, p. 40.

▼ **Figure 2.15 Lake Baikal, Russia** Lake Baikal is the world's deepest lake, at 1,615 meters (5,300 feet—over a mile), and contains about 20 percent of all the freshwater on Earth—more than North America's five Great Lakes combined.

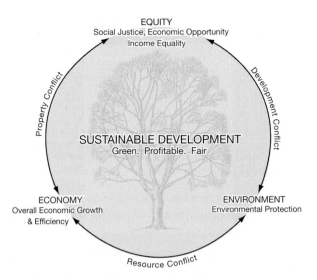

EQUITY
Social Justice, Economic Opportunity
Income Equality

Property Conflict

Development Conflict

SUSTAINABLE DEVELOPMENT
Green. Profitable. Fair

ECONOMY
Overall Economic Growth
& Efficiency

ENVIRONMENT
Environmental Protection

Resource Conflict

▲ **Figure 2.16 Sustainability** There are three key aspects of sustainability in the long run—the physical environment, equity, and economic efficiency—and there are tensions between each of these.

APPLY YOUR KNOWLEDGE

1. How is climate change the result of human activity? Think about the example of Lake Baikal in your answer and how industry, governmental policy, and management affect the natural environment.

2. Give an example of an environmental concern that affects your home region and suggest how it relates to issues of economic development and social equity.

Health Issues The increased intensity of international trade and travel has also heightened the risk and speed of the spread of disease. A striking example of the health risks associated with increasing interdependency has been the spread of the MERS (Middle East Respiratory Syndrome) virus. First reported in 2012 in Saudi Arabia, MERS is a coronavirus, in the same family as the severe acute respiratory syndrome (SARS) virus that caused an international epidemic in 2003 infecting around 8,000 people and killing more than 700 of them before it was stopped. By 2014, MERS cases had been reported in a dozen other countries.

Today, health care professionals are concerned that a new strain of influenza virus is likely to result in an influenza pandemic. A **pandemic** is an epidemic that spreads rapidly around the world with high rates of illness and death. Entirely new flu strains develop several times each century. Because no one has a chance to develop immunity to a new flu strain, it can spread rapidly and widely—and especially so in today's globalized and highly interconnected world. Similarly, there is serious concern about the possibility of epidemics in the human population resulting from zoonotic diseases (diseases originating with other species, e.g., anthrax, avian flu, ebola, West Nile virus).

Security Issues As sociologist Ulrich Beck has pointed out, the high degree of interdependence that is now embedded in a globalizing and highly interconnected world has brought about all sorts of security issues. In traditional societies, the risks faced by individuals and groups were associated mostly with hazards generated by nature (disease, flood, famine, etc.), along with socially determined hazards such as invasion and conquest and regressive forms of thought and culture. The industrial societies of the nineteenth and twentieth centuries, with more powerful technologies and weaponry, faced still more hazards, but they were mostly local and regional in nature.

Contemporary society, Beck points out, is characterized by another set of hazards, many of them uncontrollable and with a global reach. Examples include climate change as a result of human activity; the spread of weapons of mass destruction (i.e., nuclear and biological warfare); the risk of accidents involving radiation or contamination by radioactivity from nuclear fuel or nuclear waste; the risk of epidemics in the human population resulting from zoonotic diseases; the risk of epidemic disease in food animals (such as the devastating outbreak of foot-and-mouth disease, which affects cattle and sheep, in parts of northwestern Europe in 2001); and the risk of catastrophic instability in global financial markets.

Overall, Beck argues, we are moving toward a **risk society,** in which the significance of wealth distribution is being eclipsed by the distribution of risk and in which politics—both domestic and international—is increasingly about avoiding hazards. As a result, knowledge—especially scientific knowledge—becomes increasingly important as a source of power, while science itself becomes increasingly politicized—as, for example, in the case of global warming. Increased awareness of the multiple risks faced by places and regions has focused attention on the concept of resilience. **Resilience** is the ability of people, organizations or systems to prepare for, respond, recover from and thrive in the face of hazards. The goal is to ensure the continuity and advancement of economic prosperity, business success, environmental quality, and human well-being, despite external threats. In a globalized world, only the most resilient places and regions will remain economically competitive and attractive for business growth, and capable of adapting to continually changing conditions. Resilience requires the capacity to reorganize, and the ability to create and sustain the capacity to learn and adapt to change. Both are attributes that are more widespread in core regions than in semiperipheral and peripheral regions.

APPLY YOUR KNOWLEDGE

1. What is a risk society and how is the role of knowledge a tool of power?

2. Use the Internet to find the degree of international disparity in rates of infant mortality—the number of deaths of infants under 1 year of age per 1,000 live births. (*Hint:* Good sources are the World Bank, http://data.worldbank.org/, and the United Nations Development Programme, http://hdr.undp.org/en/statistics/.) Suggest why some countries still have high rates of infant mortality.

2.3 Window on the World

America's Drowning Atlantic Seaboard

In 2012, Hurricane Sandy tracked up the Atlantic coast of the United States, leaving a trail of damage estimated at $65 billion. The storm surge caused especially severe problems in New Jersey and New York (**Figures 2.B, 2.D**), and focused attention on the vulnerability of coastal communities. Twenty-three of the most populous counties in the United States are located on the coast, most of them on the Atlantic seaboard. Their vulnerability to Atlantic storms and hurricanes is rapidly increasing because of climate change and, in particular, sea level rise. Since 1980, the 600-mile stretch of Atlantic coastline between Cape Hatteras, North Carolina and Boston, Massachusetts, has experienced a rise in sea level that is between three and four times the global average. The consensus among scientists is that a significant rise in sea level—a meter or more—will occur in the next 80–90 years. In 2009, the U.S. Global Change Research Program had detailed the vulnerabilities of the mid-Atlantic coast of the United States,[1] noting that:

- rates of relative sea level rise in the mid-Atlantic region were higher than the global average and generally ranged between 0.1 and 0.2 inches per year;

- many tidal wetlands were already on the decline, in part from rising sea levels; and

- if sea level rises 39 inches (one meter) in the next century, most wetlands will be lost and many narrow barrier islands may disintegrate.

[1] James G. Titus, Eric K. Anderson, Donald R. Cahoon, Stephen Gill, Robert E. Thieler, and J. S. Williams, "Coastal Sensitivity to Sea-Level Rise: A Focus on the Mid-Atlantic Region," US Environmental Protection Agency, 2009.

▲ **Figure 2.B** Damage from Hurricane Sandy in the Staten Island borough neighborhood of Oakwood in New York, a month after the storm hit in October 2012.

▲ Figure 2.C A row of condemned homes in Nags Head, North Carolina, in the Outer Banks.

The impacts are likely to intensify the problems that coastal areas already face, including shoreline erosion (**Figure 2.C**), flooding, water pollution, damage to homes and infrastructure, and damage to natural ecosystems. Maine is preparing for a rise in sea level of up to 2 meters by 2100, and Delaware is developing contingency plans for a rise of 1.5 meters. Increasing resilience against the impacts of sea level rise and climate change means investing heavily in infrastructure and imposing planning controls on residential development. But some local governments, influenced by real estate interests and property owners and fearful of discouraging investment, have passed resolutions against sea level rise policies. In North Carolina, where a state panel of engineers and scientists concluded that the sea will likely rise 39 inches by 2100, the legislature passed a bill in 2012 placing a 4-year ban on acknowledging rising sea levels when considering coastal development within the state. At risk are more than 30,500 homes and other buildings, including some of the state's most expensive real estate. But pressure from developers meant that, rather than using any projection at all, the legislature elected to wait until a new study is completed in 2016.

1. What are some of the impacts of sea level rise on coastal regions?

2. Why did North Carolina politicians pass laws that refuse to acknowledge sea level rise, even after the panel of scientists and engineers did extensive studies? What are some other issues about the environment where you see a clash between politicians and experts? Do an Internet search to find another example of this type of debate.

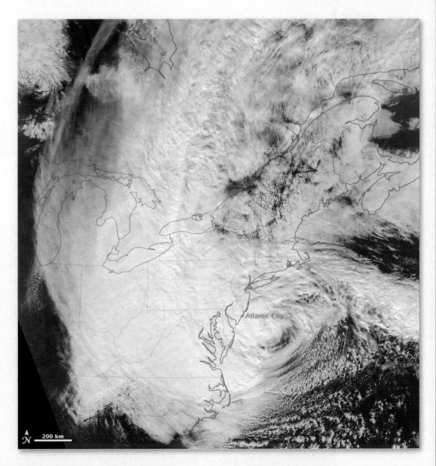

▲ Figure 2.D A satellite view of Hurricane Sandy on October 29, 2012

2.4 Spatial Inequality   Core-Periphery Disparity

One of the most striking outcomes of contemporary globalization is the consolidation of the core of the world-system. The core is now a close-knit triad of the geographic centers of North America, the European Union, and Japan. These three geographic centers are connected through three main circuits, or flows, of investment, trade, and communication: between Europe and North America, between Europe and the Far East, and among the regions of the Pacific Rim.

Globalization, although incorporating more of the world more completely into the capitalist world-system, has intensified differences in prosperity between the core and the periphery.

According to the United Nations Development Program, the gap between the poorest fifth of the world's population and the wealthiest fifth increased more than threefold between 1965 and 2012. Some parts of the periphery have almost slid off the economic map. In 55 countries, per capita income actually fell during the 1990s. In sub-Saharan Africa, economic output fell by one-third during the 1980s and stayed low during the 1990s and 2000s, so that people's standard of living there is now, on average, lower than it was in the early 1960s. In 2010, the fifth of the world's population living in the highest-income countries had:

- 75 percent of world income (the bottom fifth had just 1 percent); and

- 84 percent of world export markets (the bottom fifth had just 1 percent).

While 3 billion people around the world struggle to live on less than U.S.$2.00 a day, the world's billionaires—only 1,000 or so people—were together worth U.S.$3.5 trillion (equivalent to more than 5 percent of world GDP). OECD countries (the Organization for Economic Cooperation and Development, an association of 30 industrialized countries), with 19 percent of the global population, control more than 75 percent of global trade in goods and services and consume 86 percent of the world's goods. **Figure 2.E** shows the distribution of global population (the brighter the dot, the more the people) and their national average annual income. The disparities are the legacy of the historical geography of the world-system described in this chapter.

Such enormous differences lead many people to question the equity of the geographical consequences of globalization. The concept of **spatial justice** is important here because it requires us to consider the distribution of society's benefits and burdens at different spatial scales, taking into account both variations in people's need and in their contribution to the production of wealth and social well-being.

Many people, nations, and ethnic groups around the world feel marginalized, exploited, and neglected as a result of the quickening pace of change. Across much of the peripheral world, the perception of injustice has been brewing for a long time. Resentment at past colonial and imperial exploitation has been compounded as the more affluent places and regions of the world have become increasingly dependent on the cheap labor and resources of the periphery and as transnational businesses have displaced the traditional economic and social practices of peripheral and semiperipheral regions under the banner of modernization. Thinking about spatial justice is an important aspect of the "geographical imagination" described in Chapter 1 and is a recurring theme in the remainder of this book.

---

1. Why are the wealth disparities more extreme between the core and periphery since 1965 and with increasing globalization? Why do you think the gap has widened?

2. Can you find an example of a spatial justice project, law, incentive, or the like that has attempted to balance economic disparity in a city, town, country, or neighborhood?

▼ **Figure 2.E** Where the world's people live, by economic status. Data source: Oak Ridge National Laboratory, World Bank. (David Whitmore, John Grimwade / National Geographic)

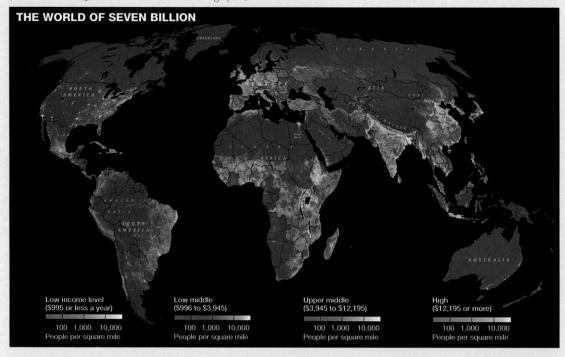

Westernization and Cultural Imperialism

Political scientist Benjamin Barber characterizes the cultural tensions associated with contemporary globalization as "Jihad versus McWorld." McWorld is shorthand for the pop culture and shallow materialism that is part of Western, capitalist modernization. **Jihad** is shorthand for cultural values that are underpinned by religious fundamentalism, traditional tribal allegiances, and opposition to Western materialism. (Note that the term *jihad* properly refers to a struggle waged as a religious duty on behalf of Islam.) Neither "Jihad" nor "McWorld" is conducive to a healthy democracy or civil society, argues Barber, while tensions between the two make for potentially volatile situations.

At the heart of these tensions is a marked disillusionment with the West, especially within traditional Islamic societies. Across much of the world, modernization is now taken to mean Westernization (**Figure 2.17**) and, more specifically, Americanization. Most Americans think of modernization as necessary and good but many other people see it as the cause of their exploitation. In most peripheral countries, only a minority can enjoy Western-style consumerism and the impoverished majority is acutely aware of the affluence of the core countries. While the gap between rich and poor countries has actually been widening for several decades, the U.S. aid budget—already low compared to the aid budgets of other developed countries—has been declining substantially. The United States, as a result, tends to be easily portrayed as a swaggering superpower, rigging the world-system to serve its own interests and doing relatively little by way of economic or humanitarian aid.

The current phase of globalization thus involves a distinctive new geopolitical element that has been described as the "new imperialism": the imperialism of the United States, the world's only superpower. Although Americans do not like to think of their country as territorially aggressive or exploitative, the "war on terror" and invasion of Afghanistan and Iraq following the al-Qaeda attacks in 2001 are widely interpreted elsewhere in the world as an exercise in imperialism, motivated in large part by a desire for military control over global oil resources. This interpretation of the United States as the instigator of a new imperialism has been reinforced by military threats against Iran, North Korea, Syria, and the Islamic State, deployment of special forces around the globe, use of "extraordinary rendition" (the apprehension or kidnapping of suspects followed by their transfer to countries known to employ harsh interrogation techniques or torture), and unilateral rejection of international environmental treaties and international aid agreements. What is less widely discussed in the world's newspapers is that this new imperialism is also viewed by some academics as the result of a highly competitive global economic environment in which the United States is no longer able to achieve superiority through innovation, product design, productivity, and marketing and so has had to resort to military intervention.

APPLY YOUR KNOWLEDGE

1. What do you think would be a way of mitigating the negative effects of both the "Jihad" and "McWorld" worldviews? (Think about the role of technology, industry, global income disparity, climate change, and environmental resources.)

2. Find a story in a national newspaper that addresses an issue associated with contemporary globalization. Provide three examples from the article that illustrate the increasing interdependence of place and region.

◀ **Figure 2.17 Westernization** Indonesian women talk in front of the billboard of a new store of the Swedish company H&M in a U.S.-style mall in Jakarta, Indonesia.

FUTURE GEOGRAPHIES

The globalization of the capitalist world-system involves processes that have been occurring for at least 500 years. But since World War II, world integration and transformation have been remarkably accelerated and dramatic. How will the forces of broadening global connectivity—and the popular reactions to them—change the fates and fortunes of world regions whose current coherence owes more to eighteenth- and nineteenth-century European colonialism than to the forces of integration or disintegration in the twenty-first century? Optimistic futurists stress the potential for technological innovations to discover and harness new resources, to provide faster and more effective means of transportation and communication, and to make possible new ways of living. Space and place, we are led to believe, will be transcended by technological fixes.

Pessimistic futurists stress the finite nature of Earth's resources, the fragility of its environment, and population growth rates that exceed the capacity of peripheral regions to sustain them. Such doomsday forecasting scenarios include irretrievable environmental degradation, increasing social and economic polarization, and the breakdown of law and order. The sort of geography associated with these scenarios is also rarely explicit, but it usually involves the probability of a sharp polarization between the haves and have-nots at every geographical scale.

Fortunately, we don't have to choose between the two extreme scenarios of optimism and pessimism. Using our geographical imagination, we can suggest a more grounded outline of future geographies. To do so, we must first glance back at the past. Then, looking at present trends and using what we know about processes of geographic change and principles of spatial organization, we can begin to map out the kinds of geographies that the future most probably holds.

Looking back at the way that the geography of the world-system has unfolded, we can see that a fairly coherent period of economic and geopolitical development occurred between the outbreak of World War I (in 1914) and the collapse of the Soviet Union (in 1989). Some historians refer to this period as the "short twentieth century." In this short century, the modern world was established, along with its now familiar landscapes and spatial structures: from the industrial landscapes of the core to the unintended metropolises of the periphery; from the voting blocs of the West to the newly independent nation-states of the South.

Today, much of the established familiarity of the modern world and its geographies seems to be disappearing. We have entered a period of transition, triggered by the end of the Cold War in 1989 and rendered more complex by the geopolitical and cultural repercussions of the terrorist attacks of September 11, 2001, and the global financial "meltdown" of 2008. Obviously, we cannot simply project our future geographies from the landscapes and spatial structures of the past. At the same time, we can only guess at some aspects of the future. But we can draw some conclusions from a combination of existing structures and budding trends. We have to anticipate how the shreds of tradition and the strands of contemporary change might be rewoven into new landscapes and new spatial structures. Among the relative certainties over the next decade or so are the increasing power and influence of East and Southeast Asia, with its enormous population (**Figure 2.18**) and its untapped natural resources; a general

▼ Figure 2.18 The uneven distribution of the world's population

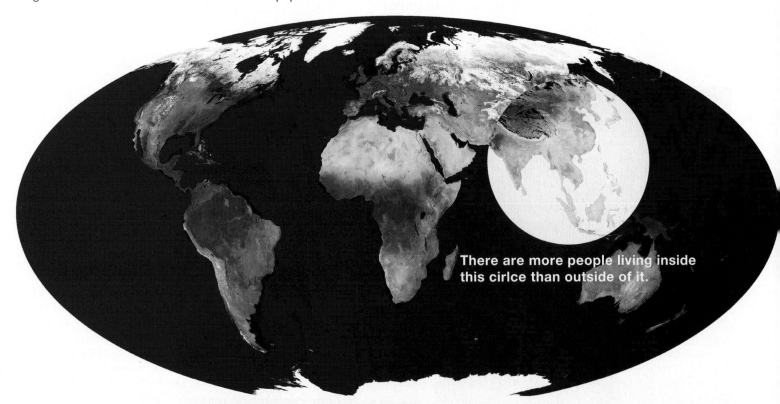

There are more people living inside this cirlce than outside of it.

shift in relative wealth and economic power from the West to the East; an increased probability of conflict in the Middle East; and an intensification of problems resulting from pressure on food, water, and energy resources. Among the key uncertainties are the speed of climate change, the resolution of the Arab-Israeli conflict, and the effectiveness of alternative energy technologies. In subsequent chapters, we examine these and related issues in detail.

APPLY YOUR KNOWLEDGE

1. What is the middle ground between the optimistic and pessimistic futurist perspectives?

2. What are specific examples of some of the "relative certainties": the shift of wealth from West to East, rising power of Asia, conflict in the Middle East, and pressure on resources? How do these global issues directly impact your geography?

CONCLUSION

Places and regions everywhere carry the legacy of a sequence of major changes in world geography. The evolution of world geography can be traced from the prehistoric hearths of agricultural development and human settlement, through the trading systems of the precapitalist, preindustrial world, to the modern world. The foundations of the modern world are industrialization, colonization, and the international market economy. Today, these foundations can be seen in the geography of the Information Age, a geography that provides a global context for places and regions.

Today's world is highly integrated. Places and regions have become increasingly interdependent, linked through complex and rapidly changing commodity chains orchestrated by transnational corporations. Using new technology systems that allow for instantaneous global telecommunications and flexible patterns of investment and production, these corporations span the world. This integration blurs some national and regional differences as the global marketplace brings about a dispersion of people, tastes, and ideas. The overall result, though, has been an intensification of differences between the core and the periphery. Within this new global context, local differences in resource endowments remain and people's territorial impulses endure. Many local cultures continue to be resilient or adaptive. Fundamental principles of spatial organization also continue to operate.

The emergence of globalization—with its transnational architectural styles, dress codes, retail chains, and popular culture and ubiquitous immigrants, business visitors, and tourists—seems as if it might inevitably impose a sense of placelessness and dislocation, a loss of territorial identity, and an erosion of the distinctive sense of place associated with certain localities. Yet the common experiences associated with globalization are still modified by local geographies. The structures and flows of globalization are variously embraced, resisted, subverted, and exploited as they make contact with specific places and specific communities. In the process, places and regions are reconstructed rather than effaced. Often this involves deliberate attempts by the residents of a particular area to create or re-create territorial identity and a sense of place. Human geographies change, but they don't disappear.

LEARNING OUTCOMES REVISITED

- **Summarize the distinctive stages of the evolution of the modern world-system.**

  Premodern geographies were organized around minisystems and regional empires. The modern world-system was established over a long period that began in the late fifteenth century. More and more peoples around the world have become exposed to one another's technologies and ideas since the fifteenth century. Different resources, social structures, and cultural systems resulted in quite different pathways of development, however. Some societies were incorporated into the new, European-based international economic system faster than others; some resisted incorporation; and some sought alternative systems of economic and political organization.

- **Analyze how and why the new technologies of the Industrial Revolution helped bring about the emergence of a global economic system.**

  The new technologies of the Industrial Revolution brought about the emergence of a global economic system that reached into almost every part of the world and into virtually every aspect of people's lives. New transportation technologies triggered successive phases of geographic expansion, allowing for an intensive period of external colonization and imperialism. The core of the world-system (Europe) grew to include the United States and Japan, while most of the rest of the world was systematically incorporated into the capitalist world-system as a dependent periphery.

- **Examine the changing patterns of interdependence among different world regions.**

  Each place and region carries out its own particular role within the competitive world-system. Because of these different roles, places and regions are dependent on one another. The development of each place affects, and is affected by, the development of many other places. Since the seventeenth century, the world-system has been consolidated, with stronger economic ties between countries. It has also been extended, with all the world's countries eventually becoming involved to some extent in the interdependence of the capitalist system and the consequent flows of resources, capital, goods, ideas, and people among places and regions.

■ **Compare the three tiers that constitute the modern world-system.**

Today, the world-system is highly structured and is characterized by three tiers: core regions, semiperipheral regions, and peripheral regions. The core regions of the world-system are those that dominate trade, control the most advanced technologies, and have high levels of productivity within diversified economies. Peripheral regions are characterized by dependent and disadvantageous trading relationships, by primitive or obsolescent technologies, and by undeveloped or narrowly specialized economies with low levels of productivity. Semiperipheral regions are able to exploit peripheral regions but are themselves exploited and dominated by the core regions. This three-tiered system is fluid, providing a continually changing framework for geographical transformation within individual places and regions.

■ **Explain how the growth and internal development of the world's core regions could take place only with the foodstuffs, raw materials, and markets provided by the colonization of the periphery.**

Peripheral regions were originally developed and exploited to provide the raw materials for industrializing regions and the food supplies for their rapidly growing populations. In the eighteenth and nineteenth centuries, the industrial core nations embarked on the inland penetration of the world's midcontinental grassland zones to exploit them for grain or stock production. At the same time, as the demand for tropical plantation products increased, most of the tropical world came under the political and economic control—direct or indirect—of one or another of the industrial core nations. For these peripheral regions, European overseas expansion meant political and economic dependency.

■ **Identify an example of each of the four key issues caused by globalization—environmental, health, core-periphery disparity, and security issues.**

Many of the important issues facing modern society are the consequences—intended and unintended—of human modifications of our physical environment. In addition, the increased intensity of international trade and travel has also heightened the risk and speed of the spread of disease. Globalization has also intensified differences in prosperity between the core and the periphery, and contemporary society is characterized by new vulnerabilities and hazards, many of them uncontrollable and with a global reach.

KEY TERMS

capitalism (p. *36*)
climate change (p. *54*)
colonialism (p. *43*)
colonization (p. *33*)
commodity chain (p. *50*)
comparative advantage (p. *45*)
core regions (p. *43*)
division of labor (p. *45*)
environmental determinism (p. *49*)

ethnocentrism (p. *49*)
external arena (p. *40*)
globalization (p. *50*)
Gross Domestic Product (GDP) (p. *50*)
hearth areas (p. *32*)
hegemony (p. *45*)
hinterland (p. *37*)
imperialism (p. *43*)
import substitution (p. *41*)

jihad (p. *59*)
law of diminishing returns (p. *33*)
leadership cycles (p. *44*)
minisystem (p. *32*)
neocolonialism (p. *49*)
pandemic (p. *55*)
peripheral regions (p. *43*)
plantation (p. *41*)
resilience (p. *55*)

risk society (p. *55*)
semiperipheral regions (p. *43*)
Slash-and-Burn (p. *32*)
Spatial Justice (p. *58*)
Sustainability (p. *54*)
Transnational Corporations (p. *50*)
Urbanization (p. *35*)
World-Empire (p. *33*)
World-System (p. *37*)

REVIEW & DISCUSSION

1. Discuss in your group whether the United States can still be considered a core country (see definitions on p. 45). Develop three reasons in support of and against your arguments. What statistical evidence supports your argument? Also consider which countries might be moving into the core or into the semiperiphery or periphery? What evidence could you find to support such transitions?

2. Two major events have had significant impacts on places and regions, namely the droughts in Russia in the summer of 2010 and the Egyptian revolution of February 2011. Discuss in your group how these events are related to globalization. Provide two examples for each case.

3. Consider the core, the semiperiphery, and the periphery, not from the perspective of states, but from the viewpoint of transnational corporations. Research two corporations and find out where their corporate headquarters are located, develop a list of what they produce, and, if possible, find out where their products come from. Take into consideration any raw materials and manufacturing that are involved. For example, your group might want to consider cell phone companies. Where are the companies located? Are they in the core, semiperiphery, or periphery? Where do they get the material to make cell phones, within the core, semiperiphery, or periphery? And finally where are the cells phones manufactured? Once you have compiled this information, develop a world map displaying your information and the ways the transnational corporations' products move between the core, semiperiphery, and periphery.

UNPLUGGED

1. The present-day core regions of the world-system, shown in Figure 2.10 (p. 44), are those that dominate trade, control the most advanced technologies, have high levels of productivity within diversified economies, and enjoy relatively high per capita incomes. What statistical evidence can you find in your local library to support this characterization? Consider and list what core countries produce. (*Hint:* Look for data in annual compilations of statistics published in annual reviews, such as the Encyclopaedia Britannica's *Yearbook* and in the annual reports of organizations, such as the United Nations Development Programme [UNDP], the World Bank, and the World Resources Institute, and consider per capita incomes and how they differ between countries). How could differences in trade be understood from these publications?

2. The idea of an international division of labor is based on the observation that different countries tend to specialize in the production or manufacture of particular commodities, goods, or services. In what product or products do the following countries specialize: Bolivia, Ghana, Guinea, Libya, Namibia, Peru, and Zambia? You will find the data you need in a good statistical yearbook, such as the Encyclopaedia Britannica's *Yearbook*, or in a world reference atlas or economic atlas. Take one product that you found and use in your everyday life and reflect on the process it underwent to come into your possession.

3. Consider the division of labor in your place of work (either as a student or outside the university). How are tasks divided and who is responsible for what types of work? Is there an apparent division of labor (see p. 45)? If so, list three examples of how work is divided in your workplace. Think of who does what. And consider why certain tasks are divided differently.

DATA ANALYSIS

The Take
http://goo.gl/Z0ENpf

In this chapter we have looked at globalization, the legacy of imperialism, income disparity, commodity chains, climate change, risk societies, and other elements of our globalized world. Terms like "Jihad" or "McWorld" construct a world of extremes, but are there other alternatives to social, economic, and environmental problems created by the global capitalism? Consider the worker-run factories in Argentina that sprang up after the collapse of their deregulated, privatized economy. View the documentary *The Take* (2007) (http://www.thetake.org), which looks at worker-run factories in Argentina and the alternative economies that have developed. Think about these questions while viewing:

1. Why did the Argentine economy collapse in 2001?

2. How does the worker-run factory system work?

3. Who began the movement and how do those workers take care of their community?

4. How has the state responded to the worker-run factories? What is the role of violence in economic revolutions? Think about who has something to lose when these alternative economies work.

5. Research the current state of the Argentinian economy. Have the worker-run factories worked? Why or why not?

6. Finally, what other alternative economies operate successfully on the planet? Take a look by conducting an Internet search on terms like collective economies, entrepreneurs in the developing world, women-run businesses around the globe, worker rights, and ethical businesses,. Do you think alternative economies can destabilize global income disparities? Can a different system of profit and production increase the quality of life for more people?

MasteringGeography™

Looking for additional review and test prep materials? Visit the Study Area in MasteringGeography™ to enhance your geographic literacy, spatial reasoning skills, and understanding of this chapter's content by accessing a variety of resources, including **MapMaster** interactive maps, Videos, *In the News* RSS feeds, flashcards, web links, self-study quizzes, and an eText version of *Human Geography*.

LEARNING OUTCOMES

■ *Explain* why populations change, where those changes occur, and what the implications of population change are for the future of different places around the globe.

■ *Identify* the two most important factors in population dynamics, birth and death, and how they shape population characteristics.

■ *Analyze* how geography is a powerful force in the incidence of health and disease.

■ *Demonstrate* how the movement of populations is affected by both push and pull factors, and explain how these factors are key to understanding new settlement patterns.

■ *Describe* the challenges of providing for the world's growing population with adequate food and safe drinking water, as well as a sustainable environment.

3

GEOGRAPHIES OF POPULATION AND MIGRATION

China, the world's most populous nation, has also been the most aggressive when it comes to controlling its size through legislative means. In 1979, China introduced a policy requiring couples from China's ethnic Han majority to have only one child. The law continues to be in effect, though there are exceptions such as would-be parents who are also both single children. There are also special benefits for parents who conform to the one-child policy. They receive longer maternity leave, for example. The policy was meant to stabilize a rapidly growing population beginning in the late 1970s and it has been successful in preventing some 250 million births.

And yet, the policy has not been an absolute success. One negative outcome of the one-child policy has resulted in an unbalanced sex-ratio as many Chinese parents have exercised a preference for sons—especially after 1986 when ultrasounds and abortions became more widely available—and chose to have sex-selective abortions to ensure that choice (until 1994 when prenatal screening was banned). Due in large part to the one-child policy, new technologies and cultural preferences, the National Bureau of Statistics reports that by 2020, Chinese men between the ages of 20 and 40 will outnumber women by 24 million. In a country where there is, at least numerically, a huge number of eligible, marriageable age men, heterosexual women would seem to be in an enviable position. And yet, many women in their twenties and early thirties appear to be rejecting marriage or are at least putting it off resulting in what China's State Council calls a "threat to social stability."

In China, young women who are not married by the time they are 27 are *shengnu*, literally "left-over" women. Yet contrary to what the term might suggest, leftover women are highly educated, progressive in their thinking and have well-paying and interesting professional employment. In theory, such women should be seen as highly attractive marriage partners, but many of them report they can't find men who are as accomplished as they are and who want to support their continued career success.

The *shengnu* "problem" is not just a concern for the government, however, as parents, academics and businesses are conceptualizing their failure to marry as having the potential to generate significant social and economic effects from driving down real estate prices on homes to luring married men into affairs. Even popular culture reflects the concern around *shengnu* where television shows like *The Price of Being a Shengnu, Go, Go, Shengnu,* and *Even Shengnu Get Crazy* capture a national preoccupation with encouraging smart, successful heterosexual women to find a man. The real problem may lie in sexism or, more specifically in men wishing to be the dominant partner in a marriage. "This leads to a phenomenon in which A-grade men marry B-grade women, B-grade men marry C-grade women, and C-grade men marry D-grade women. Only A-grade women and D-grade men can't find partners," observes Nin Lin, host of a popular Shanghai match-making television show. It may be that official, social and cultural pressures on *shengnu* are having an effect that will itself have other effects: in 2013 by age 35, 90 percent of Chinese women were married. And yet, National Health and Family Planning Commission figures indicate that marriages between couples in their thirties are the most "fragile" meaning they are more likely than marriages between couples in their twenties, forties or otherwise to end in divorce.

Population geographers are interested in situations like the *shengnu* because they illustrate the great variety that exists across the world, and even across cities and neighborhoods, in the forces that shape birth, death and human movement. How do the *shengnu* in China compare to similar 20-something women in the United States? Or how do the same women in New York City compare to those in rural South Africa where the AIDS epidemic is four times more likely to affect women between the ages of 15 and 24 than men in the same age cohort because of poverty, violence against women, cultural factors that promote intergenerational sex, and a whole host of other factors? In this chapter, we explore these sorts of questions and others at the same time that we discuss the tools that demographers—population experts including geographers—use to comprehend population differences and similarities that exist across countries and even across town.

THE DEMOGRAPHER'S TOOLBOX

Demography, the study of the characteristics of human populations, is an interdisciplinary undertaking. Geographers study population to understand the areal distribution of Earth's peoples. They are also interested in the reasons for, and the consequences of, the distribution of populations from the international to the local level. Using many of the same tools and methods of analysis as other population experts, geographers think of population in terms of the places that populations inhabit. As well they consider populations in terms of the way that places are shaped by populations and in turn shape the populations that occupy them.

Censuses and Vital Records

Population experts rely on a wide array of instruments and institutions to carry out their work. Government entities, schools, and hospitals collect information on births, deaths, marriages, migration, and other aspects of population change (**Figure 3.1**). The most widely known instrument for assessing the state of the population is the census, a survey first developed to obtain information for tax collection but used increasingly for all sorts of applications from predicting election results to shaping economic forces.

The Census   A **census** is a count of the number of people in a country, region, or city. Undertaking a census and establishing an accurate count, however, are not simple. Most censuses are also directed at gathering other information about people, such as previous residences, number of people in a household, and income. Many countries comprehensively assess the characteristics of their national populations every 10 years. In the United States, for example, the Bureau of the Census has surveyed the population every 10 years since 1790. The information gathered is used to apportion seats in the U.S. House of Representatives, as well as to redistribute federal tax funds and other revenues to states, counties, and cities.

Vital Records   In addition to the census, population experts employ other data sources to assess population characteristics. One such source is **vital records**, which report births, deaths, marriages, divorces, and the incidence of certain infectious diseases. These data are collected and recorded by city, county, state, and other levels of government.

▼ **Figure 3.1 Gay marriage registration, San Francisco**   The demise of the Defense of Marriage Act in 2013 resulted in the opportunity for gay couples to marry legally in 19 states including California.

In combination with the census data, vital records help show the way the population of an area—whether country, region, or city—is changing. Schools, hospitals, police departments, prisons, and other public agencies, such as the U.S. Bureau of Citizenship & Immigration Services and international groups like the World Health Organization, also collect demographic statistics that are useful to population experts.

Limitations of the Census

Censuses are extremely expensive and labor-intensive undertakings for any governmental jurisdiction, and as a result, they occur infrequently (**Figure 3.2**). In many less-developed countries, governments are not always able to finance a decennial census such as the comprehensive surveys undertaken in more developed countries like the United States, France, or Japan. For example, Cambodia conducted its first complete census in 1962, another in 1998, and its most recent one in 2008. Liberia conducted its first census in 24 years in 2008.

The incompatibility of census dates (caused by the different collection dates) makes comparisons between, among, and within countries quite difficult. For example, the United States conducted its last decennial census on April 1, 2010. China conducted a comparable decennial census in November 2010, its sixth nationwide census. The difference in collection times between the two censuses makes comparisons difficult, especially because processes such as international migration can have a significant impact on the actual numbers.

Despite the massive and costly efforts to count population around the world, no census is entirely comprehensive. All censuses tend to underrepresent nonmainstream kinds of households. For example, it has never been possible in the United States to explicitly identify gay households, though researchers used the category of same-sex couple households as a proxy for a gay household. The problem has been, of course, that lots of individuals of the same sex may share an address, many of whom may be heterosexual including unrelated same-sex roommates, elderly siblings, or divorced friends. The point is that while the census is the best approximation of the national population, it has its limitations.

In the United States, because federal revenue-sharing formulas as well as the apportionment of Congressional seats are tied to population numbers, political officials from many of America's large and medium-sized cities care deeply about the accuracy of population numbers. Undercounted communities stand to lose significant financial support for local schools, roads, and health services, whereas overcounted communities can reap more grants than warranted. The census is central to the way national goverments operate and its population gets access to key governmental services.

APPLY YOUR KNOWLEDGE

1. Describe three things the census provides that shape government practices and decision making.

2. Give an example, other than from the United States, of how the census provides more than just a straightforward counting of a population.

▲ **Figure 3.2 Biometric census taking in India, 2011** India became the first country in the world to collect photographs and fingerprints of all of its population as part of its decennial census exercise. Populations in many western countries have refused to participate in a biometric census count as they are reluctant to have the government possess such information about them.

POPULATION DISTRIBUTION AND COMPOSITION

Because human geographers explore the interrelationships and interdependencies between people and places, they are interested in demography. Population geographers bring to demography a special perspective—the spatial perspective—that emphasizes description and explanation of the "where" of population distribution, patterns, and processes.

The U.S. Racial Dot Map

http://goo.gl/CvzZjx

For instance, the seemingly simple fact that as of autumn 2014 the world was inhabited by nearly 7.28 billion people is one that geographers like to think about in a more complex way. This number is undeniably large and increasing overall with each passing second, but its most important aspect for geographers is the unevenness of population from region to region and from place to place. Equally important are the implications and impacts of these differences. Looking at population numbers, geographers ask themselves two questions: Where are these populations concentrated, and what are the causes and consequences of such a population distribution?

Population Distribution

Many geographic reasons exist for the distribution of populations throughout the globe. As the world population density map demonstrates (**Figure 3.3**), some areas of the world are heavily inhabited, others only sparsely. Some areas contain no people whatsoever. Bangladesh and the Netherlands, for example, have high population densities throughout. Egypt, on the other hand, displays a pattern of especially high population concentrations along the coasts and the Nile

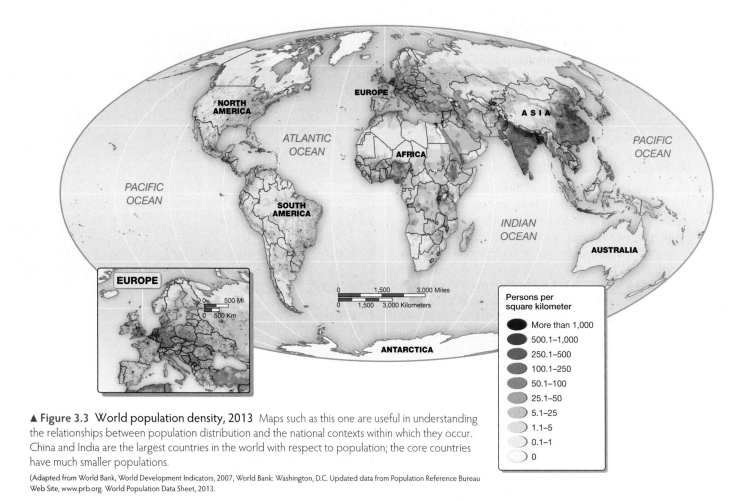

▲ **Figure 3.3** World population density, 2013 Maps such as this one are useful in understanding the relationships between population distribution and the national contexts within which they occur. China and India are the largest countries in the world with respect to population; the core countries have much smaller populations.

(Adapted from World Bank, World Development Indicators, 2007, World Bank: Washington, D.C. Updated data from Population Reference Bureau Web Site, www.prb.org. World Population Data Sheet, 2013.

River but a relatively low population density elsewhere. The sparsest inland population densities in the world occur across the mountains and rolling plateaus of Mongolia. These examples illustrate how environmental and physical factors are important influences on population distributions and concentrations.

Degree of accessibility, topography, soil fertility, climate and weather, water availability and quality, and type and availability of other natural resources are some of the factors that shape population distribution. Other factors are also crucial—first and foremost are a country's political and economic experiences and characteristics. For example, the high population concentrations along Brazil's Atlantic coast date back to the trade patterns set up during Portuguese colonial control in the sixteenth and seventeenth centuries. Another important factor is culture as expressed in religion, tradition, or historical experience. One of the key reasons cities like Medina and Mecca in the Middle East, possess relatively large population concentrations is because they are Islamic sacred sites. **Table 3.1** lists population estimates in terms of continental distributions. Asia is far and away the most populous continent. Running a distant second and third are Africa and Europe.

The population clusters that take shape across the globe have a number of physical similarities. Almost all of the world's inhabitants live on 10 percent of the land. Most live near the edges of landmasses, near the oceans or seas, or along rivers with easy access to a navigable waterway. Approximately

TABLE 3.1   World Population Estimates by Continents, 2013

Continent	Number of Inhabitants (in millions)	% of Total Population
Asia	4,302	60
Africa	1,100	15
Europe	740	10
Latin America and the Caribbean	606	8.5
North America	352	0 5
Oceania	38	0.5
Total	7,138	100

Source: Population Reference Bureau Web Site, www.prb.org. *World Population Data Sheet,* 2013.

90 percent live north of the equator, where the largest proportion of the total land area (63 percent) is located. Finally, most of the world's population lives in temperate, low-lying areas with fertile soils.

Population numbers are significant not only on a global scale. Population concentrations within countries, regions, and even metropolitan areas are also important. For example, much of the population of North Africa is distributed along the coastal areas where most of the large cities are, as well as along the Nile River in Egypt. In Australia, much of the population is clustered along the coast (**Figure 3.4**).

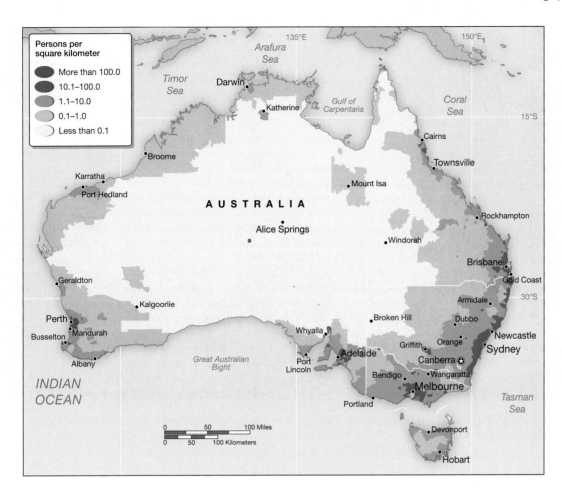

◄ **Figure 3.4 Population density, Australia, 2013** This map shows that population densities are highest along the coasts and lowest in the interior of the country where harsh desert and grasslands prevail.

Population Density

Another way to explore population is in terms of **density**, a numerical measure of the relationship between the number of people and some other unit of interest expressed as a ratio. Crude density is probably the most common measurement of population density. **Crude density**, also called **arithmetic density**, is the total number of people divided by the total land area. The metropolitan area of Bangkok, Thailand, is a very populous Asian city containing over 14.5 million residents, with a population density of approximately 6,200 persons per square kilometer (about 16,200 persons per square mile), which is about 2.5 times the density of Los Angeles (**Figure 3.5**).

The limitation of the crude density ratio—and hence the reason for its "crudeness"—is that it is one-dimensional. It tells us very little about the variations in the relationship between people and land. For that, we need other tools for exploring population density, such as nutritional density or agricultural density.

■ **Nutritional density** is the ratio between the total population and the amount of land under cultivation in a given unit of area.

■ **Agricultural density** is the ratio between the number of agriculturists—people earning their living or subsistence from working the land—per unit of farmable land in a specific area.

■ **Health density** can be measured as the ratio of the number of physicians to the total population (**Figure 3.6**).

Population Composition

In addition to exploring patterns of distribution and density, population geographers also examine population in terms of composition—that is, the subgroups that constitute it. Understanding population composition enables geographers to gather important information about population dynamics. For example, knowing the composition of a population in terms of the total number of males and females, number and proportion of senior citizens and children, and number and proportion of people active in the workforce provides valuable insights into the ways in which the population behaves now and how it might behave in the future. The following Spatial Inequality feature provides a snapshot of population segregation in the United States and the effect segregation has on that population's life chances.

▲ **Figure 3.5 Bangkok's population density** Bangkok, Thailand is a crowded city with a very high population density which leads to some of the worst traffic in the world as well as poor air quality from slow moving vehicles.

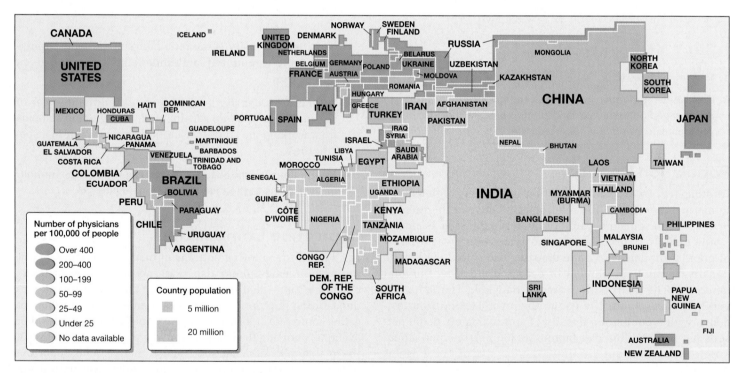

▲ **Figure 3.6 Health care density** Another measure of population density is reflected in this map, which shows the number of people per physician in the total population. Most of the core countries as well as Cuba, the former Soviet Union and some Central Asian and South American countries have the highest ratio of doctors to overall population. Most of the continent of Africa, except South Africa and Egypt, Libya, and Morocco in North Africa, has the lowest ratio, reflecting another dimension of core-periphery inequality.

(*Source:* Adapted from H. Veregin (ed.), *Goode's World Atlas*, 22nd ed. Chicago, IL: Rand McNally & Co, 2010, p. 55. Updated data from NationMaster.com, http://www.nationmaster.com/graph/hea_phy_per_1000_peo-physicians-per-1-000-people; World Development Indicators database: http://data.worldbank.org/data-catalog/world-development-indicators.)

3.1 Spatial Inequality   Population Segregation in the United States

As a country, the United States is more racially and ethnically diverse than ever before. And in metropolitan areas, statistics indicate that black-white segregation has been on a slow but steady decline since the 1970s (**Figure 3.A**). Of the largest U.S metro areas, only New York has become more segregated (**Figure 3.B**). But if we look more carefully at black-white residential occupation, we can discern a more complicated story about segregation. Using the metro area of Atlanta as an illustration, for example, it's clear from **Figure 3.C** that urban blacks are actually becoming more isolated from whites. This is because Latinos and Asians are living in the neighborhoods where whites previously lived and whites have moved even further away from blacks. have moved into previously white areas. As such, the residential distance between blacks and whites has actually increased. The result is actually increased segregation between blacks and whites, which is a problem because it drives down the economic growth of cities, as witnessed in the declared bankruptcy of the city of Detroit in 2013. Important, as well, is the fact that urban segregation also affects the wider region within which the city is situated.

Why is it that the effects of segregation ripple out from their sources? As reported in *The Atlantic*, it's because segregation affects the life chances of minorities and their limited futures mean less growth and prosperity for the regions in which they are anchored. How does segregation affect places and regions?[1] The following is a list of the most clear-cut ways that segregation affects economic status.

- **Education**: Each year spent in a desegregated school increases black students' average annual earnings after graduation by about

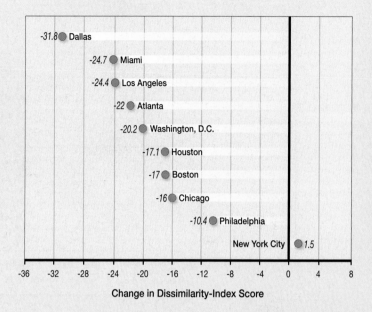

▲ **Figure 3.B Declining segregation in U.S. cities** This graphic shows that there have been significant declines in black-white segregation in most of the largest U.S. cities, except for New York.

5 percent. Schools in segregated neighborhoods are often poorly funded due to a limited property tax base, which affects the quality of teachers who can be recruited and the resources that can be invested in the schools.

- **Incarceration**: Black students who attend desegregated schools starting in elementary school are 16.9 percent less likely to be in prison by the age of 30 as compared with black students who attend segregated schools. A poor reading level at the end of the fourth grade is one of the best predictors of a future prison term. Underfunded schools in low-income minority neighborhoods often produce underprepared students.

- **Health**: Full black-white residential integration would lower the black infant mortality rate by at least two deaths per 1,000 live births. And older male adults in racially segregated high-crime neighborhoods have a 31 percent higher chance of developing cancer than their counterparts in safer, less segregated areas. Hospitals serving segregated black neighborhoods have less technology and fewer medical specialists than those for a white population of equivalent socioeconomic status. As well, black segregated neighborhoods have more fast food outlets and fewer supermarkets and recreational facilities.

- **Pollution**: Residents of minority neighborhoods experience 5 to 20 times more exposure to pollution than people who live in predominantly white neighborhoods. Due to high housing costs and historical discrimination, low-income minority neighborhoods are often clustered around industrial sites, truck routes, ports, and other air pollution and toxin hotspots.

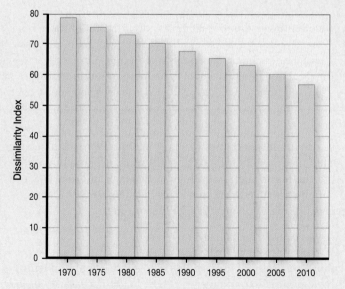

▲ **Figure 3.A Black-white segregation in the United States, 1970–2010** Residential segregation, as the graphic shows, appears to be in decline.

[1]Adapted from: Emily Badger, "The Real Cost of Segregation—in One Big Chart," *The Atlantic*, August 14, 2–13, p. 29.

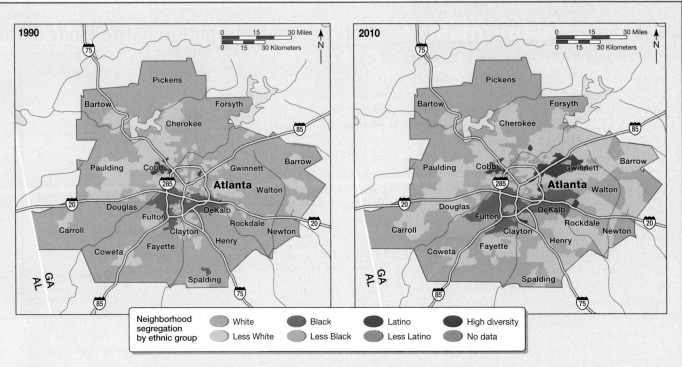

▲ **Figure 3.C Declining segregation** The trend of declining segregation is hidden by the urban blacks who are actually becoming more isolated as Latinos and Asians are occupying neighborhoods that were once dominated by white families.

1. Low standardized test scores on reading level for fourth grade children is one of the indicators that private prison investment groups use when determining how many prisons to build in a region. Explore reading level test data in your local school system, which can be found through your state's department of education website or through your local school districts' websites. Compare a school with low fourth grade reading score to one with high scores. Then find data from the U.S. census on the demographics of these two areas. Do the data in your locale link poor reading scores to low income areas; high scores to high income areas? Is there a racial dimension to the data? Why or why not might that be the case?

2. Using newspaper reports on the Internet, identify incidents of protests against local pollution in any city in the United States. From carefully reading a sample of articles about the problem, determine what the demographic composition of that neighborhood is as well as the socioeconomic status of the residents. What are some reasons why pollution sources are located in this neighborhood?

Age-Sex Pyramids

Areal distributions are not the only way that demographers portray population distributions. To display variations within particular subgroups of a population or with respect to certain descriptive aspects, such as births or deaths, demographers also use bar graphs displayed both horizontally and vertically. The most common way for demographers to graphically represent the composition of the population is an **age-sex pyramid**, which is a representation of the population based on its composition according to age and sex. An age-sex pyramid is actually a bar graph displayed horizontally. Ordinarily, males are portrayed on the left side of the vertical axis and females on the right. Age categories are ordered sequentially from the youngest at the bottom of the pyramid to the oldest at the top. By moving up or down the pyramid, one can compare the opposing horizontal bars in order to assess differences in frequencies for each age group.

Age-sex pyramids allow demographers to identify changes in the age and sex composition of populations. For example, an age-sex pyramid depicting Germany's 2000 population clearly revealed the impact of the two world wars, especially with regard to the loss of large numbers of males of military age and the deficit of births during those periods (**Figure 3.7**). Demographers call population groups like these cohorts.

Population Cohorts

A **cohort** is a group of individuals who share a common temporal demographic experience. A cohort is not necessarily based only on age, however. Cohorts may be defined based on criteria such as time of marriage or time of graduation.

In addition to revealing the demographic implications of war or other significant events, age-sex pyramids can provide information necessary to assess the potential impacts that growing or declining populations might have. The shape of an age-sex pyramid varies depending on the proportion of people in each age cohort. The pyramid for the peripheral countries, shown in **Figure 3.8a**, reveals that many dependent children,

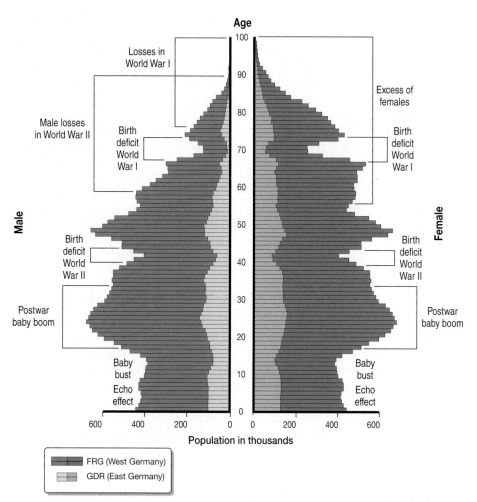

Age

Losses in World War I

Male losses in World War II

Birth deficit World War I

Male

Birth deficit World War II

Postwar baby boom

Baby bust

Echo effect

Excess of females

Birth deficit World War I

Female

Birth deficit World War II

Postwar baby boom

Baby bust

Echo effect

Population in thousands

FRG (West Germany)

GDR (East Germany)

◄ **Figure 3.7 Population of Germany, by age and sex, 2007** Germany's population profile is that of a core country that has passed through the postwar baby boom and currently possesses a low birthrate. It is also the profile of a country whose population has experienced the ravages of two world wars.

(*Source:* Adapted from J. McFalls, Jr., "Population: A Lively Introduction," 5th ed., *Population Bulletin, 62*(1), 2007, p. 20.)

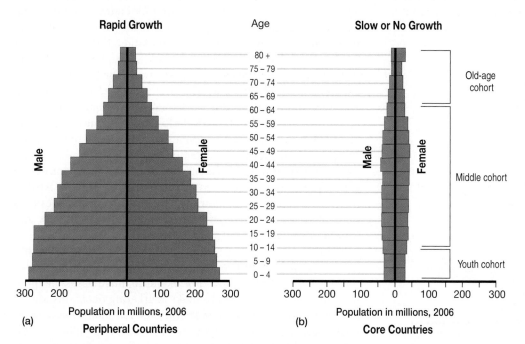

Rapid Growth   Age   **Slow or No Growth**

Old-age cohort

Middle cohort

Youth cohort

Male   **Female**   **Male**   **Female**

Population in millions, 2006   Population in millions, 2006

(a) **Peripheral Countries**   (b) **Core Countries**

◄ **Figure 3.8 Population pyramids of (a) peripheral and (b) core countries** Population pyramids vary with the age and sex structure of the population being depicted. We can derive important information about the population growth rates of different countries over time by analyzing changes in the numbers of people in each sex and age category.

ages 0 to 14, exist relative to the rest of the population. The considerable narrowing of the pyramid toward the top indicates that the population has been growing very rapidly in recent years. The shape of this pyramid is typical of peripheral countries with high birthrates and low death rates.

Serious implications are associated with this type of pyramid. First, in the absence of high productivity and wealth, resources are increasingly stretched to their limit to accommodate even elemental schooling, nutrition, and health care for the growing number of children.

Furthermore, when these children reach working age, a large number of jobs will have to be created to enable them to support themselves and their families. Also, as they form their own families, the sheer number of women of childbearing age will almost guarantee that the population explosion will continue. This will be true unless strong measures are taken, such as intensive and well-funded birth-control campaigns, improved education, and outside opportunities for women, as well as modifications of cultural norms that place a high value on large family size.

In contrast, the pyramid for the core countries (see **Figure 3.8b**) illustrates the typical shape for a country experiencing a slow rate of growth. Most countries in the core such as Denmark, Spain, and the United States are experiencing birthrates that are at or below replacement level. Thus, the pyramid is very columnar, hardly a pyramid at all. People are equally distributed among the cohorts, though the base is perceptibly narrower. In all these countries, however, high levels of production and wealth, combined with low birthrates, translate into a generally greater capacity to provide not only high levels of health, education, and nutrition but also jobs, as those children grow up and join the workforce. Visualizing Geography 3.1 provides some insight into educational trends among young people. (Their possible effect on the future of wealth and level of production suggest that globalization will continue to be significant.) It is important to note that age-sex pyramids can be constructed at any level from the national to the neighborhood. **Figure 3.9** is an age-sex pyramid from Tompkins County, New York. Cornell University is located there.

A critical aspect of the population pyramid is the **dependency ratio**, which is a measure of the economic impact of the young and old on the more economically productive members of the population. In order to assess this relation of dependency in a particular population, demographers divide the total population into three age cohorts, sometimes further dividing those cohorts by sex. The **youth cohort** consists of those members of the population who are less than 15 years of age and generally considered to be too young to be fully active in the labor force. The **middle cohort** consists of those members of the population aged 15 to 64, who are considered economically active and productive. Finally, the **old-age cohort** consists of those members of the population aged 65 and older, who are considered beyond their economically active and productive years. By dividing the population into these three groups, it is possible to obtain a measure of the dependence of the young and old upon the economically active and the impact of the dependent population upon the independent (see **Figures 3.8a** and **3.8b**).

The Effect of Population Cohorts

Countries with populations that contain a high proportion of old people face unique challenges. This is a situation most core countries will soon be facing as their "baby boom" generation ages. The **baby boom** generation includes those individuals born between 1946 and 1964. A considerable amount of a country's resources and energies will be necessary to meet the needs of a large number of people who may no longer be contributing in any significant fashion to the creation of the wealth necessary for their maintenance. There might also be a need to import workers to supplement the relatively small working-age population.

Similarly, knowing the number of women of childbearing age in a population, along with other information about their status and opportunities, can provide valuable information about the future growth potential of that population. For example, populations in core countries like Denmark, which has a small number of women of childbearing age relative to the total population size but with high levels of education, socioeconomic security, and wide opportunities for work outside the home, will generally grow very slowly if at all. Peripheral countries like Kenya, on the other hand, where a large number of women of childbearing age have low levels of education and socioeconomic security and relatively few employment opportunities, will continue to experience relatively high rates of population growth, barring unforeseen changes. The variety that exists within a country's population shapes the opportunities and challenges it must confront nationally, regionally, and locally.

Understanding population composition, then, not only tells us about the future demographics of regions but is also quite useful in the present. For example, businesses use population composition data to make marketing decisions and to decide where to locate their operations. For many years, these businesses used laborious computer models to help target their markets. With the development of geographic information systems (GIS), however, this process has been greatly simplified. Assessing the location and composition of particular populations is known as **geodemographic analysis**.

Figure 3.10 shows the present state and potential future impact of the baby boom cohort, the largest population cohort in

Age

▲ Figure 3.9 Population pyramid for Tompkins County, New York, 2010 Cornell University, is in the town of Ithaca, which is part of Tompkins County. The extended bars for 15–19 and 20–24 age ranges can be easily explained by the predominance of young people in this college town.

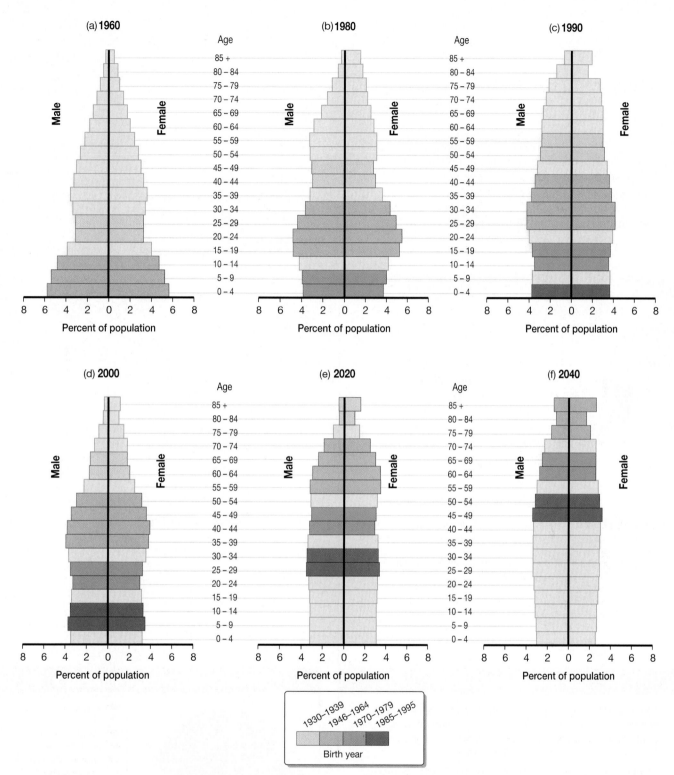

▲ **Figure 3.10 Population pyramids of U.S. baby boomers, 1960–2040** Population pyramids may be constructed based on estimates of the future. In this series of pyramids, we can clearly see the progression of the baby boomers, shown in orange, up the pyramid as their cohort ages. Note how the "pyramid" becomes a column by the year 2040 as birthrates remain below death rates for each cohort. Note also the significantly higher number of women compared with men in the oldest age group for this same year, reflecting the statistical tendency for women to live longer than men.

(*Source:* Adapted from "Twentieth Century U.S. Generations" *Population Bulletin*, *64*(1), 2009, p. 5.)

U.S. demographic history. This series of pyramids illustrates how the configuration changes as the boomers age. The narrower column of younger people rising below the boomer cohort in these pyramids reveals the biggest problem facing this population: a significantly smaller cohort moving into its main productive years having to support a growing cohort of aging and decreasingly productive boomers. Congressional fights over Social Security, the Affordable Care Act, and Medicare funding are only the tip of the iceberg with regard to this problem.

3.2 Visualizing Geography

Education Abroad

Study abroad is a phenomenon that is gaining increasing popularity in U.S. institutions of higher education. Whereas twenty years ago very few college age students left the U.S. to enjoy a period of time in a foreign educational experience, today nearly 10% of U.S. undergraduates have studied abroad before they graduate.

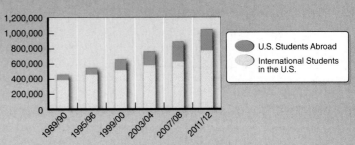

Figure 3.A.1 International Students in the U.S. and U.S. Students Abroad, 1989-2012

3.1 International students studying in the U.S.

World total of international students studying in the U.S. 2011/2012 = **764,495**

Figure 3.1.1 U.S. Destinations for International Students

The **top 10 states** together host **61%** of all international students. The **top 3 states**, California, New York, and Texas, host **32%**.

Top ten places of origin of international students in the U.S. (2011/2012):

China 29%	Japan 2%
India 12%	Vietnam 2%
South Korea 9%	Mexico 2%
Saudi Arabia 5%	Turkey 1%
Canada 3%	Other places of origin: 32%
Taiwan 3%	

3.2 Economic impact

The 819,644 international students who enrolled in U.S. postsecondary institutions in 2012/2013 contributed about $24 billion dollars to the U.S. economy which translates into the support of around **313,000 jobs**.

Net Contribution to U.S. Economy by Foreign Students and their Families: **$23,954,000,000**

For every **seven** international students enrolled in a U.S. institution of higher learning...

...**three** U.S. jobs are created or supported by spending that is spread in sectors across the spectrum that includes accommodation, dining, transportation, communication (especially telecommunication), and health care and insurance.

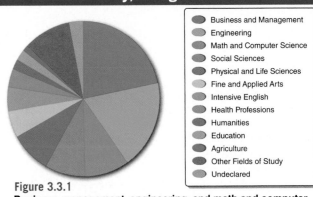

CANADA 1%

Mexico

LATIN AMERICA 16%

3.3 Fields of study, foreign students in the U.S.

- Business and Management
- Engineering
- Math and Computer Science
- Social Sciences
- Physical and Life Sciences
- Fine and Applied Arts
- Intensive English
- Health Professions
- Humanities
- Education
- Agriculture
- Other Fields of Study
- Undeclared

Figure 3.3.1
Business management, engineering, and math and computer science are the top three fields of study for international students in the U.S.

3.4 U.S. students studying abroad

It's important to appreciate that just as U.S. students are travelling outside of the country to experience education in another place, undergraduate and graduate students from other parts of the world are increasingly coming to the United States for an educational experience. Nearly half of all international students who come to the U.S. to study come from just three countries: China, India and South Korea. And in contrast to the international educational exposure U.S. students are seeking, foreign students coming to the United States are likely to be studying business management, engineering and math and computer science. These international movements of individuals between the ages of 18 and 30 have lots of effects that include the culture and politics but most especially the economy.

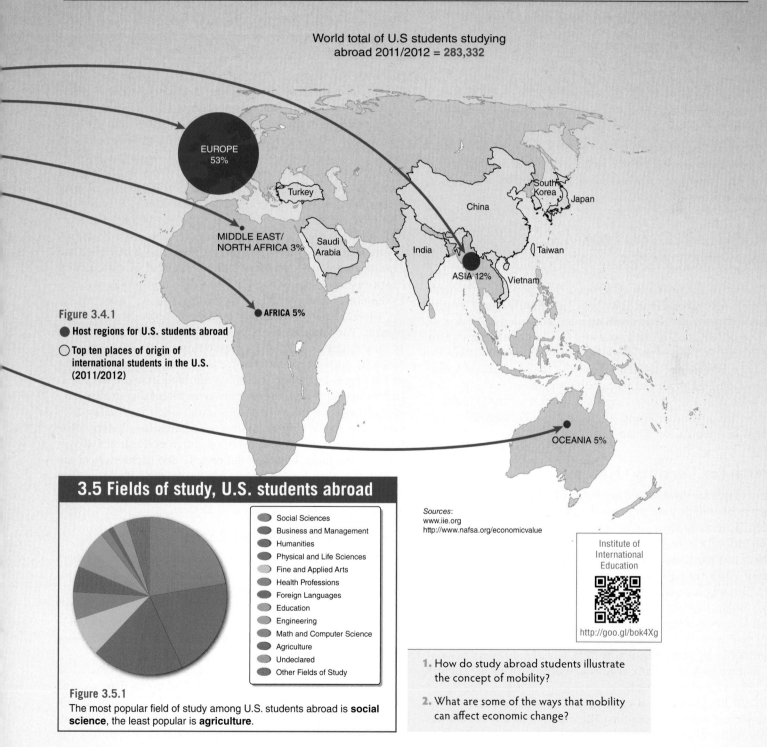

World total of U.S students studying
abroad 2011/2012 = 283,332

Figure 3.4.1

● **Host regions for U.S. students abroad**

○ **Top ten places of origin of international students in the U.S. (2011/2012)**

Sources:
www.iie.org
http://www.nafsa.org/economicvalue

Institute of
International
Education

http://goo.gl/bok4Xg

3.5 Fields of study, U.S. students abroad

- Social Sciences
- Business and Management
- Humanities
- Physical and Life Sciences
- Fine and Applied Arts
- Health Professions
- Foreign Languages
- Education
- Engineering
- Math and Computer Science
- Agriculture
- Undeclared
- Other Fields of Study

Figure 3.5.1

The most popular field of study among U.S. students abroad is **social science**, the least popular is **agriculture**.

1. How do study abroad students illustrate the concept of mobility?

2. What are some of the ways that mobility can affect economic change?

TABLE 3.2 Baby Boomer Population Structure

Year	% of Population Who Are Baby Boomers	Age
1990	30	25–44
2000	20	35–59
2020	15	55–79
2040	7	75–85+

The baby boom demographically dominated the last half of the twentieth century, but its influence will wane by the end of the first half of the new one (**Table 3.2**).

APPLY YOUR KNOWLEDGE

1. Why do researchers divide the population of a country into youth, middle, and old-age cohorts? What do these categories indicate about the potential of a country's population?

2. Using the rate of natural increase for both China and India—with the first and second largest populations in the world—identify which of these two countries adds more people to its population annually?

POPULATION DYNAMICS AND PROCESSES

In order to arrive at an understanding of population growth and change, experts look first at two significant factors: fertility and mortality. Birth and death rates are important indicators of a region's level of development and its place within the world economy. To understand population growth overall, however, they must also look at the movement of the population. A simple equation for calculating population growth is $G = B - D + (I - E)$, where G (growth) equals B (births) minus D (deaths) plus I (immigration) minus E (emigration). We look at each of these key population dynamics in turn.

Birth (or Fertility) Rates

The **crude birthrate (CBR)** is the ratio of the number of live births in a single year for every thousand people in the population. The crude birthrate is indeed crude, because it measures the birthrate in terms of the total population and not with respect to a particular age-specific group or cohort. For example, as of 2013, the CBR for the entire population of Canada was 11. But when we break that large population into subsections we find birthrates varying across age groups quite significantly. For Canadian women aged 20 to 24 the CBR was 45.7; for 25 to 29 year olds, it was 95.2; for 30 to 34 year olds, it was 105.9; and 35 to 39 year olds, it was 52.3. Clearly, differences exist when we look at specific groups and especially at age cohorts at their reproductive peak.

Although the level of economic development is a very important factor shaping the CBR, other, often equally important, influences also affect it. In particular, it may be heavily influenced by the demographic structure of the population, as graphically suggested by age-sex pyramids. In addition, an area's CBR is influenced by women's educational achievement, religion, social customs, and diet and health, as well as by politics and civil unrest. Most demographers also believe that the availability of birth-control methods is critically important to a country's or region's birthrate. A world map of the CBR (**Figure 3.11**) shows high levels of fertility in most of the periphery of the world economy and low levels of fertility in the core. The highest birthrates occur in Africa, the poorest region in the world.

Total Fertility Rate

The crude birthrate is only one indicator of fertility and in fact is somewhat limited in its usefulness, revealing very little about the potential for future fertility levels. Two other indicators formulated by population experts—the total fertility rate and the doubling time—provide more insight into the potential of a population. The **total fertility rate (TFR)** is a measure of the average number of children a woman will have throughout the years that demographers have identified as her childbearing years, approximately ages 15 through 49 (**Table 3.3**). Whereas the CBR indicates the total number of births per 1,000 people in a given year, the TFR is a more predictive measure that attempts to portray what birthrates will be among a particular cohort of women over time. A population with a TFR of slightly higher than 2 has achieved replacement-level fertility. This means that birthrates and death rates are approximately balanced and there is stability in the population.

Closely related to the TFR is the doubling time of the population. The **doubling time**, as the name suggests, is a measure of how long it will take the population of an area to grow to twice its current size. A country whose population increases at 1.8 percent per year will have doubled in about 40 years. In fact, world population is currently increasing at this rate. By contrast, a country whose population is increasing 3.18 percent annually will double in only 22 years—the doubling time for Kenya. Birthrates and the population dynamics we can project from them, however, tell us only part of the story of the potential of the population for growth. We must also know the death (mortality) rates.

Death (or Mortality) Rates

Countering birthrates and shaping overall population numbers and composition is the **crude death rate (CDR)**, the ratio of the number of deaths in one year to every thousand people in the population. As with crude birthrates, crude death rates often roughly reflect levels of economic development—countries with low birthrates generally have low death rates (**Figure 3.12**).

Although often associated with economic development, CDR is also significantly influenced by other factors. A demographic structure with more men and elderly people, for example, usually means higher death rates. Other important influences on mortality include health care availability, social class, occupation, and even place of residence. Poorer groups in the population have

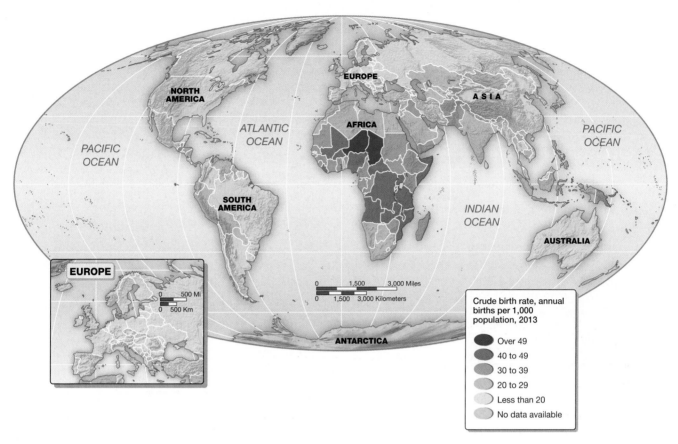

▲ **Figure 3.11** World crude birthrates, 2013   Crude birthrates and crude death rates are often indicators of the levels of economic development in individual countries. For example, the CBR of Australia offers a stark contrast to that of Ethiopia, a very poor and underdeveloped peripheral country.

(*Source:* Data from Population Reference Bureau, *World Population Data Sheet,* 2013.)

TABLE 3.3   Highest and Lowest TFR for the Top and Bottom Five Countries, 2013

Country	Total Fertility Rate (TFR)
Niger	7.6
Chad	7.0
Somalia	6.8
Democratic Rep. of Congo	6.3
Angola	6.3
Bosnia-Herzegovina	1.2
Taiwan	1.3
Moldova	1.3
Poland	1.3
Portugal	1.3

Source: Population Reference Bureau, *World Population Data Sheet,* 2013.

higher death rates than the middle class. In the United States, coal miners have higher death rates than schoolteachers and urban areas often have higher death rates than rural areas. The difference between the CBR and CDR is the rate of:

- **natural increase**—the surplus of births over deaths (**Figure 3.13**), or

- **natural decrease**—the deficit of births relative to deaths.

Infant Mortality and Life Expectancy

Death rates can be measured for sex and age cohorts; one of the most common measures is the **infant mortality rate**. This figure reflects the annual number of deaths of infants under 1 year of age compared to the total number of live births for that same year. The figure is usually expressed as the number of deaths during the first year of life per 1,000 live births. The infant mortality rate has been used by researchers as an important indicator both of the adequacy of a country's health care system and of the general population's access to health care. Global patterns show that infant mortality rates are high in the peripheral countries of Africa (68) and Asia (35) and low in the more developed countries of Europe (5) and North America (6) (**Figure 3.14**). Generally, the core's low rates reflect adequate maternal nutrition and the wider availability of health care resources and personnel.

However, when patterns are examined at the level of countries, regions, and cities, infant mortality rates are not uniform. In the United States, for example, blacks as well as other ethnic minorities in urban and rural areas, suffer infant mortality rates that are twice as high as the national average. In east central Europe, Bulgaria has a 7.8 per-1,000 infant mortality rate, while Czechia has a rate of 2.6 per 1,000. In Israel, the infant mortality rate per 1,000 is 3.5 and in the neighboring Palestinian territories, it is 20.

When war or political strife is introduced into the equation, the infant mortality rate skyrockets, as the Palestine example

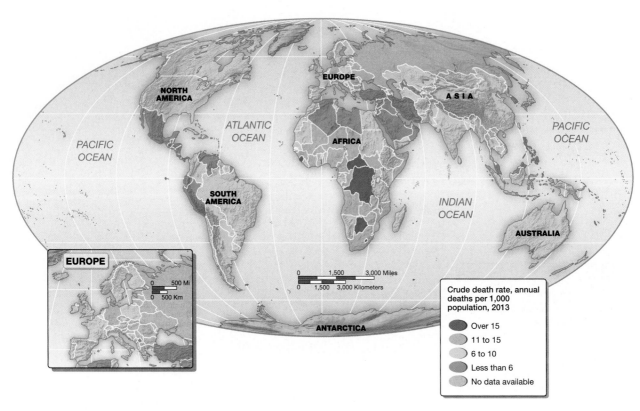

▲ **Figure 3.12 World crude death rates, 2013**   The global pattern of crude death rates varies from crude birthrates. Most apparent is that the difference between highest and lowest crude death rates is relatively smaller than the difference for crude birthrates, reflecting the impact of factors related to the middle phases of the demographic transition.

(*Source:* Data from Population Reference Bureau, *World Population Data Sheet,* 2013.)

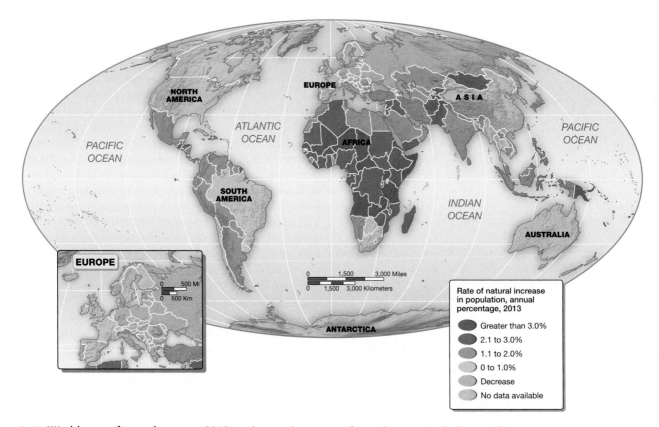

▲ **Figure 3.13 World rates of natural increase, 2013**   As the map shows, rates of natural increase are highest in sub-Saharan Africa, the Middle East, and parts of Asia, as well as parts of South and Central America. Europe, the United States, and Canada, as well as Australia and parts of central Asia and Russia have slow to stable rates of natural increase.

(*Source:* Data from Population Reference Bureau, *World Population Data Sheet,* 2013.)

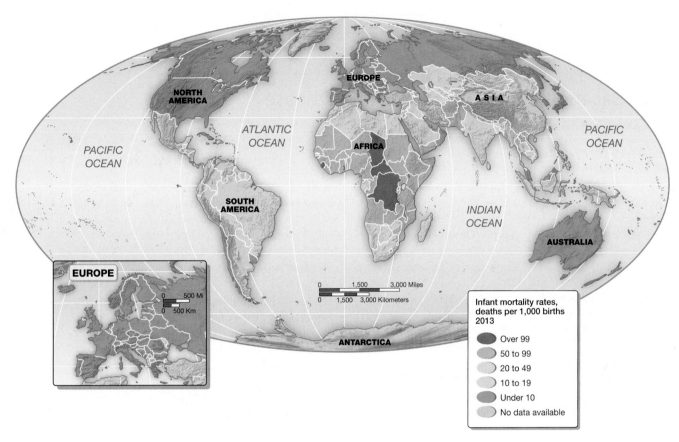

▲ **Figure 3.14** World infant mortality rates, 2013 The geography of poverty underlies the patterns shown in this map and allows us to analyze the linkages between population variables and social conditions. Infant mortality rates generally parallel crude death rates, with sub-Saharan Africa generally reporting the highest rates. These rates reflect a number of factors, including inadequate or completely absent maternal health care as well as poor nutrition for infants.

(*Source:* Data from Population Reference Bureau, *World Population Data Sheet*, 2013.)

illustrates. More dramatic, however, is that during the height of the Iraq war, Iraq had a rate of 84 infant deaths per 1,000 live births. Today the rate is 28 with the rate for surrounding countries such as Saudi Arabia 16 and Oman 9. Yemen, a poor country that is also experiencing high levels of civil strife has a rate of 72. Global patterns often mask regional, local, and even neighborhood variation in mortality rates for both infants and other population cohorts.

Related to infant mortality and the crude death rate is **life expectancy**, the average number of years a newborn infant can expect to live. Not surprisingly, life expectancy varies considerably from country to country, region to region, and even from place to place within cities and among different classes and racial and ethnic groups. In the United States, a child born in 2013 can expect to live 79 years. If we begin to specify the characteristics of that infant by sex and race, however, variation emerges. An African American male born in 2013 has a life expectancy of 69.5 years, while a white female born that same year can expect to live, on average, 81 years.

Another key factor influencing life expectancy is epidemics, which can quickly and radically alter population numbers and composition. In our times, epidemics can spread swiftly over great distances, largely because people and other disease carriers can travel from one place to another rapidly. Epidemics can have profound effects, from the international to the local level, and reflect the increasing interdependence of a shrinking globe. Geographers are particularly interested in understanding the ways that epidemics, disease, and health in general

may affect different population groups in different ways and have a greater or lesser impact on different localities.

Medical and Health Geographies

Medical geography is a subarea of the discipline that specializes in understanding the spatial distribution and extent of disease and illness as well as health care access and delivery. This perspective tends to start from the existence of disease and then uses quantitative approaches, such as disease mapping, to understand its spatial distribution and diffusion. In contrast, **health geography** starts with health as an initial condition and focuses on the dynamic relationship between health, people, and place. For example, health geographers consider race, class, ethnicity, and gender and how it affects or is affected by place (**Figure 3.15**). Whereas medical geographers explore the spatial manifestation of disease change—old diseases are returning (tuberculosis), new diseases are emerging (Ebola) and other diseases are spreading beyond their previous boundaries (dengue fever)—health geographers look to understand how social inequalities, environmental factors, or significant transformations in economies such as changing agricultural practices—from traditional rice crops to GMO (genetically modified organism) crops in India—can lead to health inequalities for particular populations. Health geographers are also interested in alternative medicine and what it means to have healthy places in which to live (**Table 3.4**). The **Geography Matters**

3.3 Geography Matters

The Global and the Local in Health and Disease

by Vincent J Del Casino Jr, University of Arizona

If you ask a health geographer what the greatest threats are to global health today, she would have to answer you by starting with, "It depends." That's because health is a product of where people live and the health-related conditions one finds in those particular places. For example, a late 2013 UN report provided figures showing that in that year, Mexico surpassed the United States as the world's most obese country.[1] **Obesity** is a medical condition in which excess body fat may cause health complications, leading to reduced life expectancy (**Figure 3.D**). The UN report's data were surprising because Americans have held the title as the world's most obese population for a decade. And, the difference between the two countries is quite small: roughly one-third of the adult population in both countries is considered obese. Obesity—often found in higher rates among poorer populations—is complicated by diabetes and related conditions and leads to hundreds of thousands of deaths each year in Mexico and the United States.

But let's look at the flip side of weight-related health issues in the world's rich and middle-income countries. When we do, what we learn is that while the poorest adult citizens in some of the world's wealthier countries are dying from obesity-related diabetes, in low-income countries, impoverished children and mothers are dying from malnutrition or suffering severe long-term ill-health from chronic under-nutrition. Both are examples of what experts call the double burden of malnutrition: the case of both over and under-nutrition in a national population (**Figure 3.E**). Remarkably, in both cases and despite being very different diseases, researchers find that inexpensive, good quality, readily available nutritious food in adequate quantity can be a powerful intervention for recovery and prevention.

These two threats to human health are examples of **non-communicable diseases** (NCDs). Non-infectious or not transmissible among people, the four main types of NCDs are cardiovascular disease, cancer, respiratory disease, and diabetes. Combined, these NCDs kill more than 36 million people each year across the globe. Many of these diseases are tied to certain behaviors, including tobacco use, physical inactivity, and unhealthy diet, as well as the harmful effects of alcohol consumption. However, some of these diseases are also tied to one's genetics and family history.

Recognizing that NCDs are now the leading cause of death globally should not distract us from communicable diseases, especially the most worrying ones. These are the diseases made increasingly popular in "medical thrillers," including those that jump from non-humans to humans, or are influenzas of one sort or another. *The Happening* (2008), about a plant-based virus, *Outbreak* (1995) based on the Ebola virus, or *The Stand* (1994) about an influenza virus, are popular illustrations of communicable diseases. The most recent film in this genre, said by the CDC to be "highly accurate," is *Contagion* (2011) in which a bat-pig virus (MEV-1) is transmitted to an American businesswoman visiting Hong Kong setting off a global pandemic. These movies illustrate how our interconnected world can facilitate the rapid and dramatic spread of communicable diseases across vast global spaces. Indeed, the story lines in these movies parallel recent international outbreaks, such as Severe Acute Respiratory Syndrome (SARS), which spread through cities, such as Singapore in Southeast Asia and Toronto in Canada, at very rapid rates in 2003.

These epidemics—both those portrayed in film and in the real life experience of SARS or even more recently with the Ebola outbreak in western Africa—reflect serious concerns about human-non-human disease communication and occupy the attention of national agencies, such as the U.S. Centers for Disease Control and Prevention (CDC) and international ones like the World Health Organization (WHO). Known as **zoonotic diseases** or **zoonoses**—transmissible from animals to humans—they include diseases familiar in the global North, including swine fever, mad cow disease, rabies, West Nile virus, dengue fever, and Hantavirus, diseases known in the global South, including river blindness (Onchocerciasis), sleeping sickness (trypanosomiasis), yellow fever, and Lassa fever (**Figure 3.F**). Why the map of infectious diseases appears to be changing is not completely understood. But there

Mobile Phones for Health
http://goo.gl/9Ng8p3

Height (ft)

Weight (lbs)	4'9"	4'11"	5'1"	5'3"	5'5"	5'7"	5'9"	5'11"	6'1"	6'3"
154	33	31	29	27	26	24	23	22	20	19
165	36	33	31	29	28	26	24	23	22	21
176	38	36	33	31	29	28	26	25	23	22
187	40	38	35	33	31	29	28	26	25	24
198	43	40	37	35	33	31	29	28	26	25
209	45	42	40	37	35	33	31	29	28	26
220	48	44	42	39	37	35	33	31	29	28
231	50	47	44	41	39	36	34	32	31	29
243	52	49	46	43	40	38	36	34	32	30
254	55	51	48	45	42	40	38	35	34	32
265	57	53	50	47	44	42	39	37	35	33
276	59	56	52	49	46	43	41	39	37	35
287	62	58	54	51	48	45	42	40	38	36
298	64	60	56	53	50	47	44	42	39	37
309	67	62	58	55	51	48	46	43	41	39
320	69	60	60	57	53	50	47	45	42	40

BMI	Weight category
18.5 - 24.9	Normal weight
25.0 - 29.9	Overweight
30.0 - 34.9	Obesity
35.0 - 39.9	Severe obesity
over 40	Morbid obesity

▲ **Figure 3.D Obesity and the Body Mass Index** Obesity is calculated through the Body Mass Index, a weight-to-height ratio, determined by dividing one's weight by the square of one's height.

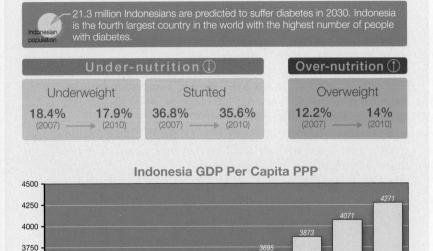

21.3 million Indonesians are predicted to suffer diabetes in 2030. Indonesia is the fourth largest country in the world with the highest number of people with diabetes.

Indonesian population

Under-nutrition ⬇		Over-nutrition ⬆
Underweight	Stunted	Overweight
18.4% → 17.9%	36.8% → 35.6%	12.2% → 14%
(2007) (2010)	(2007) (2010)	(2007) (2010)

Indonesia GDP Per Capita PPP

Year	Value
2005	3014
2006	3141
2007	3267
2008	3425
2009	3580
2010	3695
2011	3873
2012	4071
2013	4271

▲ **Figure 3.E The double burden of malnutrition in Indonesia** An accumulating body of evidence is suggesting that when GDP per capita PPP improves, obesity and diet-related non-communicable diseases may increase, even when there is an existing high level of under-nutrition.

is considerable agreement among health professionals and natural scientists that climate change may be in large part responsible for the changing conditions that support the emergence of diseases in new places. Warming climates have been linked to the wider spread of dengue fever, malaria, and cholera because these diseases—and the mosquitoes that transmit them—depend upon higher temperatures to survive. The same is true for the ticks that carry Lyme disease—their geographic range, limited by temperature, is expanding globally northward as air temperatures are rising.

One of the most positive responses to the new geographies of disease is the concomitant rise in the use of technology to combat disease, from large-scale geographic surveillance systems to mobile phones, electronic media, and innovative and inexpensive devices. For example, mobile phones delivering medical and public health information supported by mobile devices, known as mHealth, support the improvement of health care delivery around the globe. These improvements may include recording home births in rural Africa for remote monitoring by medical personnel (thereby including those new lives into the local health care system) or can enable the delivery of nutritional information and weight loss interventions for obese people.

1. What are the different ways geographers contribute to the study of health and disease?

2. What are the challenges to stemming the migration of non-communicable diseases in an era of a highly interactive, global world?

Safe Drug Delivery

http://goo.gl/0CxSF0

▼ **Figure 3.F Emerging zoonotic hotspots** This map identifies zoonotic events over the last 72 years in red with recent events identified in yellow. Notice that western Europe and the United States are hotspots. What isn't so obvious from the data is that peripheral countries are starting to catch up (*The Scientist*, 2012, http://www.the-scientist.com/?articles.view/article No/32325/title/Maps-Show-Animal-Disease-Hotspots/ (accessed 3/23/14). Original map by IOZ and published in ILRI Report, *Mapping of Poverty and Likely Zoonoses Hotspots,* 2012.)

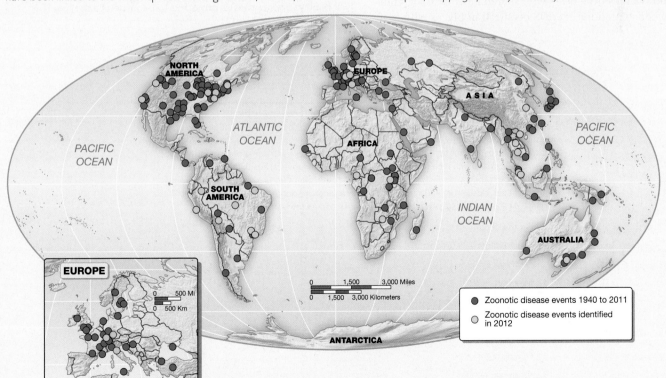

▲ **Figure 3.15 Environmental health protest, Brooklyn** These young men from the Williamsburg neighborhood are calling for the closure of Radiac, a nuclear waste and hazardous waste storage facility two blocks away from a public school.

feature investigates how and why non-communicable diseases and communicable diseases are changing across the globe and in particular countries and populations.

No discussion of health geographies would be complete without recognition of the most widespread and well-known epidemic of our times, HIV/AIDS (human immunodeficiency virus/acquired immunodeficiency syndrome). The disease is a serious problem in regions ranging from Southeast Asia to sub-Saharan Africa and also affects certain populations in many core countries of Europe as well as the United States. In the United States, for example, HIV/AIDS first arose largely among male homosexuals and intravenous drug users who shared needles. Geographically, early concentrations of AIDS occurred in places with high concentrations of these two subpopulations. It had perhaps the most severe impact in inner-city areas but cropped up in every region of the United States, appearing in the male and female heterosexual population. The rate for blacks being much greater than the rate for whites. African American males continue to bear the greatest burden of HIV/AIDS infection.

The pattern of the disease is different, however, in central Africa, where it is overwhelmingly associated with heterosexual, nondrug users and affects both sexes equally. The geographical diffusion of HIV/AIDS in Africa has occurred along roads, rivers, and coastlines, all major transportation routes associated with regional marketing systems. The central African nations, including Congo (formerly Zaire), Zambia, Uganda, Rwanda, and the Central African Republic, have been hard-hit. In sub-Saharan Africa, however, an estimated 22.5 million people are infected with the virus (**Figure 3.16**). The worst impact is in urban areas, though no area has been immune to the disease's spread.

Although no cure yet exists, four decades after the first reported cases of AIDS we do have the technological and medical understanding to substantially alter the HIV epidemic at a country level regardless of the mode of transmission of HIV.

APPLY YOUR KNOWLEDGE

1. Describe some of the ways life expectancy is shaped by geography, that is, how does where a person is born, lives, or works shape how long they are likely to live?

2. Using the most recent *World Population Data Sheet*, rank the 10 countries with the largest projected populations for both 2025 and 2050 (from largest to the smallest). Which country's (or countries') population is projected to drop out of the top 10 by 2050? Which country (or countries) is projected to be added to the top 10?

TABLE 3.4 Top Ten U.S. States in Well-Being

Gallup-Healthways produces an annual well-being index that surveys more than 176,000 Americans. Survey participants answer questions about six measures of well-being: life evaluation (how people feel they're doing currently, and how they expect to do in five years); emotional health (how many positive and negative emotions they experienced in the past day); work environment (how happy they are at work and with their relationship with their bosses); physical health (including obesity and illnesses); healthy behaviors (how well people eat and how often they exercise); and basic access to food, shelter, clean water and medical care. The rankings are out of a possible 100 points. West Virginia (61.4), Kentucky (63.0) and Mississippi (63.7) score at the bottom of the list.

1. North Dakota – 70.4
2. South Dakota – 70.0
3. Nebraska – 69.7
4. Minnesota – 69.7
5. Montana – 69.3
6. Vermont – 69.1
7. Colorado – 68.9
8. Hawaii – 68.4
9. Washington – 68.3
10. Iowa – 68.2

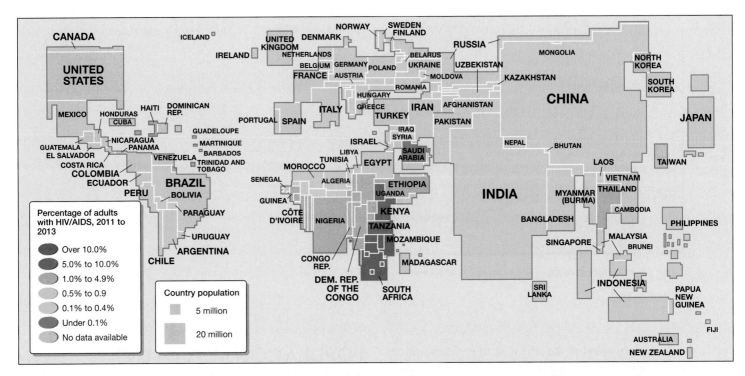

▲ **Figure 3.16 HIV Infection, 2012** HIV infections are concentrated in the periphery and semiperiphery, with most of the people with HIV/AIDS living in Asia and sub-Saharan Africa. Compare the cartograms for Africa with those of North America, Europe, and China. Campaigns in some African countries, especially in Zimbabwe and Uganda, have helped to stem the number of new cases and ultimately the number of deaths.

(*Source:* Adapted from H. Veregin (ed.), *Goode's World Atlas,* 22nd edition, Chicago, IL: Rand McNally & Co, 2010, p. 54. Updated data from *CIA World Factbook* (percentage of adults (15–49) living with HIV/AIDS. Estimated for year 2012, https://www.cia.gov/library/publications/the-world-factbook/fields/print_2155.html)

Demographic Transition Theory

Many demographers believe that fertility and mortality rates are directly tied to the level of economic development of a country, region, or place. Pointing to the history of demographic change in core countries, they contend that many of the economic, political, social, and technological transformations associated with industrialization and urbanization lead to a demographic transition.

A **demographic transition** is a model of population change in which high birth and death rates are replaced by low birth and death rates. Once a society moves from a preindustrial economic base to an industrial one, population growth slows. According to the demographic transition model, the slowing of population growth is attributable to improved economic production and higher standards of living brought about by better health care, education, and sanitation.

As **Figure 3.17** illustrates, the high birth and death rates of the preindustrial phase (Phase 1) are replaced by the low birth and death rates of the industrial phase (Phase 4) only after passing through the critical transitional phase of steady birth-rates and falling death rates (Phase 2) and then more moderate rates (Phase 3) of natural increase (increase through birth, not migration). This transitional phase of rapid population growth is the direct result of early and steep declines in mortality at the same time that fertility remains at levels characteristic of a place that has not yet been industrialized. Importantly, a new phase (Phase 5) appears to be emerging that is shown with a question mark on the model. This phase, which we are witnessing in a few countries in the developed world, has death rates slightly exceeding birth rates, leading to a slow but overall decline in population. Countries in this phase include Germany and Italy.

The Demographic Trap

Some demographers have observed that many peripheral and semiperipheral countries appear to be stalled in the transitional phase—caught in a "demographic trap." Despite a sharp decline in mortality rates, most peripheral countries retain relatively high fertility rates. The reason for this lag in declining fertility rates relative to mortality rates is that while societies have developed new and more effective methods for fighting infectious diseases, social attitudes about the desirability of large families are only recently changing.

Although the demographic transition model is based on actual birth and death statistics, many population geographers and other population experts question its applicability to all places and all times. Although the model adequately describes the history of population change in the core countries of the world, it appears less useful for explaining the demographic histories of countries and regions in the periphery. Its significance has therefore been contested. Among other criticisms is that industrialization—which, according to the theory, is central to moving from Phase 2 to Phases 3 and 4—is seldom domestically generated in the peripheral countries. Instead, foreign investment seems to drive peripheral industrialization. As a result, the features of demographic change, such as higher

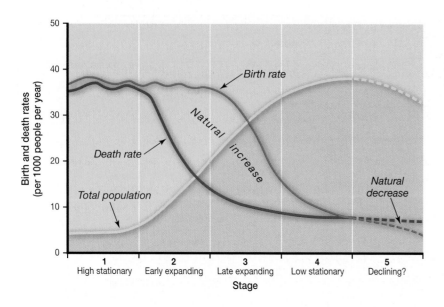

◄ **Figure 3.17 Demographic transition model** The transition from a stable population based on high birth and death rates to one based on low birth and death rates. A country is expected to progress in clearly defined stages. Population experts disagree about the usefulness of the model.

living standards, witnessed in the core countries, where industrialization was largely a result of internal capital investment, have not been as dramatic in many peripheral countries.

Education, Women, and Demographic Transformation

One factor that very clearly leads to demographic change is the education of women. Demographers have shown that the education of women is likely to increase the pace of the transition from Phase 2 to Phase 3 because of its impact on fertility and mortality. It has been statistically demonstrated that better educated women have fewer and healthier children and have better health themselves as compared with women who have little or no education. Studies undertaken by population geographers and others have found that the first generation of mothers with education are effecting a complex process in the transformation of the attitudes of whole communities with respect to reduced family size. Still, there are significant differences among and within regions with respect to fertility rates. In general, the data show that in Africa, and parts of Asia, literacy rates for girls are low and fertility rates remain high; whereas in Latin America and the Caribbean region, by contrast, universal literacy and equal education rates for men and women have been achieved and fertility rates have declined. In short, education, and in particular the education of women, plays a very important role in the demographic change (**Figure 3.18**).

APPLY YOUR KNOWLEDGE

1. Provide an example of why a country might be concerned about its national population being too small. How could that affect its population characteristics?

2. Japan and Denmark both have low birthrates. Investigate one of these countries or any other that has a low birthrate and provide a discussion of how it is addressing this issue.

POPULATION MOVEMENT AND MIGRATION

In addition to the population dynamics of death and reproduction, the third critical influence on population is the movement of people from place to place. Individuals may make far-reaching international or intraregional moves, or they may simply move from one part of a city to another. For the most part, mobility and migration reflect the interdependence of the world-system. For example, global shifts in industrial investment result in local adjustments to those shifts as populations move or remain in place in response to the creation or disappearance of employment opportunities.

Mobility and Migration

One way to describe such movement is by the term **mobility** which is the ability to move from one place to another, either permanently or temporarily. Mobility may be used to describe a wide array of human movement, such as a daily commute from suburb to city or a move from one side of the globe to the other. Mobility is a term that is meant to capture the capacity of an individual or a population to move.

The second way to describe population movement is in terms of an actual action, not just a capacity to take an action. The term used for this movement is **migration**, a long-distance move to a new location. Migration involves a permanent or temporary change of residence from one neighborhood or settlement to another. Moving from a particular location is defined as **emigration**, also known as out-migration. Moving to a particular location is defined as **immigration** or in-migration. For example, a Moroccan who moves to Paris emigrates from Morocco and immigrates to France. This type of move, from one country to another, is termed **international migration**. Moves may also occur within a particular country or region, in which case they are called **internal migration**. Both permanent and temporary changes of residence occur for many reasons

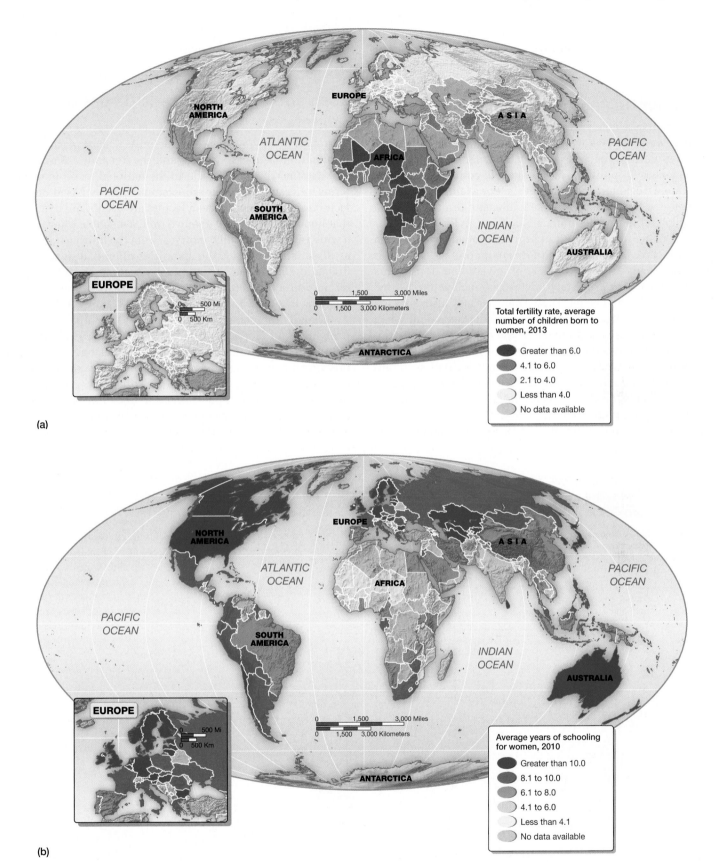

(a)

(b)

▲ **Figure 3.18   The connection between women's fertility and education**   These two maps show an important correlation between (a) female education and (b) fertility rates. It is important to appreciate as well that fertility rates decrease as well when both parents are educated.

but most often involve a desire for economic betterment or an escape from adverse political conditions, such as war or oppression.

Documenting and Explaining Migration

Governments are concerned about keeping track of migration numbers and rates, as well as the characteristics of the migrant populations, because these factors can have profound consequences for political, economic, and cultural conditions on national, regional, and local levels. For example, a peripheral country that has experienced substantial out-migration of highly trained professionals, such as the Philippines or Ukraine, may find it difficult to provide needed services such as health care. Benefiting from low labor costs are countries that have received large numbers of low-skilled in-migrants willing to work for extremely low wages, such as the United States, Germany, and Israel. These countries may also face considerable social stress in times of economic recession, when unemployed citizens begin to blame the immigrants for "stealing" their jobs or receiving welfare benefits.

Demographers have developed several calculations of migration rates. The in-migration and out-migration rates provide the foundation for gross and net migration rates for an area under study. **Gross migration** refers to the total number of migrants moving into and out of a place, region, or country. **Net**

migration refers to the gain or loss in the total population of that area as a result of the migration.

Push and Pull Factors and Types of Migration

Migration rates, however, provide only a small portion of the information needed to understand the dynamics of migration and its effects from the local to the national level. In general terms, migrants make their decisions to move based on push factors and pull factors. **Push factors** are events and conditions that impel an individual to move from a location. They include a wide variety of possible motives, from the idiosyncratic, such as dissatisfaction with the amenities offered at home, to the dramatic, such as war, economic dislocation, or ecological deterioration. **Pull factors** are forces of attraction that influence migrants to move to a particular location (**Figure 3.19**).

Usually the decision to migrate is a combination of both push and pull factors, and most migrations are voluntary. In **voluntary migration** an individual chooses to move. Where migration occurs against the individual's will, push factors can produce **forced migration**. Oftentimes, the decision to migrate is a mixed one reflecting both forced and voluntary factors. Forced migration (both internal and international) remains a critical problem in the contemporary world.

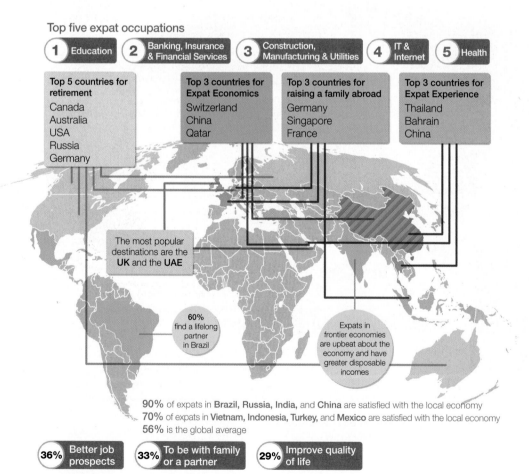

◄ Figure 3.19 The U.K. expatriate journey, 2013 An expatriate, or "expat," leaves his or her country of birth and moves, mostly permanently, to another country. This infographic provides a helpful description of pull factors in the migration decision of people from all over the world. It shows that jobs, family reunion, and quality of life are roughly equally important in the decision to move.

Top five expat occupations

1 Education
2 Banking, Insurance & Financial Services
3 Construction, Manufacturing & Utilities
4 IT & Internet
5 Health

Top 5 countries for retirement
Canada
Australia
USA
Russia
Germany

Top 3 countries for Expat Economics
Switzerland
China
Qatar

Top 3 countries for raising a family abroad
Germany
Singapore
France

Top 3 countries for Expat Experience
Thailand
Bahrain
China

The most popular destinations are the **UK** and the **UAE**

60% find a lifelong partner in Brazil

Expats in frontier economies are upbeat about the economy and have greater disposable incomes

90% of expats in **Brazil, Russia, India,** and **China** are satisfied with the local economy
70% of expats in **Vietnam, Indonesia, Turkey,** and **Mexico** are satisfied with the local economy
56% is the global average

36% Better job prospects
33% To be with family or a partner
29% Improve quality of life

Top three reasons for moving

These migrants may be fleeing a region or country for many reasons, but some of the most common are war, famine (often war-induced), life-threatening environmental degradation or disaster, or governmental coercion or oppression. And while **refugees**—individuals who cross national boundaries to seek safety and asylum—are a significant global problem, **internally displaced persons (IDPs)**—the number of individuals who are uprooted within the boundaries of their own country because of conflict or human rights abuse—is also growing globally (see **Window on the World:** Internal Displacement).

APPLY YOUR KNOWLEDGE

1. What distinguishes migration from mobility?

2. Using yourself as an example, identify three push and pull factors that will shape your decision about where you will live when you complete higher education. Do you suspect you will migrate after graduation? If so, will you be doing it out of necessity or choice? Please provide specific reasons for both the pull and push factors.

International Voluntary Migration

Migration does not always involve force or even a permanent change of residence. Voluntary migration can occur for any number of reasons, such as high wage differentials between places, better experience and job opportunities elsewhere, family links abroad, or local underemployment or unemployment conditions (**Figure 3.20**). But migrants are not free to simply travel wherever they wish as significant laws about entry and staying govern migration in every country in the world. Some countries, such as the United States, have relatively open borders, meaning they are more willing than other countries to accept migrants, though migrants are regulated through strict laws and quota systems. Japan, by contrast, has some of the strictest immigration laws in the world. Japan's modern preference for cultural and geographical isolation can be traced back to the seventeenth century. Recently, however, the Japanese government announced plans to introduce an immigration system that awards visas to skilled foreign professionals. The reason for this change in sentiment is the expectation that Japan will need workers to fill the gaps caused by its shrinking population.

Labor Migration

Temporary labor migration has long been an indispensable part of the world economic order and has at times been actively encouraged by governments and companies alike. Sending workers abroad is an important economic strategy for many peripheral and semiperipheral countries; it lessens local unemployment and also enables workers to send substantial amounts of money to their families at home. This arrangement helps supplement the workers' family income and supports the dominance of the core in global economic activities. Receiving countries which have developed programs to facilitate the immigration of what have sometimes been called **guest workers**—laborers given temporary visas to work for limited periods of time—gain many of the benefits of working abroad that they otherwise would not have at home. These include a higher salary and possibly better working conditions. It also might, however, occur under highly difficult working conditions such as estrangement of parents from

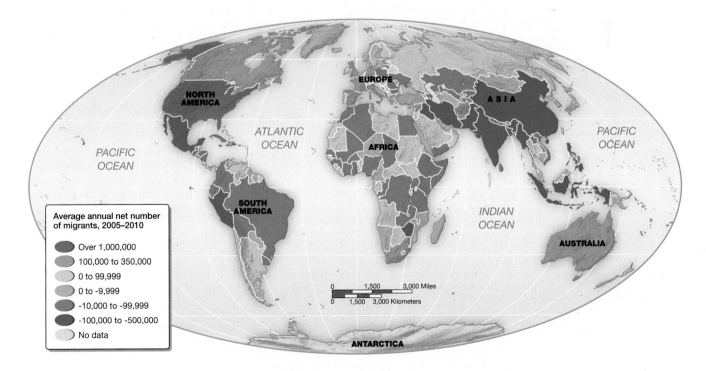

▲ **Figure 3.20 International voluntary migration, 2005-2010** The net migration shown here is a measure of the residual of voluntary out-migration and in-migration.

(*Source:* United Nations Dept. of Economic and Social Affairs, Population Division, World Population Prospects, *2012.* http://esa.un.org/unpd/wpp/Excel-Data/migration.htm.)

3.4 Window on the World

Internal Displacement

While the global humanitarian community has paid close attention to refugee populations for decades, the problem of internally displaced persons began to draw international attention only around 20 years ago. Across the globe, at the end of 2012—the latest year for which reliable data are currently available—there were nearly 30 million internally displaced persons (IDPs)—individuals who were uprooted within their own countries because of civil conflict or human rights violations, sometimes by their own governments.

War and Conflict

Most IDPs are the victims of internal conflict with their numbers more than twice as high as the global refugee population. It is also frequently the case

that the plight of IDPs is worse than that of the refugees. This is because the IDPs' governments are either unable or unwilling to protect them or provide the assistance they have a right to expect. It is also the case that the international community has not secured the resources needed to help them.

Figure 3.G provides a snapshot of the number of IDPs in and around Syria. By early 2013, that number had risen dramatically for some countries, like Syria—torn by civil-war—which has been experiencing the displacement of 1 family every 60 seconds. IDPs continue to also rise in Colombia, where guerrillas and paramilitaries target civilian populations through arbitrary killings, looting, and destruction of property. Their aim is to depopulate rural areas for political and economic gains and to control or regain strategic territories. Conflict continues to plague the Democratic Republic of Congo

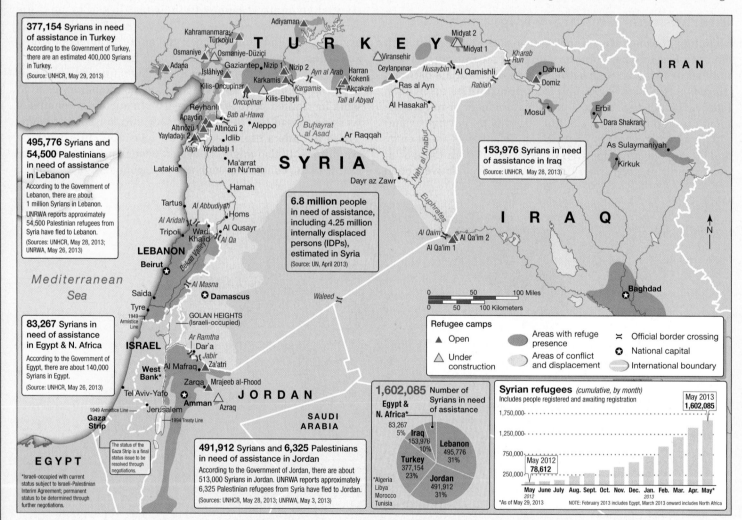

▲ **Figure 3.G Internally displaced and refugee Syrians, 2013** This map shows the scattering of Syrians due to the civil war there. In addition to those that have been pushed into other parts of the Syria, the map also shows those who are living in refugee camps on Syria's borders with Lebanon, Jordan, and Iraq. http://reliefweb.int/map/syrian-arab-republic/syria-numbers-and-locations-refugees-and-idps-29-may-2013

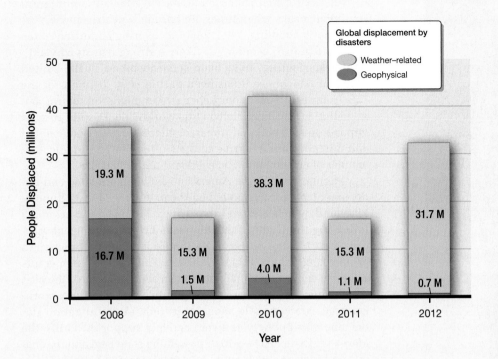

◄ **Figure 3.H** Global displacement due to disasters This graph suggests extreme years of displacement punctuated by ones that are relatively quiet where weather related disasters appear to affect more people than geophysical ones. http://www.internal-displacement.org/global-figures

resulting in further population dislocation due continued violence in the eastern, mineral-rich part of the country, despite the end of the war in 2003.

The largest increase in IDPs in 2012 was in the Middle East and North Africa where the after effects of the "Arab Spring"—popular uprisings against entrenched government leaders—are still being felt. Sparked by the first anti-government protest in the region in Tunisia, other countries there followed suit with protests erupting (and in some cases persisting) in Egypt, Algeria, Yemen, Jordan, Syria, Kuwait, Saudi Arabia, and Bahrain. These protests often brought government retaliation so that by the end of 2012, there were almost six million IDPs in the region, a rise of 20 percent over 2011.

Disaster and Climate-Induced Displacement

While IDPs are most typically driven by state or civil violence from their homes, a second force for displacement is climate-related and geographical hazards. The Norwegian Refugee Council provides a helpful typology for understand the forces driving disaster-induced displacement.[2]

- Sudden onset of hazard events, for example, floods and cyclones.

- Environmental degradation and/or slow onset hazard events, for example, drought.

- Armed (environmental) conflict/violence, for example, conflict over shrinking natural resources.

Natural disasters, often the result of climate-related events, are an increasing cause of internal displacement. The number of recorded natural disasters has doubled from approximately 200 to over 400 per year over the past two decades. By some estimates, the number of people displaced by natural disasters is twice that of those forced to flee conflict. Disasters and the resulting environmental degradation can be direct causes of displacement, or they may indirectly affect displacement through environmental conflicts.

The UN projects that 150 million people will lose their homes and become climate-related IDPs by the year 2050. The state of climate- and geophysical-related IDPs in countries across the globe for 2012 is shown in **Figure 3.H**. It is important in recognizing the effects of climate change on IDPs that climate change does not necessarily create a new type of IDP, rather it increases the frequency and severity of weather events that may lead to displacement.

---

1. What are some causes and locations of internal displacement in the world today?

2. Why do you think that internal displacement is an issue that has only gotten attention in the last 20 years? How might climate change affect internal displacement in the next 50 years?

Climate Change Impacts on Health

Economic historians have shown that anti-immigrant sentiment peaks when times are hard and declines when the economy is booming. The slowly recovering condition of the U.S. (and global) economy is likely a strong reason why anti-immigrant feelings have been so widespread in the United States of late where it has been particularly difficult for the U.S. Congress to pass comprehensive immigration reform. Immigration is also a conflicted issue worldwide. Despite popular opposition and political impasse, migrants, both documented and not, continue to arrive in the United States and affect all manner of culture, politics, economy, and social life.

Migrants from Latin American—both documented and undocumented—for example, have been embraced in Baltimore, Maryland, which is actively working to boost its local economy following the flight of middle class households to the suburbs. In 2011, the mayor of Baltimore launched a program to bring 10,000 families back to the city and she sees immigrants as key to that program. To facilitate a repopulation of the city, she's signed an order prohibiting police and other city workers from asking people about their (federal) immigration status. She also publicly and very actively supported a bill—the Maryland Dream Act—which passed the state Legislature. That bill provides undocumented migrants access to instate tuition rates and state financial aid, a very strong incentive for families with aspirations for upward mobility to settle in Baltimore and other parts of Maryland.

Contrast the welcoming situation for immigrants in Maryland with that of Arizona where in April 2010, state law S.B. 1070, was passed. The law augments federal immigration law by requiring certain noncitizens to register with the U.S. government and makes it illegal for a noncitizen to be in Arizona if he or she is not carrying the required documents. It also restrains state or local government entities from restricting enforcement of federal immigration laws and cracks down on individuals sheltering, hiring, and transporting undocumented workers. In 2012, the U.S. Supreme Court nullified three of

▲ **Figure 3.21 Foreign caregivers** Foreign women are often employed as caregivers for middle and upper class children. Shown here is a Filipina woman who is a nanny for her young European charge in Hong Kong.

their children left behind in the country of origin as well as problems with employers that are not appealable through legislation that only protects citizens, etc. (**Figure 3.21**). It should also be pointed out that labor migration is very likely to improve the economic conditions of the employer as well as the community in which he or she lives.

In the United States, the most controversial form of international voluntary migration is of **undocumented workers**, those individuals who arrive in the country without official entry visas and are considered by the government to be in the country illegally (**Figure 3.22**). Most of these workers come to the United States from Latin America (the largest share from Mexico). Other migrants who come to the United States with appropriate visas but then fail to return to their country of origin when their visas expire are also considered to be undocumented. Since the United States was founded, the immigration of working-age individuals has been crucial to its growth; it has also been a periodic source of conflict.

▼ **Figure 3.22 Marching for immigration reform in Texas** Hundreds of individuals and families march to the state capitol in Austin to push for immigration reform that would allow amnesty to undocumented people already living in the United States.

the four provisions of the Arizona law because they effectively usurped federal policy or interfered with federal enforcement. The Court left in place the provision that allowed law enforcement to investigate a person's immigration status.

An increasingly important category of international migrants is **transnational migrants**, so called because they set up homes and/or work in more than one country. Sometimes these migrants are low-paid workers, as in the case of many migrants from the former Soviet Union who have taken up jobs in the unskilled sectors of manufacturing, agriculture, and the service economy in Europe. Other transnational migrants occupy the higher end of the socioeconomic spectrum, such as the Hong Kong Chinese, who have established substantial property and business investments and residences in Canada while maintaining their citizenship status in Hong Kong as well. Transnational migrants seek destinations all over the world and it is difficult to think of any country where transnational migrants have not appeared.

Amenity Migration

While labor migration is the predominant form of international migration, other forms of voluntary international migration also occur. One form, gaining increasing popularity among the baby boom cohort in the United States and Europe is **amenity migration**. This is a form a migration in which the migrant seeks not necessarily employment, but cultural, environmental, or social benefits in a new country or city.

Baby boom amenity migrants from the United States have sought out foreign retirement destinations that are welcoming and where health care is excellent and living costs are low.

▼ **Figure 3.23 The city of San Miguel de Allende** This Mexican city is home to many U.S. retirement migrants who have flocked to the city because of a wide range of amenities including lower health care and living expenses.

Costa Rica, Panama, and Mexico, just to name a few, have become attractive destinations for U.S. retirement migrants in search of a package of amenities. In Mexico, the most popular municipalities for these migrants are Chapala, Mexicali, Los Cabos and San Miguel Allende (**Figure 3.23**). Though this type of amenity migration is small and still growing, the effects on the migrants as well as the communities in which they live are significant. While the migrants bring their retirement dollars to these locations and contribute to local economic development and job creation, they benefit by having access to private rather than public health care, larger houses than in the United States, low-cost domestic help, and lower, and sometimes, no taxes in the countries to which they migrate.

International Forced Migration

Forced migration is a worldwide phenomenon (**Figure 3.24**). As described above, an example of example of international forced migration is Syrians fleeing their country as it erupted into civil war (**Figure 3.25**). Another more historically prolonged example is the Lebanese and the Kurds who have also been scattered widely throughout the globe because of war and civil strife. The 1975–1990 civil war and more recent violence in Lebanon have resulted in a particularly large number of emigrations. The failure of the Kurds to establish an autonomous state in the early twentieth century led to their being split among Iran, Iraq, and Turkey, with a small minority in Syria. Many Kurds have moved to different parts of the region or left it altogether because of military aggression, persecution, and the failure of repeated attempts to establish a Kurdish state.

Other prominent examples of international forced migration include the migration of Jews from Germany and Eastern Europe following the First World War and preceding the Second World War as well as the deportation of Armenians from the Ottoman Empire after the First World War. **Figure 3.26** shows the scattering of Palestinians from their homeland since the establishment of the state of Israel in 1948. The refugees have been fleeing ongoing violence as well as discrimination and land dispossession.

In other parts of the world, though, refugee populations are pouring across borders and creating difficult conditions for national governments. The wave of popular uprisings against decades-old dictatorships that first began in Tunisia in early 2011 has sparked the largest movement of migrants the world has seen since the Second World War. Many of these refugee populations have sought safe passage north to Europe (especially Italy, Spain, France, Greece, and Turkey) as well as south to adjacent counties in Africa, south of the Sahel.

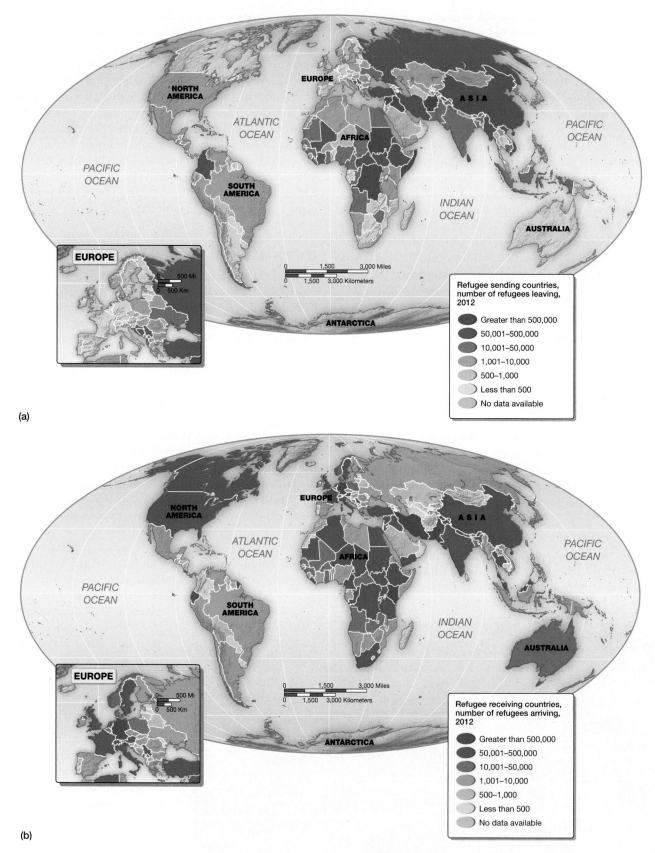

(a)

(b)

▲ **Figure 3.24 Refugee-sending and refugee-receiving countries** Shown are the sending countries (a), those whose internal situations propelled people to leave, as well as those countries to which they fled (b), sometimes for short periods on route to somewhere else or eventually home, but sometimes permanently. What is perhaps most distressing about this graphic is that refugee populations have increased over the last decade, and almost exclusively so in the periphery.

(Data from United Nations High Commission for Refugees, *UNHCR Statistical Yearbook 2012*, http://www.unhcr.org/cgibin/texis/vtx/home/opendocPDFViewer.html?docid=52a723f89&query=2013%20yearbook)

▲ **Figure 3.25 Syrian refugees crossing into Turkey** The civil war in Syria has caused a massive exodus of the population. Many refugees have fled to Turkey as well as Lebanon, Jordan and Iraq as well as internally within Syria.

Internal Voluntary Migration

One way to understand the geographical patterns of migration is to think in terms of waves of migration. In the United States, for example, three important and overlapping waves of internal migration over the past two centuries altered the population geography. As with a great deal of migration activity, these three major migrations were tied to broad-based political, economic, and social changes.

First Wave The first internal migration wave began with colonization and increased steadily through the twentieth century. This wave had two parts. The first was characterized by the large movement of people from the settled eastern seaboard into the interior of the country. This westward expansion, which began spontaneously during the British colonial period—in blatant disregard of British restrictions on such expansion—became official settlement policy after the American Revolution. The federal government encouraged migration over the course of more than a century as part of the country's expansionist strategies. The second part consisted of massive rural-to-urban migration associated with industrialization, especially occurring during the mid-nineteenth to the early twentieth centuries. **Table 3.5** illustrates how, between 1860 and 2000, the United States was transformed from a rural to an urban society as industrialization created new jobs and as increasingly redundant numbers of agricultural workers (along with foreign immigrants) moved to urban areas to work in the manufacturing sector.

Second Wave The second great migration wave, which began early in the 1940s and continued through the 1970s, was the massive and very rapid movement of mostly African Americans out of the rural South to cities in the South, North, and West. Although African Americans had formed considerable populations in cities such as Chicago and New York before the onset of this wave, the mechanization of cotton picking pushed additional large numbers of black Americans out of the rural areas. Tenant cotton picking was a major source of livelihood for African Americans in the Deep South until mechanization reduced the number of jobs available. At the same time, pull factors attracted them to the large cities. In the early 1940s, large numbers of jobs in the defense-oriented manufacturing sector became available as other urban workers joined the war effort. This second wave of migration can be seen as part of a wider pattern of rural-to-urban migration among agricultural workers as industrialization spread globally.

Third Wave The third internal U.S. migration wave began shortly after Second World War and continues into the present. Following the end of the war and directly related to the impact of governmental defense policies and activities on the country's politics and economy, the region of the United States lying below the thirty-seventh parallel, also known as the Sunbelt, emerged as a migration destination. Between 1950 and 1990, this region, which includes 15 states and extends from North Carolina in the east to southern California in the west and Florida and Texas to the south, experienced a 97.9 percent increase in population. Beginning in the late 1970s and early 1980s, the West also began to grow dramatically. At the same time that the West and the South were booming, the Midwest and Northeast, known variously as the Snowbelt, Frostbelt, or Rustbelt, declined in population.

The most compelling explanation for the large-scale population shift characteristic of the third migration wave was the pull of economic opportunity. Rather than being invested in upgrading the aged and obsolescent urban-industrial areas of the Northeast and Midwest, venture capital was invested in Sunbelt locations, where cheaper land, lower labor costs, and the absence of labor union organizations made the manufacturing and service-sector activity more profitable. By the 1990 census, there was a decrease in the amount of in-migration to the Sunbelt, but the geography of the U.S. population at the end of the twentieth century was nevertheless almost diametrically opposed to that of 150 years ago.

Internal Forced Migration

One of the best-known forced migrations in the United States is the "Trail of Tears," a tragic episode in which nearly the entire Cherokee Nation was forced to leave its once treaty-protected Georgia homelands for Oklahoma. Approximately 17,000 Cherokees were forced to march across the continent in the early 1830s, suffering from drought, food scarcity, bitterly cold weather, and sickness along the way. By some estimates at least a quarter of the Cherokees died as a result of the removal.

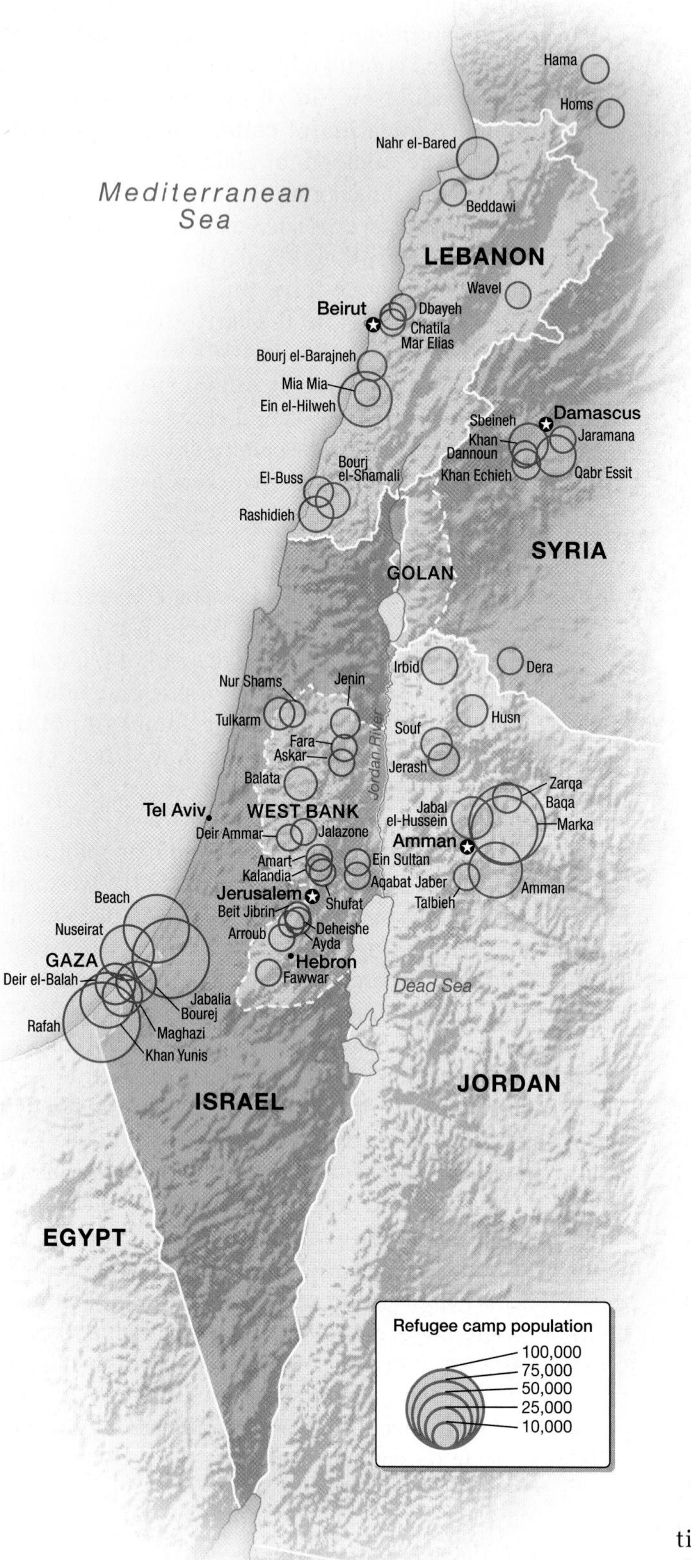

Map labels

Mediterranean
Sea

Hama
Homs
Nahr el-Bared
Beddawi
LEBANON
Wavel
Beirut
Dbayeh
Chatila
Mar Elias
Bourj el-Barajneh
Mia Mia
Ein el-Hilweh
Sbeineh
Khan
Dannoun
Khan Echieh
Damascus
Jaramana
Qabr Essit
El-Buss
Bourj el-Shamali
Rashidieh
GOLAN
SYRIA
Irbid
Dera
Nur Shams
Jenin
Tulkarm
Souf
Husn
Fara
Askar
Jerash
Balata
Zarqa
Baqa
Marka
Tel Aviv
WEST BANK
Jabal el-Hussein
Deir Ammar
Jalazone
Amman
Amart
Kalandia
Ein Sultan
Aqabat Jaber
Amman
Beach
Jerusalem
Beit Jibrin
Shufat
Talbieh
Nuseirat
Arroub
Deheishe
Ayda
GAZA
Hebron
Deir el-Balah
Fawwar
Dead Sea
Jabalia
Bourej
Maghazi
Rafah
Khan Yunis
JORDAN
ISRAEL
EGYPT

Jordan River

Refugee camp population

- 100,000
- 75,000
- 50,000
- 25,000
- 10,000

▲ **Figure 3.26** Palestinian refugees in the Middle East

During the 1948 war, 700,000 Palestinians were either expelled or fled and hundreds of Palestinian towns were depopulated and destroyed.

(Adapted from *The Guardian*, October 14, 2000, p. 5.)

Body text

Placed within the national and international context, the movement of Native American populations during the nineteenth century can be seen as a response to larger political and economic forces. European populations were At the same time, European populations were migrating to the United States, and the national economy was on the threshold of an urban-industrial revolution. The eastern Native American populations posed an obstacle to economic expansion, which was dependent upon geographic expansion. Growing Anglo American prosperity, it was believed, had to be secured by taking Indian land.

Today, many native people in Alaska are being forced to migrate from their ancestral homes, this time because of climate-induced changes. Alaska has warmed twice as fast as the global average during the past half-century, resulting in sea ice diminishing, sea level rising, and the land itself experiencing temperature increases. These changes have lead to accelerated rates of erosion and flooding, which is damaging and destroying infrastructure and preventing local people from pursuing their traditional livelihoods. While the state and federal governments have been spending millions of dollars to mitigate these problems, in many places they are insufficient and community relocation is the only option available that can protect them from accelerating climate change (**Figure 3.27**).

Forced Migration Due to Climate Change

http://goo.gl/tzrAru

Other, recent examples of internal forced migration are provided by China and South Africa. In the late 1960s and 1970s, as part of the Cultural Revolution, the government of China forcibly relocated 10 to 17 million of its citizens to rural communes in order to enforce Chinese Communist dogma and to ease pressures arising from high urban unemployment. The policy has since been disavowed, but the effects on an entire generation of Chinese youth were profound. Another example what took place in South Africa between 1960 and 1980, when apartheid policies forced some 3.6 million blacks to relocate to government-created homelands, resulting in much suffering and strife. Indeed, civil war, ethnic conflict, famine, deteriorating economic conditions, and political repression have produced an extraordinary series of internal forced migrations in several sub-Saharan countries as well as elsewhere throughout the globe.

These forced migrations, both internal and international in scope, become particularly significant in light of changing population. Forecasts predict that 80 percent of the world population increase in the next decade will take place within the poorest countries of the world. Many of these countries have some of the highest rates of forced migration. The combination bodes ill for these countries' prospects for economic and political improvement.

TABLE 3.5   United States Rural to Urban Population Change, 1860–2010

Year	1860	1870	1880	1890	1900	1910	1920	1930	1940	1950	1960	1970	1980	1990	2000	2010
Total U.S. population (in millions)	31.4	38.5	50.1	62.9	76.2	92.2	106	123.2	132.1	151.3	179.3	203.2	226.5	248.7	281.4	308.75
Total urban population (in millions)	6.2	9.9	14.1	22.1	30.2	42	54.2	69.1	74.7	96.8	125.2	149.6	167	187	222.3	249.25
% Urban	19.8	25.7	28.2	35.1	39.6	45.6	52.2	58.1	58.5	64	69.9	73.6	73.7	75.2	78.9	80.7
% Increase in total population	35.6	22.6	30.2	25.5	21	21	15	16.2	7.3	14.5	18.5	13.4	11.4	9.8	13.1	9.7
% Increase in urban population	35.6	22.6	30.2	20.5	21	21	15	16.2	7.3	14.5	18.5	13.4	11.4	9.8	18.8	12.1

Source: United States Census Bureau.

APPLY YOUR KNOWLEDGE

1. Summarize the three waves of internal voluntary migration that occurred in the United States.

2. Identify an environmental or economic issue in a country that interests you. Summarize two ways that this issue corresponds to push and pull factors that affect migration to and from that country.

POPULATION DEBATES AND POLICIES

One big question occupies the agenda of population experts studying world population trends today: How many people can Earth sustain without depleting or critically straining its resource base? The relationship between population and resources, which lies at the heart of this question, has been a point of debate among experts since the early nineteenth century.

Population and Resources

The debate about population and resources originated in the work of an English clergyman named Thomas Robert Malthus (1766–1834), whose theory of population relative to food supply established resources as the critical limiting condition upon population growth. Importantly, many critics have emerged to challenge him.

Malthusian Theory   Malthus's theory was published in 1798 in a famous book called *An Essay on the Principle of Population*. In this tract, Malthus sets up two important postulates:

- Food is necessary to the existence of human beings.

- The passion between the sexes is necessary and constant.

It is important to put the work of Malthus into the historical context within which it was written. Revolutionary changes—prompted in large part by technological innovations—had occurred in English agriculture and industry and were

◀ Figure 3.27
Kivalina, Alaska
Low-lying coastal Native Alaskan villages like this one are in danger of inundation as climate change is causing sea level rise.

eliminating traditional forms of employment faster than new ones could be created. This condition led to a fairly widespread belief among wealthy members of English society that a "surplus" of unnecessary workers existed in the population. The displaced agriculturists began to be a heavy burden on charity, and so-called "Poor Laws" were introduced to regulate begging and public behavior.

In his treatise, Malthus insisted that "the power of the population is indefinitely greater than the power of the earth to produce subsistence." He also believed that if one accepted this premise, a natural law would follow; that is, the population would inevitably exhaust food supplies. Malthus's response to this imbalance was to advocate for the creation of laws to limit human reproduction, especially among poor people.

Critiques of Malthus

Malthus was not without his critics, and influential thinkers such as William Godwin, Karl Marx, and Friedrich Engels disputed his premises and propositions. Godwin argued that "there is no evil under which the human species cannot labor, that man is not competent to cure." Marx and Engels were in general agreement that technological development and an equitable distribution of resources would solve what they saw as a fictitious imbalance between people and food.

In the mid-twentieth century Ester Boserup, an agricultural expert who championed the role of women in economic development, challenged Malthus's pessimistic assumptions about the relationship between population and resources. In contrast to the Malthusian argument that an insurmountable ceiling on food production acted as a natural limit on population growth, she countered that agricultural intensity rises with population density. She believed that households oriented their production to household requirements and not to the market and were therefore adopted more productive technologies in order to meet their needs. In short, producers intensify production in order to maintain an adequate food supply. Boserup's ideas on agricultural intensification became prominent among international development agencies for several decades in the mid-twentieth century and are still applicable today.

Taking a similar though more structural approach to the population-resource issue as Boserup did, geographer David Harvey has shown that adherence to Malthus's approach results in a doomsday conclusion about the limiting effect of resources on population growth. But by following Marx's approach, quite different perspectives on, and solutions to, the population-resources issue can be generated. These solutions are based on human creativity and socially generated innovation, which allow people to overcome the limitations of their environment.

Neo-Malthusians and Others Today

Neo-Malthusians—people today who share Malthus's perspective—predict a population doomsday. They believe that growing human populations the world over, with their potential to exhaust Earth's resources, pose the most dangerous threat to the environment. Although they acknowledge that the people of the core countries consume the vast majority of resources, they and others argue that only strict demographic controls everywhere will solve the problem, even if they require severely coercive tactics.

A more moderate approach argues that people's behaviors and governmental policies have a much greater impact on the condition of the environment and the state of natural resources than population size in and of itself. Proponents of this approach reject casting the population issue as a biological one in which an ever-growing population will inevitably create ecological catastrophe. They also reject framing it as an economic issue in which technological innovation and the sensitivities of the market will regulate population increases before a catastrophe can occur. Rather, they see the issue as a political one—one that governments have tended to avoid dealing with because they lack the will to redistribute wealth or the resources to reduce poverty, the latter condition being strongly correlated with high fertility.

The question of whether too many people exist for Earth to sustain bedeviled population policymakers and political leaders for most of the second half of the twentieth century. This concern led to the formation of international agencies that monitor and often attempt to influence population change. It also led to the organizing of a series of international conferences that attempted to establish globally applicable population policies. The underlying assumption of much of this policymaking, which has continued into the twenty-first century, is that countries and regions have a better chance of achieving improvement in their level of development if they can keep their population from outstripping the supply of resources and jobs.

Population Policies and Programs

Contemporary concerns about population—especially whether too many people exist for Earth to sustain—have led to the development of international and national policies and programs. A **population policy** is an official government strategy designed to affect any or all of several objectives, including the size, composition, and distribution of population. The implementation of a population policy takes the form of a population program. Whereas a policy identifies goals and objectives, a program is an instrument for meeting those goals and objectives. Most of the international population policies of the last three decades have attempted to reduce the number of births worldwide.

Figure 3.28 provides a graph of the recent history of world population growth and a reasonable projection of future growth. By the year 2050, the world is projected to contain nearly 9 billion people. In comparison, over the course of the entire nineteenth century, fewer than a billion people were added to the population. Over the next century, population growth is predicted to occur almost exclusively in Africa, Asia, and Latin America, while Europe and North America will experience very low, and in some cases zero,

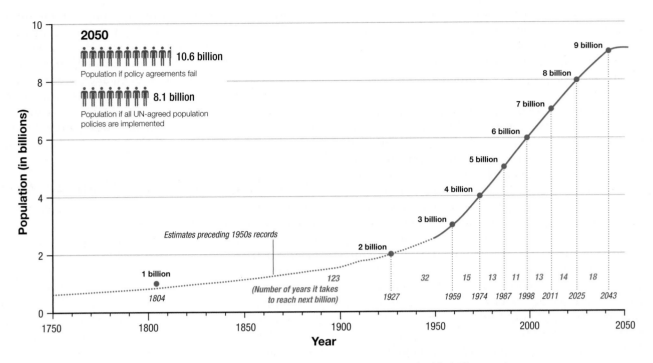

▲ **Figure 3.28** World population growth, 1750–2050   Since 1960, the world population has doubled. The annual number of people added to the world population is 80 million per year. Population projections indicate a high of 10.6 million and a low of 8.1 billion, depending on the implementation of U.N. population policies.

population growth. The differing rates of natural increase listed in **Table 3.6** illustrate this point. At the turn of the twentieth century, the core contained 32 countries with zero population growth. By contrast, the periphery contained 39 countries with rates of natural increase of 3.0 or more. A sustained rate of natural increase of 3.0 per year means a population will double in approximately 24 years, which would be the case for.

International Population Policies

Since 1954, the United Nations has sponsored international conferences at 10-year intervals to develop population policy at the global level. Out of each conference emerged explicit population policies aimed at lowering fertility rates in the periphery and semiperiphery. Significantly, all the world population conferences have recognized that the history, social and cultural practices, development level and goals,

TABLE 3.6   Global Demographic Indicators, 2011

This table illustrates the substantial population pressures emanating from the global periphery, which contained roughly four times the population of the core with a rate of natural increase also nearly four times as great. While Asian countries possess the largest proportion of the world's population, their rate of natural increase is considered moderate and their level of economic development—measured by their GNI PPP—higher than the average overall for the periphery.

Region	Population Mid-2011 (millions)	Natural Increase (annual %)	Birthrate (per 1,000 pop.)	Death Rate (per 1,000 pop.)	Life Expectancy at Birth (years)
World	7,137	1.2	20	8	70
More Developed	1,246	0.1	11	10	78
Less Developed	5,891	1.4	22	7	69
Africa	1,100	2.6	37	11	59
Asia	4,302	1.1	18	7	71
Latin America and Caribbean	606	1.1	18	8	75
Europe	740	0.0	11	11	74
North America	352	0.4	12	8	79

Source: Population Reference Bureau Web site, www.prb.org. *World Population Data Sheet,* 2013.

and political structures for countries and even regions within countries are highly variable and that one rigid and overarching policy to limit fertility will not work for all. Whereas some programs and approaches will be effective for some countries seeking to cut population numbers, they will be fruitless for others.

For instance, China's family-planning policy of one child per household appears to be effective in driving down the birthrate and dampening the country's overall population growth. Because so little data are available on China, however, it is not clear whether the policy is operating in the same fashion throughout the country. For instance, some population experts believe an urban bias exists, and while the policy is adhered to in the cities, it is disregarded in the countryside. Interestingly, in the wake of the May 2008 Sichuan earthquake—with a death toll of approximately 90,000 including children caught in collapsed school buildings—the Chinese government suspended the policy. There is increasing discussion throughout China that the one-child policy should be abandoned for the entire country as the Chinese become increasingly urbanized and the fertility rate is dropping in spite of the policy, not because of it.

Family-planning regulations include setting a legal age for marriage, offering incentives to couples to have only one child, and mandating increasingly severe disincentives for couples who have larger families. In India, family-planning policies offering free contraceptives and family-planning counseling have resulted in lowering the birthrate. Other countries, such as Sri Lanka, Thailand, and Cuba, as well as the Indian state of Kerala, have lowered their birthrates not through regulation of family size, but by increasing access to social resources such as health care and education, particularly for women. More equality between men and women inside and outside the household is also believed to have a significant impact on reducing fertility.

Today, over 92 percent of all countries support family-planning programs and contraceptives, either directly through government facilities or indirectly through support of nongovernmental activities, such as nonprofit family-planning associations. Despite the pervasiveness of government support for contraceptive methods, however, some 215 million women would like but currently do not have ready access to safe and effective modern means of contraception. The reasons why supply is not keeping up with demand are complex, but they involve several key factors: the increasing cost of providing contraception commodities; the cost of providing clinicians, facilities, counselors, and education materials; and the uncertain financial support of international donor organizations such as the U.S. Agency for International Development (USAID) (the single largest donor of contraceptives globally) and the UN Population Fund for family-planning programs in poor countries.

Recognizing the central importance of economic development to improving the lives of women, men, and children throughout the world, international agencies are increasingly turning to sustainable economic development as a way

of limiting births and ensuring an improved quality of life through the reduction of poverty. A host of related international conferences have been held to address this theme, the most central one being the UN Millennium Summit. The goals, listed in **Table 3.7**, are aimed at reducing poverty by improving economic development through aid, trade, and debt relief and through the enhancement of democratic governance institutions and careful attention to the impact of development on the environment.

The UN Millennium Development Goals

The UN Development Program (UNDP) Millennium Development Goals (MDGs) are based on a partnership between core and peripheral countries. With respect to governance transformations, for instance, UNDP is working with an oil company and Amnesty International in Venezuela to provide the country's judges with a comprehensive understanding of human rights laws, regulations, and issues. With respect to the environment, UNDP is working with farmers in Ethiopia by supporting the planting and marketing of traditional crops and, in the process, strengthening the country's Biodiversity Research Institute and encouraging farmers to create biodiversity banks, while the crops make their incomes more secure.

The goals and targets are based on the UN Millennium Declaration, and the UN General Assembly has approved them as part of the Secretary General's road map toward implementing the declaration. UNDP worked with other UN departments, funds, and programs, the World Bank, the International Monetary Fund, and the Organization for Economic Cooperation and Development to identify over 40 quantifiable indicators to assess progress.

The eight major MDGs are aimed at enabling peripheral countries to achieve core economic standards of wealth and prosperity while recognizing that preexisting conditions will have to be taken into account to construct a place-specific development path. As the goals imply, enabling more sustainable economic development worldwide is seen as a way of also shaping population growth and the quality of life for populations in the periphery. It is also a way of opening up new markets for core products and services and extending the capitalist world-system.

TABLE 3.7 Millennium Development Goals (MDGs)

Goals and Targets
Goal 1: Eradicate extreme poverty and hunger
Goal 2: Achieve universal primary education
Goal 3: Promote gender equality and empower women
Goal 4: Reduce child mortality
Goal 5: Improve maternal health
Goal 6: Combat HIV/AIDS, malaria, and other diseases
Goal 7: Ensure environmental sustainability
Goal 8: Develop a Global Partnership for Development

Progress on the MDGs

Each year the UN publishes a report outlining progress in each of the UN Millennium goals. The report published in 2013 states that: "… despite the impact of the global economic and financial crisis … several important targets have or will be met by 2015, assuming continued commitment by national governments, the international community, civil society, and the private section."[3] While the report is detailed and we are only providing the indicators of progress here, it is important to point out that sustained efforts are still required to address environmental sustainability, maternal and child health, and gender-based inequalities as they affect women's opportunities to shape their own lives.

Positive outcomes

- The proportion of people living in extreme poverty has been halved at the global level

- Over 2 billion people gained access to improved sources of drinking water

- Remarkable gains have been made in the fight against malaria and tuberculosis

- The proportion of slum dwellers in the cities of the developing world is declining

- Lowered debt burden and improved trading climate are widening the options for developing countries

- The hunger reduction target is within reach

Negative factors

- Environmental sustainability is under severe threat, new level of global cooperation is needed

- While big gains have been made in child survival, more must be done to meet our obligations to the youngest generation

- Most maternal deaths are preventable, but progress in this area is falling short

- Access to antiretroviral therapy and knowledge about HIV prevention must expand

- Too many children are still denied their right to primary education

- Gains in sanitation are impressive—but not good enough

- There is less aid money overall, with the poorest countries most adversely affected

- The poorest children are most likely to be out of school

- Gender-based inequalities in decision-making power persist

[3]United Nations, The Millennium Development Goals Report 2013. http://www.un.org/millenniumgoals/pdf/report-2013/mdg-report-2013-english.pdf (accessed 3/19/14)

APPLY YOUR KNOWLEDGE

1. What are the key issues in global population policies today?

2. Using the most recent UN Millennium Development Goals Report, identify one country where any one of the goals has been successful and one where that same goal has not. What other differences between these two countries that might explain why one has been more successful than the other.

FUTURE GEOGRAPHIES

In autumn 2014, the world contained about 7.28 billion people. The population division of the UN Department of Social and Economic Affairs projects that the world's population will continue to increase by 1.2 percent annually to mid-century when the 2014 average U.S. college freshman will be 50 years old. The distribution of this projected population growth is noteworthy: Over the next several decades, population growth is predicted to occur overwhelmingly in regions least able to support it. Just six countries will account for half the increase in the world's population, including Bangladesh, China, India, Indonesia, Nigeria, and Pakistan. Meanwhile, Europe and North America are expected to experience very low and in some cases zero population growth. Collectively, the periphery will grow 58 percent, as opposed to 2 percent for the core regions through 2050. The periphery will account for 99 percent of the expected increment in world population in this period.

U.S. and World Population Clock
Census.gov

Nevertheless, it is increasingly clear that human population will not continue to grow indefinitely, even in countries with already high populations. Demographic transition theory has prompted some population analysts to suggest that many of the economic, political, social, and technological transformations associated with continued urbanization and industrialization across the globe will shift the world's population from a period of high growth to one of low growth, or even population decline. This will be due to falling fertility. Current total fertility of 3.11 children per woman will fall to 2.04 by mid-century—just below replacement level but still above the current rate in core regions.

In addition to the number of people who will populate the planet in the coming decades, a second issue is one more appropriately situated at the individual level. How old can each of us expect to live and what will be the quality of our lives as we age?

Much has been written in the popular press about expected improvements in human longevity with attention paid especially to those adults who are already past the century mark. The term, "blue zone"—areas of the world where people live measurably longer lives—suggests that there is a geography to long life. For those of us who wish to live to 100, 110, even 120 in the twenty-first century, there is much to be learned from "longevity hotspots" and how they can help people live longer lives.

Using data generated from these genetically isolated communities as well as powerful genomic technologies and basic molecular research, scientists are exploring what makes places like Okinawa (Japan), Sardinia (Italy), Nicoya (Costa Rica), Icaria (Greece), Vilcabamba (Ecuador), and Seventh-day Adventists in Loma Linda, California, support healthier and longer lives than the rest of the population. **Figure 3.29** shows a Venn diagram based on the common characteristics among the long-living residents of Loma Linda, Okinawa, and Sardinia.

Blue Zones

Bluezones.com

APPLY YOUR KNOWLEDGE

1. Will today's 18-year-olds be able to extend their lives to the ages of these populations? Will they do this through technology or through similar lifestyle choices?

2. Will the environments in which they live constrain their ability to achieve such longevity or will they help transform their worlds in order to allow a new pathway to a longer life?

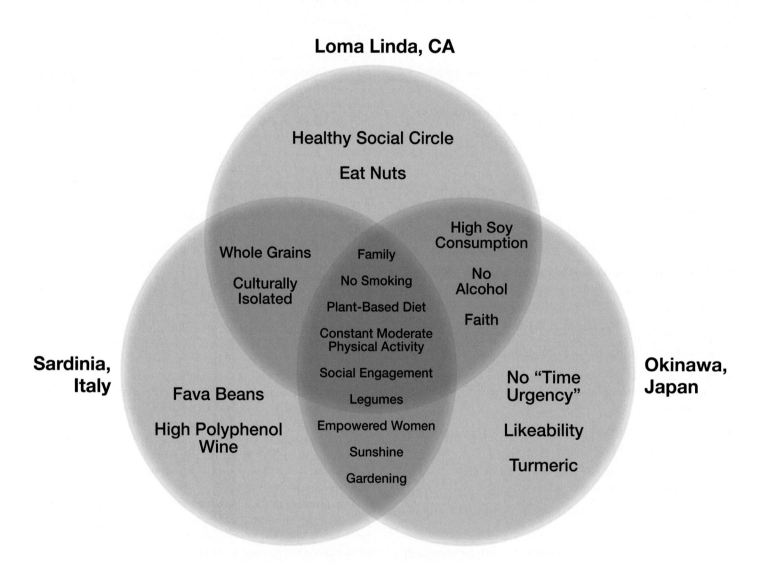

▲ Figure 3.29 Longevity factors in Loma Linda, California, Okinawa, Japan and Sardinia, Italy This Venn diagram shows the various factors thought to contribute to longevity among the 80-year-olds and older residents of these cities. Where the factors overlap in the darkest part of the diagram indicates the factors most frequently shared across the group.

■ CONCLUSION

The geography of population is directly connected to the complex forces that drive globalization. And since the fifteenth century, the distribution of the world's population has changed dramatically as the capitalist economy has expanded, bringing new and different peoples into contact with one another and setting into motion additional patterns of national and regional migrations.

When capitalism emerged in Europe in the fifteenth century, the world's population was experiencing high birthrates, high death rates, and relatively low levels of migration or mobility. Four hundred years later, birth, death, and migration rates vary—sometimes quite dramatically—from region to region, with core countries experiencing low death and birthrates and peripheral and semiperipheral countries generally experiencing high birthrates and fairly low death rates. Migration rates vary within and outside the core. These variations may be seen as reflections of the level and intensity of political, economic, and cultural connectedness between core and periphery and are difficult to predict.

The example of once colonized peoples migrating to their former ruling countries in search of work provides insights into the dynamic nature of the world economy. The same can be said of U.S. migrants who in the 1970s and 1980s steadily left their homes in the Northeast and Midwest to take advantage of the employment opportunities that were emerging in the Sunbelt: the Southeast, Southwest, and West. Both examples show the important role that people play in acting out the dynamics of geographic variety.

LEARNING OUTCOMES REVISITED

■ **Describe why populations change, where those changes occur, and what the implications of population change are for the future of different places around the globe.**

Population geographers bring to demography a special perspective—the spatial perspective—that emphasizes description and explanation of the "where" of population distribution, patterns, and processes. The distribution of population is a result of many factors, such as employment opportunities, culture, water supply, climate, and other physical environment characteristics. Geographers explore these patterns of distribution and density, as well as population composition in order to comprehend the complex geography of populations. Understanding the reasons for and implications of variation in patterns and composition provides geographers with insight into population change and the potential impacts that growing or declining populations might have.

■ **Identify the two most important factors in population dynamics, birth and death, and how they shape population characteristics.**

In order to arrive at a different understanding of population growth and change, experts look first at two significant factors: fertility and mortality. Birth and death rates are simple but central indicators of a place's level of economic development. Fertility and mortality rates provide key insights into how well a country, region, or city is able to provide for its population, especially in terms of income, education, and health care. Population geographers consider life expectancy, immigration, and emigration as well, which also affect population dynamics.

■ **State how geography is a powerful force in the incidence of health and disease.**

All people, things and events either occupy or happen in a particular place. Health and disease are no exception. Medical and health geographers help us to understand how a place can support healthy bodies or how it constitutes the condition for disease and poor health. They also provide maps and spatial explanations for the state of disease, such as HIV/AIDS, and health, such as the power of the "blue zones."

■ **Demonstrate how the movement of population is affected by both push and pull factors and explain how these factors are key to understanding new settlement patterns.**

In general terms, migrants make their decisions to move based on push factors and pull factors. Remember that push factors are events and conditions that impel an individual to move from a location. Pull factors are forces of attraction that influence migrants to move to a particular location. Mobility is the capacity to move from one place to another, either permanently or temporarily. Migration, in contrast, is an actual long-distance move to a new location. Permanent and temporary changes of residence can occur for a variety of reasons. Striving for economic betterment or escaping from adverse political conditions, such as war or oppression, are the most frequent causes. Push factors can produce forced migration, but it is usually the case that the decision to migrate reflects both push and pull factors.

■ **Describe the challenges of providing for the world's growing population with adequate food and safe drinking water, as well as a sustainable environment.**

A moderate response to the question "How can the global economy provide the world's growing population with adequate food and safe drinking water, as well as a sustainable environment?" rejects casting the population issue as a biological one in which an ever-growing population will inevitably create ecological catastrophe. It also rejects framing it as an economic issue in which technological innovation and the sensitivities of the market will regulate population increases before a catastrophe can occur. Importantly, the response to this question is more convincingly understood as a political one. Yet governments across the globe tend to avoid dealing with the population-resource problem because they lack the will to redistribute wealth or the resources to reduce poverty. This leaves the burden on citizens to organize themselves to force government to address the problem and change their own behaviors to lessen its effects.

KEY TERMS

age-sex pyramid (p. 72)
agricultural density (p. 69)
amenity migration (p. 93)
arithmetic density (p. 69)
baby boom (p. 74)
census (p. 66)
cohort (p. 72)
crude birthrate (CBR) (p. 78)
crude death rate (CDR) (p. 78)
crude density (p. 69)
demographic transition (p. 85)
demography (p. 66)
density (p. 69)
dependency ratio (p. 74)
doubling time (p. 78)
emigration (p. 86)
forced migration (p. 88)
geodemographic analysis (p. 74)

geography matters (p. 81)
health density (p. 69)
health geography (p. 81)
gross migration (p. 88)
guest workers (p. 89)
immigration (p. 86)
infant mortality rate (p. 79)
internally displaced persons (IDPs) (p. 89)
internal migration (p. 86)
international migration (p. 86)
life expectancy (p. 81)
medical geography (p. 81)
middle cohort (p. 74)
migration (p. 86)
mobility (p. 86)
natural decrease (p. 79)
natural increase (p. 79)
net migration (p. 88)

non-communicable disease (p. 82)
nutritional density (p. 69)
obesity (p. 82)
old-age cohort (p. 74)
population policy (p. 98)
pull factors (p. 88)
push factors (p. 88)
refugees (p. 89)
total fertility rate (TFR) (p. 78)
transnational migrant (p. 93)
undocumented workers (p. 92)
vital records (p. 66)
voluntary migration (p. 88)
youth cohort (p. 74)
zoonotic disease (p. 82)
zoonoses (p. 82)

REVIEW & DISCUSSION

1. Consider the population concentration in your town or city and some of its characteristics. Look up the census and vital records. Pay particular attention to birthrates. Compare 1950 with 2010 census data. What changed with respect to birthrates? What might be one reason for the change or lack of change? Now compare the data with birth data from another city or town. Is there a difference? If so, list two reasons why there might be a difference. If there is not a significant change, why might this be the case?

2. Pick two major cities within the United States and consider them from the perspective of migration. What have been the gross migration patterns of people in and out of the cities? How has migration changed over the last 20 years? List three reasons for these migration patterns.

3. Do an Internet search and find a current example each of refugees and internally displaced people. Compare and contrast these different categorizations. What similarities do these groups share? What are their primary differences? Develop a list of two reasons for each situation. Also, consider how refugees and internally displaced people have changed the population characteristics of their given place. For example, have these factors had an effect on birthrates or infant mortality rates? Please be as specific as possible in your answer by citing data that support your conclusion.

UNPLUGGED

1. The geographic distribution of population is a result of many factors, such as employment opportunities, culture, water supply, climate, and other physical environment characteristics. Look at the distribution of population in your state or province. Is it evenly distributed, or are the majority of people found in only a few cities? List three factors that influence where people live in your state or province.

2. Immigration is an important factor contributing to the increase in the population of the United States. Chances are your great-grandparents, grandparents, parents, or even you immigrated to, or migrated within, your country of residence. Construct your family's immigration or migration history. Identify push and pull factors influencing your family's decision to immigrate to or migrate within your country. Have you had to move within your lifetime? If so, please write a paragraph describing your decision.

3. During the 1990s, an interesting example of return migration occurred. Some areas of the United States that had previously lost large numbers of African Americans began to regain them. Compare data from the 1980 census with data from the 2010 census for the southeastern region of the United States. Which states experienced an increase of African American migrants? What are some of the characteristics of this migrant stream, and why might it have occurred? What impact might this have on electoral issues?

4. Every few years, UNESCO publishes a data book on global refugee statistics indicating both sending and receiving countries, among other variables. Locate the most recent data

book and identify one country that has been a large sender of refugees and the country that has been the largest receiver of those refugees. The data book provides information not only on the numbers of refugees but on their age, gender, and other variables. Discuss some of the demographic implications for both countries if the refugee population was not to be allowed to return to the sending country. How could this affect population policy in both countries?

DATA ANALYSIS

How critical is the link between women's education and fertility? In this chapter we have seen that fertility rates are highest in Africa and Asia. At the same time, these two regions have had "outstanding gains" in women's education in the last two decades, according to the United Nations. Charting demographic trends, policymakers and scholars have argued that when women are more educated, fertility rates drop. Furthermore, women's education creates higher standards of living for women, men and children, resulting in healthier populations globally. Take a deeper look at this relationship between fertility and education through the U.N. report,
The World's Women, 2010 (http://unstats.un.org/unsd/demographic/products/Worldswomen/WW2010pub.htm).

To begin, find Figure 3.7, "Distribution of population by sex and the highest level of education attained, 1995–2007." This is a population pyramid comparing sex and education levels around the world. What do you observe?

As you read the report, compare the graphics to find answers to these questions:

1. Where in the world is there a "surplus" of men?

2. What are the literacy rates for women in those same areas?

3. According to the United Nations, there are 57 million more men in the world than women and these numbers are concentrated in the lower age groups. What factors create these statistics?

4. Who lives longer—men or women?

5. How does that difference in longevity. impact the data you are finding?

6. Why do you think these numbers vary around the world?

Finally, do an Internet search to find out how education initiatives are working in some of the different countries in Africa and Asia. Look at non-governmental organizations like Amnesty International, OxFam International, Human Rights Watch and the United Nations office on women, called UN Women. Why do you think some education programs work better than others?

Mastering Geography™

Looking for additional review and test prep materials? Visit the Study Area in MasteringGeography™ to enhance your geographic literacy, spatial reasoning skills, and understanding of this chapter's content by accessing a variety of resources, including **MapMaster** interactive maps, Videos, *In the News* RSS feeds, flashcards, web links, self-study quizzes, and an eText version of *Human Geography*.

LEARNING OUTCOMES

■ *Describe* how people and nature form a complex relationship in which nature is understood as both a physical realm and a social construct.

■ *Compare* the many views of nature operating both historically and in society today, from the traditional Western approach to the radical left and contemporary ecotheological approaches.

■ *Explain* how extensive human settlement, colonization, and accelerated capitalist development have transformed nature in unprecedented ways.

■ *Describe* the drivers and impacts of climate change and state your position with respect to it within current debates.

■ *Summarize* how the globalization of the capitalist political economy has affected the environment so that environmental problems have become increasingly global in scope.

■ *Evaluate* the ways sustainability has become a predominant approach to global economic development and environmental transformation.

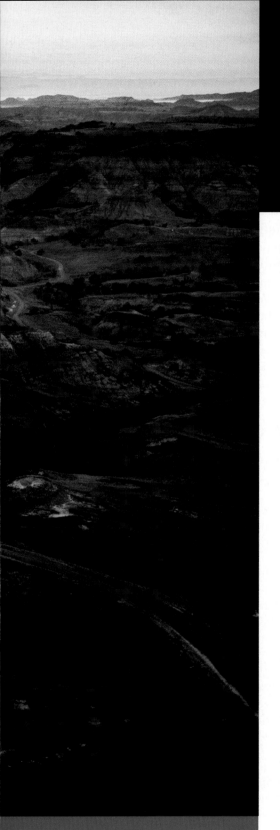

PEOPLE AND NATURE

4

In late October 2013, the Elsipogtog First Nation in New Brunswick, Canada formed a road blockade to stop an energy company from conducting shale oil and gas exploration on their land without their consent. The blockade was formed by approximately 300 tribal people as well as local community members in reaction to the announcement by SWN Resources Canada, a subsidiary of the Houston-based Southwestern Energy Company, that they would soon commence shale gas exploration in First Nations' territory.

The response to the blockade was swift and aggressive with over 100 Royal Canadian Mounted Police launching a paramilitary assault on the group, brandishing rifles and accompanied by attack dogs. Forty participants were arrested and the blockade was broken up as SWN reconsidered how to proceed. Since that initial event, the Elsipogtog have been joined by another First Nations' group, the Mi'kmaq, to carry on their objections to proposals by SWN to conduct hydraulic fracking on the shale sands that lie below their native lands. The tribes are arguing that SWN is violating treaty rights and fear for the effects fracking will have on their land and water as well as the people and animals that depend on those resources.

Groups across Canada, the United States, and the world have emerged to protest fracking, fearing that irreparable environmental damage will result from this extreme mining process. As of summer 2014, over 420 communities in the United States alone have passed measures to prohibit fracking on lands near or in their jurisdictions.

Induced hydraulic fracturing or "fracking," is a mining process designed to extract previously inaccessible energy resources from subsurface geological formulations. In fracking, highly pressurized fluid—usually a combination of water, sand, and chemicals—is injected into a wellbore to create small fractures in deep-rock formations. The production of the fractures allows oil and gas to escape the location where they are trapped in underground shale formations and rise up the wellbore to the surface and thus become more readily available for capture.

▲ Induced hydraulic fracturing, Watford City, North Dakota. New drilling techniques including fracking, shown here, are leading to a boom in oil production in this state.

Fracking in Canada and the United States is presented as an alternative to North American dependence on fossil fuels—particularly oil—found in the Middle East and other areas of the globe. Thus fracking represents not just a technology meant to support continued economic growth, but also a geopolitical strategy to reduce dependence on foreign oil and to augment domestic supply and demand tensions. Unfortunately, fracking holds the potential to produce extensive and dramatic environmental damage that can result in toxic landscapes for both humans and animals, including poisoned water resources, denuded soil, and radioactive pollution affecting fish, wildlife, and human water systems.

An example of these negative environmental effects is the 2011 Chesapeake Energy Corporation fracking accident that blew out a gas well in northeastern Pennsylvania. The explosion resulted in fracturing fluid escaping from the well and flowing into the Susquehanna River and down into the Maryland's Chesapeake Bay. The Environmental Protection Agency has identified more than two dozen highly toxic and hazardous chemicals, including methanol, glutaraldehyde, formaldehyde, 2-butoxyethanol and hydrochloric acid in the fluid that fracking pumps into the earth. In addition to the toxic and hazardous chemicals, along with water, that are pumped into the wells to create the pressure necessary to release oil and gas from the shale reserves, fracking waste fluids also contain contaminants picked up from deep within Earth, most notably heavy metals, volatile organic compounds, salty brine, and radioactive materials.

Besides the environmental damage that can result from onsite fracking, disposing of fracking's radioactive waste is another aspect of environmental concern. In March 2014, a Pennsylvania-based drilling company attempted to deposit two roll-off boxes of sludge in a landfill in Chartiers, Pennsylvania, but tripped radioactivity alarms. Turned away from the Pennsylvania landfill, the company then hauled the radioactive sludge to a landfill in Bridgeport, West Virginia, which lacked a radiation detector, since state law there does not require one. In fact, U.S. states have a diverse spectrum of laws regarding radioactive waste suggesting that some states are more likely to become depositories for radioactive fracking waste than others, even states that have no active fracking occurring within their jurisdiction.

While fracking is a widely used response to decreasing access to more readily available oil and gas and therefore offers a pathway to energy independence, at least temporarily, for countries like the United States, Canada, China, South Africa, and Australia, as well as parts of North Africa and Europe, it does not come without significant costs. Even as popular opposition to fracking is growing globally, more and more wells are being opened. Yet because fracking also provides jobs, resistance to it is complicated by the opportunity for many who might not otherwise have access to employment.

In the chapter that follows, we examine the nature–society relationship by looking first at different approaches to it. We then examine how changing conceptions of nature have translated into very different uses of and adaptations to it. We conclude the chapter with an examination of sustainable development as a way of addressing global environmental problems and a discussion of the new institutional frameworks and activist organizations that are emerging to promote sustainability.

NATURE AS A CONCEPT

As discussed briefly in Chapter 2, a simple model of the nature–society relation is that nature, through its grand force and subtle expressions, limits or shapes society. This model is known as environmental determinism. A second model posits that society also shapes and controls nature, largely through technology and social institutions. This second model, explored in this chapter, emphasizes the complexity of nature–society interactions. In this section, we explore the relationship between people, nature, and technology to understand how humans have shaped the natural world and what those human impacts mean for Earth's future.

The Earth Summits

Interest in the relationship between nature and society has experienced a resurgence as the scope of environmental problems over the last 50 years has widened to include not only those that are locally or regionally contained, but also those that have implications for the whole planet. The single most dramatic manifestation of this interest occurred in the summer of 1992, when more than 100 world leaders and 30,000 other participants attended the second Earth Summit in Rio de Janeiro (the first Earth Summit was held in Stockholm in 1972). The central focus of the 1992 agenda was to ensure a sustainable future for Earth by establishing treaties on global environmental issues, such as climate change and biodiversity, by raising public awareness of the need to integrate environment and development.

The signatories to the 1992 Earth Summit conventions created the Commission on Sustainable Development to monitor and report on implementation. Ten years later, Earth Summit 2002 (also known as Rio+10) was held in Johannesburg, South Africa. Its central aim was to assemble leaders

◀ **Figure 4.1 Rio+20 protests**
The marchers shown here are part of the People's Summit for Social and Environmental Justice who are directing their desire for sustainable development to the Rio+20 participants.

from government, business, and NGOs to agree on a range of measures toward achieving sustainable development at the national, regional, and international levels. At this summit, the need to enhance the integration of sustainable development in the activities of all relevant United Nations agencies, programs, and funds was highlighted.

The Rio+20—a celebration of the initial conference held in Rio in 1992—occurred in June 2012 (**Figure 4.1**). Also known as Earth Summit 2012, its goals are to secure renewed political commitment to sustainable development, to assess progress toward internationally agreed goals on sustainable development, and to address new and emerging challenges. Rio+20 focused on two specific themes: a green economy in the context of poverty eradication and sustainable development, and an institutional framework for sustainable development.

Using Agenda 21 as its focal point, the outcome document from Earth Summit 1992, Rio+20 sought affirmation for the political commitments made at past Earth Summits and set the global environmental agenda for the next 20 years. Agenda 21 was considered a revolutionary document in 1992 in that it not only popularized the term *sustainable development,* but also created an agenda that guided the next 20 years of government activities and policies toward improving the environment by increasing economic development in the world's periphery. **Sustainable development** seeks a balance among economic growth, environmental impacts, and social equity.

The main outcome of Rio+20 was a nonbinding document, "The Future We Want," in which the heads of state of the 192 governments attending the conference renewed their political commitment to sustainable development. They also declared their commitment to the promotion of a sustainable future by reaffirming previous action plans like Agenda 21.

Some very important changes have occurred since the first summit in Stockholm 40 years ago. One change has been the emergence of international institutions to facilitate and monitor environmental improvements. Another has been real progress on the global phaseout of leaded gasoline. A third has been increasing scientific and popular interest in global environmental issues. Perhaps most significant has been the proliferation of concern for sustainability, especially as it has broadened out from being a conference theme to a set of practices being taken up by governments, corporations, public and private organizations, and individuals. Among these new practices and objectives are environmental restoration, green urban development, the use of alternative energy technologies (such as wind, geothermal, solar, and nuclear energy), sustainable tourism, "greening" of the commodity supply chain, sustainable agriculture, promotion of **biomass**—the total mass of organisms in a given area of volume—and biofuels, green real estate development, green government practices, the emergence and growth of carbon markets, the transitioning of towns toward sustainability, the construction of sustainable buildings, and "green" marketing.

APPLY YOUR KNOWLEDGE

1. What is the role of international summits in balancing economic development and environmental impacts?

2. Do an Internet search for additional insight into the recent Rio+20 conference. What was the outcome of the conference and who attended (developed nations, underdeveloped, specific regions)? Did the attending states issue any resolutions to mitigate environmental problems?

Alternative Solutions to Environmental Problems

In the past, technology was viewed as the obvious solution to most environmental problems, but today's technological progress seems often to aggravate rather than to solve such problems (**Figure 4.2**). As a result, researchers and activists have begun to ask different questions and to abandon the assumption that technology is the *only* solution.

For instance, philosopher and environmental activist Vandana Shiva has led a broad-based and increasingly global movement advocating changes in agriculture and food production practices (**Figure 4.3**). In 1982, she established the Research Foundation for Science, Technology and Ecology, and in 1991, she founded Navdanya, a social movement with the goal of protecting the diversity and integrity of living resources, especially promoting the use and preservation of native seeds, organic farming, and fair trade.

In India, Navdanya has successfully conserved more than 5,000 crop varieties, created awareness of the hazards of genetic engineering, and defended people whose food rights are under threat from globalization and biopiracy. A controversial practice with many conflicting definitions, **biopiracy** is the commercial exploitation of biological materials, that is living matter, including anything from genetic cell lines to plant or animal substances, and patenting that material without

▲ **Figure 4.3 Vandana Shiva** An Indian environmental activist who was trained as a physicist and calls herself an ecofeminist and philosopher. Shiva is also a leading figure in the global solidarity movement known as the alter-globalization movement

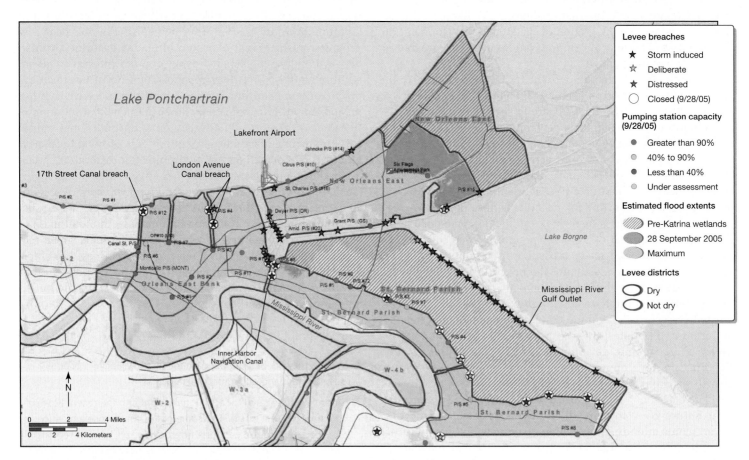

▲ **Figure 4.2 Levees system, New Orleans, Louisiana** Levees in New Orleans are an engineering feat keeping the mighty Mississippi River from flooding the city that sits below the normal river channel. The system failed during Hurricane Katrina when the storm surge both overtopped and in some cases breached the levees leading to both loss of life and extensive property damage.

compensating the peoples whose knowledge revealed the value of them or on whose territory it was discovered.

Navdanya and other movements like it have influenced environmental experts, including a number of geographers, to conceptualize nature not as something apart from humans but as inseparable from us. These experts believe that nature and questions about the environment need to be considered in conjunction with society, which shapes our attitudes toward nature and how we identify sources of and solutions to environmental problems. Such an approach—thinking of nature and society as interactive components of a complex system—enables us to ask new questions and consider alternatives to our current practices with respect to nature. One such practice, is described in **Box 4.1**, "Geography Matters: The Political Ecology of Carbon Offsets."

APPLY YOUR KNOWLEDGE

1. Explain how human beings and nature are part of a complex system. Think about the influence of social policies, consumer habits, technology, and governments on the environment.

2. What are some of the ways your community interacts with the natural world around your city? List at least three relationships that you observe and discuss whether those relationships are sustainable for both people and nature.

Nature and Society Defined

The central concepts of this chapter—nature and society—have very specific meanings. Although we discuss the changing conceptions and understandings of nature in some detail, we hold to one basic conception here, that **nature** is a social creation as much as it is the physical universe that includes human beings. Therefore, understandings of nature are the product of different times and different needs. Nature is not only an object, it is a reflection of society in that the philosophies, belief systems, and ideologies people produce shape the way we think about and use nature. The relationship between nature and society is two-way: Society shapes people's understandings and uses of nature at the same time that nature shapes society and its many manifestations. The amount of shaping by society is dependent to a large extent on the state of technology and the constraints on its use at any given time.

Society is the sum of the inventions, institutions, and relationships created and reproduced by human beings across particular places and times. Society's relationship with nature varies from place to place and among different social groups. The relationship between society and nature is usually mediated through technology. Knowledge, implements, arts, skills, and the sociocultural context—all are components of technology. If we accept that all of these components are relevant to technology, then we can provide a definition that has three distinguishable, though equally important, aspects. **Technology** is defined as

- physical objects or artifacts (e.g., the plow);

- activities or processes (e.g., steelmaking);

- knowledge or know-how (e.g., biological engineering).

This definition recognizes tools, applications, and understandings equally as critical components of technology. The manifestations and impacts of technology can be measured in terms of concepts, such as level of industrialization and per capita energy consumption.

The definitions provided in this section reflect current thinking on the relationship between society and nature. For centuries, humankind, in response to the constraints of the physical environment, has been as much influenced by prevailing ideas about nature as by its realities. In fact, prevailing ideas about nature have changed over time, as evidence from literature, art, religion, legal systems, and technological innovations makes abundantly clear.

The Complex Relationship of Environmental Impacts, Population, and Affluence

Concern about global environmental changes has given rise to a recent attempt to conceptualize the relationship between society and the environment based on the premise that individual societal changes can be both subtly and dramatically related to environmental changes. A formula now widely used for distinguishing the sources of social impacts on the environment is $I = PAT$; it relates human population pressures on environmental resources to a society's level of affluence and access to technology. More specifically, the formula states that $I = PAT$, where I (impact on Earth's resources) is equal to P (population) times A (affluence, as measured by per capita income) times T (a technology factor). According to this formula, the differential impact on the environment of two households' food consumption in two different places would equal the number of people per household times the per capita income of the household times the type of technology and energy used in producing foodstuffs for that household.

Each of the variables in the formula—population, affluence, and technology—is complex. For example, with regard to population numbers, it is generally believed that fewer people on the planet will result in fewer direct pressures on resources. Some argue, however, that increased world population is quite desirable, since more people means more labor coupled with more potential for the emergence of innovation to solve present and future resource problems. Clearly, there is no simple answer to the question of how many people are too many.

Affluence also cannot simply be assessed in terms of "less is better." Certainly, increasing affluence—a measure of per capita consumption multiplied by the number of consumers and the environmental impacts of their technologies—is a drain on Earth's resources and a burden on Earth's ability to absorb waste. Yet how much affluence is too much is difficult to determine. Furthermore, evidence shows that the core countries, with high levels of affluence, are more effective than the poor countries of the periphery at protecting their environments. Unfortunately, core countries often do so by exporting

4.1   Geography Matters

The Political Ecology of Carbon Offsets

By Tracey Osborne, University of Arizona

Do you know what your carbon footprint is—and what you can do about it? Your carbon footprint depends on where you live, your level of consumption, and the carbon intensity of your activities. Many of your daily activities—such as driving, heating and cooling your home, even what you eat—produce greenhouse gas emissions. Add all these calculations together, and they account for your carbon footprint. Global greenhouse emissions are changing the climate, warming the planet, and causing extreme weather events such as droughts and storms. You can reduce your personal emissions by using less energy in your home, car pooling more often, or flying less often—but you can also choose to "offset" them by paying to reduce emissions somewhere else.

Some companies, individuals, countries, and even music bands are using "carbon offsets" to voluntarily compensate for their emissions or to meet obligations under the Kyoto Protocol climate change treaty (see **Figure 4.A**). Because the atmosphere is globally connected, emission reductions made anywhere can contribute to reducing the risk of climate change. Carbon offsets represent a type of global trade, allowing you to buy emission reduction "offset credits" from projects anywhere in the world that reduce greenhouse gas emissions. Projects might improve the efficiency of an existing power plant in China or install wind turbines in India. Because forests absorb carbon and removing or degrading them releases it, reforestation can be a popular offset option. Often implemented in developing countries where deforestation is on the rise and the costs of projects tend to be lower, forest offsets are considered to be low cost solutions to climate change (see **Figure 4.B**).

Geographers who study forest carbon offsets have argued that while these projects offer a host of opportunities for climate policy and development, trading these carbon offset credits has many downsides. One problem is what geographers have called "the commodification of nature" where

valued components of the natural environment—carbon, biodiversity, forests, clean water—are measured, priced, sold, and traded using market logic. According to geographer Kathleen McAfee, when we commodify nature, we reduce diverse ecosystems to their exchangeable components and abstract nature from its broader social and ecological context. This can result in the dispossession of some local people of the right to use or profit from their land, and the reinforcement of unequal power relations, which may marginalize local communities. These issues are evident in the example of a carbon forest project in southern Mexico.

Corn, Cattle, and the Carbon Market in Southern Mexico

Established in the mid-1990s, Scolel Té (which means "the tree that grows" in a Mayan language) is located in the Mexican state of Chiapas. One of the earliest examples of forest-based carbon offsets, the project originated in an effort to reverse deforestation and generate local sustainable development benefits for small landholders, partly by selling carbon offsets to companies and individuals abroad. The project involves the participation of small farmers, many of whom are indigenous and engage in semi-subsistence agriculture of corn and beans, coffee production, forest management, and in some cases cattle ranching on communally owned land. By protecting forests and planting trees the project stores and sequesters carbon as trees mature. In return, the farmers are provided with a series of payments generated through the carbon market and the sale of offsets (see **Figure 4.C**).

The Scolel Té project has mainly targeted subsistence farmers who will change their land use at lower costs than the large-scale cattle ranchers whose expanding pastures are the leading driver of tropical deforestation in Mexico (as in much of Latin America—see Figure 2). Some communities, such as the

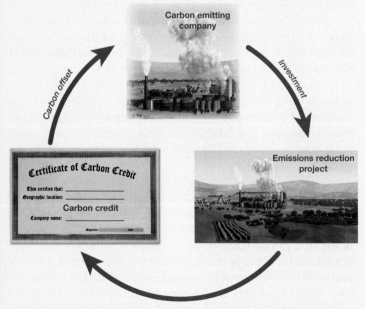

▲ **Figure 4.A The carbon offset cycle** Shown here is the process by which emissions made in one place can be offset in another.

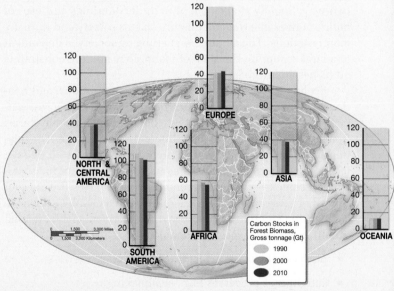

▲ **Figure 4.B Forest offsets** Forests absorb carbon and are often a popular offset option.

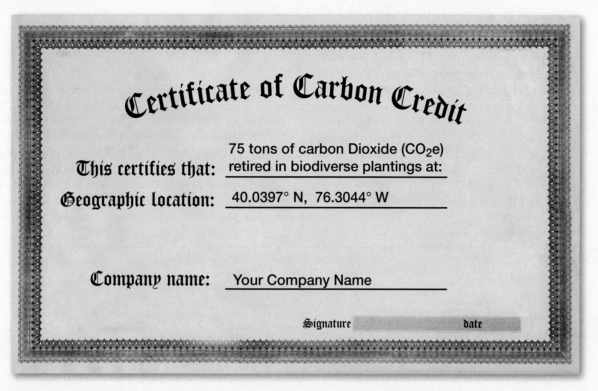

▲ Figure 4.C Certificate of Carbon Credit Entities who pay others to compensate for their emissions are provided with certificates like this one.

indigenous Chol Mayans, are concerned that selling offsets will constrain their use of forests for subsistence or traditional social and cultural uses.

There are ecological concerns associated with the carbon project as well. Because carbon payments are low and irregular, many participants in the tropical lowlands plant valuable timber species such as mahogany and tropical cedar. The expected revenue from timber is a major financial driver for participation in the project because carbon producers are allowed to cut timber after trees mature, provided they replant following harvesting. However, some have argued that the primary focus on carbon sequestration with fast-growing trees has produced timber plantations without sufficient attention to biodiversity.

Reducing Emissions from Deforestation and Forest Degradation

The experience of carbon forest projects like Scolel Té provide important lessons for a new global initiative to reduce climate change by protecting forests called REDD+. REDD+ pays governments and landowners for Reducing Emissions from Deforestation and Forest Degradation (REDD) in developing countries and includes conservation, sustainable forest management, and the enhancement of carbon stocks (the plus +). While some groups see REDD+ as a solution to both climate change and the sustainability of forests, others are opposed to it. Local people and some international activists are raising concerns about what REDD+ may mean for the rights of local people and the biodiversity of ecosystems. Will large areas of land be set aside for climate protection? Will the traditional and indigenous owners and users have access to them, and will they be paid fairly for the offsets? Should timber plantations—which lack biodiversity—be included? These are all important questions under debate with regards to REDD+.

Case studies of forest carbon projects indicate the need for caution with regard to the further commodification of forest carbon. It may be better to pursue other non-market strategies to protect forests, reduce greenhouse gas emissions, and generate local benefits—such as taxing carbon emissions where they are released or funding energy efficiency and renewable energy technologies for transport and electricity. Whatever approach we take, it is clear that effective and immediate action on climate change is urgently important.

1. Go onto the Climate Cool Network Web site, http://coolclimate. berkeley.edu/ click on Calculators & Maps tab above and select Household Calculator. Calculate the average carbon footprint for someone in your area or your personalized footprint based on your activities. What actions could you take to reduce your carbon emissions?

2. What are the key social and ecological problems caused by the commodification of carbon in the Scolel Té carbon forestry project in Mexico?

3. What might be alternative non-market strategies or mechanisms for mitigating global climate change?

Web sites:
http://coolclimate.berkeley.edu/
Diana Liverman's blog in REDD+ series on the Public Political Ecology Lab (PPEL) Web site http://ppel.arizona.edu/?p=248

Technological Waste http://goo.gl/5deoF6

their noxious industrial processes and waste products to peripheral countries (**Figure 4.4**). By exporting polluting industries and the jobs that go with them, however, core countries may also be contributing to increased affluence in the receiving countries. Given what we know about core countries, such a rise fosters a set of social values that ultimately leads to better protection of the environment in a new place. It is difficult to identify just when environmental consciousness goes from being a luxury to a necessity. The role of affluence in terms of environmental impacts is, in short, like population, difficult to assess.

Not surprisingly, the technology variable is no less complicated. Technologies affect the environment in three ways, through

- the harvesting of resources;
- the emission of wastes in the manufacture of goods and services; and
- the emission of waste in the consumption of goods and services.

A technological innovation can create access to an existing resource that was otherwise not available for use. In addition, technology can sometimes be a solution and sometimes a problem. Both principles can be seen in the case of fracking that opened this chapter. Producing this energy creates hazards at the same time that it threatens other important resources such as safe water supply.

It is therefore clear that increases in human numbers, in levels of wealth, and in technological capacity are key components of social and economic progress that have had an extremely complex

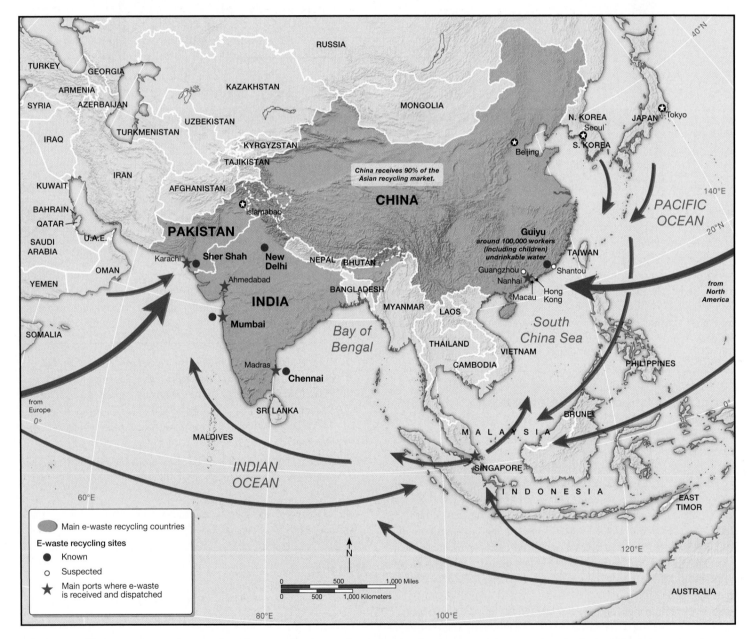

▲ **Figure 4.4 Electronic waste** This map shows the major illegal waste shipment routes for electronic garbage produced in Europe, North America and Australia. It also indicates key trans-shipment points in Asia. (*Source:* Basel Action Network, Silicon Valley Toxics Coalition, Toxics Link India, SCOPE (Pakistan), Greenpeace China, 2002.)

impact on the environment. In the last 100 years, this complexity has come to be seen as a triple-barreled threat to the quality of the natural world and the availability and quality of environmental resources. Before we look more carefully at the specific impacts of populations, affluence, and technology on nature, we need to look first at how differing social attitudes toward nature shape the human behaviors that are a basis for *I = PAT*.

Nature–Society Interactions

The concept of *adaptation* to the natural environment is part of the geographical subfield of cultural ecology most closely associated with the work of Carl Sauer and his students. **Cultural ecology** is the study of how human society has adapted to environmental challenges like aridity and steep landscapes through technologies such as irrigation and terracing and organizing people to construct and maintain these systems. These adaptations can be seen clearly in the rice terraces of Southeast Asia or the canals and reservoirs of the southwestern United States. More recent adaptations include the use of biotechnology and agricultural chemicals to increase agricultural production and the development of new pharmaceuticals to cope with diseases. Indeed, it has been argued that climate adaptation measures—reducing fossil-fuel consumption, adopting more sustainable farming practices, and recycling—are a form of cultural ecology. We discuss these and other such adaptations throughout this chapter.

Before the 1980s, the principles of cultural ecology did not include the political dimensions of ecological questions, but since then cultural ecologists have moved away from a strict focus on particular cultural groups' relationship with the environment, placing that relationship within a wider context instead. The result is political ecology, the merging of political economy with cultural ecology. **Political ecology** stresses that human–environment relations can be adequately understood only by relating patterns of resource use to political and economic forces, particularly the way that national states and capitalist economic practices can lead to, as well as exacerbate, environmental degradation.

Political ecology is understood to have three fundamental assumptions:

1. The costs and benefits associated with environmental change are distributed unequally because changes to the environment are not themselves homogenous.

2. This unequal distribution of environmental change inevitably reinforces or reduces existing social and economic inequalities.

3. The unequal distribution of environmental change results in the unequal distribution of costs and benefits affecting preexisting inequalities and altering power relationships.

APPLY YOUR KNOWLEDGE

1. Reflect on the *I = PAT* formula and determine what factors you think are most significant for environmental impact: population, affluence, or technology. When do richer countries create more impact and when do poorer countries? (Hint: Compare sizes of population with rates of consumption).

2. List three examples of cultural ecology from your own community or campus. (Hint: How specifically has your town or campus adapted to environmental challenges over the last 20 years?)

ENVIRONMENTAL ATTITUDES AND PHILOSOPHIES IN THE UNITED STATES

As we mentioned at the beginning of this chapter, nature is a construct that is shaped by social ideas, beliefs, and values. As a result, different societies and different cultures have different views of nature. In the contemporary world, views of nature are dominated by the Western (also known as Judeo-Christian) tradition that understands humans to be superior to nature. In this view, nature is something to be tamed or dominated. But other views of nature have emerged that depart dramatically from the dominant view. These include the environmental philosophies that became popular in the nineteenth and early twentieth centuries and the more radical political views of nature that gained prominence in the late twentieth century. Among the latter are approaches based on ecotheology—including Christianity, Hinduism, Islam, and Judaism—which reject the long-standing Western tradition. **Figure 4.5** provides a snapshot of changing attitudes in the United States toward the environment over nearly three decades.

Gallup's Environmental Polling
http://goo.gl/PGVWb1

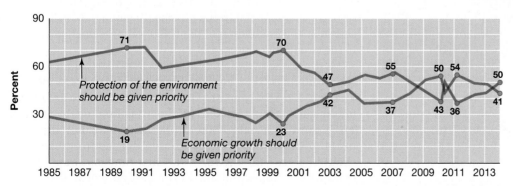

▲ **Figure 4.5** Environmental attitudes, Gallup Poll, 2014 This table shows changing attitudes toward the environment since 1985. Note that protecting the environment appears again to becoming important for many Americans. (*Source:* Adapted from Swift, Art "Americans again pick environment over economic growth" Gallup, March 20, 2014)

[1]Adapted from Bryant, Raymond L. and Sinead, Bailey. 1997. *Third World Political Ecology*. Abingdon, Oxon: Routledge.

Classical Environmental Philosophies

Henry David Thoreau (1817–1862), an American naturalist and activist, perhaps best illustrates a view that began to emerge in the mid-nineteenth century in the United States—the incorporation of North American Indian conceptions of nature into approaches to ecology. Thoreau lived and studied the natural world around the town of his birth, Concord, Massachusetts. He is most famous for his book *Walden*, which chronicles the two years he spent living and observing nature in solitude in a house he built at Walden Pond (**Figure 4.6**). Thoreau represented

I do not propose to write an ode to dejection, but to brag as lustily as chanticleer in the morning, standing on his roost, if only to wake my neighbors up. — Page 92.

▲ **Figure 4.6** Frontispiece of Henry David Thoreau's, *Walden*, **1854** The author's objective was to appreciate nature by removing himself from the rapidly industrializing and urbanizing world around him.

a significant alternative to the "humans-over-nature" approach that characterized his times. Many people regard him as the originator of a U.S. environmental philosophy.

Thoreau was impressed with the power of nature. He emphasized the interrelatedness of the natural world, where birds depended upon worms, fish depended upon flies, and so on along the food chain. Most notably, however, Thoreau regarded the natural world as an antidote to the negative effects of technology on the landscape and the American character. Concord was just 32 km (20 miles) west of Boston and an equal distance south of the booming mill towns of Lowell and Lawrence. Although he spent his life in a more or less rural setting, the Industrial Revolution was in full force all around Thoreau, and he was keenly aware of its impacts. His research on the animals and plants that surrounded Concord was an attempt to reconstruct the landscape as it had existed before colonization and massive European immigration.

Thoreau's ideas embraced European notions of **romanticism**, a philosophy that emphasized the interdependence of humans and nature. In direct revolt against those who espoused a Judeo-Christian understanding of nature, the romantics believed that *all* creatures—human and otherwise—were infused with a divine presence that commanded respect and that humans were not exceptional in this scheme. Rather, human divinity issued from humble participation in the natural community.

A branch of American romanticism known as **transcendentalism** also influenced understandings of nature during the early 1800s. Transcendentalism was espoused most eloquently by a Unitarian minister turned poet and philosopher, Ralph Waldo Emerson, a neighbor and contemporary of Thoreau. It encouraged people to attempt to rise above nature and the limitations of the body to the point where the spirit dominates the flesh and a mystical and spiritual life replaces a primitive and savage one.

Thoreau and Emerson are two of the most important influences on contemporary American ideas about the human–nature relationship. Another major influence on U.S. environmentalism was George Perkins Marsh, a native Vermonter, who in 1864 wrote a treatise entitled *Man and Nature, or Physical Geography as Modified by Human Action* (heavily revised and republished in 1874 as *The Earth as Modified by Human Action*). The first work to suggest that human beings are significant agents of environmental change, it is considered one of the most important advances in geography, ecology, and resource management in the nineteenth century. Marsh's ideas served as the foundation of the U.S. environmental movement in the twentieth century.

Early in the twentieth century, writers like Gifford Pinchot and politicians like Theodore Roosevelt drew on the ideas of Thoreau, Emerson, and Marsh to advocate the wise use of natural resources and the conservation of natural environments. Their view that nature should be conserved has survived to the present. **Conservation** holds that natural resources should be used thoughtfully and that humans should serve as stewards, not exploiters, of the natural world. Conservation implies responsibility to future generations as well as to the natural world itself in the utilization of resources. The writings of all these individuals eventually helped to inspire a wide range of environmental organizations, including the

▲ **Figure 4.7 African wildlife and land conservation** Nature conservancy focuses on involving local communities in conservation efforts intended to sustain the habitat of some of the world's most iconic wildlife.

Nature
Conservancy
Interactive
Conservation
Maps

http://goo.gl/1BF4ht

Environmental Defense Fund, World Watch Institute, the Sierra Club and the Nature Conservancy (**Figure 4.7**).

Those who espouse a more radical approach to nature see the conservation approach as too passive to be truly effective in protecting the environment. Such individuals believe that conservation leaves intact the political and economic system that drives the exploitation of nature. They believe that nature is sacred and should be preserved, not used at all. This more extreme position, **preservation**, advocates that certain habitats, species, and resources should remain off-limits to human use, regardless of whether the use maintains or depletes the resource in question.

CONTEMPORARY ENVIRONMENTAL PHILOSOPHIES

Environmental ethics is a philosophical perspective that prescribes moral principles as guidance for our treatment of nature. From this perspective, society has a moral obligation to treat nature according to the rules of moral behavior that exist for human beings. An aspect of environmental ethics that has caused a great deal of controversy is the idea that animals, trees, rocks, and other elements of nature have rights in the same way that humans do. If the moral system of our society insists that humans have the right to a safe and happy life, then, it is argued, the same rights should be extended to nonhuman nature.

Ecofeminism holds that patriarchy—a system of social ideas that values men more highly than women—is at the center of our present environmental malaise. Because patriarchy has equated women with nature, it has promoted the subordination and exploitation of both. The many varieties of ecofeminism range from nature-based spirituality oriented toward a goddess to more political approaches that emphasize resistance and opposition to the dominant models that devalue what is not male. Some ecofeminists are also environmental ethicists. Not only a movement of the core, ecofeminism has also been widely embraced in the periphery, where women are primarily responsible for the health and welfare of their families in environments that are being rapidly degraded. The unifying objective in all of ecofeminism is to dismantle the patriarchal biases in Western culture and replace them with a perspective that values social, cultural, and biological diversity.

Deep ecology, which shares many points of view with ecofeminism, is an approach to nature revolving around two key components: self-realization and biospherical egalitarianism. Self-realization embraces the view that humans must learn to recognize they are part of the nonhuman world. As a form of egalitarianism, deep ecology insists that Earth, or the biosphere, is the central focus of all life and that all components of nature, human and nonhuman, deserve the same respect and treatment. Deep ecologists, like environmental ethicists, believe that there is no absolute divide between humanity and everything else and that a complex and diverse set of relations constitutes the universe. The belief that all things are internally related should enable society to treat the nonhuman world with respect and not simply as a source of raw materials for human use.

The **environmental justice** (EJ) movement considers the pollution of neighborhoods by, for example, factories and hazardous-waste dumps to be the result of a structured and institutionalized inequality that is pervasive in both the capitalist core and the periphery. EJ activists see their issues as distinct from those of middle-class mainstream groups like the Sierra Club or even Greenpeace. For environmental justice activists, their struggles are uniquely rooted in their economic status. Thus, theirs are not quality-of-life issues, such as whether any forests will be left for recreation, but sheer economic and physical survival. As a result, the questions raised by EJ activists involve the distribution of economic and political resources. Such questions are not easily resolved in courts of law, but speak of more complicated issues such as the nature of racism and sexism and of capitalism as a class-based economic system. Like ecofeminism, the environmental justice movement is not restricted to the core. Indeed, poor people throughout the world are concerned that the negative impacts of economic development consistently affect them more than the rich (**Figure 4.8**).

Ecotheology calls for a reevaluation of the Western relationship to nature. The term came into prominence in the late twentieth century—mainly in Christian circles, though it has since spread across religious boundaries—in association with the scientific field of ecology. Within religious circles, there is a fear that science may not be capable of inspiring the changes in behavior necessary to thwart continuing environmental destruction. Ecotheologists also argue that capitalist political

▲ **Figure 4.8 International Campaign for Justice in Bhopal** Thirty years after a Union Carbide plant in Bhopal, India, accidentally leaked a toxic gas into the community, the activist group, International Campaign for Justice in Bhopal, is still seeking reparation for death, disabilities, and persistent illness caused by the leak that is believed to have caused nearly 4,000 deaths and over half a million injuries.

and economic institutions actively contribute to environmental degradation. In their view, it has become necessary, therefore, to address the current environmental crisis through belief systems that will overcome the inadequacies of humanly created institutions. Ecotheology recognizes the value of other creatures and God's intent for the cosmos as the basis for developing ethical models that take into account politics, economics, and practical issues in the quest for intelligent environmental policies.

All of these contemporary philosophies of nature attest to a growing concern over the environmental effects of globalization. Acid rain, deforestation, the disappearance of species, nuclear accidents, and toxic waste have all been important stimuli for newly emerging philosophies about the relationships between society and nature within a globalizing world. While none of these philosophies is a panacea, each has an important critique to offer. More than anything, however, each serves to remind us that environmental crises are not simple, and simple solutions will not suffice.

APPLY YOUR KNOWLEDGE

1. Choose three of the environmental philosophies (classic and contemporary) discussed, define them, and analyze why you think they are the most relevant to your views of human-nature relationships.

2. Provide three examples of how Thoreau's ideas about nature can be understood in terms of late-twentieth-century environmental philosophies such as environmental justice and ecotheology. (Hint: What is Thoreau saying about nature and what are the different environmental movements' views of nature? How are they similar?) Apply the same analysis to Marsh or Emerson.

GLOBAL CHANGE IN THE ANTHROPOCENE

In this section, we build on our understanding of attitudes and practices toward nature by considering Nobel prize-winning atmospheric chemist Paul Crutzen's characterization of our current era. He coined the term **anthropocene** to signal the current geological era in Earth's history in which the extent of human activities from agriculture to industrialization to urbanization have had far-reaching effects on Earth's ecosystems. An **ecosystem** is a community of different species interacting with each other and with the larger physical environment that surrounds it. We begin our discussion with European colonial expansion as it introduced new forms of agriculture, industry, and accelerated the development of cities throughout the globe. We end the section with a discussion of global climate change as a manifestation of the cumulative effects of human resource exploitation of Earth's atmosphere today.

European Colonial Expansion

The history of European expansion provides a powerful example of how a society possessing new environmental attitudes was able to radically transform nature. These new attitudes drew upon a newly emerging science and its contribution to technological innovation as well as the capitalist political and economic system that was being consolidated across Europe. This history also provides a helpful foundation for understanding the attitudes and practices that have helped produce some of our biggest environmental challenges today.

Initially, European expansion was internal—largely contained within continental boundaries. The most obvious reason for expansion was population increase: from 36 million in 1000 to over 44 million in 1100, nearly 60 million in 1200, and about 80 million by 1300 (**Figure 4.9**). As population increased, more land was brought under cultivation. In addition, more forests were cleared for agriculture, more animals were killed for food, and more minerals and other resources were exploited for a variety of needs. Forests originally covered upward of 90 percent of Western and Central Europe. At the end of the period of internal expansion, however—around 1300—the forested area amounted to only 20 percent.

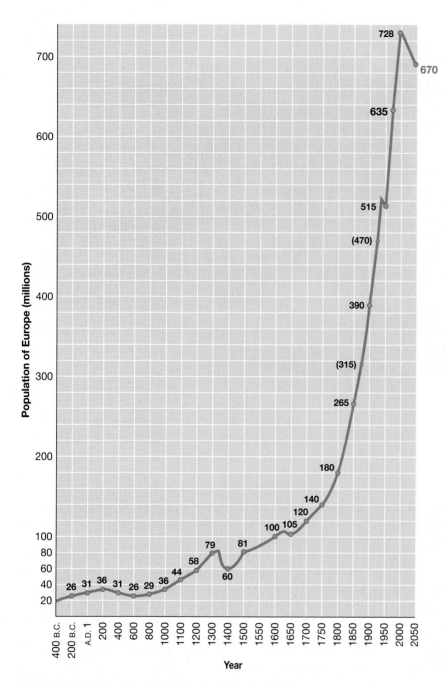

◄ **Figure 4.9 Population growth in Europe** This graph shows growth and change in the European population from 400 B.C. to A.D. 2050. The growth in the European population has been especially dramatic in the last 500 years as a result of capitalist globalization. The increase in human numbers at the beginning of the 1500s gave an important push to exploration and colonization beyond the confines of Europe. The dip in the graph from 1300 to 1500 is partially explained by the bubonic plague epidemic and food shortages. The dip around the mid-twentieth century shows the effect of the two world wars. European population is expected to continue to decline over the next four decades. (Updated from C. McEvedy and R. Jones, *Atlas of World Population History*. London: Allen Lane, 1978, Fig. 1.2, p. 18.)

ultimately meant increased wealth and power for a new class of people—the merchants—as well as for the aristocracy.

Over the centuries, Europe came to control increasing areas of the globe. The two examples we discuss in the next sections of the chapter illustrate how the introduction of the European people, ideologies, technologies, plant species, pathogens, and animals changed the environments into which they were introduced and also the societies they encountered.

Disease and Depopulation in the Spanish Colonies

Historians generally agree that European colonization of the New World was eventually responsible for the greatest loss of human life in history. Historians also agree that the primary reason for that loss was disease. New World populations, isolated for millennia from the Old World, possessed immune systems that had never encountered some of the most common European diseases. **Virgin soil epidemics**—in which the population at risk has no natural immunity or previous exposure to the disease within the lifetime of the oldest member of the group—were common in the so-called Columbian Exchange. Though in this case mostly one-way, the **Columbian Exchange** was the interaction between the Old World (Europe) and the New World (the Americas) initiated by the voyages of Columbus. Diseases such as smallpox, measles, chicken pox, whooping cough, typhus, typhoid fever, bubonic plague, cholera, scarlet fever, malaria, yellow fever, diphtheria, influenza, and others were unknown in the pre-Columbian New World.

Scholars refer to the phenomenon of near genocide of native populations as **demographic collapse**. The ecological effect of the population decline caused by the high rates of mortality was the transformation of many regions from productive agricultural land to abandoned land. Many of the Andean terraces, for example, were abandoned and dramatic soil erosion ensued. And large expanses of cleared land eventually returned to forest in areas such as the Yucatán.

In the fifteenth century, Europe underwent its second phase of expansion. This phase was external and not only changed the global political map but also launched a period of environmental change that continues to this day. European external expansion—colonialism—was the response to several impulses, ranging from self-interest to altruism. Europeans were fast running out of land, and as we saw in Chapter 2, explorers were being dispatched by monarchs to conquer new territories, enlarge their empires, and collect tax revenues from new subjects. Many of these adventurous individuals were also searching for fame and fortune or avoiding religious persecution. Behind European external expansion was also the Christian impulse to bring new souls into the kingdom of God. Other forces behind European colonialism included the need to expand the emerging system of trade, which

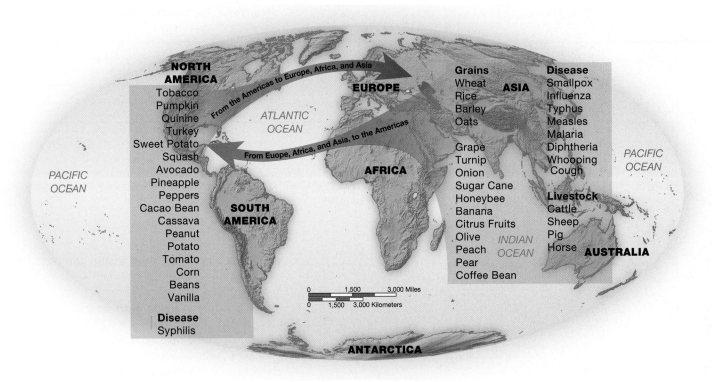

▲ **Figure 4.10 The Columbian Exchange** The encounter between the Old World and the New World that followed from Columbus's voyages enabled significant interactions between social and environmental systems. Most beneficial was the exchange of crops that improved nutrition on both sides of the Atlantic; most destructive was the carrying of diseases from Europe to the Americas causing massive loss of life in the New World.

Old World Plants and Animals in the New World A second case study of the environmental effects of European colonization involves the introduction of Old World plants and animals into the New World, and vice versa. The introduction of exotic plants and animals into new ecosystems is called **ecological imperialism**, a term now widely used by geographers, ecologists, and other scholars of the environment. The interaction between the Old and the New Worlds resulted in both the intentional and unintentional introduction of new crops and animals on both sides of the Atlantic (**Figure 4.10**). Europeans brought from their homelands many plants and animals that were exotics, that is, unknown to American ecosystems. For example, the Spanish introduced wheat and sugarcane, as well as horses, cattle, and pigs. These introductions altered the environment, particularly as the emphasis on select species led to a reduction in the variety of plants and animals that constituted local ecosystems. Inadvertent introductions of hardy exotic species included rats, weeds such as the dandelion and thistle, and birds such as starlings, which crowded out the less hardy indigenous species. As with the human population, the indigenous populations of plants, birds, and animals had few defenses against European plant and animal diseases and were sometimes seriously reduced or made extinct through contact.

Contact between the Old and the New Worlds was, however, an exchange—a two-way process—and New World crops and animals as well as pathogens were likewise introduced into the Old World, sometimes with devastating effects. Corn,

potatoes, tobacco, cocoa, tomatoes, and cotton were all brought back to Europe; so was syphilis, which spread rapidly throughout the European population.

Importantly, contacts between Europe and the rest of the world, though frequently violent and exploitative, were not uniformly disastrous. There are certainly examples of contacts that were mutually beneficial. The largely favorable ones were mostly knowledge-based or nutritional. Columbus's voyages (**Figure 4.11**) added dramatically to global knowledge, expanding understanding of geography, botany, zoology, and other rapidly growing sciences.

The encounter also had significant nutritional impacts for both sides by bringing new plants to each. European colonization, although responsible for the extermination of hundreds of plant and animal species, was also responsible for increasing the types and amounts of foods available worldwide. It is estimated that the Columbian Exchange may have tripled the number of cultivable food plants in the New World. It certainly enabled new types of food to grow in abundance where they had never grown before, and it introduced animals as an important source of dietary protein. The advantages of having a large variety of food plants are myriad. For instance, if one crop fails, another more than likely will succeed because not all plants are subject to failure from the same set of environmental conditions.

Before the Columbian Exchange, the only important sources of animal energy in the New World were the llama and the dog. The introduction of the horse, the ox, and the donkey

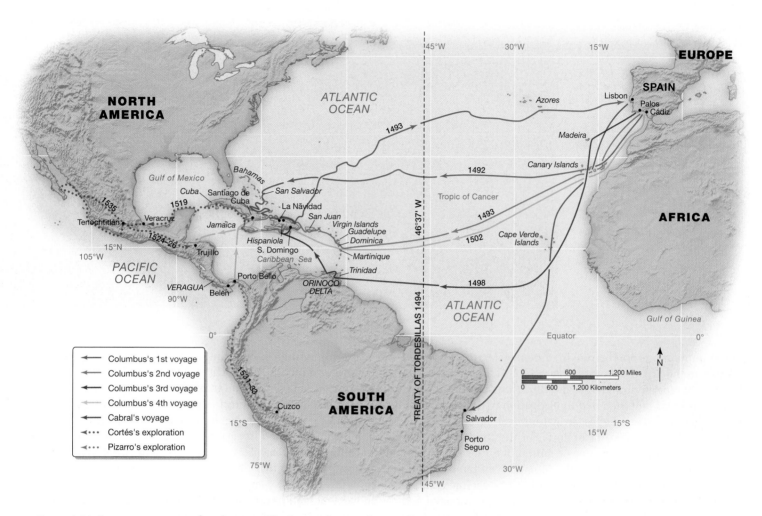

▲ **Figure 4.11 European voyages of exploration** The lines on this map illustrate the voyages and missions of Columbus, Pizarro, Cabral, and Cortés. Departing from Portugal and Spain, Columbus encountered several of the islands of the Caribbean as well as the coastal area of present-day Honduras and Venezuela.

(Adapted from *The Penguin Atlas of the Diasporas*, by G. Chaliand and J. P. Rageau, translated by A. M. Berrett. Translation copyright © 1995 by G. Chaliand and J. P. Rageau.)

created a virtual power revolution in the New World. These animals also provided fibers and, after death or slaughter, hides and bones to make various tools, utensils, and coverings. Most significant in its environmental impact, however, was the ox. Land that had escaped cultivation because the indigenous digging sticks and tools were unable to penetrate the heavy soil and matted root surface became workable with an ox-drawn plow. The result was that the indigenous form of intensive agricultural production (small area, many laborers) was replaced by extensive production (large area, fewer laborers). This transformation was not entirely without negative impacts, such as soil destabilization and erosion.

Indigenous People and their Pre-Columbian Environments
When discussing the impact of exchange, it is important to acknowledge native New World peoples and their relationships with the environment. The popular image of indigenous peoples living in harmony with nature, having only a minimal effect on their environment, is flawed. In reality, different groups had different impacts, and it is erroneous to conflate the thousands of groups into one romanticized

caricature. In New England, for example, prior to European contact, some groups hunted for wild game and gathered wild foods. More sedentary types, living in permanent and semi-permanent villages, cleared and planted small areas of land. Hunter-gatherers were mobile, moving with the seasons to obtain fish, migrating birds, deer, wild berries, and plants. Agriculturalists planted corn, squash, beans, and tobacco and used a wide range of other natural resources. The economy was a fairly simple one based on personal use or on barter (e.g., trading corn for fish). The idea of a surplus was foreign: People cultivated or exploited only as much land and resources as they needed to survive. Land and resources were shared, without concepts such as private property or land ownership. Fire was used to clear land for planting as well as for hunting. Although vegetation change did occur, it was minimal and not irreversible.

Native peoples of South and Central America altered their environment as well, though perhaps in more dramatic ways. The Aztecs of Mexico and the Incas of Peru had developed complex urban civilizations—many of them more sophisticated and highly populated than any European city—dependent on

▶ **Figure 4.12** Tenochtitlán, circa **1500** This famous contemporary painting by Luis Covarrubias of the capital city of the Aztecs, Tenochtitlán, illustrates the existence of dense social, cultural, and political activity in the core of the city with agricultural fields on the periphery, particularly to the north. Agricultural goods were also imported from the area surrounding the capital beyond the shores of the lake. With a population of 200,000, it was roughly twice the size of Seville, then Spain's largest city. When Cortés came upon the capital he noted that "... in Spain there is nothing to compare with it."

dense populations and employing intensive agricultural techniques on the urban outskirts (**Figure 4.12**). These groups were responsible for environmental modifications through cultivation techniques that included the irrigation of dry regions and the terracing of steep slopes. Irrigation over several centuries results in the salinization of soils. In the lowland tropics, intensive agricultural practices resulted in widespread deforestation as people cut and set fire to patches of forest, planted crops, and then moved on when soil fertility declined. Unlike in North America, a surplus was key to the operations of both societies, as tribute by ordinary people to the political and religious elite was required in the form of food, animals, labor, or precious metals. The construction of the sizable Inca and Aztec empires required the production of large amounts of building materials in the form of wood and mortar. Concentrated populations and the demands of urbanization meant that widespread environmental degradation existed prior to European contact.

APPLY YOUR KNOWLEDGE

1. How did the interaction between the Old World and the New World, the Columbian Exchange, affect human and natural systems in both Europe and the Americas? Name one legacy of the exchange that we are still experiencing.

2. Conduct online or library research to find three plants and animals in your own region that, in your opinion, are examples of ecological imperialism.

Global Climate Change

In the 400 years since the Columbian Exchange occurred, Earth has been dramatically transformed through human adaptation to the environment as well as exploitation of its resources. In some cases, the human use of nature has resulted in degradation or pollution. For example, overcultivation of steep slopes

has resulted in erosion of the soil needed for subsequent agricultural production, and the use of toxic agricultural chemicals has caused the contamination of rivers and lakes.

The Industrial Revolution produced a dramatic growth in the emissions of waste material to land, water, and the atmosphere and resulted in serious air pollution and health problems in many areas. These emissions and the associated pollution persist to this day. **Figure 4.13** provides a snapshot of how the last 200 years of burning fossil fuels for heat, light, movement, and improvement of daily life has conveyed carbon dioxide emissions into the atmosphere. We talk at length about these effects in the next major section of the chapter; here we wish to focus specifically on human-induced (also called anthropogenic) change on Earth's atmosphere by examining the complexities of climate change.

What Is Climate Change?

Climate change is defined by the Intergovernmental Panel on Climate Change (IPCC) as "a change in the state of the climate that can be identified (e.g., using statistical tests) by changes in the mean and/or the variability of its properties, and that persists for an extended period, typically decades or longer. It refers to any change in climate over time, whether due to natural variability or as a result of human activity."

Scientists are largely in agreement that global and regional climates have varied over geological time, largely due to slight changes in the tilt of Earth's axis in its orbit around the sun or associated changes in the amount of solar radiation reaching Earth. They also recognize that climate has changed over time because of large-scale events such as volcanic eruptions. The kind of climate change Earth is currently experiencing, however, is largely due to human activity that is altering the composition of the atmosphere. Of greatest concern is **global warming**, an increase in world temperatures and change in climate associated with increasing levels of carbon dioxide (CO_2), methane and other trace gases (**Figure 4.14**). These so-called

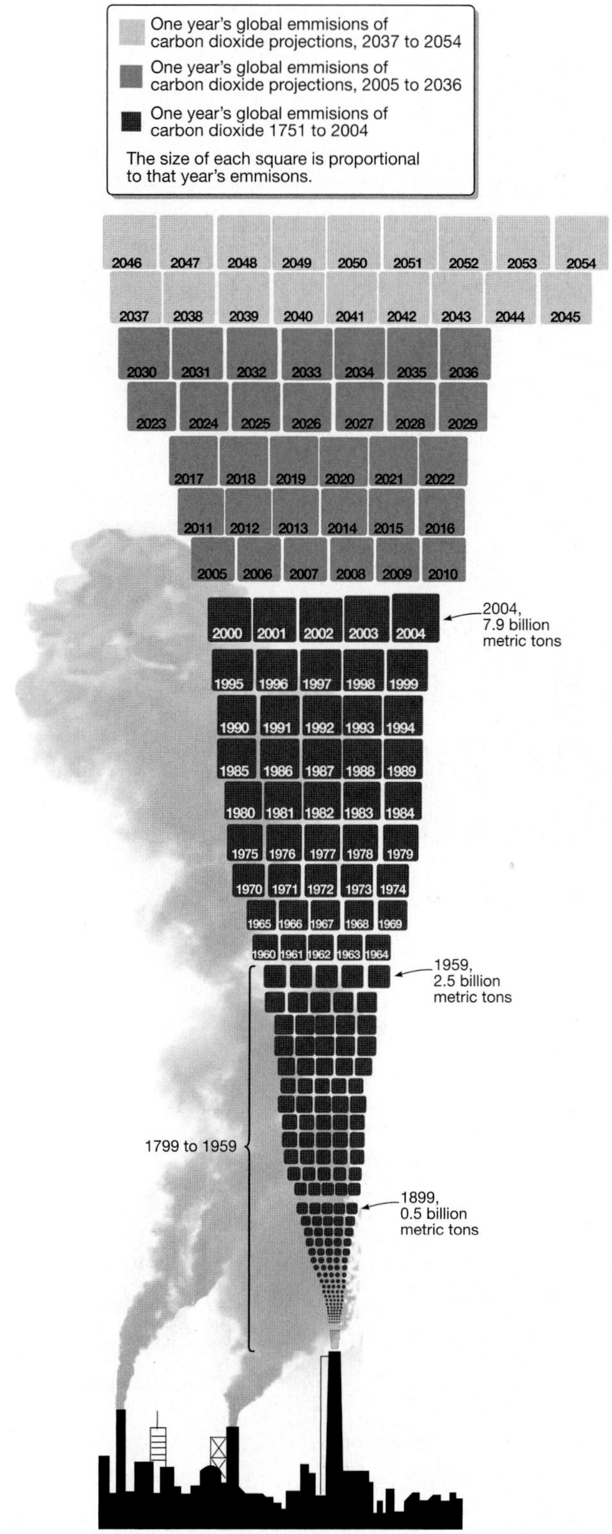

The size of each square is proportional to that year's emmisons.

- One year's global emmisions of carbon dioxide projections, 2037 to 2054
- One year's global emmisions of carbon dioxide projections, 2005 to 2036
- One year's global emmisions of carbon dioxide 1751 to 2004

2004, 7.9 billion metric tons

1799 to 1959

1959, 2.5 billion metric tons

1899, 0.5 billion metric tons

▲ **Figure 4.13 Global emissions of carbon dioxide** Each square represents one year's global emissions of carbon dioxide, measured by the weight of carbon it contains. (*Source:* Adapted from *New York Times*, December 16, 2007. http://www.nytimes.com/interactive/2007/12/16/weekinreview/20071216_EMISSIONS_GRAPHIC.html, accessed June 22, 2014.)

greenhouse gases (GHGs) result from human activities such as the burning of fossil fuels, cement production, and deforestation. Greenhouse gases act to trap heat within the atmosphere, resulting in its warming as well as that of Earth's surface.

How Do We Know Climate Is Changing?

It's important to appreciate that climate change is not an assertion based on a speculative theory or a political agenda. The evidence for it is based on facts, many of which are identified next.

- Temperature increase is widespread over the globe and is greater at higher northern latitudes. In the past 100 years, average Arctic temperatures have increased at almost twice the global average rate.

- Global average sea level rose at an average rate of 1.8 (1.3 to 2.3) mm per year over 1961 to 2003 and at an average rate of about 3.1 (2.4 to 3.8) mm per year from 1993 to 2003.

- Satellite data since 1978 show that annual average Arctic sea ice extent has shrunk by 2.7 (2.1 to 3.3) percent per decade, with larger decreases in summer of 7.4 (5.0 to 9.8) percent per decade. Mountain glaciers and snow cover have declined on average in both hemispheres.

- Since 1900, precipitation has increased significantly in eastern parts of North and South America, northern Europe, and northern and central Asia, whereas precipitation declined in the Sahel, the Mediterranean, southern Africa, and parts of southern Asia. Globally, the area affected by drought has likely increased since the 1970s.

- Average Northern Hemisphere temperatures during the second half of the twentieth century were very likely higher than during any other 50-year period in the last 500 years and likely the highest in at least the past 1,300 years.

What Are Some Projected Impacts of These Changes?

The effects of climate change are not only expected to happen in the future, many of them are already occurring.

- Based on a range of models, it is likely that future tropical cyclones (typhoons and hurricanes) will become more intense, with larger peak wind speeds and more heavy precipitation associated with ongoing increases of tropical sea-surface temperatures.

- Crop productivity is projected to increase slightly at mid- to high latitudes for local mean temperature increases of up to 1° to 3°C, depending on the crop, and then decrease beyond that in some regions. At lower latitudes, especially in seasonally dry and tropical regions, crop productivity is projected to decrease for even small local temperature increases (1° to 2°C), which would increase the risk of hunger.

- Coasts are projected to be exposed to increasing risks, including coastal erosion, due to climate change and sea level rise. A rising sea level would be disastrous for some countries. About 70 percent of Bangladesh, for example, is at sea level, as is much of Egypt's most fertile land in the Nile delta (**Figure 4.15**).

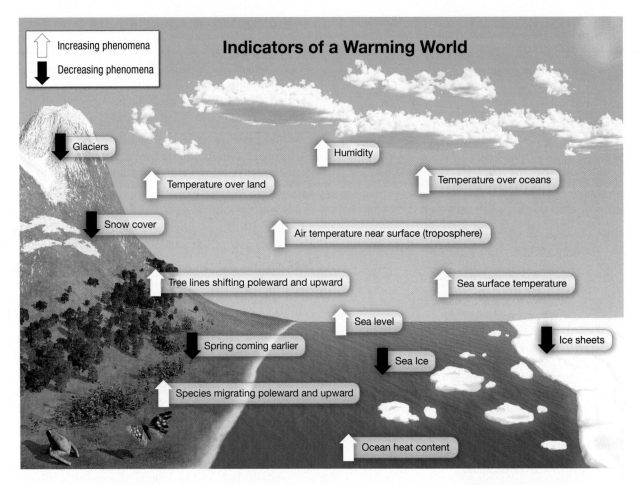

▲ **Figure 4.14 Major indicators of climate change** This diagram shows the major observed indicators of climate change, including higher temperatures over land and oceans, higher humidity, and higher sea levels.

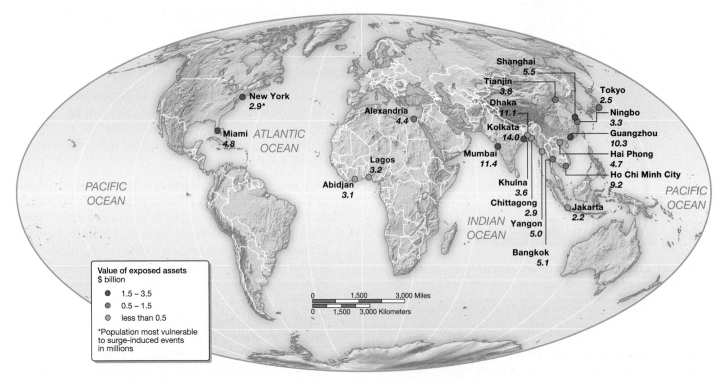

▲ **Figure 4.15 Effect of rising sea level on major cities** This graphic shows the largest cities vulnerable to rising sea waters by the year 2070.

(*Source:* http://phys.org/news/2012-11-sea-levels-faster-ipec.html, accessed June 22, 2016.)

- The health status of millions of people is projected to be affected through, for example, increase in malnutrition; increased deaths, diseases, and injury due to extreme weather events; increased burden of diarrheal diseases; increased frequency of cardiorespiratory diseases due to higher concentrations of ground-level ozone in urban areas related to climate change; and the altered spatial distribution of some infectious diseases.

- Global climate change is causing sea levels to rise as polar ice caps melt which will greatly affect coastal areas across the globe. This effect is already occurring in northern polar regions as well as among small islands in the southern hemisphere.

- Population growth patterns and the changing geography of economic development allow us to predict with some confidence that the air and water pollution generated by low-income countries will more than double in the next 10 to 15 years as they become more industrialized.

- Climate change is already exacerbating current stresses on water resources from population growth and economic and land-use change, including urbanization.

- The resilience of many ecosystems is already compromised in some places and is likely to be exceeded this century by an unprecedented combination of climate change, associated disturbances (e.g., flooding, drought, wildfire, insects, ocean acidification), and other global change drivers (e.g., land-use change, pollution, fragmentation of natural systems, overexploitation of resources).

The causes and consequences of these global climate changes vary considerably by where in the world they are unfolding. For example, the industrial countries have higher carbon dioxide emissions. Increased carbon dioxide emissions are contributing to rising temperatures through the trapping of heat in Earth's atmosphere. In order to survive in many of the world's peripheral regions, the rural poor are often impelled to degrade and destroy their immediate environment by cutting down forests for fuelwood, leading to the destruction of forests, which help to cool Earth's surface (**Figure 4.16**). Thus, both the core and the periphery are contributing to the problem of global climate change in different, but significant, ways. Moreover, environmental problems are becoming inseparable from processes of demographic change, economic development, and human welfare and are becoming increasingly enmeshed in matters of national security and regional conflict.

What's to Be Done?

There are many ways that individuals, organizations, and governments can begin to address the effects of climate change as described earlier, from using compact fluorescent light bulbs, buying local produce, and using reusable water bottles instead of disposable ones; developing and expanding local food resources; and supporting clean energy sources such as solar and wind power.

One increasing popular strategy for reducing GHGs is emissions trading. Already an established practice, emissions trading allows governments to regulate the amount of emissions produced in aggregate by setting the overall cap but allowing corporations the flexibility of determining how and where the emissions reductions will be achieved. Corporations that want to limit their emissions are allocated allowances, with each allowance representing a ton of the relevant emission, such as carbon dioxide. Corporations can emit in excess of their allocation of allowances by purchasing allowances from the market. Similarly, a company that emits less than its allocation of allowances can sell its surplus allowances.

The Climate Change Controversy

Despite overwhelming consensus from the international scientific community, a small number of critics—known as climate change deniers or skeptics—do not accept that Earth's climate is being changed by human activities. These individuals are generally not climate scientists and they do not publish in

▶ **Figure 4.16 Deforestation in the state of Para, Brazil** Shown here is a section of the Jamanxim National Forest where illegal clear-cutting has occurred. Large areas of deforestation like this one are not uncommon in the Amazonian rainforest.

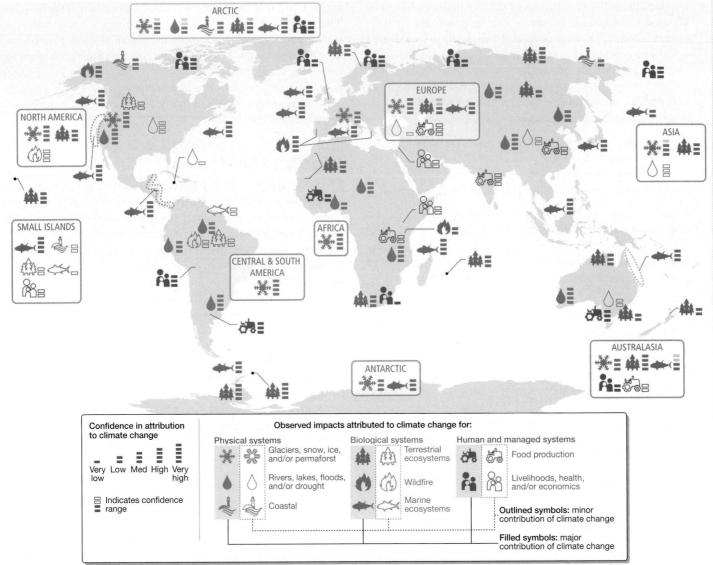

▲ **Figure 4.17 Climate change impacts on earth systems** The Fifth IPCC Report provides helpful information on the already observed impacts of climate change, such as on physical systems, biological systems, and human and managed systems, as shown here. (*Source:* http://ipcc-wg2.gov/AR5/press-events/press-kit/, accessed June 22, 2014.)

peer-reviewed scientific journals or engage in debate at conferences on climate science. Yet many of them are influential individuals as members of Congress or as chief officers of large corporations and as such have access to news media that feel compelled to offer a balanced perspective on the problem. The objective of this group of climate skeptics is to delay or prevent action by government to address climate change such as initiatives to push for alternative energy sources or tighter regulations for high-polluting energy like coal. Their approach has been to insist that global warming is a theory, not a fact, and as such, little or no regulatory action is needed.

A comprehensive study of 928 articles published on climate change in peer-reviewed scientific journals over a ten-year period found that not one of the articles disagreed with the scientific consensus that human-induced climate change is occurring. To gain a better understanding of the arguments advanced by climate change deniers click on the following link to an article in Scientific American that addresses them. It provides fact-based evidence to rebut climate change skepticism.

The IPCC's *Fifth Assessment Report: Mitigation of Climate Change* (2014) provides helpful data about the state of the world's climate and efforts to mitigate the effects of global warming in a format that is easily accessible to a lay audience. The report understands mitigation as an effort to stabilize greenhouse gases in the atmosphere to prevent the dangerous effects of human-induced change. It also recognizes that mitigation must be achieved in such a way that ecosystems will not be fatally threatened, food production will not be compromised, and economic development will occur in a sustainable manner (**Figure 4.17**).

Response to Climate Change Denialists

http://goo.gl/wdlYga

IPCC Fifth Assessment Report

http://goo.gl/QEqZCm

ENERGY, LAND-USE CHANGE AND ENVIRONMENT

Certainly the most central and significant technological breakthrough of the Industrial Revolution was the discovery and utilization of fossil fuels: coal, oil, and natural gas. Although the very first factories in Europe and the United States relied on waterpower to drive machinery, hydrocarbon fuels provided a more constant, dependable, and effective source of power. A steady increase in power production and demand since the beginning of the Industrial Revolution has been paralleled, not surprisingly, by an increase in resource extraction and conversion. **Box 4.2**, "Spatial Inequality: Energy Consumption and Production" provides insight into the global variation that exists in energy use.

Energy Needs and Environmental Impacts

The burning of home heating oil, along with the use of petroleum products for fuel in internal combustion engines, also launches harmful chemicals into Earth's atmosphere—causing air pollution and related health problems. The production and transport of oil have resulted in oil spills and substantial pollution of water and ecosystems. Media images of damage to seabirds and mammals after tankers have run aground and spilled oil have shown how immediate the environmental damage can be. Indeed, the oceans are acutely affected by the widespread use of oil for energy purposes. There is no better example of the damage of oil spills to the environment than the accident that occurred when British Petroleum's *Deepwater Horizon* oil rig exploded and caught fire on April 20, 2010, in the Gulf of Mexico, 400 km (250 miles) offshore of Houston.

The explosion killed 11 people and injured at least two dozen more when a blowout preventer, intended to check the release of oil, failed to activate. As the well continued to leak an estimated 40,000 barrels of oil a day before it was effectively capped, oil slicks spread across the Gulf from Florida to Texas, contaminating the environment and seriously damaging the Gulf coastal fisheries and tourism economy, especially in Louisiana, Mississippi, Alabama, and northern Florida. When the well was finally capped three months after the explosion, it is estimated that a total of 4.9 million barrels, or 205.8 million gallons, of oil had been released into the environment. As the world's largest accidental oil spill to date, it will continue to contaminate or damage marine and terrestrial life—both human and nonhuman (**Figure 4.18**). Despite the huge cost of the Gulf oil spill cleanup (estimated at $40 billion) to British Petroleum, the company registered a 17 percent profit in its 2011 first-quarter earnings.

Natural gas is one of the least noxious of the hydrocarbon-based energy resources because its combustion is relatively clean. Now supplying nearly one-quarter of global commercial energy, natural gas is predicted to be the fastest-growing energy source in this century. Reserves are still being discovered, with Russia holding the largest amount—about one-third of the world's total (**Figure 4.19**). While regarded as a preferred alternative to oil and coal, natural gas is not produced or consumed without environmental impacts. The risk of explosions at natural gas conversion facilities is significant; groundwater contamination and leakages and losses of gas from distribution systems contribute to the deterioration of Earth's atmosphere, as well as increasingly, some of the unintended effects of fracking as a form of natural gas mining.

▶ **Figure 4.18** Oil spill caused by the explosion of the *Deepwater Horizon rig* Pictured is an aerial view of crude oil on May 24, 2010, in Elmer Island, Louisiana. Skimmer ships, floating containment booms, anchored barriers, and sand-filled barricades were assembled along shorelines as chemical dispersants were used to attempt to prevent hundreds of miles of beach and wetlands from the spreading oil. In addition, underwater plumes of dissolved oil not visible on the surface and a 210 kilometer (80 square mile) "kill zone" surrounding the blown well were reported by scientists.

4.2 Spatial Inequality   Energy Consumption and Production

At present, the world's population relies most heavily for its energy needs on nonrenewable energy resources that include fossil fuels and nuclear energy, as well as renewable resources such as biomass, solar, hydroelectric, wind, and geothermal power. The distribution and production of the world's energy resources as well as its consumption, however, are deeply spatially uneven (**Figure 4.D** & **4.E**). Currently, 30 percent of the world's oil supplies are

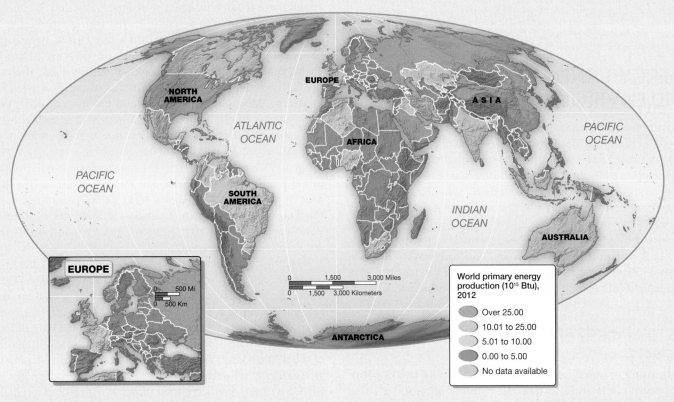

▲ **Figure 4.D** **Global energy production, 2012**   This map provides a picture of energy production around the world. The United States is the largest producer.   (*Source:* Adapted from US Energy Information Administration http://www.eia.gov/cfapps/ipdbproject/IEDIndex3.cfm#.)

located in the Middle East, and most of the coal is mined from the Northern Hemisphere, with China accounting for 41 percent of world production, followed by the United States at 19 percent, and the former Soviet Union at 8 percent. Nuclear reactors are a phenomenon of the core regions of the world as well. For example, France has 59 nuclear reactors operated by government-owned Electricité de France, supplying the country with over 430 billion kilowatt hours of electricity per year, which is 78 percent of the total generated there.

The consumption of energy is also spatially unequal as it relies on the ability to pay as well as the ability to afford the mitigation costs of its environmental impacts. In one year, global energy consumption is equal to about 1.3 billion tons of coal. What is most remarkable is that this is four times what the global population consumed in 1950 and 20 times what it consumed in 1850. The affluent core regions of the world far outstrip the peripheral regions in energy consumption. With nearly four times the population of the core regions, the peripheral regions account for less than one-third of global energy expenditures. Yet consumption of energy in the peripheral regions is rising quite rapidly as globalization spreads industries, energy-intensive consumer products such as automobiles, and energy-intensive agricultural practices into regions of the world where they were previously unknown.

It is projected that within the next decade or so, the peripheral regions will become the dominant consumers of energy-dependent products from appliances to homes to vehicles.

Finally the effects of energy production and consumption are also unevenly experienced. Every stage of the energy conversion process—from discovery to extraction, processing, and utilization—has an impact on the physical landscape as well as the populations who live on or near that landscape. In the coalfields of the world, from the U.S. Appalachian Mountains to western Siberia, mining results in a loss of vegetation and topsoil, in erosion and water pollution, and in acid and toxic drainage. It also contributes to cancer and lung disease in coal miners. Coal burning, is associated with relatively high emissions of environmentally harmful gases, such as carbon and sulfur dioxide, most directly affects populations living close to coal plants, especially in countries with limited environmental regulations.

---

1. Why is energy consumption on the rise for peripheral countries?

2. Compare the effects of nuclear, coal, and oil energy consumption on the social and natural environments. Is one fuel source more destructive than the others? Why?

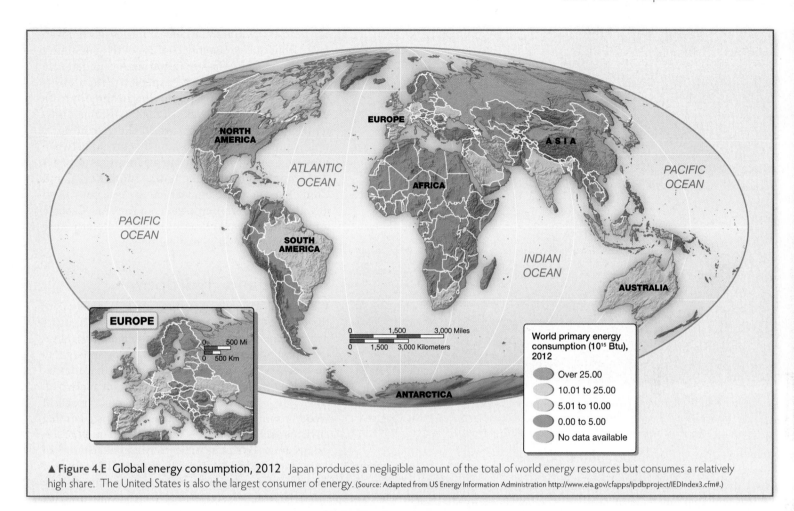

▲ **Figure 4.E** Global energy consumption, 2012 Japan produces a negligible amount of the total of world energy resources but consumes a relatively high share. The United States is also the largest consumer of energy. (*Source:* Adapted from US Energy Information Administration http://www.eia.gov/cfapps/ipdbproject/IEDIndex3.cfm#.)

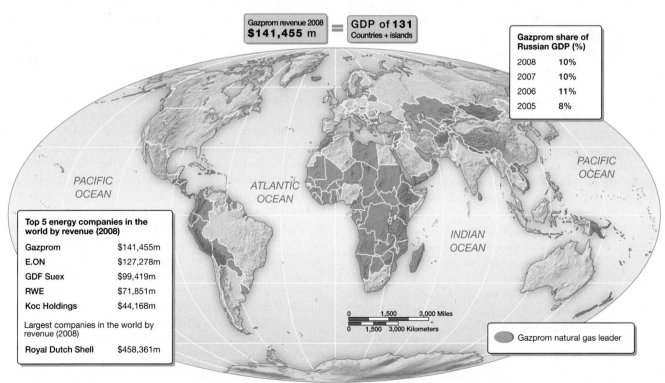

▲ **Figure 4.19** Gazprom natural gas reserves, 2013 Almost three-quarters of the world's natural gas reserves are located in the Middle East and Eurasia. Together, Russia, Iran, and Qatar account for about 55 percent of the world's natural gas reserves. This figure shows the dominance of the Russian gas company, Gazprom, with extraction sites in South America, Africa, the Middle East and Asia. (*Source:* http://www.arcticgas.gov/can-gas-liquids-technology-get-traction, accessed June 22, 2014.)

▲ **Figure 4.20 Disaster at Fukushima nuclear power plant** Shown in the background is the black smoke from the burning power plant caused by an earthquake that generated a tsunami that struck the power facility causing reactor explosions.

At the midpoint of the twentieth century, nuclear energy was widely promoted as a clearly preferable alternative to fossil fuels. It was seen by many as the answer to the expanding energy needs of core countries, especially as the supply of uranium worldwide was thought to be more than adequate for centuries of use. Nuclear energy was also regarded as cleaner and more efficient than fossil fuels. Although nuclear war was a pervasive threat, and there were certainly critics of nuclear energy even in the early years of its development, the civilian "atomic age" was widely seen as a triumphant technological solution to the energy needs of an expanding global economic system.

It was when serious accidents at nuclear power plants began to occur—such as those at Windscale in Britain and Three Mile Island in the United States, Chernobyl in Russia, and now Fukushima in Japan—that the voices of concerned scientists and citizens were raised in opposition. Incontestable evidence points to the need to address the problems associated with nuclear energy production, such as how to ensure nuclear reactor safety and how to safely dispose of nuclear waste (which remains radioactive for tens of thousands of years). After the devastating accidents and meltdowns, many countries decided to go nuclear-free, such as Sweden, Denmark, Finland, and New Zealand. In light of the recent meltdown at the Fukushima plant in Japan, the safety of nuclear energy is again being publicly debated. France, for example, which derives 75 percent of its electricity from nuclear energy power, saw protests against its continued use emerge in the spring of 2011. Other opposition to nuclear power occurred across Europe in Germany, Bulgaria, Italy, and the United Kingdom as well as in the United States, Asia, and Latin America.

And yet, while some core countries have moved away from nuclear energy because of the possibility of environmental disaster in the absence of fail-safe reactors, many more—and especially populous—developing countries are moving in the opposite direction, even in the wake of the Fukushima disaster (**Figure 4.20**). Given that world energy consumption is predicted to increase by 57 percent between now and 2030, it is not surprising that rapidly developing countries like India, South Korea, and China have growing nuclear energy programs. And developing Eastern European countries like Czechia and Romania, as well as Russia and Belarus, are continuing to invest in nuclear power.

Biomass and Hydropower

While nuclear power problems are still largely confined to the core, the periphery is not without its energy-related environmental problems. Because a large proportion of populations in the periphery rely on biomass for their energy needs, as the populations have grown, so has the demand for fuelwood. One of the most immediate environmental impacts of wood burning is air pollution, but the most alarming environmental problem is the rapid depletion of forest resources. With the other conventional sources of energy (coal, oil, and gas) being too costly or unavailable to most peripheral households, wood or other forms of biomass—any form of material that can be used as fuel, such as animal waste, livestock operation residues, and aquatic plants—is the only alternative. The demand for fuelwood has been so great in many peripheral regions that forest reserves are being rapidly used up (**Figure 4.21**).

Fuelwood depletion is extreme in the highland areas of Nepal, as well as in the Bolivian and Peruvian Andes Mountains. The clearing of forests for fuelwood in these regions has led to serious steep-slope soil erosion. In sub-Saharan Africa, where 90 percent of the region's energy needs are supplied by wood, overcutting of the forests has resulted in denuded areas, especially around rapidly growing cities. And although wood gathering is usually associated with rural life, it is not uncommon for city dwellers as well to use wood and other forms of biomass to satisfy their household energy needs. In Niamey, the capital of Niger, the zone of overcutting is expanding as the city itself expands. It is estimated that city dwellers in Niamey travel from 50 to 100 kilometers (31 to 62 miles) to gather wood. The same goes for inhabitants of Ouagadougou in Burkina Faso, where the average haul for wood is also over 50 kilometers. And fuelwood use in Asian and South American highland regions as well as in sub-Saharan Africa is expected to continue to increase substantially (34 percent by some estimates) over the next decade, at least.

Hydropower Hydroelectric power, also known as hydropower, was once seen as a preferred alternative to the more obviously environmentally polluting and nonrenewable fossil-fuel sources. The wave of dam building that

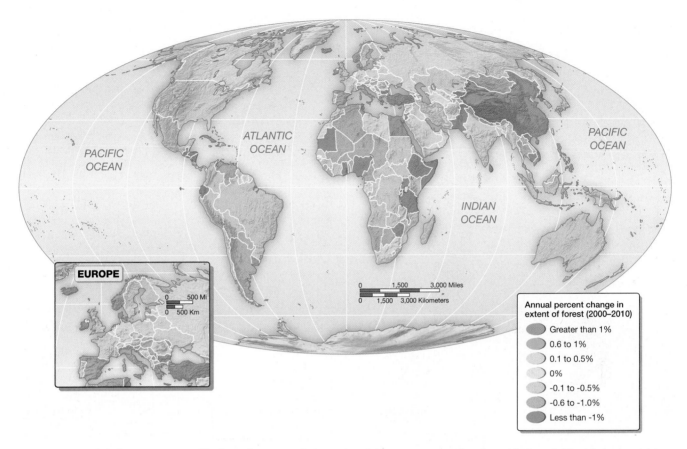

▲ **Figure 4.21** **Global consumption of fuelwoods** Firewood, charcoal, and dung are considered traditional fuels, and although their availability is decreasing, dependence upon them is increasing. Dependence on traditional sources of fuel is especially high in the periphery, where, in Africa, for example, they are the most important energy source for cooking and heating. Wood and charcoal, although renewable sources, are replenished very slowly. Acute scarcity will be a certainty for most African households in the twenty-first century. (*Source:* Data from UN Food and Agricultural Organization, *State of the World's Forests 2011*, Annex, Table 4, Rome, 2011.)

occurred throughout the world over the course of the twentieth century improved the overall availability, quality, cost, and dependability of energy as well as harnessed water resources for food production, generation, flood control, and domestic use (**Figure 4.22**). Unfortunately, however, dams built to provide hydroelectric power (as well as water for irrigation, navigation, and drinking) for the burgeoning cities of the core and to encourage economic development in the periphery and semiperiphery have also had profound negative environmental impacts. Among the most significant of these impacts are changes in downstream flow, evaporation, sediment transport and deposition, mineral quality and soil moisture, channeling and bank scouring, and aquatic biota and flora, as well as the development of conditions threatening to human health.

Furthermore, the construction of dams dramatically alters terrain, often with serious consequences. For example, clearance of forests for dam construction often leads to large-scale flooding. Felled trees are usually left to decay in the impounded waters, which become increasingly acidic. The impounded waters can also incubate mosquitoes, which carry diseases such as malaria. The effects on human populations are also significant as dam projects often require the relocation of long-settled populations. These populations are often indigenous and their displacement disrupts dense social networks, generations-long connections to the land, as well as established livelihoods.

Despite these problems and in response to the rising cost of fossil fuels, the use of hydroelectricity is expected to continue to expand over the next two decades by 1.9 percent per year, particularly where its use is supported by government policies and incentives, such as in Asia and South America. The construction of the Three Gorges Dam, as well as several other large-scale hydroelectric projects in China, illustrates this trend (**Figure 4.23**). In contrast, most of the increase in energy production in core countries is expected to be in renewable resources, as shown in Box 4.3 "Visualizing Geography: Renewable Energy Resources."

Energy-Related Pollutants One reason hydroelectric power continues to be appealing in the periphery is that it produces fewer atmospheric pollutants than fossil fuels. Indeed, coal and gas power stations as well as factories, automobiles, and other forms of transportation are largely responsible for the increasingly acidic quality of Earth's atmosphere. Although it is true that people as well as other organisms naturally produce many gases, including oxygen and carbon

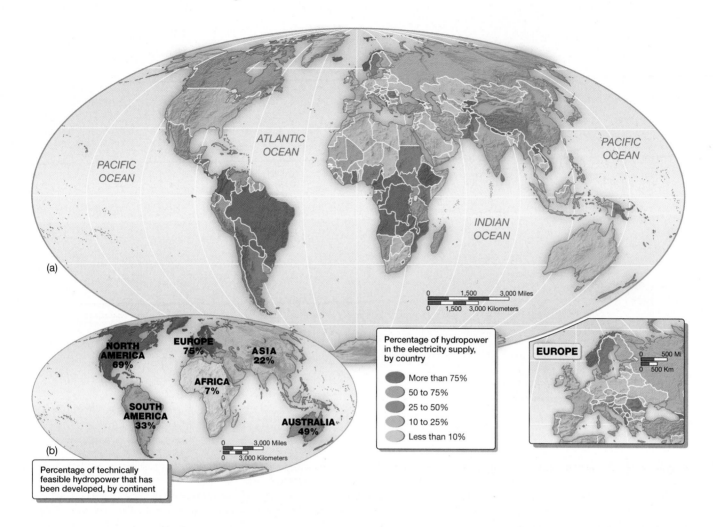

▲ **Figure 4.22 Percentage of hydropower in the electricity supply by country** Although the great dam-building era for core countries is now largely completed, many peripheral countries, in a bid to participate more actively in the world economy, are building dams. Only a few countries are almost exclusively dependent on the hydropower produced from dams. Given the increasing need for electricity by rapidly developing peripheral countries, hydropower is becoming a more attractive energy option for many of them because of the uncertain supply of oil in the future. (*Source:* U.S. Energy Information Administration, U.S. Department of Energy)

dioxide, increasing levels of industrialization and motor vehicle use have contributed to the destablization of the long-term balance of such gases, leading to serious atmospheric pollution. Sulfur dioxide, nitrogen oxides, and hydrocarbons, among other gases, which are released into the atmosphere from motor vehicle exhaust, industrial processes, and power generation (based on fossil fuels), are increasing the level of acids in the atmosphere. If these gases reach sufficient concentrations and are not effectively dispersed in the atmosphere, acid rain can result.

Acid rain or acid deposition, as it is known scientifically, is the wet deposition of acids upon Earth through the natural cleansing properties of the atmosphere. The term *acid rain* also includes acid mists, acid fogs, and smog. Acid rain occurs as the water droplets in clouds absorb certain gases that later fall back to Earth as acid precipitation. **Figure 4.24** illustrates the widespread nature of acid emissions. The problem first emerged as a health problem in the industrial countries of the Northern Hemisphere, especially the United Kingdom, Scandinavia, the

United States, and Canada, in the 1950s and eventually came to be seen as an environmental concern as well. Acid deposition poisons soils and water bodies that become too acidic to support life. In urban areas, acid rain corrodes marble and limestone structures, affecting important buildings such as the Parthenon in Athens and St. Paul's Cathedral in London as well as others in Europe. By the late 1990s, however, acid deposition became far less of a problem in Europe and North America as decades of environmental regulations and cooperative agreements diminished the release of of chlorflourocarbons (CFCs) into the atmosphere. The problem has not gone away entirely, however, and is emerging with great negative effect in Asia, especially China, where few governmental limits on industrial pollution are in place. In fact, because of rapid industrial growth, China has some of the world's most polluted cities and rivers. The steady emissions of sulfur dioxide from unregulated factories and vehicles has resulted in one-third of China's territory being affected by acid rain, posing a major threat to soil water and food safety.

▲ **Figure 4.23 Three Gorges Dam, China** (a) In May 2006, the world's largest dam was completed in Yichang, central China's Hubei Province. The NASA bird's eye view of the dam shows the dramatic difference in the width of the river on both sides of the dam. (b) The axis of the dam is 2,309 meters (1.4 miles), the longest in the world. More than 1.3 million people have been relocated to make way for the dam and its reservoir. (c) The inset map shows the 477 km (297 mi) extent of the reservoir that is impounded behind the dam..

133

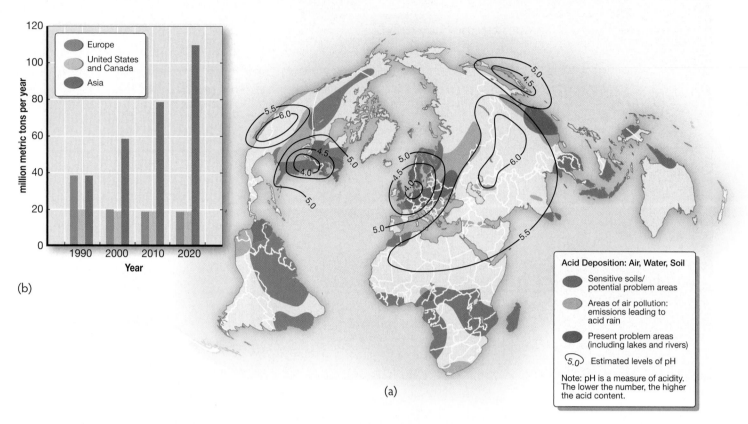

(b)

(a)

▲ **Figure 4.24 Global acid emissions** (a) Acid emissions affect various elements of the natural and the built environment. In some parts of the world, the damage to soils is especially severe. In others, acid emissions cause serious air pollution. (b) Most industrialized countries have cut sulfur dioxide emissions to help mitigate the damage of acid rain to ecosystems. But the acid rain problem is not yet solved, as it is emerging as a major problem in the developing world, especially in parts of Asia and the Pacific where energy use has surged. (*Source:* Adapted from J.L. Allen, *Student Atlas of Environmental Issues*, Guilford, CT: Duskin/McGraw-Hill: 1997, p. 45; World Resources Institute, "Acid Rain: Downpour in Asia," *World Resources 1998–1999*, 1998; The International Energy Agency.)

APPLY YOUR KNOWLEDGE

1. What is the relationship between energy consumption and environmental degradation? When answering, consider issues such as pollution, contamination, resources depletion, explosions and spills.

2. Take a closer look at some of the most significant nuclear disasters of the last 40 years by researching one of these sites: Windscale in Britain; Three Mile Island in the United States; Chernobyl in Russia; or Fukushima in Japan. What is different about a nuclear disaster from other types of energy plant accidents? What are the effects to humans, animals, and the environment? Will the disaster affect future generations and if so, how?

Land Use and Environmental Impacts

In addition to industrial pollution and steadily increasing demands for energy, the environment is also being dramatically affected by land use. The clearing of land for fuel, farming, grazing, resource extraction, highway building, energy generation, and war all have significant impacts. Geographers classify land into five categories: forest, cultivated land, grassland, wetland, and areas of settlement. Geographers also speak of land-use change as occurring in either of two ways: conversion or modification. *Conversion* is the wholesale transformation of land from one use to another (e.g., the conversion of forest to settlement). *Modification* is an alteration of existing land use (e.g., a grassland overlaid with railroad line or a forest thinned and not clear-cut).

One of the most dramatic impacts is loss or alteration of forest cover, which humans have cleared for millennia to make way for cultivation and settlement. Forests are cleared not only to obtain land to accommodate population increases but also to extract the vast timber resources they contain. The forested area of the world has been reduced by about 8 million square kilometers (about 3 million square miles) since preagricultural times. Rapid clearance of the world's forests has occurred through logging, settlement, and agricultural clearing or through fuelwood cutting around urban areas. **Figure 4.25** shows the global extent of deforestation in recent years.

Forests The permanent clearing and destruction of forests, **deforestation**, is currently occurring most alarmingly in the world's rain forests. The UN Food and Agricultural Organization has estimated that we are destroying rain forests globally at the rate of 0.40 hectare (1 acre) per second. Today, rain forests cover less than 7 percent of the land surface, half of what they covered only a few thousand years ago.

Destruction of the rain forests, however, is not just about the loss of trees, a renewable resource that is being eliminated

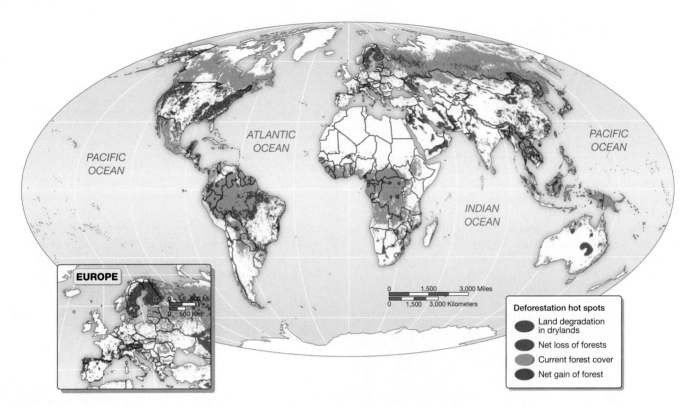

▲ **Figure 4.25** **Global deforestation** Since agriculture emerged about 10,000 years ago, human activities have diminished the world's forest resources by about 25 percent. Whereas forests once occupied about one-third of Earth's surface, they now take up only one-quarter. Forests filter air and noise pollution, provide a habitat for wildlife, and slow down water runoff, helping to recharge streams and groundwater. They also influence climate at local, regional, and global levels. (*Source:* World Resources Institute, http://images.wri.org/sdm-gene-02-deforestation.jpg or http://www.wri.org/image/view/10981/_original.)

more quickly than it can be regenerated. It is also about the loss of the biological diversity of an ecosystem. This also translates into the potential loss of biological compounds that may have great medical value. In addition, the destruction of rain forests can lead to the destabilization of the oxygen and carbon dioxide cycles of the forests, which may have long-term effects on global climate.

Much of the destruction of the South American rain forests is the result of peripheral countries' attempts at economic development. In the Bolivian Amazon rain forest, the introduction of coca production has become an important source of revenue for farmers in the region and has led to the removal of small tracts of forest. The pressure of economic development persists in the region such that net rain forest and forest land more generally are continuing to decline in Central and South America at a faster rate than in the 1990s.

Great geographical variability exists with respect to human impacts on the world's forests. For most of the core regions, the forests that were once cleared have been replaced by regeneration. For most of the periphery, in contrast, clearance has accelerated to such an extent that one estimate shows a 50 percent reduction in the amount of forest cover since the early 1900s.

Cultivated Lands Cultivation is another important component of global land use, which we will deal with extensively in Chapter 9. In this section, we briefly cover one or two points about the environmental impacts of cultivation that are particularly pertinent to our current discussion. During the past 300 years, the land devoted to cultivation has expanded globally by 450 percent. In 1700, the global stock of land in cultivation took up an area about the size of Argentina. Today, it occupies an area roughly the size of the entire continent of South America. While the most rapid expansion of cropland since the mid-twentieth century has occurred in the peripheral regions, the amount of cropland has either held steady or been reduced in core regions. The expansion of cropland in peripheral regions is partly a response to growing populations and rising levels of consumption worldwide. It is also due to the globalization of agriculture (see Chapter 9), with some core-region production having been moved to peripheral regions.

The phenomenon of corporations and rich governments investing in the agricultural land of peripheral countries is often called a "land grab" in the popular press. While it is impossible to know exactly how much peripheral land is being sold to investors, because land deals are often secret, there are reliable estimates. The Land Coalition, a nongovernmental organization, estimates that by 2011 almost 80 million hectares had been subject to some sort of negotiation with a foreign investor, more than half in Africa (**Figure 4.26**). To appreciate just how much land this is, 80 million hectares is more than all the farmland of the United Kingdom, France, Germany, and Italy combined. Big investors are countries with concerns about feeding their own people that purchase farmland abroad

4.3 Visualizing Geography

Renewable Energy Resources

It is important to realize that alternatives exist to fossil fuels, hydroelectric power, and nuclear energy. Energy derived from the sun, the wind, Earth's interior (geothermal sources), and the tides, is clean, profitable, and dependable (Figure 4.1.1). Japan, the US, and Germany all have solar energy production facilities that are cheap and nonpolluting. The production of energy from geothermal and wind sources has also been successful in a few locations around the globe. Italy, Germany, Iceland, the US, Mexico, Denmark, and the Philippines all derive some of their energy production from geothermal or wind sources (Figure 4.4.1).

While the sources here are not likely to be reliable for large scale consumers like cities or regions, there is some optimism that they can be effective at the individual or household level.

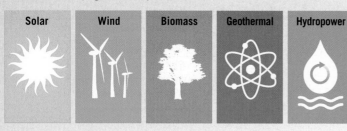

Figure 4.1 Common types of renewable energy

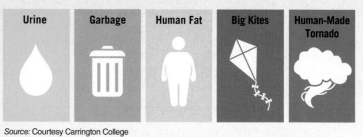

Source: Courtesy Carrington College

Figure 4.2 Some unusual renewable energy resources

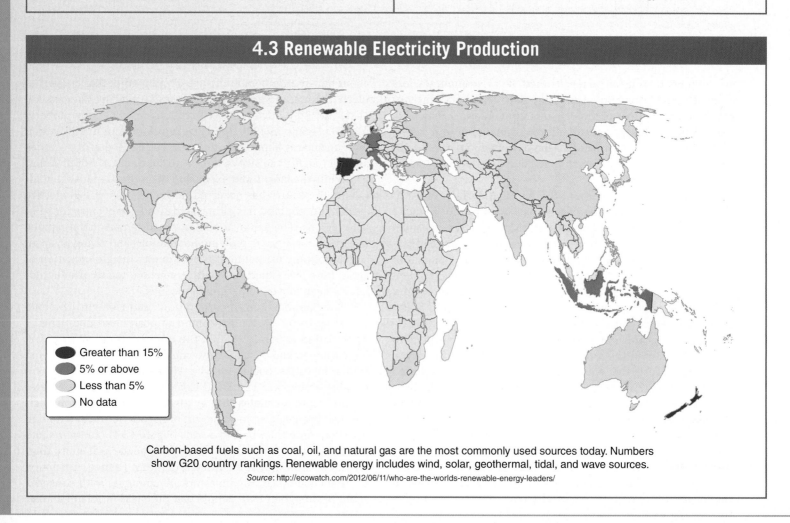

4.3 Renewable Electricity Production

Greater than 15%
5% or above
Less than 5%
No data

Carbon-based fuels such as coal, oil, and natural gas are the most commonly used sources today. Numbers show G20 country rankings. Renewable energy includes wind, solar, geothermal, tidal, and wave sources.

Source: http://ecowatch.com/2012/06/11/who-are-the-worlds-renewable-energy-leaders/

4.4 Hybrid and Electric Vehicles

Most of the world's energy consumption is for transportation. Hybrid (HEV) and plug-in electric vehicles (PEV) are limited due to the high cost of this technology. City and county governments in the US and large corporations like Federal Express (FedEx) are converting their transportation fleets to hybrids. Hybrid and electric automobiles are also very popular in Europe and the United Kingdom. By mid-2011, sales of new hybrid and electric cars were growing faster than sales of conventional vehicles in the US, the largest vehicle market in the world. The growing hybrid and electric automobile market is being bolstered by the fleet market, which every year sees new commitments to all-electric delivery vehicles.

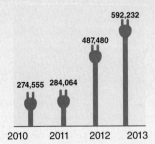

Figure 4.4.1 Sales of HEV and PEV vehicles 2010-2013
Source: www.electricdrive.org

 At the end of 2013 there were **55 HEV and PEV models** available from over 20 manufacturers. Most American car buyers don't know that various goverment incentives exist for electric-car ownership.

 In August 2013 there were **130,767** electric vehicles on the road, by August 2014 there were **246,426**, an **89% increase**.

In 2011, about 20% of the world's energy was used for transportation.

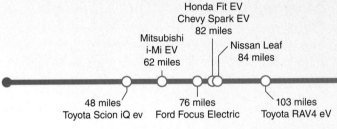

Figure 4.4.2 Driving ranges of some electric passenger cars.
http://wallstcheatsheet.com/automobiles/top-10-electric-vehicles-with-the-longest-driving-range.html/?a=viewall
http://www.autotrader.com/research/article/best-cars/221440/top-10-evs-with-the-longest-range.jsp

4.5 Energy Efficiency

A significant change in energy consumption is greater efficiency of use. In the US, for example, average household energy use dropped between 1978 and 2010. However, the increased use of electronic devices—cell phones, televisions, computers—is partly offsetting these greater efficiencies. Energy consumption to fuel the increasing use of these devices is expected to double over the next 10 years (Figure 4.3.1).

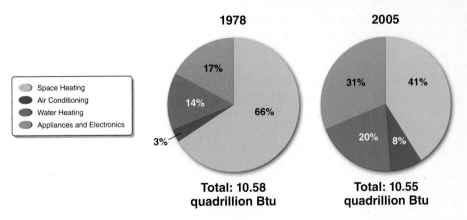

Figure 4.5.1 Energy consumption in U.S. homes 1978 and 2005 *Source*: U.S. Energy Information Administration, 1978 and 2005 Residential Energy Consumption Survey

Renewable Energy Institute International

http://goo.gl/mZd0Mn

1. In what ways is renewable energy key to sustainable development?

2. What are some of the limitations of renewable energy? Choose three sources and describe how the issues of availability and cost are central to renewable energy.

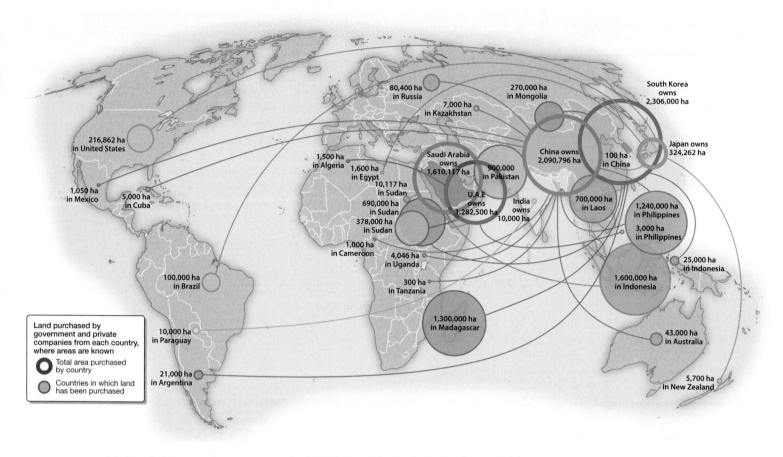

▲ Figure 4.26 Global land grab This graphic illustrates both public and private investment in land in Africa, North and South America, Asia, and Australia. The shades shown on the filled circles corresponds to the country or the national origin of the company that has purchased the land. For example, the 324,262 hectares of that has been purchased by the goverment of Japan or Japanese companies is show in orange lines connecting it to Brazil, The United States, Egypt, China, and New Zealand. (*Source:* Copyright Guardian News & Media Ltd. 2008. Reprinted with permission.)

as a guarantee of future food supplies. China is the largest of these investors, buying or leasing twice as much land as anyone else, usually in Africa. The most worrying aspect of these land grabs is the high levels of corruption among the buyers and low levels of benefit for the sellers.

Grasslands

Distinct from agricultural land, grasslands are also used productively the world over, either as rangeland or pasture for livestock grazing. Most grasslands are found in arid and semiarid regions that are unsuitable for farming due to lack of water or poor soils (**Figure 4.27**). Some grasslands, however, occur in more rainy regions where tropical rain forests have been removed. Others are at the mid-latitudes, such as the tall- and short-grass prairies of the central United States and Canada. Approximately 68 million square kilometers (26 million square miles) of global land surface is currently taken up by grasslands.

Human impacts on grasslands are largely of two sorts. The first is the clearing of grasslands for other uses, most frequently settlement. As the global demands for beef production have increased, the use of the world's grasslands has intensified. Widespread overgrazing of grasslands has led to acute degradation. In its most severe form, overgrazing has led to desertification.

Desertification is the spread of desert conditions resulting from deforestation, overgrazing, and poor agricultural practices, as well as reduced rainfall associated with climatic change. Until recently, one of the most severe examples of desertification had been occurring in the Sahel region of Africa. The degradation of the grasslands bordering the Sahara Desert has not been a simple case of careless overgrazing by thoughtless herders, however. Severe drought, recurrent famine, and the breakdown of traditional systems for coping with disaster have all combined to create increased pressure on fragile resources, resulting in a loss of grass cover and extreme soil degradation since the 1970s. However, increasing evidence is suggesting that between 1982 and 2000 the desertification in this region has been declining (**Figure 4.28**). Scientists believe they are seeing signals that the Sahara Desert and surrounding regions are actually **greening**—adding biomass including grasses as well as trees—due to increasing rainfall. The increased rainfall may be the result of global climate change, which is leading to warmer temperatures. Hotter air can hold more moisture, which in turn creates more rain. Aerial photographs as well as ground studies have confirmed the greening phenomenon. The most optimistic projection of the various climate models that attempt to better understand the impacts of the greening is that the rains could continue

▶ **Figure 4.27 African grasslands** Also known as savannas, grasslands include scattered shrubs and isolated small trees and are normally found in areas with high-to-average temperatures and low-to-moderate precipitation. They occur in an extensive belt on both sides of the equator. African tropical savannas, such as the one pictured here in Kenya, contain extensive herds of hoofed animals, including gazelles, giraffes, zebras, wildebeests, antelopes, and elephants.

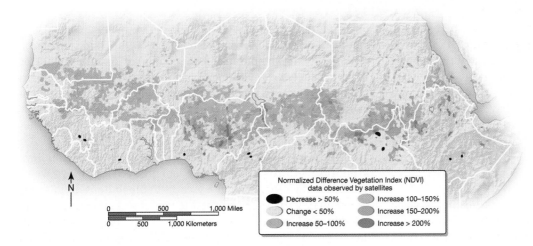

Normalized Difference Vegetation Index (NDVI)
data observed by satellites

Decrease > 50%   Increase 100–150%
Change < 50%   Increase 150–200%
Increase 50–100%   Increase > 200%

◀ **Figure 4.28 Greening of the Sahel** Desertification has been a significant problem in many parts of the world, particularly in Africa south of the Sahara Desert. Overgrazing of fragile arid and semiarid rangelands and deforestation without reforestation have been the chief causes. Recently, however, there is growing evidence of increasing biomass production in this region as a result of increasing rainfall. This figure is based on trend analyses of time series over the Sahel region from 1982 to 1999. (*Source:* NOAA AVHRR NDVI-data from 1982 to 1999.)

to revitalize drought-ravaged regions, providing new grazing lands for farming communities.

Wetlands The wetlands category covers swamps, marshes, bogs, peatlands, and the shore areas of lakes, rivers, oceans, and other water bodies. Wetlands can be associated either with saltwater or freshwater. Most of Earth's wetlands are associated with the latter (**Figure 4.29**). The human impacts on wetland environments are numerous. The most widespread has been the draining or filling of wetlands and their conversion to other land uses, such as settlement or cultivation. One reliable estimate places the total area of the world's wetlands at about 8.5 million square kilometers (3.3 million square miles), with about 1.5 million square kilometers (0.6 million square miles) lost to drainage or filling. For example, Australia has lost all of its original 20,000 square kilometers (7,740 square miles)

of wetlands to conversion. For the last 400 years or so, people have regarded wetlands as nuisances and as sources of disease. In core countries, technological innovation made modification and conversion of wetlands possible and profitable.

In San Francisco, California, the conversion of wetlands in the mid-nineteenth century allowed speculators and real estate developers to extend significantly the central downtown area into the once marshy edges of San Francisco Bay. The Gold Rush in the Sierra Nevada sent millions of tons of sediment down the rivers into the bay, filling in its marshland and reducing its nearshore depth. It is estimated that in 1850 the San Francisco Bay system, which includes San Pablo Bay as well as Suisun Bay, covered approximately 315 square kilometers (about 120 square miles). One hundred years later, only about 125 square kilometers (about 50 square miles) remained. By the 1960s, the conversion and modification of the wetlands (as

◀ **Figure 4.29 Mangrove in Thailand** Mangroves grow in saline coastal habitats in the tropics and subtropics and act as protective zones against high-energy wave action that would otherwise cause erosion.

well as the effects of pollution pouring directly into the bay) had so dramatically transformed water quality and the habitats of fish, fowl, and marine life that the viability of the ecosystem was seriously threatened. Since then, restoration activities have been undertaken and parts of the bay have returned to something approximating their former state, but large parts are heavily urbanized and cannot be restored.

Water Use and Access

One of the most crucial elements in the relationship between people and their environment is water. We can't live without it and all of our economic practices depend on it. For instance, it takes 150 gallons of water to produce one newspaper; 1,000 gallons for one quart of orange juice, 2,500 gallons for 1 pound of beef, and 40,000 gallons of water for one new car. The water embedded in the production of the food or other things we consume is referred to as **virtual water** (**Figure 4.30**).

Without abundant water, people in the core could not live the high-quality lives they currently enjoy; without clean water, those in the periphery die. It's therefore important to question what our lives would be like if the price of water was as volatile as, for example, oil. The price of oil has risen and fallen repeatedly due to civil unrest in the Middle East most recently, but also to the power of 12 countries—known as the Organization of Petroleum Exporting Countries (OPEC)—to set oil production schedules and prices. Privatized water could be subject to similar conditions and price effects when private companies set the price for water and not governments.

Water rates in most municipalities around the world are subsidized since water is absolutely essential to life and the "real" cost of water—acquisition, treatment, delivery, waste management—can be prohibitive. Moreover, subsidies exist

not just for households but also for agriculture and industries. And yet, while people in the core enjoy artificially low water costs because the government subsidizes the price of water, in the periphery populations have been feeling the effects of escalating water prices for nearly a decade, and in some places far longer. The *2013 UN Human Development Report* states that 1.1 billion people in peripheral countries do not have adequate access to water, with 2.6 billion lacking basic sanitation. Almost two-thirds of the people lacking access to clean water survive on less than $2 a day. Access to piped household water is unequal, with 85 percent of the wealthiest 20 percent of the population having adequate access, compared to a meager 20 percent of the poorest 20 percent of the population. Perhaps most disturbing is the statistic that 1.4 million children will die each year from lack of access to safe drinking water and sanitation.

Experts around the world have begun to talk about a global "water crisis," which is occurring not because there is a shortage, but because water as a public good is rapidly turning into a privatized commodity. That is, the privatization of water (rather than the provision of water through publicly owned utilities)—touted as a way of bringing equitable access and efficiency to all water users—has instead led to increased prices and accessibility problems for poor and marginalized peoples not only in the global periphery but in peripheral areas of many core countries as well (**Figure 4.31**).

Nearly two decades ago, many governments—national as well as local—looked to the privatization of water as a way of unburdening themselves of a relatively expensive service in response to the public's demand for smaller government. Corporations—mostly multinational ones—began buying up municipal water providers and offering water provision based on profit-and-loss considerations.

+ how much did *your* textbook cost?

12 glasses of beer
20 gal/glass

0.91 loaf of bread
264 gal/loaf

120 days worth of daily water needs
2 gal/day per person

KNOX · MARSTON
HUMAN GEOGRAPHY
Places & Regions in Global Context
7th Edition

240 gallons

8 loads of laundry
30 gal/load

120 toilet flushes
2 gal/flush

7 cups of coffee
34 gal/cup

12 showers
20 gal/shower

◄ **Figure 4.30 Virtual water** This graphic provides an illustration of how much water is used for even the most ordinary products like a textbook. It translates the 240 gallons used to produce the book into other product equivalences. (Source: Based on Diagram by Soo Lee http://ssb2012.wordpress.com/2012/09/18/virtual-waterdiagrams/.)

International water companies such as Paris-based Suez, the largest water company, Veolia (formerly Vivendi), also French and the world's second-largest water giant, and RWE, a German utility conglomerate, are competing for water supplies across the world. Since 2002, Lexington, Kentucky, for instance, has been fighting to regain control of its local water utility from RWE. The water privatization in New Orleans was predicted to be the largest private water contract in the nation when it was proposed in 2000. Vivendi (as it was then known) and Suez vied to profit from control of Crescent City's (California) water supply. The deal was predicted to be worth more than $1.5 billion over the next 20 years. Ultimately, the city was able to defeat the takeover.

Reports abound across the United States and the globe of the private purchase of local water supplies and resulting rate hikes, negative economic impacts, inadequate customer service, and harm to natural resources. The bottom line is that water—clean, drinkable water—like lots of other natural resources, is becoming increasingly scarce and an attractive investment opportunity. One of the most significant challenges of the twenty-first century will be how to ensure the wide accessibility of adequate supplies of quality drinking water for the world's people.

In response to the global water privatization movement, individuals as well as governments and nongovernmental organizations are increasingly arguing that access to safe and adequate supplies of water should be seen as a human right and that large corporations are not capable of guaranteeing that right. It is important to be aware that the conflict over who should provide safe water is occurring in cities and towns throughout the core as well as the periphery. Besides issues of access and public safety,

▼ **Figure 4.31 Protest against water privatization** Shown here are activists from a non-governmental organization registering their objection to the move across the globe to privatize water. They are marching outside the Third World Water Forum, held in Japan.

many of those opposing water privatization argue that conflicts over water are really about fundamental questions of democracy. In particular, if water is no longer a public good, who will make the decisions that affect our future access to it and who will be denied access? This question is an especially significant one as climate change is expected to have dramatic impacts on water quality and quantity throughout the world.

APPLY YOUR KNOWLEDGE

1. What is the "global water crisis"? Think about the change from water as a public good to a private commodity.

2. Conduct an Internet search to find a core country and a peripheral country that share a common environmental concern relating to threats to forests, grasslands, or wetlands. Compare and contrast the different ways core and peripheral countries are addressing the same issue.

THE STATE OF THE GLOBAL ENVIRONMENT

The combustion of fossil fuels, the destruction of forest resources, the damming of watercourses, and the massive change in land-use patterns brought about by the pressures of globalization—industrialization being the most extreme phase—contribute to environmental problems of enormous proportions. It is now customary to speak of the accumulation of environmental problems we, as a human race, experience as global in dimension. Very little, if anything, has escaped the embrace of globalization, least of all the environment.

In fact, no other period in human history has transformed the environment as profoundly as the anthropocene. While we reap the benefits of a modern way of life, it is important to recognize that these benefits have not been without cost. Fortunately, the costs have not been accepted uncritically. Over the last two to three decades, responses to global environmental problems have been on the increase as local groups have mobilized internationally.

Global Environmental Governance

The increasing importance of flows and connections—economic, political, social, and cultural—means that contemporary globalization has resulted in an increasingly shrinking world. In addition to allowing people and goods to travel farther faster and to receive and send information more quickly, a smaller world means that political action has also become connected across the globe. It can now move beyond the confines of the state into the global political arena, where rapid communications enable complex supporting networks to be developed and deployed, facilitating interaction and decision making. Good examples can be seen in the protests that occurred at the World Trade Organization (WTO) meeting in Seattle, Washington, in 1999; at the Group of 8 (G8) summit in Genoa, Italy, in 2001; at the Asian Pacific Economic Cooperation summit meeting in Bangkok in 2003; around the UN climate negotiations conference in Copenhagen, Denmark, in 2009; and again during the UN climate conference in Cancún, Mexico, in 2010. Telecommunications as well as Facebook, Twitter, and other social media sites enable protest leaders to organize and deploy demonstrators from interested groups all over the world. Such political protests reflect attempts to match the political reach of institutions like the WTO, the International Monetary Fund (IMF), the World Bank, and the United Nations.

One indication of the expanding influence and geographical extent of popular political groups is the growth of environmental organizations whose purview and membership are global. These organizations have emerged in response to the global impact of such contemporary environmental problems as fisheries depletion, global warming, genetically modified seeds, and the decline in global biodiversity. Since the 1990s, these groups—ranging from lobbying organizations and nongovernmental organizations (NGOs) to direct-action organizations and political parties like the Green Party in Europe and drawing on distinctive traditions and varying levels of resources—have become an important international force. Increasingly, agreements and conventions protecting biodiversity are being created, and not a moment too soon. The decline in the diversity of simple foodstuffs, like lettuce, potatoes, tomatoes, and squash, occurred most dramatically over the course of the twentieth century. For instance, in 1903 there were 13 known varieties of asparagus; by 1983 there was just one, a decline of 97.8 percent. There were 287 known varieties of carrots in 1903; this number today is just 21, a fall of 92.7 percent. A decline in the diversity of foodstuffs means that different resistances to pests inherent in these different varieties have also declined, as have their different nutritional values and tastes. It should be pointed out, however, that a growing local food movement in the core countries is working hard to recover some of these lost varieties **(Figure 4.32)**.

Moreover, new sources of medicine may be lost not only because of deforestation in tropical forests but also because of the decline in indigenous languages, cultures, and traditions. Recognizing that many indigenous people have extensive knowledge of local plants and animals and their medicinal uses, the Convention on Biological Diversity that emerged from the Rio Summit in 1992 is attempting to protect global biodiversity by preserving and protecting these cultures and traditions. Traditional knowledge and practices are being lost as globalization homogenizes languages and draws more and more people into a capitalist market system. The UN Environment Program devotes a great deal of its energies to biological and cultural diversity. Even the WTO has begun to recognize the value of indigenous knowledge and the promise of biodiversity through its advocacy of intellectual property rights of both corporations and indigenous peoples. For the latter, this means protection against **bioprospecting**, the scientific or commercial practice of searching for a useful application, process, or product in nature, often in extreme environments such as deserts, rain forests, and cold places like the Arctic and Antarctic.

▲ **Figure 4.32 Seed Savers exchanges and saves heirloom seeds**   This map shows all the locations around the United States where heirloom seeds are available for purchase.

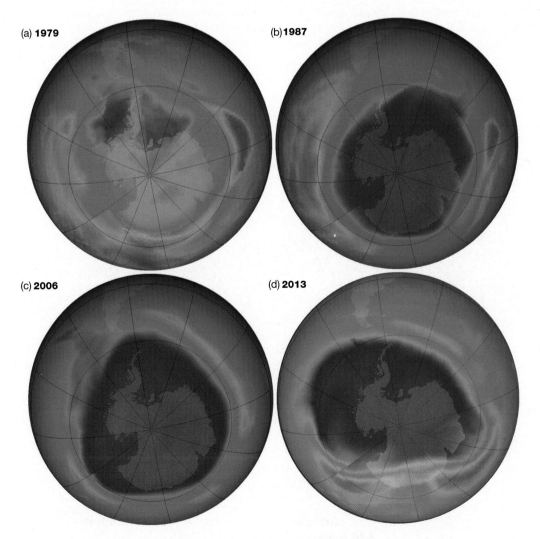

▲ **Figure 4.33 Antarctic ozone hole**   This series of enhanced satellite images over time shows the once growing ozone hole over Antarctica diminishing. The Montreal Protocol is credited with reversing the trend and avoiding an environmental disaster.

4.4 Window on the World

Ecotourism

At its most basic, **ecotourism** is responsible travel that conserves or improves the local environment as well as supports the well-being of the people who inhabit it. Ecotourism is also known as sustainable tourism and its goals, as defined by the leading sustainable tourism organization, the International Ecotourism Society, are provided in the following declaration:

The elements of ecotourism include environmental conservation that is accomplished through including local communities in decision-making as well as inviting tourism that is premised on the principle of sustainability. The bottom line in making tourism ecologically sound is to support the participation of partners—both communities and tourists—who operate in environmentally sustainable ways."

The Movement

The ecotourism movement, now 20 years old, has established guidelines, standards, policies, and a global agenda meant to standardize practices to assure that all ecotourism programs adhere to a high level of performance on a range of activities, including conservation, community capacity building, and knowledge and understanding transfer between tourists and local communities. The United Nations declared 2002 the International Year of Ecotourism to help bring NGOs, governments, and industry together to support the growth and development of sustainable tourism practices around the globe. Ecotourism experiences are varied and depend very much on the activities and environmental and social conditions of the destination (**Figure 4.F**).

The Experience

The *New York Times* publishes an ecotourism guide as part of its very popular Travel feature. It lists a wide range of ecotourism opportunities on its Web site that range from archaeology through to beaches, cruises, gay and lesbian destinations, spas, surfing and wildlife encounters in sites across the planet. One such trip includes the highly endangered Galapagos Islands, a chain of 19 remote volcanic islands that have mostly been converted to national parkland 600 miles off the coast of Ecuador. There visitors can drift with sea turtles and swim with sea lions and observe all sorts of birdlife.

▼ **Figure 4.F Growth of shark ecotourism** Ecotourism centers on observing the behaviors of sharks through scubadiving, boating, and the use of shark cages. The industry currently generates more than $314 million but projections for 2034 indicate it could double. (*Source:* Based on Craig Layman, "Shark Ecotourism to Double" http://appliedecology.cals.ncsu.edu/absci/2013/05/shark-ecotourism-to-double/)

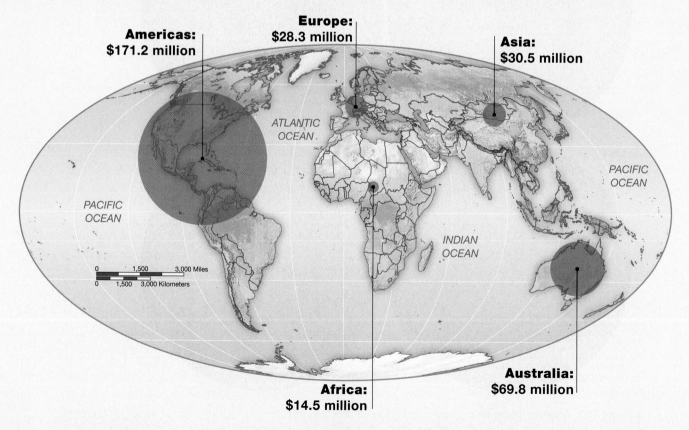

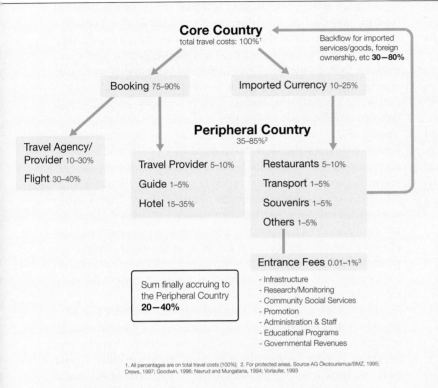

Core Country
total travel costs: 100%[1]

Backflow for imported
services/goods, foreign
ownership, etc **30–80%**

Booking 75–90%

Imported Currency 10–25%

Peripheral Country
35–85%[2]

Travel Agency/
Provider 10–30%

Flight 30–40%

Travel Provider 5–10%

Guide 1–5%

Hotel 15–35%

Restaurants 5–10%

Transport 1–5%

Souvenirs 1–5%

Others 1–5%

Entrance Fees 0.01–1%[3]

- Infrastructure
- Research/Monitoring
- Community Social Services
- Promotion
- Administration & Staff
- Educational Programs
- Governmental Revenues

Sum finally accruing to
the Peripheral Country
20–40%

1. All percentages are on total travel costs (100%); 2. For protected areas. Source AG Ökotourismus/BMZ, 1995;
Drews, 1997; Goodwin, 1996; Navrud and Mungatana, 1994; Vorlaufer, 1993

▲ **Figure 4.G Income from ecotourism** This graph shows that revenues generated by ecotourism largely return to core countries. Peripheral countries that are the sites of ecotourism receive expenditures of between 20 and 40%.

Visitors are not expected to do much more than adhere to strict rules that prevent the introduction of nonnative species to the islands and contribute money to support the preservation of this UNESCO World Heritage site. A more engaged ecotourism experience can include a wide range of activities from working with a local community in Costa Rica to helping fund and construct a school for local children to developing microenterprises—bakeries, sustainable fishing tours, basket-making for tourist consumption—to participating in animal rescue programs (**Figure 4.G**).

The Criticisms

As ecotourism has grown in popularity, so too have criticisms of its ethos and practices. Some point out that the term *ecotourism* is itself an oxymoron as tourism in general depends upon increasing air travel and the greenhouse gas emissions that accompany it and contributes to global warming and climate change. The problem, many critics of ecotourism point out, is that the regulation of ecotourism is poorly implemented or nonexistent since helicopter tours of the Grand Canyon and cruises to Antarctica can be categorized as ecotourism along with camping and canoeing. But the list of problems that some see as endemic to ecotourism is long:

- Revenues bypass the local communities experiencing the tourism.
- Tour operators within natural areas do not adhere to ecotourism principles.
- Local communities are forced to leave their homes to ensure a more "natural" experience for tourists.
- Government resources are redirected from social programs to environmental ones that don't generate positive impacts for both.
- Tourist infrastructure creates conflict over preexisting land uses.
- Effects of increasing ecotourism are too great for local water-treatment and sanitation facilities.

It must be appreciated that ecotourism is a consumer-centered economic activity that might be seen, by its very nature, to be contrary to sustaining the environment.

Solutions

Many of the extreme ecotourism critics would argue that ecotourism should mean no tourism! Since this extreme response is unlikely to be adopted, there are other ways that ecotourism can more closely adhere to its basic principles. These require that tourists be better educated about their impacts, even those who believe they have none, such as hikers, who contribute to soil impaction and erosion and plant damage. These include:

- Assuring the economic benefits of ecotourism are spread so that local communities are significant beneficiaries (**Figure 4.H**).
- Making sure that management plans are in place so that local natural resources are sustained.
- Informing ecotourists of the local social organization and cultural practices and how they can be respectful visitors.

There are moves currently within the ecotourism industry to create tour guide training programs, so that tourists can become directly informed about the impacts of their visit; establish national and international accreditation standards that rate programs on their level of environmental commitment; and establish national ecotourism certification programs. Add to these measures routine and formal environmental impact assessments at the site and there could be a strong form of accountability in place. While all of these measures are controversial in one way or another, at least one of them has already been put into place in a handful of countries.

1. What are the principles of ecotourism and how do these contrast with typical tourism?
2. Describe a responsible ecotourism activity from the perspective of the local population.

▲ **Figure 4.H Ecotourism, San Ignacio Lagoon, Baja, Mexico** Here an ecotourist reaches out to touch a gray whale who use the lagoon as their calving grounds.

Clearly, global environmental awareness is on the rise from both ends of the political spectrum, including the conservative (such as the WTO) and the progressive (such as programs devoted to preserving genetic diversity in seed strains). This increasing awareness is directly responsible for the staging of global environmental conferences like Rio in 1992, Kyoto in 1997, and the signing of the Montreal Protocol in 1987, a treaty meant to halt the growing hole in the ozone layer over Antarctica. Over 25 years ago, scientists and policymakers unveiled what the United Nations calls "the most successful treaty in UN history." Nearly 200 nations signed the Montreal Protocol on Substances that Deplete the Ozone Layer and thereby agreed to ban the use of chlorine- and bromine-based chemicals (particularly chlorofluorocarbons, or CFCs) that destroy atmospheric ozone. It is widely accepted that this international agreement likely saved the world from an environmental crisis, while setting an example for how to develop and implement environmental policy (**Figure 4.33**).

TED Talk on Biodiversity

http://goo.gl/VPhOfs

APPLY YOUR KNOWLEDGE

1. What are some ways that environmental awareness has become a global, not national, perspective? Do you think the stakes are the same for core and peripheral countries?

2. Conduct an Internet or library search to find two different environmental policies that demonstrate cooperation on an environmental problem across state boundaries. How successful have they been? Provide statistical data to support your conclusion.

Environmental Sustainability

The interdependence of economic, environmental, and social problems, often located within widely different political contexts, means that some parts of the world are ecological time bombs. Most environmental threats are greatest in the world's periphery, where daily environmental pollution and degradation amount to a catastrophe that will continue to unfold in slow motion in the coming years. One approach to this problem that has emanated from core countries is ecotourism, also known as sustainable tourism. We examine this well established but still growing form of alternative tourism in **Box 4.4**, "Window on the World: Ecotourism."

In the peripheral regions there is simply less money to cope with environmental threats. The poverty endemic to peripheral regions also adds to environmental stress. In order to survive, the rural poor are constantly impelled to degrade and destroy their immediate environment, cutting down forests for fuelwood and exhausting soils with overuse. In order to meet their debt repayments, governments generate export earnings by encouraging the harvesting of natural resources. In the cities of the periphery, poverty encompasses so many people in such concentrations as to generate its own vicious cycle of pollution, environmental degradation, and disease. Even climate change, an inherently global problem, seems to pose its greatest threats to poorer, peripheral regions.

A more benign relationship between nature and society has been proposed under the principle of *sustainable development,* a term that incorporates the ethic of intergenerational equity, with its obligation to preserve resources and landscapes for future generations. Sustainable development involves employing ecological, economic, and social measures to prevent environmental degradation while promoting economic growth and social equality. Sustainable development insists that economic growth and change should occur only when the impacts on the environment are benign or manageable and the impacts (both costs and benefits) on society are fairly distributed across classes and regions. This means finding less-polluting technologies that use resources more efficiently and managing renewable resources (those that replenish themselves, such as water, fish, and forests) to ensure replacement and continued yield. In practice, sustainable development policies of major international institutions, such as the World Bank, have promoted reforestation, energy efficiency and conservation, and birth control and poverty eradication programs to reduce the environmental impact of rural populations.

But cities and towns throughout the world have also launched policies and practices meant to foster sustainability at the local level. These include reorganized municipal waste programs, where cities collect organic waste and turn it into rich and usable soil, or remove some of the most problematic elements of it to keep it from entering into landfills or even converting organic waste to energy. The recycling networks are not only local but can span the globe so that material deemed as waste in an affluent country can be exported to a less affluent one where it takes on new value in a different context. Indeed, the global transfer of obsolete goods is big business so that in 2008, the transfer of scrap and other waste material exceeded the value of all semi-conductors and electronic goods exported by the United States to China (**Figure 4.34**).

There is also an emerging movement known as "transition towns." The practices undertaken in this movement are based on groups of active citizens who wish to improve the resilience of their communities in the face of threats like climate change and even economic uncertainty. These citizens work collaboratively to find local solutions. As their Web site states, transition towns:

> … succeed by regeneratively using their local assets, innovating, networking, collaborating, replicating proven strategies, and respecting the deep patterns of nature and diverse cultures in their place. Transition initiatives work with deliberation and good cheer to create a fulfilling and inspiring local way of life that can withstand the shocks of rapidly shifting global systems.[3]

APPLY YOUR KNOWLEDGE

1. What is sustainable development?

2. Find an example of a "transition town" on the Internet. Where is the town located and what are some of the policies they have enacted toward sustainable practices and preparation for climate change threats?

[3]http://transitionus.org/transition-town-movement, accessed August 8, 2014.

▲ **Figure 4.34 Waste Containers** Recyclable paper and metal, half-processed forest products, and other raw materials as well as agricultural products account for most of the U.S. goods that are shipped to China and other ports in northeast Asia. Those that come from China to the United States are loaded with manufactured commodities such as clothing, shoes, toys, furniture, appliances and other household goods.

FUTURE GEOGRAPHIES

The continued expansion of the global economy and the globalization of industry will undoubtedly boost the overall demand for raw materials and energy and continue to shape the relationship between people and nature. While the extraction of raw materials will be important in the future, the issue of most concern, by far, will be energy resources. World energy consumption has been increasing steadily. As the periphery is industrialized and its population increases further, the demand for energy will expand even more rapidly. Basic industrial development tends to be highly energy-intensive. The International Energy Agency, assuming (fairly optimistically) that energy in peripheral countries will be generated in the future as efficiently as it is in developed ones now, estimates that developing-country energy consumption will more than double by 2025, lifting total world energy demand by almost 50 percent. Unless peripheral countries are able to limit the degradation associated with energy use, the globe will continue to feel the negative public health and environmental effects of air, water, and terrestrial pollution.

Despite the threat to people and the environment, industrialization geared to meeting the growing worldwide market for consumer goods, such as automobiles, air conditioners, refrigerators, televisions, and household appliances, will continue. Without higher rates of investment in exploration and extraction than at present, production will be slow to meet the escalating demand.

The past 25 years have seen a growing public awareness of how continued globalization will affect the world. Increasingly, citizens, nongovernmental organizations, and environmental policymakers are expressing concern over the negative outcomes of rapid and enduring global economic growth. However, because growth is so critically tied to improving the lives of poor people around the world, governments are reluctant to limit it. The response from the global community, hammered out during international meetings, through academic publications, and in response to social protest, is to link globalization to governmental cooperation across states with the assumption that global challenges will require international political and economic cooperation.

APPLY YOUR KNOWLEDGE

1. What energy resources do you think should be exploited to establish and maintain growth industries in peripheral countries?

2. What do you see as the most viable response to growing energy demands in the United States?

▲ **Figure 4.35 Korangi Town in coastal Pakistan.** This photograph taken from the International Space Station shows the contrast between the highly urbanized and industrialized Korangi area and the dense mangrove forests and waterways of the Indus River Delta.

■ CONCLUSION

The relationship between people and nature is to a great extent mediated by institutions and practices, from technology to religious beliefs. In this chapter, we have seen how the nature–society relationship has changed over time and how the globalization of the capitalist world economy has had a more widespread impact on attitudes and practices than any cultural or economic system that preceded it.

The expansion of European trade, followed by colonization and eventually industrialization, broadcast worldwide the belief that humans should take their place at the apex of the natural world. The Western attitude toward nature as manifested by the capitalist economic system is the most pervasive shaper of nature–society interactions today.

In addition to exploring the history of ideas about nature and contemporary environmental philosophies and organizations in the United States, this chapter has also shown that people and nature are interdependent and that events in one part of the global environmental system affect conditions in the system elsewhere. Finally, we have shown that events in the past have shaped the contemporary state of society and nature.

In short, as economies have globalized so has the environment. We can now speak of a global environment in which not only the people but also the physical environments where they live and work are linked in complex and essential ways.

Along with the recognition of a globalized environment have come new ways of thinking about global economic development. Sustainable development, one of these new ways of thinking, has come to dominate the agenda of international institutions, as well as environmental organizations, as the new century unfolds.

LEARNING OUTCOMES REVISITED

■ **Describe how people and nature come to form a complex relationship in which nature is understood as both a physical realm and a social construct.**

Nature is not only an object, it is a reflection of society in that the philosophies, belief systems, and ideologies people produce shape the way we think about and employ nature. Society is the sum of the inventions, institutions, and relationships created and reproduced by human beings across particular places and times. The relationship between nature and society is two-way: Society shapes people's understandings and uses of nature at the same time that nature shapes society.

■ **Compare the many views of nature operating both historically and in society today, from the traditional Western approach to the radical left and contemporary ecotheological approaches.**

In the contemporary world, views of nature are dominated by the Western (also known as Judeo-Christian) tradition that understands humans to be superior to nature. In this view, nature is something to be tamed or dominated. But other views of nature have emerged that depart dramatically from the dominant view. These include the environmental philosophies that became popular in the nineteenth and early twentieth centuries and the more radical political views of nature that gained prominence in the late twentieth century. Among the latter are approaches based on ecotheology, which reject the long-standing consumption-based Western tradition.

■ **Explain how extensive human settlement, colonization, and accelerated capitalist development have transformed nature in unprecedented ways.**

In the fifteenth century, Europe initiated territorial expansion that changed the global political map and launched dramatic environmental change. Europeans were fast running out of land, and explorers were dispatched to conquer new territories, enlarge their empires, and collect tax revenues from new subjects. European people, ideologies, technologies, plant species, pathogens, and animals changed the environments into which they were introduced and the societies they encountered. The reverse was also true as the societies they encountered changed Europe. European colonization ultimately helped to stimulate rapid industrialization and urbanization, both of which have had profound effects on the global environment.

■ **Describe the drivers and impacts of climate change and state your position with respect to it within current debates.**

The scientific evidence for global climate change is both extensive and widely accepted and includes increasing temperatures, sea level rise, glacial melt, species migration, and persistent drought, among other signals. The impacts of climate change are already being felt widely through increasing numbers of tropical storms and accompanying storm surges as well as ecosystem degradation. Projected impacts include coastal land submersion and accompanying forced population migration. Climate skepticism can easily be countered by reference to scientific evidence.

■ **Summarize how the globalization of the capitalist political economy has affected the environment so that environmental problems have become increasingly global in scope.**

No other transition in human history has had the impact on the natural world that industrialization and urbanization have. The combustion of fossil fuels, the destruction of forest resources, the damming of watercourses, and the massive change in land-use patterns brought about by the pressures of globalization contribute to environmental problems of enormous proportions. Geographers and others use the term **global change** to describe the combination of political, economic, social, historical, and environmental problems with which human beings across Earth must currently contend.

■ **Evaluate the ways that sustainability has become a predominant approach to global economic development and environmental transformation.**

Sustainable development involves employing ecological, economic, and social measures to prevent environmental degradation while promoting economic growth and social equality. Sustainable development insists that economic growth and change should occur only when the impacts on the environment are benign or manageable and the impacts (both costs and benefits) on society are fairly distributed across classes and regions. This means finding less-polluting technologies that use resources more efficiently and managing renewable resources (those that replenish themselves, such as water, fish, and forests) to ensure replacement and continued yield.

KEY TERMS

acid rain (p. *132*)
anthropocene (p. *118*)
biopiracy (p. *110*)
biomass (p. *109*)
bioprospecting (p. *142*)
climate change (p. *122*)
Columbian Exchange
    (p. *119*)
conservation (p. *116*)
cultural ecology (p. *115*)

deep ecology (p. *117*)
deforestation (p. *134*)
demographic collapse
    (p. *119*)
desertification (p. *138*)
ecofeminism (p. *117*)
ecological imperialism
    (p. *120*)
ecosystem (p. *118*)
ecotheology (p. *117*)

ecotourism (p. *144*)
environmental ethics (p. *117*)
environmental justice (p. *117*)
global change (p. *149*)
global warming (p. *122*)
greenhouse gases (GHGs)
    (p. *123*)
greening (p. *138*)
nature (p. *111*)
political ecology (p. *115*)

preservation (p. *117*)
romanticism (p. *116*)
society (p. *111*)
sustainable development
    (p. *109*)
technology (p. *111*)
transcendentalism (p. *116*)
virgin soil epidemics
    (p. *119*)
virtual water (p. *140*)

REVIEW & DISCUSSION

1. Consider the relationship between industrialization and physical geography by examining mountain top removal in Appalachia. Visit the Beehive Collective graphic campaign entitled "The True Cost of Coal" at http://www.beehivecollective.org/english/coal.htm. Describe the environmental philosophies that the Beehive Collective most closely adheres to. (*Hint:* Is it the collective part of the environmental justice or ecofeminist movements? Or a combination of a variety of different theories?) Explain your choice. Be specific in the reasons you give by citing examples from the organization's Web site. List four examples of how the graphics on the site depict the complex relationship between nature and society.

2. Colleges and universities have become much more involved in sustainability over the last decade. To know where your institution fits into the larger college and university sustainability movement, visit the College Sustainability Report Card Web site at http://www.

greenreportcard.org/about/faq and determine how your institution is rated. Discuss how it compares to other institutions and how your institution might improve or how sustainability practices engaged in at your institution could be transferred to other colleges or universities.

3. As you know from this chapter, Gallup, Inc. has conducted a national survey of environmental attitudes in the United States every year since 1986. Visit their Web site at http://www.gallup.com/poll/1615/Environment.aspx and investigate the kinds of questions asked as well as the results from the survey. Selecting what you feel are the most pertinent questions for your region of the country, design your own questionnaire and administer it to students on your campus, and then analyze the survey results. How do the attitudes of your fellow students compare with those of the nation? Determine why they might the same or different. Be aware that Gallup does categorize responses by age category.

UNPLUGGED

1. Many communities produce an index of environmental stress, in essence a map of the toxic sites in a city or region. Plot a rudimentary map like this of your own community using the local phone book as a data source. Consult the Yellow Pages to identify the addresses of environmentally harmful and potentially harmful businesses, such as dry-cleaning businesses, gas stations, automotive repair and car-care businesses, aerospace and electronic manufacturing companies, agricultural supply stores, and other such commercial enterprises where noxious chemicals may be produced, sold, or applied. Alternatively consider your campus and where its most toxic sites are. Your school directory is the ideal place to start.

2. Read a natural history of the place where your college or university is located. What sorts of plants and animals dominated the landscape there during the Paleolithic

period? Do any plants or animals continue to survive in altered or unaltered forms from that period? What new plants have been introduced and how extensive are they? What sorts of tending and maintenance do these new species require and how have they changed the landscape? Finally, with the information you have gathered, create a map of the paths that these specific plants have taken to arrive in your region.

3. Colleges and universities are large generators of waste, from plain-paper waste to biomedical and other sorts of waste, that can have significant environmental impacts. Identify how your campus handles this waste stream and how you, as a member of the academic community, contribute to it. Where does the waste go when it leaves the school? Is it locally deposited? Does it go out of state?

DATA ANALYSIS

America's
Climate
Refugees

http://goo.gl/iletzz

In this chapter, we have examined the interactions between society and nature from the perspective that humans shape the natural world, be it with technology, population growth or decline, the exchange between cultures (seen in ecological imperialism), energy consumption, and/or global awareness of climate change. To analyze more deeply how human intervention impacts the environment and the effects of climate change on certain communities, look at the Alaska Native village of Newtok with this interactive site: "America's Climate Refugees" at: http://www.theguardian.com/environment/interactive/2013/may/13/newtok-alaska-climate-change-refugees

Watch and read the three different multimedia sections: 1) "America's first climate refugees"; 2) An undeniable truth?; and 3) "It's happening now… The village is sinking." As you listen to the video stories, read the reports and analyze the maps to answer these questions:

1. Why is Newtok sinking?

2. How is climate change a direct threat to the Newtok community's existence?

3. Why is the Arctic, particularly Alaska, at risk for the early effects of climate change? How have Alaskan politicians responded to climate change?

4. Look at the map on the page, "The at risk list" (http://www.theguardian.com/environment/interactive/2013/may/14/alaska-villages-frontline-global-warming). Why do Alaska's indigenous communities face some of the most detrimental effects from climate change? Think about both geographic location and economic realities. Do you see any aspects of ecological imperialism at play?

5. Watch the video story, "Life in Newtok is Difficult" (in the "It's happening now… The village is sinking" section, http://www.theguardian.com/environment/interactive/2013/may/15/newtok-safer-ground-villagers-nervous). Why are goods so expensive? What resources does the community have in comparison to the cost of living?

6. What are some of the proposals to save Newtok? How feasible are those solutions?

7. How effective was the Charles family's "great escape" from Newtok to Metarvik (http://www.theguardian.com/environment/interactive/2013/may/13/alaskan-family-newtok-mertarvik)?

8. With all of this information, how is climate change affecting certain communities more than others? Do you think this is fair? How complicit are all countries in creating climate change? Should there also be a global response to solve these issues?

MasteringGeography™

Looking for additional review and test prep materials? Visit the Study Area in MasteringGeography™ to enhance your geographic literacy, spatial reasoning skills, and understanding of this chapter's content by accessing a variety of resources, including **MapMaster** interactive maps, Videos, *In the News* RSS feeds, flashcards, web links, self-study quizzes, and an eText version of *Human Geography*.

LEARNING
OUTCOMES

- *Describe* how place and space shape culture and, conversely, how culture shapes place and space.

- *Compare* and contrast the different ways that contemporary approaches in cultural geography invoke the role played by politics and the economy in establishing and perpetuating cultures and cultural landscapes.

- *Understand* the ways that cultural differences—especially gender, class, sexuality, race, and ethnicity—are both products of and influences on geography, producing important variations within, as well as between, individuals and groups.

- *Explain* the conceptual changes that are taking place in cultural geography that include actor-network theory and non-representational theory.

- *Demonstrate* how globalization does not necessarily mean that the world is becoming more homogeneous and recognize that in some ways, globalization has made the local even more important than before.

5

CULTURAL GEOGRAPHIES

Thailand is considered to have one of the highest, if not *the* highest, number of transgendered people worldwide.[1] **Transgender** is the term that refers to a person whose self-identity does not conform to conventional notions of the male or female gender. There is no official count, but knowledgeable scholars estimate the number to be close to 200,000. This number would not come as a surprise to visitors to Thailand and might even be rejected as an underestimate given kathoeys ubiquity. Indeed, surf the Internet and you'll find kathoeys repeatedly touted as an interesting and exciting feature of nightlife in the city.

Kathoey is the Thai term that refers largely to male-to-female transgender. Kathoeys, or "lady boys" as they refer to themselves and are referred to by others, are everywhere in Bangkok but also live openly in rural villages across the country. Importantly, many Thai people perceive kathoeys as not comfortably fitting into the category of transgender and prefer to recognize them as belonging to a third gender—that is they are neither male nor female—while others, including many kathoeys themselves, view them as either a kind of man or a kind of woman. Equally important is the fact that while certainly marginalized, as are most sexual minorities around the world, in Thailand, the kathoey enjoy wide acceptance and even national fame.

Scholars have been drawn to understanding kathoeys because of the fact that they live and move through Thai society fluidly and seemingly with little harassment. Survey research on the ubiquity of kathoeys in Thai society has shown that both children and adults can often identify at least one kathoey in their school or village. In short, it appears that there are a large number of males in Thailand who decide to make a dramatic transition from one gender to another and researchers believe that, among other reasons, geographic factors—that is, the particular characteristics of Thailand, and especially its cultural characteristics—help explain this phenomenon.

▲ These transgendered men are participants in the Miss Tiffany's Universe beauty pageant, the winner of which gets a chance to join the world's longest-running kathoey review.

[1]This description of kathoeys is adapted from S. Winter (2010) "Why are there so many kathoey in Thailand?", http://web.hku.hk/~sjwinter/TransgenderASIA/paper_why_are_there_so_many_kathoey.htm, accessed June 25, 2014; and Winter, S. and Udomsak, N. "Male, Female and Transgender: Stereotypes and Self in Thailand. *International Journal of Transgenderism*, *6*(1), 2002.

The first one has to do with spiritual beliefs: in Judeo-Christian tradition, sex and gender are rather strictly defined in terms of human anatomy. In Thailand, however, which is a largely Buddhist country, gender is often defined in terms of the social roles males or females perform in the home or in the public sphere. At the same time, sex roles are correlated not with the sexual organs one has, but with what one does with them.

Importantly, as in most other cultures, though Thais also operate through stereotypes of male and female behaviors, the differences between them in terms of those personal characteristics, is actually smaller than elsewhere. In short, males and females are seen as far more similar to each other than in other cultures and, moreover, women are very highly revered in Thailand. The result is though physically the transition from male to female is a significant one, psychologically it is less so in a country where families pay a bride price to marry off their sons, not their daughters.

Finally, Thai Buddhism is a highly tolerant belief system that recognizes homosexuality and transgender as the result of karma, a misdeed or transgression from a past life. As such, the current life of a kathoey is determined by the consequences of their past lives and thus they deserve acceptance and understanding and not scorn and rejection as anyone can be affected by karma.

Kathoeys provide an important window into the complexity of culture, particularly at the intersection with sexuality (including sexual role and sexual anatomy) and gender (including gender identity and gender presentation). It helps us see that though sexuality and gender may appear to be the same worldwide, there are actually dynamic contextual factors that underlie the expression of significant differences, despite apparent similarities. In this chapter, as shown here in the kathoey example, we explore culture as a process that is not understandable outside of the geographical context in which it operates.

CULTURE AS A GEOGRAPHICAL PROCESS

Anthropologists, geographers, and other scholars who study culture, such as historians and cultural studies specialists, agree that culture is a complex concept. Over time, our understanding of culture has been changed and enriched. A simple definition of culture is that it is a particular way of life, such as a set of skilled activities, values, and meanings surrounding a particular type of practice. Scholars also describe culture in terms of classical standards and aesthetic excellence in, for example, opera, ballet, or literature.

Culture

The term *culture* also describes the range of activities that characterize a particular group, such as working-class culture, corporate culture, or teenage culture. Although all these understandings of culture are accurate, for our purposes they are incomplete. Broadly speaking, **culture** is a shared set of meanings that is lived through the material and symbolic practices of everyday life. Our understanding in this book is that culture is not something that is necessarily tied to a place and thus a fact to be discovered. Rather, we regard the connections among people, places, and cultures to be emerging and always evolving creations that can be altered sometimes in subtle and at other times in more dramatic ways. The "shared set of meanings" can include values, beliefs, practices, and ideas about family, childhood, race, gender, sexuality, and other important identities or other strong associations such as vegetarianism or a devotion to Star Trek **(Figure 5.1)**. These values, beliefs, ideas, and practices are routinely subject to reevaluation and redefinition and can be, and very frequently are, transformed from both within and outside a particular group.

In short, culture is a dynamic concept that revolves around and intersects with complex social, political, economic, and even historical factors. For much of the twentieth century, geographers, like anthropologists, have focused most of their attention on material culture, as opposed to its less tangible symbolic or spiritual manifestations. This understanding of culture is part of a longer tradition within geography and other disciplines. We will look more closely at the development of the cultural tradition in geography in the following section, in which we discuss the debates surrounding culture within the discipline.

▼ **Figure 5.1 Star Trek fans** These young people are attending the 'Destination Star Trek' event in Frankfurt, Germany dressed up as characters from the television show.

Like agriculture, politics, and urbanization, globalization has had effects on culture. Terms such as *world music* or *global connectivity* are a reflection of the sense that the world seems a smaller place now as people everywhere are sharing aspects of the same culture through the widespread influence of television, the Internet, and other media. Yet, as pointed out in Chapter 2, although powerful homogenizing forces are certainly at work, the world has not become so uniform that place no longer matters. With respect to culture, just the opposite is true. Place matters more than ever in the negotiation of global forces, as local forces confront globalization and translate it into unique place-specific forms (**Figure 5.2**).

▼ **Figure 5.2** St Patrick's Day in Japan. The wearing of green and a traditional march is also characteristic of St. Patrick's Day in Tokyo. The marchers are playing a traditional Irish whistle or feadóg stáin.

Geography and Culture

The place-based interactions occurring between culture and global political and economic forces are at the heart of cultural geography today. **Cultural geography** focuses on the way space, place, and landscape shape culture at the same time that culture shapes space, place, and landscape. As such, cultural geography demarcates two important and interrelated parts. Culture is the ongoing process of producing a shared set of meanings and practices, while geography is the dynamic context within which groups operate to shape those meanings and practices and in the process form an identity and act. Geography in this definition can be a space that is as small as the body and as large as the globe.

A Two-Way Relationship

A contemporary example of this two-way relationship between geography and culture can be seen in the widespread popularity of online social networks designed to connect friends, family, co-workers, and people with similar interests. The most popular online social networking website Facebook—with around 1.23 billion active monthly users worldwide—allows members to share photos, blogs, music, and videos, and a host of other personal information, as well as organize events, join groups, or seek relationship partners. In 2014, Facebook celebrated its tenth birthday. The social media site is increasingly important not just as a location for social connection but also for other forms of connection from economics to politics. For instance, Facebook and other social media played an important role in the Arab Spring and continue to be used in political ways around the world. Obama's first presidential election was thought to have been successful in part for its use of social media.

And while Facebook and other virtual networks have revolutionized the way people communicate with each other, there is far more to the Internet than connecting with old friends and making new ones. Perhaps the most powerful force on the Internet now is something known as the "Internet of things," called this because of the power of the Internet to connect objects, that is things like cars, refrigerators, and home security systems, across time and space. Connectivity is obviously a powerful economic force as investors see the Internet of things as a new "industrial revolution." But it's also a powerful cultural force as it transforms the way we, as cultural beings, relate to these connected objects and the relationships they generate with other things and people through their connectivity. **Figure 5.3** is an illustration of one kind of connectivity that the Internet of things will enable. The figure shows how the Internet's ability to connect things across space can make our lives less subject to unexpected disruption thus making life less frustrating and more efficient economically.

What it doesn't show is the how increasing connectivity across things also creates inequalities and challenges to fundamental rights such as privacy. Yes, having all your monthly bills get consolidated into one that is immediately registered and paid by your bank is highly convenient. But it also means

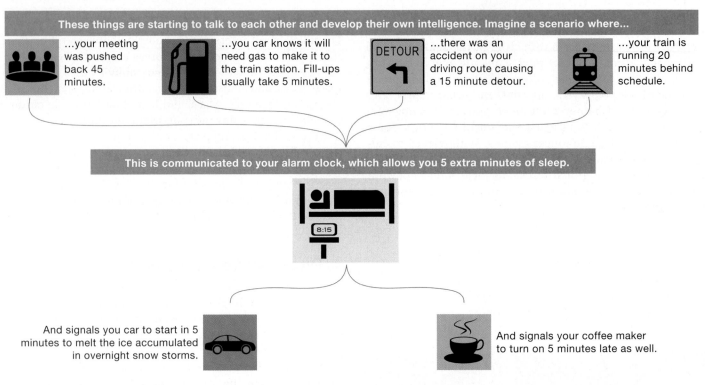

▲ **Figure 5.3 The Internet of things** Experts who study the effect of present and evolving technologies on the future believe that one of the biggest influences on the way we live our lives in the next 20 years is the internet of things.

that powerful corporations come to know a lot about you that you may not want them to. Moreover, as the Internet becomes increasingly connected across all of our things in some parts of the world, other parts will not be able to be so connected as they can't afford the cost of the "smartness" of these things. Experts who study the impact of technology on society, culture, and ethics have reservations about the kinds of changes that the Internet of things will bring about.

But it's not just that the Internet of things will affect issues of privacy; it's also that the things themselves will change our cultural practices too, such as dating, physical intimacy, and marriage. An obvious example is the way the Internet allows couples in long-distance relationships to connect in their own private shared space. The app "Couple" has a feature called ThumbKiss™, which allows distant couples to send a kiss when each of the couple touches the same spot on a smartphone that makes their smartphones vibrate simultaneously.

In the United States, marriage ceremonies and receptions using drones and bots are gaining popularity. As drones fly over the wedding party taking videos from above they communicate with bots on the dance floor that are capturing conversation and snapping candid photographs. The film *Her* (2013), directed by Spike Jonze and starring Joaquin Phoenix, attempted to address the issue of how the Internet of things, and computers more generally, has the power to dramatically affect some of the most basic cultural institutions of contemporary life such as companionship and love.

Traditions of Cultural Geography

Before we proceed any further in our discussion of culture, it is important to discuss a significant difference between our view and that of more traditional cultural geographers. Many introductory human geography texts divide culture into two major categories: folk and popular culture. **Folk culture** is seen by specialists as the traditional practices of small groups, especially rural people with a simple lifestyle (compared with modern, urban people), such as the Amish in Pennsylvania or the Roma (also known as Gypsies or Travelers) in Europe, who are seen as homogeneous in their belief systems and practices. **Popular culture**, by contrast, is viewed by some cultural geographers as the practices and meaning systems produced by large groups of people whose norms and tastes are often heterogeneous and change frequently, often in response to commercial products. Hip-hop would be seen by these theorists as an example of popular culture,

In this text, we do not divide culture into categories. We see culture as an overarching process that is shaped by and shapes politics, the economy, and society and cannot be neatly demarcated by reference to the number of characteristics or degree of homogeneity of its practitioners. We see culture as something that can be enduring as well as newly created, but always influenced by a whole range of interactions as groups maintain, change, or even create traditions from the material of their everyday lives. For us, there is no purpose served in categorically differentiating between hip-hop and Hinduism, as both are significant expressions of culture and both are of interest to geographers (**Figure 5.4**).

▲ **Figure 5.4 Tamil hip-hop** The Hindu music duo Adhithya Venkatapathy and Jeeva of HipHop Tamizha are the pioneers of Tamil hiphop.

APPLY YOUR KNOWLEDGE

1. How is culture a dynamic concept? Think about how culture is a process and all the different factors that shape culture.

2. How do you define cultural geography? Identify three aspects of your own culture and the ways that place and space have shaped it.

BUILDING CULTURAL COMPLEXES

Geographers focus on the interactions between people and culture and among space, place, and landscape. One of the most influential geographers was Carl Sauer, who taught at the University of California, Berkeley, in the early to mid-twentieth century. Sauer was largely responsible for creating the "Berkeley School" of cultural geography. He was particularly interested in trying to understand the material expressions of culture by focusing on their manifestations in the landscape **(Figure 5.5)**.

Cultural Landscape

Sauer's interest in culture and geography came to be embodied in the concept of the **cultural landscape,** a characteristic and tangible outcome of the complex interactions between a human group—with its own practices, preferences, values, and aspirations—and its natural environment. The concept of the cultural landscape is illustrated in Box 5.1 Visualizing Geography: UNESCO World Heritage Sites. Sauer differentiated the cultural landscape from the natural landscape. He emphasized that the former was a "humanized" version of the latter, such that the activities of humans resulted in an identifiable and

▶ **Figure 5.5 Masai village, Kenya** The cultural landscape, as defined by Carl Sauer, reflects the way that cultural and environmental processes come together to create a unique product as they do in the small village pictured here, where herding is the major occupation. The village is enclosed by thorny brambles and branches harvested from the surrounding area. Within the enclosure, the dwellings are arranged in a unique circular pattern, with the animal pens in the middle of the settlement for easy observation by the residents.

5.1 Visualizing Geography

UNESCO World Heritage Landscapes

The United Nations Educational, Scientific and Cultural Organization's (UNESCO) World Heritage Program provides an excellent example of the concept of cultural landscape and how it is being adopted and adapted today. The mission of the UNESCO is to "encourage the identification, protection and preservation of cultural and natural heritages around the world considered to be of outstanding value to humanity."

5.1 UNESCO World Heritage Sites

Based on a convention adopted in 1974 and ratified by 191 countries, the World Heritage Program works with countries around the globe to nominate, protect, manage, raise public awareness and foster international cooperation in the conservation of the world's most significant cultural and natural heritage landscapes (**Figure 5.1.1**). There are over 1000 properties that have been accepted for World Heritage status.

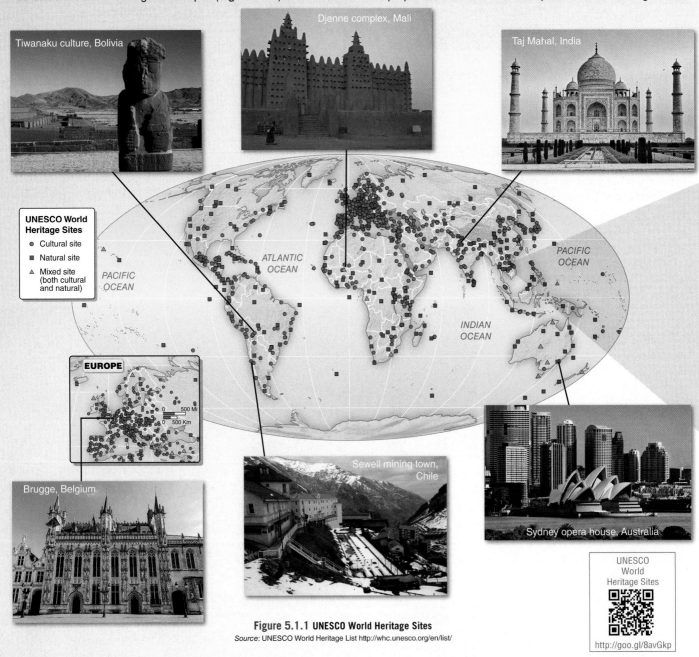

UNESCO World Heritage Sites
- Cultural site
- Natural site
- Mixed site (both cultural and natural)

Tiwanaku culture, Bolivia

Djenne complex, Mali

Taj Mahal, India

Brugge, Belgium

Sewell mining town, Chile

Sydney opera house, Australia

Figure 5.1.1 UNESCO World Heritage Sites
Source: UNESCO World Heritage List http://whc.unesco.org/en/list/

UNESCO World Heritage Sites

http://goo.gl/8avGkp

5.2 Recently Added Sites

In summer 2014, UNESCO added twenty new sites to the list including the village of Battir in Palestine (**Figure 5.2.1**). The village of Battir is especially rewarding as both Israel and Palestine collaborated to protect this property, despite the contention between the two countries in and around the West Bank.

Figure 5.2.1 Battir village, Palestine

5.3 Selection Criteria

The criteria for inclusion of a humanly produced landscape as a World Heritage property reflects many of the concerns and elements Sauer specified as central to his concept of the cultural landscape. The complete list of cultural and natural criteria can be found at http://whc.unesco.org/en/criteria/

UNESCO criteria

☑ Represent a masterpiece of human creative genius

☑ Exhibit an important interchange of human values, over a span of time or within a cultural area of the world, on developments in architecture or technology, monumental arts, town-planning or landscape design

☑ Bear a unique or at least exceptional testimony to a cultural tradition or to a civilization which is living or which has disappeared

☑ Outstanding example of a type of building, architectural or technological ensemble or landscape which illustrates significant stages in human history

Figure 5.3.1 Angkor complex, Cambodia
This site meets four of the UNESCO criteria as shown.

From the UNESCO committee's decision:

"Khmer architecture evolved largely from that of the Indian sub-continent, from which it soon became clearly distinct as it developed its own special characteristics, some independently evolved and others acquired from neighboring cultural traditions. The result was a new artistic horizon in oriental art and architecture".

The influence of Khmer art as developed at Angkor was a profound one over much of South-east Asia and played a fundamental role in its distinctive evolution.

The Khmer Empire of the 9th-14th centuries encompassed much of South-east Asia and played a formative role in the political and cultural development of the region. All that remains of that civilization is its rich heritage of cult structures in brick and stone.

Khmer architecture evolved largely from that of the Indian sub-continent, from which it soon became clearly distinct as it developed its own special characteristics, some independently evolved and others acquired from neighboring cultural traditions. The result was a new artistic horizon in oriental art and architecture.

---

1. If culture is dynamic, why is it important to preserve cultural landscapes that are considered of heritage quality? What can landscape preservation teach us about human-spatial interactions?

2. Look up the UNESCO criteria to preserve a heritage cultural landscape such as those in Figure 5.1.1. Compare the UNESCO criteria to Sauer's definition of cultural and natural landscapes. What are the similarities and differences between the two?

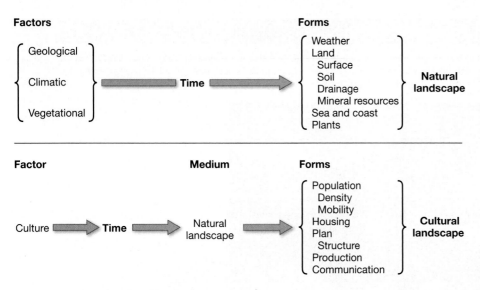

▲ **Figure 5.6 Sauer's cultural landscape** This figure summarizes the ways that natural and cultural landscapes are transformed. Physical and climatic factors shape the natural landscape. The results of cultural factors are cultural forms, such as population distributions and patterns of housing. Over time, people—through culture—reshape the natural landscape to meet their needs. (Adapted from C. Sauer, "The Morphology of Landscape," in J. Leighly (ed.), *Land and Life: Selections from the Writings of Carl Ortwin Sauer.* Berkeley and Los Angeles: University of California Press, 1963, pp. 315–350.)

understandable alteration of the natural environment. **Figure 5.6** lists the differences between a natural and a cultural landscape. While the original diagram from Sauer's work is obviously static, the arrow of time can be complicated by all sorts of events such as, for the natural landscape, earthquakes and volcanic eruptions, or slower weathering processes brought about by wind or water. The cultural landscape component of the diagram should also be understood to recognize that time is active and ongoing and processes like industrialization or political revolution and war can dramatically shape both the natural and human landscape.

Critique of Sauer's Cultural Landscape

Sauer's ideas about the effect of cultural practices on the physical landscape are a little less than a century old and have been challenged and critiqued by geographers over the years. The most prominent challenge was leveled by James Duncan who rejected the implicit notion of culture upon which Sauer built the cultural landscape concept. Duncan argued that Sauer had adopted what he called a "superorganic" notion of culture in this conceptualization. What he meant was that for Sauer culture seemed to be something that was somehow exterior to humans, a kind of well that humans drew from to make sense of the world. Duncan and most contemporary geographers reject this view and see culture, as we do in this book, as a set of meanings and practices that are produced as humans go about living their lives.

For roughly five decades, interest in culture within the field of U.S. geography largely followed Sauer's important work. His approach to the cultural landscape was ecological, and his many published works reflect his attempts to understand the myriad ways that humans transform the surface of Earth. In his own words:

The cultural landscape is fashioned from a natural landscape by a cultural group. Culture is the agent, the natural area is the medium, the cultural landscape is the result. Under the influence of a given culture, itself changing through time, the landscape undergoes development, passing through phases, and probably reaching ultimately the end of its cycle of development. With the introduction of a different—that is an alien—culture, a rejuvenation of the cultural landscape sets in, or a new landscape is superimposed on remnants of an older one.[2]

European Approaches to Culture and Place

In Europe, geographers interested in human interactions with the landscape took slightly different approaches. For example, in Great Britain the approach to understanding the human imprint on the landscape was termed *historical geography,* while in France it was conceptualized as *genre de vie.* **Historical geography,** very simply defined, is the geography of the past. Its most famous practitioner was H. C. Darby, who attempted to understand "cross sections" or sequences of evolution, especially of rural landscapes. ***Genre de vie,*** a key concept in Vidal de la Blache's approach to cultural geography in France, referred to a functionally organized way of life characteristic of a particular culture group. *Genre de vie* centered on the livelihood practices of groups that were seen to shape physical, social, and psychological bonds **(Figure 5.7)**. Although emphasizing some landscape components over others or giving a larger or smaller role to the physical environment, all of these approaches placed the cultural landscape at the heart of their study of human-environment interactions.

H. C. Darby most successfully implemented his historical approach to cultural geography and landscape by developing a geography based on the *Domesday Book,* a key historical document. William the Conqueror ordered the *Domesday* compiled in 1085 so he could have a list of his spoils of war. The book provides a rich catalog of the ownership of nearly every tract of land in England. It includes the names of landholders in each county; the manors they held and their values; the names of their subtenants; the names of many boroughs and details of their customs; the number of freemen, sokemen (freemen landholders), unfree peasants, and slaves on each manor; and the resources of each manor. Geographers like Darby found such data invaluable for reconstructing the political, economic, and social forces that shaped past landscapes.

Vidal de la Blache emphasized the need to study small, homogeneous areas to uncover the close relationships that exist between people and their immediate surroundings. He constructed multifarious descriptions of preindustrial France that demonstrated how the various *genres de vie* emerged

[2]C. Sauer, "The Morphology of Landscape," in J. Leighly (ed.), *Land and Life: Selections from the Writings of Carl Ortwin Sauer.* Berkeley, CA: University of California Press, 1964, pp. 315–350.

▲ **Figure 5.7 Market gardens in Corsica** This image shows a rural setting in Corsica, an island nation in the Mediterranean, where commercial agriculture is being undertaken. Farming is a way of life—a *genre de vie*—that we can read from the landscape where extensive cultivated fields and isolated farmhouses constitute key elements.

from the possibilities and constraints posed by local physical environments. Subsequently, he wrote about the changes in French regions brought on by industrialization, observing that regional homogeneity was no longer the unifying element. Instead, the increased mobility of people and goods produced new, more complex geographies where previously isolated *genres de vie* were integrated into a competitive industrial economic framework. Anticipating the widespread impacts of globalization, de la Blache also recognized how people in various places struggled to cope with the big changes transforming their lives.

APPLY YOUR KNOWLEDGE

1. What are the differences between American and European approaches to studying human interactions in the landscape?

2. Reflecting on Sauer's words on how culture shapes landscape, think of one example where you have seen this take place—it might be a neighborhood, an international border, or regional space.

Cultural Traits

Geographers also examine specific aspects of culture, ranging from single attributes to complex systems. One simple aspect of culture of interest to geographers is the idea of special traits, which include such things as distinctive styles of dress, dietary habits, and styles of architecture (**Figure 5.8**). A **cultural trait** is a single aspect of the complex of routine practices that

constitute a particular cultural group. The use of chopsticks is an example of a cultural trait. Chopsticks are a common eating implement across China as well as other parts of Asia from Vietnam to Japan. As this example illustrates, cultural traits are not necessarily unique to one group, and understanding them is only one aspect of the complexity of culture (**Figure 5.9**).

Another example of a cultural trait is a **rite of passage**. These are acts, customs, practices, or procedures that recognize key transitions in human life—birth, menstruation, and other markers of adulthood, such as sexual awakening and marriage. Rites of passage are not uncommon among many of the world's cultures. Some non-Western cultures, for example, send adolescent boys away from the village to experience an ordeal—ritual scarring or circumcision, for example—or to meditate in extended isolation on the new roles they must assume as adults (**Figure 5.10**).

Cultural Complexes and Regions

Cultural traits always occur in combination. The combination of traits characteristic of a particular group is known as a **cultural complex**. Referring back to our example of chopsticks, a cultural complex in which they figure prominently would include other elements such as culinary preferences, foods associated with holidays or special events, and activities such as toasts and drinks or prayers to celebrate those events.

Another concept key to traditional approaches in cultural geography is the cultural region. Although a cultural region may be quite extensive or very narrowly described and even

▼ **Figure 5.8 Tuareg men in Niger** The Tuareg men are also known as the Blue Men of the Sahara because of their practice of wearing distinctive blue robes and veils. Tuareg women do not wear veils but men do when they reach maturity.

5.2 Geography Matters

Shaping Place Through Fact and Fiction, Practice and Representation

By Dydia DeLyser, California State University, Fullerton

Contemporary cultural geographers study both practices and representations—that is, they are interested both in people's actual actions *and* in how those actions are talked about and written about. Because human beings communicate significantly (though not exclusively) through language, both what we do and what we say are important. Cultural geographers are interested in understanding how people make meaning in their daily lives through complex, entangled webs of practices and representations. And one interesting thing about the way people make meaning is that those meaningful places, objects, and stories do not have to be factual.

How Art Shapes Place

Cultural geographers studying literature and film, for example, explore how works of fiction and film help shape place identities and stimulate cultural flows. Consider how the recent *Lord of the Rings* and *Hobbit* films, both shot on locations across New Zealand, have stimulated tourism to New Zealand, complete with scene-by-scene location guides. While it's possible that some Hobbit tourists believe they are actually visiting "Middle Earth" (or wish they were), more likely they simply enjoy seeing the "actual" places where scenes from a favorite film were shot. It's also common for tourists to buy souvenirs that remind them of favorite films or works of fiction. Souvenirs associated with the Harry Potter novels include Gryffindor key rings, Hogwarts crest mugs, and Voldemort wand/pens and are available at Oxford University in the UK where scenes from the novels were filmed; there the Cathedral Shop gift store now offers fiction-inspired items alongside school-branded clothing and recordings of the University's choir. Such fictional worlds can stimulate more than tourism and souvenir purchases—they can actually shape place identity (**Figure 5.A**).

This was the case with the nineteenth-century novel *Ramona* written by Helen Hunt Jackson. Jackson's 1884 novel sought to draw readers' attention to the plight of the California Mission Indians who had been denied treaty rights, and whose lands were being stolen, crops and homes pillaged; they endured assaults and murder by incoming white settlers who desired native lands. Jackson knew her topic was politically sensitive, so she disguised it in romance: the half-Indian Ramona is raised by an adoptive mother on an opulent rancho where California's beneficent climate creates bloom and abundance. Ramona falls in love and elopes with the son of an Indian Chief; in her new life she endures all the suffering of her people: white invaders steal their lands and her husband is murdered before Ramona's eyes—real-life incidents Jackson took from her research and applied to the life of her fictional character.

The novel was a tremendous success, but it failed to draw attention to native issues, gaining attention instead for its romance and lavish landscapes. Jackson's was the first novel written about southern California, so when a tremendous tourism-and-real-estate boom began there in 1885, thousands of tourists prepared for their travels by reading *Ramona*. Soon clever boosters were selling Ramona-related souvenirs and developing *Ramona*-related tourist attractions: by 1925 there were two ranchos that were known as the "Home of Ramona," as well as sites called "Ramona's Marriage Place," "Ramona's Birthplace" and even "Ramona's grave." Southern California's past and present landscapes were being understood through the lens of fiction (**Figure 5.B**).

Scholars scoffed at *Ramona* tourists who seemed to have confused fact with fiction, and been so easily relieved of their cash for crass curios like *Ramona* salt-and-pepper shakers, *Ramona* letter openers, *Ramona* postcards, and *Ramona* souvenir teaspoons (**Figure 5.C**). Both the tourists and the things they bought were dismissed as insignificant in the shaping of southern California's *Ramona*-inspired landscapes.

The Significance of Souvenirs

But were these representations (the novel and the souvenirs) along with the tourist practices

▲ Figure 5.A Tourists watch as a Harry Potter fan tries to enter Platform 9 3/4 at King's Cross Station, London, England.

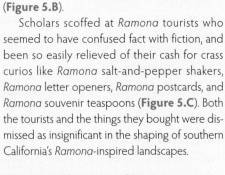

▲ **Figure 5.B** "La Casa de Estudillo," house in Old Town San Diego, real house but imagined by readers of Helen Hunt Jackson's "Ramona" to have been site of Ramona's marriage.

and the practices of tourists at those locales have to be understood in tandem; together representations and practices reveal the complex webs of fiction and fact from which we construct our realities.

1. Look around you both at home and in stores. What souvenirs do you find? What do they remind you of? What geographies—of landscapes, emotions, peoples, and travels—do these material objects recall for you or for their collectors?

2. How else do we connect fact and fiction in our daily lives? Think of an example of something significant and influential that is nevertheless not really "real." How does this connection between fact and fiction influence you? Does it shape what you wear or eat, desire or disdain, aspire to become? Whatever its influence, consider also how it influences the landscapes, fact of fiction that are part of your life.

(visiting *Ramona*-related places) really so insignificant? Tourists and souvenirs are difficult to study because of their rapid geographical dispersion: the tourists return home, and the souvenirs are brought home or gifted elsewhere. So geographers have observed and interviewed tourists at tourist sites, and then followed them and their purchased souvenirs back home to understand what these objects mean in their new contexts. What they have learned is that these objects serve as reminders of places visited, stories and histories engaged there, and personal experiences shared there. Each souvenir serves to link people, through travel, to places and understandings of those places, in personally meaningful ways, creating what cultural geographers call "intimate spatialities of social memory"— personal geographies of connection to place and history.

We may sneer or smile at this proliferation of souvenirs. But their existence, and the details of their materiality is revealing for cultural geographers. Tourists used their souvenirs to connect themselves and their travels to southern California and the story of Ramona— like one woman who visited Ramona's Marriage Place on her honeymoon and wrote on a postcard: "In this building is the chapel and altar where Ramona married Allessandro [sic]," and, noting that a wishing well was full of coins, wrote, "I made wish too! But I won't tell—." Her hopes for her future became entwined with the fictional story of Ramona and the factual landscapes of southern California. Because postcards, when mailed are also cancelled and dated, careful research reveals that, in the case of every Ramona-related tourist attraction, the tourists came first, *before* the boosters. It was the tourists who shaped southern California's Ramona landscapes. And today's tourists may be shaping landscapes in New Zealand and Britain, where landscapes of fiction, souvenirs,

▲ **Figure 5.C** Souvenirs from the 19th-century novel *Ramona* proliferated all over southern California between 1885 and the 1950s. At "Ramona's Marriage Place," a prominent tourist attraction in San Diego, a dedicated curio shop sold dozens of different kinds of souvenirs. Anything from a covered-wagon lamp, to a sterling-silver matchbox cover, from salt-and-pepper shakers to seeds from the garden, all branded "Ramona's Marriage Place." (*Souvenirs from the collection of Dydia DeLyser*)

▲ **Figure 5.9 A Chinese toddler using chopsticks** Usually made of wood for everyday use, chopsticks are an example of a cultural trait that characterizes eating practices in China. Originating in China, chopstick use diffused beyond that country's border to include Vietnam, Korea and Japan.

discontinuous in its extension, it is the area within which a particular cultural system prevails. A **cultural region** is an area where certain cultural practices, beliefs, or values are more or less practiced by the majority of the inhabitants. Cultural regions should be seen as key parts of a cultural complex and not simply the setting where cultural complexes occur.

▼ **Figure 5.10 A coming-of-age ceremony, South Korea** The coming-of-age ceremony for girls in South Korea is held every May 15 in Seoul. There, young women who are turning 20, participate in dances and a citywide celebration that is meant to remind them of the responsibilities they face as adults. In South Korea, people who turn 60 are also ritually celebrated. Called *hwan-gap*, this celebration is significant because it marks the day on which an individual has completed a full zodiacal cycle. But, more importantly, it is also celebrated, because to live to be 60 is seen as a great accomplishment. In the past, most people in South Korea died before their 60th birthday.

Returning to our chopstick example, when combined with the cultural complex of food, holidays, family and related practices and undertaken in China, they demonstrate how a cultural system is at work, though its specific manifestations vary by the cultural region in which they are occurring.

CULTURAL SYSTEMS

Broader than the cultural complex concept is the cultural system, a collection of interacting components that, taken together, shape a group's collective identity. A **cultural system** includes traits, territorial affiliation, and shared history, as well as other, more complex elements, such as language and religion. In a cultural system, it is possible for internal variations to exist in particular elements at the same time that broader similarities lend coherence. The phrase, "American culture" is a reference to the concept of a cultural system. This system includes shared territory and history, a standardized education and a shared language (or more precisely a national language) as well as other cultural elements such as a predominant form of family organization, political norms like the two-party system, and even attachment to certain national sports activities or even television shows. Of course, not everyone in the United States actually has the same history nor do all Americans speak the national language of English or enjoy baseball or live as a nuclear family (**Figure 5.11**). The point is that cultural systems are about broad similarities at the national level and more particular geographic variation at the regional and local level where the local level may refer to a city, a neighborhood or even a household.

Culture and Society

Through its influence on social organization, culture has an important impact. Social categories like kinship, gang, or generation, or some combination of categories, can figure more or less prominently, depending upon geography. Moreover, the salience of these social categories may change over time as the group interacts with people and forces outside of its boundaries.

For example, countries in the Middle East and North Africa are as culturally dynamic as any other region of the globe. Their society is shaped by cultural ties and meaning systems that highlight gender, tribe, nationality,

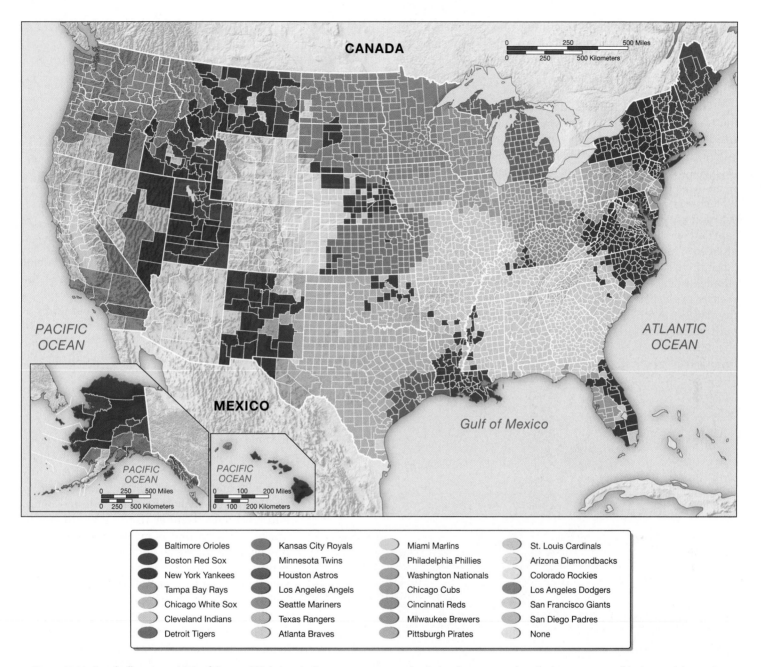

Legend:
- Baltimore Orioles
- Boston Red Sox
- New York Yankees
- Tampa Bay Rays
- Chicago White Sox
- Cleveland Indians
- Detroit Tigers
- Kansas City Royals
- Minnesota Twins
- Houston Astros
- Los Angeles Angels
- Seattle Mariners
- Texas Rangers
- Atlanta Braves
- Miami Marlins
- Philadelphia Phillies
- Washington Nationals
- Chicago Cubs
- Cincinnati Reds
- Milwaukee Brewers
- Pittsburgh Pirates
- St. Louis Cardinals
- Arizona Diamondbacks
- Colorado Rockies
- Los Angeles Dodgers
- San Francisco Giants
- San Diego Padres
- None

▲ **Figure 5.11 Baseball regions, United States** While baseball teams are associated with their home cities, their fan base can extend far beyond the city limits. Notice that the New York Yankees and the Boston Red Sox, are highly popular in Alaska and Hawaii, thousands of miles away from their east coast homes.

kinship, and family. Global media technologies, such as satellite television and the Internet, are increasingly penetrating the region, however, and the potential for new social forms to emerge and old ones to be reconfigured is increasing. Although the predominant forms of social organization in the region have persisted for hundreds of years, it would be incorrect to assume that both subtle and dramatic changes within these forms have not already occurred.

Kinship is a form of social organization that is particularly central to the culture system of the Middle East and North Africa. **Kinship** is normally thought of as a relationship based on blood, marriage, or adoption. This definition needs to be expanded, however, to include a *shared notion of relationship* among members of a group. The point is that not all kinship

relations are understood by social groups to be exclusively based on biological or marriage ties. While biological ties, usually determined through the father, are important in the Middle East, they are not the only important ties that link individuals and families.

Kinship is such a valued relationship for expressing solidarity and connection that it is often used to assert a feeling of group closeness and as a basis for identity even where no "natural" or "blood" ties are present. This notion of kinship might also be seen in America among fraternity "brothers," sorority "sisters," gang members, and even police officers and firefighters who feel a strong familial bond with co-workers. In the Middle East and North Africa, kinship is even an important factor in shaping the spatial relationships of the home, as well as outside the home, determining who can interact with whom and under what circumstances. This is especially the case in the interaction of gender and kinship, where women's and men's access to public and private space is sharply differentiated.

The idea of the tribe is also central to understanding the sociopolitical organization of the Middle East and North Africa, as well as other regions of the world. Although tribally organized populations appear throughout the region, the tribe is not a widespread form of social organization. Generally speaking, a **tribe** is a form of social identity created by groups who share a set of ideas about collective loyalty and political action (**Figure 5.12**). The term *tribe* is a highly contested concept and one that should be treated carefully. For instance, it is often seen as a negative label applied by colonizers to suggest a primitive social organization throughout Africa. Where it is adopted in the Middle East and North Africa, however, tribe is seen as a valuable element in sustaining modern national identity. Tribes are grounded in one or more expressions of social, political, and cultural identities created by individuals who share those identities. The result is the formation of collective loyalties that result in primary allegiance to the tribe.

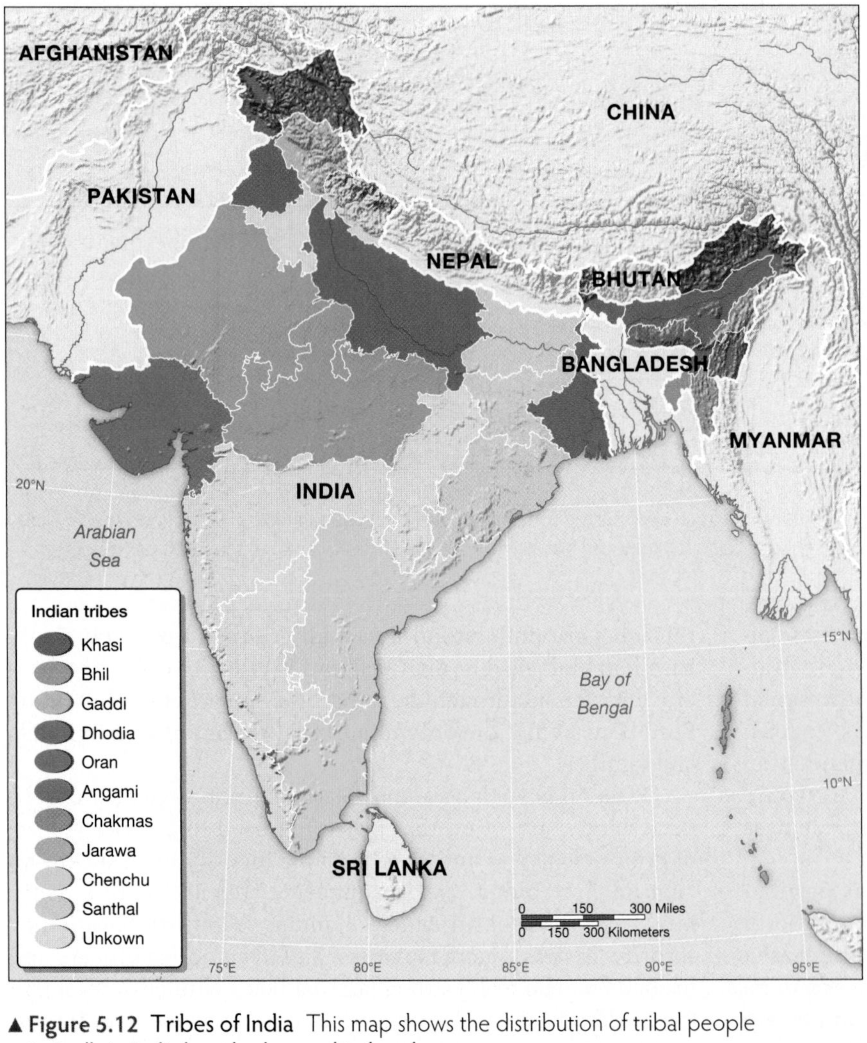

APPLY YOUR KNOWLEDGE

1. What is the difference between kinship and a tribe? Identify two other places in the world where tribal relationships are key to the culture?

2. Create a kinship map of your family. Once you have done this, identify a kinship system from the Middle East or North Africa. Compare and contrast that system with your own. What aspects of that kinship organization are similar to yours? What aspects are different and why?

CULTURE AND IDENTITY

In addition to exploring cultural forms, such as religion and language, and movements, such as **cultural nationalism**—the belief that a nation shares a common culture—geographers have increasingly begun to ask questions about other forms of identity. This interest largely has to do with certain long-established and some more recently self-conscious cultural groups that are beginning to use their identities to assert political, economic, social, and cultural claims.

Sex and Gender

In standard dictionaries of the English language **sex** is consistently defined as the biological and physiological characteristics that differentiate males and females at birth, based on bodily characteristics such as anatomy, chromosomes, and hormones. **Gender** in the same reference books is a term that is meant to differentiate biology from the social and cultural distinction between the sexes. Sex as a category is meant to specify man and woman, whereas gender categories would more appropriately be masculine and feminine.

This current distinction between sex and gender has been criticized as misleading, however, by both social and biological scientists. The reason for this is that it implies that the behavior of an individual can be separated out into being shaped by either biological or cultural factors. The critics would respond that biological manifestations of sex are crosscut with psychosocial and cultural variables so much so that there is no way to determine that the differences between males and females are due exclusively to biology or to culture (**Figure 5.13**).

Indian tribes

- Khasi
- Bhil
- Gaddi
- Dhodia
- Oran
- Angami
- Chakmas
- Jarawa
- Chenchu
- Santhal
- Unkown

▲ **Figure 5.12 Tribes of India** This map shows the distribution of tribal people principally in India but also beyond its borders.

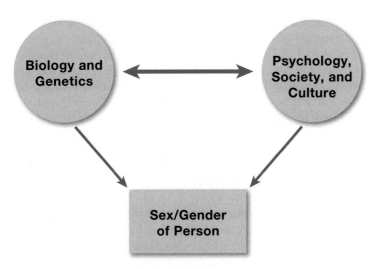

▲ **Figure 5.13 Biological-Psychosocial-Phenotype** This diagram shows the mutual influences on sex and gender.

Our approach in this book is understanding that sex and gender operate along three lines:

- Sex and gender are not categories that should be separated—they very clearly shape each other—because they are deeply connected categories of being.

- There are a multitude of sexes in between the two extremes of male and female and one's sex need not align with one's biology.

- Gender is most effectively understood as a socially constructed performance that uses codes of femininity and masculinity to operate in the world.

We turn next to gender identity, relinquishing our focus on sex. Our reason for this is that in both scholarship and in popular treatments, the term *gender* has come to be used as the preferred term for the complex relationship between sex and gender. Because of that, we discuss gender in its multifaceted and sometimes confusing aspects as well as its relevance to geographic concerns.

Gender Identity As illustrated in the chapter opener, gender is an identity that has captured the attention not only of cultural geographers but is widely recognized in popular culture as well. Indeed, gender identity is so widely acknowledged in everyday life that in 2013, Facebook added more than 50 custom categories to their gender option. Facebook, recognizing that the simple male/female dichotomy was inadequate in capturing the multiplicity of genders that reside in the space between them, took the rather radical step of opening up the category to 58 options.

This move by Facebook is a fundamental recognition that is a key feature of all human identity marking and defining humans as "sexed" beings. It is also an acknowledgment that gender is not something people essentially are (because of a given set of physical characteristics), but something people do (something we enact by the way we present our bodily selves to the world) and something we understand ourselves to be (**Figure 5.14**).

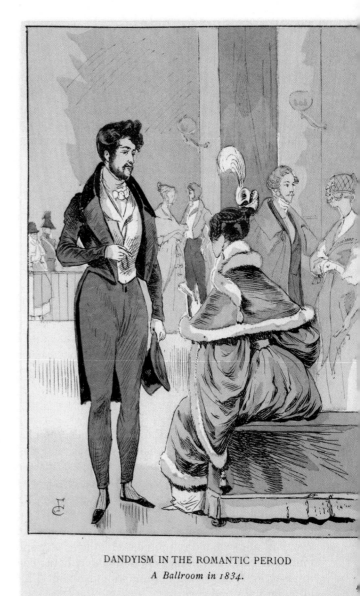

DANDYISM IN THE ROMANTIC PERIOD
A Ballroom in 1834.

▲ **Figure 5.14 Masculinity and the dandy in the nineteenth century** Across much of Europe and the United Kingdom, the nineteenth century dandy, pictured here, was considered the height of masculinity. He was concerned with physical appearance and a cult of the self. The man on the left is dressed in a frock.

Research on the gendered body in geography recognizes that space is central to its understanding. For instance, where transgendered identities can be performed plays a central role in who occupies those spaces, how they are occupied, and even what gender identity people might be performing if they occupy one space and not another. But issues of gender and space go beyond the performance of identity to the emergence of new political cultures that are being constructed to protect the rights of people who don't identify as simple male or female in particular national and international spaces.

What Facebook's Gender Identity Means

http://goo.gl/cNZrVV

Feminism and Gender

Whereas the previous section on gender has focused on gender as a sexual identity that extends far beyond the simple binary or male or female, gender has also been a category that feminist geographers and others have used to understand inequality between men and women. Much has been written about this uneven relationship. As used by feminists, gender implies a socially created difference in power between males and females that gives an advantage to the former over the latter and is not biologically determined but socially and culturally created (Box 5.2 Spatial Inequality: The Global Gender Gap). Gender interacts with other forms of identity and can intensify power differences among and between groups.

APPLY YOUR KNOWLEDGE

1. What do you think about the UN statement that, "No society treat its women as well as its men"? Why do you think that is? How do you think that can change?

2. Chose one of the ten examples of global inequality cited earlier and do an Internet search to find out more about that statement. Where in the world are people trying to change those statistics? How?

Feminists use the word intersectionality to characterize what happens when other identities such as race, class, age, or ability also have an effect on power dynamics. **Intersectionality** is a recognition of the ways that different forms or systems of oppression, domination, or discrimination overlap. Black feminism advanced the original example of intersectionality by arguing that being black and female are inseparable identities that intersect and reinforce each other. Geographers have found the concept of intersectionality convincing but add that intersecting forms of oppression are played out differently in different parts of the world.

For example, although gender differences play an important part in shaping social life for men and women in the Middle East, as elsewhere around the globe, there is no single Islamic, Christian, or Jewish notion of gender that operates exclusively in the region. Many in the West have formed stereotypes about the restricted lives of Middle Eastern women because of the operation of rigid Islamic traditions (**Figure 5.15**). It is important to understand, however, that these do not capture the great variety in gender relations that exists in the Middle East and North Africa across lines of class, generation, level of education, and geography (urban versus rural origins), among other factors.

Gender and Class

In South Asia, gender is greatly complicated by class such that among the poor, women bear the greatest social and economic burden and the most suffering. Generally speaking, South Asian society—India, Pakistan, Afghanistan, Nepal, Bhutan, Sri Lanka, Maldives and Bangladesh—is intensely patriarchal, though the form that patriarchy takes varies by region and class. The common denominator among the poor throughout South Asia is that women not only have the constant responsibilities of motherhood and domestic chores but also have to work long hours in informal-sector occupations. In many poor communities, 90 percent of all production occurs outside of formal employment, more than half of which is the result of women's efforts. In addition, women's property rights are curtailed, their public behavior is restricted, and their opportunities for education and participation in the waged labor force are severely limited.

The picture for women in South Asia, the Middle East, and elsewhere is not entirely negative, however, and one of the most significant developments has been the increasing education of girls. Because women's education is so closely linked to improvements in economic development, more and more countries are investing in their education. As more and more women become educated, they have fewer children and greater levels of economic independence and political empowerment.

Ethnicity

Ethnicity is another area in which geographers explore cultural identity. **Ethnicity** is a socially created system of rules about who belongs to a particular group based upon actual or perceived commonalities, such as language or religion. A geographic focus on ethnicity is an attempt to understand how it shapes and is shaped by space and how ethnic groups use space with respect to mainstream culture as well as other ethnic groups. For cultural geographers, territory is also a basis for ethnic group cohesion (see Chapter 10 for more on territory). For example, cultural groups—ethnically identified or otherwise—may be spatially segregated from the wider society in ghettos, ethnic enclaves, homelands, and tribal areas.

▼ **Figure 5.15** Turkish women in Berlin, Germany   Turks have migrated in large numbers to Germany due to a guest worker program. These women are veiled, in contrast to the women in the crowd behind them. Yet they are clearly out in public where they are dancing in celebration of May Day, the festival to welcome spring weather.

5.3 Spatial Inequality   The Global Gender Gap

In 1990, the United Nations published the first of its annual Human Development Reports. The report analyzes how economic growth and human development are inextricably tied and provides statistics about changes in both over time as well as suggestions for how to improve them (**Figure 5.D**). Since 1990, the report has taken the position that women are at a structural disadvantage compared to men and in its 1997 report, stated baldly, "No society treats its women as well as its men." While the differences between women's and men's pay in the developed world is a common topic of discussion and concern (where in the United States for every $1 men earn, women earn 77 cents), in the developing world, women experience deep deprivation, exploitation, and harm. The following are ten examples of gender inequality globally.[3]

1. Women everywhere experience a gender wage gap whether in the developed or the developing world.

2. Women in many parts of the world experience limited mobility from not being allowed to drive on public roads to refusing to go out by themselves at night for fear of attack or rape.

3. One in every three women around the world is likely to be beaten, coerced into sex, or otherwise abused sometime in her lifetime.

4. In some countries, a male child is more valuable than a female child and parents who don't want a girl may either abort the fetus or kill the child after birth.

5. In some countries, women are legally prohibited from owning land.

6. According to the United Nations, women do two-thirds of the world's work, receive ten percent of the world's income and own one percent of the means of economic production.

7. Women have more limited access to health care than men while one women dies in childbirth every minute of every day.

8. Forced marriages and the lack of legal access to divorce limits many women's life chances.

9. Despite making up half the global population, women hold only 15.6 percent of elected seats in national parliaments or congresses.

10. Women make up more than two-thirds of the world's illiterate adults.

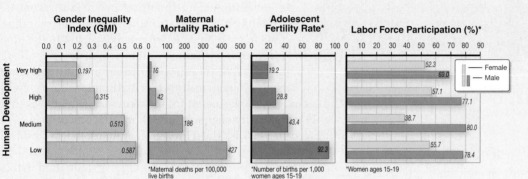

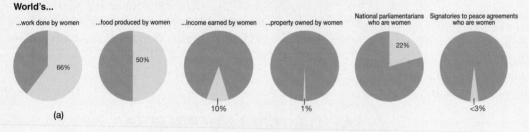

(a)

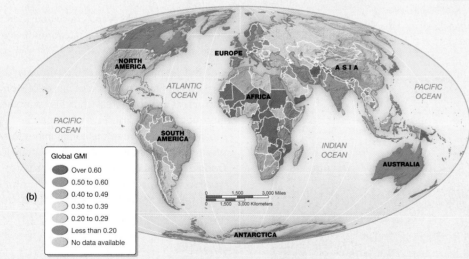

(b)

1. What is the gender gap?

2. In what ways would a narrowing of the gender gap improve the lives of women around the globe?

◄ **Figure 5.D The geography of the global gender gap** Shown in this graphic are (a) key indicators as well as (b) a map of the gender inequality index globally.

[3]Adapted from Molly Edmonds, 2014, "Examples of Gender Inequality around the World", http://www.discovery.com/tv-shows/curiosity/topics/examples-gender-inequality-around-world.htm (accessed June 29, 2014).

▲ **Figure 5.16 Uyghur protest** Uyghur residents protest in the city of Urumqi, following bloody riots in the capital of China's Xinjiang region. 156 people died and over 1,400 people were arrested.

In China, for example, 91 percent of the population is Han. There are 55 different ethnic groups in addition to the Han. These are mostly residual groups of indigenous people such as the Miao, the Dong, Li, Naxi, and Qiang, who are economically disadvantaged and found in relatively remote border regions, removed from central authority in Beijing. Tensions exist between the dominant Han and several of the larger minority groups such as the Tibetans and the Uyghurs. The latter are a Turkic Muslim minority group living in western China. They demand greater control over their territory to practice self-governance and to follow their own cultural traditions and not those of the mainstream (Han-dominated) society. The Uyghur ethnic identity is fragmented, however. Some Uyghur support a pan-Islamic vision, exemplified in the East Turkestan Islamic Movement; others support a pan-Turkic vision, as in the East Turkestan Liberation Organization; and a third group—the East Turkestan independence movement—promotes a "Uyghurstan" state. As a result, Uyghurs do not speak with one voice and members of each of these groups have committed violence against other Uyghurs whom they view as too assimilated to Chinese or Russian society or not religious enough (**Figure 5.16**).

The term ethnicity comes from the Greek *ethnos,* which in classical times was understood as "the people." But as time passed and the term migrated, its meaning changed so that in England, from the fifteenth until the mid-nineteenth century, ethnic came to mean, heathen, foreign or pagan. In the nineteenth century United States, which was experiencing massive immigration then, ethnicity came to mean a racial, cultural, or national minority group and served as a replacement for the term *race,* which was thoroughly emotionally charged through its attachment to slavery, outlawed in 1865 by the Thirteenth Amendment to the U.S. Constitution.

In its contemporary usage, ethnicity is widely understood as a cultural category constituted through shared history and often through language, religion, and an attachment to a particular place or homeland (**Figure 5.17**). And though there are many examples of ethnic groups who share a cohesive sense of community and belonging, not all do. As with any identity, variation exists in practice.

APPLY YOUR KNOWLEDGE

1. Define all the different ways geographers analyze cultural identity: sex and gender, and ethnicity. Are there other categories that you think are critical in assessing cultural identity?

2. One indicator of ethnicity and territory is the geographic distribution of places of worship across an urban area. Churches, temples, mosques, meeting houses, and chapels are usually located close to the communities they serve. Using a spatial visualization program like Google Maps, map your town or city's places of worship, and speculate on what this distribution might tell you about the local ethnic geography.

Race

Geographers also use prevailing ideas and practices with respect to race to understand places. **Race** is a problematic classification of human beings based on skin color and other

▼ **Figure 5.17 Arab Street neighborhood** This area in central Singapore contains many Arabic shops, restaurants, cafes and residences. Like Chinatowns in North American cities, it is an ethnic place within a larger more diverse urban space.

physical characteristics. **Racialization** is the practice of creating unequal castes based on the norm of whiteness. Biologically speaking, no such thing as race exists within the human species. Yet consider the categories of race and place that correspond to "African American" and "ghetto." Powerful Western ideas about race have enabled the emergence and perpetuation of segregated spaces in many North American cities and elsewhere. In this and other cases, the visible characteristics of hair, skin, and bone structure made race into a category of difference that was (and still is) widely accepted and often spatially expressed.

The mainstream approach views neighborhood as a spatial setting for systems of affiliation more or less chosen by people with similar skin color. Cultural geographers have overturned this approach to see neighborhoods as spaces that affirm the dominant society's sense of identity. Place continues to be a mechanism for creating and preserving local systems of racial classification and for containing geographical difference within defined geographical confines. The homelands of South Africa and the dismantling of apartheid there also illustrate the interaction of race and place on a much larger scale.

Unexamined in the kinds of perceptions and practices that create racialized places as different as Arab Street in Singapore and the homelands of South Africa is the taken-for-granted notion of whiteness. Whiteness is seen to be the norm or the standard against which all other visible differences are compared. But whiteness, as geographers and other scholars have shown, is itself a category of difference that depends on visible distinctions, not biological ones, and is always constructed in relation to other categories (**Figure 5.18**). Recently, researchers in the humanities and social sciences have begun to "denaturalize" whiteness by investigating the ways whiteness has been constructed as a social category in different periods and places as well as the ways whiteness operates in particular sites such as classrooms, on the street, and in boardrooms. It is only when we begin to challenge the naturalness

of whiteness and the spaces that enable it that we can truly begin to undo racist practices.

Whiteness, Blackness, and Rap Music

To appreciate the complexity of the relationship between whiteness and blackness in the United States two examples provide insight: one is about space and race in the form of racial desegregation and the other is about blackness as an object of white fascination and appropriation. Consider these two facts: the racial integration of public K-12 schools in the United States, successfully achieved within a decade of court-mandated integration in the 1960s and 1970s, is increasingly being dismantled so that now many schools, all over the country but especially in the South, are as segregated as they were in the 1950s. Contrast the deterioration of racial integration with the second fact of the huge popularity of black rap music among white audiences who are the largest consumers of recordings by black rap artists. At the same time that white teenagers appropriate rap music, they also demonstrate their fascination with African American culture through the outfits they wear and the language—both verbal and nonverbal—they adopt. And yet, they are highly unlikely to attend high school with more than a handful of black youth. How can these two contrary phenomena—"blackophobia" or fear of blacks by whites and "blackophilia" or adulation of blacks by whites—exist simultaneously in the United States?

Cultural theorists contend that these two seemingly contradictory facts can only be understood in relation to each other. They assert that the tendency to embrace black culture by consuming rap and other African American artifacts is related to the destabilization of what it means to be white in U.S. society today as the country moves from majority white, to no racial majority, thus decentering whiteness in the process. So just as white youth are embracing black culture and appropriating it for their own use, they (and their parents who make decisions about where to send them to high school) are refusing to

◄ **Figure 5.18 Melbourne, Australia** High school and college students are protesting against racist land rights legislation that discriminates against aboriginal people.

support policies or practices that challenge the foundations of their white privilege.[4] **White privilege**—advantages that accrue to white people beyond what is commonly experienced by people of color—and **white supremacy**—the belief that white people are superior to other races—sit at the heart of the racist practices described here.

Race is a highly complicated category of difference and its expression in daily life varies from place to place. Geographers probe the foundations of these spatial differences.

Geographies of Disability

While it might seem that disability is a far reach from racism, it's not. The kind of discrimination that people with disabilities face is premised on a similar belief that the able-bodied are "normal" and superior to the disabled who possess a lesser social status. **Disability** is a physical or mental condition that limits a person's movements, senses, or activities. Disability—both in its physical and mental form—has begun to be studied fairly widely by geographers who are particularly interested in the way that disabled people encounter the spaces in which they must conduct their daily lives. Studies that examine how the built environment can "dis-able" people are common, their goal being to provide research to shape public policies that address these disabling arrangements.

Besides studying the ways that the environment creates barriers for persons with disability, other work by geographers explores the ways persons with disabilities experience space, what their perceptions of the environment are, and how they create communities and form identities (**Figure 5.19**). The formation of a common cause and identity has been critical to the political movements that persons with disabilities have formed, demanding recognition of their human right to an accessible and safe environment in which to live and work. One success from this activism has been the United Nations Convention on the Rights of Persons with Disabilities. A key tenet of the convention is, "Reaffirming the universality, indivisibility, interdependence and interrelatedness of all human rights and fundamental freedoms and the need for persons with disabilities to be guaranteed their full enjoyment without discrimination."[5]

Children's Geographies and Geographies of Childhood

Another focus of cultural geography is the spaces and places of the lives of youth and children, known as **children's geographies**. In addition to recognizing that attunement to culture means recognizing differences in race, gender, sexuality and ethnicity, it also means appreciating that children—including teenagers, youth, and other individuals—who are largely dependent on adults for their health, welfare, and safety also

[4]This discussion is drawn from B. Yousman (2003) "Blackophilia and Blackophobia: White Youth, the Consumption of Rap Music, and White Supremacy," *Communication Theory*, *13*(4): 366–391.

[5]Preamble, part c from the Convention on the Rights of Persons with Disabilities, 2006. http://www.un.org/disabilities/convention/conventionfull.shtml (accessed 27 June 2014).

▲ **Figure 5.19 Disability and space** Our urban spaces are premised on a norm of the able-bodied, such as staircases, which are impossible to negotiate when in a wheelchair.

operate in and through geography in particular ways. The study of children by cultural geographers focuses on how children understand the world around them and how they negotiate it. It also explores how adult-centered accounts and experiences of the world exclude the different and equally complex experiences and accounts of children.

Geographers' recognition that children experience oppression and deprivation in ways that are different from adults is a reflection of the fact that in the United States, 22% of all children live in families with incomes below the poverty level. In this situation, the effects of poverty on children mean that their ability to learn is affected and social, emotional, and behavioral problems are a result. The effects of poverty on adults, all though no less profound, are not the same.

Geographies of childhood are distinct from children's geographies in that the former reflect an interest by researchers

▲ **Figure 5.20 Free play and environmental exploration** This image shows girls involved in what is considered by some parents to be risky behavior that could lead to bodily harm through falling. An emerging movement around children's play encourages them to become more actively involved in experiencing and exploring their environment.

in how childhood as a stage of life is conceived. Researchers recognize that childhood is not a natural category of the life cycle but one that has been "invented" by society. They also insist upon the need to appreciate children as possessing the capacity to act and think autonomously and to affect and interact with the world around them in unique ways (**Figure 5.20**). For example, geographers have explored how economic processes such as deindustrialization affect children through the kinds of futures they imagine for themselves.

APPLY YOUR KNOWLEDGE

1. Looking at the geographies of race, disability, and children—what is the role of power in each of these areas of study? (*Hint:* Seemingly unrelated, each study area analyzes how one idea of "normal" became so powerful in society. Why is that? What is problematic about a concept of "normal"?)

2. Do you see other examples like "blackophobia" and "blackophilia" in your everyday life—where different groups are simultaneously vilified and fetishized? How does this cultural schizophrenia contribute to ideas of white supremacy?

EMERGENT CULTURAL GEOGRAPHIES

Since the arrival of the new century, cultural geography has experienced, again, a transformation in the way its practitioners think about the relationship between people and their worlds. These new ways of conceptualizing culture and space are still developing but they hold the promise of opening up a wide range of previously ignored aspects of daily life to advance our attempts at explanation.

Actor-Network Theory

Perhaps the most important influence on cultural geography in the twenty-first century has been **actor-network theory** (ANT), which is actually less of a theory and more of an orientation. Actor-network theory views the world as composed of "heterogeneous things," including humans and nonhumans and objects.

What makes ANT so interesting is that the approach attributes to nonhumans and objects as much force in the composition of social life as humans have. Rather than elevating humans as the superior species that determines all social practice and action, ANT recognizes that humans coexist with nonhumans (who may be other living species or inert objects) in a network that includes all sorts of social and material bits and pieces. Things are just as important to social life as are humans in an actor network.

An example of an actor network is a family. According to ANT, a family is both a network *and* an actor that hangs together and for certain purposes acts as a single entity. A family is connected through all sorts of objects: cell phones, houses, automobiles, family dinners, the family pet, and so on. Nonhumans help constitute the fabric of "family" and enable it to persist, though its contours are always changing. It is not only the humans who make the family what it is, it is also the nonhumans that help the assemblage to come together as a recognizable and coherent entity. Agency—the ability to make things happen—resides not only in the human members of the assemblage but also in the nonhuman (the house, the dog, the barbeque grill, the flat-screen television) (**Figure 5.21**).

Recall through **Figure 5.3**, how the Internet of things is shaping wake up time, when the coffee gets made, when the car gets started, the time of departure for the train station, and so on. The point of ANT is not that all these things are connected so much that humans are not solitary agents in the world but rather that human behaviors need to be understood as shaped and often driven by objects. The Internet of things is an actor-network where all sort of objects and nonhuman others come together to shape how we operate in the world on a very basic level.

Consider the cell phone as an example. This mobile device connects people to family and friends, nearly instantaneously and certainly in ways that a landline never could. The materiality of that object—its "thingness"—is not inert; it has an active forcefulness that can produce significant outcomes such that we are often at a loss without them or especially empowered with them. In a way, cell phones have become objects that make us do things we might not or could not otherwise.

Non-Representational Theory

One of the criticisms of ANT has been that it treats humans as undifferentiated and lacking in affect and emotion as they go about the business of social life. In attempting to go further into what it means to be human and particularly into what is present in human experience, cultural geographers have become increasingly interested in non-representational theory. In brief,

◀ **Figure 5.21 An actor-network**
This image shows a family gathered together but also being occupied by various electronic devices which act to both connect them as well as distract them from each other.

non-representational theory (NRT) understands human life as a process that is always unfolding, always becoming something different, even if only slightly so. It recognizes that much of this becoming occurs outside of conscious thought. Because much of human existence is precognitive (decisions are made before the conscious self is aware of them), NRT is interested in those moments of indeterminacy when events emerge that produce new orderings that may persist or give way to older, more settled ones. These are moments we have all experienced, when something happens that no one expected and yet in which everyone participated in enabling the new ordering to emerge. This focus on indeterminacy means that NRT's task is a difficult one in that it attempts to attend to things that words (as representations) cannot express.

Geographer Ben Anderson has used non-representational theory to explore memory and music. His work is concerned with how the process of listening to music produces both remembering and forgetting within the context of ordinary living. He describes a young woman (whom he interviewed) listening to music as background to her getting ready for work. As she sits flipping through a magazine and eating breakfast with a Frank Sinatra tune playing in the background, she suddenly hears her mother (who died several years before) singing along to the music. It's a moment from her past; her mother would frequently sing along to all music, and this song was one of her favorites. And in that moment, the young woman is transported from her apartment back in space and time to her family home. The woman notes that when this happened she got "shivery feelings really suddenly."[6] Anderson identifies that "shivery feeling" as **affect**—emotions that are embodied reactions to the social and physical environment. Affect is also about the power of these emotions to result in or enable action.

Others have described how hearing their national anthem produces an almost indescribable feeling, even when a person does not consider him or herself patriotic or is uncomfortable in general with displays of nationalism (**Figure 5.22**). And yet, many of these same people also describe a kind of sudden "shivery feeling" that involuntarily occurs when the music begins and the whole stadium stands in respect and recognition of being a part of something larger than the individual.

NRT is keenly interested in events in which things suddenly shift and something involuntary occurs. And it is particularly interested in events when whole groups experience something not anticipated in advance. These are moments that hold significant political potential. For example, in Tunisia in the spring of 2011, the self-immolation of a fruit seller ignited a city's anger at their political dispossession, launching a political event that spread across the Middle East.

Emotional Geographies

In addition to an interest in affect and NRT, geographers also explore emotion. Those who study emotion differentiate it from affect by specifying that emotion is personal feeling based on one's biography or a social context. Emotion is a response to an internal state of being. Affect by contrast is a nonconscious experience of intensity; it is always outside of conscious awareness. Facial expressions or tone of voice, for instance, have the capacity to transmit affect. Once transmitted, emotions register the ways our bodies and our minds react to those expressions. We may become excited or afraid as we process the intensity of those expressions. Our emotional response depends on who we are and what our social conditioning has been.[7]

[6]B. Anderson, "Recorded Music and Practices of Remembering," *Social and Cultural Geography*, *5*(1), 2004, pp. 3–18.

[7]Eric Shouse, 2005, "Feeling, Emotion, Affect," *Affect: a Journal of Media and Culture*, *8*(6), http://journal.media-culture.org.au/0512/03-shouse.php (accessed 28 June 2014).

◀ **Figure 5.22 U.S. flag raised at Ground Zero, New York City** A flag can be a powerful material stimulus, generating a strong and sometimes indescribable response that is shared by all those viewing it. The U.S. flag raised at Ground Zero is an example of such a moving material object.

Geographers are particularly interested in how emotions are tied to places. How different environments, for instance, produce different emotional responses. Because emotions are a central component of being human they possess the potential to both shape and be shaped by geography. In addition, emotions can both emerge from as well as produce spatial orders and thus geographers are interested in understanding how emotions are part of the various elements that make up what it's like to live in a place or visit or operate in a particular space (**Figure 5.23**).

Materialism

Both ANT and non-representational theory share an interest in matter and materiality. **Materialism** emphasizes that the material world—its objects and nonhuman entities—is at least partly separate from humans and possesses the power to affect humans. A materialist approach is about attempting to understand the ways that specific properties of material things affect the interactions between humans and nonhuman entities. What materialism adds to NRT and ANT is a commitment to understanding the material world as it unfolds in unpredictable ways. Both theories point to real physical and mental entities as significant to our attempts to explain the world. And they both recognize that humans are not separate from or in any way superior to the world of things. The aim of these new ways of thinking about cultural geography is to rethink how material objects work in the world so that they come to be seen as "lively"—having real force and intensity in the world.

When seen in terms of cultural geography, the materialism that is at the center of these new ways of thinking directs our

▶ **Figure 5.23 Genocide Museum, Rwanda** Shown here are photographs of individuals who died in the genocide in Rwanda. The museum is meant to be a space that operates through emotion by calling on the visitor to see the event not as an abstraction but something that happened to real human beings.

▲ **Figure 5.24 Statue of Bussa, Barbados** Statues such as this one of Bussa, an African-born slave who led a rebellion against British forces, are a material reminder to Barbadians of one of their heroes who led 400 freedom fighters in opposition to their enslavement.

attention to how cultural beliefs and values gain permanence and power through material form. Buildings, symbols, commodities, or rituals—the twin towers of the World Trade Center, a military uniform, a diamond ring, a Thanksgiving meal—shape us and affect the way we are recognized by others (**Figure 5.24**).

APPLY YOUR KNOWLEDGE

1. What is the difference between "affect" and "emotion"? Can you think of an example when you have been in a particular space and experienced "affect"?

2. Identify an object (other than a cell phone, which has already been discussed) that has a powerful influence over the way you operate in the world. How does that object work on you? How does it affect your relationships to others? Are there particular times and places in which the object is effective or ineffective? Why might that be?

GLOBALIZATION AND CULTURAL CHANGE

Anyone who has ever traveled between major world cities will have noticed the many familiar aspects of contemporary life in settings that until recently were thought of as being quite different from one another. Airports, offices, and international hotels have become notoriously alike, and their similarities of architecture and interior design have become reinforced by the near-universal dress codes of the people who frequent them. For example, the business suit, especially for males, has become the norm for office workers throughout much of the world. Jeans, T-shirts, and cell phones, meanwhile, have become the norm for young people, as well as those in lower-wage jobs.

Americanization and Globalization

It is these commonalities—as well as others, such as the same automobiles, television shows, popular music, food, and global brands like McDonalds, Coca-Cola, and Apple—that provide a sense of familiarity to core travelers abroad. From the point of view of cultural nationalism, the "lowest common denominator" of this familiarity is often seen as the culture of fast food and popular entertainment that emanates from the United States. Popular commentators have observed that cultures around the world are being Americanized (see Chapter 2). This process represents the beginnings of a single global culture based on material consumption, with the English language as its medium.

There is certainly some evidence to support this point of view, not least in the sheer numbers of people around the world who view *American Idol,* drink Coca-Cola, and eat in McDonald's franchises or similar fast-food chains. Largely through consumer goods, U.S. culture is increasingly embraced by local entrepreneurs around the world. It seems clear that U.S. products are consumed as much for their symbolism of a particular way of life as for their intrinsic value. McDonald's burgers, along with Coca-Cola, Hollywood movies, rock music, and NFL and NBA insignia, have become associated with a lifestyle package that features luxury, youth, fitness, beauty, and freedom.

It is important to recognize, however, that U.S. products often undergo changes when they travel across the globe. For instance, L'Oreal, a Paris-based international corporation and the leading mass-market producer of cosmetics, skin- and hair-care products, and fragrances, is very aware of the need to vary its products for their consumption in non-French markets. Its website provides insight into the wide variety of skin types, hair types, and treatment preferences across the world. For instance, one of their pages is devoted to understanding "When the Diversity of Types of Beauty Inspires Science," thus linking their research aims with a sense that beauty is a category that varies throughout the world. On a linked page entitled "Expert in Skin and Hair Types Around the World" is a color chart displaying 66 skin tones called "A New Geography of Skin Color." L'Oreal believes this mapping of skin color allows the company to adapt their cosmetics to the expectations of different national consumers. *Box 5.4, Window on the World: Geographies of Plastic Surgery,* provides a contrast to L'Oreal's aim to cater to differences in cultural norms of beauty by showing how the increasing popularity of plastic surgery around the world appears to be premised on narrow, largely Western notions of beauty.

L'Oreal

http://goo.gl/6MnafK

APPLY YOU KNOWLEDGE

1. Why has the "culture of beauty"—which used to be very specific to different places of the world—become more Westernized? What do you think might be lost as a result of a "global standard of beauty"? Can anything be gained?

2. What is the relationship between "beauty" and "power"? Do an Internet search on ideals of beauty in different cultures, and compare those images to global beauty products and fashion. What are some of the links among physical beauty, consumption, and social power? For example, why is a Louis Vuitton bag a symbol of status not just in France, but also in Mexico, Dubai, Algeria, and Japan? Think about how "looking rich" or "looking white" creates a niche market *and* influences social power.

The same might be said of other cultural products emanating from the U.S. entertainment industry. Today the entertainment industry is the leading source of foreign income in the United States, with a trade surplus of over $25 billion. The originals of over half of all the books translated in the world (more than 25,000 titles) are written in English, the majority of which are produced by U.S. publishers. In terms of international flows of everything from mail and phone calls to press-agency reports, television programs, radio shows, and movies, a disproportionately large share originates in the United States. What this huge export of U.S. entertainment products suggests is that the market is being saturated by U.S. cultural norms to the disadvantage of local ones (**Figure 5.25**).

Yet neither the widespread consumption of U.S. and U.S.-style products nor the increasing familiarity of people around the world with global media and international brand names adds up to the emergence of a single global culture. Rather, what is happening is that processes of globalization are exposing the world's inhabitants to a common set of products, symbols, myths, memories, events, cult figures, landscapes, and traditions. People living in Tokyo or Tucson, Turin or Timbuktu, may be perfectly familiar with these commonalities without necessarily using or responding to them in uniform ways. It is also important to recognize that cultural flows take place in all directions, not just outward from the United States. Think, for example, of European fashions in U.S. stores; of Chinese, Indian, Italian, Mexican, and Thai restaurants in U.S. towns and cities; and of U.S. and European stores selling exotic craft goods from the periphery.

A Global Culture?

The answer to the question of whether there is a global culture must therefore be "no," or at least there is no indisputable sign of it yet. While people around the world share an increasing familiarity with a common set of products, symbols, and events (many of which originate in the U.S. culture of fast food and popular entertainment), these commonalties are configured in different ways in different places, rather than constituting a single global culture. The local interacts with the global, often producing hybrid cultures. Sometimes traditional local cultures become the subject of global consumption; sometimes it is the other way around.

◀ **Figure 5.25 Top feature films, 2013** U.S. films are highly popular around the world and help to boost the profits of U.S. production companies. Some critics worry that their popularity undermines the stability of foreign production companies and their ability to offer their own cultural products to their national audiences.

TOP 10 FEATURE FILMS

#	Film	Studio	Gross
1	*Frozen*	Disney	$1,274,219,009
2	*Iron Man 3*	Marvel Studio	$1,215,439,994
3	*Despicable Me 2*	Universal / Illumination	$970,761,885
4	*The Hobbit: The Desolation of Smaug*	Warner Bros. / New Line / MGM	$958,366,855
5	*The Hunger Games: Catching Fire*	Lionsgate	$864,565,663
6	*Fast and Furious 6*	Universal	$788,679,850
7	*Monsters University*	Disney / Pixar	$743,559,607
8	*Gravity*	Warner Bros.	$716,392,705
9	*Man of Steel*	Warner Bros. / Legendary	$668,045,518
10	*Thor: The Dark World*	Marvel Studios	$644,783,140

5.4 Window on the World

Geographies of Beauty and Plastic Surgery

The phrase, "beauty is in the eye of the beholder" has been credited to a range of individuals from Plato to Hume. This is no doubt because the sentiment is one that has been repeated down through the ages: beauty is subjective, so the assessment of what is beautiful is up to the observer. But it's also possible to contend that beauty is in geography, in the cultural ideals of beauty that vary across the globe. For instance, in Mauritania, a large woman is considered beautiful among some urban groups (**Figure 5.E**). Mauritania is a country that routinely experiences famine and a stout woman, especially a wife, signals social standing and wealth in this context. Full-figured women are also highly prized in Fiji. Though the country is not prone to drought, social life there revolves around family and eating. There lavish meals are a cultural artifact of the uncertainly that is central to a traditional subsistence-based economy where an extended spell of rains, for example, can eliminate a season's harvest.

In India, light skin is another signal of the relationship between beauty and wealth for both men and women (**Figure 5.F**). In a country that straddles the equator where the sun is constant and human skin is subject to harsh exposure and darkening, it is a sign of wealth to be able to stay indoors and avoid any kind of work that requires constant exposure to the sun. And in many Muslim countries, where women cover their hair or both their hair and bodies, the eyes become the focus of the female face and are said to be the source of a woman's beauty. Numerous other examples could be offered but the point is that for a very long time, these cultural standards of beauty have persisted. And yet, globalization and the circulation of beauty norms from the West are starting to displace these local norms with dangerous consequences in some cases.

One example of this is illustrated in **Figure 5.G** that documents the rise in plastic surgery globally. Surgical alteration is one way to achieve the beauty norms that are otherwise impossible to achieve through skin-care, dental retainers, or other less invasive procedures. For example South Korea has the highest per capita incidence of plastic surgery in Asia where one in five women between the ages of 19 and 49 has undergone a procedure. Three highly popular procedures are rhinoplasty (a nose job), blepharoplasty (eyelid lifts) and jaw reconstruction, all designed to produce a more Caucasian facial presentation. China and Japan also have very high rates of plastic surgery with the same procedures among the most popular. And before this appears to be a fixation of young girls, plastic surgery is also rising among young Asian men.

The point is that the variation of regional understandings of beauty is changing rapidly as Western images of supermodels and other "beautiful" celebrities circulate well beyond the West through television and the Web. The implications of these European-American models of beauty entering into "foreign" cultural contexts, where other models have persisted for generations, can be quite profound, disturbing, and even disastrous for these communities. An example comes from Fiji where researchers have been studying the effect of television on cultural norms from when it was introduced in the early 1990s.

What these researchers from Harvard Medical School have found is that in 1995, when the study began, there were no reported eating disorders among girls in Fiji. By 1998, however, after a steady diet of soap operas and seductive commercials, 11.3 percent of the adolescent girls reported purging at least once as a form of weight loss. The allure of Western representations of beauty on television had a profound effect on body image and dieting practices there, where only positive body images attached to robust women (and men) had once prevailed and no one even knew what dieting was. The result has been a new culture of dissatisfaction with one's own body and a desire to remake it as a vehicle for economic success. Most worrying, 25 percent of the girls surveyed about their body image

▲ **Figure 5.E Mauritanian norms of female beauty** In Mauritania, middle and upper class women in many urban areas are considered beautiful when they they're more full-figured.

reported their negative view of their bodies had led to suicidal thoughts in the previous year.

Cultural geographies of beauty reveal not only how ideals have emerged organically, connected to the particularities of a place, but also how foreign ideals have been introduced with little connection to the already established ideals at work in that place. In all cases, it seems that transforming the face and body are expected to have profound effects on future happiness and success, though there is no research as yet that unequivocally supports such a view except in cases where plastic surgery has been used to address significant disfigurement.

---

1. How is it that we can say that norms of beauty are a geographical phenomenon?

2. How has the spread of capitalism "displaced" beauty norms and what are some of the effects of this displacement?

▲ **Figure 5.F** **Skin-lightning advertisement** The desired to lighten one's skin exists in many places around the world and for men as well as women. This product is from India.

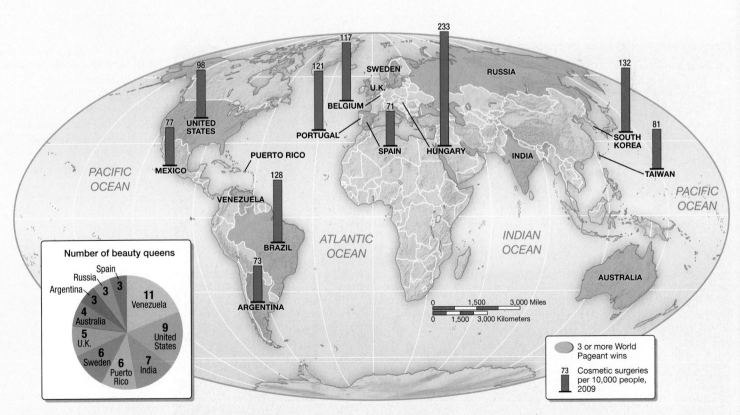

▲ **Figure 5.G** **Plastic surgery worldwide** This infographic provides a snapshot of the growing global popularity of plastic surgery and its relationship to cultural alignments with Western ideals of female beauty.

An illustration of the absence of a homogenous global culture is **world music**, the musical genre defined largely by the surge of non–English-language recordings released in the United Kingdom and the United States during the 1980s. The term is one employed primarily by the media and record stores, and it includes such diverse sources as Tuvan throat singers and Malian griots. (Throat singing is a performance style originating in south-central Russia where the vocal sounds are projected from the throat, allowing two to four tones to be simultaneously produced by the singer. Griots were ancient praise singers of West Africa and also important advisors to royalty.)

There are at least two major views on the effect of globalization on indigenous musical productions. The first emphasizes how the Western music industry has enabled indigenous music to be more widely disseminated and therefore more widely known and appreciated. This position sees local roots mixing with Western popular musical styles, with a hybrid sound resulting. The second view worries that the influence of Western musical styles and the Western music industry have transformed indigenous musical productions to the point where their authenticity has been lost and global musical heterogeneity diminished. Despite their fundamental disagreement, holders of both positions recognize that world music has enabled cultural diversity to flourish and hope that indigenous performers will be able to resist the power of the Western music industry to homogenize their work and that at the same time hybrid forms will emerge that are satisfying to a wide audience (**Figure 5.26**).

APPLY YOUR KNOWLEDGE

1. Why is homogeneity detrimental to world cultures? In what ways might the idea of "hybridity" be an alternative?

2. Can you think of other examples that either support or deny the idea of a "global culture"? What are some ways different cultures have resisted western influences and maintained their original music, fashion, food, and ideas? How might some of these cultures affect Western culture and change an "Americanization" of the globe?

FUTURE GEOGRAPHIES

In 2005, the United Nations made history by adopting the Convention on the Protection and Promotion of the Diversity of Cultural Expressions (CDPGE). The convention "is a legally binding international agreement that ensures artists, cultural professionals, practitioners, and citizens worldwide can create, produce, disseminate, and enjoy a broad range of cultural goods, services, and activities, including their own."[8] This UN convention is just one of the ways that

◄ **Figure 5.26 Bossa nova dancers** Both a musical style and a dance, bossa nova (Portugese for "new beat") originated in Brazil in the late 1950s. It faded in popularity in the 1960s and has experienced a resurgence today where it has been merged with four wall country line dancing now performed as a group, rather than in couples, as originally performed.

[8]UNESCO, http://www.unesco.org/new/en/culture/themes/cultural-diversity/2005-convention/the-convention/ (accessed June 29, 2014).

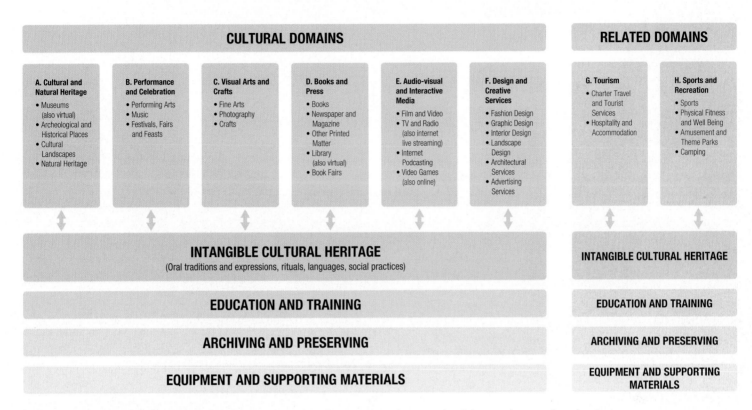

CULTURAL DOMAINS

A. Cultural and Natural Heritage
- Museums (also virtual)
- Archeological and Historical Places
- Cultural Landscapes
- Natural Heritage

B. Performance and Celebration
- Performing Arts
- Music
- Festivals, Fairs and Feasts

C. Visual Arts and Crafts
- Fine Arts
- Photography
- Crafts

D. Books and Press
- Books
- Newspaper and Magazine
- Other Printed Matter
- Library (also virtual)
- Book Fairs

E. Audio-visual and Interactive Media
- Film and Video
- TV and Radio (also internet live streaming)
- Internet Podcasting
- Video Games (also online)

F. Design and Creative Services
- Fashion Design
- Graphic Design
- Interior Design
- Landscape Design
- Architectural Services
- Advertising Services

RELATED DOMAINS

G. Tourism
- Charter Travel and Tourist Services
- Hospitality and Accommodation

H. Sports and Recreation
- Sports
- Physical Fitness and Well Being
- Amusement and Theme Parks
- Camping

INTANGIBLE CULTURAL HERITAGE
(Oral traditions and expressions, rituals, languages, social practices)

INTANGIBLE CULTURAL HERITAGE

EDUCATION AND TRAINING

EDUCATION AND TRAINING

ARCHIVING AND PRESERVING

ARCHIVING AND PRESERVING

EQUIPMENT AND SUPPORTING MATERIALS

EQUIPMENT AND SUPPORTING MATERIALS

▲ **Figure 5.27 The cultural economy** Culture is not only of significance to peoples across the globe, it is also something that the global community is interested in supporting, preserving, and showcasing.

governing entities are actively working to encourage and protect culture and creativity in a rapidly changing world (**Figure 5.27**). Given this growing commitment to recognizing and proactively appreciating the relevance of cultural diversity to the global community, it seems reasonable to predict that culture—in both its material as well as less tangible manifestations—will continue to be the focus of formal attention and support well into the future.

Importantly, connections across global space are making it increasingly possible for distant groups to share their ideas, cultural practices, and products in a way that helps both circulate and sustain them. Geographers who work with culture as their research focus are beneficiaries of these trends as they seek to understand how culture is produced and maintained and how it morphs as it is practiced and shared across distances that once would have prohibited contact.

The UN CDPGE is just one example of how cultural diversity matters in a globalizing world. One hundred and thirty-three countries as well as the European Union have ratified the CPDCE pledging to protect and foster creativity and cultural diversity. With this broad commitment to cultural diversity well-established across the globe, there appears to be, at the very least, a clear recognition of its present and future importance to social and economic development, environmental sustainability, peace, and security.

Interestingly, at the same time that nations across the globe are reaching out to protect cultural diversity, cultural majorities in the United States are collapsing. As we saw in Chapter 3, predominantly black urban neighborhoods are receiving large influxes of Hispanics at the same time that white rural communities are changing as Asian and Hispanic migrants are moving there as well. The U.S. Census projects that white Americans will soon be a large minority among many other populous minorities. It appears that in the next decade or so, Americans will all be cultural minorities as there will be no majority cultural group. Scholars have stated that there are no clear precedents in human history for such a culturally diverse place as the United States is soon to become.

APPLY YOUR KNOWLEDGE

1. What does cultural diversity mean in the context of the UN Convention on the Protection and Promotion of the Diversity of Cultural Expressions?

2. Go the U.S. Census website and identify which ethnic groups, along with white Americans, will be "large minority" groups? What effect do you believe that the absence of one majority population might have on American culture more generally?

▲ **Figure 5.28 Rainbow Bridge, Arizona** This satellite image shows in the upper right, the dramatic and mysterious Rainbow Bridge, 88 meters (290 feet) tall and 84 meters (275 feet) wide, spanning Rainbow Creek. The bridge figures significantly in several Native American cultures that have inhabited, and still do, the region around Rainbow National Monument, of which the bridge is a part.

■ CONCLUSION

Culture is a complex and exceedingly important concept within the discipline of geography. A number of approaches exist to understanding culture. It may be approached through a range of elements and features, from single traits to complex systems. Cultural geography recognizes the complexity of culture and emphasizes the roles of space, place, and landscape and the ecological relationships between cultures and their environments (**Figure 5.28**). It distinguishes itself from other disciplinary approaches, providing unique insights that reveal how culture shapes the worlds we live in at the same time that the worlds we inhabit shape culture.

Cultural geography is a diverse subfield that incorporates three general approaches. The first, traditional cultural geography, is a reflection of the work of Carl Sauer, who understood landscape as the definitive unit of geographic study. The second approach is centered on identity, ideology, power, and

meaning. Gender and sexuality, race and ethnicity, as well as age and disability are frequent empirical objects of this approach. Most recently, cultural geography has been influenced by what has come to be known as the "non-representational" approach. Cultural geographers who are non-representationalists reject what they see as static views of space inherent in more conventional approaches and turn instead to an empirical focus on embodied practices and dynamic processes and how best to comprehend them and their effects without sacrificing their dynamism.

Cultural geographers continue to embrace all of these approaches in their aims to understand how culture is the product of humans' relationships with the world around them, a world that is alive with symbols, artifacts, practices, and discourses.

LEARNING OUTCOMES REVISITED

■ Describe how place and space shape culture and, conversely, how culture shapes place and space.

A simple understanding of culture is that it is a particular way of life, such as a set of skilled activities, values, and meanings surrounding a particular type of economic practice. Geographers understand culture to be shaped by the places in which people live and make meaning from their lives. This means that social relations, politics, and the economy all play a role in the production of cultural practices by different groups in different places.

■ Compare and contrast the different ways that contemporary approaches in cultural geography interpret the role played by politics and the economy in establishing and perpetuating cultures and cultural landscapes.

Culture is not something that is necessarily tied to a place and thus a fact waiting to be discovered. Rather, the connections among people, places, and cultures are social creations that can be altered by new impulses from the economy or politics, for example, and are therefore always changing, sometimes in subtle and other times in more dramatic ways. As a result, a particular ethnic landscape may change dramatically after only a decade as the economy improves or declines and members of the group have access to additional or fewer resources that then shape their homes, vehicles, businesses, and so on.

■ Understand the ways that cultural differences—such as gender, class, sexuality, race, and ethnicity—are both products of and influences on geography, producing important variations within, as well as between, individuals and groups.

Like most social scientists, geographers understand that cultural groups are not homogeneous. All women are not alike anymore than all working-class people are. Where people live can have an important impact on their sexual identity, for instance, when they are living in a place that is homophobic.

■ Explain the conceptual changes that are taking place in cultural geography that include actor-network theory and non-representational theory.

Over the last decade, cultural geography has experienced a dramatic change in the way its practitioners think about the relationships between peoples and their worlds. These new ways of conceptualizing culture and space are still developing, but they focus on the importance of objects and material practices and how they shape the ways we experience and conduct our daily lives.

■ Demonstrate how globalization does not necessarily mean the world is becoming more homogeneous, and recognize that in some ways, globalization has made the local even more important than before.

While globalization is undoubtedly reshaping the world and bringing different cultural groups closer together than they have ever been previously, there is no conclusive evidence that globalization leads to cultural homogenization. Instead, globalization seems to be a differential process, which means that it is deployed differently in different places and experienced and responded to differently by the people who live in those places.

KEY TERMS

actor-network theory (p. *173*)
affect (p. *174*)
children's geographies (p. *172*)
cultural complex (p. *161*)
cultural geography (p. *155*)
cultural landscape (p. *157*)
cultural nationalism (p. *166*)
cultural region (p. *164*)

cultural system (p. *164*)
cultural trait (p. *161*)
culture (p. *154*)
disability (p. *172*)
ethnicity (p. *168*)
folk culture (p. *156*)
gender (p. *166*)
genre de vie (p. *160*)

historical geography (p. *160*)
intersectionality (p. *168*)
kinship (p. *165*)
materialism (p. *175*)
non-representational theory (p. *173*)
popular culture (p. *156*)
race (p. *170*)
racialization (p. *171*)

rite of passage (p. *161*)
sex (p. *166*)
transgender (p. *153*)
tribe (p. *166*)
white privilege (p. *172*)
white supremacy (p. *172*)
world music (p. *180*)

REVIEW & DISCUSSION

1. Make a list of movies that members of your group have viewed recently. Pick two of the films on which to conduct research. If possible, watch the trailers for each film. While watching the trailers, identify three different aspects of culture that are depicted. In terms of cultural geography, how do the aspects of culture you identified interact with space, place, and landscape?

2. Variations in gender identities are often the result of the different spaces in which they are enacted. In your group, discuss what this means by creating a list of how gender might be performed in different spaces by members of your group. For instance, in what spaces are you likely to project a highly feminized or masculinized identity?

In what spaces might you project an identity that is not traditionally feminine or masculine? How does the space—for example, a venue for job interview, a date, a sports competition, or a place of worship—shape your gender performance?

3. Find a description of a coming-of-age ceremony in any part of the world other than the United States, in your library or on the Internet. Summarize that description and then compare it to one you have either experienced directly or observed in the United States. What are the differences and the similarities between your experience and the one you read about? What might be some of the reasons for these?

UNPLUGGED

1. Using *Billboard Magazine* (a news magazine of the recording industry), construct a historical geography of the top 20 singles over the last half century to determine how different regions of the country (or the globe) have risen and fallen in terms of musical significance. Determine an appropriate interval for sampling—3 to 5 years is generally accepted. You may use the hometowns of the recording artists or the headquarters of the recording studios as your geographic variables. Once you have organized your data, answer the following questions: How has the geography you have documented changed? What might be the reasons for these changes? What do these changes mean for the regions of the country (or globe) that have increased or decreased in terms of musical prominence?

2. Ethnic identity is often expressed spatially through the existence of neighborhoods or business areas dominated by members of a particular group. One way to explore the spatial expression of ethnicity in a place is to look at newspapers over time. Look at ethnic change in a particular neighborhood over time by using your library's holdings of local or regional newspapers. Examine change over at least a four-decade period. To do this, identify an area of the city

in which you live or some other city for which your library has an extensive newspaper collection. Trace the history of an area you know is now occupied by a specific ethnic group. How long has the group occupied that area? What aspects of the group's occupation of that area have changed over time (school, church, or sports activities, or the age of the households)? If different groups have occupied the area, what might be the reasons for the changes?

3. College and university campuses generate their own cultural practices and ideas that shape behaviors and attitudes in ways that may not be so obvious at first glance. Observe a particular practice that occurs routinely at your college or university. (For example, fraternity and sorority initiations are important rituals of college life, as are sports events and class discussions.) Who are the participants in this practice? What are their levels of importance? Are there gender, age, or status differences in those who carry out this practice? What are the time and space aspects of the practice? Who controls its production? What are the intended outcomes? How does the practice contribute to or detract from the maintenance of order in the larger culture?

DATA ANALYSIS

UNESCO -
Interactive
Map

http://goo.gl/po0inP

In this chapter, we have looked at culture as a dynamic process, integral to our identities and interactions in space, and as a complex category of study for geographers with notions of "affect," "emotion," and "materialism." We have also investigated how global economic forces can homogenize culture and how that "Westernization" can be both internalized and resisted by non-Western cultures. Part of the resistance to homogenizing culture is preserving traditions and the unique differences all over the world, and for every culture that sacred spaces. Why do we humans signify certain spaces to symbolize our cultural notions, and why do those spaces often conjure unique experiences?

Look at the UNESCO World Heritage Site and go to the interactive map at http://whc.unesco.org/en/interactive-map/

1. Scroll over a region of the world and choose one of each of the following: a) a cultural site; b) a natural site; and 3) a site in danger. List the name of the site, what it is, and where in the world it is located.

2. Analyzing the map, are there more sites in certain areas of the world? Why might that be?

3. Next, review the latest "inscribed properties" at http://whc.unesco.org/en/newproperties/. Choose one and list what the site is, where it is located and why it is culturally significant. What do you think about these spaces?

4. Go to the UNESCO resource manual, *Managing Cultural World Heritage* at http://whc.unesco.org/en/resourcemanuals/ and skim the section, "Context: Managing World Heritage" (pp. 12–27) Then answer these questions:

   What is cultural heritage? Why is it of international importance to preserve it?

5. Finally, what place in the world do you think has cultural significance and why should it be preserved for future generations? This can be any place you know or seek to know, constructed by humans or natural; it can be the best place to skateboard, a favorite place on a trail, or a movie theatre. Why would you preserve it? What is it about that space that shapes human identity?

UNESCO -
Resource
Manuals

http://goo.gl/SqW6iK

MasteringGeography™

Looking for additional review and test prep materials? Visit the Study Area in MasteringGeography™ to enhance your geographic literacy, spatial reasoning skills, and understanding of this chapter's content by accessing a variety of resources, including **MapMaster** interactive maps, Videos, *In the News* RSS feeds, flashcards, web links, self-study quizzes, and an eText version of *Human Geography*

LEARNING OUTCOMES

- *Describe* how language both reflects and influences the way different groups understand and interpret the world.

- *Compare* and contrast different forms of communication, including standard language, slang, dialects, social media, and nonverbal modes of expression.

- *Interpret* how different geographies impact the spread or preservation of language and how different groups use language to give or change a place's meaning.

- *Describe* the global distribution of the world's religions—how they developed in specific regions and how they proliferated around the world.

- *Recognize* the difference between religions and religious movements around the world, and analyze the impacts of both on political and social life.

- *Interpret* the importance of space to religion in pilgrimages and sacred spaces in every culture.

186

6

LANGUAGE, COMMUNICATION, AND BELIEF

As an infant—as young, perhaps, as 8 months old—you had to determine the internal structure of a system that possesses tens of thousands of individual elements. Each of the elements is derived from the same collection of materials and combined into larger units. Those units can be put together into an infinite set of combinations, although only a limited set of those joined units are correct within the context of the system. How does an infant proceed? Fortunately, we tend to learn this system effortlessly: The system is language, and it is composed of words, sounds, and sentences.

But now imagine that you're a deaf child, 6 or 7 years old. You have reached this age not fully understanding what it means to be deaf. Imagine how much more difficult the mission of acquiring language will be for you. Of course, there will not necessarily be sounds involved in forming your language, but there must be something else to take the place of sound that will allow you to communicate the words and the sentences you wish to convey.

Imagine further in this already challenging scenario that it's 1970 and you live in Managua, Nicaragua, and there are no teachers at your school who know sign language. What is perhaps even more remarkable than the capacity of the hearing infant's ability to comprehend and eventually use language is the capacity of a group of deaf children, assembled in a collective but without the aid of a sign language instructor, to develop their own language so they're able communicate with each other.[1]

These children developed the Nicaraguan Sign Language. It is a unique example of how language emerges and becomes populated with a structure, words, and sentences. The deaf children created the language, not with the help of their teachers or their parents or any other adults but through their interactions with each other. Independently, they constructed a natural sign language that contains the kinds of grammatical regularities that are key to all languages. And, since the

[1]Adapted from J. R. Saffran, A. Senghas, and J. S. Trueswell, 2001, *The Acquisition of Language by Children*, Proceedings of the National Academy of Sciences, 98, 23: npn.

1970s, the language has continued to develop and change as each new generation has learned it from the previous one while adding their own structure and content.

For geographers, one particularly interesting part of this story is the way the deaf children came to develop spatial devices to signal differences. A hand signal establishes directional coordinates such as *here* and *there* or *near* and *far* or *left, right,* and *center.* And spatial coordinates are as key to the Nicaraguan Sign Language as they are to any language. Where something is happening or has happened, or how the language incorporates space, matters. As with all languages, in the Nicaraguan case, while new generations of deaf children were figuring out the code that was being passed along to them from an older one, they were also adding their own words and understandings and improving upon the language by drawing from their internal learning abilities. Language, like culture more generally, is dynamic.

This story of Nicaraguan language inventors tells us much about the human need to communicate. It also tells us that geographical context is central to communication whether through sign language or the spoken word. In this chapter we more closely examine language and communication as well as belief systems to reveal the extent and type as well as the geographical distribution and spatial significances of all three.

▼ **Figure 6.1 Pranāma** In some parts of India, pranāma, or touching feet, is an expected gesture of respect made by a young person to an older one.

GEOGRAPHIES OF LANGUAGE

Language and religion are two key elements in a cultural system. Their dynamism is anchored in communication within a group of speakers and adherents. Both are also markers of individual and collective identity and their geographies—where particular languages are spoken or particular religions are practiced. In the first section of this chapter we explore language and communication as interdependent processes. In the second section we turn to belief. In both sections we explore not only how language, communication, and religion work but also how they are distributed across a range of spaces and how space itself shapes their practice. We address ourselves to the organization, diffusion, and strength or decline of both language and religion as well as explore less-conventional forms of both.

What Is Language?

Language is a means of communicating ideas or feelings by way of a conventionalized system of signs, gestures, marks, or articulate vocal sounds (**Figure 6.1**). In short, communication is symbolic, based on commonly understood meanings of signs or sounds. It is an important aspect of culture, and without it, cultural accomplishments would be lost. As well, we would have a hard time even staying alive. In fact, language is so central to almost everything we do as humans that it's difficult to

imagine what it would be like not to have it. What if we didn't have names for things—food, help, you and me—and what if we weren't able to make statements or ask questions?

Linguists, the people who study language, believe that without concepts—words that contain meaning—it would be difficult to think. We would still experience things like lightness or cold or pain, but without the concept of pain, we couldn't communicate our condition to others so that we might get some relief. The importance of concepts to thought doesn't mean that without them it would be impossible to speak or even think; it just means that certain kinds of thinking can happen only because we have language. For instance, the Amondawa people of the Amazon rain forest have no concept of time. They have no calendars, watches, or mechanisms that measure or count the passage of time. There are only divisions of day and night and rainy and dry seasons. No one in the community has an age; instead they change their names to reflect their stage in the life cycle.

Language both reflects and influences the way different groups understand and interpret the world around them by providing each group with often distinct concepts and vocabulary as the example of the Amondawa illustrates. Every language contains the elements necessary to communicating all kinds of different thoughts. Yet what makes language so interesting is that those thoughts can be conveyed in very different ways depending on the language. Additionally, all languages have a structure that includes rules of grammar, specialized vocabularies, and systems of pronunciation.

Within standard languages, regional variations, known as dialects, exist. **Dialects** feature pronunciation, grammar, and vocabulary that are place-based. The term is most often associated with regional speech patterns. It is often said that dialects reflect the everyday lives of its speakers as they encounter new people and things. **Standard language** (also known as **official language**) is made in government offices, in the courts, and in the schools and is less alive and more artificial, though it, of course, also changes.

In the next section we will examine the distribution of languages at the global, regional, and local levels. These different distributions provide us with a sense of who is communicating how, where, and with whom through standard languages. It will also help us to appreciate where some languages are gaining ground—bringing in new speakers in places where it was not previously dominant or even spoken at all—while others are losing it as fewer individuals are learning or using the language that was previously dominant.

APPLY YOUR KNOWLEDGE

1. What is the difference between a "dialect" and a "standard language"?

2. What dialects are spoken in the United States? Has your own dialect changed as you've traveled, gained employment, gone to college? Why do you think that might be? Why are "professional spaces" the places where a more standardized form of language is required?

The World of Language

The distribution of languages can tell us much about the changing history of human geography and the impact of the movement of people, ideas, capital, and technology globally. Language experts generally agree that there are about 6,900 different languages in the world today. Very large languages tend to have millions or billions of speakers, while very small languages have fewer than 100 speakers. English and Chinese are among the very large global languages with billions of speakers. Jeru, a language of the Andaman Islands south of Burma, and Dothraki and High Valyrian, the invented languages of *Game of Thrones,* each currently have fewer than 30 fluent speakers and are considered very small languages.

Figure 6.2 provides a global snapshot of language distribution. Of course the scale is too small to see much within-country variation let alone regional or local variation. **Figure 6.3** provides a more refined view of language variation with one very large country, India. And **Figure 6.4** is a graphic map of New York City that shows the languages most often used with Twitter. Although the 2010 U.S. Census indicates that New York is the most linguistically diverse city in the world, the Twitter map shows only the tweets in the major languages spoken there; it is estimated that as many as 800 different languages are actually spoken.

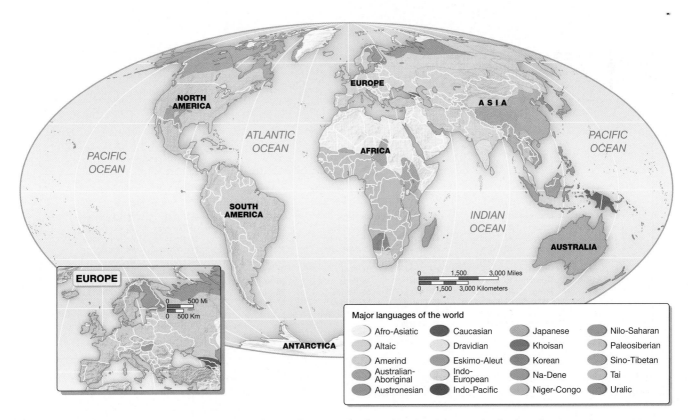

▲ **Figure 6.2 World distribution of major languages and major language families** Classifying languages by family and mapping their occurrence across the globe provide a range of insights into human geography. For example, we may begin to understand something about population movements across broad expanses of space and time.

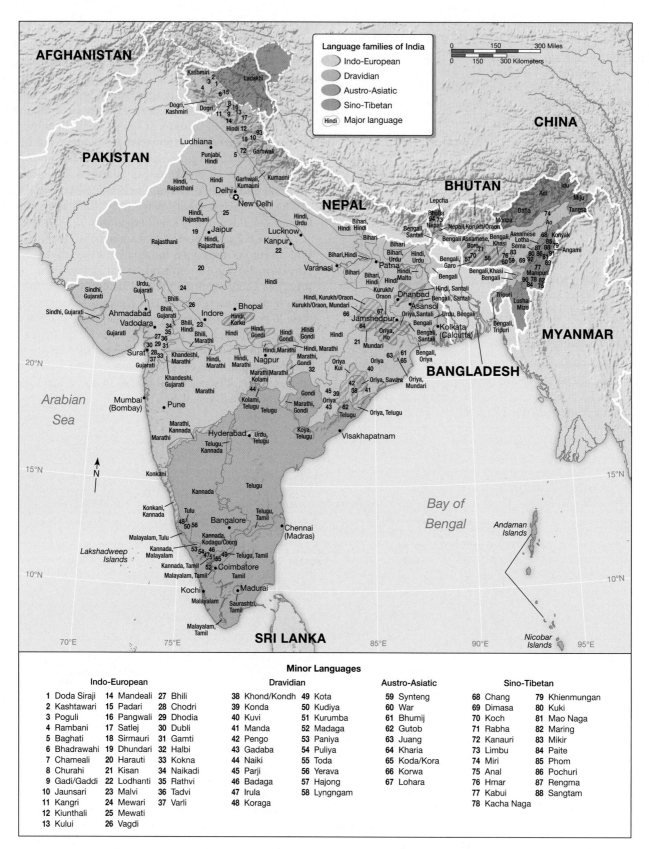

Language families of India
- Indo-European
- Dravidian
- Austro-Asiatic
- Sino-Tibetan
- (Hindi) Major language

Minor Languages

Indo-European		Dravidian	Austro-Asiatic	Sino-Tibetan	
1 Doda Siraji	14 Mandeali	27 Bhili	38 Khond/Kondh	59 Synteng	68 Chang	79 Khienmungan
2 Kashtawari	15 Padari	28 Chodri	39 Konda	60 War	69 Dimasa	80 Kuki
3 Poguli	16 Pangwali	29 Dhodia	40 Kuvi	61 Bhumij	70 Koch	81 Mao Naga
4 Rambani	17 Satlej	30 Dubli	41 Manda	62 Gutob	71 Rabha	82 Maring
5 Baghati	18 Sirmauri	31 Gamti	42 Pengo	63 Juang	72 Kanauri	83 Mikir
6 Bhadrawahi	19 Dhundari	32 Halbi	43 Gadaba	64 Kharia	73 Limbu	84 Paite
7 Chameali	20 Harauti	33 Kokna	44 Naiki	65 Koda/Kora	74 Miri	85 Phom
8 Churahi	21 Kisan	34 Naikadi	45 Parji	66 Korwa	75 Anal	86 Pochuri
9 Gadi/Gaddi	22 Lodhanti	35 Rathvi	46 Badaga	67 Lohara	76 Hmar	87 Rengma
10 Jaunsari	23 Malvi	36 Tadvi	47 Irula		77 Kabui	88 Sangtam
11 Kangri	24 Mewari	37 Varli	48 Koraga		78 Kacha Naga	
12 Kiunthali	25 Mewati		49 Kota			
13 Kului	26 Vagdi		50 Kudiya			
			51 Kurumba			
			52 Madaga			
			53 Paniya			
			54 Puliya			
			55 Yerava			
			56 Yerava			
			57 Hajong			
			58 Lyngngam			

▲ **Figure 6.3 Language map of India** It is possible to appreciate from this map the persistence of the linguistic complexity of India. A country that did not suppress regional languages when it became independent, India possesses hundreds of distinct languages.

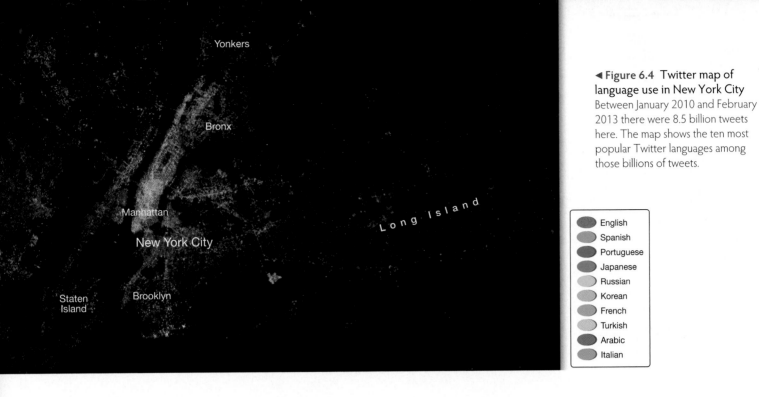

Yonkers

Bronx

Manhattan

New York City

Long Island

Staten Island

Brooklyn

◄ Figure 6.4 Twitter map of language use in New York City Between January 2010 and February 2013 there were 8.5 billion tweets here. The map shows the ten most popular Twitter languages among those billions of tweets.

- English
- Spanish
- Portuguese
- Japanese
- Russian
- Korean
- French
- Turkish
- Arabic
- Italian

The Garifuna Example Garifuna is one among hundreds of New York City languages, but it is mainly spoken in Honduras and Belize, with smaller numbers of speakers in Nicaragua and Guatemala. Garifuna is a member of the Arawakan language family and may have been brought to the Island of St. Vincent, in the Lesser Antilles, from the Orinoco River Valley in present-day Venezuela as well as from coastal Guyana, Suriname, French Guiana, and parts of the island of Trinidad. Arawakan was the largest indigenous language family in South America and the Caribbean at the time of European contact.

The Garinagu people (those who speak Garifuna) are largely descended from West Africans who were en route to South America having been pressed into slavery. Historical records suggest that because of a shipwreck, they ended up in St. Vincent. In addition to Arawak, Garifuna language contains a high number of words from Carib, another native language of South America from people who also originated along the northern coast of South America and the southern West Indies. Garifuna also contains contributions from a number of West African languages including Yoruba, Ibo, and Ashanti from present-day Benin, Ghana, Nigeria, Mali, Sierra Leone, and the Democratic Republic of Congo. Finally, Garifuna includes words and speech patterns from a number of European languages due, of course, to the arrival of Columbus in 1492. In fact, the Arawak were the first New World people Columbus and his crew encountered. Because Garifuna is influenced by so many other languages, it is known as a **fusion language** (**Figure 6.5 a & b**).

Over the centuries, the Garinagu experienced prosperity as well as war and imprisonment. They were eventually banished from St. Vincent and fled to Honduras, Belize, Nicaragua, and Guatemala. In the nineteenth and twentieth centuries, the Garinagu began migrating in large numbers to New York, Los Angeles, Chicago, New Orleans, and other U.S. cities in search of better employment and education opportunities. Today, classes in Garifuna are being offered to the grandchildren and great-grandchildren of these migrants in an effort to keep the language alive among ethnic Garinagu who are more likely to speak Spanish or English.

One notable aspect of the Garifuna language is its relationship to music, through which the oral traditions of the culture have been maintained. The language is not only spoken and written, but also sung, in its modern form. Its musical form has been described by the United Nations Educational, Scientific and Cultural Organization (UNESCO) as "a veritable repository of the history and traditional knowledge of the Garifuna [sic], such as cassava-growing, fishing, canoe-building and the construction of baked mud houses." In addition to illustrating the complexity of language change and distribution and the multiple forces at work that effect these, Garifuna also provides an example of the relevance of the different ways language is communicated. Before looking more closely at the geography of language and the impacts of globalization upon the changing distribution of languages, it is necessary to become familiar with some of the basics on how language is organized.

APPLY YOUR KNOWLEDGE

1. How is the Garifuna language also a story of human migration and power relations between people? Think about slavery, colonialism, cultural exchange, and the effect of music and dance on language. What do these human stories tell you about language development?

2. What is another language that has developed from the cultural mixing that comes from colonialism or forced migration? Come up with either a historical or current example.

191

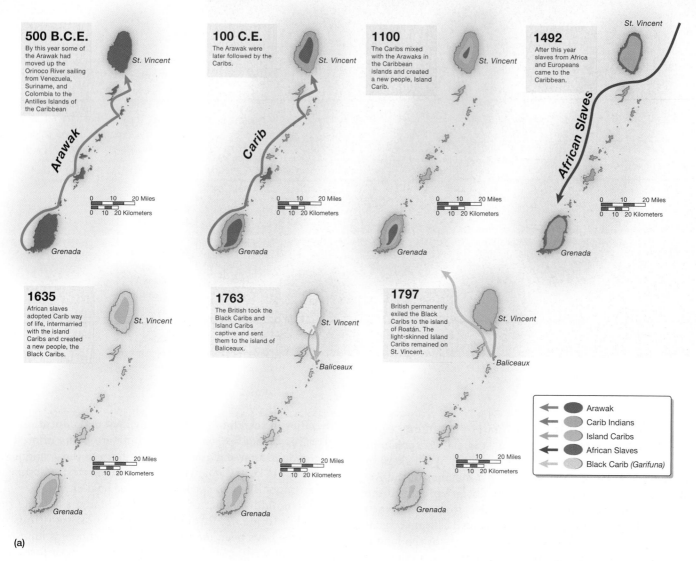

500 B.C.E.
By this year some of the Arawak had moved up the Orinoco River sailing from Venezuela, Suriname, and Colombia to the Antilles Islands of the Caribbean

St. Vincent

Arawak

Grenada

0 10 20 Miles
0 10 20 Kilometers

100 C.E.
The Arawak were later followed by the Caribs.

St. Vincent

Carib

Grenada

0 10 20 Miles
0 10 20 Kilometers

1100
The Caribs mixed with the Arawaks in the Caribbean islands and created a new people, Island Carib.

St. Vincent

Grenada

0 10 20 Miles
0 10 20 Kilometers

1492
After this year slaves from Africa and Europeans came to the Caribbean.

St. Vincent

African Slaves

Grenada

0 10 20 Miles
0 10 20 Kilometers

1635
African slaves adopted Carib way of life, intermarried with the island Caribs and created a new people, the Black Caribs.

St. Vincent

Grenada

0 10 20 Miles
0 10 20 Kilometers

1763
The British took the Black Caribs and Island Caribs captive and sent them to the island of Baliceaux.

St. Vincent

Baliceaux

Grenada

0 10 20 Miles
0 10 20 Kilometers

1797
British permanently exiled the Black Caribs to the island of Roatán. The light-skinned Island Caribs remained on St. Vincent.

St. Vincent

Baliceaux

Grenada

0 10 20 Miles
0 10 20 Kilometers

Legend:
- Arawak
- Carib Indians
- Island Caribs
- African Slaves
- Black Carib (*Garifuna*)

(a)

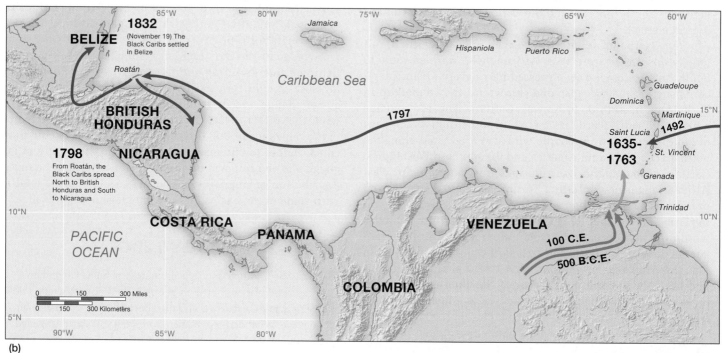

1832
(November 19) The Black Caribs settled in Belize

BELIZE

Roatán

BRITISH HONDURAS

1798
From Roatán, the Black Caribs spread North to British Honduras and South to Nicaragua

NICARAGUA

COSTA RICA

PACIFIC OCEAN

PANAMA

COLOMBIA

VENEZUELA

Jamaica

Caribbean Sea

Hispaniola   *Puerto Rico*

Guadeloupe

Dominica

Martinique   15°N

Saint Lucia   **1492**

1635–1763

St. Vincent

Grenada

Trinidad

1797

100 C.E.
500 B.C.E.

0 150 300 Miles
0 150 300 Kilometers

(b)

▲ **Figure 6.5 Garifuna language migration and influences** This timeline (a) and map (b) provides insight into the movement of the people who are now known as the Garinagu. The various colors shown in the legend help to demonstrate how the language changed over time and the ways it was influenced by other languages until the period before the expulsion of the Garinagu from St. Vincent.

LANGUAGE RELATIONSHIPS AND DYNAMICS

To delineate the relationships among and between the thousands of languages spoken across the globe, language specialists classify them largely with respect to their connections. Most languages are related to one or more other languages so that the thousands of languages that exist globally can be classified into 90 different language families, six of which are considered major language families. Families are further divided into branches and groups (**Figure 6.6**), and their relationships are graphically represented as trees.

Language Family

A **language family** is a collection of individual languages believed to be related in their prehistoric origin. About 50 percent of the world's people speak a language that belongs to the Indo-European family. Although part of a family, a **language branch** is a collection of languages that possesses a definite common origin but have split into individual languages. A third category used to show the structure of language is a **language group**, a cluster of individual languages that are part of a language branch, have shared a common origin in the recent past, and have relatively similar grammar and vocabulary. Spanish, French, Portuguese, Italian, Romanian, and Catalan are a language *group,* classified under the Romance *branch* as part of the Indo-European language *family.* Some language families contain more than 1,000 languages and others contain only a few.

Language Trees

A **language tree** is a representation of the relationships of languages to each other. The tree model is meant to mimic a family tree with languages substituted for real family members. The form of the language tree is a node-link diagram that contains branch points, or nodes, from which the daughter languages—offspring of older languages—descend by different links. The nodes are **proto-languages**, also known as common languages. In conceptualizing the descent of language, linguists intend to show that a linked language is an offshoot established through a process

▼ **Figure 6.6** Iranian language tree Iranian languages are spoken in Iran, Afghanistan, Tajikistan, and parts of Iraq, Turkey, Pakistan, and scattered areas of the Caucasus Mountains. There are 87 Iranian languages, shown in black letters with living languages underscored by green lines. The dialects are shown in grey letters. The extinct languages are shown with brown lines.

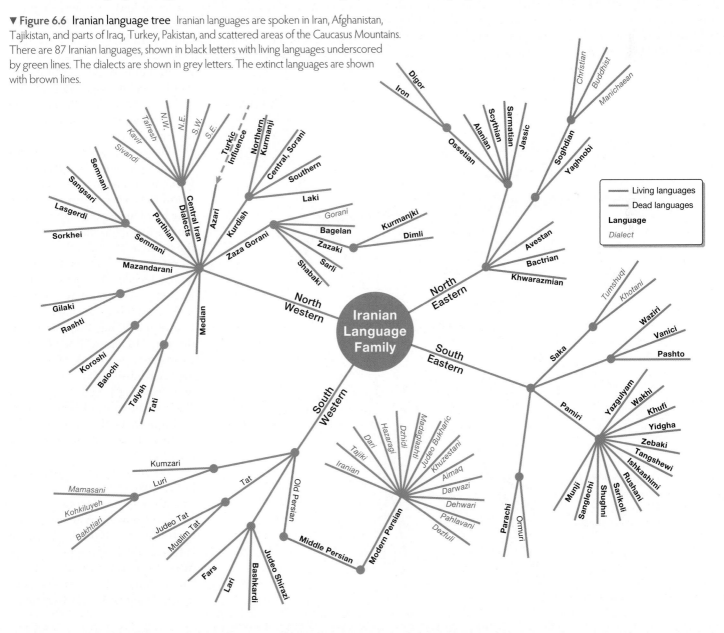

of gradual modification over time. According to the family tree model, languages are thought to have a kind of genetic relationship to those languages in which they share a line of descent.

If you were to use a map in conjunction with the tree model so that the language at the end of each node was connected to the corresponding country, you'd see that geography is central to understanding language transformation, you'll see that geography is often central to understanding its transformation and the location of its speakers. Take the node just below North Germanic in **Figure 6.7**. Old Norse, Old Swedish, and Old Danish were produced by **linguistic drift**, a process of random change inherent to all languages. Though the speakers would appear to be geographically close in modern terms, in the period of language divergence, communication would have gradually become more difficult as people migrated or somehow became disconnected over time from their original community.

Before a language can change, speakers must adopt new words, sentence structures, and sounds; spread them through the community; and transmit them to the next generation. It is widely accepted that children are usually the most active agents of language change. When learning the language of their parents, they internalize the language differently and, especially with advances in communication media over time, form and disseminate a variation of that language.

The Icelandic Example

A contrasting example of language change to Garifuna is Icelandic, one of the most unchanged and purest in the world. It is the **mother tongue**—a language that a person has learned from birth or the first few years of life—of about 320,000 people, most of whom live in Iceland. Icelandic, a direct descendant of Old Norse, is what the Vikings would have spoken (between about 800 C.E. and 1050 C.E.). Icelandic belongs to the Western Scandinavian language group. Of all the languages in the Old Norse branch, Icelandic has changed the least in the past 1,000 years.

This relative absence of change is because of the somewhat remote location of the Icelandic people, who were able to maintain their language because there were few external forces of change; as well, it was culturally dominant; with respect to the other languages they encountered. Icelandic has changed so little that contemporary Icelandic speakers could have a completely intelligible written correspondence with a Viking—a 1,000-year-old person! That is because the spelling, semantics (the meaning of a word or phrase), and word order vary only slightly. To appreciate how amazing this is, consider the experience of reading Shakespeare in Early Modern English, the language in which he wrote. Some of you will have had that experience already in a course and know what a struggle it is. Most students need a translation sheet to accomplish this.

Recall that Early Modern English was in use between 1450 C.E. and 1600 C.E. whereas Old Norse is much older, in use between 800 C.E. and 1200 C.E..

Old Norse is part of the Indo-European language family. It follows the Germanic branch and the North Germanic group as well as daughter languages, including Icelandic. English is also a branch of the Germanic family, though it is an offshoot of the West Germanic branch (**Figure 6.7**). Some common English words that have come to us from Old Norse include awe (as in awesome!), bug (the insect, not the verb "to annoy"), fog (which, incorrectly, we think of as thoroughly English, as in London fog), hell, trust, ugly, and many more. The reason we find these words in English is that the Vikings colonized eastern and northern England.

When speakers of different but related varieties of languages are able to understand each other, the languages are said to be **mutually intelligible**. Mutually intelligible languages are frequently geographically close. Those languages that are part of the Modern North Germanic

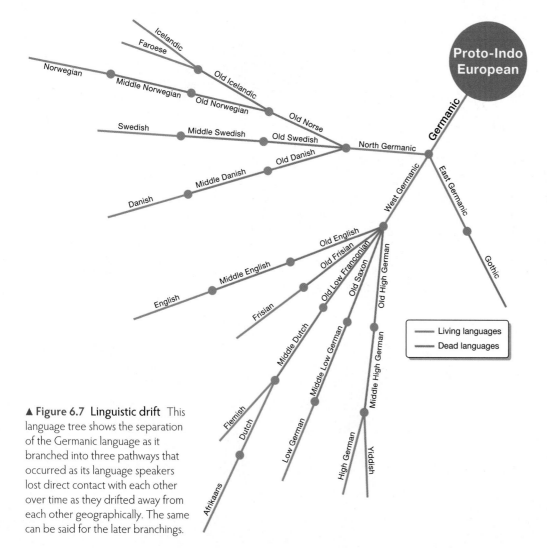

▲ **Figure 6.7 Linguistic drift** This language tree shows the separation of the Germanic language as it branched into three pathways that occurred as its language speakers lost direct contact with each other over time as they drifted away from each other geographically. The same can be said for the later branchings.

language branch—including Faroese, Norwegian, Danish, and Swedish as well as Icelandic—are mutually intelligible.

APPLY YOUR KNOWLEDGE

1. How does geography affect language development and/or preservation? Think about the Icelandic example and compare it to how other Germanic languages have developed in different parts of the world.

2. Can you find other examples of a language that has been preserved as well as one that has developed through spatial interactions?

Ultraconserved Words Linguists believe that most words disappear from the human vocabulary after 8,000–9,000 years. This occurs because of the adoption of new words as well as **linguistic weathering**—a sort of "wearing out" of words. But recently, linguists have found that about a dozen words in use today are about 15,000 years old. These are words that have survived while the languages of which they were a part have died out. The following speech assembles all the words linguists believe have survived for 150 centuries:

> You, hear me! Give this fire to that old man. Pull the black worm off the bark and give it to the mother. And no spitting in the ashes!

Where these ultraconserved words originated geographically is the subject of much controversy. There is some agreement among experts, however, that they are from a language that existed in places where people survived the last glacial period around 20,000–25,000 years ago. This area is thought to be in Eurasia.

Official Language (Standard Language) Linguistic experts often think of colonialism as the force behind the erasure of local languages. A closer look, however, reveals that the modern state and capital are the institutions likely to be the force behind language loss through the promotion of official language. Recall that official language is the language in which the government, including the courts, the legislature, and the administrative branch, conducts its business. To ease the way for economic growth and official government business, such as tax collection, a governing body requires a reduction in the number of languages existing in any territory. For example, as early as the sixteenth century, the highly centralized French State began to actively discourage the use of regional languages and dialects in official transactions. Following the overthrow of the

monarchy in France in the late eighteenth century, the new French republic more actively advanced a policy intended to establish unity among the various provinces by suppressing the regional languages. The multiplicity of languages was seen as a barrier to stable democracy and egalitarianism.

The argument for such a policy was that free people, meaning those who were no longer subjects of the monarch, must speak the same language (north-central, or Parisian French) to unify France and promote democracy as a way of life. Otherwise how could the people create and operate a government if they could not speak to each other? As a result, the regional languages and dialects of France went into a decline, hastened by official government policies spanning an extended period from Napoleon to de Gaulle. **Figure 6.8** shows the distribution of regional languages and dialects in France before the French Revolution.

In the early 1980s, the French government, in a reversal of nearly 400 years of regional language and dialect suppression, reassessed its official language policies. Responding to the emergence of the European Union and fearing the erosion of French culture more generally, the French government now regards regional languages as a treasure. In the early 1990s, the French government began financing bilingual education in

▼ **Figure 6.8 Regional languages and dialects of France** French is the official language of France, but several regional languages are also spoken as a secondary language after French. The government is required by law to communicate to the population in French.

state schools in regions with indigenous languages. Now, the revival of regional languages is viewed by the government as a way of moderating the homogenizing forces of globalization.

Language Hearths

Traditional approaches in geography have identified the source areas of the world's languages and the paths of diffusion of those languages from their places of origin. Carl Sauer labeled the geographical source of certain cultural practices as "cultural hearths." Cultural hearths, discussed in Chapter 5, are the geographic origins of innovations, ideas, or ideologies. Language hearths are a subset of cultural hearths; they are the source areas of languages. **Figure 6.9** depicts the hearth area of the Indo-European language family which possesses over 3 billion speakers.

Place Names One way that language and geography come together is in **place names (toponyms)**. The names are a kind of living record of the previous occupants of a place and their ways of doing things. They also can reflect migrants' longings for their homeland, a group's attitudes toward the environment, as well as a range of other emotions including humor, revenge, cynicism, and even reverence. Mostly, however, place names are a reflection of the kinds of economic activities that once occurred there such as the Meatpacking District in New York City or the ubiquitous Bank Street. They can also reflect colonial insensitivities in a new land.

In the United States, for example, many Native American groups have appealed to the U.S. Board of Geographic Names to have the word "squaw" removed from over 800 U.S. geographic landforms such as Squaw Peak, Squaw Lake, and Squaw Mesa. These groups argue that in a range of Indian languages, squaw is a pejorative term for vagina, though there is controversy about the derivation of the word among linguists. The U.S. Board has chosen not to make any changes despite the many appeals. In contrast, the city of Phoenix was moved by the appeal from local Indigenous people to change the iconic landform Squaw Peak (originally Squaw Tit Peak) within its urban boundaries to Piestewa Peak (**Figure 6.10**).

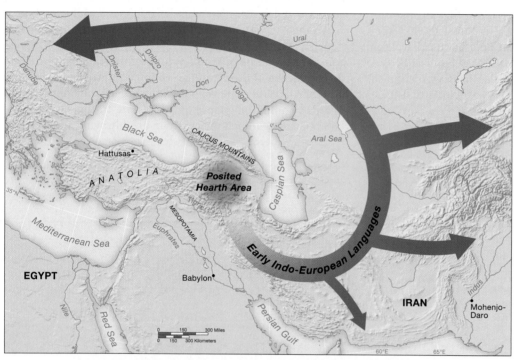

▲ **Figure 6.9 Hearth area of the Indo-European language family** Language, like other manifestations of culture, are thought to have originated in an area and then spread, carried most often by migrating populations.

As place names, these words remind us of how powerful language can be—even a single word. It is interesting that while the federal government has been reluctant to remove the word *squaw,* derived from the Algonquin (Indian) language, when it is used as a toponym on federal lands, a city experiencing that toponym on a daily basis was far more sensitive to its offensiveness and able to remove it from the urban landscape at the same time that it was able to commemorate a Native American woman—Lori Piestewa, a member of the Hopi nation who died in combat in the Iraq war—in a respectful and honorable way.

▲ **Figure 6.10 Piestewa Peak, Phoenix, Arizona** Lori Piestewa was the first Native American woman in history to die in combat while serving with the U.S. military. Piestewa Peak was named after her to honor her bravery.

Identity and Language

India presents an interesting illustration of a country where state-mandated language standardization is relatively minimal compared to countries such as Germany, where 95 percent of the population speaks German. Over 1,600 languages are spoken in India representing four major language branches. The Indo-European family, introduced by Aryan herdsmen who migrated from central Asia between 1500 and 500 B.C.E., is prevalent in the northern plains region, Sri Lanka, and the Maldives. This language family includes Hindi, Bengali, Punjabi, Bihari, and Urdu. Munda languages are spoken among the tribal peoples who still inhabit the more remote hill regions of peninsular India. Dravidian languages (which include Tamil, Telugu, Kanarese, and Malayalam) are spoken in southern India and the northern part of Sri Lanka. Finally, Tibeto-Burmese languages are scattered across the Himalayan region.

When India became independent in 1947, the boundaries of many of the country's constituent states were established on the basis of language, resulting in no single language being spoken or understood by more than 40 percent of the people. Efforts have been made since independence to establish Hindi, the most prevalent language, as the national language. This move, however, has been resisted by many of the states within India, whose political identity is now closely aligned with a different language.

The **lingua franca**—a common language among speakers whose native languages are different—in India is English, which serves as the link between its states and regions but is spoken by fewer than 6 percent of Indians. As in other former British colonies in South Asia, English is the language of higher education, the professions, and national business and government. Hindi and English are the official languages of the country, but English is more powerful than Hindi in terms of its wider global reach, offering opportunity for economic or social mobility.

Most children who attend school are taught only their local language, which inevitably restricts their prospects. A guard, sweeper, cook, or driver who speaks only Hindi or Urdu will likely do the same work all his life. In contrast, those who can speak English—by definition the upper and middle classes—are able to practice their profession or do business in any region of their country or in most parts of the world. India illustrates the decline or endangerment of local languages globally as more and more of the population comes to understand that learning English is their ticket to a better life (see Box 6.1: "Geography Matters: Language Revival").

Where the *majority (or officially powerful) language* provides access to economic opportunities, information, cultural products, social status, power, prestige, and other benefits and the *minority language,* however widely practiced, mainly serves to maintain ties to the past, the elderly, and the local community, the latter is likely eventually to be lost. The loss occurs over several generations, as the minority language is relevant in more and more restricted contexts and for fewer and fewer significant purposes. **Figure 6.11** shows endangered language hotspots around the world.

National Geographic Enduring Voices

http://goo.gl/vJ2XWF

APPLY YOUR KNOWLEDGE

1. What is the role of the state in creating an official language? What do you think is gained in this process? What is lost?

2. Think of an example of a toponym that is controversial, provocative, or simply funny. How is this place name linked to the community who inhabits the space? Is there a story behind the name that brings the community together or perhaps alienates some of them? How might toponyms reflect social relationships of gender, race, ethnicity, class, and culture?

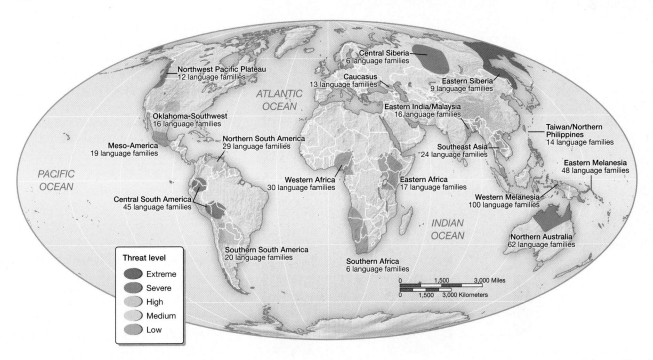

▲ **Figure 6.11 Endangered language hotspots globally** Experts estimate that in 2100 only 50 percent of the language spoken around the world today will still be spoken then. Languages have died or evolved for centuriesa but their disappearance is occurring more quickly than ever before.

6.1 Geography Matters

Language Revival

By Paul C. Adams, University of Texas at Austin

Around the world, "big" languages—like English, Spanish, Russian, and Chinese—are gaining speakers, while "small" languages—like Inupiat, Q'anjob'al, Kildin Sami, and Kangjia—lose speakers. This cultural erosion process is due to the combined effects of government policies, new communication technologies, and the countless choices made by people as they go about their daily lives. Language decline is so common that when a small language undergoes a revival, scholars pay attention.

Signs of Linguistic Revival

How do we know when a language is reviving? In recent years it has become common for organizations, political parties, churches, and academic associations in New Zealand to refer to their home as "Aotearoa New Zealand."

Driving through the Four Corners region of the American southwest, it is possible to catch news broadcasts in the Navajo language. In Wales, English is now supplemented with Welsh in school curricula, the media, and public space. The change is obvious in Quebec, where French is the predominant language on billboards, shop windows, and road signs (see **Figure 6.A**).

A language revival may involve an increase in the number of people who speak a language, but more often (and more importantly) it involves improvements in a language group's *ethnolinguistic vitality*. This term, coined in 1977 by a team with international expertise in linguistics and psychology,[i] refers to three interconnected and crucial aspects of a language: status, demography, and institutional support. When the outlook for a minority language improves, it is through any—or all—of the following: improved public perception of the language, growing numbers of speakers, and better incorporation of the language into media, schools, and government.

What Languages Are Threatened?

Many of the world's threatened languages are minority languages. This means they are a mother tongue for well under half of a particular country's population. Often these are indigenous languages, like Native American and First Nations languages in North America, the Aboriginal languages in Australia, and Māori in New Zealand.

In some cases the majority language in a country can be threatened with a loss of status and then display a comeback. During the Soviet era, citizens in Eastern Europe were expected to speak Russian, effectively making Russian the region's *lingua franca*. But in the post–Soviet era, state leaders in Eastern Europe have worked to reestablish the ethnolinguistic vitality of their various state languages. In 2007, for example, Ukraine's Constitutional Court ruled that all foreign language films must be dubbed or subtitled in Ukrainian, symbolically demoting Russian to the same subordinate status as any other foreign language.

◀ **Figure 6.A** The office of the French language (*L'office de la langue française*) has gone to great lengths to erase the English influence from the landscape in Quebec. Stop signs say "ARRÊT," although in France they say "STOP." Note the Quebec flag in the background and the term "guichet automatique" in the window rather than "ATM."

◀ **Figure 6.B** Basque Country (*Euskal Herria*) claiming political autonomy from France and Spain. Photo courtesy of Scott Larsen. Giles, H., Bourhis, R. Y., & Taylor, D. M. (1977). Towards a theory of language in ethnic group relations. *Language, Ethnicity and Intergroup Relations*, ed. H. Giles, pp. 307–348. London: Academic.

Linguistic Contestation

Where are the battles fought for the status of endangered languages? One place is in the streets. Hundreds of people have lost their lives to bombings and assassinations related to the independence struggle of *Euskal Herria*, or "the land of the Basque language," a region at the western end of the Pyrenees (see **Figure 6.B**). In striking contrast, only one life has been lost in Quebec's struggle for *souveraineté*. There, the battle has been fought in the Parliament of Canada, the National Assembly of Quebec, and the Canadian Supreme Court. Quebec's Law 101 was particularly important as it made French the official language of government, schools, businesses, and public places. This law demoted English from the position of official language. Across the Atlantic, Latvian leaders took inspiration from the success of Law 101 in Canada, and passed a similar law establishing Latvian as the official state language of Latvia while demoting Russian.[ii]

Annual festivals and gatherings are also important settings for the revitalization of endangered languages. These are particularly important for groups that are widely dispersed or split by national borders. For example, the Riddu Riddu Riđđu festival, held every summer since 1993 in northern Norway, brings together the indigenous Sami people of Norway, Sweden, Finland, and Russia. This festival includes Sami language classes, yoiking (an indigenous oral tradition), youth camp, literature, and storytelling.

Why Protect Endangered Languages?

From the perspective of a speaker of one of these languages, the stakes are clear: protecting a language preserves the heart of a culture. For others, the reasons may be a bit more abstract. The European Union offered protection to Europe's minority languages with a 2001 resolution stating that: "all European languages are equal in value and dignity from the cultural point of view and form an integral part of European culture and civilization." The United Nations has passed a number of resolutions aimed at protecting minority languages as a foundation for civil rights and as a way of safeguarding humanity's "intangible cultural heritage."

Competing with English

How do these changes relate to the rise of English as an international lingua franca? Careers in science and technology, international business, medicine, or public policy increasingly require fluency in English. While English speakers have the luxury of being understood all over the world, the rise of English complicates the preservation of endangered languages. The task of learning English cuts into the time that the speakers of these endangered languages could spend transmitting their languages to the next generation. The necessity of learning English also reduces the time spent perfecting skills to meet the local standards of literacy in national languages other than English. It is not unusual for college-bound students in minority language regions to be trilingual, but mastering the lingua franca of English can take a toll on the ethnolinguistic vitality of many minority languages.

1. What does it mean to say that a language is endangered? How can endangered languages be revived?

2. Why do regional languages experience endangerment and what does it matter if they die out?

[ii]Schmid, C., Zepa, B., & Snipe, A. (2004). Language policy and ethnic tensions in Quebec and Latvia. *International Journal of Comparative Sociology, 45*(3–4), 231–252.

COMMUNICATION

Language lives through practice, and without that language slowly dies. There are cases, such as in France, where languages on the brink of extinction have been brought back to life. Language revival, as the Geography Matters box discusses, happens because the standard practices that constitute language—speaking, writing, reading, and so on—are supported through government programs or a concerted, organized effort by a language community that puts the language back into use.

The most common and original way that language is communicated is verbally through speech. Most of us are surrounded by speech inside the womb and as infants. Typically, this early exposure to the spoken word is the pathway for our entrance into our own language community. Geographers are interested not only in where different languages are spoken—whether across the globe or across town—but also in the question of how that language is understood and deployed by speakers and how children's language acquisition varies across class and geography.

Geographies of Language Learning Researchers note that in white middle-class homes in Australia, norms and expectations about how children participate in social situations are different from those in the nearby region of rural and lower-class Papua New Guinea. In the Australian homes, mothers believe that infants possess intentionality and have the ability to express it, whether in simple sounds or full words. When teaching children to speak there, mothers engage in intense face-to-face interactions and treat their children as competent speaking partners. In parts of Papua New Guinea, however, mothers believe they must speak for their children and so turn them to face outward toward other members of the household. When those members of the household address the infant, their mothers respond by speaking for them in a high-pitched voice that is different from their own speaking voice. Importantly, both groups' children learn to speak their language effectively despite the different ways they are introduced to them. These examples demonstrate that children acquire language in culturally explicit and implicit ways.

Research also has shown that language acquisition is key to a child's future success in school. In the United States, extensive studies of children who have limited speaking interactions in early language acquisition have shown them to have a reduced vocabulary and inhibited intellectual development. This situation can occur in households in the United States where a single, working parent has limited time to spend with a growing infant. The communication of language and the acquisition of language-speaking skills vary on any number of axes, not only around the world but even within the same city or town.

Writing and Reading Language is also communicated through the written word. Books, newspapers, advertisements, text messages, greeting cards, tweets, instagrams, posts, e-mails, and Web pages, all are the products of written language and require an ability to read letters, words, sentences, and even images through a particular language structure. Written communication depends not only on the ability to relay information using an understood language system but also on an equivalent ability to read that information. Increasingly, of course, electronic translation software is enabling a kind of mutual intelligibility, even between languages that have nothing in common.

Those unable to read or write are said to be **illiterate**. Access to written information is often central to social and economic success, so the limitations of literacy can have a profound impact on an individual. Illiteracy is a problem both in poor countries in the periphery with limited access to schooling as well as in wealthy countries in the same spaces where literacy among middle- and upper-income populations is very high. Box 6.2: "Spatial Inequality: Geographies of Literacy" illustrates the geography of this worldwide problem, which tends to affect females more than males.

The Story of Writing The story of writing parallels the story of language but does not have the same plot, and its cast of characters and key events come from different places in human history. The Sumerians—who lived in Mesopotamia, which is effectively modern-day Iraq—are believed to be one of the earliest people to establish a writing system, known as **cuneiform**, named for the wedge shape of its letters. The Sumerians' writing system was originally a series of scratches in soft clay, done first by hand and later with a stylus. These scratches were **pictograms**, pictures meant to represent words. Early Sumerian cuneiform was comprised of approximately 1,500 pictograms, each of which represents a thing or an idea or even a sound, so that a picture of a bird is the same as the word for bird. While the Sumerians were among the first to develop a written language, other cultures developed their own, including the Egyptians with their hieroglyphic inscriptions, a written language also based on pictograms, but more abstract than cuneiform (**Figure 6.12**).

Researchers believe that the first alphabets were developed in Egypt. That alphabet was soon improved by the

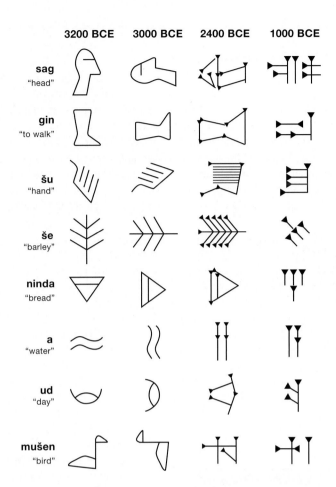

	3200 BCE	3000 BCE	2400 BCE	1000 BCE
sag "head"				
gin "to walk"				
šu "hand"				
še "barley"				
ninda "bread"				
a "water"				
ud "day"				
mušen "bird"				

▲ **Figure 6.12 Pictograms** An example of early writing, pictograms represent a thing or an idea. Pictured here are both the sound and the pictogram representing a word. As you can see, the pictograms become more abstract over time.

▲ **Figure 6.13 Roman graffiti** In Pompeii, graffiti is often of gladiators as shown here. This graffiti was found on the walls of a house.

Eskaat el nizam!" ("The people/want/to topple the regime!"). Considered to be the spark that ignited the Syrian Revolution in March 2012, these boys were arrested and later tortured by soldiers enforcing a law against antigovernment speech.

APPLY YOUR KNOWLEDGE

1. Where and how did writing first develop in the world?

2. How do you think technology has changed writing? Think about content, formality, new spellings, and modes of expression (emoticons, photos with texts, sound). How have these different types of writing changed language and symbology? For example, the sign #, used to signify "number", has recently come to signify a "hashtag" (a metadata symbol that allows grouping of messages) through media like Twitter and has taken on a whole new conceptual meaning in all forms of writing.

Phoenicians, traders in the circum-Mediterranean Sea who created a network of city-states along the coasts they visited in Africa, Spain, and Sicily. The Phoenicians were responsible for replacing the pictograms with an alphabet of consonants that were easy to learn, write, and change. The Greeks introduced vowels to the alphabet. The Latin alphabet, introduced by the Romans (by way of the Etruscans), is the same one used in this book, except that the Romans had only 23 letters—omitting the letters J, U, and W.

Early writing was done on clay, stone, wax tablets, wood, metal, papyrus, parchment, or vellum, and the writing implement was a reed, chisel, quill, and eventually a broad nib pen. Writing was also done on walls. The word **graffiti**, refers to the inscriptions—largely figure drawings—scratched on walls in ancient Rome (**Figure 6.13**). Then grafitti was meant to convey philosophical thoughts or advance political rhetoric. It was also occasionally used to publicly communicate love for another. Today, grafitti is also used to express love and politics but is also frequently used to mark territory. A recent tragic illustration of the political power of graffiti is the messages painted by 15 schoolboys between the ages of 10 and 15 on the walls of buildings in Daara, Syria. They wrote: "As-Shaab/Yoreed/

Slang Because of its liveliness, language is often adapted for different purposes or new language is developed by subsets of language users such as teenagers or people who live in separate communities or who work in special sorts of jobs. **Slang** is language that consists of nonstandard words and phrases and is a common occurrence among most languages. It's also known as patios, argot, and jargon. It's possible that you are part of a subset of language users who have invented, or at the very least transmitted, nonstandard words or phrases within your own language, whether that's English or something else. As mentioned, young people are frequently the most active agents of change in a language, introducing not only new words but also new forms of language use. Today, much slang comes from rap music and gets deployed through the lyrics of tunes.

6.2 Spatial Inequality   Geographies of Literacy

At a very basic level, **literacy** is the ability to read and write. Being able to read and write allows us to determine more readily the course of our lives as we push beyond simply comprehending language and reproducing it to transforming who we are and what we are able to do in the world. UNESCO defines literacy as the "ability to identify, understand, interpret, create, communicate and compute, using printed and written materials associated with varying contexts. Literacy involves a continuum of learning in enabling individuals to achieve their goals, to develop their knowledge and potential, and to participate fully in their community and wider society."[2]

In the United States, one of the wealthiest countries on Earth, there are 44 million people—many of whom are incarcerated—who are **functionally illiterate**. Functional illiteracy means that an individual's reading and writing skills are inadequate to manage daily living or hold down a job that requires reading skills beyond a basic level. **Figure 6.C** is a variation

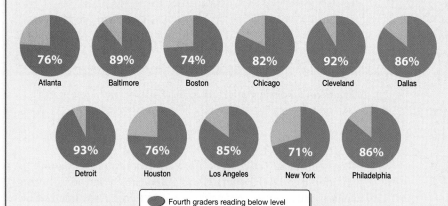

76%	89%	74%	82%	92%	86%
Atlanta	Baltimore	Boston	Chicago	Cleveland	Dallas

93%	76%	85%	71%	86%
Detroit	Houston	Los Angeles	New York	Philadelphia

◀ **Figure 6.C School-to-prison pipeline** Illiteracy among U.S. youth is the result of a combination of initial conditions but the result is often prison. One of the largest illiterate populations in the country is the prison population.

⬤ Fourth graders reading below level
⬤ Fourth graders reading at or above level

◀ **Figure 6.D Race, literacy and prison** Fourth-grade reading level is one of the measures private prison firms use to determine how large to build their prisons. Illiteracy and crime are highly related.

FROM SCHOOL TO PRISON

STUDENTS OF COLOR FACE HARSHER DISCIPLINE AND ARE MORE LIKELY TO BE PUSHED OUT OF SCHOOL THAN WHITES.

 40% OF STUDENTS **EXPELLED** FROM U.S. SCHOOLS EACH YEAR ARE BLACK.

 70% OF STUDENTS INVOLVED IN "IN-SCHOOL" **ARRESTS** OR REFERRED TO LAW ENFORCEMENT ARE BLACK OR LATINO.

 3.5 X BLACK STUDENTS ARE THREE AND A HALF TIMES MORE LIKELY TO BE **SUSPENDED** THAN WHITES.

 2 X BLACK AND LATINO STUDENTS ARE TWICE AS LIKELY TO **NOT GRADUATE** HIGH SCHOOL AS WHITES.

68% OF ALL MALES IN STATE AND FEDERAL **PRISON** DO NOT HAVE A HIGH SCHOOL **DIPLOMA.**

FROM FOSTER CARE TO PRISON

YOUTH OF COLOR ARE MORE LIKELY THAN WHITES TO BE PLACED IN THE FOSTER CARE SYSTEM.

 50% OF CHILDREN IN THE **FOSTER CARE SYSTEM** ARE BLACK OR LATINO.

 30% OF FOSTER CARE YOUTH ENTERING THE **JUVENILE JUSTICE SYSTEM** ARE PLACEMENT-RELATED BEHAVIORAL CASES. (e.g. RUNNING AWAY FROM A GROUP HOME.)

 25% OF YOUNG PEOPLE LEAVING FOSTER CARE WILL BE **INCARCERATED** WITHIN A FEW YEARS AFTER TURNING 18.

 50% OF YOUNG PEOPLE LEAVING FOSTER CARE WILL BE **UNEMPLOYED** WITHIN A FEW YEARS AFTER TURNING 18.

70% OF INMATES IN CALIFORNIA **STATE PRISON** ARE FORMER **FOSTER CARE** YOUTH.

THE COLOR OF MASS INCARCERATION AND LITERACY LEVELS

 BLACK OR LATINO **61%** OF INCARCERATED POPULATION

VS

 BLACK OR LATINO **30%** OF U.S. POPULATION

LITERACY LEVELS, U.S. ADULTS

Percent

⬤ Completely illiterate
⬤ Functionally illiterate

General population   Incarcerated adults

[2]"The Plurality of Literacy and Its Implications for Policies and Programs." UNESCO Education Sector Position Paper, 2004, p. 13.

on a widely broadcast and particularly worrying trend in the United States that a grade four reading level is the best predictor of a person ending up in prison **Figure 6.D**.

Data from the U.S. Departments of Justice and Education show that 85 percent of all juveniles who come into the court system are functionally illiterate; 60 percent of adult prison inmates are also illiterate.

At the global level, illiteracy is a particular problem for women (**Figure 6.E**). Of the 776 million adults who are illiterate worldwide, two-thirds are women. Achieving literacy is still a significant challenge for some of the world's peripheral countries where children don't stay in school long enough to secure basic reading and writing skills. Often girls never get any schooling at all and so lose out on the opportunity to establish basic literacy skills that would serve to improve their social and economic chances in life. This is because girls face discrimination and social stigma and are expected to take on care-giving duties and household responsibilities rather than attend school.

---

1. What is the difference between illiteracy and functional illiteracy?

2. Why is illiteracy higher for women than for men around the globe?

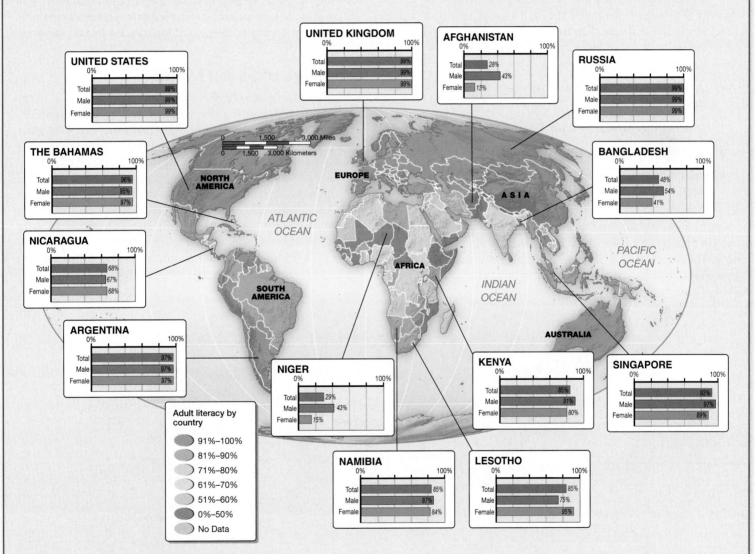

▲ **Figure 6.E Gender literacy gap** Women are more likely to be illiterate than men as girls are less likely to be educated than boys. And yet, increasing education for girls is tied to increased economic prosperity.

Since you're probably already pretty familiar with American slang, slang from Britain may be interesting as the words are also in English, but they may not have migrated across the ocean to the United States. Words such as *bare, cray, dench,* and *well jel* are common slang among English teenagers. Bare means lots of or very or exceedingly, as in "he's got bare cash." Cray, somewhat obviously, is derived from crazy and is a cross-over from rappers Jayzee and Kanye West. In England its use is really popular. And a return word from England to the United States, thanks to rapper Lethal Bizzle, is dench, meaning "remarkably good" in homage to Judi Dench, the acclaimed British actress. Well jel means very jealous, in a positive sort of way, as in when someone has a great new pair of jeans and her complimenting friend says, "I'm well jel!"

It's not only youth who create slang. Here are some slang words that were introduced through classic literature from Dickens to Shakespeare that are not only still in use today but are also quite popular. Included in the short list is the author and the date of first appearance:

- high (as in under the influence of drugs or alcohol, Thomas May, 1627)

- hang out (to spend time with, Charles Dickens, 1836)

- crib (as in dwelling place, William Shakespeare, 1623)

- unfriend (the severing of a friendship, Thomas Fuller, 1659).

Vocal Fry It's not just words, however, that characterize youth language use. **Vocal fry** is the practice of speaking in the lowest voice register to produce a popping or creaky sound at a very low frequency. It is a voice pattern that characterizes how many college-age U.S. women, currently speak. Vocal fry occurs at the end of sentences and can be almost indiscernible to the uninitiated. Among other young people, however, this slow fluttering of the voice at the end of a sentence producing a kind of rattling sound is widely recognized and appreciated. If you don't know what it sounds like, listen to Kim Kardashian or Zooey Deschanel on an audio file.

Vocal fry communicates hipness to a wider like-minded group of young women; to potential employers it signals something else. Research is now showing that women who practice vocal fry—which until recently was considered a voice disorder—are passed over for jobs. In a 2014 study, employers indicated they were less likely to say they'd want to hire the person with the fry voice, mostly because they found them to be less trustworthy. When making hiring judgments, employers preferred a normal voice 86 percent of the time for female speakers.

Language of Social Media

Social media is among the most powerful forces in communication today. It is most widely used by young people, allowing them to convey messages instantly across a wide collection of "friends," fans, and others—from businesses to universities to people in situations of strife or conflict—letting them know what is happening where they are. It is well known that social media as a communication mechanism has affected, at a very basic level, words and sentence construction. But geographers are also interested in who uses it, where they are, and in what ways they are using it. **Figure 6.14** is a map of which countries use which social networks and how many registered users are there.

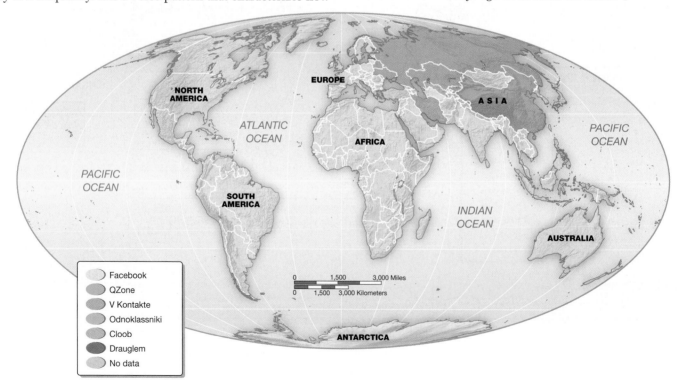

▲ **Figure 6.14 World map of social networks** Social networking is a mainstay of contemporary society around the world. Noted here are the top social networking sites in all the major countries. Facebook dominates global social networking, but there are certainly regional social networks such as VKontakte, which dominates Russia, and Qzone, which dominates China, where Facebook is less dominant.

10 Levels of Intimacy in Today's Communication

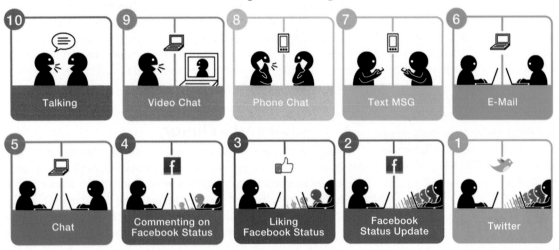

◀ **Figure 6.15 Intimacy and communication** Contemporary communication has truly diminished the level of intimacy involved between two interactants. The most intimate form of communication is face-to-face and the least is sending a tweet, where the message is significantly limited to 140 characters.

Communication is also a spatial practice that can take place at a distance or in close proximity. A letter operates at a distance while a one-to-one conversation is immediate. **Figure 6.15** depicts the different levels of intimacy involved in today's communication universe. What is missing from this graphic, however, are nonverbal forms of communication that, without a doubt, convey a language all their own.

Sign Language

Nonverbal signals account for two-thirds of all human communication. This shouldn't be surprising as the aphorism "actions speak louder than words" must surely indicate. But what if someone communicated with you using only body signals and eye contact? There is a branch of research that studies the ways tacit signals like these can be traced back to behaviors that were once critical for human survival. What's interesting about these facial and body language signals—pursed lips, crossed arms, an averted gaze—is that we still employ them today, largely subconsciously. Importantly, these signals can be read by others.

Researchers believe that our subconscious minds may register all sorts of information that our conscious minds do not access or no longer know what to do with. The television drama "Lie to Me" is premised on body and facial language with its chief protagonists being experts at reading human subconscious communication with respect to truth, fear, shame, surprise, sadness, anger, and other emotions. The show is based on the work of a practicing clinical psychologist who is himself a leading expert on lie detection. As in the show, he observes the facial expressions and bodily gestures and posture to determine when someone is lying. His research indicates that our facial expressions are innate and impossible to conceal. We communicate these micro-expressions of the language of emotion while not being aware that we're doing so.

Sensual language

There are also cultures that have no written form of communication and so, in addition to the spoken word, express themselves in other ways, often through art, song, and dance. For example, the Aboriginal people of Australia map their landscapes through painting, dance, and traditional songs. These are also known as songlines or dreaming tracks; a knowledgeable person can traverse land by repeating the words of a particular song and thus locate the landmarks and other natural sites that are described in the song. And by singing the songs sequentially, the Aboriginal people are able to move easily across vast distances in the Australian interior.

Art, and not just Aboriginal art, is also a form of communication as are dance and music. Each form contains symbols or elements meant to tell the viewer something about the performance or artwork. The modern Russian ballet, Swan Lake, for example, is as full of symbolism, body language, and communication as the Aboriginal songlines, so long as you understand the language of classical music and ballet.

Music and dance are special forms of language communication because they have the power to move us in conscious and subconscious ways. Think of the songs you love that make you want to dance, or the anthems that energize soldiers to go into battle. Music therapy among Alzheimer's patients has been successful in enabling middle- to late-stage cases—where language has deteriorated and patients are heading into speechlessness—to begin to communicate again through singing.

APPLY YOUR KNOWLEDGE

1. List three examples of nonverbal communication used around the world. How important is nonverbal communication in social relations? Why do you think we often inflate the importance of words over other ways of communicating?

2. What are five slang terms you and your colleagues use regularly? List them and trace the roots of their development? Are they from movies, music, and pop culture? Are they foreign words that people from different backgrounds are sharing? Did you read this word in social media, a news article, or a book? How does your own community of friends embrace new words and change language in the process?

RELIGIOUS GEOGRAPHIES

Religions are belief systems and sets of practices that incorporate the idea of a power higher than humans. When belief systems and associated rituals are systematically arranged and formally established, they are referred to as **organized religions**. Although religious affiliation is on the decline in some parts of the world's core regions, religious beliefs are intensifying in other parts of the world, working as a powerful shaper of daily life, from eating habits and dress codes to coming-of-age rituals and death ceremonies. And, like language, religious beliefs and practices change as new interpretations are advanced or new spiritual influences are adopted.

Global Distribution of the World's Religions

The most important influence on religious change has been conversion from one set of beliefs to another. From the Arab invasions following Muhammad's death in 632, to the Christian Crusades of the Middle Ages, to the onset of globalization in the fifteenth century, religious missionizing—propagandizing and persuasion—as well as forceful and sometimes violent conversion have been key elements in changing geographies of religion. Especially in the 500 years since the onset of the Columbian Exchange (see Chapter 4), conversion of all sorts has escalated throughout the globe. Since 1492, traditional religions have become dramatically dislocated from their sites of origin: not only through missionizing and conversion but also by way of diaspora and emigration. Whereas missionizing and conversion are deliberate efforts to change the religious views of a person or peoples, diaspora and emigration involve the involuntary and voluntary movement of peoples who bring their religious beliefs and practices to their new locales.

Migration and Religion

Diaspora is the spatial dispersion of a previously homogeneous group. The processes of global political and economic change that led to the massive movement of the world's populations over the past five centuries have also meant the dislodging and spread of the world's many religions from their traditional sites of practice. Religious practices have become so spatially mixed that it is a challenge to present a map of the contemporary global distribution of religion that reveals more than it obscures. This is because the globe is too gross a level of resolution to portray the wide variation that exists among and within religious practices. **Figure 6.16** identifies the contemporary distribution of what religious scholars consider to be the world's major religions because they contain the largest number of adherents. As with other global representations, the map is useful in that it helps to present a generalized picture.

Figure 6.17 identifies the source areas of four of the world's major religions and their diffusion from those sites over time.

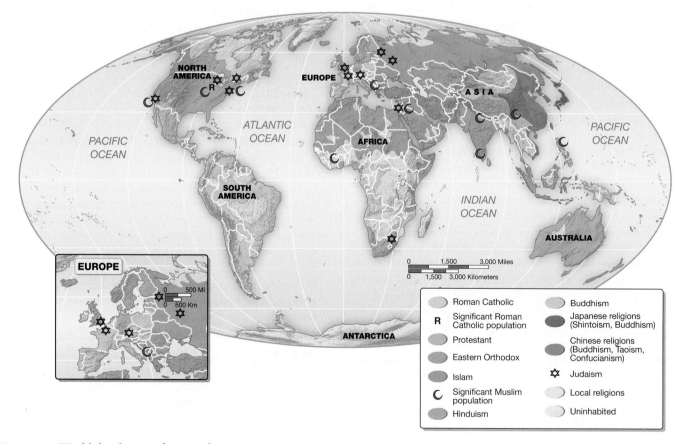

▲ **Figure 6.16 World distribution of major religions** Most of the world's peoples are members of these religions. Not evident on this map are the local variations in practice, as well as the many religions with relatively few adherents around the world.

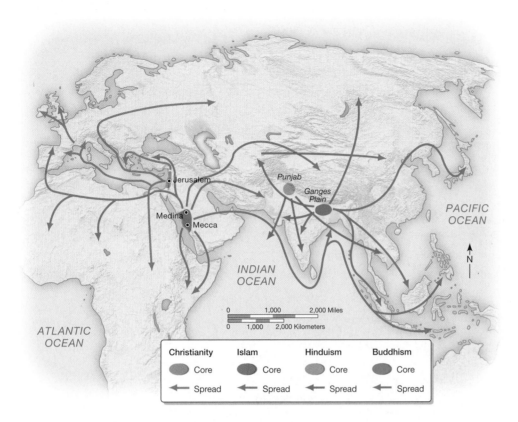

◀ **Figure 6.17** **Origin areas and diffusion of four major religions** The world's major religions originated in a fairly small region of the world. Judaism and Christianity began in present-day Israel and Jordan. Islam emerged from western Arabia. Buddhism originated in India, and Hinduism in the Indus region of Pakistan. The source areas of the world's major religions are also the cultural hearth areas of agriculture, urbanization, and other key aspects of human development.

The map illustrates how the world's major religions originated and diffused from two fairly small areas of the globe. The first, where Hinduism and Buddhism (as well as Sikhism) originated, is an area of lowlands in the subcontinent of India drained by the Indus and Ganges rivers (Punjab on the map). The second area, where Christianity and Islam (as well as Judaism) originated, is in the deserts of the Middle East.

Hinduism was the first religion to emerge, among the peoples of the Indo-Gangetic Plain, about 4,000 years ago. Buddhism and Sikhism evolved from Hinduism as reform religions, with Buddhism appearing around 500 B.C.E. and Sikhism developing in the fifteenth century. It is not surprising that Hinduism helped to produce new religions because India has long been an important cultural crossroads. As a result, ideas and practices originating in India spread rapidly. At the same time, other ideas and practices were brought to India from far-flung places and were absorbed and translated to reflect Indian needs and values.

For example, Buddhism emerged as a branch of Hinduism in an area not far from the Punjab. At first a very small group of practitioners surrounding Prince Siddhartha Gautama, the founder of the religion, were confined to northern India. Slowly and steadily Buddhism dispersed to other parts of India and was carried by missionaries and traders to China (100 B.C.E.–200 C.E.), Korea and Japan (300–500 C.E.), Southeast Asia (400–600 C.E.), Tibet (700 C.E.), and Mongolia (1500 C.E.) (see **Figure 6.18**). And as Buddhism spread, it developed many regional forms; Tibetan Buddhism, for example, is distinct from Japanese Buddhism.

The World's Major Religions

The world's most widespread religious belief systems are codified and institutionalized as organized religions. Nevertheless, it is important to recognize that each of the world's religions contains all sorts of variation. For example, Christians may be Catholics or Protestants, and even within these large groups there exists a great deal of variation.

Hinduism, the oldest religion, is codified in the Veda and the Upanishads, its sacred scriptures, but *Hinduism* is a term that broadly describes a wide array of sects. Hindus strive, through the practice of yoga, adherence to Vedic scriptures, and devotion to a personal guru, for release from repeated reincarnation. Hinduism's most central deities are the divine trinity, representing the cyclical nature of the universe: Brahma the creator, Vishnu the preserver, and Shiva the destroyer.

An offshoot of Hinduism, Buddhism was founded by Gautama—known as the Buddha (Enlightened One)—in southern Nepal. The central tenets of Buddhism are that meditation and the practice of good religious and moral behavior can lead to nirvana, the state of enlightenment. Before reaching nirvana, however, one must experience repeated lifetimes that are good or bad depending on one's actions (karma). There are four noble truths in Buddhism: (1) existence is a realm of suffering; (2) desire, along with the belief in the importance of one's self, causes suffering; (3) achievement of nirvana ends suffering; and (4) nirvana is attained only by meditation and by following the path of righteousness.

Like Hinduism and Buddhism, Sikhism advocates the pursuit of salvation through disciplined, personal meditation in

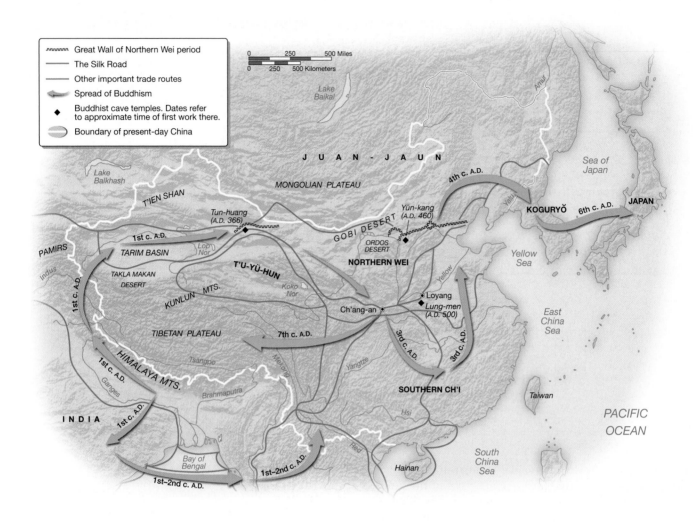

▲ **Figure 6.18 Spread of Buddhism** Buddhism diffused from its source area in India to China and then from China on to Korea and Japan. Commercial routes, like the Silk Road, were important vectors for the spread of the religion from India to China. Missionaries were responsible for the spread of Buddhism from China to Korea and Japan.

(*Source:* Adapted from C. Schirokauer, *A Brief History of Chinese and Japanese Civilizations*, 2nd ed. Florence, KY: Wadsworth, 2005.)

the name and message of God, who is regarded not as a person but as the Universe itself.

Christianity, Islam, and Judaism all developed among the Semitic-speaking people of the deserts of the Middle East. And like the Indo-Gangetic religions, these three religions are related. Although Judaism is the oldest of the three, it is the least widespread. Judaism originated about 4,000 years ago, Christianity about 2,000 years ago, and Islam about 1,300 years ago. Judaism developed out of the cultures and beliefs of Bronze Age peoples and was the first monotheistic (belief in one God) religion. Although Judaism is the oldest monotheistic religion, and one that spread widely and rapidly, it is numerically small because it does not seek converts.

Christians and Jews, along with Muslims and Bahá'í and Druze followers, share the same source as Abrahamic religions, which means that they all claim a strong connection to the prophet Abraham and the source area of the Middle East. Although sharing the same origin, Judaism is also based on the teachings of the prophets Isaac and Jacob as well as Abraham. Jews espouse belief in a God who leads his people by speaking

through these prophets. They believe that God's word is revealed in the Hebrew Bible (or Old Testament), especially in that part known as the Torah, which contains a total of 613 biblical commandments, including the Ten Commandments, explicated in another part of the Hebrew Bible known as the Talmud. Jews believe that the human condition can be improved, that the letter and the spirit of the Torah must be followed, and that a Messiah will eventually bring the world to a state of paradise.

Christianity developed in Jerusalem among the disciples of Jesus of Nazareth, who announced that he was the Messiah expected by the Jews. The religion is based on the life and teachings of Jesus as written up by various authors in the New Testament of the Christian Bible. It is the world's largest religion, with 1 to 2 billion adherents. Christians believe that Jesus was the son of God and the Messiah prophesied in the Old Testament, a scripture that is also significant to Judaism. As a central figure in all Christian religions, Jesus is revered as a teacher, the model of a pious life, the manifestation of God, and most importantly a savior of humanity who suffered, died, and was resurrected to bring about human salvation from sin. Christians believe that upon his death, Jesus ascended into heaven; that at some point he will return to judge the living and the dead; and that he will grant life everlasting to those who have followed his teachings. As Christianity moved east and south from its hearth area, its diffusion was advanced by missionizing and by imperial sponsorship and military crusades. The diffusion of Christianity in Europe is illustrated in **Figure 6.19**.

Islam is an Arabic term that means "submission," specifically submission to God's will. A **Muslim** is a member of the community of believers whose duty is obedience and submission to the will of God. Islam recognizes the prophets of the Old and New Testaments of the Christian Bible, but to Muslims, Muhammad is the last prophet and God's messenger on Earth. The Qur'an, the principal holy book of the Muslims, is considered the word of God as revealed to Muhammad by the Angel Gabriel beginning in about 610 C.E. There are two fundamental sources of Islamic doctrine and practice: the Qur'an and the Sunna. Muslims regard the Qur'an as communicated by God to Muhammad. The Sunna is not a written document but a set of practical guidelines to behavior. It is effectively a body of traditions that are derived from the words and actions of the Prophet Muhammad. Like Christianity, Islam was for centuries routinely spread by forced conversion, to save souls but also often for the purposes of political control. Inevitably, adherents of the two religions have clashed, often at geographic flashpoints, and sometimes with bloody and violent results.

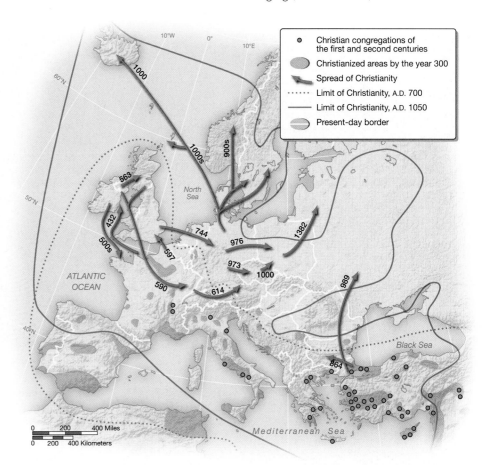

▲ **Figure 6.19 Spread of Christianity in Europe** Christianity spread through Europe largely because of missionary efforts. Monks and monasteries were especially important as hubs of diffusion in the larger network. Monasteries and other sorts of religious communities are indicated by dots on the map. The shaded areas are places where Christian converts dominated by the year 300.
(*Source:* Adapted from C. Park, *Sacred World*. New York: Routledge, 1994).

APPLY YOUR KNOWLEDGE

1. What are the world's major religions? Where did they develop? Why is the concept "diaspora" important to the proliferation of world religions?

2. How is religion like language in terms of its diffusion and potential for change? Why do you think the world's major religions are oriented around men and not women?

Established Churches and Religions

The relationship between organized religions and states varies a great deal. The First Amendment US Constitution states that "Congress shall make no law respecting an establishment of religion, or prohibiting the free exercise thereof." This separation of church and state is intended to ensure that individuals do not have to fear government interference in their manner of worship. There is a similar separation in other countries,

including Australia, France, India, and Singapore. In contrast, some countries have **established churches**: recognized by law as the official church of the state. Catholicism, for example, is the official religion in Argentina and Costa Rica, the Eastern Orthodox Church of Greece in Greece, the Anglican Church of England in the United Kingdom, and so on. Similarly, many Muslim-majority countries have established a specific form of Islam as a state religion: Sunni Islam in Afghanistan, Malaysia, and Saudi Arabia, for example, and Shi'a Islam in Iran.

Church and State in England

The influence of established churches and state religions often carries important implications for many aspects of life and can have all sorts of geographic implications and outcomes. In the United Kingdom, for example, 26 bishops of the established Church of England—the "Lords Spiritual"—sit in parliament's unelected upper chamber, the House of Lords. They are now a minority in the House of Lords, but at one time bishops, archbishops, and abbots were a majority, and the House of Lords itself was much more powerful. As a result, the Church

◄ **Figure 6.20** Cross Bones burial grounds in Southwark, London The repository for thousands of deceased prostitutes, most of whom had operated under license to the Bishops of Winchester. In the nineteenth century, Cross Bones became the haunt of body-snatchers seeking specimens for the anatomy classes at nearby Guy's Hospital. The burial ground was finally closed in 1853 as it was "completely overcharged with dead" and a danger to public health. The railings have become an impromptu memorial, adorned with messages and mementoes, ribbons, flowers, and other totems.

▲ **Figure 6.21** Planted church In 1711 Parliament had passed an Act for Building Fifty New Churches across London: part of a Tory strategy to maintain the Church of England as an arm of the state. The Act allocated money from the Coal Tax to the commission for Building Fifty New Churches, and the commission's design guidelines laid stress on imposing exteriors and impressive entrances for the new churches. Only 12 churches were ever built under the Act—Christ Church was one of them. The architect was Nicholas Hawksmoor.

and its leadership acquired a great deal of power, wealth, and influence. In medieval England, for example, parts of many of the bishoprics (the territories under the leadership of a bishop) functioned as civil jurisdictions with legal powers. Sometimes this led to exploitation. For several centuries the Bishops of Winchester enriched themselves by licensing prostitutes in the part of south London that they controlled. The bishops nevertheless did not allow the women to be buried on consecrated ground. Prostitutes ended up in Cross Bones burial grounds (**Figure 6.20**), along with paupers and other unlicensed sinners.

In the eighteenth century, the influence of the established Church was imprinted on the townscape and social geography of London as parliament passed a succession of Acts to pay for new churches. Worried that the population of the rapidly growing city was becoming increasingly nonconformist, and with an influx of migrants with different religious affiliations, Parliament sought to secure London's burgeoning suburbs for the established Church by dominating new neighborhoods with impressive Anglican churches (**Figure 6.21**).

Islam as a State Religion Where Islam is the state religion, various interpretations of *sharia* law—based directly on precepts set forth in the Qur'an and the Sunna—control not only people's private comportment but also their public behavior, including their dress, diet, education, and entertainment. As a result, *sharia* law is a fundamental influence on the economic, social, cultural, and political geographies of Islamic countries. As we shall see in Chapter 12, for example, Islam's emphasis on personal privacy and virtue, on communal well-being, and on the inner essence of things rather than on their outward appearance has shaped the layout and design of traditional Islamic cities.

Sharia is one aspect of religious law that dictates the practices and behaviors of its adherents. Box 6.3: "Visualizing

Geography: Belief Systems and Restrictions on Behavior" provides a wider view of the ways that religion shape human behaviors as well as spirituality.

Religiosity

It is important to remember that not everyone adheres to a religious faith of any kind, organized or otherwise. Globally, an estimated 18 percent of the population is unaffiliated with any religion. There is, however, a great deal of variation. Surveys have shown that in many countries a majority of people say that they believe there is a god and that religion is important in their personal lives. Religiosity is highest in countries like Indonesia, Tanzania, Pakistan, and Nigeria, where more than 90 percent of the population believe there is a god and say that religion is important to them. In contrast, barely half of the citizens of the European Union believe there is a god. Fewer than 30 percent of the population in Denmark, France, and the Netherlands believe there is a god, while in the Czech Republic, Estonia, and Sweden the figure is less than 20 percent (**Figure 6.22**). Within each country there is also a great deal of variation by region and even by neighborhood in the same city. Data from the United Kingdom's census of 2011, for example, show that in some districts in London,

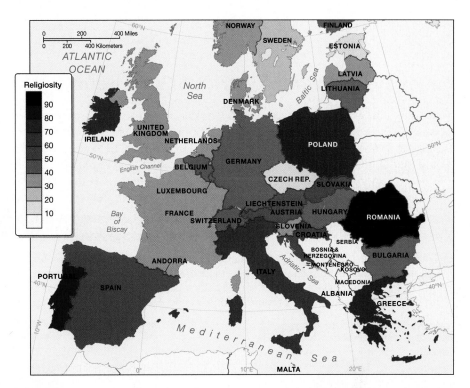

▲ **Figure 6.22 Religiosity in Europe** The map shows the percentage of people in each country who believe there is a god.

fewer than 10 percent of the population have no religion while in others the figure exceeds 40 percent (**Figure 6.23**). Similar variations exist across other UK cities.

▲ **Figure 6.23 Religiosity in London** The percentage of residents with 'no religion' varies a great deal from one district to another.

6.3 Visualizing Geography

Belief Systems and Restrictions on Behavior

Religions and belief systems can place restrictions on human behavior including diet, sexual behavior, gender roles, education, speech and health care. For example, Christians promote abstinence until marriage, which are monogamous, heterosexual unions. All of the major world religions—Judaism, Christianity, Islam, Hinduism, and Buddhism—have strong ideas about menstruation and its "negative effect" on women, leading to prohibitions of physical intimacy, cooking, attending places of worship, and even having women live separately from men at that time (http://ispub.com/IJWH/5/2/8213). The followers of these different belief systems choose to participate in these beliefs, based on personal choice or cultural influences.

6.1 Government Restrictions on Religion Around the World

When governments codify religious beliefs into state laws, these restrictions evolve into laws, enforceable by the state. For example, on Shabbat in Israel, all business and public transportation is shut down in observation of the holy day. Moreover, states that adhere to one form of beliefs can vilify or outlaw other religions, making practices adhered to by followers a social taboo or even illegal. Secular states, like many in Western Europe, restrict religious expression in spaces that are considered state spaces.

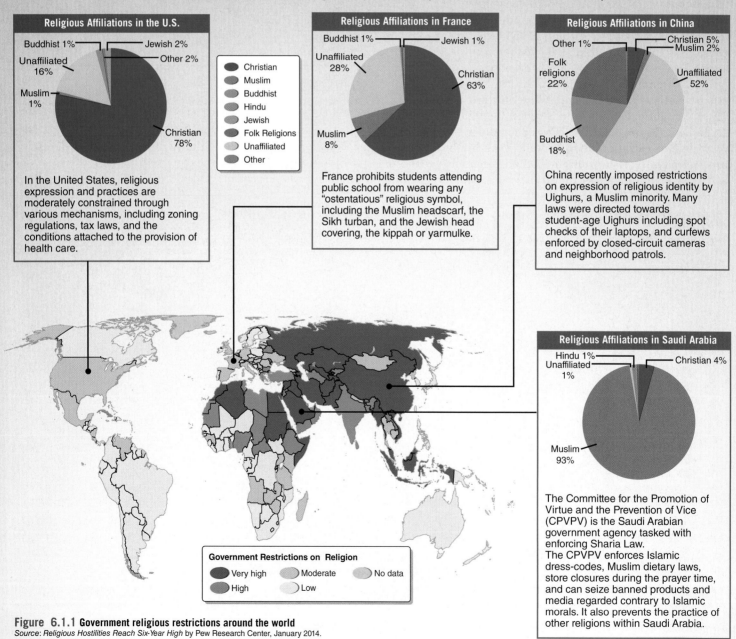

Religious Affiliations in the U.S.

Buddhist 1%
Jewish 2%
Other 2%
Unaffiliated 16%
Muslim 1%
Christian 78%

Legend:
- Christian
- Muslim
- Buddhist
- Hindu
- Jewish
- Folk Religions
- Unaffiliated
- Other

In the United States, religious expression and practices are moderately constrained through various mechanisms, including zoning regulations, tax laws, and the conditions attached to the provision of health care.

Religious Affiliations in France

Buddhist 1%
Jewish 1%
Unaffiliated 28%
Christian 63%
Muslim 8%

France prohibits students attending public school from wearing any "ostentatious" religious symbol, including the Muslim headscarf, the Sikh turban, and the Jewish head covering, the kippah or yarmulke.

Religious Affiliations in China

Other 1%
Christian 5%
Muslim 2%
Folk religions 22%
Unaffiliated 52%
Buddhist 18%

China recently imposed restrictions on expression of religious identity by Uighurs, a Muslim minority. Many laws were directed towards student-age Uighurs including spot checks of their laptops, and curfews enforced by closed-circuit cameras and neighborhood patrols.

Religious Affiliations in Saudi Arabia

Hindu 1%
Unaffiliated 1%
Christian 4%
Muslim 93%

The Committee for the Promotion of Virtue and the Prevention of Vice (CPVPV) is the Saudi Arabian government agency tasked with enforcing Sharia Law.
The CPVPV enforces Islamic dress-codes, Muslim dietary laws, store closures during the prayer time, and can seize banned products and media regarded contrary to Islamic morals. It also prevents the practice of other religions within Saudi Arabia.

Government Restrictions on Religion
- Very high
- High
- Moderate
- Low
- No data

Figure 6.1.1 Government religious restrictions around the world
Source: Religious Hostilities Reach Six-Year High by Pew Research Center, January 2014.

6.2 Social Hostilities Involving Religion

Figure 6.2.1 Government restrictions and social hostilities involving religion in the world's 25 most populous countries
The Social Hostilities Index (SHI) measures acts of religious hostility by private individuals, organizations or groups in society. The SHI includes 13 measures of social hostilities such as religion-related armed conflict or terrorism, mob or sectarian violence, harassment over attire for religious reasons or other religion-related intimidation or abuse.
Source: Religious Hostilities Reach Six-Year High by Pew Research Center, January 2014.

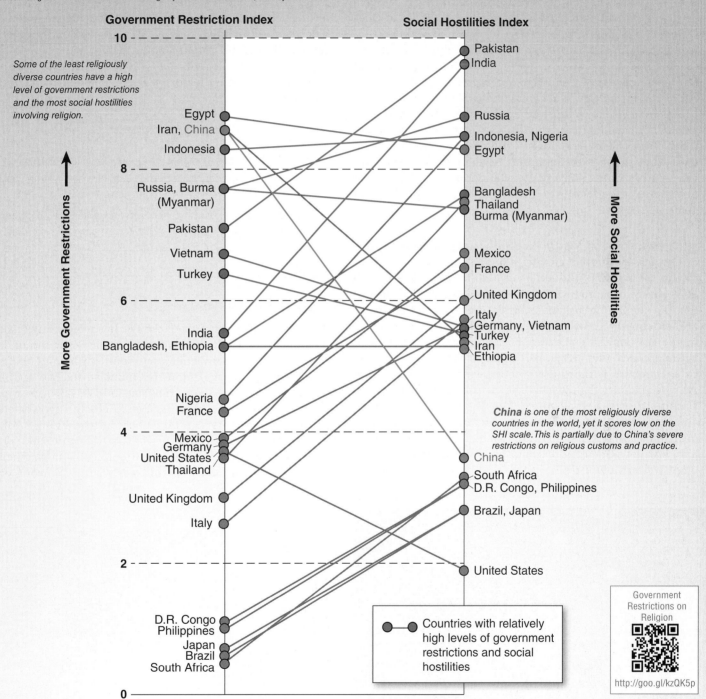

Some of the least religiously diverse countries have a high level of government restrictions and the most social hostilities involving religion.

China is one of the most religiously diverse countries in the world, yet it scores low on the SHI scale. This is partially due to China's severe restrictions on religious customs and practice.

Countries with relatively high levels of government restrictions and social hostilities

Government Restrictions on Religion
http://goo.gl/kzQK5p

1. What do you think has more influence on human behavior around the world: religious beliefs or state laws? Think about the examples in the text and ones you have read or seen in the news, your hometown, around your campus, the way laws are debated. What is the most powerful influence--church or state? Why?

2. Look up a country not discussed in the text and find out the history of religious movements in that place. For example, why are countries in South America predominantly Catholic? What religions are practiced in the Pacific islands? Where happened to ancient religions like Zoroastrianism? Compare the religious traditions to current government laws about human behavior--reproductive rights, dress, religious adherence in state spaces, sacred spaces. What do you see?

Overall, religiosity is decreasing in most of Europe. In the United Kingdom, for example, only 14 percent of respondents to a 1963 Gallup Poll described themselves not believing in a god. But a poll undertaken in 2012 for the Church of England found that more than 40 percent of respondents identified themselves as not believing in a god. One outcome is that many of the country's churches stand empty, or have been sold off and converted to apartments, galleries, or cafes (**Figure 6.24**). In France, regular attendance at Mass was already down to less than 40 percent of the Catholic population in 1961 but is now down to less than 10 percent. In several European countries, a shortage of recruits to religious seminaries has meant that some communities have had to recruit priests from Africa or Latin America. On the other hand, religiosity has been increasing in many Islamic countries (where it has always been high), Russia, and Eastern European countries since the fall of communism (where it had been very low), and in China since the relaxation of strict communist principles under the leadership of Deng Xiaoping in the 1980s and 1990s.

Wealth and Religiosity   Generally, religion plays a strong role in less-developed countries but a much less important role in the lives of individuals in high-income countries. A 2008 survey of more than 24,000 individuals in 23 countries found a very strong negative correlation between the wealth of their country, measured in terms of standardized purchasing power, and the percentage of people saying that religion is important to them (**Figure 6.25**). The clear exception to the overall pattern is the United States, which is a much more religious country than its degree of prosperity would suggest. Despite its wealth, the United States is in the middle of the global pack when it comes to the importance of religion. Indeed, on this question, the United States is closer to considerably less-developed nations such as India and Brazil than to other Western countries.

It is a matter of speculation as to why this may be so. Some observers suggest that it is because Americans feel far less secure in terms of their personal well-being than would be expected given the overall wealth of the country. Other wealthy countries have established well-developed welfare states and more comprehensive health care systems that serve to redistribute wealth and provide a safety net for the poor, reducing feelings of insecurity and anxiety that, it is argued, often provide a fertile ground for religion. Nevertheless, the cultural exceptionalism of the United States, together with its distinctive demographic profile and relatively brief history, means that there are many other factors that may explain the country's religiosity.

APPLY YOUR KNOWLEDGE

1. Where in the world is religion part of everyday life? What parts of the world are experiencing a decline in religious affiliation? What factors explain this regional phenomenon?

2. In looking at the relationship between wealth and religion, where does the United States stand? What can explain this exception? In what areas of U.S. public and private life has religion remained important?

Other Belief Systems

Along with the adherents of the world's major religions and the hundreds of millions of atheists and agnostics around the world, there are many others with quite different belief systems. Among them are animism, Confucianism, Taoism, Shintoism, Mormonism, Zoroastrianism, and Jansenism, as well as Voodoo, Wicca, Rastafarianism, and Scientology, created by science-fiction writer L. Ron Hubbard. In addition, most social systems generate **cults**: systems of veneration and devotion among a relatively small group of people having beliefs or practices regarded by outsiders as weird or sinister, or, at

▲ **Figure 6.24 Adaptive re-use of a church** An arts and crafts center in a converted Methodist church in Lynton, England.

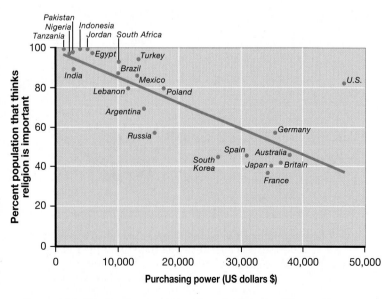

▲ **Figure 6.25 Graph of national wealth and religiosity** The wealthier the country, the less important religion is to people. The United States is an exception.

▲ **Figure 6.26 Norse mythology** A statue of a troll in the Holmenkollen district of Oslo, Norway.

least, weirder than some of the belief systems that are legally accepted as religions. One of the best-known examples of a cult from the United States was the People's Temple, widely publicized because of the mass suicide of 913 of its members who drank Kool-Aid laced with poison in their Jonestown, Guyana, settlement in 1978.

But more significant than the small and ephemeral cults that characterize modern societies are the legacies of traditional belief systems and mythologies: grand narratives that helped to give life meaning and that helped people make sense of the world. Many of them once had the standing of a religion before being eclipsed by the missionizing of Christianity or Islam. While no longer regarded as religions, their legacy is often an important part of people's cultural heritage and identity. A good example is Norse mythology, associated with the Vikings of Scandinavia and Denmark and the Germanic peoples of Northern Europe between 800 and 1000 C.E. Norse mythology was a highly sophisticated and complex amalgam of gods, goddesses, giants, poems, and sagas. While few Norwegians, Swedes, or Danes may believe today in the existence of a warrior-god Thor; of Freya, the goddess of beauty and fertility; of elves; or of trolls (**Figure 6.26**), Norse mythology is a significant element of Scandinavian folk culture and contemporary landscapes, place names, literature, and cultural identity.

GLOBALIZATION, COMMUNICATION, AND RELIGION

Changing geographies of belief systems are a significant outcome of globalization. As disparate belief systems come into contact, in some cases causing tension and even violent conflict, new religions are introduced among populations dominated by adherents of a different one. An excellent illustration of the global forces behind the changing geography of religious faith is the Columbian contact with the New World. Before Columbus and later Europeans reached the continents of North and South America, the people living there practiced, for the most part, various forms of animism and related rituals. They viewed themselves holistically, as one part of the wider world of animate and inanimate nature. Shamanism, in which spiritually gifted individuals are believed to possess the power to control preternatural forces, is one important aspect of the belief system that existed among Native American populations at the time of European contact (**Figure 6.27**).

Cargo cults are another example of the impact of modernity and globalization on traditional belief systems. **Cargo cults** involve the belief that certain ritualistic acts will lead to a bestowing of material wealth ("cargo"). They are chiefly associated with the Melanesian Islanders of the Pacific Ocean, where traditional societies were disrupted by the arrival during World War II of military forces—first Japanese, then American—with their aircraft, equipment, food, soft drinks, and cigarettes. After the military bases closed and the personnel departed, some charismatic leaders on remote Melanesian Islands sought to lure back the outsiders' goods by carving model aircraft and jeeps, lighting signal flares, and performing parade-ground style drills with wooden rifles.

▲ **Figure 6.27 San Xavier del Bac Mission near Tucson, Arizona** The influence of missionaries on native populations was profound, as indigenous rituals became mixed with Christian ones. Even today in missions such as San Xavier, native peoples practice Christianity along with components of their native belief system.

Perhaps what is most interesting about recent trends in the geography of religion is how during the colonial period religious missionizing and conversion flowed from the core to the periphery. In the postcolonial period, however, an opposite trend has occurred. For example, the fastest-growing religion in the United States today is Islam, and it is in the core countries where Buddhism is making the greatest numbers of converts. Recent Popes have been the most widely traveled leaders in Roman Catholic history. And the same can be said for the Dalai Lama, who, as the spiritual leader of Tibetan Buddhists and the head of the Tibetan government-in-exile, has been a tireless world traveler. The Popes' efforts have been mostly directed at maintaining Roman Catholic followers and attempting to dissuade their conversion to other religions, such as evangelicalism in the United States and Latin America. The Dalai Lama promotes conversion to Buddhism; he carries its message to new places, especially in the core, and advocates sovereignty for Tibet, a region once independent but currently controlled by China.

The Influence of Modern Communications

Another impact of globalization upon religious change occurs through the electronic media. One example is the rise of television evangelism, or *televangelism,* which has led to the conversion of large numbers of people to *Christian fundamentalism,* a term popularly used to describe strict adherence to Christian doctrines based on a literal interpretation of the Bible. Christian televangelism is well established in the United States and also widespread in other countries, including Brazil, Argentina, and Chile, as well as India, Kenya, and China. (And Muslim televangelism is just beginning to emerge.) The Christian reference to *fundamentalism* derives from a late-nineteenth- and early-twentieth-century transdenominational Protestant movement that opposed the accommodation of Christian doctrine to modern scientific knowledge, especially concerning origin and evolution of life on Earth. It began with the publication of a series of pamphlets between 1910 and 1915 entitled "The Fundamentals: A Testimony to the Truth."

Today, Christian fundamentalism is growing stronger in the United States with megachurches becoming popular sites of worship (**Figure 6.28**). The conventional definition of a megachurch involves Protestant congregations that share several distinctive characteristics: 2,000 or more persons in attendance at weekly worship, a charismatic senior minister, a 7-day-a-week congregational community, and multiple social and outreach activities. In 1970 there were about ten such churches in the United States; by 2014 there were more than 1,650. They have been called the "soul of the exurbs" and it is certainly the case that the most sprawling metropolitan regions—in the South, Southwest, and West—have

▼ **Figure 6.28 Megachurch, Louisville, Kentucky** Churches with congregations of 2,000 or more are widely regarded as megachurches. There are about 1,300 such churches in the United States.

6.4 Window on the World

Fashion Veiling

In the Netherlands, Islamic fashion researcher Annelies Moors explored the trend of young women, children of immigrant guest workers, moving toward wearing headscarves. Though it is important to appreciate that not all Muslim women wear headscarves, these young women, who are usually better educated than their mothers, have taken an interest in Islamic dress in response to an increasing commitment to their religion. In exploring this phenomenon, Moors interviewed three Muslim women to understand their reasons for these choices. It should be appreciated that while all the women identify as Muslim and are recognizable as such by their dress, their outfits are distinctive, reflecting their personal differences.

Feride (who is in her mid-thirties and is the daughter of working-class Turkish immigrant parents) is most comfortable in elegant suits, combinations of floor-length skirts and well-cut jackets of high-quality materials, often brought from Turkey.

Lisa (a 22-year-old college student whose father is from Pakistan and mother is a Hindu from Suriname, who converted to Islam) describes her style as casual, sporty, urban, and cool. She usually wears jeans with a tunic or a blouse over them (a headscarf), and "always a hoodie, combined with a cool bag and Nikes."

Malika (a 24-year-old college student whose parents were Moroccan immigrants) wears an outfit that shows the least variety. She wears a long, loose, all-covering dress, made by a seamstress who specializes in such outfits, and combines this with a three-quarter-length, all-enveloping veil (khimar); sometimes she also wears gloves. Underneath, however, she wears very fashionable styles, including brand-name jeans.

Although Feride considers herself very much a religious person, she also explained that she had to grow into wearing these styles of covered dress. "You have your own personal taste, and also fashion plays a role. If long splits are fashionable, you buy skirts with long splits. … Covering is a form of worship, and you are not really supposed to draw attention to yourself or to make yourself beautiful. That is a thin line."

Malika began wearing her veil almost a year ago. Although to an outsider this shift in clothing may look like a huge change, she said, "I do not think it is that much of a change. I still love fashion, but now it is halal [permissible according to Islamic law]. I simply wear a long dress over my jeans. Sometimes people who know me from before say what a pity, all those curls, what happened, you were always so free. Then I say that was my appearance, but what you see is not the same as what is inside. I had always been practicing."

Lisa likes brand names: "G-Star, for instance. The quality is much better." She often goes to the smaller stores where they have different brands. "But for simple things, I go to Vero Moda or H&M." Thinking about what she finds Islamically acceptable, she says, "I would not say everything except the face and the hands, but I do cover my chest and the section between my waist and my knees. Also, it should not be too tight or too short. If I wear skirts, it is a long skirt, or I wear trousers underneath. I wear such styles with my headscarf, because of [my] love for God. It is not that it makes you less attractive, because your eyes or your face can also do that."

The stories these very different women tell about themselves and their personal choices indicate that traditional cultural traits, such as dressing the female body consistent with religious strictures, are never static but vary, even within the same location (**Figure 6.F**).

1. What is fashion veiling?
2. How does veiling among Muslim women reflect both religious beliefs as well as popular culture?

Source: Annelies Moors, 2009, "Islamic Fashion in Europe: Religious Conviction, Aesthetic Style and Creative Consumption," *Encounters*, 1:175–200. Reprinted by permission of *Encounters* and the author.

▲ **Figure 6.F   Three examples of Islamic dress** (a) Traditional; (b) a more liberal interpretation in London; (c) a mixture of styles in Singapore.

the most megachurches. Megachurch facilities are typically designed to draw in every aspect of life, with an active presence in social media, round-the-clock activities that include aerobics classes, social services, fast food, bowling alleys, sports teams, aquatic centers with Christian themes, multimedia bible classes, and customized apps that provide sermon notes, sermon audio, and listings of the church calendar and events.

While communication technologies, mass media, and social media can accelerate missionizing and reinforce religious precepts, they can also work in the opposite direction. Movies, TV shows, music, and the Internet rapidly spread trends and fashions around the globe, and some of them, inevitably, clash with the traditions and interpretations of one religion or another. In some cases, religious traditions are compromised. In others, they can be modified. Take, for example, the tradition of Muslim women wearing headscarves or veils when they are out in public. The Muslim holy book, the Qur'an, instructs them, as well as men, to be modest in their public comportment. Some Muslim women wear headscarves to demonstrate their piety and publicly identify themselves as people who adhere to Muslim religious beliefs. For them, covering one's head is an expression of religious conviction. But being religious doesn't mean a Muslim woman must also be unfashionable. And just because many Muslim women wear headscarves (or other forms of more modest dress such as full-length skirts and long sleeves) does not mean that they all dress the same. In fact, fashionable Muslim dress has grown increasingly popular (see Box 6.4: "Window on the World: Fashion Veiling").

Geographers Banu Gökariksel and Anna Secor have studied the veiling fashion industry in Turkey, a particularly interesting place to explore because of the over 200 firms operating there. With retail outlets in the Middle East, Europe, Asia, and North America, these firms clearly have a global reach.

Turkish veiling-fashion designers visit the annual fashion shows in Paris to find inspiration for their new styles in emerging Western trends of fabric, color, cut, and style. Veiling-fashion is little different from Western fashion in that it is part of a larger system of production, consumption, and changing cultural tastes.

APPLY YOUR KNOWLEDGE

1. Why do the three women, Feride, Lisa, and Malika, wear veils? Are their differences in their reasoning? Similarities? How do their personal experiences affect their choices?

2. Locate online the video entitled "Lifting the Veil: Muslim Women Explain Their Choice" (http://www.npr.org/2011/04/21/135523680/lifting-the-veil-muslim-women-explain-their-choice). Before watching, write down your current understanding and knowledge of the hijab. After viewing the segment, describe how your impressions changed or stayed the same in response to the content of the program.

Reaction: Islam and Cultural Nationalism

Through the media of film and television, many aspects of U.S. culture travel widely outside its borders. Although many of these are welcomed abroad, many others are not. France, for example, has been fighting for years against the "Americanization" of its language, food, and films. Nations can respond to the homogenizing forces of globalization and the spread of U.S. culture in any number of ways. Some groups attempt to seal themselves off from undesirable influences. Other groups attempt to legislate the flow of foreign ideas and values. Most recently, Muslim countries have been especially resistant to the effects of Western culture propelled by globalization on their beliefs and practices. The movement, known as **cultural nationalism**, is an effort to protect regional and national cultures from the homogenizing impact of globalization, especially from the penetrating influence of U.S. culture.

The Islamic world includes very different societies and regions, from Southeast Asia to Africa. Muslims comprise over 85 percent of the populations of Afghanistan, Algeria, Bangladesh, Egypt, Indonesia, Iran, Iraq, Jordan, Pakistan, Saudi Arabia, Senegal, Tunisia, Turkey, and most of the independent republics of central Asia and the Caucasus (including Azerbaijan, Turkmenistan, Uzbekistan, and Tajikistan). In Albania, Chad, Ethiopia, and Nigeria, Muslims make up 50 to 85 percent of the population. In India, Burma (Myanmar), Cambodia, China, Greece, Slovenia, Thailand, and the Philippines, significant Muslim minorities exist. After Christianity, Islam possesses the next largest number of adherents worldwide—about 1 billion. The map in **Figure 6.29** shows the relative distribution of Muslims throughout Europe, Africa, and Asia as well as the heartland of Islamic religious practice.

One of the most widespread cultural forces in the world today is Islamism, more popularly, although incorrectly, known as Islamic fundamentalism. Whereas *fundamentalism* is a general term that describes the desire to return to strict adherence to the fundamentals of a religious system, **Islamism** is an anticolonial, anti-imperial political movement. In Muslim countries, Islamists resist core, especially Western, forces of globalization—namely modernization and secularization. Not all Muslims are Islamists, although Islamism is the most militant movement within Islam today.

The basic intent of Islamism is to create a model of society that protects the purity and centrality of Islamic precepts through the return to a universal Islamic state—a state that would be religiously and politically unified. Islamists object to modernization because they believe the corrupting influences of the core place the rights of the individual over the common good. They view the popularity of Western ideas as a move away from religion to a more secular (nonreligious) society. Islamists desire to maintain religious precepts at the center of state actions, such as introducing principles from the sacred law of Islam into state constitutions.

Islamism—a radical and sometimes militant movement—should not be regarded as synonymous with the practices of Islam, any more generally than Christian fundamentalism is with Christianity. Islam is not a monolithic religion, and even

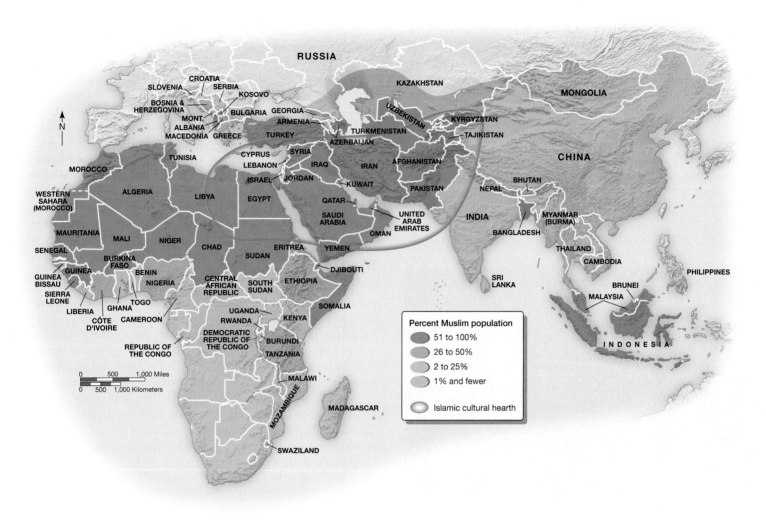

▲ **Figure 6.29 Muslim world** Like the Spanish colonial effort, the rise and growth of Muslim colonization were accompanied by the diffusion of the colonizers' religion. The distribution of Islam in Africa, Southeast Asia, and South Asia that we see today testifies to the broad reach of Muslim cultural, colonial, and trade activities. But the heart of the Muslim culture remains the Middle East, the original cultural hearth. It is also in this area that Islamism is most militant.

(*Source:* Adapted from D. Hiro, *Holy Wars.* London: Routledge, 1989.)

though all adherents accept the basic pillars of the faith, specific practices vary according to the different histories of countries, nations, and tribes. Some expressions of Islam allow for the existence and integration of Western styles of dress, food, music, and other aspects of culture, while others call for the complete elimination of Western influences.

APPLY YOUR KNOWLEDGE

1. What is Islamism? How does it differ from the practices of Islam? Where in the world does Islamism have the most traction? Why?

2. Look at the news headlines this week and find examples of both Islamism and Christian fundamentalism around the world. What did you find? How were the movements incorporated into other aspects of social life: politics, economy, fashion, reproduction, gender, education, and so on? How do those influences compare to the "Americanization" of social life around the globe?

RELIGION AND LANDSCAPE

The long-standing power of organized religions and established churches, together with the importance of religion in many people's lives, means that religion has been an important shaper of landscape in almost every part of the world. The most obvious examples, perhaps, are the cathedrals, churches, mosques, and temples that in many cases stand out in high relief amid their immediate surroundings (**Figure 6.30**). Less visible, but even more influential in shaping many of the world's landscapes, are the land holdings and investments of religious institutions. The Catholic Church, for example, owns not only the 110 acres of land in the middle of Rome that constitutes the Vatican City but also roughly 177 million more acres of property—agricultural, residential, and commercial, as well as ecclesiastical—around the world. The Catholic Church is also the wealthiest single institution in the world, with an estimated U.S.$150 billion in *annual* spending, investment, and lobbying. The impact of that level of expenditure on local landscapes and human geographies is significant. In the United States, most of the expenditure is directed toward health care and to Catholic colleges and universities.

In the United Kingdom, the established church has property and investment assets worth more than £5.3 billion (U.S.$9 billion), including a huge "shoppertainment" center in northern England and industrial parks, residential estates, and vacant land temporarily converted to parking lots in many cities. In the eighteenth and early nineteenth centuries, the Church had much more land in London, and grew rich on the rents of its tenants. Much of the housing owned by

▲ **Figure 6.30 Visibility** Parish church spires, like this one in Chamery, in the Champagne region of France, were visible from long distances.

▲ **Figure 6.31 Bedford Estate, London** The enormous size of their holdings gave the owners of the Great Estates the opportunity to exercise an exceptional degree of architectural and urban design control.

the Church, however, was substandard, and the Ecclesiastical Commissioners of the Church attracted widespread criticism for deriving so much of their profits from impoverished tenants. When the railways needed to purchase land to bring their tracks and termini to central London in the 1840s, the commissioners were only too happy to rid themselves of the worst of their embarrassing slum properties while securing healthy compensation. Nevertheless, the Church remained one of the biggest "slumlords" in London into the mid-twentieth century. In the 1960s the Ecclesiastical Commissioners sold off many of their slum properties, thereby removing at least some of the grounds for shame. At the same time it raised capital that the commissioners needed to enter into partnerships with property companies and begin the renovation and rebuilding of its better-managed estates.

Without doubt the single most profound influence on London's landscapes and geography, however, dates from a much earlier time. Before the sixteenth-century Protestant Reformation that led to the eventual adoption of the Anglican Church as the established religion, the Catholic Church and its representatives owned a great deal of property in the city and huge tracts of land just beyond it. Before Henry VIII's Dissolution of the Monasteries, there had been 23 religious houses in the city itself, while beyond the walls was a ring of hospitals, abbeys, priories, convents, and ecclesiastical palaces that effectively constituted a medieval green belt. Their confiscation by the Crown gave London room to grow, and the allocation of most of the undeveloped monastic land to a select group of courtiers was a windfall privatization that established London's Great Estates, which were later to be developed as the distinctive residential districts of London's West End (**Figure 6.31**).

Sacred Spaces and Pilgrimages

Sacred spaces are physical settings recognized by individuals or groups as worthy of special attention because they are the sites of special religious experiences and events. They do not occur naturally; rather, they are assigned sanctity through the values and belief systems of particular groups or individuals. Geographer Yi-Fu Tuan insists that what defines the sacredness of a space goes beyond the obvious shrines and temples. Sacred spaces are those that rise above the commonplace and interrupt ordinary routine.

In almost all cases, sacred spaces are segregated, dedicated, and hallowed sites that are maintained as such generation after generation. Believers—including mystics, spiritualists, religious followers, and pilgrims—recognize sacred spaces as being endowed with divine meaning. The range of sacred spaces includes sites as different as an elegant and elaborate temple in Cambodia, Angkor Wat, sacred to Hindus (**Figure 6.32**), and the Black Hills of South Dakota, the sacred mountains of the Lakota Sioux.

Often, members of a specific religion are expected to journey to especially important sacred spaces to renew their faith or to demonstrate devotion. A pilgrimage is a journey to a

◀ **Figure 6.32 Angkor Wat, Cambodia** Built for the King Suryavarman II in the early twelfth century, Angor Wat was dedicated to the Hindu god Vishnu. In the late thirteenth century, this state temple became a sacred site for Buddhists, which continues to this day.

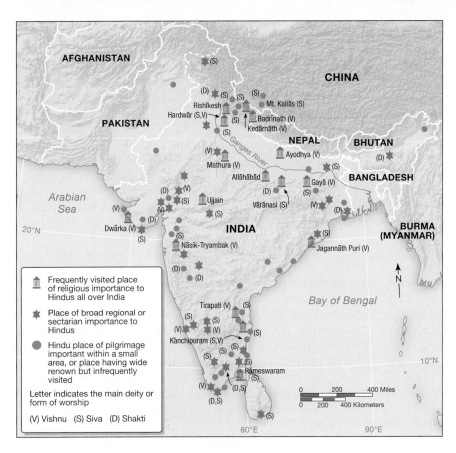

◀ **Figure 6.33 Sacred sites of Hindu India** India's many rivers are holy places within the Hindu religion, so sacred sites are located along the country's many riverbanks. Shrines closer to the rivers are regarded as holier than those farther away.

(*Source*: Adapted from Ismail Ragi al Farugi and David E. Sopher, *Historical Atlas of the Religions of the World*. New York: Macmillan, 1974.)

sacred space, and a pilgrim is a person who undertakes such a journey. In India many of the sacred pilgrimage sites for Hindus are concentrated along the seven sacred rivers: the Ganges, the Yamuna, the Saraswati, the Narmada, the Indus, the Cauvery, and the Godavari. The Ganges is India's holiest river, and many sacred sites are located along its banks (**Figure 6.33**). Hindus visit sacred pilgrimage sites for a variety of reasons, including seeking a cure for sickness, washing away sins, and fulfilling a promise to a deity. Perhaps the most well-known pilgrimage is the **hajj**, the obligatory once-in-a-lifetime journey of Muslims to Mecca. For one month every year, the city of Mecca in Saudi Arabia swells from its base population of 150,000 to over 1,000,000 as pilgrims from all over the world journey to fulfill their obligation to pray in the city and receive the grace of Allah. **Figure 6.34** shows the principal countries that send pilgrims to Mecca.

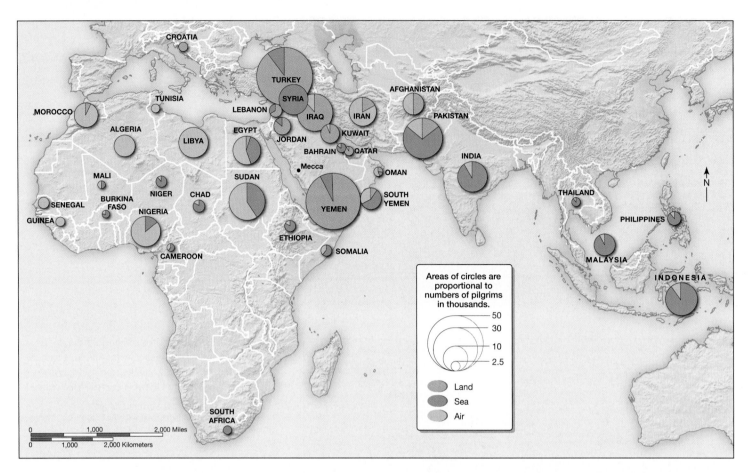

▲ **Figure 6.34 Source areas for pilgrims to Mecca** Islam requires that every adult Muslim perform the pilgrimage to Mecca at least once in a lifetime. This obligation is deferred for four groups of people: those who cannot afford to make the pilgrimage; those who are constrained by physical disability, hazardous conditions, or political barriers; slaves and those of unsound mind; and women without a husband or male relative to accompany them. The pattern of actual pilgrimages to Mecca (which is located close to the Red Sea coast in Saudi Arabia) suggests a fairly strong distance-decay effect, with most traveling relatively short distances from Middle Eastern Arab countries. More distant source areas generally provide smaller numbers of pilgrims, though Indonesia and Malaysia are notable exceptions.

(*Source:* After C. C. Park, *Sacred Worlds.* London: Routledge, 1994, p. 268.)

Pilgrimages to sacred sites are made all over the world, including Christian Europe. The most visited sacred site in Europe is Lourdes, at the base of the Pyrenees in southwest France, not far from the Spanish border (**Figure 6.35**). Another sacred site that attracts pilgrims throughout the world is the city of Jerusalem, and the Holy Land more generally, which is visited by Jews, Orthodox, Catholics, Protestants, Christian Zionists, and followers of many other religions. Jerusalem certainly has few rivals for the title of the most sacred city in the world, possessing as it does an unmatched Christian, Jewish, and Islamic history. The history and the map of the city reflect the histories of the various empires that dominated and were succeeded by new and yet more powerful empires (**Figure 6.36**).

After the break-up of the Ottoman Empire in the aftermath of World War I, a Mandate for Palestine was established by the League of Nations, whereby Palestine would become a Protectorate, administered by Britain, that would include a home for the Jewish people. British policy was that Jerusalem should be an international city with no one state claiming it as entirely its own. The creation of Israel as a national state in 1949 put an end to that policy. Today, Jerusalem is a highly contested city

as Palestinians, Christians, Muslims, and Jews fight for control of it. Nationalist Israelis maintain that Jerusalem will be the "eternal and undivided capital" of Israel, while Palestinians believe that Jerusalem should be the future capital of the Palestinian state. Peace accords agreed in Oslo, Norway, in 1992 led to a joint declaration that hinted at the possibility of negotiating the future of Jerusalem. But despite further negotiations in subsequent years, Jerusalem remains entirely controlled by Israel, with little prospect of a just and peaceful resolution of the issue.

An example of the ongoing contest for control is the continuing dispute over the Dome of the Rock, which was constructed between 688 and 691 C.E. Muslims claim the Dome as a asacred site (**Figure 6.37**). Yet the Dome sits on a site sacred to the Jews, the Temple Mount, the site where the Great Temple and Second Temple were built and later destroyed. Indeed, the Dome is believed to enclose the sacred rock upon which Abraham prepared to sacrifice his son, according to Jewish tradition, and according to Islamic tradition, is the same rock from which the Prophet Muhammad launched his spirit on a heavenly visit. Also located at the Temple Mount is the Al-Aqsa Mosque, a central sacred site to Muslims.

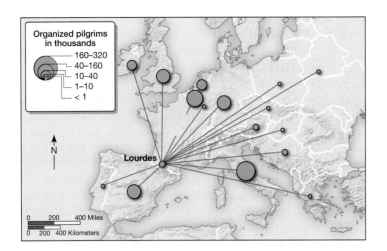

◀ **Figure 6.35 Source areas for pilgrims to Lourdes** This map shows the points of origin of European group-organized pilgrims to Lourdes in 1978. These represent only about 30 percent of all pilgrims to Lourdes, most of whom travel to the shrine on their own. Improved transportation (mainly by train) and the availability of organized package trips have contributed to a marked increase in the number of pilgrims visiting. Many of the 5 million pilgrims who visit the town each year do so in the hope of a miraculous cure for medical ills at a grotto where the Virgin Mary is said to have appeared before 14-year-old Bernadette Soubirous in a series of 18 visions in 1858.

(*Source:* After C. C. Park, *Sacred Worlds.* London: Routledge, 1994, p. 284.)

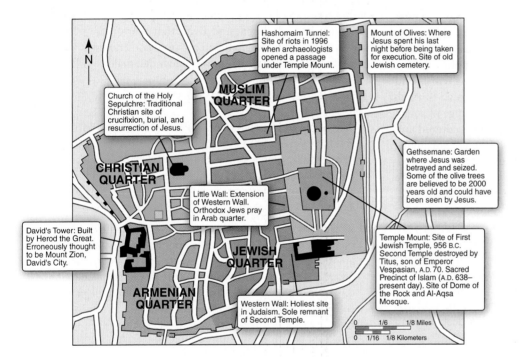

◀ **Figure 6.36 Jerusalem, the Holy City** This map of Jerusalem demarcates the main sections of the city. Over the many years of the Israeli-Palestinian peace process, numerous proposals have been advanced about how to divide the city to satisfy the wishes of both Palestinians and Israelis. The disposition of Jerusalem is one of the major issues in the ongoing conflict.

(*Source:* Redrawn from *The Guardian,* 14 October 2000, p. 5.)

◀ **Figure 6.37 Dome of the Rock** Located in Haram Al-Shaif, or Temple Mount, the Dome of the Rock sits in the part of Jerusalem that is technically neither Jewish nor Muslim, but both. The Muslims want this entire holy site. The Israelis also want it because the Wailing Wall, sacred to Jews, occupies part of the Temple Mount.

APPLY YOUR KNOWLEDGE

1. What is a sacred space? How does geographer Yi-Fu Tuan define a sacred space?

2. For Muslims the Dome of the Rock is holy, and for the Lakota, the Black Hills are a sacred space. While the Dome is a human-made sanctuary on a holy rock, the Black Hills are not, showing the importance of both built and natural environments as sacred. What are sacred places for you? For your family? Are all of these spaces religious? Or are they sacred for other reasons? What does it mean to be in a sacred space? How are your activities or speech different in those spaces?

FUTURE GEOGRAPHIES

As this chapter has shown, language and religion are dynamic components of culture and the ways that communication is changing is having profound effects on both. It seems reasonable to assume, then, that the future of language, communication, and belief will be one of accelerating change as new words and ways of speaking, reading, and writing will get created through the complex interactions of people using a wide range of social and other media. Religion likewise will be steadily altered and adapted to new conditions as it has been since it emerged among settled populations hundreds of thousands of years ago.

What appears to be occurring at the beginning of the twenty-first century is that religion will remain an influential force globally, although secularization, especially in the core countries, will continue. New forms of belief and worship will continue to emerge in the twenty-first century at the same time that established religions will grow and change as its adherents migrate, become modernized, and the religious establishment responds to these transformations. For instance, Buddhism is an increasingly popular religion affecting Western ideas of science and psychology. In contrast, Protestantism appears to be in decline and its leaders must consider how best to address this change if it is to remain a key part of twenty-first-century life.

Like religion, language is also changing. Whereas many languages, very small in terms of the number of speakers, are in decline, others are experiencing fortification as more and more speakers are being added. English is a language that is experiencing dramatic growth so much so that experts expect its dominance will only grow. Migration, globalization, popular culture, and the World Wide Web, in combination, support English as the key to the world's business, politics, and society. With so many speakers, however, English itself is bound to continue to change as it has for centuries. At the same time that English is likely to remain an ascendant language for years to come, languages already experiencing decline will most likely continue to do so, especially as the native speakers of those languages migrate or die. This has been the case, for example, with a number of Native American languages such as Apalachee, Atakapa, and Eyak that were ultimately replaced by English and are now considered extinct with the last native speaker now deceased.

Future of World Religions

http://goo.gl/2JBezx

APPLY YOUR KNOWLEDGE

1. Why is English growing in dominance around the world? What do you think the effects of extinct languages are? What do we gain by having less languages around the world? What do we lose? Think about not only the actual languages but concepts and ideas that are preserved in languages that do not necessarily translate into English. Do you think languages should be preserved?

2. What changes in religion have you noticed in your own life? Compare your beliefs to your parents, your grandparents, and great-grandparents. What do you observe over time and space in terms of social changes, world events, and the rate of communication across the world to share ideas? How do you think religion will continue to change for your own children?

▇ CONCLUSION

Language and religion are two key elements of cultural systems: markers of individual and collective identity. As such, they are important dimensions of human geographies. They are, in many contexts, closely interdependent. The diffusion of religion has carried linguistic traits into new areas, while the diffusion of language has facilitated the spread of certain religious beliefs and practices. Language and religion are also implicated in other aspects of human geographies: Many territorial conflicts, in particular, involve linguistic and religious dimensions.

Language both reflects and influences the way different groups understand and interpret their environment, while different groups use language to give or change a place's meaning. The names given to places often reflect the physical environment, local histories, or economic functions, but sometimes in ways that are unsuited to contemporary sensibilities, so that place names change, just as language itself changes.

The distribution of languages and belief systems can tell us much about the changing history of human geography and the impact of the movement of people, ideas, capital, and technology globally. Language and religion both developed in specific hearth regions and diffused around the world in various ways. Languages have "drifted" in syntax and meaning through time and space, while belief systems have splintered and mutated.

Today there is great variation within countries and even within individual cities and regions in terms of dialect and, sometimes, the preferred language. Similarly, there are significant variations in both religious adherence and religiosity.

Social media and new forms of communication are rapidly altering some aspects of language and facilitating the spread of certain aspects of language around the globe. At the same time, globalization prompts people to seek to reaffirm the distinctiveness of places and cultures. One outcome of this has been the revival of minority languages; another has been the growth of religious nationalism. More generally, modern communications systems and social media have contributed both to the propagation of religious fundamentalism and to changes and challenges to religious beliefs and precepts.

LEARNING OUTCOMES REVISITED

- **Describe how language both reflects and influences the way different groups understand and interpret the world.**

  Language is a means of communication and all languages have structures that include rules of grammar, specialized vocabularies, and systems of pronunciation. Around the world, there are both very large and very small languages that link speakers, with New York City being the most linguistically diverse city in the world. Language is also a result of social, economic, and political phenomenon like imperialism, migration, or slavery where people groups are moved from one region of the world to another and develop new ways of communicating, based on these experiences and the dynamic nature of their lives.

- **Compare and contrast different forms of communication be it standard language, slang, dialects, social media, and nonverbal modes of expression.**

  Languages are organized into different families, which are rooted in their prehistoric origins and then are further categorized into branches and groups. States have been influential in formalizing language and promoting that use in courts, legislature, administration, and business. However, people adapt language and speech patterns every day through various modes of expression including creating new words and expressions of slang, responding to technology through social media, and developing regional variations called dialects. Human communication is also greatly dependent on nonverbal modes like dance, visual cues, sound, gestures, and sign language.

- **Interpret how different geographies impact the spread or preservation of language and how different groups use language to give or change a place's meaning.**

  Through global communication and business, the dominance of English is growing, while other languages, like those of Indigenous peoples in the Western Hemisphere, are becoming extinct from nonuse. Geography can affect language loss and preservation through the lack of migration in certain areas. Isolated spaces like Iceland can preserve older forms of their language because the rugged landscape and cold climate have discouraged migration from other people who bring new concepts/pronunciations to the language. The relationship between language and space can also be seen in how various communities signify places with toponyms. Place names reflect the values of a community, the collective use of that space, or honor a significant figure. Additionally, toponyms can reveal the racial, class, and gender disparities in society, and change as people resist the use of stereotypes in the landscape.

- **Describe the global distribution of the world's religions—how they developed in specific regions and how they proliferated around the world.**

  The four major world religions—Hinduism, Buddhism, Christianity, and Islam—have emerged from specific places on the globe but have spread through diaspora. Hinduism began in India, which has always been an important cultural crossroads. This unique geography allowed ideas to mix and migrate from the site of origin, and Buddhism later developed as an offshoot of Hinduism in Nepal. Christianity and Islam were both born in the Middle East, another region with waves of human migration and cultural sharing. Once states assumed these beliefs as the official religions, their impact on law, culture, language, and social relationships intensified.

- **Recognize the difference between religions and religious movements around the world and analyze their impacts on political and social life.**

  Compare these movements to the global movement of "Americanization" and regional cultural nationalism. Religions are belief systems and sets of practices that incorporate the idea of a power higher than humans. When these beliefs are formalized through a system of rituals, they are organized religions. Religious movements occur when people interpret the teachings of a religion and harness those beliefs in a social and political movement. For many people around the world, there is a response to the homogenizing effects of "Americanization" on culture and a desire to protect communities from "outside influences." Islamism is one such political movement and does not include all Muslims. Militant Islamists resist core, Western forces like secularization and modernization, and seek to create a fundamentalist society.

- **Interpret the importance of space to religion in pilgrimages and sacred spaces in every culture.**

  Sacred spaces are physical settings recognized by a community of people as significant to religious and/or spiritual events. They are assigned this value by a group of people and can be as diverse as the Dome of the Rock in Jerusalem for Muslims or the Black Hills in South Dakota for the Lakota people—one is a temple on a sacred rock and other is the natural environment of the hills without any human-made structure. Geographer Yi-Fu Tuan defines sacred spaces as those that rise above the commonplace and interrupt ordinary routine. Pilgrimages are journeys to these sacred spaces.

KEY TERMS

cargo cults (p. *215*)	Islamism (p. *218*)	Muslim (p. *209*)
cults (p. *214*)	language (p. *188*)	mutually intelligible (p. *194*)
cultural nationalism (p. *218*)	language branch (p. *193*)	official language (p. *189*)
cuneiform (p. *200*)	language family (p. *193*)	organized religions (p. *206*)
dialects (p. *189*)	language group (p. *193*)	pictograms (p. *200*)
diaspora (p. *206*)	language hearth (p. *196*)	place name (toponym) (p. *196*)
established churches (p. *209*)	language tree (p. *193*)	proto-languages (p. *193*)
fusion language (p. *191*)	lingua franca (p. *197*)	religion (p. *206*)
functional illiteracy (p. *202*)	linguistic drift (p. *194*)	sacred spaces (p. *220*)
graffiti (p. *203*)	linguistic weathering (p. *195*)	slang (p. *203*)
hajj (p. *221*)	literacy (p. *202*)	standard language (p. *189*)
Islam (p. *209*)	mother tongue (p. *194*)	vocal fry (p. *204*)

REVIEW & DISCUSSION

1. Conduct an Internet search on a small or endangered language not discussed in the text. Where is it spoken and by how many speakers? In what language family/branch/group is this language a member? Why is it endangered? Compare the phenomenon of language loss and preservation outlined in this chapter to the processes in the language you research. Do you notice any similarities? Are there any efforts to revitalize or preserve the language?

2. As discussed, nonverbal communication is a critical component of how human beings interact. But how does nonverbal communication allow us to communicate with *non-humans*? Consider the relationships people have with their pets or working animals by looking up articles on how animals communicate with people. Think about your own nonverbal communication with animals. (Hint: Look at the subject of "anthropomorphizing." Do animals pick up on our social cues and mimic them, or is it a reciprocal relationship? In other words, how much do we adopt animal cues? How do you think nonverbal human-to-animal communication impacts social space? Consider stories of how horses and elephants heal from trauma or how dogs hold memories of loved ones, etc.)

3. Research the religious organizations in your area. Do you live near any of these spaces? Are certain religions more widely represented than others? List all of the religious centers in your area and examine their role in the community. For example, do followers offer social services or host special events in your neighborhood? Do any of these organizations work with each other; that is, are there any interfaith dialogues directed toward shared social or environmental issues? Are some of these spaces involved in politics—either directly by supporting candidates and policy or indirectly, such as serving as a polling place? Are some centers wealthier than others? Older than others? Why do you think?

4. Reconsider your notions about a particular religion or faith. What is a system of belief that you don't know much about, but have stereotypes or preconceptions? Make a list of four conceptions you have about this faith and then do research on those four aspects to get the full story. For example, why do both Muslim and Orthodox Jewish women cover their hair? Why is the Ba'hai faith headquartered in Haifa, Israel, but founded by a Persian? Why did the Quakers flee England? Why is Mary the Mother of Jesus so revered by Catholics? Who is I'itoi and why is he important to the Tohono O'odham?

UNPLUGGED

1. Draft a language tree based on your background. Using a language family map as your guide, construct a tree or visual graph of the languages you speak back to your ancestors. Think about where they come from, what languages they were born into, the languages they adopted, and the languages of their spouses. How far back do you go to non-English as the language spoken at home? None or one generation? Twelve or more generations? What does this tell you about language acquisition and human migration?

2. The speed and access of cell phone communication has become a critical component in diverse political movements—from the Arab Spring to immigration protests in Tucson, AZ, to the police and activists of Ferguson, MO. Choose an event where cell phone communication served to connect the events that transpired in one space to a national, or even global, audience. How did a network of communication contribute to this social or political event? Do you think the outcome or processes would be different without "the whole world watching"? How do cell phones change the landscape of communication?

3. Consider Yi-Fu Tuan's theory on sacred spaces and write an essay on a space sacred to you. How does he define sacred spaces and how is that different (or the same) as a place of worship? What is your sacred place and why is it special? Does it hold an important memory? Do you go to this place to mediate or to celebrate? Is this place secret or shared with others (like a family vacation space)?

DATA ANALYSIS

In this chapter we have looked at both language and religion and the effect geography has on both particularly through the movement of people and spiritual beliefs and the relationships between people and places, especially sacred one. Let's now look at how both concepts can be manipulated in geopolitical spaces.

First, analyze how "Islam" became synonymous with "terrorism." Think about how both "Islamist" militant groups falsely represent the religion and how anti-Islamic politicians stereotype all Muslims. Read Professor Juan Cole's "Top Ten Ways Islamic Law Forbids Terrorism."

Islamic Law Forbids Terrorism

http://goo.gl/Lcn7EN

1. List at least five ways Islam prohibits terrorism.

2. Why do you think many militant groups like Boko Haram and Al-Qaeda present themselves as pious Muslims?

3. How do misrepresentations affect perceptions of Islam to outsiders of that religion?

4. Discuss another example of a religion that has been misrepresented and how different followers attempt to educate outsiders.

Second, read how the *New York Times* (NYT) debated using the word *torture* in their reports, partially through the influence of the U.S. Justice Department under both Presidents Bush and Obama. Go to "The Executive Editor on the Word 'Torture.'"

The Word 'Torture'

http://goo.gl/fheQ8m

5. What is "torture"?

6. Why is it hard to define, or do you think it is hard to define?

7. Do you think the NYT censored its own journalists by not using the word *torture* when writing about C.I.A. methods? Why or why not?

8. How does politics intersect with communication to push certain agendas? Do you think there is a way to report political news without using manipulative language?

Using what you learned from both of these stories, answer these last two questions:

9. Language can be used to bring the world together, but also to separate people based on beliefs and perceptions of those beliefs. Do you use the terms *torture* and *terror* when talking about politics today? Define what you mean when you use them and challenge yourself to be specific. Think about the difference between religion and politics, and how they intersect.

10. Complex spiritual, political, and social issues around the world are often represented as stereotypes or sound bytes. Discuss one recent news story where you think language was used to manipulate public opinion. Share how you resisted making assumptions and drew your own conclusions.

MasteringGeography™

Looking for additional review and test prep materials? Visit the Study Area in MasteringGeography™ to enhance your geographic literacy, spatial reasoning skills, and understanding of this chapter's content by accessing a variety of resources, including **MapMaster** interactive maps, Videos, *In the News* RSS feeds, flashcards, web links, self-study quizzes, and an eText version of *Human Geography*.

LEARNING
OUTCOMES

- *State* how environment shapes people and how people shape environments.

- *Describe* how place making stands at the center of issues of culture and power relations and is a key part of the systems of meaning through which humans make sense of the world.

- *Define* the concept of territoriality both as an instinctive process and as the result of socio-political systems.

- *Identify* how different cultural identities and status categories influence the ways people experience and understand landscapes, as well as how they are shaped by—and able to shape—landscapes.

- *Describe* how codes signify important information about landscapes, a process known as semiotics.

- *Analyze* the effects of global consumption on cultural landscapes as they range from homogenization to preservation, and describe how a place can become a commodity.

INTERPRETING PLACES AND LANDSCAPES

The average tourist usually seeks out landscapes that are beautiful or famous. Think: Yosemite National Park, the Eiffel Tower, the Taj Mahal, or the Serengeti Plains. These sites are thought to evoke grandeur, natural beauty, and awe. In addition to being attracted to splendid or famous landscapes, tourists have also been keen to visit the more prosaic landscapes of everyday life, such as the floating villages of Cambodia, the night markets of Chang Mai, Thailand, and the horse farms of Kentucky blue grass country. Tourists spend billions of dollars each year to visit these landscapes both grand and ordinary.

A new trend in tourism, however, is one that is becoming most commonly known as anti-tourism. Anti-tourism deliberately seeks to avoid the sublime and even the commonplace in order to know more about the underbelly of our world; i.e. the dirty, dangerous, the prohibited, and even the toxic landscapes that have been made by human hands but hidden away, abandoned, or just not attractive or interesting to the mainstream traveller for one reason or another.

Anti-tourists are interested in getting away from the fully inclusive tour that installs the visitor in relaxing, luxurious surroundings. They prefer to explore the unsavory stuff that the tourism bureaus don't want visitors to know about. For example, a tour that has been offered for four years, "The Poison Cauldron," enables tourists to visit one of New York City's most toxic landscapes: Newton Creek. Located along the banks of Newton Creek are the waste transfer and petroleum districts of North Brooklyn, a borough of New York City. With the help of a local historian, tourists are guided along a three-mile trek through Greenpoint, the northernmost neighborhood in Brooklyn that faces the financial district of Manhattan across the East River.

Greenpoint was farmland during the early colonial period. In the nineteenth century it became industrialized with rope factories, lumberyards, shipbuilding, glasswork, and foundry activities. In the twentieth century, when the industrial activities had mostly ceased, Greenpoint became the site for a huge incinerator facility, a sewage treatment plant, an electroplating company, and a transfer point for municipal waste collected from New York City. In 1950, Newton Creek, a small tributary of the East River, experienced what was then the worst oil spill in U.S. history, which has yet to be completely ameliorated.

In short, Newton Creek is hardly the sort of landscape that would attract the average tourist. And yet, walking tours (closed-toed shoes highly recommended) of The Poison Cauldron run every Saturday afternoon commencing at DUKBO, an area Down Under the Kosciuszko Bridge Onramp. Why visit a toxic site where you learn about how the bottom of Newton Creek is coved by a 15–20 foot layer of "black mayonnaise," that smells like petroleum and contains 150 years of industrial waste, human excrement, petroleum residue, and coal tar? The answer is that the mostly young people who take the tour are seeking a form of fun that deviates from the usual, the tame, and the predictable. They want an experience that isn't "prettified" and that few others would be interested in having.

Other forms of anti-tourism that are directed at experiencing unconventional landscapes include "urban exploration" and "dark tourism." Urban exploration, also called urbex or UE, focuses on the investigation of infrastructure, including abandoned ruins such as unused grain elevators, factories, fallout shelters, or schools—or built landscapes like sewers, storage sites, and utility tunnels that are normally off-limits or have restricted access. Urban explorers thrive on the physical danger and possibility of arrest and punishment that often accompany the "infiltration" of these sites.

In contrast, dark tourism involves travel to sites that are associated with death or tragedy such as castles, battlefields, slave quarters and dungeons, as well as sites of disasters including Ground Zero in New York, after the 9/11 disaster or the Auschwitz concentration camp in Poland. Both dark tourism and urban exploration are forms of anti-tourism that push the boundaries of human relationships to landscapes as they reject the kind of experience that celebrates and reinforces the desirability of made-for-tourists sites and embrace experiences that peel back the complexity and contradictory aspects that constitute all such places.

In this chapter, we examine those complex relationships to place and landscape, attempting to unravel the complexity and understand the way it's generated and understood and the effects it has on people. We will also look at how places and landscapes are interpreted differently by different groups of people and why those differences matter.

BEHAVIOR, KNOWLEDGE, AND HUMAN ENVIRONMENTS

In addition to attempting to understand how the environment shapes and is shaped by people, geographers also seek to identify how it is perceived and understood by people. Asserting that there is interdependence between people and places, geographers explore how individuals and groups acquire knowledge of their environments and how this knowledge shapes their attitudes and behaviors. Some geographers focus their research on natural hazards as a way of addressing environmental knowledge, while others try to understand how people ascribe meaning to landscape and places. In this chapter, we consider the key geographical concepts of place, landscape, and space and explore the ways in which people understand them, create them, and operate within them. Our goal is to understand how individuals and groups experience their environments, create and struggle within places, and find meaning in the landscapes they create.

In their attempts to understand environmental perception and knowledge, geographers share a great deal with other social scientists, but especially with psychologists. Human cognition and behavior are at the center of psychology. What makes environmental knowledge and behavior uniquely geographical is their relation to both the environmental context and the humans who struggle to understand and operate within it. Much of what we as humans know about the environment we live in is learned through direct and indirect experience. Our environmental knowledge is also acquired through a filter of personal and group characteristics, such as race, gender, stage of the life cycle, religious beliefs, and where we live (**Figure 7.1**).

▼ **Figure 7.1 People's Climate March in New York** Demonstrators hold signs on a street next to Central Park in response to the U.N. Climate Summit held there on September 23, 2014.

For instance, children have interesting and distinct relationships to the physical and cultural environment. How do children acquire knowledge about their environments? How do boys and girls differ in the ways they learn about and negotiate their environments? What kind of environmental knowledge do children acquire, and how do they use it? What role do cultural influences play in the process? What happens when larger social, economic, and environmental changes take place?

Geographer Cindi Katz conducted research in rural Sudan to find answers to these questions. Working with a group of 10-year-old children in a small village, she sought to discover how they acquired environmental knowledge. What she also learned was how the transformation of agriculture in the region changed not only the children's relationships to their families and community but also their perceptions of nature.[1] In this Sudanese village, as in similar communities elsewhere in the periphery, children were important contributors to subsistence activities, especially planting, weeding, and harvesting. The villagers were strict Muslims and thus had stringent rules

[1]C. Katz, *Growing Up Global: Economic Restructuring and Children's Everyday Lives*. Minneapolis: University of Minnesota Press, 2004.

about what female members of the community were allowed to do and where they were allowed to go. Many of the subsistence activities that required leaving the family compound were customarily delegated to male children. Within the traditional subsistence culture, boys predominated in all agricultural tasks except planting and harvesting and were responsible for herding livestock as well. Many boys (and occasionally girls) were also responsible for fetching water and helping to gather firewood. Both boys and girls collected seasonal foods from lands surrounding the village. Work and play were often mixed together, and play, as well as work, provided a creative means for acquiring and using environmental knowledge and for developing a finely textured sense of the home area (**Figure 7.2**).

What happens when the agricultural production system is thoroughly changed, as it was in this village when irrigated cash-crop cultivation was introduced? Through an international development scheme, with the financial assistance of outside donors, a Sudanese government project transformed the agriculture of the region from subsistence livestock raising and cultivation of sorghum and sesame to cultivation of irrigated cash crops such as cotton. The new cash-crop regime, which required management of irrigation works, application of

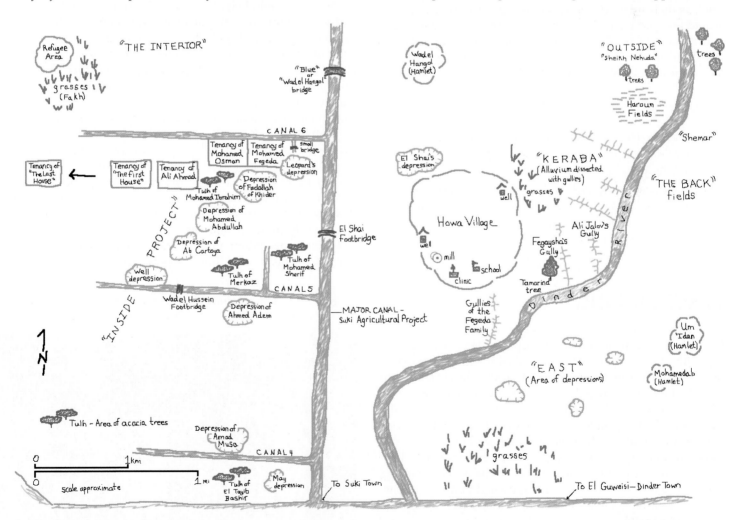

▲ **Figure 7.2 Shepherd's map** This map, drawn by a 10-year-old Sudanese boy, shows the area over which the sheep are herded. It illustrates the detailed environmental knowledge Sudanese children possess of the landscape that surrounds their village. The village is an Islamic one, and norms determine the kinds of tasks in which boys and girls can participate. Only boys are allowed to tend sheep, which requires a particular environmental knowledge about grazing areas and water availability.

(*Source:* Image courtesy of C. Katz.)

fertilizers, herbicides, and pesticides, and more frequent weeding, required children as well as adults to work longer and harder. Parents were often forced to keep their children out of school because many of the tasks had to be done during the school term.

Destruction of local forests required children to range farther afield to procure fuelwood and to make gathering trips more frequently. Soon wealthier households began buying wood rather than increasing the demands on their children or other household members. For the children of more marginalized households, the selling of fuelwood, foods, and other items provided a new means for earning cash to support their families but also placed increasing demands on their energies and resourcefulness and changed their whole experience of their world.

For the children of this village, globalization (in the form of the transition to cash-crop agriculture) altered their relationships with their environments and with their futures. The kinds of skills the children had learned for subsistence production were no longer useful for cash-crop production. As they played less and worked longer hours in more specialized roles, their experience of their environment narrowed. At the same time, as their roles within the family changed, they attended school less and learned less about their world through formal education. As a result, their perceptions of their environment transformed, along with their values and attitudes toward the landscape and the place they knew as home.

APPLY YOUR KNOWLEDGE

1. Describe the difference in the lifestyle, work, play, and education regime for young people in the rural Sudan example, when the community moved into cash crop cultivation.

2. Scrutinize how globalization has shaped the environment you operate in as a student. How has it shaped you—the buildings, the people, the climate, the social life, et cetera—and how do you shape it by interacting with it?

PLACE-MAKING

Places are socially constructed—given different meanings by different groups for different purposes. Most people identify with places as part of their personal identity, drawing on particular images and particular histories of places in order to lend distinctiveness to both their individuality and their sense of community. But identifying with place may also imply the exclusion of other people and the stereotyping of other places. People often reinforce their sense of place and of who they are by contrasting themselves with places and people they feel are very different from them. Seen in this context, place-making stands at the center of issues of culture and power relations, a key part of the systems of meaning through which we make sense of the world.

There is another important dimension to the relationship between places and their inhabitants, and it has a lot to do with identity. We each "manage" several distinct roles or identities at once. This is particularly characteristic in urban environments because of the physical and functional separation of the "audiences" to which different roles are addressed: family, neighbours, co-workers, club members, and so on. People are consequently able to present very different "selves" in different socio-spatial contexts. The city, with its wide range of different roles and identities, becomes a "magic theatre, " or an "emporium of styles, "—the anonymity afforded by the ease of slipping from one role to another often facilitates the emergence of unconventional behavior.

Territoriality

Some social scientists believe that wanting to have a place where you feel you belong is a natural human attribute, part of a strong territorial instinct. Humans, it is argued, have an innate sense of territoriality, just like many other species. The concept of **territoriality** refers to the persistent attachment of individuals or peoples to a specific location or territory. The concept is important to geographers because it can be related to fundamental place-making forces.

The specific study of people's sense of territoriality is part of the field of **ethology**, the scientific study of the formation and evolution of human customs and beliefs. The term also refers to the study of the behavior of animals in their natural environments. According to ethologists, humans carry genetic traits resulting from our species' need for territory. Territory provides a source of physical safety and security, a source of stimulation (through border disputes), and a physical expression of identity. These needs add up to a strong territorial urge that can be seen in people's behavior: claims to space in reading rooms or on beaches, for example, and claims made by gangs to neighborhood turf (**Figure 7.3**). Ethologists argue that the territorial urge also can be observed when people become frustrated because of overcrowding. They become stressed and, in some circumstances, begin to exhibit aggressive or deviant behavior. Ethologists and environmental psychologists link crowding to everything from vandalism and assault to promiscuity, listlessness, and clinical depression.

▼ Figure 7.3 **Graffiti as territorial markers** Graffiti are used by neighborhood gangs to establish and proclaim their identity.

Proxemics While such claims are difficult to substantiate, as is the whole notion that humans have an inborn sense of territoriality, the idea of territoriality as a product of *culturally* established meanings is supported by a large body of scientific evidence. Some of this evidence comes from the field of **proxemics**, the study of the social and cultural meanings that people give to personal space. These meanings make for unwritten territorial rules that can be seen in the microgeography of people's behavior. It has been established, for example, that people develop unwritten protocols about how to claim space. One common protocol is simply regular use (think of students' habits in classroom seating arrangements). Another is through the use of spatial markers such as a newspaper or a towel to fix a space in a reading room or on a beach. There are also bubbles, or areas, of personal space that we try not to invade (or allow to be invaded by others). Varying in size and shape according to location and circumstance, these bubbles tend to be smaller in public places and in busier and more crowded situations; to be larger among strangers and in situations involving members of different social classes; and to vary from one social class or cultural group to another.

Social and cultural needs On larger spatial scales, territoriality is mostly a product of political relations and cultural systems. This aspect of territoriality underpins a great deal of human geography. All social organizations and the individuals who belong to them are bound at some scale or another through formal or informal territorial limits. Many organizations—nations, corporations, unions, clubs—actually claim a specific area of geographic space to be under their influence or control. In this context, territoriality can be defined as any attempt to assert control over other people, resources, or relationships over a specific geographic area. Territoriality also is defined as any attempt to fulfill socially produced needs for identity, defense, and stimulation. Territoriality covers many phenomena, including the property rights of individuals and private corporations; the neighborhood covenants of homeowners' associations; the market areas of commercial businesses; the heartlands of ethnic or cultural groups; the jurisdictions of local, state, and national governments; and the reach of transnational corporations and supranational organizations.

Territoriality thus provides a means of meeting three social and cultural needs:

- The regulation of social interaction
- The regulation of access to people and resources
- The provision of a focus and symbol of group membership and identity

Territoriality also gives tangible form to power and control but does so in a way that directs attention away from the personal relationships between the controlled and the controllers. In other words, rules and laws become associated with particular spaces and territories rather than with particular individuals or groups. Finally, territoriality allows people to create and maintain a framework through which to experience the world and give it meaning. Bounded territories, for example, make it easier to differentiate "us" from "them."

Street Art

Street art has become widespread as an element of place making. In some localities, it can be a major component of community identity. In San Francisco's Latino Mission district, for example, the streets are lined with the highest concentration of mural art in the world—more than 500 public art pieces within a 30-block radius (**Figure 7.4a**). The powerful association between street art and place has been noticed by business. The visual identity system for Coca-Cola's 2014 World Cup advertising campaign, for example, used the work of São Paulo street artist Speto (**Figure 7.4b**), while Pepsi-Cola's Art of Football campaign also had a street-art flavor.

Street art has also become an instrument of social commentary and satire. The most famous exponent is the British artist Banksy, who began "bombing" walls in Bristol, England, during the 1990s with his stenciled art (**Figure 7.4c**) and has since bombed cities as far apart as Vienna, San Francisco, Barcelona, Paris, and Detroit. A similar approach has been followed by the Iranian street artist Black Hand—dubbed the "Iranian Banksy" by some media outlets—whose daring work on the walls and buildings of Tehran have gone viral on Facebook and Twitter. In 2014 he painted an image of a young woman wearing Iran's soccer national team's jersey and raising a bottle of dish-washing soap (called *Jaam*, which means championship cup in Farsi). The image (**Figure 7.4d**) refers to the political culture that encourages women to stay at home, as in the ban against women attending sporting events.

APPLY YOUR KNOWLEDGE

1. What is territoriality? Give an example on both small and large scales. What are the three social and cultural needs that territoriality satisfies?

2. Describe the relationship between the field of ethology and the concept of territoriality. List and evaluate three examples that you experience in everyday life of proxemics as a territorializing force (beyond the classroom). How are they an expression of power and culture?

People and Places, Insiders and Outsiders

Places are constantly under social construction as people respond to the opportunities and constraints of their particular locality. As the famous American critic and historian Lewis Mumford put it, "in the state of building at any period one may discover, in legible script, the complicated process and changes that are taking place within civilization itself."[2] But places are never simply straightforward mirrors or neutral containers of social processes. They are created by specific sets of people, and they draw their distinctive character from the people that inhabit them. As social groups occupy places, they gradually impose themselves on their environment,

[2]L. Mumford, *The Culture of Cities*. New York: Harcourt, Brace and World, 1938, p. 403.

▲ **Figure 7.4** (a) San Francisco's Mission District and its community-based Mission Muralismo movement Balmy Alley houses 37 public art pieces by over three dozen community muralists, and serves as an important cultural hub for the Mission (b) Banksy art, London (c) Pepsi-Cola World Cup branding by street artist Jaz (d) Black Hand street art, Iran.

modifying and adjusting it, as best they can, to suit their needs and express their values. Yet at the same time people themselves gradually accommodate both to their physical environment and to the people around them. There is thus a continuous two-way process, what geographer Edward Soja called a **sociospatial dialectic**, in which people create and modify places while at the same time being conditioned in various ways by the spaces in which they live and work. People are constantly modifying and reshaping places, and places are constantly coping with change and influencing their inhabitants.

Places are both centers of meaning for people and the frameworks for their actions and behavior. It is important to remember that places are constructed by their inhabitants from their own subjective point of view and that they are simultaneously constructed and seen as an external "other" by outsiders. A neighborhood, for example, is both an area of special meanings to its residents as well as an area containing houses, streets, and people that others may view from an outsider's ("decentered") perspective.

As we saw in Chapter 1, a key concept is that of the *life-world*, the taken-for-granted pattern and context through which people conduct their day-to-day lives without having to make it an object of conscious attention. People's familiarity with one another's vocabulary, speech patterns, dress codes, gestures, and humor, and with shared experiences of their

7.1 Spatial Inequality   Outsider Art

In Chandigarh, India, a one-of-a-kind sculpture installation called the *Rock Garden* was created by Nek Chand Saini. Nek Chand was a road inspector in the 1950s, while the city was being redeveloped as the first "planned city in India" under the influence of Swiss architect Le Corbusier. Nek Chand collected debris from the construction sites and began building his "magic kingdom" on government owned land. This angered city officials who wanted to destroy Nek Chand's garden for violating laws established under Le Corbusier's "beautiful city" plan. Public opinion, however, sided with Nek Chand and he was able to quit his job and devote his life to creating the *Rock Garden* (http://nekchand.com/background). Today, Nek Chand's *Rock Garden* hosts over 5,000 visitors a day and people continue to build his vision. You can take a virtual tour of the garden with these panoramic photos (http://nekchand.com/gallery/interactive-panoramas) and this video (http://vimeo.com/67800249).

Nek Chand's *Rock Garden* is an example of what many call visionary or "outsider art." Unlike fine art, which is taught in schools, galleries and museums, "outsider art" is produced by people from the fringes of society. Nek Chand was from a lower caste and had no formal art education, his methods were self-taught and he used debris as his material—long before recycling was a hip thing to do. Nek Chand created a space that undermined the "beautiful city" plan and created an "outsider" landscape deep within an urban area. A place that many outsiders from around the world continue to seek.

"Outsider art" is a term coined in 1922 by Dr. Hans Prinzhorn for art produced by psychiatric patients, but grew to encompass all art outside the fine art world. Dr. Prinzhorn articulated a boundary between the "inside" art in a society, and art outside the mainstream. Insider art or fine art is work which reinforces social ideas about beauty and form, philosophies of life and death, the techniques we prize, and/or the artists we choose to venerate as "masters." With outsider art, artists are laborers, farmers, housewives, retirees, disabled, poor, and often are members of ethnicities and groups underrepresented in fine art institutions. Because they lack formal training, they develop original techniques and use non-conventional materials—anything from toothpicks to tattoos to motorcycle parts. Outsider art is highly individualized and intuitive, but a common theme for many is creating utopias like the *Rock Garden*. Drawing from their experiences outside the mainstream, these artists seek to create alternative landscapes.

1. What makes outsider art "outside"? What is this form of expression outside of? What are other cultural expressions that are "outside" the mainstream forms? What does this tell you about cultural and social categories?

2. Go to the American Visionary Art Museum website at http://www.avam.org. Look at the current exhibitions and then compare those works and artists with a more mainstream art museum website like the National Gallery of Art in Washington, D.C. or the Metropolitan Museum of Art in New York City. What are the differences you notice? Does one space have more resonance to you? Why or why not? Finally, what are the art spaces around you? Who are the artists in your city? Are they inside or outside your community?

▲ **Figure 7.A** Nek Chand's Rock Garden, Chandigarh, India.

physical environment, often carries over into people's attitudes and feelings about themselves and their locality and to the symbolism they attach to that place. When this happens, the result is a collective and self-conscious "structure of feeling": a sociocultural frame of reference generated among people because of the experiences and memories that they associate with a particular place.

Experience and Meaning

The interactions between people and places raise some fundamental questions about the meanings that people attach to their experiences: How do people process information from external settings? What kind of information do they use? How do new experiences affect the way they understand their worlds?

What meanings do particular environments have for individuals? How do these meanings influence behavior? How do people develop and modify their sense of a place, and what does it mean to them? The answers to these questions are by no means complete. It is clear, however, that people filter information from their environments through neurophysiological processes and draw on personality and culture to produce cognitive images of their environment, representations of the world that can be called to mind through the imagination (**Figure 7.5**). Cognitive images are what people see in the mind's eye when they think of a particular place or setting.

Cognitive images both simplify and distort real-world environments. Research has suggested, for example, that many people tend to organize their cognitive images of particular parts of their world in terms of several simple elements (**Figure 7.6**):

Paths: The channels along which they and others move; for example, streets, walkways, transit lines, canals.

Edges: Barriers that separate one area from another; for example, shorelines, walls, railroad tracks.

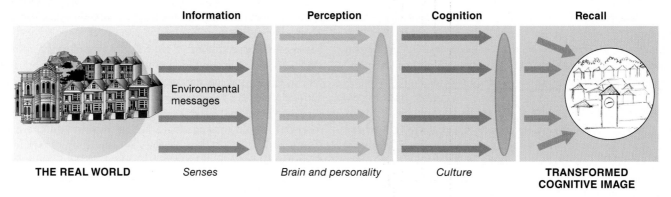

▲ **Figure 7.5 The formation of cognitive images** People form cognitive images as a product of information about the real world, experienced directly and indirectly, and filtered through their senses, brain, personality, and the attitudes and values they have acquired from their cultural background.

(*Source:* Adapted from R. G. Golledge and R. J. Stimpson, *Analytical Behavior Geography.* Beckenham, UK: Croom Helm, 1987, Fig. 3.2, p. 3.)

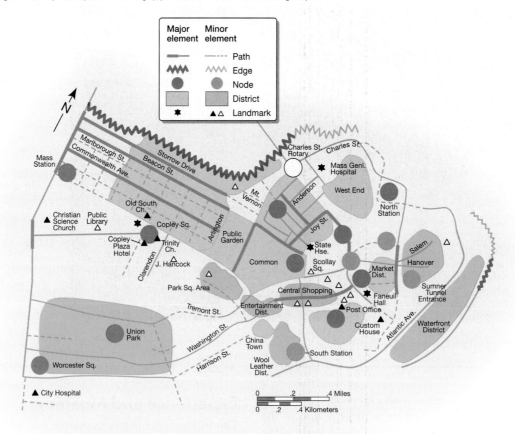

▲ **Figure 7.6 Cognitive image of Boston** This map was compiled by Kevin Lynch from interviews with a sample of Boston residents. Lynch found that the residents of Boston tended to structure their cognitive images of the city with the same elements. He produced ingenious maps, such as this one, to demonstrate the collective "mental map" of the city, using symbols of different boldness or color to indicate the proportion of respondents who had mentioned each element.

(*Source:* Adapted from K. Lynch, *The Image of the City.* Cambridge, MA: M.I.T. Press, 1960, p. 146.)

Districts: Areas with an identifiable character (physical and/or cultural) that people mentally "enter" and "leave"; for example, a business district or an ethnic neighborhood.

Nodes: Strategic points and foci for travel; for example, street corners, traffic junctions, city squares.

Landmarks: Physical reference points; for example, distinctive landforms, buildings, or monuments.

Individual landscape features may function as more than one kind of cognitive element. A freeway, for instance, may be perceived as both an edge and a path in a person's cognitive image of a city. Similarly, a railroad terminal may be seen as both a landmark and a node.

Distortions in our cognitive images are partly the result of incomplete information. Once we get beyond our immediate living area, we know few spaces in complete detail. Yet our worlds—especially those of us in developed countries that are directly tied to global networks of communication and knowledge—are increasingly large in geographic scope. As a result, these worlds must be conceived, or understood, with limited direct stimuli. We have to rely on fragmentary and often biased information from other people, from books, magazines, television, and the Internet. Distortions in cognitive images are also partly the result of our own biases. What we remember about places, what we like or dislike, what we think is significant are all functions of our personalities, our experiences, and the cultural influences to which we have been exposed.

APPLY YOUR KNOWLEDGE

1. What is the "sociospatial dialectic", referred to by geographer Edward Soja? Give an example of both people reshaping space, and places influencing their inhabitants.

2. Use the five elements noted in this section of the chapter to map out your image of your college campus. What are the key paths, edges, districts, nodes, and landmarks that form your cognitive image?

Images and Behavior

Cognitive images are compiled, in part, through behavioral patterns. Environments are "learned" through experience. Meanwhile, cognitive images, once generated, influence behavior. Via this two-way relationship, cognitive images are constantly changing. Each of us also generates—and draws on—different kinds of cognitive images in different circumstances.

Elements such as districts, nodes, and landmarks are important in the kinds of cognitive images that people use to orient themselves and to navigate within a place or region. The more of these elements an environment contains—and the more distinctive they are—the more legible that environment is to people and the easier it is to get oriented and navigate. In addition, the more firsthand information people have about their environment, and the more they are able to draw on secondary sources of information, the more detailed and comprehensive their images are.

This phenomenon is strikingly illustrated in **Figure 7.7**, which shows the collective image of Los Angeles as seen by the residents of three different neighborhoods: Westwood, an affluent neighborhood; Avalon, a poor, inner-city neighborhood; and Boyle Heights, a poor, immigrant neighborhood. The residents of Westwood have a well-formed, detailed, and comprehensive image of the entire Los Angeles basin. At the other end of the socioeconomic spectrum, residents of the African-American neighborhood of Avalon, near Watts, have a vaguer image of the city, structured only by the major east–west boulevards and freeways and dominated by the gridiron layout of streets between Watts and the city center. The Spanish-speaking residents of Boyle Heights—even less affluent, less mobile, and more isolated by language—have an extremely restricted image of the city. Their world consists of a small area around Brooklyn Avenue and First Street, bounded by the landmarks of city hall, the bus depot, and Union Station.

The importance of these images goes beyond people's ability simply to navigate around their environments. The narrower and more localized people's images are, for example, the less they will tend to venture beyond their home area. Their behavior becomes circumscribed by their cognitive imagery. People's images of places also shape particular aspects of their behavior. Research on shopping in cities, for example, has shown that customers do not necessarily go to the nearest store or to the one with the lowest prices; they are influenced by the configuration of traffic, parking, and pedestrian circulation within their imagery of the environment. The significance of this clearly has not been lost on the developers of shopping malls, who always provide extensive space for free parking and multiple entrances and exits.

In addition, shopping behavior, like many other aspects of behavior, is influenced by people's values and feelings. A district in a city, for example, may be regarded as attractive or repellent, exciting or relaxing, fearsome or reassuring, or, more likely, a combination of such feelings. As is the case with all cognitive imagery, such images are produced through a combination of direct experience and indirect information, all filtered through personal and cultural perspectives. Images such as these often exert a strong influence on behavior. Returning to the example of consumer behavior, one of the strongest influences on shopping patterns relates to the imagery evoked by retail environments—something else that has not escaped the developers of malls, who spend large sums of money to establish the right atmosphere and image for their projects.

Shopping behavior is one narrow example of the influence of place imagery on behavior. Additional examples can be drawn from every aspect of human geography and at every spatial scale. The settlement of North America, for example, was strongly influenced by the changing images of the Plains and the West. In the early 1800s, the Plains and the West were perceived as arid and unattractive, an image reinforced by early atlases. When the railroad companies wanted to encourage settlement in these regions, they changed people's image of the Plains and the West with advertising campaigns that portrayed them as fertile and hospitable regions. The images associated with different regions and localities continue to shape settlement patterns. People draw on their cognitive imagery, for example, in making

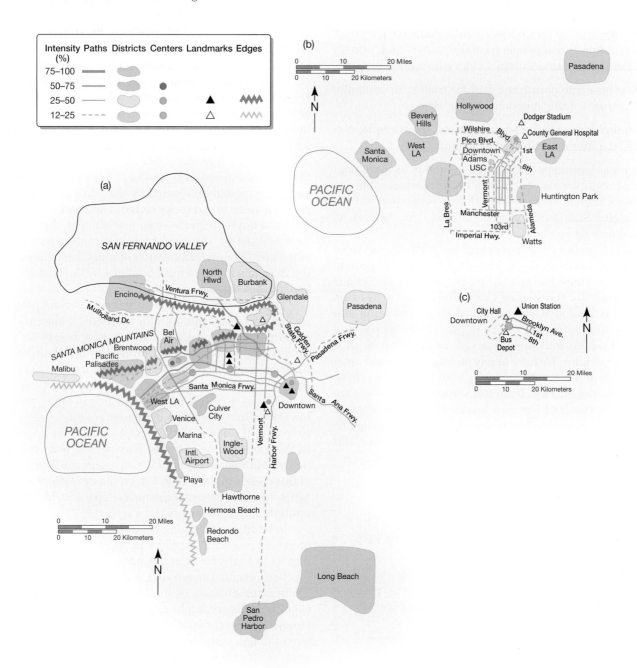

▲ **Figure 7.7 Images of Los Angeles** These images, as seen by residents of (a) Westwood, (b) Avalon, and (c) Boyle Heights are drawn to the same scale. The intensity of particular elements in the urban images of residents of different areas is measured by the percentage of residents who recall a particular element as being part of their own mental map of the city.

(*Source:* Adapted from P. Orleans, in R. Downs and D. Stea (eds.), *Image and Environment.* Chicago: Aldine, 1973, pp. 120–122.)

decisions about migrating from one area to another. **Figure 7.8** shows the composite image of the United States held by a group of Virginia Tech students, based on the perceived attractiveness of cities and states as places in which to live.

Another example of the influence of cognitive imagery on people's behavior is the way that people respond to environmental hazards, such as floods, droughts, earthquakes, storms, and landslides, and come to terms with the associated risks and uncertainties. Some people attempt to change the unpredictable into the knowable by imposing order where none really exists (resorting to folk wisdom about weather, for example),

while others deny all predictability and take a fatalistic view. Some tend to overestimate both the degree and the intensity of natural hazards, while others tend to underestimate them.

Finally, one aspect of cognitive imagery is of special importance in modifying people's behavior: the sentimental and symbolic attributes ascribed to places. Through their daily lives and the cumulative effects of cultural influences and significant personal events, people develop bonds with places. They do this simultaneously at different geographic scales: from the home, through the neighborhood and locality, to the national state. The tendency for people to do this has been

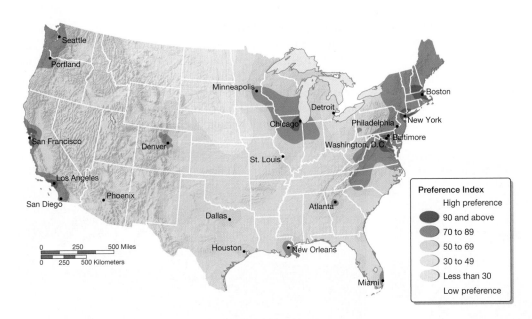

◄Figure 7.8 **Preference map of the United States** This isoline map illustrates collective preferences for cities in the coterminous United States as places in which to live and work, as expressed by architecture students at Virginia Tech in 1996. It is a generalization based on the scores the students gave to the 150 largest cities in the country.

called topophilia. **Topophilia** literally means "love of place." Geographers use the term to describe the complex emotions and meanings associated with particular places that, for one reason or another, have become significant to individuals. Most people have a home area, hometown, or home region for which they have a special attachment or sense of identity and belonging.

APPLY YOUR KNOWLEDGE

1. What is cognitive imagery? How do cognitive images direct to "learn" an environment? How is cognitive imagery built differently by each of us?

2. Place-making involves a wide range of cultural, social, and psychological processes. Choose a specific location in your town or university and elaborate on three ways that people make sense of this place through territoriality, proxemics, or cognitive images. Assess how age, race, gender, or class might influence this meaning making.

LANDSCAPES AS HUMAN SYSTEMS

Landscape is a term that means different things to different people. For some, the term brings to mind the design of formal gardens and parks, as in landscape architecture. For others, landscape signifies a bucolic countryside or even the organization of residences and public buildings. For still others, landscape calls to mind the artistic rendering of scenery, as in landscape painting.

Since 1925, when Carl Sauer advocated the study of the *cultural landscape* as a uniquely geographical pursuit, new generations of geographers have been expanding the concept. The fact that different people comprehend the landscape differently is central to the **humanistic approach** in geography. This approach places the individual—especially individual values, meaning systems, intentions, and conscious acts—at the center of analysis. As the Sudanese example given earlier

in the chapter suggests, children's perceptions of their worlds are different from those of their parents, and girls may see their world differently from boys, even in the same family.

Environmental perception and its close relative, behavioral geography, are interdisciplinary, drawing together geographers, landscape architects, psychologists, architects, and others. Professionals in these disciplines investigate what preferences people have in landscapes, how they construct cognitive images of their worlds, and how they find (or fail to find) their way around in various settings. A focus on the perceptions of individuals is an important counterweight to the tendency to talk about a social group or society more generally. Nevertheless, some critics argue that humanistic research has limited utility because individual attitudes and views do not necessarily add up to the views held by a group or a society.

One alternative to the humanistic approach explores both the role of larger forces, such as culture, gender, and the state, and the ways in which these forces enhance or constrain individuals' lives. Much recent cultural geographical work, therefore, conceptualizes the relationship of people and the environment as interactive, not one-way, and emphasizes the role that landscapes play in shaping and reinforcing human practices. This most recent conceptualization of landscape is more dynamic and complex than the one Carl Sauer advanced, and it encourages geographers to look outside their own discipline—to anthropology, psychology, sociology, and even history—to fully understand its complexity.

The Social Meaning of Landscapes

Geographers study vernacular (ordinary) landscapes because they reflect the distinctive attributes of particular places or regions. They also study "landscapes of power," such as clusters of corporate tower blocks, and "landscapes of despair," such as homeless encampments and **derelict landscapes**. The latter are landscapes that have experienced abandonment, misuse, disinvestment, or vandalism. Additionally, geographers study symbolic landscapes because they reflect certain values or ideals—either those intended by the builders or financiers of particular places or those perceived by other groups. Some individual buildings and structures are so powerfully symbolic that they come to stand for entire cities: the Eiffel Tower in Paris, the Colosseum in Rome, and the Sugarloaf in Rio de Janeiro, for example (**Figure 7.9**). However, generic landscapes of different kinds are most interesting to geographers.

Ordinary Landscapes Some landscapes come to symbolize entire nations or cultures. Quite ordinary landscapes can be powerfully symbolic because they are understood as being a particular kind of place. The stereotypical New England townscape, for example (**Figure 7.10a**), is widely taken to represent not just a certain type of regional architecture but the best that Americans have known "of an intimate, family-centered,

▲ **Figure 7.9 Famous landmarks** Some cities are immediately recognizable because of certain buildings, landmarks, or cityscapes that have come to symbolize them. These examples—(a) the Colosseum in Rome, (b) the Opera House in Sydney, (c) the Sugarloaf in Rio de Janeiro, and (d) the River Thames in London— are known worldwide.

▲ **Figure 7.10 Ordinary landscapes** Some ordinary cityscapes are powerfully symbolic of particular kinds of places. (a) The New England village and (b) the Main Street of Middle America are in this category, so much so that they have been taken as symbolizing the United States.

God-fearing, morally conscious, industrious, thrifty, democratic *community*."[3] A second ordinary townscape with powerful symbolic connotations is the typical Main Street of Middle America (**Figure 7.10b**). It is "middle" in several respects: between the frontier to the west and the cosmopolitan seaports to the east, between agricultural regions and industrial metropolises, between affluence and poverty. It has come to symbolize places with a balanced community, populated by property-minded, law-abiding citizens devoted to free enterprise and a certain kind of social morality. Another everyday landscape that looms large in the national and international popular imagination is the American highway. US Route 66, connecting Chicago to Los Angeles along more than 2,400 miles of roadway, is the classic example, and its quintessentially American signage and automobile-oriented buildings (See Box 7.2: "Visualizing Geography: Route 66") have become iconic of an era of optimism, movement, new opportunities, and indomitable spirit. John Steinbeck, in his famous novel *The Grapes of Wrath*, proclaimed U.S. Highway 66 the "Mother Road" because hundreds of thousands of migrants took it to California to escape the despair of the Oklahoma Dust Bowl in the 1930s. In 1946 Nat King Cole recorded the song—Route 66—that became a pop classic a generation later when it was recorded by Chuck Berry (in 1961) and the Rolling Stones (1964). Since the early 1990s, a loose coalition of federal agencies such as the National Park Service and the United States Forest Service, state entities, and private individuals has taken steps to interpret, preserve, and commemorate the highway in other ways.

A more recent example is provided by the landscapes of contemporary American suburbia. They are landscapes of bigness and ostentation, characterized by packaged developments and simulated settings. These conservative utopias of themed and fortified subdivisions of private master-planned developments (**Figure 7.11**) reflect a presumed reciprocity between size and social superiority.

The key point is that ordinary landscapes, as geographers such as Don Mitchell have established, are instruments of social and cultural power that naturalize political-economic structures as if they were simply given and inevitable. As powerful complexes of signs, they perform vital functions of social regulation. Many of the landscapes of contemporary American suburbia, for example, have naturalized an ideology of competitive consumption and disengagement from civic affairs.

Landscapes of National Identity

As we saw in Chapter 1, landscapes can become a way of picturing a nation, and some landscapes are powerfully symbolic of national identity: the classical Tuscan landscape (**Figure 1.13**), the West of Ireland (**Figure 1.14**), and the rural landscapes by which England is popularly imagined (**Figure 1.15**). Other examples include Mount Rushmore in the United States, the bulb fields of the Netherlands, and the fjords of Norway. National identity is, however, something that is constantly challenged, negotiated, and modified. In this context, some places take on particular importance for their iconography and the events that come to be associated with them. Tahrir Square, in Cairo, Egypt, for example, became familiar around the world during the "Arab Spring" of 2011, when the Square became the principal arena for contesting the fundamental values of the nation. In Britain, Trafalgar Square has become the multi-layered arena in which national identity has been both asserted and challenged (see Box 7.3: "Window on the World: Trafalgar Square and British National Identity").

[3]D. W. Meinig (ed.), *The Interpretations of Ordinary Landscapes: Geographic Essays.* New York: Oxford University Press, 1979, p. 27.

▼ **Figure 7.11 Vulgaria** The dominant theme in upscale residential development in the United States is size and ostentation. This example is from McLean, Virginia.

7.2 Visualizing Geography

Route 66

Between 1916 and 1925 the U.S. Congress created the National Highway System Program, which provided for the construction of the first public highways throughout the country. The official designation of the highway that would connect Chicago to Los Angeles with over 2,400 miles of roadway was Route 66. The diagonal configuration of Route 66 was also significant to the growing trucking industry, which by 1930 had come to rival the railroad for preeminence in U.S. shipping.

7.1 Historic Route 66

2,451 miles (3,945 km) through **8 states** and **3 time zones**

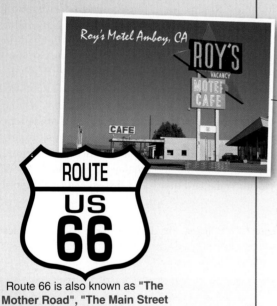

Route 66 is also known as **"The Mother Road"**, **"The Main Street of America"** and **"The Will Rogers Highway"**.

The first McDonald's restaurant was built along Route 66 in San Bernardino, California in 1945.

1926	1938	1940s	1950s
Route 66 was commissioned on November 26, 1926, connecting as many sections of existing road as possible. The final stretches of the road were constructed during the height of the Depression by thousands of unemployed young men who were put to work in a federal program.	The road was completed and its surface continuously paved from east to west.	Route 66 was an important part of the U.S. effort in World War II. The War Department needed improved highways to achieve rapid mobilization during the war and to promote national defense after the war. Because of its all-weather capability, Route 66 helped to facilitate the single greatest wartime mobilization in the history of the nation.	The end of the war found U.S. residents more mobile than ever before, and Route 66 helped to facilitate a new migration of easterners to the West, as well as to promote automobile travel as a tourist experience.

Sources: National Historic Route 66 Federation, http://www.historic66.com/, U.S. National Park Service. World Monuments Fund

7.2 Route 66 today

A 2012 economic impact study found that $132 million per year is spent in communities along historic route 66.

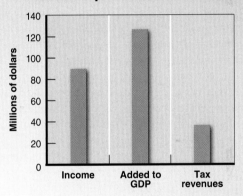

Income from Route 66

$38 million

$94 million

○ Tourism related to historic places

● Heritage preservation

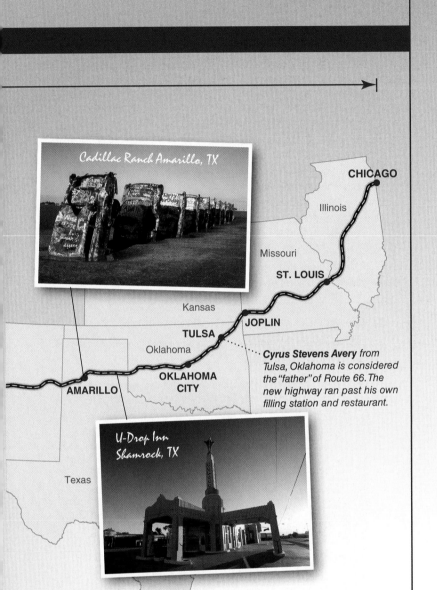

Cadillac Ranch Amarillo, TX

CHICAGO

Illinois

Missouri

ST. LOUIS

Kansas

JOPLIN

TULSA

Oklahoma

OKLAHOMA CITY

AMARILLO

Cyrus Stevens Avery from Tulsa, Oklahoma is considered the "father" of Route 66. The new highway ran past his own filling station and restaurant.

U-Drop Inn Shamrock, TX

Texas

National Impact of Route 66 Tourism

Millions of dollars

Income | Added to GDP | Tax revenues

Parts of Route 66 are still drivable today although it has been replaced by various interstate highways. Current maps do not include old Route 66.

1960s

By 1960 road travel and commerce were so important that Route 66 had become inadequate to the task of moving people and goods eastward and westward.

1970

By 1970 large parts of Route 66 had been bypassed or replaced by a new national interstate highway system.

1985

Route 66 was officially decommissioned on June 27, 1985.

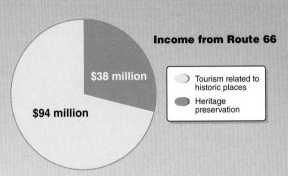

Route 66 Travel Itinerary

http://goo.gl/czCDMZ

1. List at least three ways Route 66 contributed to U.S. economy, culture and identity. How can a highway system change people's relationship with land, with each other and with new spaces? How do we change when we can travel longer distance in shorter spans of time? When we interface with more people and places? How did Route 66 create a specific cultural landscape?

2. What was the last road trip you took? Where did you go? What did you notice about the changing landscapes as you traveled? What would be your dream road trip?

7.3 Window on the World

Trafalgar Square and British National Identity

Trafalgar Square in London was developed at the height of British imperial power in the nineteenth century. As the country's collective imagination has grappled with historic events it has subsequently become an iconic public space of immense symbolic significance.

The square was heavily freighted with symbolic meaning from the start. Named after the decisive naval victory over the French at the 1805 Battle of Trafalgar, and with Admiral Nelson's Column (**Figure 7.B**) at its center, it also featured the equestrian statue of the absolutist monarch Charles I.

The present layout of the square (**Figure 7.C**) derives from designs by Charles Barry in the 1840s. His design featured four plinths for statuary (one in each corner of the square), and two fountain basins that were intended not only for decoration but also to hinder too large a crowd from gathering. This was a time when national identity was closely bound up with the fact of empire. Two of Barry's four plinths are occupied by statues of army generals, both of whom were prominently involved in British imperial activity in South Asia. A third plinth carries an equestrian statue of George IV in Roman dress. The fourth plinth was left vacant. Later, the busts of three World War I Royal Navy admirals—Jellicoe, Beatty, and Cunningham—were added to the square.

The square's prominence as the 'heart of empire' and its everyday accessibility to both workers and visitors have also made it an important gathering place for political protest and, therefore, a site of challenge and resistance: a place of contested meanings. Significant demonstrations have been held in the square in support of female suffrage, labor movements, and nuclear disarmament; and against fascism and unemployment.

On the other hand the square has also been the focal point of national celebrations, both planned and spontaneous: VE Day and VJ Day at the end of the Second World War, for example; New Year celebrations; and the celebration of national sporting victories. Recent programming of events in the square has emphasized Britain's—and especially London's—cosmopolitanism, celebrating the Chinese New Year, Saint Patrick's Day, Diwali, Vaisakhi, and Disability Rights, for example. It is also now routinely used for pop-cultural and educational events: all of which, given the Square's symbolic importance, contribute to the evolving discourse on national identity.

Under Mayor Ken Livingstone in the 2000s the square was remodeled as part of London's "World Squares for All" program. The Fourth Plinth, which had been vacant since the 1840s, has since been used for a series of sculptures and installations that have been deliberately provocative of issues of identity, tolerance, diversity, history, and modernity (**Figure 7.D**).

1. Why was the Battle of Trafalgar significant to Britain?
2. What is a place that you consider to be of special symbolic significance in your own country?

▲ **Figure 7.C** Trafalgar Square

▲ **Figure 7.B** Nelson's column

▲ **Figure 7.D** Fourth plinth with Marc Quinn's sculpture of Alison Lapper Pregnant

Landscape and Art   Landscape and art has been of interest to geographers for a long time beginning with the sketches done by the artists who accompanied the great explorers like Alexander von Humboldt. The field sketches produced by artists onboard ships, such as Captain Cook's *Endeavor* helped people "back home" to develop a geographical imagination of places and peoples far distant from them. As well, this type of landscape art in the service of science and empire was often understood to foreground accuracy and authenticity, being valued as a "true" representation of what was seen by these "on the spot" artists.

More recently, landscape paintings by Romantic artists J.W.M. Turner and John Constable have captured the attention of physical geographers who study them to understand weather and climate during the eighteenth century in England. Today, the earth structures built by land artists such as Robert Smithson, or the purposeful walks and landscape performances of artists such as Richard Long and Ana Mendiata, have prompted geographers to use this art as empirical material upon which to build their thinking about landscape, especially issues of power and politics, as well as enduring concerns with questions of nature, culture, and the environment. Oftentimes these works encourage us to ponder on our own relationships with the landscape, to reflect on the materiality of the stone, earth, and vegetation and the animate forces (weather, organic growth) that are part of it, to explore a multi-sensual engagement with landscape that is about feeling, smelling and touching, not just seeing.

Over the centuries the sense of landscape art as an information source, has been challenged by critics who worry that landscape representations such as the Romantic art hid the realities of life in the places depicted in the images. For example, the English rural scenes, known as "picturesques", painted by John Constable portrayed the eighteenth century countryside as a calm, contented, ordered place (**Figure 7.12**). Produced for rich landowners, Constable chose not to depict the rural unrest that had spread from Europe to England during this period that led to riots and hay burning in protest over ongoing conditions of hardship and starvation.

Geographers continue to explore landscape art for its potential to offer commentary and different possibilities for humans' relationship with the environment. Indeed some go so far as to claim that the sorts of environmental encounters art works create have the potential to foster an environmental ethics that is increasingly important in a world threatened by climate change.

Funerary Landscapes   Burial practices provide a different perspective on landscape and provide some interesting insights on human geography. Funerary landscapes are deliberately created and highly organized cultural landscapes. For countless generations, humans have disposed of their dead in the landscape, and the ways they have done it can tell us a lot about the socio-cultural ideals. Sociologist W. Lloyd Warner was one of the first to develop this perspective, in his 1959 book on symbolism in American cemeteries.[4] Warner pointed

[4]W. Lloyd Warner, *The living and the dead: a study of the symbolic life of Americans.* New Haven: Yale University Press, 1959.

▲ **Figure 7.12   The Cornfield**   This painting by John Constable (1776–1837), is a representation of the Romantic tradition.

out that cemeteries are symbolic replicas of the living community that express many of a community's basic beliefs and values. The social and status structures that organize the living community, he argued, are reflected and expressed in the forms and arrangement of cemeteries. Thus there are "good" and "bad" neighborhoods in cemeteries as well as in cities and towns. Similarly, sections of many of the cemeteries he analyzed were marketed exclusively to particular ethnic or religious groups, while exclusionary clauses kept out people of color altogether.

As social structures and cultural ideals change over the course of time, so do funerary landscapes evolve. Early in United States' history, relatively small and slow-growing communities tended to bury their dead in churchyards or, in the rural South, in family burial grounds. Many followed the ancient Christian practice of burials with the head of the body to the west, feet to the east, thus facing the presumed direction from where Christ is supposed to appear again (an orientation, incidentally, favored by pre-Christian communities for an entirely different reason: facing sunrise). This typically leads to a logical spatial arrangement of burial plots in rows, and a formal, rectilinear, layout (**Figure 7.13a**).

By the early decades of the nineteenth century the beginnings of industrialization and the growth of urban populations led to a significant cultural shift: the so-called **American Renaissance**, rooted in the literary works of Ralph Waldo Emerson and Henry David Thoreau and propagated by

Walt Whitman and Herman Melville. Thoreau, a disciple of Emerson's, popularized the idea of nature as a spiritual wellspring for city dwellers in his book *Walden* (1854). It led to the popularization of pastoral and picturesque settings, and one of the first manifestations of these ideals came in "rural" cemeteries like Laurel Hill in Philadelphia, PA, Green-Wood in Brooklyn, NY, Evergreen in Portland, ME, and Forest Hill, in Madison, WI. Rural cemeteries were consciously designed to provide sanctuary, solitude, quiet, adornment, and beauty. The social ideals of the time were expressed in spaciousness and garden-like settings with shrubs, lawns, and (symbolically) evergreen trees (**Figure 7.13b**).

By the late nineteenth century, the romanticism of the American Renaissance had been eclipsed by a more aggressive and materialistic ethos. Cemeteries changed accordingly, with elaborate decorative tombstones and monuments intended to communicate the wealth, affluence, and status of the deceased (**Figure 7.13c**). Early in the twentieth century, the influence of the City Beautiful movement, with its ideals of order and rationalized efficiency, led to simpler and more conservative

funerary landscapes, and by the end of the twentieth century many cemeteries had become so rationalized that tombstones and monuments had almost disappeared from park-like landscapes (**Figure 7.13d**).

In recent decades the cost of land and of burials has led to an increase in cremation as an American funerary practice. Meanwhile, another socio-cultural shift is beginning to be reflected in funerary landscapes. Increasing environmental awareness has led to a growing trend for " green" burials. According to the Casket and Funeral Association of America, more than 91,000 cubic yards of hard wood, 90,000 tons of steel, 2,700 tons of copper and bronze, and 800,000 gallons of embalming fluid are used in traditional burials every year. Cremation also takes its toll on the environment: it can take up to four hours at temperatures ranging from 760 to 1,150 degrees Celsius to fully cremate a body. This is roughly equivalent to the energy required to drive a car 4,800 miles. With an increasing focus on minimizing people's carbon footprint during life, some are turning to natural burial practices as a way of demonstrating their eco-credentials, using natural coffins or caskets

(a)

(c)

(b)

(d)

▲ **Figure 7.13** **Funerary landscapes** (a) New England church and churchyard, in Barnstead, New Hampshire, rectilinear layout; (b) Laurel Hill Cemetery in Philadelphia, Pennsylvania, a rural-style cemetery; (c) Oakland Cemetery, Atlanta, Georgia, with an elaborate Victorian landscape; (d) Golden Gate National Cemetery in San Bruno, California, a lawn-park cemetery.

to enable the body to decompose naturally, returning its nutrients to the ground in woodland, meadowland, and open field settings.

CODED SPACES

A dynamic and complex approach to understanding landscape is based on the conceptualization of **landscape as text**, by which we mean that, like a book, landscape can be read and written by groups and individuals. This approach departs from traditional attempts to systematize or categorize landscapes based on the different elements they contain. The landscape-as-text view holds that landscapes do not come ready-made with labels on them. Rather, there are "writers" who produce landscapes and give them meaning, and there are "readers" who consume the messages embedded in landscapes. Those messages can be read as signs about values, beliefs, and practices, though not every reader will take the same message from a particular landscape (just as people may differ in their interpretation of a passage from a book).

In short, landscapes both produce and communicate meaning, and one of our tasks as geographers is to interpret those meanings. In order to interpret or read our environment, we need to understand the language in which it is written. We must learn how to recognize the signs and symbols that go into the making of landscape. The practice of writing and reading signs is known as **semiotics**.

Semiotics asserts that innumerable signs are embedded or displayed in landscape, space, and place, sending messages about identity, values, beliefs, and practices. These signs may have different meanings for those who produce them and those who read, or interpret, them. Some signs are so subtle as to be recognizable only when pointed out by a knowledgeable observer; others may be more readily available and more ubiquitous. For example, semiotics enables us to recognize that by the way they dress, people, including college students, send messages about who they are and what they value. For some of us, certain types, such as the "skater," the "goth," or the "emo," are readily identifiable by their clothes, hairstyle, or footwear, by the books they carry, or even by the food they eat.

Commercial Spaces

Semiotics, however, is not only about the signs that people convey by their mode of dress. Messages are also deployed through the landscape and embedded in places and spaces. Consider the very familiar landscape of the shopping mall. Although there is certainly a science to the size, scale, and marketing of a mall based on demographic research as well as environmental and architectural analysis, there is more to the mall than these concrete features. The placement and mix of stores and their interior design, the arrangement of products within stores, the amenities offered to shoppers, and the ambient music all combine to send signals to the consumer about style, taste, and self-image (**Figure 7.14**). Called by some "palaces of consumption," malls are complex semiotic sites, directing important signals

▲ **Figure 7.14** Aeroville shopping mall, Paris

not only about what to buy but also about who should shop there and who should not.

And yet, as much as Americans seem to enjoy shopping, a great many express disdain for shopping and the commercialism and materialism that accompany it. Shopping is a complicated activity about which people feel ambivalence. It is not surprising, therefore, that developers have promoted shopping as a kind of tourism. The mall is a "pseudoplace" meant to encourage one sort of activity—shopping—by projecting the illusion that something else besides shopping (and spending money) is actually going on.

The South Coast Plaza in Orange County, California, is a highly popular retail center in the United States, and with almost 3 million square feet of enclosed space covering 128 acres, it is one of the largest malls in the country. It is the most profitable too, boasting over $1.5 billion annually in sales. The mall contains 250 retail outlets, including luxury goods stores like Gucci, Versace, Chanel, Tiffany, Jimmy Choo, and Cartier as well as more popular, less pricey upscale stores like Bloomingdales and Nordstrom and finally middlebrow stores like Macy's and Sears. The latter stores anchor the mall at its outside corners, while the more expensive and luxurious shops and boutiques occupy interior locations. Thus the central stretches of the mall convey signs of affluence and luxury appealing to upper-class patrons while the periphery is more oriented to necessity and practicality for middle- and lower-middle-class patrons.

However complex the messages that malls send, one focus is consistent across class, race, gender, age, ethnicity, and other cultural boundaries: consumption, a predominant aspect of globalization. Indeed, malls are the early twenty-first century's spaces of consumption, where just about every aspect of our lives has become a commodity. Consumption—or shopping—defines who we are more than ever before, and what we consume sends signals about who we want to be. Advertising and the mass media tell us what to consume, equating ownership of products with happiness, a good sex life, and success in general. Within the space of the mall these signals are collected and resent. The architecture and design of the mall are an important part of the semiotic system shaping our choices and molding our preferences. As architectural historian Margaret Crawford writes:

> All the familiar tricks of mall design—limited entrances, escalators placed only at the end of corridors, fountains and benches carefully positioned to entice shoppers into stores—control the flow of consumers through the numbingly repetitive corridors of shops. The orderly processions of goods along endless aisles continuously stimulate the desire to buy. At the same time, other architectural tricks seem to contradict commercial consideration. Dramatic atriums create huge floating spaces for contemplation, multiple levels provide infinite vistas from a variety of vantage points, and reflective surfaces bring near and far together. In the absence of sounds from the outside, these artful visual effects are complemented by the "white noise" of MUZAK and fountains echoing across enormous open courts.[5]

Malls, condominium developments, neighborhoods, university campuses, and any number of other possible geographic sites possess codes of meaning. By thinking about their complex makeup and the individuals and groups they seek to influence, it is possible to interpret them and understand the implicit messages they contain.

APPLY YOUR KNOWLEDGE

1. What is semiotics and how does it help us to understand important information about landscapes?

2. Apply what you just learned about codes and provide a description of the systems of signification that operate in your neighborhood. Think about different spaces in your neighborhood, how it is organized, who the population is, where the common spaces are, what levels of surveillance are operating, et cetera. What do these signs tell you about the "text" of your neighborhood landscape?

GLOBALIZATION AND PLACE-MAKING

The spread of Modernity to peripheral regions can be seen as part of globalization. These globalization processes have not only brought about a generalization of forms of industrial production, market behavior, trade, and consumption but reinforced and extended the commonalities among places. Three factors are especially important in this context.

First, mass communications media have created global culture markets in print, film, music, television, and the Internet. Indeed, the Internet has created an entirely new *kind* of space—cyberspace—with its own "landscape" (or technoscape) and its own embryonic cultures. The instantaneous character of contemporary communications has also made possible the creation of a shared, global consciousness from the staging of global events such as the Olympic Games and the World Cup. Mass communications media have also facilitated the development of celebrity culture, which has its own associations with specific places (see Box 7.4 "Geography Matters: The Geography of Stardom: Celebrity and Place"). Second, mass communications media have diffused certain values and attitudes toward a wide spectrum of sociocultural issues, including citizenship, human rights, child rearing, social welfare, and self-expression. Third, international legal conventions have increased the degree of standardization and level of harmonization not only of trade and labor practices but of criminal justice, civil rights, and environmental regulations.

These commonalities have been accompanied by the growing importance of material consumption within many cultures. Increasingly, people around the world are eating the same foods, wearing the same clothes, and buying the same consumer products. Yet the more people's patterns of consumption converge, the more fertile the ground for countercultural movements. The more transnational corporations undercut the authority of national and local governments to regulate economic affairs, the greater the popular support for regionalism. The more universal the diffusion of material culture and lifestyles, the more local and ethnic identities are valued. The faster the information highway takes people into cyberspace, the more they feel the need for a subjective setting—a specific place or community—they can call their own. The faster the pace of life in search of profit and material consumption, the more people value leisure time. And the faster their neighborhoods and towns acquire the same generic supermarkets, gas stations, shopping malls, industrial estates, office parks, and suburban subdivisions, the more people feel the need for enclaves of familiarity, centeredness, and identity. The United Nations Center for Human Settlements (UNCHS) notes:

> In many localities, people are overwhelmed by changes in their traditional cultural, spiritual, and social values and norms and by the introduction of a cult of consumerism intrinsic to the process of globalization. In the rebound, many localities have rediscovered the "culture of place" by stressing their own identity, their own roots, their own culture and values and the importance of their own neighborhood, area, vicinity, or town.[6]

Going Slow

One example of the impulse for people to recover a sense of place is provided by the slow city (Cittaslow) movement. The Cittaslow movement is a grassroots response to globalization and is closely related to the longer-established and better-known Slow Food movement. The aims of the two movements

---

[5]M. Crawford, "The World in a Shopping Mall," in M. Sorkin (ed.), *Variations on a Theme Park.* New York: Noonday Press, 1992, p. 14.

[6]United Nations Center for Human Settlements, *Global Report on Human Settlements 2001.* London: Earthscan, 2001, p. 4.

are different but complementary: In broad terms, both organizations are in favor of local, traditional cultures, a relaxed pace of life, and conviviality. Both are a response to the increasing pace of everyday life associated with the acceleration of money around local, national, and global circuits of capital. Both are hostile to big business and globalization, though their driving motivation is not so much political as ecological and humanistic. **Slow food** is devoted to a less hurried pace of life and to the true tastes, aromas, and diversity of good food (**Figure 7.15**). The movement also serves as a rallying point against globalization, mass production, and the kind of generic fast food represented by U.S.-based franchises like McDonald's, Burger King, Pizza Hut, Taco Bell, and Kentucky Fried Chicken. By 2011, the slow food movement, based in Bra, near Turin in northern Italy, had established convivia (local branches) in more than 100 countries, with over 100,000 members worldwide. Its campaigns cover a range of specific causes, from protecting the integrity of chocolate to promoting the cultivation of traditional crop varieties and livestock breeds and opposing genetically engineered foods.

The slow city movement was formed in October 1999, when Paolo Saturnini, mayor of Greve-in-Chianti, a Tuscan hill town, organized a meeting with the mayors of three other municipalities (Orvieto, Bra, and Positano) to define the attributes that might characterize a *città lente*—slow city. At their founding meeting in Orvieto, the four mayors committed themselves to a series of principles that included working toward calmer and less polluted physical environments, conserving local aesthetic traditions, and fostering local crafts, produce, and cuisine. They also pledged to use technology to create healthier environments, to make citizens aware of the value of more leisurely rhythms to life, and to share their experience in seeking administrative solutions for better living. The goal is to foster the development of places that enjoy a robust vitality based on good food, healthy environments, sustainable economies, and the seasonality and traditional rhythms of community life.

In South Korea, economic success and modernization took place very rapidly. Large business conglomerates (called *chaebols*) have been dominant, and entrepreneurial opportunities for small firms have been limited. Local traditions and cultures have vanished from many places because of the rapid change from a predominantly rural and agricultural society to a highly modern, internationally connected, and globalized society. There is, however, a growing recognition of the value of history, culture, traditions, and local arts and crafts. "Slowing down" has also caught on with stressed city dwellers. In Samjicheon, a South Korean Cittaslow town (**Figure 7.16**), there are workshops where participants can learn how to make traditional dishes such as kimchi and practice traditional handwriting, papermaking, or textile dyeing. A special market for local products, arts, and crafts is held every other week near the town's Cittaslow visitor centre and a daily local market in the newer part of the village allows farmers to sell their produce. Meanwhile, the town has restored its traditional gravel pathways and creeks, and introduced strict building regulations to ensure that new buildings conform to the town's historic character.

Places as Objects of Consumption

In much of the world, people's enjoyment of material goods now depends not just on their physical consumption or use. It is also linked more than ever to the role of material culture as a social marker. A person's home, automobile, clothing, reading, viewing, eating, and drinking preferences, and choice of vacations are all indicators of that person's social distinctiveness and sense of style. This pressures individuals to continuously search for new and distinctive styles. The wider the range of foods, products, and ideas from around the world—and from past worlds—the greater the possibilities for establishing such styles.

Given that material consumption is so central to the repertoire of symbols, beliefs, and practices of modern

▼ **Figure 7.15 Slow food festival, Bra, Italy** The international Slow Food organization is headquartered in the town of Bra, where a festival featuring traditional and artisan cheeses is held every other year.

▼ **Figure 7.16 Samjicheon, South Korea** One of the first officially designated 'slow cities' in East Asia.

7.4 Geography Matters

The Geography of Stardom: Celebrity and Place

By Elizabeth Currid-Halkett, University of Southern California

Is Celebrity Placeless?

Celebrity is everywhere and nowhere. In his book, *The Image*, the social critic Daniel Boorstin observed that "[Celebrities] help make and … publicize one another. Being known primarily for their well-knownness, celebrities intensify their celebrity images simply by becoming widely known for relations among themselves." One aspect of celebrity status involves being photographed repeatedly with other celebrities. This process is why stars—whether Marilyn Monroe, Madonna, Paris Hilton or Angelina Jolie—become such iconic and powerful images around the world (**Figure 7.E**).

Yet celebrity photos are taken in particular places. Because celebrity culture and its stars appear in print, television, movies, and the internet, we don't realize that those images are photographed on streets, in restaurants, and outside of houses in real places. Like other cultural industries – fashion, art, film, music – celebrity is a global industry but also requires physical places and real logistics to produce. Andy Warhol's Marilyn silkscreens are recognized around the world, plastered on coffee mugs and t-shirts and college dorm walls. But Warhol himself made these silkscreen images in his New York City art studio called The Factory. Same goes for Prada's latest handbags (designed and made in Milan, Italy) or Jay-Z's latest album—we

may hear "Empire State of Mind" in Tokyo, but the work of writing and producing the song came from a recording studio in New York City. Celebrity may seem placeless but, like technology in Silicon Valley or finance on Wall Street, it requires people, resources, and networks that are located in particular cities around the world. In fact, it requires the backdrop of the cultural industries to generate celebrity culture and its people and events: film awards, art exhibitions, fashion runway shows, and theaters.

The Practicality of a Seemingly Nebulous Industry

What makes a star a star is hard to define. The celebrity industry and celebrity culture feel ephemeral and intangible. The same type of ambiguity emerges in other cultural industries. For example, why Andy Warhol or Picasso and not someone else? Why Bob Dylan and not another aspiring musician? Because culture is about taste, why one person becomes successful—or attains what sociologists call **symbolic capital**— is not always clear. However, in studying cultural industries we learn that beneath the aesthetics there is a practical scaffolding . Warhol needed paints and canvases to produce his work, a gallery to show it, a dealer to sell it, and art critics to write about it and tell the world about his talent. Similarly, Jennifer Aniston needs

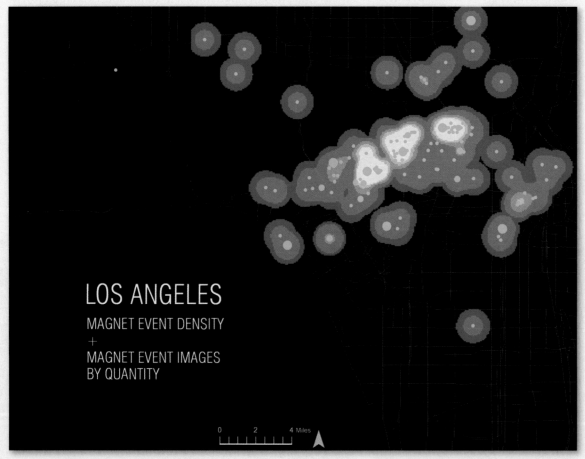

◀ **Figure 7.E The geography of celebrity:** Map of where celebrity events occur in Los Angeles using photographic data to geocode and map them.

LOS ANGELES

MAGNET EVENT DENSITY
+
MAGNET EVENT IMAGES
BY QUANTITY

0   2   4 Miles

a TV show or movies to be filmed in, a stylist to make her look beautiful, a trainer to keep her in shape, an agent to negotiate her deals, a lawyer to read her contracts, a photographer to take her pictures, and the media to publish photos. The TV show needs a back lot, a studio, and skilled producers and directors, most of which are behind the scenes.

Celebrity is a practical matter that relies on resources and people to get things done. This concentration of skilled labor (people), resources and related businesses in one place are what economists and geographers call *agglomeration economies* or *industrial clusters*. Often these concepts are applied to more conventional industries like automobiles or apparel manufacturing: Think about Detroit's auto industry or Silicon Valley's technology hub – they may produce more tangible products, but the underlying need for extensive resources and talented people are the same.

While agglomeration economies help support an industry they also make those specific places that have such clusters the only places in the world where the industry truly thrives—which is why Detroit dominated the automobile industry for so long and New York City's Wall Street remains one of the few global capitals of finance. Similarly, only a few places in the world produce celebrity culture: Los Angeles, London, and New York are the epicenters of stardom—homes not just to the stars but to the industries (fashion, art, film and music), people (producers, directors, art dealers) and media who create them (**Figure 7.F**).

▼ **Figure 7.F The world cities of celebrity** Social network graph displaying the connections across cities based on the stars that frequent them (data attained from photographs, who is in them and where they were taken).

How Do We Study the Importance of Celebrity?

As intangible as celebrity culture seems to be, it also possesses something most industries do not: worldwide interest and media following. We can understand celebrity culture through the very photos that uphold the celebrity industry, the very material that Boorstin said reaffirmed their "known for their well-knownness" quality.

Like other sectors, the celebrity industry can be studied through basic industrial and economic analysis. Recall that agglomeration economies offer a high concentration of practical resources and people needed for the success of a particular industry. From a research perspective, these supporting jobs show up in government employment and establishment data.

There is, however, something unique about the celebrity industry that sets it apart from other industries. The visual nature of celebrity allows us to study it in innovative ways – photographs, blogs, and newspaper stories become *data* in a way that is not possible with medicine, finance, technology, and even fashion. The stars and their events, fleeting as they are, give us tangible evidence of the celebrity industry and its economic and social importance around the world and in a few key cities. After studying it, one observes celebrity industry is not so placeless at all.

---

1. How do you see the celebrity industry as similar or different to other industries in general? How about to other cultural industries?

2. How do you think social media has reframed older versions of celebrity culture (e.g. Hollywood)?

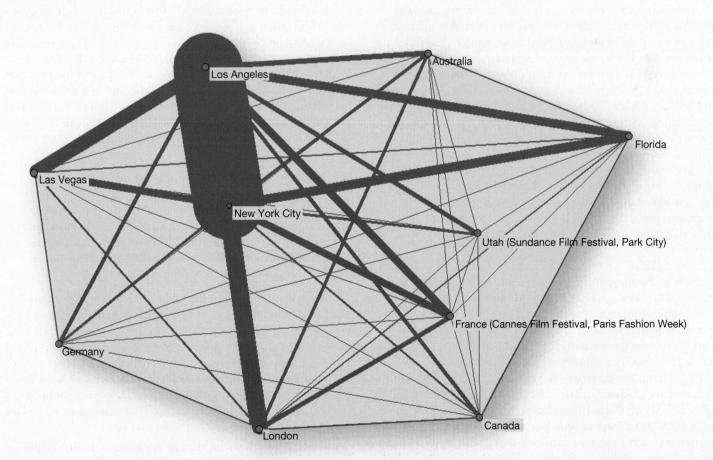

societies, the "culture industries"—advertising, publishing, communications media, and popular entertainment—have become important shapers of spaces, places, and landscapes. Because the symbolic meanings of material culture must be advertised (in the broadest sense of the word) in order to be shared, advertising (in its narrower sense) has become a key component of contemporary culture and place-making. In addition to stimulating consumer demand, advertising has always had a role in teaching people how to dress, how to furnish a home, and how to signify status through groupings of possessions.

In the 1970s and 1980s, however, the emphasis in advertising strategies shifted away from presenting products as newer, better, more efficient, and more economical to identifying them as the means to self-awareness, self-actualization, and group stylishness. Increasingly, products are advertised in terms of their association with a particular lifestyle rather than in terms of their intrinsic utility. Many of these advertisements deliberately draw on international or global themes, and some entire advertising campaigns (for Coca-Cola, Benetton, and American Express, for example) have been explicitly based on the theme of cultural globalization. Many others rely on stereotypes of particular places or kinds of places (especially exotic, spectacular, or "cool" places) in creating the appropriate context or setting for their product. Images of places therefore join with images of global food, architecture, pop culture, and consumer goods in the global media marketplace. Advertisements both instruct and influence consumers about products and also about spaces, places, and landscapes.

One result of these trends is that contemporary cultures rely much more than before not only on material consumption but also on *visual* and *experiential* consumption: the purchase of images and the experience of spectacular and distinctive places, physical settings, and landscapes. Visual consumption can take the form of magazines, television, movies, sites on the World Wide Web, tourism, window shopping, people-watching, or visits to galleries and museums. The images, signs, and experiences that are consumed may be originals, copies, or simulations.

The significance of the increased importance of visual consumption for place-making and the evolution of landscapes is that settings such as theme parks, shopping malls, festival marketplaces, renovated historic districts, museums, and galleries have all become prominent as centers of cultural practices and activities.

Examples of the re-creation and refurbishment of historic districts and settings are widespread and have become a mainstay of the "heritage industry." This industry, based on the commercial exploitation of the histories of peoples and places, is now worldwide, as evidenced by the involvement of the United Nations Educational, Scientific, and Cultural Organization (UNESCO) in identifying places for inclusion on World Heritage lists. In countries like the United Kingdom, with a high density of historic districts and settings, place marketing relies heavily on the heritage industry. In the United Kingdom, more than 90 million people pay to view about 650 historic properties each year, and millions more visit free-of-charge heritage sites such as cathedrals, field monuments, remote

ruins, and no longer useful industrial waterways. In 2010, more than 200 million tourists visited designated heritage sites in the United Kingdom, spending about $53 billion on entry fees, retail sales, travel, and hotel accommodations.

A key question when it comes to heritage lists is *"Whose heritage?"* The buildings chosen to represent a community's heritage will almost inevitably exclude or alienate some members of that community. Herbert Gans, a prominent American urban sociologist and critic, wrote an article for the New York Times in 1975 in which he attacked the New York Landmarks Commission for what he called the rewriting of New York's architectural history. He was concerned that the only buildings that the Commission tended to designate were the stately mansions of the rich and buildings designed by famous architects, so that the Commission was simply preserving the elite portion of the architectural past. More recently, urban historian Delores Hayden analyzed heritage buildings in Los Angeles that had been selected by panels of experts and found that most were not even remotely representative of the largely black and Hispanic populations of the communities in which they were located.

The architectural elite does not always prevail, however. In London's East End a 1972 housing project designed by acclaimed architects Peter and Alison Smithson was scheduled for demolition in 2014. *Building Design* magazine, backed by star architects like Zaha Hadid and Richard Rogers, ran a campaign to get the project officially listed in order to save it from destruction. English Heritage, the official body responsible for listing protected buildings, declined, noting that while Robin Hood Gardens might be admired by architects for its sculptural qualities, it had failed as a place for human beings to live. More than seventy-five percent of the project's residents had supported the demolition when consulted by the local public authority.

One important consequence of the heritage industry is that urban cultural environments are vulnerable to a debasing and trivializing "Disneyfication" process. The United Nations Human Settlements Programme (UN-Habitat) *Cities in A Globalizing World - Global Report on Human Settlements 2001* notes:

> The particular historic character of a city often gets submerged in the direct and overt quest for an international image and international business. Local identity becomes an ornament, a public relations artifact designed to aid marketing. Authenticity is paid for, encapsulated, mummified, located, and displayed to attract tourists rather than to shelter continuities of tradition or the lives of its historic creators.[7]

As a result, contemporary landscapes contain increasing numbers of inauthentic settings—what geographer David Harvey has called the "degenerative utopias" of global capitalism. These are as much the product of contemporary material and visual culture as they are of any cultural heritage. Particularly influenced by images and symbols derived from movies, advertising, and popular culture, examples include the re-creation of the Wild West in the fake cowboy town of Old

---

[7]United Nations Center for Human Settlements, Global Report on Human Settlements 2001. London: Earthscan, 2001, p. 4.

Tucson in Arizona; the Bavarian village created in Torrance, California; and the simulation of other cities and tourist hot spots in Las Vegas, Nevada (**Figure 7.17**). The town of Helen, Georgia, boasts a full-scale reconstruction of a Swiss mountain village, complete with costumed staff and stores selling Swiss merchandise. In China, a reproduction of the entire village of Hallstatt, Austria, has been constructed in Guangdong province (**Figure 7.18**). Hallstatt is a UNESCO World Heritage site, and its residents were not pleased to hear about the copycat version.

APPLY YOUR KNOWLEDGE

1. What is the Cittaslow movement? What are the values of the international movement and what influences are they resisting? Discuss the Cittaslow towns in Italy and Korea detailed in the text and how they maintain "slow " values.

2. How have places become objects of consumption? Define what geographer David Harvey means by the term "degenerative utopia" and think of an example not used in the text.

▲ **Figure 7.17 Venice, Las Vegas** The boundaries between the heritage industry and the leisure and entertainment industries have become increasingly blurred, and a great deal of investment has been channeled toward the creation of inauthentic "historic" settings whose characteristics owe more to movies and popular stereotypes than historic realities. Shown here is the fake canal within the Venetian Hotel in Las Vegas, imitating the cityscape of Venice, Italy.

(a)

◀ **Figure 7.18 Chinese mimicry:** (a) the village of Hallstatt, Austria, and (b) a copycat version of Hallstatt in China's southern city of Huizhou in Guangdong province.

(b)

FUTURE GEOGRAPHIES

Globalization has already brought a significant degree of homogenization of culture through the language of consumer goods. This is the material culture of the West, enmeshed in Airbus jets, CNN, music video channels, cell phones, and the Internet and swamped by Coca-Cola, Budweiser, McDonald's, GAP clothing, Nikes, iPods, PlayStations, Toyotas, Disney franchising, and formula-driven Hollywood movies (**Figure 7.19**). Furthermore, sociologists have recognized that a distinctive culture of "global metropolitanism" is emerging among the transnational elite. This is simply homogenized culture at a higher plane of consumption (French wines instead of Budweiser, molecular gastronomy and international cuisine instead of McDonald's, Hugo Boss clothes instead of Levis, Range Rovers instead of Toyotas, and so on). The members of this new culture are people who make international conference calls, make decisions and transact investments that are transnational in scope, edit the news, design and market international products, and travel the world for business and pleasure, and it is likely to grow significantly in size and influence.

The idea of the emergence of a single global culture is too simplistic, but the pervasive emphasis on material, visual, and experiential consumption means that many aspects of contemporary culture will increasingly transcend local and national boundaries. More and more of the world's population are world travelers—either directly or via the TV in their living room—so that many are knowledgeable about aspects of others' cultures. This contributes to **cosmopolitanism**, an intellectual and aesthetic openness toward divergent experiences, images, and products from different cultures.

Cosmopolitanism is an important geographic phenomenon because it fosters a curiosity about all places, peoples, and cultures, together with at least a rudimentary ability to map, or situate, such places and cultures geographically, historically, and anthropologically. It also suggests an ability to reflect upon, and judge aesthetically between, different places and societies. Optimists would speculate that this bodes well for global peace and understanding: We can, perhaps, more easily identify with people who use the same products, listen to the same music, and appreciate the same sports stars that we do. At the same time, however, focusing attention on material consumption obscures the emergence of other trends. As people's lives are homogenized through their jobs and their material culture, many of them want to revive subjectivity, reconstruct we/us feelings, and reestablish a distinctive cultural identity. One feasible outcome of this is an increased probability of cultural and territorial conflict.

APPLY YOUR KNOWLEDGE

1. What is cosmopolitanism? What are some of the positive and negative aspects of this culture?

2. Think of three examples of how globalization of culture shapes places and landscapes. Compare the values of the Cittaslow movement to globalization of culture, and analyze how homogenization can change places and landscapes.

▲ **Figure 7.19 Western pop culture** Hollywood movies dominate the schedule at the Gaumont cinema in Paris.

■ CONCLUSION

Geographers study the interdependence between people and places and are especially interested in how individuals and groups acquire knowledge of their environments and how this knowledge shapes their attitudes and behaviors. People ascribe meanings to landscapes and places in many ways, and they also derive meanings from the places and landscapes they experience. Different groups of people experience landscape, place, and space differently. For instance, the experience that rural Sudanese children have of their landscapes and the ways in which they acquire knowledge of their surroundings differ from how middle-class children in an American suburb learn about and function in their landscapes. Furthermore, both landscapes elicit a distinctive sense of place that is different for those who live there and those who simply visit.

As indicated in previous chapters, the concepts of landscape and place are central to geographic inquiry. They are the result of intentional and unintentional human action, and every landscape is a complex reflection of the operations of the larger society. Geographers have developed categories of landscape to help distinguish the different types that exist. Ordinary landscapes, such as suburban neighborhoods, are ones that people experience in the course of their everyday lives. By contrast, symbolic landscapes represent the particular values and aspirations that their developers and financiers want to impart to a larger public. An example is Mount Rushmore in the Black Hills of South Dakota, designed and executed by sculptor Gutzon Borglum. Chiseling the heads of George Washington, Thomas Jefferson, Theodore Roosevelt, and Abraham Lincoln into the granite face of the mountain,

Borglum intended to construct an enduring landscape of nationalism in the wilderness.

More recently, geographers have come to regard landscape as a text, something that can be written and read, rewritten, and reinterpreted. This concept suggests that a landscape can have more than one author, and different readers may derive different meanings from what is written there. The idea that landscape can be written and read is further supported by the understanding that the language in which the landscape is written is a code. To understand the significance of the code is to understand its semiotics, the language in which the code is written. The code may be meant to convey many things, including a language of power or of playfulness, a language that elevates one group above another, or a language that encourages imagination or religious devotion and spiritual awe.

Globalization has altered cultural landscapes, places, and spaces differently as individuals and groups have struggled to negotiate the local impacts of this widespread shift in cultural sensibilities. The shared meanings that insiders derive from their place or landscape have been disrupted by the intrusion of new sights, sounds, and smells as values, ideas, and practices from one part of the globe have been exported to another. The Internet and the emergence of cyberspace have meant that new spaces of interaction have emerged that have neither a distinct historical memory attached to them nor a well-established sense of place. Because of this, cyberspace carries with it some unique possibilities for cultural exchange. It remains to be seen, however, whether access to this new space will be truly open—or whether the Internet will become another landscape of power and exclusion.

LEARNING OUTCOMES REVISITED

■ *Investigate* how environment shapes people and how people shape environments.

People not only filter information from their environments through neurophysiological and psychological processes but draw on personality and culture to produce cognitive images of their environment—pictures or representations of the world that can be called to mind through the imagination. Thus human–environment relationship results in a variety of ways of understanding the world around us as well as different ways of being in the world as information about our environment is filtered by people.

■ *Recognize* that place-making stands at the center of issues of culture and power relations and is a key part of the systems of meaning through which humans make sense of the world.

Places are the result of a wide range of forces from economic to social. Economically, places emerge through all sorts of transactions that result from the complexities of the land market. But places are also more than just real estate. They can reflect tensions between social groups as well as harmonious interaction.

■ *Define* the concept of territoriality both as an instinctive process and the result of socio-political systems.

Territoriality is the persistent attachment of individuals or peoples to a specific location or territory. Some social scientists have argued that territoriality is a natural human instinct, impacting our sense of safety and security, our sense of stimulation and our identity. Proxemics is the study of the social and political meanings people give to space and includes protocols about space. Territoriality is in many ways a material form of power, as it regulates interaction, access to resources and people, and can be a symbol of group membership (seen in borders delineating "us" and "them").

■ *Identify* how different cultural identities and status categories influence the ways people experience and understand landscapes, as well as how they are shaped by—and able to shape—landscapes.

Among the most important relations are the cultural identities of race, class, gender, ethnicity, and sexuality. Often these identities come together in a group, and their influence in combination becomes central to our understanding of how group identity shapes space and is shaped by it.

■ *Understand* how codes signify important information about landscapes, a process known as semiotics.

To interpret our environment, we must learn how to read the codes that are written into the landscape. Landscapes as different from each other as shopping malls and war memorials can be understood in terms of their semiotics, although it is important to appreciate that even when certain landscapes have intended meanings by those who have created them, those who perceive them may make their own sense of that landscape.

■ *Analyze* the effects of global consumption on cultural landscapes from homogenization to preservation and describe how a place can become a commodity.

A shared consciousness has diffused certain values and attitudes around the globe, and the commonalities have been accompanied by the growing importance of material consumption. One of the effects of global culture is the commodification of spaces, or the "Disneyfication" in the heritage industry where regions capitalize on stereotypical imagery of their regions to "sell" to global consumers. Simultaneously, many people have responded to the homogenized global culture with counter movements including rediscovering their own "culture of place", preserving distinct cultural values and difference to global culture, and participating in countercultures like the Cittaslow movement.

KEY TERMS

Cosmopolitanism (p. *254*)
derelict landscapes (p. *239*)
ethology (p. *232*)
humanistic approach (p. *239*)

landscape as text (p. *247*)
proxemics (p. *233*)
semiotics (p. *247*)
slow food (p. *249*)

territoriality (p. *232*)
topophilia (p. *239*)

REVIEW & DISCUSSION

1. Consider the environment of your own town or campus. Conduct research exploring how the environment has been shaped or reshaped over the last 5 to 10 years. (*Hint:* You might want to research the building of new parking structures or sports complexes or consider how your university has expanded and bought more land or is not using all of its buildings.) Are there opposing perceptions regarding how your university has been shaped?

2. Conduct research to identify and explore the different elements (paths, edges, districts, nodes, and landmarks) within your campus or town. Begin by making a list of the different elements and who occupies these spaces. What are the territorial, social, and cultural markers of each space? List three of the proxemics you observe in these spaces. (*Hint:* You might start by considering the different coffee shops, bars, or restaurants at your university. Who hangs out at these different locations? List and assess the types of social interactions that are taking place. List the symbols or cultural behaviors that are associated with the location and the people who occupy it.) Once you have the information compiled, draw a territorial cognitive image map like the ones in Figures 6.5 and 6.6.

3. As a student on a campus, in what ways might your environmental perception of the campus landscape differ from that of the faculty on campus or from that of the janitorial staff who work on campus? Give three examples of how these different groups will have different perceptions of the landscape. Also provide a detailed sketch of how each group would read the landscape-as-text. For example, what is the meaning of a classroom, hallway, bathroom, or office for each group? List the different semiotics that is involved for each group.

4. Consider a place that has become a commodity. Think about your last vacation or landscape you desire to visit. How did you discover this place? What made you want to visit this place? How was it different from your home—think about the food, the language, the environment, the people, the architecture? Do an Internet search on this place to find "iconic" images of the place. What was the same about this place as your home? Did you see any evidence of globalized culture there? Did you see efforts to preserve regional identity?

UNPLUGGED

1. Write a short essay (500 words or two double-spaced pages) that describes, from your personal perspective, the sense of place that you associate with your hometown or county. Write about the places, buildings, sights, and sounds that are especially meaningful to you.

2. Draw up a list of the top 10 places in the United States in which you would like to live and work; then draw up a list of the bottom 10. How do these lists compare with the map of student preferences shown in Figure 7.8? Why might your preferences be different from theirs?

3. Compare the places where you have been an "outsider" and an "insider". Think about spaces where age, gender, religion, skill level (sports, expertise in films or chess, rock climbing), community (teams, clubs, family, political affiliation), personal experience (jobs, family circumstances, health circumstances) and interests (music preference, fashion choices, favorite TV shows, dietary choices like veganism) dictate the landscape of that space. Where were the places you were an outsider? Did you adapt to the space or did people in that space adapt to you? Did your identity change as a result? What were the spaces you were the insider? How did you react to outsiders? What did you share with other insiders? How did you all draw identity from the place you had created?

4. On a clean sheet of paper and without reference to maps or other materials, sketch a detailed map of the town or city in which you live. Does your sketch contain nodes? Landmarks? Edges? Districts? Paths? How does "your" cognitive image map compare to your "real" town or city?

DATA ANALYSIS

Project Row House
http://goo.gl/zAJMVo

In this chapter, we have seen how place making stands at the center of issues of culture and power relations. We also have discussed how different cultural identities and status categories influence the ways people experience and understand their landscapes. We have also seen how preservation of culture is an issue of power, race, class, and gender. And through the work of Edward Soja, we can analyze those relationships with landscape as part of a "sociospatial dialectic" in that we change spaces, but those places also inform our worldviews. Thinking about race, class, and art in the landscape, explore Project Row House (PRH), a community based art and culture organization in the northern Third Ward, a traditionally African-American neighborhood in Houston, Texas. Go to the Project Web site, view the art, read about the programs, and answer these questions.

1. What is PRH? Why was it started?

2. Read the information on the "About" page. What architecture is being preserved? What is the style? Who did these homes belong to?

3. Why did the artists want to work in this space? What is the relationship between the building preservation and community?

4. The "Young Mothers Residential Program" is an example of incorporating art and community action as part of a cultural landscape. What does this program do? What kind of "cognitive imagery" might this program offer to its participants, compared to living outside the program? What do you think about this alternative?

5. Go to the "Architecture" tab and read each section. How does community activist Rick Lowe differentiate his goals of housing with a developer?

6. What is "Row House CDC"? Why was it created? What does it provide?

7. What are some of the performances hosted by PRH? Look at both the "professional" artists and members of the community. How does this create a new landscape that challenges "cosmopolitanism" and at the same time celebrate the arts?

8. Watch this two-minute documentary on PRH with Rick Lowe.

   What does Rick Lowe say about "the issues" and solutions?

   Project Row House Documentary
   http://goo.gl/4uEPYI

9. What happened to the neighborhood when PRH began? What kind of gallery space is PRH?

10. What does Rick Lowe say about "shaping the world"?

MasteringGeography™

Looking for additional review and test prep materials? Visit the Study Area in MasteringGeography™ to enhance your geographic literacy, spatial reasoning skills, and understanding of this chapter's content by accessing a variety of resources, including MapMaster interactive maps, Videos, *In the News* RSS feeds, flashcards, web links, self-study quizzes, and an eText version of *Human Geography*.

LEARNING OUTCOMES

- *Scrutinize* the nature and degree of unevenness in patterns of economic development at national and international scales.

- *Compare* how different regions of the world contribute to technological innovation and how interdependent the global economy is on these technological systems.

- *Analyze* how geographical divisions of labor have evolved with the growth of the world-system and the accompanying variations in economic structure.

- *Interpret* how regional cores of economic development are created through the operation of several basic principles of spatial organization.

- *Explain* how spirals of economic development can be arrested in various ways, including the onset of disinvestment and deindustrialization.

- *Demonstrate* how globalization has resulted in patterns and processes of local and regional economic development that are open to external influences.

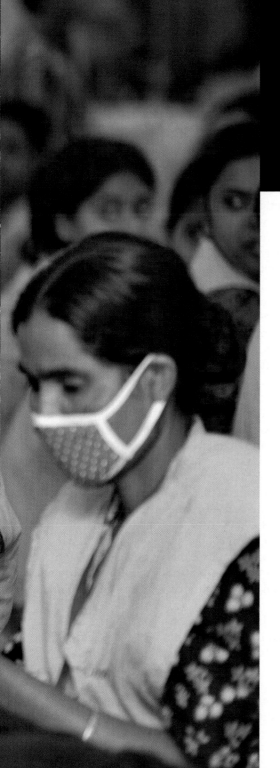

▲ Garment workers in Gajipur, Bangladesh.

8

GEOGRAPHIES OF ECONOMIC DEVELOPMENT

The garment industry in Bangladesh is a major supplier of mass-market clothing for Western retailers like Gap, H&M, Old Navy, and WalMart. In the 1970s the newly independent country had no modern industry at all, but by 2000 the garment industry was generating more than U.S. $12.5 billion in export revenue. Today Bangladesh earns $20 billion a year from garment exports. Women account for 80 percent of its 4 million employees, most of them working in terrible conditions for long hours and extremely low pay. A wide set of factors, from financial innovation to government policy, contributed to this dramatic change in the country's economic geography. It began in 1979, when Daewoo Corporation of Korea teamed up with Bangladesh's Desh Garments Ltd. to produce garments in Bangladesh for export. Desh sent 130 newly recruited, technically qualified employees to Daewoo's garment factory in Korea, where they participated in an 8-month intensive training course covering topics from sewing skills to factory management, quality control, and international procurement and marketing—skills that they then applied in the Desh factories in Bangladesh.

Within a few years, almost all the trainees had left Desh to start their own garment businesses. Some of the ex-Desh workers joined new garment factories established by affluent businessmen, while others founded trading houses, which then contributed to the proliferation of garment manufacturers by providing a variety of valuable services, including international procurement and marketing, sample making, and design reengineering. Observing Desh's good start in exporting, and subsequently the success of ex-Desh workers, highly educated people started their own garment businesses, and wealthy families actively invested in the industry. Learning from one another, they clustered in a 20-kilometer corridor just north of Dhaka, the country's capital. Learning from abroad continued as some entrepreneurs participated in training programs in Singapore, Japan, and Europe. Other countries in East Asia followed Daewoo into operation in Bangladesh and invested in training Bangladeshi workers and managers. Thus, many Bangladeshi traders and manufacturers had work experience in garment trading and production, including the experience of working in joint ventures, before starting their own businesses.

Geographers are interested in understanding the principles behind such apparently serendipitous and complex processes of economic development: from international ties to the local concentrations of industry; from the circumstances of entrepreneurship and investment to the conditions in local labor markets; and from the economic structure of countries and regions to the factors that influence the location of economic activities.

Based on *World Development Report 2013*, The World Bank, Washington, D.C., 2013, p. 117

PATTERNS OF ECONOMIC DEVELOPMENT

We often discuss economic development in terms of levels and rates of change in prosperity, as reflected in bottom-line statistical measures of productivity, incomes, purchasing power, and consumption. Increased prosperity is only one aspect of economic development, however. For human geographers and other social scientists, the term *economic development* refers to processes of change involving the nature and composition of the economy of a particular region as well as to increases in the overall prosperity of a region. These processes can involve three types of changes:

- changes in the structure of the region's economy (example, a shift from agriculture to manufacturing);

- changes in forms of economic organization within the region (for example, a shift from socialism to free-market capitalism);

- changes in the availability and use of technology within the region.

Technological Change and Economic Development

The Industrial Revolution, which began in England at the end of the eighteenth century, eventually resulted not only in the complete reorganization of the geography of the original European core of the world-system but also in an extension of the world-system core to the United States and Japan. Since then there have been several more technology systems, each opening new geographic frontiers and rewriting the geography of economic development while shifting the balance of advantages between regions. Overall, the opportunities for development created by each new technology system have been associated with distinctive economic epochs and long-term fluctuations in the overall rate of change of prices in the economy.

Beginning in the late eighteenth century, a series of technological innovations in power and energy, transportation, and manufacturing processes resulted in crucial changes in patterns of economic development. Each of these major clusters of technological innovations created new demands for natural resources as well as new labor forces and markets. The result was that each major cluster of technological innovations—called technology systems—tended to favor different regions and different kinds of places. Technology systems are clusters of interrelated energy, transportation, and production technologies that dominate economic activity for several decades at a time—until a new cluster of improved technologies evolves. What is especially remarkable about technology systems is that so far they have come along at about 50-year intervals. Since the beginning of the Industrial Revolution, we can identify four of them:

1790–1840: early mechanization based on water power and steam engines; development of cotton textiles and iron-working; development of river transport systems, canals, and turnpike roads.

1840–1890: exploitation of coal-powered steam engines; steel products; railroads; world shipping; and machine tools.

1890–1950: exploitation of the internal combustion engine; oil and plastics; electrical and heavy engineering; aircraft; radio and telecommunications.

1950–1990: exploitation of nuclear power, aerospace, electronics, and petrochemicals; development of limited-access highways and global air routes.

A fifth technology system, still incomplete, began to take shape in the 1980s with a series of innovations that are now being commercially exploited:

1990 onward: exploitation of solar energy, robotics, microelectronics, biotechnology, nanotechnology, advanced materials (e.g., fine chemicals and thermoplastics), and information technology (e.g., digital telecommunications and geographic information systems).

Each of these technology systems has rewritten the geography of economic development as it has shifted the balance of advantages between regions (see Box 8.1, "Visualizing Geography: Technological Change and Economic Development"). From the mid-1800s, industrial development spread to new regions. The growth of those regions then became interdependent with the fortunes of other regions through a complex web of production and trade.

The Unevenness of Economic Development

Geographically, the single most important feature of economic development is that it is *uneven*. At the global scale, this unevenness takes the form of core-periphery contrasts within the evolving world-system (see Chapter 2). These global core-periphery contrasts are the result of a competitive economic system that is heavily influenced by cultural and political factors. The core regions within the world-system—North America, Europe, and Japan—have the most diversified economies, the most advanced technologies, the highest levels of productivity, and the highest levels of prosperity. They are commonly referred to as *developed regions* (though processes of economic development are, of course, continuous, and no region can ever be regarded as fully developed).

Other countries and regions—the periphery and semiperiphery of the world-system—are often referred to as *developing* or *less developed*. Indeed, the nations of the periphery are often referred to as LDCs (less developed countries). Another popular term for the global periphery, developed as a political label but now synonymous with economic development in popular usage, is the *Third World*. This term had its origins in the early Cold War era of the 1950s and 1960s, when the newly independent countries of the periphery positioned themselves as a distinctive political bloc, aligned with neither the First World of developed, capitalist countries nor the Second World of the Soviet Union and its satellite countries.

Measuring Levels of Economic Development

At the global scale, levels of economic development are usually measured by economic indicators such as gross domestic product and gross national income. **Gross domestic product (GDP)** is an estimate of the total value of all materials, foodstuffs, goods, and services that are produced by a country in a particular year. To standardize for countries' varying sizes, total GDP is normally divided by total population that gives an indicator, *per capita* GDP, which is a good yardstick of relative levels of economic development. **Gross national income (GNI)** is a measure of the income that flows to a country from production wherever in the world that production occurs. For

example, if a U.S.-owned company operating in another country sends some of its income (profits) back to the United States, this adds to the U.S. GNI.

In making international comparisons, GDP and GNI can be problematic because they are based on each nation's currency. As a result, it is now common to compare national currencies based on **purchasing power parity** (PPP). In effect, PPP measures how much of a common "market basket" of goods and services each currency can purchase locally, including goods and services that are not traded internationally. When we use PPP-based currency values to compare levels of economic prosperity, we usually see lower GNI figures in wealthy countries (because of the generally higher cost of living) and higher GNI figures in poorer nations (because of the generally lower cost of living). Nevertheless, even with this compression between rich and poor, economic prosperity is very unevenly distributed across nations.

As **Figure 8.1** shows, most of the highest levels of economic development are to be found in northern latitudes (very roughly, north of 30° N), which has given rise to another popular shorthand for the world's economic geography: the "North" (the core) and the "South" (the periphery). Viewed in more detail, the global pattern of per capita GNI (measured in the "international dollars" of PPP) in 2012 is a direct reflection of the core-semiperiphery-periphery structure of the world-system. In many of the core countries of North America, northwestern Europe, and Japan, annual per capita GNI (in PPP) exceeds $30,000. The only other countries that match these levels are Australia and Singapore, where annual per capita GNI in 2012 was $43,300 and $60,110, respectively. Semiperipheral

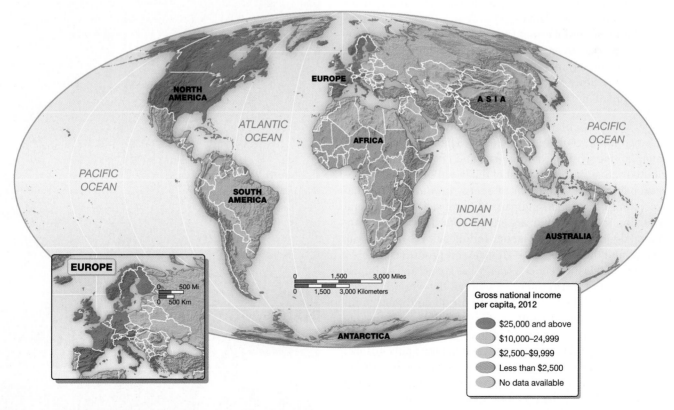

▲ **Figure 8.1 Gross National income (GNI) per capita 2012** GNI per capita is one of the best single measures of economic development.

countries such as Brazil, Russia, and Thailand have an annual per capita GNI ranging between $10,000 and $25,000. In the rest of the world—the periphery—annual per capita GNI (in PPP) is typically less than $5,000.

The gap between the highest per capita GNIs (more than $60,000 in Luxembourg, Norway and Singapore) and the lowest (less than $600 in Burundi, Eritrea, Liberia, and the Democratic Republic of the Congo) is huge. The gap between the world's rich and poor is also getting wider. In 1970, the average GNI per capita of the ten most prosperous countries in the world was 50 times greater than the average GNI per capita of the ten poorest countries. By 2012, the relative gap had increased to a factor of 70. Overall, more than 80 percent of the world's population lives in countries where income differentials are widening.

APPLY YOUR KNOWLEDGE

1. What kind of statistics besides gross national income provide an indication of international disparities in economic development?

2. Find data on one such indicator and propose two possible reasons why the variations exist in the data you found. (*Hint:* Good sources are the World Bank, http://data.worldbank.org/, and the United Nations Development Programme, http://hdr.undp.org/en/statistics/.)

Resources and Development

Current patterns of economic development are the result of many different factors. One of the most important is the availability of key resources such as cultivable land, energy, and valuable minerals. Unevenly distributed across the world are both key resources and—equally important—the *combinations* of energy and minerals crucial to economic development. A lack of natural resources can, of course, be remedied through international trade (Japan's success is a prime example). For most countries, however, the resource base remains an important determinant of development.

Energy One particularly important resource in terms of the world's economic geography is energy. The major sources of commercial energy—oil, natural gas, and coal—are unevenly distributed across the globe. Most of the world's core economies are reasonably well off in terms of energy *production,* the major exceptions being Japan and parts of Europe. Most peripheral countries, on the other hand, are energy poor. The major exceptions are Algeria, Ecuador, Gabon, Indonesia, Libya, Nigeria, Venezuela, and the Gulf states—all major oil producers.

Because of this unevenness, energy has come to be an important component of world trade. As more of the world becomes industrialized and developed, demand for oil continues to increase. Yet, as we saw in Chapter 4, it is generally agreed that the peak of oil *discovery* was passed in the 1960s and that the world started using more oil than was contained in new fields in 1981. Oil prices have risen sharply: more than 500 percent between 2000 and 2013. Oil is now the single most important commodity in world trade, making up more than 20 percent of the total by value in 2013. Newly discovered oil reserves in Canada and Brazil have eased short-term concerns about oil supplies, as has the introduction of hydraulic fracturing techniques (commonly

known as "fracking") to release hard-to-get-at reserves of oil and natural gas. But both oil exploration and fracking (**Figure 8.2**) are increasingly the focus of environmental concerns.

For many peripheral countries the cost of importing energy is a heavy burden. Consider, for example, the predicament of countries like India, Ghana, Paraguay, Egypt, and Armenia, where in 2010 the cost of energy imports amounted to more than one-quarter of the total value of exported merchandise. Few peripheral countries can afford to consume energy on the scale of the developed economies, so patterns of commercial energy *consumption* mirror the fundamental core-periphery cleavage of the world economy. In 2008, energy consumption per capita in North America was 14 times that of India, 18 times that of Mozambique, and nearly 50 times that of Bangladesh. The world's high-income countries, with 15 percent of the world's population, use half its commercial energy and 10 times as much per capita as low-income countries.

It should be noted that these figures do not reflect the use of firewood and other traditional fuels for cooking, lighting, heating and, sometimes, industrial needs. In total, such forms probably account for around 20 percent of total world energy consumption. In parts of Africa and Asia, they account for up to 80 percent of energy consumption. This points to yet another core-periphery contrast. Whereas massive investments in exploration and exploitation are enabling more of

▼ **Figure 8.2 A fracking rig at a well site in Washington, PA** Hydraulic fracturing, or "fracking", is the process of extracting natural gas from shale rock layers deep within the earth.

▲ **Figure 8.3 Deforestation** Forest clearance in Malaysia in preparation for an oil palm plantation.

serious in densely populated locations, arid and semiarid regions, and cooler mountainous areas, where the regeneration of shrubs, woodlands, and forests is particularly slow. Nearly 100 million people in 22 countries (16 of them in Africa) cannot meet their minimum energy needs even by overcutting remaining forests (**Figure 8.3**).

Cultivable Land The distribution of cultivable land is another important factor in international economic development. Much more than half of Earth's land surface is unsuitable for any productive form of arable farming, as shown in **Figure 8.4**. Poor soils, short growing seasons, arid climates, mountainous terrain, forests, and conservation limit the extent of agricultural land across much of the globe. As a result, the distribution of the world's cultivable

the developed, energy-consuming countries to become self-sufficient through various combinations of coal, oil, natural gas, hydroelectric power, and nuclear power, 1.5 billion people in peripheral countries depend on collecting fuelwood as their principal source of energy. The collection of wood fuel causes considerable deforestation. The problem is most land is highly uneven, being concentrated in Europe, west-central Russia, eastern North America, the Australian littoral, Latin America, India, eastern China, and parts of sub-Saharan Africa. Some of these regions may be marginal for arable farming because of marshy soils or other adverse conditions, while

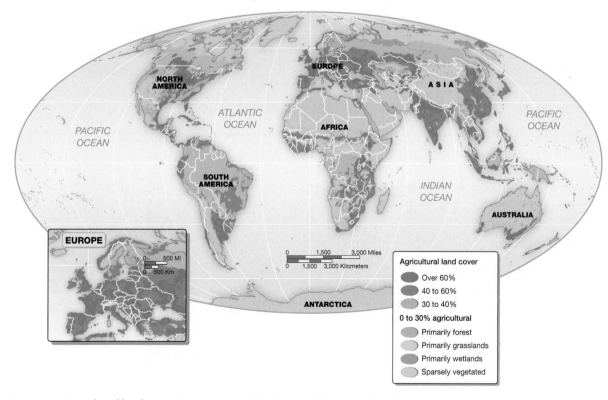

▲ **Figure 8.4 Agricultural land cover** Some countries, like the United States, are fortunate in having a broad range of cultivable land, which allows for many options in agricultural development. Many countries, though, have a much narrower base of cultivable land and must rely on the exploitation of one major resource as a means to economic development. (Adapted from *World Resources 2000–2001: People and Ecosystems.* Washington, DC: World Resources Institute, 2000, p. 57. Originally from Wood et al., 2000. The map is based on Global Land Cover Characteristics Database Version 1.2 (Loveland et al. [2000]) and USGS/EDC [1999a]. The figure is based on FAOSTAT [1999]).

8.1 Visualizing Geography

Technological Change and Economic Development

Figure 8.1.1 Technology and development timeline.

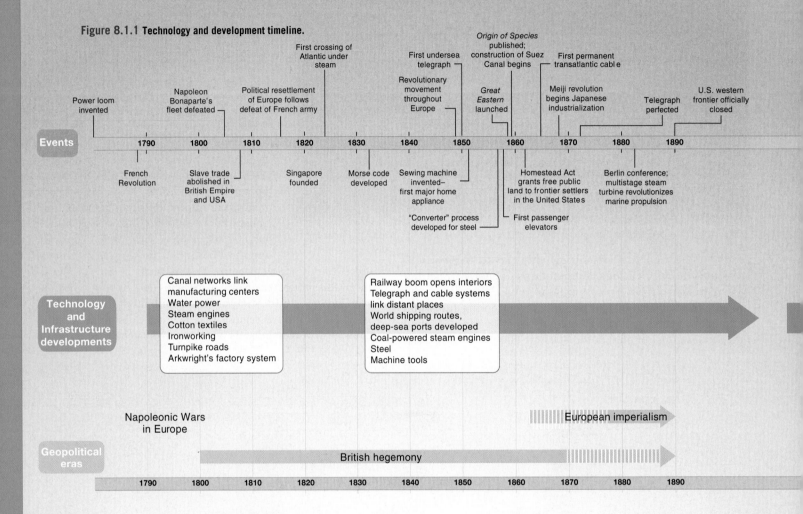

Events

Power loom invented

French Revolution

Napoleon Bonaparte's fleet defeated

Slave trade abolished in British Empire and USA

Political resettlement of Europe follows defeat of French army

First crossing of Atlantic under steam

Singapore founded

Morse code developed

Revolutionary movement throughout Europe

Sewing machine invented– first major home appliance

First undersea telegraph

"Converter" process developed for steel

Great Eastern launched

Origin of Species published; construction of Suez Canal begins

Homestead Act grants free public land to frontier settlers in the United States

First passenger elevators

First permanent transatlantic cable

Meiji revolution begins Japanese industrialization

Berlin conference; multistage steam turbine revolutionizes marine propulsion

Telegraph perfected

U.S. western frontier officially closed

1790   1800   1810   1820   1830   1840   1850   1860   1870   1880   1890

Technology and Infrastructure developments

Canal networks link manufacturing centers
Water power
Steam engines
Cotton textiles
Ironworking
Turnpike roads
Arkwright's factory system

Railway boom opens interiors
Telegraph and cable systems link distant places
World shipping routes, deep-sea ports developed
Coal-powered steam engines
Steel
Machine tools

Napoleonic Wars in Europe

European imperialism

Geopolitical eras

British hegemony

1790   1800   1810   1820   1830   1840   1850   1860   1870   1880   1890

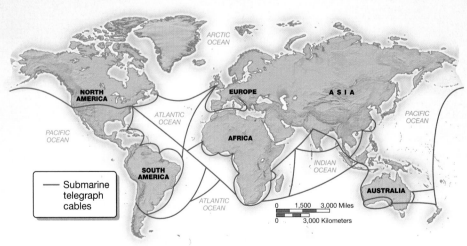

Submarine telegraph cables

Figure 8.1.2 The international telegraph network in 1900.

Figure 8.1.3 Digital technology. Your cell phone has more computing power than the computers used during the Apollo era. Left: NASA mission control in the 1970s. Right: Touch screen cell phone in use today.

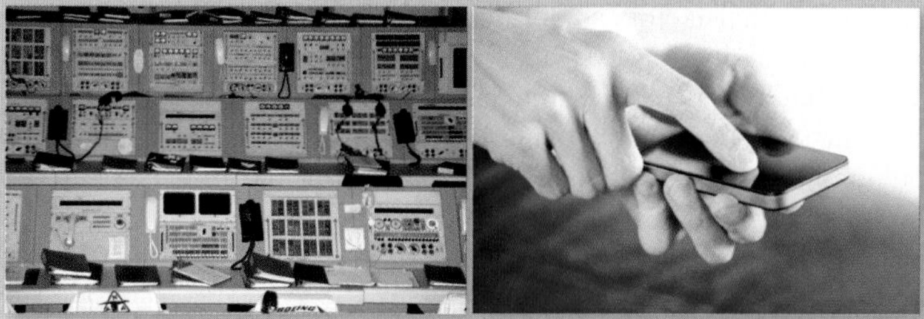

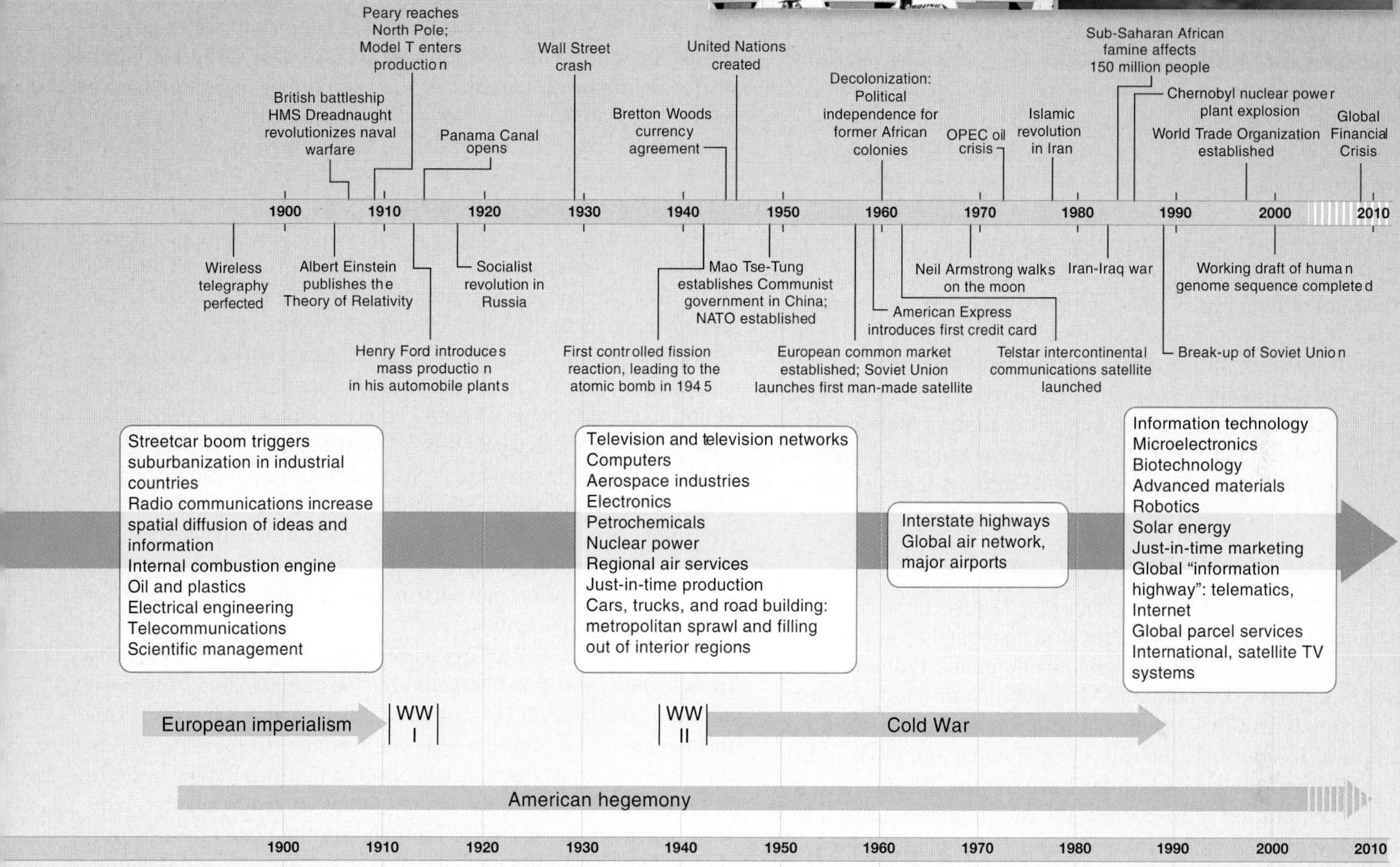

Peary reaches North Pole; Model T enters production
Wall Street crash
United Nations created
Sub-Saharan African famine affects 150 million people
British battleship HMS Dreadnaught revolutionizes naval warfare
Panama Canal opens
Bretton Woods currency agreement
Decolonization: Political independence for former African colonies
OPEC oil crisis
Islamic revolution in Iran
Chernobyl nuclear power plant explosion
World Trade Organization established
Global Financial Crisis
1900
1910
1920
1930
1940
1950
1960
1970
1980
1990
2000
2010
Wireless telegraphy perfected
Albert Einstein publishes the Theory of Relativity
Socialist revolution in Russia
Mao Tse-Tung establishes Communist government in China; NATO established
Neil Armstrong walks on the moon
Iran-Iraq war
Working draft of human genome sequence completed
American Express introduces first credit card
Henry Ford introduces mass production in his automobile plants
First controlled fission reaction, leading to the atomic bomb in 1945
European common market established; Soviet Union launches first man-made satellite
Telstar intercontinental communications satellite launched
Break-up of Soviet Union
Streetcar boom triggers suburbanization in industrial countries
Radio communications increase spatial diffusion of ideas and information
Internal combustion engine
Oil and plastics
Electrical engineering
Telecommunications
Scientific management
Television and television networks
Computers
Aerospace industries
Electronics
Petrochemicals
Nuclear power
Regional air services
Just-in-time production
Cars, trucks, and road building: metropolitan sprawl and filling out of interior regions
Interstate highways
Global air network, major airports
Information technology
Microelectronics
Biotechnology
Advanced materials
Robotics
Solar energy
Just-in-time marketing
Global "information highway": telematics, Internet
Global parcel services
International, satellite TV systems
European imperialism
WW I
WW II
Cold War
American hegemony
1900
1910
1920
1930
1940
1950
1960
1970
1980
1990
2000
2010

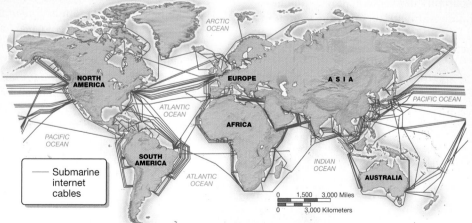

ARCTIC OCEAN
NORTH AMERICA
EUROPE
ASIA
PACIFIC OCEAN
ATLANTIC OCEAN
AFRICA
PACIFIC OCEAN
SOUTH AMERICA
INDIAN OCEAN
ATLANTIC OCEAN
AUSTRALIA
Submarine internet cables
0 1,500 3,000 Miles
0 3,000 Kilometers

Figure 8.1.4 Undersea internet cables in 2012.

Rise of Industrial America

http://goo.gl/PSsqjL

1. What are the differences in the British hegemonic era and in the American hegemonic era?

2. Where do you think technology and economic development are headed in the globalized world? Do you think there will be a new site of hegemonic influence? How do you think current world events impact development and technology?

irrigation or other factors sometimes extends the local frontier of productive agriculture. We also have to bear in mind that not all cultivable land is of the same quality. This leads to the concept of the **carrying capacity** of agricultural land: the maximum population that can be maintained in a place under current technologies and best practices, without damaging the overall productivity of that or other places.

Industrial Resources A high proportion of the world's key industrial resources—basic raw materials—are concentrated in Russia, the United States, Canada, South Africa, and Australia. The United States, for example, in addition to having 42 percent of the world's known resources of hydrocarbons (oil, natural gas, and oil shales) and 38 percent of the lignite ("brown coal," used mainly in power stations), has 38 percent of the molybdenum (used in metal alloys), 21 percent of the lead (used for batteries, gasoline, and construction), 19 percent of the copper (used for electrical wiring and components and for coinage), 18 percent of the bituminous coal (used for fuel in power stations and in the chemical industry), and 15 percent of the zinc. Russia has 68 percent of the vanadium (used in metal alloys), 50 percent of the lignite, 38 percent of the bituminous coal, 35 percent of the manganese, 25 percent of the iron, and 19 percent of the hydrocarbons (**Figure 8.5**).

Resources and Development

The concentration of known resources in just a few countries is largely a result of geology, but it is also partly a function of countries' political and economic development. Political instability in much of postcolonial Africa, Asia, and Latin America has seriously hindered their exploration and exploitation of resources. In contrast, the relative affluence and great political stability of the United States have led to a much more intensive exploration of resources. We should also bear in mind that the significance of particular resources is often tied to particular technologies. As technologies change, so do resource requirements, and the geography of economic development is "rewritten." One important example of this was the switch in

the manufacture of mass-produced textiles from natural fibers like wool and cotton to synthetic fibers in the 1950s and 1960s. When this happened, many farmers in the U.S. South had to switch from cotton to other crops.

Regions and countries that are heavily dependent on one particular resource are vulnerable to the consequences of technological change. They are also vulnerable to fluctuations in the price set for their product on the world market. These vulnerabilities are particularly important for countries whose economies are dependent on nonfuel minerals, such as the Democratic Republic of the Congo (copper), Mauritania (iron ore), Namibia (diamonds), Niger (uranium), Sierra Leone (diamonds), Togo (phosphates), and Zambia (copper).

Resources and Sustainability The ideal of **sustainable development** is one that achieves a balance among economic growth, the environmental impacts of that growth, and the fairness, or social equity, of the distribution of the costs and benefits of that growth. The importance of sustainability is cogently illustrated by the concept of an **ecological footprint**, which is a measure of the human pressures on the natural environment from the consumption of renewable resources and the production of pollution. It represents a quantitative assessment of the biologically productive area required to produce the resources (food, energy, and materials) and to absorb the wastes of an individual, city, region, or country. The ecological footprint of a country or region changes in proportion to population size, average consumption per person, and the resource intensity of the technology being used.

Humanity's footprint first grew larger than global biocapacity in the 1980s, and this overshoot has been increasing every year since. In 2006, demand exceeded supply by about 40 percent. This means that it took almost a year and five months for Earth to produce the ecological resources we used in that year. At 9.0 hectares (23.6 acres) per person, the United States currently has the sixth-largest per capita ecological footprint on the planet, just behind the United Arab Emirates, Qatar, Bahrain, Denmark, and Belgium. Other countries with extremely large ecological footprints include Australia, Canada, the Netherlands, Finland, and Sweden. Countries with the smallest ecological footprints, between 0.5 and 0.75 hectares (1.2 to 1.9 acres) per person, include Afghanistan, Bangladesh, Haiti, and Malawi.

Sustainable development means using renewable natural resources in a manner that does not eliminate or degrade them—by making greater use, for example, of solar and geothermal energy and recycled materials. It means managing economic systems so that all resources—physical and human—are used optimally. It means regulating economic systems so that the benefits of development are distributed more equitably (if only to prevent poverty from causing environmental degradation). It also means organizing societies so that improved education, health care, and social welfare can contribute to environmental awareness and sensitivity and an improved quality of life. A final and more radical aspect of sustainable development involves moving away from wholesale globalization toward increased "localization": a return to more locally based

▼ **Figure 8.5 Resource extraction** Mining potassium salt (to be used as a fertilizer) near Perm, Russia.

▲ Figure 8.6 **Promoting local economies** Increasing awareness of the benefits of locally produced foods has encouraged many supermarkets, like this one in Lugano, Switzerland, to feature local products.

economies where production, consumption, and decision making are oriented to local needs and conditions (**Figure 8.6**).

Defined this way, sustainable development sounds eminently sensible yet impossibly utopian. A succession of international summit meetings on the topic has revealed deep conflicts of interest between core countries and peripheral countries. One of the most serious obstacles to prospects for sustainable development is continued heavy reliance on fossil fuels as the fundamental source of energy for economic development. This not only perpetuates international inequalities but also leads to transnational problems such as acid rain, global warming, climatic changes, deforestation, health hazards, and, many would argue, war. The sustainable alternative—renewable energy generated from the sun, tides, waves, winds, rivers, and geothermal features—has been pursued half-heartedly because of the commercial interests of the powerful corporations and governments that control fossil-fuel resources.

A second important challenge to the possibility of sustainable development is the rate of demographic growth in peripheral countries. Sustainable development is feasible only if population

size and growth are in harmony with the changing productive capacity of the ecosystem. It is estimated that 1.2 billion of the world's 6.9 billion people are undernourished and underweight.

But the greatest single obstacle to sustainable development is the inadequacy of institutional frameworks. Sustainable development requires economic, financial, and fiscal decisions to be fully integrated with environmental and ecological decisions. National and local governments everywhere have evolved institutional structures that tend to separate decisions about what is economically rational and what is environmentally desirable. International organizations, while better placed to integrate policy across these sectors and better able to address economic and environmental "spillovers" from one country to another, have (with the notable exception of the European Union) not acquired sufficient power to promote integrated, harmonized policies. Without radical and widespread changes in value systems and unprecedented changes in political will, "sustainable development" is likely to remain an embarrassing contradiction in terms.

APPLY YOUR KNOWLEDGE

1. How sustainable do you think your lifestyle is, compared to other places in the world?

2. Calculate your ecological footprint to find out how your lifestyle compares by using this calculator on the Global Footprint Network: http://www.footprintnetwork.org/en/index.php/GFN/page/calculators/

THE ECONOMIC STRUCTURE OF COUNTRIES AND REGIONS

The *economic structure* of a country or region can be described in terms of its relative share of primary, secondary, tertiary, and quaternary economic activities (see **Table 8.1**).

Variations in economic structure—according to primary, secondary, tertiary, or quaternary activities—reflect *geographical*

TABLE 8.1   Economic Structure

Primary activities	concerned directly with natural resources	agriculture, mining, fishing, and forestry
Secondary activities	process, transform, fabricate, or assemble the raw materials derived from primary activities or that reassemble, refinish, or package manufactured goods	steelmaking, food processing, furniture production, textile manufacturing, automobile assembly, and garment manufacturing
Tertiary activities	involve the sale and exchange of goods and services	warehousing, retail stores, personal services such as hairdressing, commercial services such as accounting, advertising, and entertainment
Quaternary activities	deal with the handling and processing of knowledge and information	data processing, information retrieval, education, and research and development (R&D)

divisions of labor. Geographical divisions of labor are national, regional, and locally based economic specializations that have evolved with the growth of the world-system of trade and politics and with the locational needs of successive technology systems. They represent one of the most important dimensions of economic development. For instance, countries whose economies are dominated by **primary-sector activities** tend to have a relatively low per capita GDP. The exceptions are oil-rich countries such as Saudi Arabia, Qatar, and Venezuela. Where the **international division of labor** (the specialization, by countries, in particular products for export) has produced national economies with a large **secondary sector**, per capita GDP is much higher (as, for example, in Argentina and South Korea). The highest levels of per capita GDP, however, are associated with economies that are *postindustrial:* economies where the **tertiary** and **quaternary sectors** have grown to dominate the workforce, with smaller but highly productive secondary sectors.

As **Figure 8.7** shows, the economic structure of much of the world is dominated by the primary sector. In much of Africa and Asia, between 50 and 75 percent of the labor force is engaged in primary-sector activities such as agriculture, mining, fishing, and forestry. In contrast, the primary sector of the world's core regions is typically small, occupying only 5 to 10 percent of the labor force.

The secondary sector is much larger in the core countries and in semiperipheral countries, where the world's specialized manufacturing regions are located. In 2012, core countries accounted for almost three-quarters of world manufacturing value added (MVA). MVA is the net output of secondary industries; it is determined by adding up the value of all outputs and subtracting the value of all intermediate inputs. This share has been slowly decreasing, however. The core countries had an average annual growth rate for MVA of around 2 percent during 1990–2012, while the growth rate in the rest of the world was closer to 7 percent.

This growth has been concentrated in semiperipheral, **newly industrializing countries** (NICs). Newly industrializing countries are countries, formerly peripheral within the world-system, that have acquired a significant industrial sector, usually through **foreign direct investment**. Of the 20 biggest manufacturing countries in 2012, 7 were NICs: China, South Korea, Mexico, Brazil, India, Indonesia, and Thailand (listed here in order of importance).

In terms of individual countries, the United States remains the most important source of manufactured goods, accounting for just over 22 per cent of global MVA in 2012. Just five countries—the United States, Japan, Germany, China, and the United Kingdom—together produced over 60 percent of the world total MVA. The highly capitalized manufacturing industries of the developed countries have been able to maintain high levels of worker productivity, with the result that the contribution of manufacturing to their GDP has remained relatively high even as the size of their manufacturing labor force has decreased.

Although the United States has retained its leadership as the world's major producer of manufactured goods, its dominance has been significantly reduced. China, in particular, has experienced a dramatic increase in manufacturing

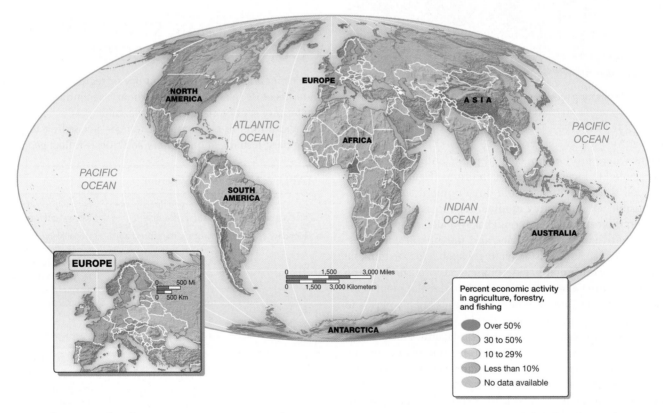

▲ Figure 8.7 **The geography of primary economic activities** Primary economic activities are those that are concerned directly with natural resources of any kind. They include agriculture, mining, forestry, and fishing. The vast majority of the world's population, concentrated in China, India, Southeast Asia, and Africa, is engaged in primary economic activities. This map shows the percentage of the labor force in each country that was engaged in primary employment in 2012.

(Source: World Development Indicators 2014, THE WORLD BANK)

▲ **Figure 8.8 Manufacturing in South Korea** Employees attach protective film to printed circuit board panels on the production line at the Seil Electronics Co. factory in Incheon, South Korea.

production, achieving annual average growth rates during the 1970s, 1980s, 1990s, and 2000s of about 8 percent, 11 percent, 14 percent, and 9 percent, respectively (see Box 8.2, "Window on the World: China's Economic Development"). South Korea has also enjoyed a spectacular increase in manufacturing production (**Figure 8.8**), achieving annual average growth rates of almost 18 percent in the 1960s, 17 percent in the 1970s, 12 percent in the 1980s, more than 7 percent in the 1990s, and around 6 percent in the 2000s. More recently, other Pacific Rim NICs, such as Malaysia and Thailand, have experienced rapid growth in manufacturing production.

These shifts are part of a globalization of economic activity that has emerged as the overarching component of the world's economic geography. As we shall see, it has been corporate strategy, particularly the strategies of large **transnational corporations** (TNCs), which has created this globalization of economic activity. Transnational corporations are companies that participate not only in international trade but also in production, manufacturing, and/or sales operations in several countries.

The tertiary and quaternary sectors are significant only in the most affluent countries of the core. Globalization has meant that knowledge-based activities have become a critical aspect of economic development, resulting in the rapid growth of quaternary industries.

Knowledge and Economics

For the world's core economies, knowledge has become more important than physical and human resources in determining levels of economic well-being. More than half of the GDP of major core countries is based on the production and distribution of knowledge. In the United States, more workers are engaged in producing and distributing knowledge than in making physical goods.

For the world's peripheral economies, lack of knowledge—along with a limited capacity to absorb and communicate knowledge—is an increasingly important barrier to economic development. Poor countries have fewer resources to devote to

research, development, and the acquisition of information technology. They also have fewer institutions for providing high-quality education, fewer bodies that can enforce standards and performance, and only weakly developed organizations for gathering and disseminating the information needed for business transactions. As a result, economic productivity tends to fall relative to the performance of places and regions in core economies, where new knowledge is constantly generated and rapidly and effectively disseminated. All these problems are compounded by gender inequalities (see Box 8.3: "Spatial Inequality: Gender and Economic Development").

APPLY YOUR KNOWLEDGE

1. Why do you think certain places in the world specialize in one sector over others?

2. Consider a product that you own, such as a T-shirt, a pair of shoes, or your smartphone and map out the product's development through the primary, secondary, and tertiary activities. Think about the entire process of your product's development: Who designed the product? From which country are the raw materials sourced? Who creates that actual object, that is, who comprises the labor force that sews, builds or assembles your object?

International Trade

The geographical division of labor on a world scale means that the geography of international trade is very complex. One significant reflection of the increased economic integration of the world-system is that global trade has grown much more rapidly over the past few decades than global production. Between 1990 and 2012, the average annual growth rate of the value of world exports was twice that of the growth of world production and several times greater than that of world population growth.

The fundamental structure of international trade is based on a few **trading blocs**—groups of countries with formalized systems of trading agreements. Most of the world's trade takes place within four trading blocs:

- Western Europe, together with some former European colonies in Africa, South Asia, the Caribbean, and Australasia;

- North America, together with some Latin American states;

- the countries of the former Soviet Union; and

- Japan, together with other East Asian states and the oil-exporting states of Saudi Arabia and Bahrain.

Nevertheless, a significant number of countries exhibit a high degree of **autarky** from the world economy. That is, they do not contribute significantly to the flows of imports and exports that constitute the geography of trade. Typically, these are smaller, peripheral countries, such as Bolivia, Burkina Faso, Ghana, Malawi, Samoa, and Tanzania.

8.2 Window on the World

China's Economic Development

Under the leadership of Deng Xiaoping (1978–1997), China embarked on a thorough reorientation of its economy, dismantling Communist-style central planning in favor of private entrepreneurship and market mechanisms and integrating itself into the world economy. Saying that he did not care whether the cat was black or white as long as it caught mice, Deng established a program of "Four Modernizations" (industry, agriculture, science, and defense) and an "open-door policy" that allowed China to be plugged in to the interdependent circuits of the global economy. As a result, China has completely reorganized and revitalized its economy. Agriculture has been decollectivized, with Communist collective farms modified to allow a degree of private profit-taking. State-owned industries have been closed or privatized, and centralized state planning has been dismantled in order to foster private entrepreneurship.

In the 1980s and early 1990s, when the world economy was sluggish, China's manufacturing sector grew by almost 15 percent each year. For example, almost all of the shoes once made in South Korea or Taiwan are now made in China. More than 60 percent of the toys in the world, accounting for more than $10 billion in trade, are also made in China. Since 1992, China has extended its open-door policy, permitted foreign investment aimed at Chinese domestic markets, and normalized trading relationships with the United States and the European Union. In 2001, China was admitted to the World Trade Organization, allowing the country to trade more freely than ever before with the rest of the world.

China's increased participation in world trade has created an entirely new situation within the world economy. Chinese manufacturers, operating with low wages, have imposed a deflationary trend on world prices for

manufacturers. The Chinese economy's size makes it a major producer, and its labor costs stay flat year after year because there is an endless supply of people who will work for 60 cents an hour. Meanwhile, the rapid expansion of consumer demand in China has begun to drive up commodity prices in the world market. Overall, China's economy is already the third largest in the world after those of the United States and Japan.

Nowhere have China's "open-door" policies had more impact than in South China, where the Chinese government has deliberately built upon the prosperity of Hong Kong, the former British colony that was returned to China in 1997. The coastline of South China provides many protected bays suitable for harbors and a series of ports, including Quanzhou, Shantou, Xiamen, and, on either side of the mouth of the Zhu Jiang (Pearl River), Macão and Hong Kong. These ports made possible South China's emergence as a core manufacturing region by providing an interface with the world economy. The established trade and manufacturing of Macão—a Portuguese colony that was returned to China in 1999—and Hong Kong provided another precondition for success.

When Deng established his "open-door" policy, a third factor kicked in: capital investment from Hong Kong, Taiwan, and the Chinese diaspora. By 1993, more than 15,000 manufacturers from Hong Kong alone had set up businesses in neighboring Guangdong Province, and a similar number established subcontracting relationships, contracting out processing work to Chinese companies. Today, the cities and special economic zones of South China's "Gold Coast" provide a thriving export-processing platform that has driven double-digit annual economic growth for much of the past two decades. The population of Shenzhen (**Figure 8.A**) has grown from just 19,000

▲ **Figure 8.A** **Shenzhen** The city of Shenzhen, just across the border from the Special Administrative Region of Hong Kong.

▲ **Figure 8.B New affluence** China's economic boom has led to a rapid increase in the size of the country's middle class, up nearly 25 percent since 2008, from 65 to 80 million people.

▲ **Figure 8.C Real estate boom** Developers plan to replace all of the small fishing villages around Sanya Harbor in Hainan province, China, with new luxury buildings.

industry enormous sums in research and development and licensing fees, while saving the country even greater sums in imports.

Foreign investors, meanwhile, have been keen to develop a share of China's rapidly expanding and increasingly affluent market. The automobile market is particularly attractive to Western manufacturers. Volkswagen was the first to establish a presence in China, in 1985. By 2003, Volkswagen had claimed around 40 percent of China's annual production of almost 4 million cars and light trucks. General Motors, in partnership with Shanghai Automotive Industry Corporation, has a 10 percent market share. Other foreign manufacturers operating in China include Honda, Toyota, Nissan, and, most recently, BMW and Mercedes-Benz.

Overall, most foreign investment in China comes from elsewhere within East Asia. Japan, Taiwan, and South Korea, having developed manufacturing industries that undercut those of the United States, now face deindustrialization themselves through the inexorable process of "creative destruction" (see page 280). More than 10,000 Taiwanese firms have established operations in China, investing an estimated $150 billion. Pusan, the center of the South Korean footwear industry that in 1990 exported $4.3 billion worth of shoes, is full of deserted factories. South Korean footwear exports are down to less than $700 million, while China's footwear exports have increased from $2.1 billion in 1990 to $44 billion in 2012. Several Japanese electronics giants, including Toshiba Corp., Sony Corp., Matsushita Electric Industrial Co., and Canon, Inc., have expanded operations in China even as they have shed tens of thousands of workers at home. Olympus manufactures its digital cameras in Shenzhen and Guangzhou. Pioneer has moved its manufacture of DVD recorders to Shanghai and Dongguan.

Toshiba's factory in Dalian illustrates the logic. Toshiba is one of about 40 Japanese companies that built large-scale production facilities in a special export-processing zone established by Dalian in the early 1990s with generous financial support from the Japanese government and major Japanese firms. By shifting production of digital televisions from its plant in Saitama, Japan, in 2001, Toshiba cut labor costs per worker by 90 percent.

in 1975 to 15 million in 2012. Such growth has generated a substantial middle class with significant spending power (**Figure 8.B**); it has also created significant inflation, especially in real estate values (**Figure 8.C**).

Much of China's manufacturing growth has been based on a strategy of import substitution (see page 274). In spite of China's membership in the World Trade Organization (which has strict rules about intellectual property), a significant amount of China's industry is based on counterfeiting and reverse engineering (making products that are copied and then sold under different or altered brand names) and piracy (making look-alike products passed off as the real thing). Copies of everything from DVDs, movies, designer clothes and footwear, drugs, motorcycles, and automobiles to high-speed magnetic levitation (maglev) cross-country trains save Chinese

1. What factors in China's "open-door policy" changed the economy, and how?

2. Alongside economic growth, China continues to face human rights accusations, especially of labor conditions. Analyze the economic boom alongside the human factors by looking at human rights organization reports on China on such websites as *Amnesty International, Human Rights Watch,* and *Oxfam International.* How can nations balance economic growth with high standards of human life?

8.3 Spatial Inequality   Gender and Economic Development

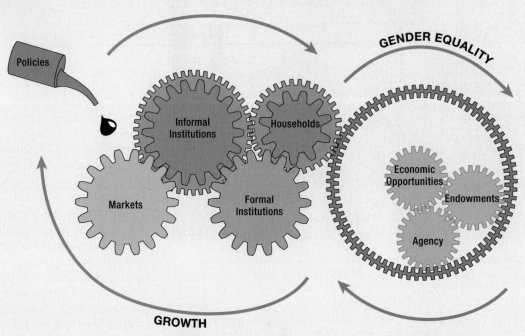

▲ **Figure 8.D** Gender outcomes result from interactions between households, markets, and institutions.
Source: World Bank, World Development Report 2012.

As we saw in Chapter 2, there are wide international disparities in education, health, and economic opportunity in terms of gender. Gender equality matters intrinsically, because the ability to live the life of one's own choosing is a basic human right and should be equal for everyone, regardless of gender. But gender equality also matters because greater gender equality contributes to economic development (**Figure 8.D**). The World Bank's *World Development Report* for 2012 covered this topic in detail.

World Development Report
http://goo.gl/KZu52k

The report concludes that gender equality can enhance economic efficiency and improve other development outcomes in three ways:

- Removing barriers that prevent women from having the same access as men to education, economic opportunities, and productive inputs can generate broad gains in productivity.

- Improving women's absolute and relative status feeds many other development outcomes, including those for their children.

- When women and men have equal chances to become socially and politically active, make decisions, and shape policies, this leads over time to more representative and more inclusive institutions and policy choices and thus to a better development path.

Eliminating barriers that prevent women from working in certain occupations or sectors would reduce the productivity gap between male and female workers by one-third to one-half and increase output per worker by 3 to 25 percent across a range of countries. Increasing the share of household income controlled by women, either through their own earnings or cash transfers, changes spending in ways that benefit children in terms of both life expectancy and educational attainment. Women's collective agency can be transformative for society. It can shape the institutions, markets, and social norms that limit their individual agency and opportunities. Empowering women as political and social actors can change policy choices and make institutions more representative of a range of voices.

1. What are some ways gender equality benefits the economy?

2. Research the World Bank's *World Development Report* (linked above) and read the overview. Using the text and graphics, analyze: 1) the challenges to gender equality in "severely disadvantaged populations"; and 2) where gender inequalities persisted and why.

Shifts in global politics have affected the geography of trade. One important change was the breakup of the former Soviet Union. Other significant geopolitical shifts include the trend toward the political as well as economic integration of Europe and the increasing participation of China in the world economy. But perhaps the most important shift in global politics in relation to world trade has been the shift toward open markets and free trade through neoliberal policies propagated by core countries (**Figure 8.9**). **Neoliberal policies** are economic policies predicated on a minimalist role for the state that assume the desirability of free markets not only for economic organization but also for political and social life. If necessary, social goals and regulatory standards have to be sacrificed, it is argued, to ensure that business has the maximum latitude for profitability. The rising tide of economic development, the argument goes, will then trickle down to less prosperous areas. Neoliberalism has meant the deregulation of finance and industry; the shedding of many of the traditional roles of central and local governments as mediators and regulators; curbs on the power and influence of labor unions and government agencies; a reduction of public investment in the physical **infrastructure** of roads, bridges and public utilities; the creation of free trade zones, enterprise zones, and other deregulated spaces; and the privatization of government services. The overall effect has been to "hollow out" the capacity of central governments to shape economic development while forcing local and regional governments to become increasingly entrepreneurial in pursuit of jobs and revenues.

The opening of markets and the globalization of economic activity has created new flows of materials, components, information, and finished products. As a global system of manufacturing has emerged, significant quantities of manufactured goods are now imported *and* exported across much of the world through complex commodity chains (as we saw in the case of blue jeans in Box 2.2. "Visualizing Geography: Commodity Chains"); no longer do developed economies export manufactured goods and peripheral countries import them. African countries are an important exception, with many of them barely participating in world trade in manufactured goods.

One important aspect of world trade is that the smaller, peripheral partners in trading relationships are highly dependent on developed economies. An aspect of dependency, in this context, is the degree to which a country's export base lacks diversity. **Dependency** involves a high level of reliance by a country on foreign enterprises, investment, or technology. External dependence for a country means that it is highly reliant on levels of demand and the overall economic climate of other countries. Dependency for a peripheral country can result in a narrow economic base in which the balancing of national accounts and the generation of foreign exchange depend on the export of one or two agricultural or mineral resources, as is the case in Angola, Chad, the Dominican Republic, Iran, Iraq, Libya, and Nigeria, for example.

Patterns of International Debt   In many peripheral countries, debt service—the annual interest on international debts—is a significant handicap to economic development. For every $1 that developing countries receive in aid, they end up having to return $5 in debt service payments to core countries. In some countries, 20 percent or more of all export earnings are swallowed up by debt service (**Figure 8.10**).

At the root of the international debt problem is the structured inequality of the world economy. The role inherited by most peripheral countries within the international division of labor has been one of producing primary goods and commodities for which both the elasticity of demand and price elasticity are low. The **elasticity of demand** is the degree to which levels of demand for a product or service change in response to changes in price.

◄ Figure 8.9 **World Economic Forum** The opening plenary session of the 2014 World Economic Forum on East Asia, in Manila, the Philippines.

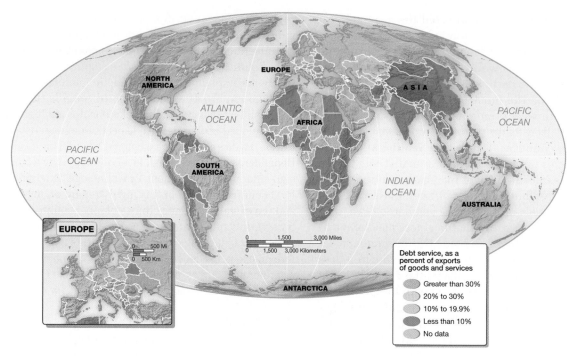

▲ **Figure 8.10 The debt crisis in 2012** In some countries, the annual interest on international debts (their "debt service") accounts for more than 20 percent of the annual value of their exports of goods and services. Many countries first got into debt trouble in the mid-1970s, when Western banks, faced with recession at home, offered low-interest loans to the governments of peripheral countries rather than being stuck with idle capital. When the world economy heated up again, interest rates rose and many countries found themselves facing a debt crisis.

(Source: World Bank, Global Development Finance.) Cocoa production Workers spread cocoa to dry in a Ghana hamlet.

Where a relatively small change in price induces a significant change in demand, elasticity is high; where levels of demand remain fairly stable in spite of price changes, demand is said to be inelastic. Demand for the products of peripheral countries in their principal markets (the more developed countries) has a low elasticity: It tends to increase by relatively small amounts in response to significant increases in the incomes of their customers. Similarly, significant reductions in the price of their products tend to result in only a relatively small increase in demand.

▼ **Figure 8.11 Cocoa production** Workers spread cocoa to dry in a Ghana hamlet.

Consider, for example, the cocoa-producing regions of West Africa (**Figure 8.11**). No matter how they improve productivity in order to keep prices low, and no matter how much more affluent their customers in core countries become, there is a limit to the demand for cocoa products. In contrast, the elasticity of demand and price elasticity of high-tech manufactured goods and high-order services (the specialties of core economies within the international division of labor) are both high. As a result, the **terms of trade** are stacked against the producers of primary goods.

The terms of trade are determined by the ratio of the prices at which exports and imports are exchanged. When the price of exports rises relative to the price of imports, the terms of trade reflect an improvement for the exporting country. No matter how efficient primary producers may become, or how affluent their customers, the balance of trade is tilted against them. Quite simply, they must run in order to stand still.

An obvious counterstrategy for peripheral countries is to attempt to establish a new role in the international division of labor, moving away from a specialization in primary commodities toward a more diversified manufacturing base. This strategy is known as **import substitution**. It is a difficult strategy to pursue, however, because building up a diversified manufacturing base requires vast amounts of start-up capital. With the terms of trade running against them, it is extremely difficult for peripheral countries to accumulate this capital; so they have to borrow.

The debt problem has led to calls for affluent lending countries to provide debt relief to some of the poorest countries. In 2005, the world's richest countries—the G8 group—agreed to write off $40 billion in debts owed by 18 of the world's poorest countries, most of them in Africa. Addressing the full magnitude of LDC debt, however, will require continued substantial and sustained efforts if the poorest countries are to break free of the crushing financial obligations of their accumulated debts.

APPLY YOUR KNOWLEDGE

1. How do economic relationships of dependency affect poverty around the globe?

2. Conduct research to determine the current debt of three different countries. How different are the debt ratios of each of the countries you selected? List two main factors contributing to each of the specific countries' debt. Do you see any evidence of the debt cycle changing?

▲ **Figure 8.12** Farmers harvesting cotton in Gujarat, India   These farmers supply fair trade cotton to supermarkets in the UK.

Fair Trade   The Fair Trade movement is a result of increasing awareness within developed countries of the weak bargaining position of many small producers at the beginning of the commodity chains that underpin the global economy. Fair Trade has become part of the "mobilization against globalization," an attempt to raise consumers' consciousness about the relationships embodied in their purchases.

The Fair Trade movement is a global network of producers, traders, marketers, advocates, and consumers focused on building equitable trading relationships between consumers and the world's most economically disadvantaged artisans and farmers. As such, it is fundamentally a strategy for poverty alleviation and sustainable development. The key principles of Fair Trade include (1) creating opportunities for economically disadvantaged producers, (2) capacity building, (3) ensuring that women's work is properly valued and rewarded, (4) ensuring a safe and healthy working environment for producers, and (5) payment of a fair price—one that covers not only the costs of production but enables production that is socially just and environmentally sound. The result is often very modest in relation to the retail price of Fair Trade goods in developed countries—only about 9 cents of the $3.10 cost of an average fairly traded 100-gram bar of chocolate goes to producers in developing countries, the rest being accounted for by core-country food processors, designers, packagers, photographers, marketing staff, advertisers, shopkeepers, and tax authorities. Nevertheless, Fair Trade demonstrably helps producers in developing countries (**Figure 8.12**) and typically involves democratic decision making over how the extra money earned is to be distributed.

Several types of Fair Trade organizations perform various roles along the commodity chains linking producers to consumers. At the beginning of the commodity chain are producer organizations—village or community groups or cooperatives, for example, often joined together under export marketing organizations. In 2014, there were more than 1,150 Fair

Trade–certified producer groups (including many umbrella bodies) worldwide. These organizations typically sell their products to a second kind of Fair Trade organization: registered importers and wholesalers in more developed countries. In 2014, these existed in about 60 countries. They in turn sell to Fair Trade retailers: "world shops" and catalog- or Internet-based retailers (Figure 8.12). In some countries, Fair Trade products are now mainstream, available in major supermarkets and independent shops and beginning to gain market share. In Switzerland, for example, Fair Trade bananas account for 20 percent of the retail market. In the United States, it is coffee that is the most important Fair Trade–certified product, accounting for up to $600 million in sales in 2009.

There are also Fair Trade organizations that are focused on labor practices, such as the Ethical Trading Initiative (ETI), which evolved from Fair Trade campaigns run by British aid organizations. ETI involves a multiagency grouping of companies, nongovernmental organizations (NGOs), and trades unions that have together devised a basic code of labor practices covering the right to collective bargaining, safe and hygienic working conditions, living wages, and a standard working week of no more than 48 hours.

Interpretations of International Patterns of Development

The overall relationship between economic structure and levels of prosperity makes it tempting to interpret economic development in distinctive stages. Each developed region or country, in other words, might be thought of as progressing from the early stages of development, with a heavy reliance on primary activities (and relatively low levels of prosperity), through a phase of industrialization and on to a "mature" stage of postindustrial development (with a diversified economic structure and relatively high levels of prosperity).

This, in fact, has been a commonly held view of economic development, first conceptualized by a prominent economist, W. W. Rostow (**Figure 8.13**).

Rostow's model, however, is too simplistic to be of much help in understanding human geography. The reality is that places and regions are now *interdependent*. The fortunes of any given place are increasingly tied up with those of many others. Furthermore, Rostow's model perpetuates the myth of "developmentalism," the idea that every country and region will eventually make economic progress toward "high mass consumption" provided that they compete to the best of their ability within the world economy. The main weakness of developmentalism is that it is simply not reasonable to compare the prospects of late starters to the experience of those places, regions, and countries that were among the early starters. For these early starters the horizons were clear: free of effective competition, free of obstacles, and free of precedents. For the late starters the situation is entirely different. Today's less developed regions must compete in a crowded field while facing numerous barriers that are a direct consequence of the success of some of the early starters.

Indeed, many writers and theorists of international development claim that the prosperity of the core countries in the world economy has been based on *under*development and squalor in peripheral countries. Peripheral countries, it is argued, could not "follow" the previous historical experience of developed countries in stages-of-development fashion because their underdevelopment (that is, exploitation) was a structural requirement for development elsewhere. The development of Europe and North America, in other words, required the systematic underdevelopment of peripheral countries. By means of unequal trade, exploitation of labor, and profit extraction, the underdeveloped countries became increasingly rather than decreasingly impoverished.

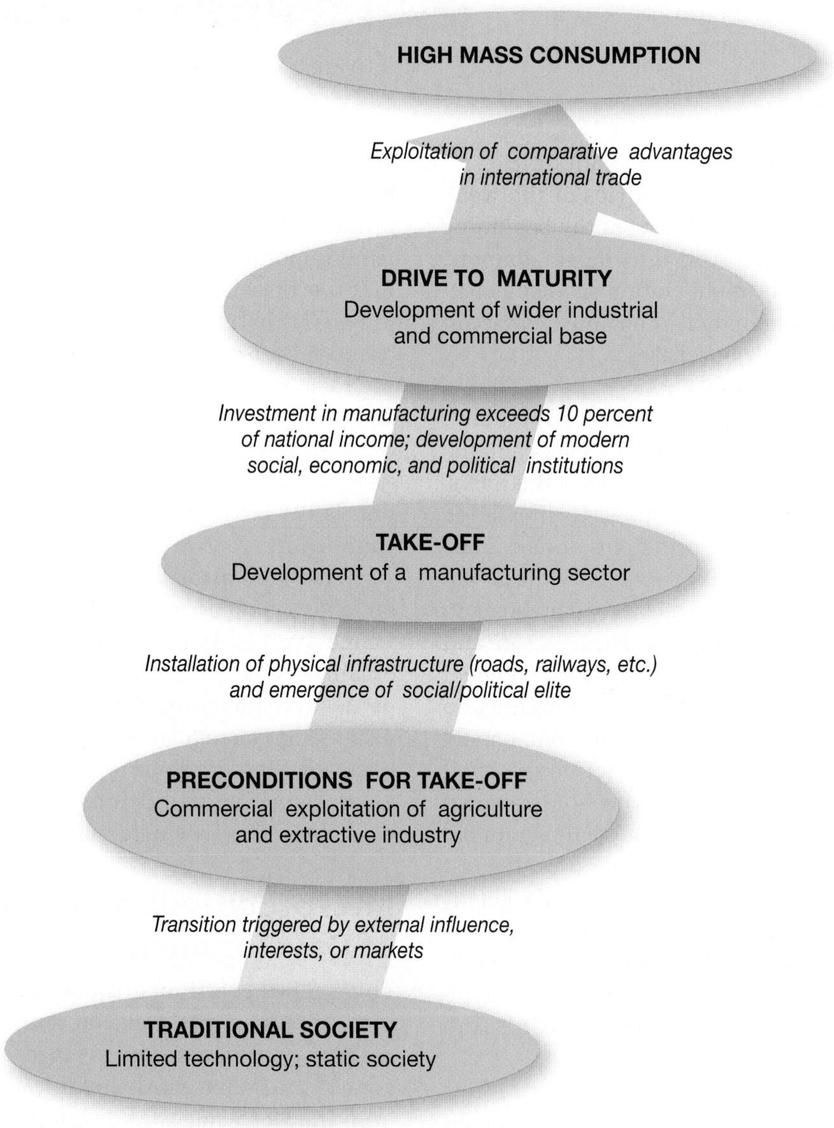

HIGH MASS CONSUMPTION

Exploitation of comparative advantages in international trade

DRIVE TO MATURITY
Development of wider industrial and commercial base

Investment in manufacturing exceeds 10 percent of national income; development of modern social, economic, and political institutions

TAKE-OFF
Development of a manufacturing sector

Installation of physical infrastructure (roads, railways, etc.) and emergence of social/political elite

PRECONDITIONS FOR TAKE-OFF
Commercial exploitation of agriculture and extractive industry

Transition triggered by external influence, interests, or markets

TRADITIONAL SOCIETY
Limited technology; static society

▲ **Figure 8.13 Stages of economic development** This diagram illustrates a model based on the idea of successive stages of economic development. Each stage is seen as leading to the next, though some regions or countries may take longer than others to make the transition from one stage to the next. According to this view, now regarded as overly simplistic, places and regions follow parallel courses within a world that is steadily modernizing. Late starters eventually make progress, but at speeds determined by their resource endowments, their productivity, and the wisdom of their people's policies and decisions.

This explanation is known as "dependency theory." It has been a very influential approach in explaining global patterns of development and underdevelopment. Dependency theory states, essentially, that development and underdevelopment are reverse sides of the same process: *Development somewhere requires underdevelopment elsewhere.* Immanuel Wallerstein's world-system theory (see Chapter 2) takes this kind of dependency into account. According to this perspective, the entire world economy is to be seen as an evolving market system with an economic hierarchy of states—a core, a semiperiphery, and a periphery. The composition of this hierarchy is variable: Individual countries can move from periphery to semiperiphery, core to periphery, and so on.

REGIONAL ECONOMIC DEVELOPMENT

Unevenness in economic development often has a regional dimension. Initial conditions are a crucial determinant of regional economic performance. Scarce resources, a history of neglect, lack of investment, and concentrations of low-skilled people all combine to explain the lagging performance of certain areas. In some regions, initial extreme disadvantages constrain the opportunities of individuals born there. A child born in the Mexican state of Chiapas, for example, has much bleaker prospects than a child born in Mexico City. The child from Chiapas is twice as likely to die before age 5, less than half as likely to complete primary school, and 10 times as likely to live in a house without access to running water. On reaching working age, he or she will earn 20 to 35 percent less than a comparable worker living in Mexico City and 40 to 45 percent less than one living in northern Mexico.

Other examples of regional inequality can be found throughout the world, and globalization has increased regional inequality within many countries. In China, for example, disparities have widened dramatically between the interior and the export-oriented regions of the coast. The transition economies of the countries of the former Soviet Union and its Eastern European satellites have registered some of the largest increases in regional inequality; some core countries—especially Sweden, the United Kingdom, and the United States—have also registered significant increases in regional inequality since the 1980s. At this regional scale, as at the global scale, levels of economic development often exhibit a fundamental core-periphery structure. Indeed, within-country core-periphery contrasts are evident throughout the world: in core countries such as France and the United States, in semiperipheral countries such as South Korea, and in peripheral countries such as Nigeria and Indonesia.

The Economics of Location

Patterns of regional economic development are historical in origin and cumulative in nature. Recognizing this, geographers are interested in **geographical path dependence**, the relationship between present-day activities in a place and the past experiences of that place. An important principle of regional economic development is the principle of initial advantage. **Initial advantage** highlights the importance of an early start in economic development. It represents a special case of external economies. Other things being equal, new phases of economic development take hold first in settings that offer external economies: existing labor markets, existing consumer markets, existing frameworks of fixed social capital, and so on. **External economies** are cost savings that result from advantages beyond a firm's organization and methods of production. These initial advantages are consolidated by **localization economies**—cost savings that accrue to particular industries as a result of clustering together at a specific location—and so form the basis for continuing economic growth.

Examples of localization economies include sharing a pool of labor with special skills or experience, supporting specialized technical schools, joining to create a marketing organization or a research institute, and drawing on specialized subcontractors, maintenance firms, suppliers, distribution agents, and lawyers. If such advantages lead to a reputation for high-quality production, localization is intensified because more producers want to cash in on the reputation. Among many examples are the electronics and software industries of Silicon Valley; recording companies in Los Angeles; Sheffield steel and steel products, especially cutlery, from England; the U.S. auto industry in Detroit; Swiss watches made in the towns of Biel, Geneva, and Neuchâtel; Belgian lace from Bruges, Brussels, and Mechelen; English worsted in Bradford and Huddersfield; Stoke-on-Trent English china; Irish linen from Athlone; and French perfume made in Grasse.

For places and regions with a substantial initial advantage, therefore, the trajectory of geographical path dependence tends to be one of persistent growth. This pattern reinforces, in turn, the core-periphery patterns of economic development found in every part of the world and at every spatial scale. That said, geographers recognize there is no single pathway to development. The consequences of initial advantage for both core and peripheral regions can be—and often are—modified. Old core-periphery relationships can be blurred, and new ones can be initiated.

APPLY YOUR KNOWLEDGE

1. What regions seem to have economic advantages in the world? Why might that be?

2. Identify an example of localized economics functioning in your own community. What industries are clustering together and how can this result in cost savings? Think about whether these industries are responding to a creative workforce, raw materials in your area, and/or demand for certain products in your community. How much of the clustering is dependent on local energies? Are there outside sources contributing to the inertia of this clustering?

Regional Economic Linkages

Regional cores of economic development are created cumulatively, through the operation of basic principles of economic geography within complex webs of *functional interdependence.* These webs include the relationships and linkages between different kinds of industries, different kinds of stores, and different kinds of offices. Particularly important here are the **agglomeration effects** associated with various kinds of local economic linkages and interdependencies. These include the cost advantages that accrue to individual firms because of their location among functionally related activities. The trigger for agglomeration effects can be any kind of economic development—the establishment of a trading port or the growth of a local industry or any large-scale enterprise, for example. The external economies and economic linkages generated by such developments are the initial advantages that can stimulate a self-propelling process of local economic development.

A number of interrelated effects come into play when new economic activity begins in an area. **Backward linkages** develop as new firms arrive to provide the growing industry with components, supplies, specialized services, or facilities. **Forward linkages** develop as new firms arrive to take the finished products of the growing industry and use them in their own processing, assembly, finishing, packaging, or distribution operations. Together with the initial growth, the growth in these linked industries helps create a threshold of activity large enough to attract **ancillary industries** and activities (e.g., maintenance and repair, recycling, security, and business services).

The existence of these interrelated activities establishes a pool of specialized labor with the kinds of skills and experience that make the area attractive to still more firms. Meanwhile, the linkages among all these firms help promote interaction between professional and technical personnel and allow the area to support R&D (research and development) facilities, research institutes, and so on, thus increasing the likelihood of local inventions and innovations that might further stimulate local economic development.

Another aspect of local economic growth results from the increase in population represented by the families of employees. Their presence creates a demand for housing, utilities, physical infrastructure, retailing, personal services, and so on—all of which generate additional jobs. This expansion, in turn, helps create populations large enough to attract an even wider variety and more sophisticated kinds of services and amenities. Last—but by no means least—the overall growth in local employment creates a larger local tax base. The local government can then provide improved public utilities, roads, schools, health services, recreational amenities, and so on—all of which serve to intensify agglomeration economies and so enhance the competitiveness of the area in attracting further rounds of investment.

Swedish economist Gunnar Myrdal, the 1974 Nobel Prize winner, was the first to recognize that any significant initial local advantage tends to be reinforced through these geographic principles of agglomeration and localization. He called the process **cumulative causation** (**Figure 8.14**). Cumulative causation refers to the spiraling buildup of advantages that occurs in specific geographic settings as a result of the development of external economies, agglomeration effects, and localization economies. Myrdal also pointed out that this spiral of local growth tends to attract people—enterprising young people, usually—and investment funds from other areas. According to the basic principles of spatial interaction, these flows tend to be strongest from nearby regions and areas with the lowest wages, fewest job opportunities, or least attractive investment opportunities.

In some cases this loss of entrepreneurial talent, labor, and investment capital is sufficient to trigger a cumulative negative spiral of economic disadvantage. With less capital, less innovative energy, and depleted pools of labor, industrial growth in peripheral regions tends to be significantly slower and less innovative than in regions with an initial advantage. This in turn tends to limit the size of the local tax base, making it difficult for local governments to furnish a competitive infrastructure of roads, schools, and recreational amenities. Myrdal called these negative impacts on a region (or regions) of the economic growth of some other region **backwash effects**. Negative impacts take the form, for example, of out-migration, outflows of investment capital, and the shrinkage of local tax bases. Backwash effects are important because they help explain why regional economic development is so uneven and why core-periphery contrasts in economic development are so common.

APPLY YOUR KNOWLEDGE

1. How does geography or location contribute to agglomeration effects in the economy?

2. Identify and research three cities that have experienced a backwash effect. List three reasons for the backwash effect in each locale.

The Modification of Regional Core-Periphery Patterns

Although very important, cumulative causation and backwash effects are not the only processes affecting the geography of economic development. If they were, the world's economic geography would be even more starkly polarized than it is now. There would be little chance for the emergence of new growth regions, like Guangdong in Southeast China, and there would be little likelihood of stagnation or decline in once-booming regions, like northern England.

Myrdal himself recognized that peripheral regions do sometimes emerge as new growth regions and partially explained them in what he called *spread* (or trickle-down) effects. **Spread effects** are the positive impacts on a region (or regions) from the economic growth of some other region, usually a core region. Growth creates levels of demand for food, consumer products, and other manufactured goods that are so high that local producers cannot satisfy them. This demand gives investors in peripheral regions (or countries) the opportunity to establish a local capacity to meet the demand. Entrepreneurs who participate are also able to exploit the advantages of cheaper land and labor in peripheral regions. If strong enough, spread effects can enable peripheral regions to develop their own spiral of cumulative causation, thus changing the interregional geography of economic patterns and flows. The economic growth of South Korea, for example, is partly attributable to the spread effects of Japanese economic prosperity.

Core-periphery patterns and relationships can also change as a result of slowdowns or reversals in the spiral of cumulative causation in core regions. The main cause of this effect is the development of **agglomeration diseconomies**. Agglomeration diseconomies are the negative economic effects of urbanization and the local concentration of industry, including the higher prices that must be paid by firms competing for land and labor; the costs of delays resulting from traffic congestion and crowded

► **Figure 8.14 Processes of regional economic growth** Once a significant amount of new industry becomes established in an area, it creates a self-propelling process of economic growth. As this diagram shows, the geographic principles of agglomeration and localization reinforce the initial advantages of industrial growth. The overall process is known as *cumulative causation*.

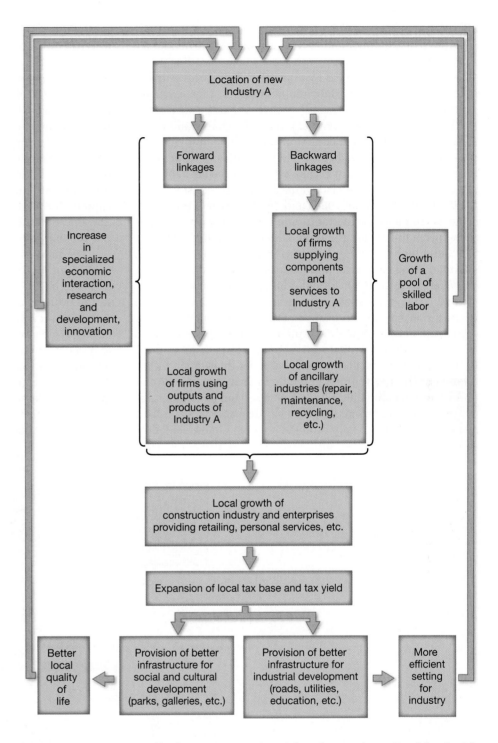

port and railroad facilities; the increasing costs of solid waste disposal; and the burden of higher taxes that eventually have to be levied by local governments in order to support services and amenities previously considered unnecessary—traffic police, city planning, and transit systems, for example.

Deindustrialization and Creative Destruction

The most fundamental cause of change in the relationship between initial advantage and cumulative causation is longer-term shifts in technology systems and the competition between states within the world-system. The innovations associated with successive technology systems generate new industries that are not yet tied down by enormous investments in factories or allied to existing industrial agglomerations. Combined with innovations in transport and communications, this creates *windows of locational opportunity* that can result in new industrial districts, with new generations of small towns or cities growing into dominant metropolitan areas through new rounds of cumulative causation.

Equally important as a factor in how core-periphery patterns change are the shifts in the profitability of old, established industries in core regions compared to the profitability of new industries in fast-growing new industrial districts. As soon as the differential is large enough, some disinvestment takes place within core regions. This disinvestment can take place in several ways. Manufacturers can reduce their wage bill by cutting back on production; they can reduce their fixed costs by closing down and selling off some of their factory space and equipment; or they can reduce their spending on research and development for new products. This disinvestment, in turn, leads to **deindustrialization** in formerly prosperous industrial core regions.

Deindustrialization involves a relative decline (and in extreme cases an absolute decline) in industrial employment in core regions as firms scale back their activities in response to lower levels of profitability (**Figure 8.15**). This is what happened to the Manufacturing Belt (sometimes called the "Rust Belt") of the United States in the 1960s and 1970s (**Figure 8.16**). It also occurred in many of the traditional industrial regions of Europe during the 1960s, 1970s, and 1980s. In France, Belgium, the Netherlands, Norway, Sweden, and the United Kingdom, manufacturing employment decreased by between one-third and one-half from 1960 to 1990. The most pronounced example of this deindustrialization has been in the United Kingdom, where a sharp decline in manufacturing employment has been accompanied by an equally sharp rise in service employment.

Meanwhile, the capital made available from disinvestment in these core regions becomes available for investment by entrepreneurs in new ventures based on innovative products and production technologies. Old industries—and sometimes entire old industrial regions—have to be "dismantled" (or at least neglected) in order to help fund the creation of new centers of profitability and employment. This process is often referred to as creative destruction, something that is inherent to the dynamics of capitalism. **Creative destruction** involves the withdrawal of investments from activities (and regions) that yield low rates of profit in order to reinvest in new activities (and new regions). The concept of creative destruction provides a powerful image, helping us understand the entrepreneur's need to withdraw investments from activities (and regions) yielding low rates of profit in order to reinvest in new activities (and, often, in new regions). In the United States, for example, the deindustrialization of the Manufacturing Belt provided the capital and the locational flexibility for firms to invest in the Sun Belt of the United States and in semiperipheral countries like Mexico and South Korea.

The process does not stop there, however. If the deindustrialization of the old core regions is severe enough, the relative cost of their land, labor, and infrastructure may decline to the point where they once again become attractive to investors. As a result, a seesaw movement of investment capital occurs, which over the long term tends to move from developed to less developed regions—then back again, once the formerly developed region has experienced a sufficient relative decline. "Has-been" regions can become redeveloped and revitalized, given a new lease on life by the infusion of new capital for new industries. This is what happened, for example, to the Pittsburgh region in the 1980s, resulting in the creation of a postindustrial

The spiral of deindustrialization

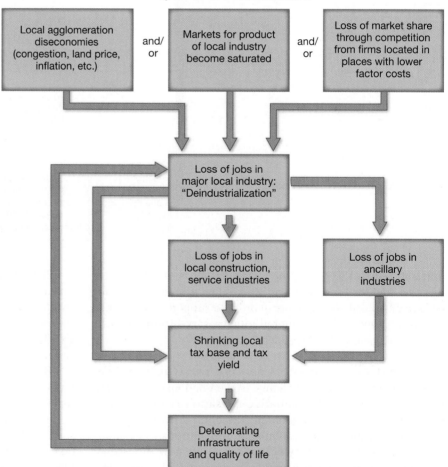

▶ **Figure 8.15 Regional economic decline** When the locational advantages of manufacturing regions are undermined for one reason or another, profitability declines and manufacturing employment falls. This can lead to a downward spiral of economic decline, as experienced by many of the traditional manufacturing regions of Europe and North America during the 1960s, 1970s, and 1980s. (Reprinted with permission of Prentice Hall, from P. L. Knox, *Urbanization*, © 1994, p. 55.)

▲ **Figure 8.16** **Deindustrialization** This abandoned Packard automobile plant in Detroit, Michigan, is testament to the downward economic spiral in what was once one of the world's most important heavy manufacturing regions.

▲ **Figure 8.17** **Congestion charges** The London congestion charge was introduced in 2003 to reduce traffic congestion in the center of the city. The standard charge is £10 ($16) for each day, for each nonexempt vehicle that travels within the zone between 7:00 am and 6:00 pm (Monday–Friday only); a penalty of between £60 ($96) and £180 ($288) is levied for nonpayment.

economy out of a depressed industrial setting. USX, the U.S. Steel group of companies, reduced its workforce in the Pittsburgh region from more than 20,000 to less than 5,000 between 1975 and 1995. These losses have been more than made up, however, by new jobs generated in high-tech electronics, specialized engineering, and finance and business services.

Government Intervention In addition to the processes of deindustrialization and creative destruction, core-periphery patterns can also be modified by government intervention. National governments realize that regional planning and policy can be an important component of strategies to stabilize and reorganize their economies, as well as to maximize their overall competitiveness. Without regional planning and policy, the resources of peripheral regions can remain underutilized, while core regions can become vulnerable to agglomeration diseconomies. For political reasons, too, national governments are often willing to help particular regions adjust to changing economic circumstances. At the same time, most local governments take responsibility for stimulating economic development within their jurisdiction, if only in order to increase the local tax base in the long term.

The nature and extent of government intervention has varied over time and by country. In some countries, special government agencies have been established to promote regional economic development and reduce core-periphery contrasts. Among the best-known examples are the Japanese MITI (Ministry of International Trade and Industry), the Italian Cassa del Mezzogiorno (Southern Development Agency, replaced in 1987 by several smaller agencies), and the U.S. Economic Development Administration. Some governments have sought to help industries in declining regions by undertaking government investment in infrastructure and providing subsidies for private investment; others have sought to devise tax breaks that reduce the cost of labor in peripheral regions. Still others have sought to deal with agglomeration diseconomies in core regions through increased taxes and restrictions on land use (**Figure 8.17**).

While each approach has its followers, one of the most widespread governmental approaches to core-periphery patterns involves the exploitation of the principle of cumulative causation through the creation of growth poles. **Growth poles** are places of economic activity deliberately organized around one or more high-growth industries. Economists have noted, however, that not all industries are equal in the extent to which they stimulate economic growth and cumulative causation. The ones that generate the most pronounced effects are known as "propulsive industries," and they have received a great deal of attention from geographers and economists who are interested in helping shape strategic policies that might promote regional economic development. In the 1920s, shipbuilding was a propulsive industry. In the 1950s and 1960s, automobile manufacturing was a propulsive industry, and today biotechnology and digital technologies are propulsive industries. The basic idea is for governments to promote regional economic growth by fostering propulsive industries in favorable locations. These locations are intended to become growth poles—places that, given an artificial start, develop a self-sustaining spiral of economic prosperity.

APPLY YOUR KNOWLEDGE

1. What is the difference between deindustrialization and creative destruction?

2. Identify two regions that have experienced deindustrialization in addition to the regions referred to in the text. Think about when industries created a "boom" of economic activity in those regions. Did these industries experience any governmental intervention post-boom?

GLOBALIZATION AND ECONOMIC DEVELOPMENT

In the past 30 years, regional core-periphery patterns have been increasingly influenced by globalization and the economic interdependence of major world regions. This globalization has been caused by four important and interrelated factors: a new international division of labor, a new technology system, the homogenization of international consumer markets, and the internationalization of finance.

■ A new international division of labor. A major wave of corporate globalization took place in the 1970s, led by manufacturing giants like General Motors and General Electric that wanted to reduce labor costs, outflank national labor unions, and increase overseas market penetration. This new international division of labor has resulted in manufacturing production shifting from the world's core regions to semiperipheral and peripheral countries as companies seek to keep production costs low by exploiting the huge differentials in wage rates around the world. Meanwhile, new specializations have emerged within the core regions of the world-system: high-tech manufacturing and **producer services** such as information services, insurance, and market research that enhance the productivity or efficiency of other firms' activities or that enable them to maintain specialized roles. These producer services industries have themselves become globalized in response to the needs of their most important clients, the global manufacturing corporations.

■ A new technology system, based on a combination of innovations, including solar energy, robotics, microelectronics, biotechnology, and digital telecommunications and information systems. This has required the geographical reorganization of the core economies. It has also extended the global reach of finance and industry and permitted a more flexible approach to investment and trade. Especially important in this regard have been new and improved technologies in transport and communications—the integration of shipping, railroad, and highway systems through containerization (**Figure 8.18**); the introduction of wide-bodied cargo jets; and the development of fiber-optic networks, communications satellites, and e-mail and information retrieval systems.

■ The homogenization of international consumer markets. Among the more affluent populations of the world, a new and materialistic international culture has taken root, in which people save less, borrow more, defer parenthood, and indulge in affordable luxuries that are marketed as symbols of style and distinctiveness. This culture is easily transmitted through the new telecommunications media, and it has been an important basis for transnational corporations' global branding and marketing of "world products" (e.g., German luxury automobiles, Swiss watches, British raincoats, French wines, American soft drinks, Italian shoes and designer clothes, and Japanese consumer electronics).

■ The internationalization of finance. The pivotal moment in the emergence of global banking and globally integrated financial markets was a "system shock" to the international economy that occurred in the mid-1970s. World financial markets, swollen with U.S. dollars by the U.S. government's deficit budgeting and by huge currency reserves held by OPEC (Organization of Petroleum Exporting Countries) after it had orchestrated a four-fold increase in the price of crude oil, quickly evolved into a new and sophisticated system of international finance, with new patterns of investment and disinvestment. Meanwhile, the capacity of computers and information systems to deal very quickly with changing international conditions added a speculative component to the

◄ **Figure 8.18 The impact of containerization on world trade** Containerization revolutionized long-distance transport by doing away with the slow, expensive, and unreliable business of loading and unloading ships with manual labor. Before containerization, ships spent one day in port for every day at sea; in the wake of containerization, they spend a day in port for every ten days at sea. By 1965, an international standard for containers had been adopted, making it possible to transfer goods directly from ship to rail to road and allowing for a highly integrated global transport infrastructure. The average container ship today holds 4,000 20-foot containers, but some are able to carry 6,000 or 8,000. Containerization requires a heavy investment in both vessels and dockside handling equipment, however. As a result, container traffic has quickly become concentrated in a few ports that handle high-volume transatlantic and transpacific trade. This photograph shows the Harem container port in Istanbul, Turkey.

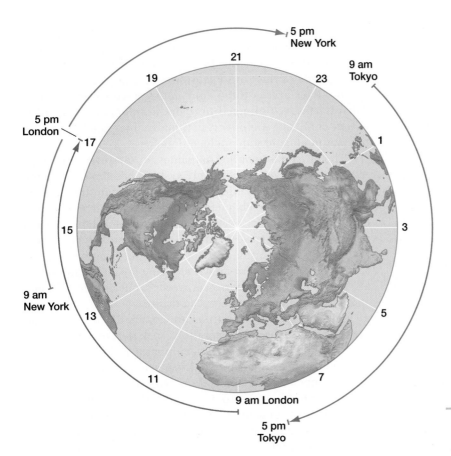

the soaring demand for housing. Few in the financial services industries fully understood the complexities of the booming mortgage market, and the various risk-assessment agencies were seriously at fault in underestimating the risks associated with these loans. Eventually, when interest rates increased, many households began to default on their monthly payments and, as the bad loans added up, mortgage lenders, in turn, found themselves in financial trouble.

APPLY YOUR KNOWLEDGE

1. How has globalization changed the core-peripheral relationships in the world?

2. Choose a semiperipheral or peripheral country and determine the ways each of the four factors of globalization have affected its economic development process. Create two lists reflecting the global economy: 1) the corporations that have factories, service centers or regional head quarters there; and 2) the international products and services that are popular in that country.

▲ **Figure 8.19** 24-hour trading between major financial markets Office hours in the two most important financial centers—New York and London—overlap one another somewhat even though the two cities are situated in broadly separated time zones. While these markets are both closed, Tokyo offices are open. This means that between them the world's three major financial centers span the globe with almost 24-hour trading in currencies, stocks, and other financial instruments.

internationalization of finance. International movements of money, bonds, securities, and other financial instruments have now become an end in themselves because they are a potential source of high profits from speculation and manipulation. The global banking and financial network handles trillions of dollars every day—90 percent of which has nothing at all to do with the traditional world economy of trade in goods and services. The nerve centers of the new system are located in just a few places—London, Frankfurt, New York, and Tokyo, in particular. Satellite communications systems and fiberoptic networks made it possible for firms to operate key financial and business services 24 hours a day around the globe (**Figure 8.19**), handling an enormous volume of transactions. Taking advantage of the relaxed controls on financial institutions resulting from neoliberal policies, mortgage lenders in core countries (and especially the United States) had been selling their mortgages on bond markets and to investment banks in order to fund

Global Assembly Lines and Supply Chains

The globalization of the world economy represents the most recent stage in a long process of internationalization. At the heart of this process are private companies that participate not only in international trade but also in production, manufacturing, and/or sales operations in several countries. Almost 80 percent of Ford's workforce, for example, is employed overseas, and foreign sales account for 55 percent of its total revenues. Over 50 percent of IBM's workforce is employed overseas, and 61 percent of its revenues are derived from foreign sales; 82 percent of the workforce of Philips, the Dutch electronics firm, is employed overseas, and 95 percent of its revenues are derived from foreign sales. Many of these transnational corporations have grown large through mergers and acquisitions and their activities span a diverse range of economic activities.

Corporations like theses that consist of several divisions engaged in quite different activities are known as **conglomerate corporations**. Altria (formerly known as Philip Morris), for example, primarily known for its tobacco products (such as Marlboro cigarettes), also controls a large group of assets in the beverage industry (including Miller Brewing) and has extensive interests in real estate, import-export, publishing, and foods (including General Foods, Tobler, Terry's, and Suchard chocolate). Nestlé, the world's largest packaged-food manufacturer, is the largest company in Switzerland but derives less than 2 percent of its revenue from its home country. Its major

U.S. product lines and brand names include beverages (Calistoga, Nescafé, Nestea, Perrier, Taster's Choice), chocolate and candy (Butterfinger, Crunch, KitKat), culinary products (Buitoni, Carnation, Libby, Maggi), frozen foods (Lean Cuisine), pet foods (Purina, Fancy Feast, Friskies, Mighty Dog), and drugs and cosmetics (L'Oreal). In addition to its 480 factories in 63 countries around the world, Nestlé operates more than 40 Stouffer hotels.

Transnationals and Globalization Transnational corporations first began to appear in the nineteenth century, but until the mid-twentieth century there were only a few. Most of these were U.S.- or European-based transnationals concerned with obtaining raw materials, such as oil or minerals, for their domestic manufacturing operations. After World War II, an increasing number of large corporations began to invest in overseas production and manufacturing operations as a means of establishing a foothold in foreign consumer markets. Beginning in the 1970s, the growth of transnational conglomerates increased sharply, not only in the United States but also in Europe, Japan, and even some semiperipheral countries. By 2012, there were more than 43,000 transnational corporations in the world. Of these, the top 147, all closely connected to one another, together dominated heart of the world economy. Many of the largest transnational corporations are now more powerful, in economic terms, than most sovereign nations. General Motors' economy is larger than Portugal's; Toyota's is larger than Ireland's; and Wal-Mart's annual sales exceed Norway's gross domestic product.

The advantages to manufacturers of a global assembly line are several:

- A standardized global product for a global market allows them to maximize economies of scale.

- A global assembly line allows production and assembly to take greater advantage of the full range of global variations in costs. With a global assembly line, labor-intensive work can be done where labor is cheap, raw materials can be processed near their source of supply, final assembly can be done close to major markets, and so on.

- A global assembly line means that a company is no longer dependent on a single source of supply for a specific component, thus reducing its vulnerability to industrial troubles and other disturbances.

- Global sourcing allows transnational conglomerates better access to local markets. For example, Boeing has pursued a strategy of buying a significant number of aircraft components in China and has therefore succeeded in opening the Chinese market to its products.

The automobile industry was among the first to develop a global assembly line. In 1976, Ford introduced the Fiesta, a vehicle designed to sell in Europe, South America, and Asia as well as North America. The Fiesta was assembled in several locations from components manufactured in an even greater number of locations. It became the first of a series of Ford "world cars" that now includes the Focus, the Mondeo, and the Contour. The components of the Ford Escort are made and assembled in 15 countries across three continents. Ford's international subsidiaries, which used to operate independently of the parent company, are now functionally integrated, using supercomputers and video teleconferences.

The other automobile companies followed suit and organized their own global assembly systems based on a common underbody platform with the flexibility to adapt the interior, trim, body, and ride characteristics to conditions in different countries. Honda, for example, has produced three distinct versions of the same car from its Accord world car platform—the bigger, family-oriented Accord for American drivers; the smaller, sportier Accord aimed at young Japanese professionals; and a shorter and narrower Accord, offering the stiff and sporty ride preferred by European drivers.

The global assembly lines are constantly being reorganized as transnational corporations seek to take advantage of geographical differences between places and regions and as workers and consumers in specific places and regions react to the consequences of globalization. Nike, the athletic footwear and clothing marketer, provides a good illustration. Nike once relied on its own manufacturing facilities in the United States and the United Kingdom. Today, however, most of its production is subcontracted to suppliers in East, South, and Southeast Asia (**Figure 8.20**). The geography of this subcontracting evolved over time in response to the changing pattern of labor costs in Asia. The first Asian production of Nike shoes took place in Japan. The company then switched most of its subcontracting to South Korea and Taiwan. As labor costs rose there, Nike's subcontracting spread across more and more peripheral countries—China, Indonesia, Malaysia, and Vietnam—in search of low labor costs. By 2006, Nike subcontractors employed more than 800,000 people in more than 680

▼ **Figure 8.20 Subcontracting** Nike factory on the outskirts of Ho Chi Minh City, Vietnam

different factories. Nike was the largest foreign employer in Vietnam, where its factories accounted for 5 percent of Vietnam's total exports. China, Indonesia, and Thailand were also major components in Nike's expanded global assembly line because of their low wage costs—around $60 per month.

Flexible Production Systems The strategies of transnational corporations are an important element in the transition from Fordism to neo-Fordism in much of the world. **Fordism** is named after Henry Ford, the automobile manufacturer who pioneered the principles involved: mass production, based on assembly-line techniques and "scientific" management, together with mass consumption, based on higher wages and sophisticated advertising techniques. In **neo-Fordism** the logic of mass production coupled with mass consumption has been modified by the addition of more flexible production, distribution, and marketing systems. This flexibility is rooted in forms of production that enable manufacturers to shift quickly and efficiently from one level of output to another and, more importantly, from one product configuration to another.

Flexible production systems involve flexibility both within firms and between them. *Within* firms, new technologies now allow a great deal of flexibility. Computerized machine tools, for example, are capable of producing a variety of new products simply by being reprogrammed, often with very little downtime between production runs for different products. Different stages of the production process (sometimes located in different places) are integrated and coordinated through computer-aided design (CAD) and computer-aided manufacturing (CAM) systems. Computer-based information systems monitor retail sales and track wholesale orders, allowing producers to reduce the costs of raw materials stockpiles, parts inventories, and warehousing through sophisticated small-batch, just-in-time production and distribution systems. **Just-in-time production** employs vertical disintegration (see p. 243) within large, formerly functionally integrated firms, such as automobile manufacturers, in which daily and even hourly deliveries of parts and other supplies from smaller (often nonunion) subcontractors and suppliers now arrive "just in time" to maintain "last-minute" and "zero" inventories. The combination of computer-based information systems, CAD/CAM systems, and computerized machine tools has also given firms the flexibility to exploit specialized niches of consumer demand so that economies of scale in production can be applied to upscale but geographically scattered markets.

The Benetton clothing company provides an excellent case-study example of the exploitation of flexible production systems within a single firm. In 1965, the Benetton company began with a single factory near Venice. In 1968, it acquired a single retail store in the Alpine town of Belluno, marking the beginning of a remarkable sequence of corporate expansion. Benetton is now a global organization with over 5,000 retail outlets in more than 120 countries and its own investment bank and financial services organizations. It achieved this growth by exploiting computers, new communications and transportation systems, flexible outsourcing strategies, and new production-process technologies (such as robotics and CAD/CAM systems) to the fullest possible extent.

Only about 400 of Benetton's employees are located in the company's home base of Treviso, Italy (**Figure 8.21**). From Treviso, Benetton managers coordinate the activities of more than 250 outside suppliers in order to stock its worldwide network of retail outlet franchises. In Treviso, the firm's designers create new shirts and sweaters on CAD terminals, but their designs are produced only for orders already in hand, allowing for the coordination of production with the purchase of raw materials. In factories, rollers linked to a central computer spread and cut layers of cloth in small batches according to the numbers and colors ordered by Benetton stores around the world. Sweaters, gloves, and scarves, knitted in volume in white yarn, are dyed in small batches by machines similarly programmed to respond to sales orders. Completed garments are warehoused briefly (by robots) and shipped out directly (via private package delivery firms) to individual stores to arrive on their shelves within 10 days of manufacture.

Sensitivity to demand, however, is the foundation of Benetton's success. Niche marketing and product differentiation are central to this sensitivity, which requires a high degree of flexibility in exploiting new product lines. Key stores patronized by trendsetting consumers are monitored closely, and many Benetton stores' cash registers operate as point-of-sale

▲ **Figure 8.21 Center of a global organization** Headquarters of the Benetton Group, Villa Minelli, Treviso, Italy

terminals so that marketing data are available to company headquarters daily. Another notable feature of the company's operations is the way that different market niches are exploited with the same basic products. In Italy, Benetton products are sold through several different retail chains, each with an image and decor calculated to bring in a different sort of customer.

Between firms, the flexibility inherent in neo-Fordism is achieved through the externalization of certain functions. One way of doing this is to reorganize administrative, managerial, and technical functions into flatter, leaner, and more flexible forms of organization that can make increased use of outside consultants, specialists, and subcontractors. This has led to a degree of vertical disintegration among firms. **Vertical disintegration** involves the evolution from large, functionally integrated firms within a given industry toward networks of specialized firms, subcontractors, and suppliers. Another route to externalization is participation in joint ventures, in the licensing or contracting of technology, and in strategic alliances involving design partnerships, collaborative R&D projects, and the like. **Strategic alliances** are commercial agreements between transnational corporations, usually involving shared technologies, marketing networks, market research, or product development. They are an important contributor to the intensification of economic globalization.

Maquiladoras and Export-Processing Zones

The governments of many peripheral and semiperipheral countries encourage the type of subcontracting carried out by big transnational corporations. These governments see participation in global assembly lines as a pathway to export-led industrialization. They offer incentives such as tax "holidays" (not having to pay taxes for a specified period) to transnational corporations. In the 1960s, Mexico enacted legislation permitting foreign companies to establish "sister factories"—*maquiladoras*—within 19 kilometers (12 miles) of the border with the United States for the duty-free assembly of products destined

for reexport (see **Figure 8.22**). By 2005, more than 3,500 such manufacturing and assembly plants had been established, employing around a million Mexican workers, most of them women, and accounting for more than 30 percent of Mexico's exports. But since 2005 hundreds of *maquiladoras* have closed, resulting in the loss of tens of thousands of jobs, as lower wage rates and better incentives in other countries—particularly China—have proved more attractive to manufacturers.

More common, and more successful, are **export-processing zones (EPZs)**: small areas where governments create especially favorable investment and trading conditions in order to attract export-oriented industries. These conditions include minimum levels of bureaucracy, the absence of foreign exchange controls, factory space and warehousing at subsidized rents, low tax rates, and exemption from tariffs and export duties. In 1985, it was estimated that there were a total of 173 EPZs around the world, which together employed 1.8 million workers. By 2006, there were about 3,500 EPZs, employing about 66 million people.[1] China alone had 164 EPZs, which together employed 40 million workers. The International Labor Organization (ILO) has criticized EPZs because very few of them have any meaningful links with the domestic economies around them, and most trap large numbers of people in low-wage, low-skill jobs.

In addition to tax incentives and EPZs, many governments also establish policies that ensure cheap and controllable labor. Sometimes countries are pressured to participate in global assembly lines by core countries and by the transnational institutions they support. The United States and the World Bank, for example, have backed regimes that support globalized production and have pushed for austerity programs that help keep labor cheap in peripheral countries. Countries pursuing export-led industrialization as an economic development strategy do not plan to remain the providers of cheap labor for foreign-based transnational corporations, however. They hope to shift from labor-intensive manufactured goods to capital-intensive, high-technology goods, following the path of semiperipheral Asian countries like Singapore and South Korea.

▼ **Figure 8.22 Maquiladora factory** The production floor at the Flextronics maquiladora in Jalisco, Mexico.

Retailing Chains and Global Sourcing

At the other end of the commodity chain from the farms, mines, *maquiladoras,* and factories of the world are the retail outlets and restaurants where the products are sold and consumed. Traditionally, retailing and food services in developed countries have been dominated by small, specialized, independent stores and local cafés, restaurants, pubs, and bars. But the logic of economic rationalization and economies of scale has displaced the traditional pattern of downtown department stores, main street shops, corner stores, and local bars with big-box superstores and national and international chains

[1] International Labor Organization, *c.* Sectoral Activities Programme, Working Paper. Geneva: International Labor Office, 2007.

of retail outlets and restaurants. Town centers once filled with a thriving mix of independent and family-owned stores now have "cloned" settings consisting of standardized supermarket retailers, fast-food chains, mobile phone shops, and fashion outlets of global conglomerates. With their cost advantages, these chains have become the economic equivalent of invasive species: voracious, indiscriminate, and often antisocial. Their big, centralized logistical operations have not only put small independent stores out of business but are driving the homogenization of consumption. The retail sector of most towns and cities is now characterized by external control. Decisions about hiring, labor policies, wages, stock, and menus are made in corporate headquarters hundreds of miles or more away.

Fast-food restaurants have become icons of this trend. McDonald's alone has some 33,000 restaurants worldwide and opens new ones at the rate of almost 2,000 each year. It is the largest purchaser of beef, pork, and potatoes and the largest owner of retail property in the world. In the United States, 40 percent of meals are eaten outside of the home, most of them at fast-food restaurants. One in four adults visits a fast-food restaurant every day. Not surprisingly, the majority of the population is overweight and the frequency of health problems associated with obesity—such as early-onset diabetes and high cholesterol—is rising rapidly. The cost of these problems to personal well-being and to health care systems is already daunting. Meanwhile, fast food's low-paying service sector has become an increasingly significant component of the economy.

Equally significant in terms of local economies, local development patterns, and global supply chains has been the success of "big-box" retail outlets such as Circuit City, Best Buy, Office Max, Home Depot, Target, and Wal-Mart. To many observers, Wal-Mart has come to symbolize the worst characteristics of globalization, including corporate greed, low wages, the decline of small-town mom-and-pop stores, and the proliferation of sweatshops in less developed countries

Supermarket chains have also become particularly influential. In the United Kingdom, for example, the top four supermarket chains—Tesco, Asda (owned by Wal-Mart), Sainsbury, and Safeway—have come to dominate the retail food market through a combination of out-of-town superstores and convenience supermarkets along high streets. As a consequence, they have killed off small general stores at the rate of one a day and specialist shops—such as butchers, bakers, and fishmongers—at the rate of 50 per week.

The centralized supply chains of supermarket chains have not only killed off small local businesses but also impacted local farmers. Supermarket chains rely on big suppliers in agribusiness. These suppliers are typically highly subsidized national and transnational firms whose global reach depends heavily on monoculture and extensive husbandry that, in turn, require the extensive use of antibiotics in animals and pesticides, fertilizer, and genetic engineering. As a result, small farmers and fishermen have been squeezed from the market. And with them many traditional local foods have disappeared or are in danger of disappearing. Meanwhile, supermarket shelves are lined with highly processed foods, out-of-season fruit and vegetables, and produce that has traveled a long way

and, often, been stored for a while. When the average North American or European family sits down to eat, most of the ingredients have typically traveled at least 1,500 miles from farm, processing, packing, and distribution, to consumption.

APPLY YOUR KNOWLEDGE

1. Describe the transnational commodity chain from production to consumption.

2. Consider three of the eateries you dine at most often in your community. Are they are locally owned or chains owned by large corporations? If they are chain outlets, were the businesses they replaced locally owned?

New Geographies of Office Employment

The globalization of production and the growth of transnational corporations have brought about another important change in patterns of local economic development. Banking, finance, and producer services are no longer locally oriented ancillary activities but important global industries in their own right. They have developed some specific spatial tendencies of their own—tendencies that have become important shapers of local economic development processes.

One of the most striking trends has been the geographic decentralization of office employment. This mainly involves "back-office" functions that have been relocated from metropolitan and business-district locations to small-town and suburban locations. Back-office functions such as record-keeping and analytical functions do not require frequent personal contact with clients or business associates. Developments in computing technologies, database access, electronic data interchanges, and telephone call routing technologies are enabling a larger share of back-office work to be relocated to specialized office space in cheaper settings, freeing space in the high-rent locations.

Some prominent examples of back-office decentralization from U.S. metropolitan areas have included the relocation of back-office jobs at American Express from New York to Salt Lake City, Fort Lauderdale, and Phoenix; the relocation of Metropolitan Life's back offices to Greenville, South Carolina, Scranton, Pennsylvania, and Wichita, Kansas; the relocation of Hertz's data-entry division to Oklahoma City, Oklahoma, Dean Witter's to Dallas, Texas, and Avis's to Tulsa, Oklahoma; and the relocation of Citibank's MasterCard and Visa divisions to Tampa, Florida, and Sioux Falls, South Dakota. Some places have actually become specialized back-office locations as a result of such decentralization. Omaha, Nebraska, and San Antonio, Texas, for example, are centers for a large number of telemarketing firms, and Roanoke, VA., has become something of a mail-order center.

Internationally, this trend has taken the form of offshore back offices. By decentralizing back-office functions to offshore locations, companies save even more in labor costs. Several New York–based life-insurance companies, for example, have established back-office facilities in Ireland, situated conveniently near Ireland's main international airport, Shannon. Insurance claim documents are shipped from New York

◀ **Figure 8.23** Outsourcing
Workers assemble air con-
ditioner units at Panasonic's
manufacturing unit at Jhajjar, in
Haryana state, India.

via Federal Express, processed in Ireland, and the results beamed back to New York via satellite or the transatlantic fiber-optic line.

The next logical step is outsourcing. The outsourcing of services is one of the most dynamic sectors of the world economy. Global outsourcing expenditures are expected to grow significantly as small and medium-sized enterprises follow the example of large transnational companies in taking advantage of low wages in semiperipheral and peripheral countries (**Figure 8.23**). Typically, international outsourcing in service industries involves the work of "routine producers" (who process data by following instructions, perform repetitive tasks, and respond to explicit procedures) rather than "symbolic analysts" (who work with abstract images, are involved with problem-identifying and problem-solving, and make decisions based on critical judgment sharpened by experience). Outsourced services range from simple business-process activities (for example, data entry, word processing, transcription) to more sophisticated, high-value-added activities (for example, architectural drawing, product support, financial analysis, software programming, and human resource services).

India has become one of the most successful exporters of outsourced service activities, ranging from call centers and business-process activities to advanced IT (information technology) services (**Figure 8.24**). More than 150 of the *Fortune* 500 companies, for example, now outsource software development to India. In the Philippines, special electronic "enterprise zones" have been set up with competitive international telephone rates for companies specializing in telemarketing and electronic commerce. Mexico, South Africa, and Malaysia have also become important locations for call centers and business-process activities.

Clusters of Specialized Offices Decentralization is outweighed, however, by the tendency for a disproportionate share of the new jobs created in banking, finance, and business services to cluster in highly specialized financial districts within major metropolitan areas. The reasons for this localization are to be found in another special case of the geographical agglomeration effects that we discussed earlier in this chapter.

Metropolitan areas such as New York City, London, Paris, Tokyo, and Frankfurt have acquired the kind of infrastructure—specialized office space, financial exchanges, teleports (office parks equipped with satellite Earth stations and linked to local fiber-optic lines), and communications networks—that is essential for delivering services to clients with a national or international scope of activity (**Figure 8.25**). These metropolitan areas have also

▼ **Figure 8.24** Globalized office work Workers at a call center in New Delhi, India, where business is outsourced from western companies.

▲ **Figure 8.25 London's financial quarter** Banking, insurance, and global financial services dominate the 'Square Mile' that forms the basis of London's world city status. Recent expansion has resulted in a clutch of new skyscrapers: from left to right, Heron Tower, 'The Cheesegrater', 'The Gherkin', and 'The Walkie-Talkie'.

established a comparative advantage both in the mix of specialized firms and expert professionals on hand and in the high-order cultural amenities (available both to high-paid workers and to their out-of-town business visitors). Above all, these metro areas have established themselves as centers of authority, with a critical mass of people in the know about market conditions, trends, and innovations. These people gain one another's trust through frequent face-to-face contact, not just in business settings but also in the informal settings of clubs and office bars. These key cities have become **world cities**—places that, in the globalized world economy, are able not only to generate powerful spirals of local economic development but also to act as pivotal points in the reorganization of global space. They are control centers for the flows of information, cultural products, and finance that collectively sustain the economic and cultural globalization of the world (see Chapter 10 for more details).

The Experience Economy and Place Marketing

As profitability becomes more challenging in agrarian, industrial, and service enterprises, so businesses have seen an opportunity to orchestrate memorable events for their customers, with memory itself becoming the "product"—the experience. It is not what is sold that characterizes the experience economy, but rather the way it is sold. Experience thus becomes a competitive advantage for products and services.

The experience economy is by no means new. Think, for example, of sports entertainment, arena rock concerts, fancy megamalls with themed entertainment, and visits to museums and galleries, as well as a great deal of traditional tourism. But businesses are increasingly combining commodities, products, and services with compelling experiences. Restaurants organize their services around particular themes (providing marketers with a new term: "eatertainment"). Shops and malls organize shows, events, or expositions ("shoppertainment"). There are now dental practices and general medical practices that double as day spas, while some manufacturers have established flagship stores that have a significant experiential component: Nike Town, SegaWorld, and Warner Studio "Villages," for example. In addition, some businesses are based on selling new kinds of experiences. Ciudad de los niños (Kids City), for example, charges an admission fee to an entire "shopping mall" where children play-shop and play-work. Similarly, American Girl Place offers a combination of attractions specifically targeted at young girls: a doll hair salon, special café, a theater featuring musicals, and shops with dolls and outfits. The American Girl Web site advertises "Café. Theater. Shops. *Memories.*"[2]

In the experience economy it is important to capitalize on places because, as we saw in Chapter 6, places have the capacity to arouse distinctive feelings and attachments. At the same time, thanks to economic and cultural globalization, places and regions throughout the world are increasingly seeking to influence the ways in which they are perceived by tourists, businesses, media firms, and consumers. As a result, places are increasingly being reinterpreted, reimagined, designed, packaged, and marketed. Through place marketing, sense of place has become a valuable commodity and culture has become an important economic activity.

Seeking to be competitive within the globalizing economy, many places have sponsored extensive makeovers of themselves, including the creation of pedestrian plazas, cosmopolitan cultural facilities, festivals, and sports and media events—what geographer David Harvey has described as the "carnival masks" and "businessmen's utopias" of global capitalism. An increasing number of places have also set up home pages on the Internet containing maps, information, photographs, guides, and virtual spaces in order to promote themselves in the global marketplace.

[2]http://www.americangirl.com/stores/brand_agplace.php

8.4 Geography Matters

How Geopolitics has Changed the World

by John Agnew, University of California, Los Angeles

Some global firms—like Starbucks, Google and Amazon—have come under fire in many countries for avoiding paying much tax on sales outside the US. Companies have long had complicated tax structures, but a recent series of stories has highlighted a number of tax-avoiding firms accused of not paying their fair share. Starbucks, for example, had sales of U.S. $675 million in the United Kingdom in 2012, but paid no corporate tax. Meanwhile, the company was able to transfer money to a Dutch sister company in royalty payments, bought coffee beans from Switzerland, and paid high interest rates to borrow from other parts of the business. Amazon, with sales in the United Kingdom of U.S. $7.7 billion in 2011, only reported a "tax expense" of U.S. $3 million. And Google's U. K. unit paid just U.S. $10.2 million to the British Treasury in 2011 on U.K. turnover of U.S. $670 million.

But everything these companies are doing is legal—it's tax *avoidance* and not tax *evasion*. Why are people outraged? Across the world, countries are experiencing deep reductions in public spending and real austerity cutting payments for public goods (schools, transportation, etc.) and human welfare (unemployment payments, food stamps, etc.). Tax avoidance is not a victimless crime. Discussions about the ethics of corporate tax avoidance are now everywhere.

This tax avoidance reflects the fact that different countries have very different rates for taxes on businesses and taxes on private incomes. Businesses have a massive incentive to minimize their taxes both at home and abroad.

Businesses might locate factories, service and distribution hubs, and regional headquarters in low-tax jurisdictions. Starbucks sources its U.K. coffee from a wholesale trading subsidiary in Switzerland. And Google operates subsidiaries in Bermuda and Ireland to take advantage of no and low taxes, respectively. Businesses also practice "transfer pricing." Transfer pricing occurs when a division of a multinational business in one country charges one of its divisions in another country for a product or a service. As a result, artificially high charges can be levied internally to siphon money from a high-tax country to a low-tax one.

How has this world in which taxes are driving these kinds of business decisions emerged? There are three major factors.

1. **Globally, the geopolitics of globalization.** The United States facilitated the opening up of the world economy. Since the end of World War II in 1945, the United States has been the world's most important country economically and militarily. U.S. economic and cultural influence around the world has been profound. After the Cold War with the Soviet Union, during which the United States championed private ownership of land and industry, the United States began removing barriers to trade such as tariffs and quotas. The United States also provided the world's main currency for world trade. U.S.-based multinational businesses and banks were important sponsors and beneficiaries of policies that opened up the world economy under U.S. government auspices. That enabled the global supply chains and massive changes in patterns of economic development discussed in this chapter.

The U.S. government stands behind these companies and their global operations.

2. **Nationally, the geopolitics of development**—the differences between states to mobilize their populations to pursue economic development and invest in public goods and infrastructure. Some governments have been more adept than others at exploiting the opportunities provided by a more open world economy. China's great success in improving its level of economic development and reducing its number of low-income people is not just due to its ability to exploit its vast pool of relatively cheap labor. The abilities to mobilize populations and create opportunities in global markets are key aspects to a country's success in the global marketplace. Mobilizing a population requires some degree of cultural homogeneity. And a population's acceptance of government legitimacy is important to foreign and domestic investors who expect political stability and minimal workplace disruptions. Accepting a government's legitimacy requires a clear sense that some economic activities (e.g., banking) need considerable government control. Investment in infrastructures such as ports, railways, and highways, and investment in public goods, such as general education and health care, also produce favorable settings for profitable private investment.

3. **Institutionally, the increasingly complex system of regulation.** Since the 1970s, world economic development has been regulated not just by governments within countries, but by increasingly influential private, quasi-public and international organizations. Private organizations such as credit-rating agencies rate the riskiness of bonds issued by businesses and governments. Their decisions are based on the claim that they have specialist knowledge and "independence" not available to governmental organizations. So when governments try to raise revenues by selling bonds, they are subject to the authority exercised by these agencies. Quasi-public organizations and their regulations also have an impact. Most national central banks have a high degree of political independence from their governments—they decide how much currency to issue and fix interest rates and exchange rates—with an eye on global markets rather than their own governments. Other regulatory effects are a result of international financial and development organizations such as the International Monetary Fund and the World Bank. These organizations support governments in difficult economic circumstances—if those governments follow policy prescriptions designed by those organizations.

Much of the world's private financial economy is increasingly moving "offshore" to avoid as much national and global regulation as possible (**Table 8.2**). To take advantage of low or nonexistent corporate and personal income taxes, many transnational businesses now incorporate in tax

TABLE 8.2   Some Examples of Tax Havens and What They Do.

Sun and shadows

Tax havens: A user's guide

Location	Population, 2010, 000	Known for	Critics say
Cayman Islands	56	World's Leading domicile for hedge funds. Also popular with big banks	Lax regulation, essentially written by its clients. Crops up in most big financial scandals
Mauritius	1,299	Close links to India. Poaching business from more-pressured European havens	Allows Indians to disguise investment in their own country as tax-advantaged foreign capital
Jersey	93	Close links to the City of London. Impenetrable off-shore trusts. No corporate/capital-gains tax	Claims that it is cleaning up its act. But Isle of Man is doing more
Luxembourg	507	Doggedly resisting European efforts to promote transparency. Second-Largest mutual-fund market after US	The "Death Star" of financial secrecy at the heart of the EU
Switzerland	7,664	Defending tax evasion as a legitimate response to "excessive" tax elsewhere. Managing a third of the world's cross-border invested wealth	Concessions offered under international pressure to end bank secrecy are mostly window-dressing
Singapore	5,086	Aggressively marketing itself as a regional hub. Like Mauritius, snatching business from Western havens	Information-sharing agreements with other countries constrained by legally enforced safeguards

Sources: Tax Justice Network: UN Population; *The Economist*

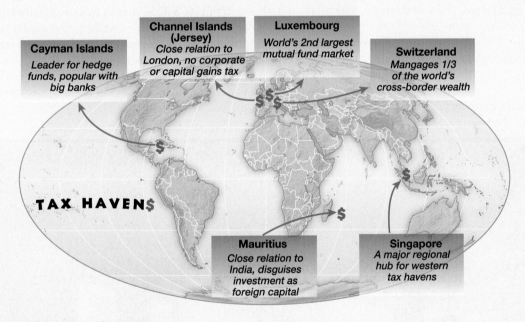

▲ Figure 8.E   Tax havens.

havens such as the Cayman Islands (**Figure 8.E**). Major global financial institutions in New York and London provide the nerve centers for this cross-jurisdictional circulation of corporate profits.

As a result of all these factors, a major crisis is brewing. The massive avoidance of taxes not only creates opportunities for "money laundering" of illegal activities, it also reduces the revenues that governments need to provide the public investments necessary for successful economic development. This is why "tax avoidance" has become a hot button political issue.

1. Why was the reduction in average tariffs on manufactured goods crucial for the growth of global supply chains?

2. Why is a relatively open and market-based financial system important for globalization?

3. Why has so much economic regulation slipped out of the hands of governments and into the hands of private and international organizations?

Central to place marketing is the deliberate manipulation of culture in an effort to enhance the appeal of places to key groups. These groups include the upper-level management of large corporations, the higher-skilled and better-educated personnel sought by expanding high-technology industries, wealthy tourists, and the organizers of business and professional conferences and other income-generating events. In part, place marketing strategies depend on promoting traditions, lifestyles, and arts that are locally rooted; in part, they depend on being able to tap into globalizing culture through new cultural amenities and specially organized events and exhibitions. Some of the most widely adopted strategies include funding for experiential settings for the arts, investment in public spaces, the re-creation and refurbishment of distinctive settings like waterfronts and historic districts, the expansion and improvement of museums (especially with blockbuster exhibitions of spectacular cultural products that attract large crowds and can be marketed with commercial tie-ins), and the designation and conservation of historic landmarks.

APPLY YOUR KNOWLEDGE

1. What is the experience economy?

2. Determine if the location you cited in the previous Apply Your Knowledge on page 31 is part of the experience economy. List two ways that manipulation of material and visual culture is used to enhance a sense of place for this location. Analyze where the tourists originate and what is the "experience" they hope to have.

FUTURE GEOGRAPHIES

As we have seen, technological breakthroughs and the availability of resources have had a profound influence on past patterns of development, and the same factors will certainly be a strong influence on future economic geographies. The expansion of the world economy and the globalization of industry will undoubtedly boost the overall demand for raw materials of various kinds, and this will spur the development of some previously underexploited but resource-rich regions in Africa, Eurasia, and East Asia.

Raw materials, however, will represent only a fraction of future resource needs. The main issue, by far, will be energy resources. World energy consumption has been increasing steadily over the recent past, and as the periphery is industrialized and its population increases, the global demand for energy will expand rapidly. Basic industrial development tends to be highly energy-intensive. The International Energy Agency, assuming (fairly optimistically) that energy in peripheral countries will be generated in the future as efficiently as it is today in core countries, estimates that developing-country energy consumption will increase dramatically, lifting total world energy demand by almost 50 percent by 2020. Peripheral and semiperipheral countries will then account for more than half of world energy consumption. Much of this will be driven by industrialization geared to meet the growing worldwide market for consumer goods, such as private automobiles, air conditioners, refrigerators, televisions, and household appliances (**Figure 8.26**).

▼ **Figure 8.26 Spreading consumerism** Customers shop for a television at an eZone retail showroom in the Jayanagae area in Bangalore, India.

Without higher rates of investment in exploration and extraction than at present, production will be slow to meet the escalating demand. Many experts believe that current levels of production in fact represent "peak oil" and that by 2020 global oil production may be only 90 percent of its current level. The result might well be a significant increase in energy prices. This would have important geographical ramifications: Companies would need to seriously reconsider their operations and force core-region households into a reevaluation of their residential preferences and commuting behavior, while peripheral-region households would be driven deeper into poverty. If the oil-price crisis of 1973 is anything to go by (after crude oil prices had been quadrupled by the OPEC cartel), the outcome could be a major revision of patterns of industrial location and metropolitan areas. Significantly higher energy costs may change the optimal location for many manufacturers, leading to deindustrialization in some regions and to new spirals of cumulative causation in others. Higher fuel costs will encourage some people to live nearer their place of work; others will be able to take advantage of telecommuting to reduce personal transportation costs.

It is also relevant to note that almost all of the increase in oil production over the next 15 or 20 years is likely to come from outside the core economies. This means that the world economy will become increasingly dependent on OPEC governments, which control over 70 percent of all proven oil reserves, most of them in the politically unstable Middle East.

Given what we know about past processes of geographic change and principles of spatial organization, it is clear that changes in transportation technology are also of fundamental importance. Consider, for example, the impact of ocean-going steamers and railroads on the changing geographies of the nineteenth century and the impact of automobiles and trucks on the changing geographies of the twentieth century. Among the most important of the next generation of transportation technologies that will influence future geographies are high-speed rail systems, smart roads, and smart cars.

Meanwhile, the impact of climate change on future economic geographies will depend a great deal on the latitude, altitude, and topography of places and regions. The 2014 report of the Intergovernmental Panel on Climate Change (IPCC) stressed that climate change will affect various energy sources and technologies differently, depending on the resources, the technological processes, and the locations (e.g., coastal regions, floodplains) involved. Climate change is also likely to influence the integrity and reliability of pipelines and electricity grids and will have both positive and negative impacts on water supply infrastructure and water demand. It will negatively affect transport infrastructure and many tourism resorts, particularly ski resorts, beach resorts, and nature resorts will be at risk. Nevertheless, the IPCC report concluded that for most economic sectors, the impact of climate change will be small relative to the impacts of other drivers such as changes in population, income distribution, technology, lifestyles, and regulation.

APPLY YOUR KNOWLEDGE

1. What are some of the geographic ramifications of the "peak oil" economy—in terms of production, use, higher cost, and so on?

2. Which alternative transportation and energy sources do you think have the most potential for dealing with the economic problems of oil dependency and declining reserves?

▼ **Figure 8.28 Cause for concern** Rising sea levels and the increasing frequency and intensity of extreme weather events threaten many coastal communities. Shown here is the town of Scheveningen, the Netherlands. In 2013 the town completed the construction of a protective sea wall beneath the sandy shore and added 2 million cubic meters (2.6 million cubic yards) of sand to the beach.

CONCLUSION

Economic development is not simply a sequential process of modernization and increasing affluence. Economic development involves not only using the latest technology to generate higher incomes but also improving the quality of life through better housing, health care, and social welfare systems, and enhancing the physical framework, or infrastructure, on which the economy rests.

Local, regional, and international patterns and processes of economic development are of particular importance to geographers. Levels of economic development and local processes of economic change affect many aspects of local well-being and so contribute to many aspects of human geography. Economic development is an important place-making process that underpins much of the diversity among regions and nations. At the same time, it is a reflection and a product of variations in natural resources, demographic characteristics, political systems, and social customs.

Economic development is an uneven geographic phenomenon. Regional patterns of economic development are tied to the geographic distribution of resources and to the legacy of the past specializations of places and regions. A general tendency exists toward the creation of regional cores with dependent peripheries. Nevertheless, such patterns are not fixed or static. Changing economic conditions can lead to the modification or reversal of core-periphery patterns, as in the stagnation of once-booming regions like northern England and the spectacular growth of Guangdong Province in Southeast China. Over the long term, core-periphery patterns have most often been modified as a result of the changing locational needs and opportunities of successive technology systems. Today, economic globalization has exposed more places and regions than ever to the ups and downs of episodes of creative destruction—episodes played out ever faster, thanks to the way that technological innovations have shrunk time and space.

At the global scale, the unevenness of economic development takes the form of core-periphery contrasts. Most striking about these contrasts today are the dynamism and pace of change involved in economic development. The global assembly line, the global office, and global tourism are all making places much more interdependent and much faster changing. Parts of Brazil, China, India, Mexico, and South Korea, for example, have developed quickly from rural backwaters into significant industrial regions. This dynamism has, however, brought with it an expanding gap between rich and poor at every spatial scale: international, regional, and local.

LEARNING OUTCOMES REVISITED

- Scrutinize the nature and degree of unevenness in patterns of economic development at national and international scales.

At the global scale, unevenness takes the form of core-periphery contrasts. Similar core-periphery contrasts exist at the regional scale. The core regions within the world-system—North America, Europe, and Japan—have the most diversified economies, the most advanced technologies, the highest levels of productivity, and the highest levels of prosperity. Other countries and regions—the periphery and semiperiphery of the world-system—are often referred to as *developing* or *less developed*. The average GNI per capita of the ten most prosperous countries in the world is 67 times greater than the average GNI per capita of the ten poorest countries. Overall, more than 80 percent of the world's population lives in countries where income differentials are widening.

- Compare how different regions of the world contribute to technological innovation and how interdependent the global economy is on these technological systems.

Technology systems are clusters of interrelated energy, transportation, and production technologies that dominate economic activity for several decades at a time—until a new cluster of improved technologies evolves. Technology systems have evolved at 50-year intervals in the world, starting with the Industrial Revolution (1790–1840) and at subsequent intervals in *1840–1890; 1890–1950; 1950–1990, and 1990 onward*. With a direct impact on economic development, technology has spread from different points of origin in each interval, and creates a web of production, trade, and communication around the world.

- Analyze how geographical divisions of labor have evolved with the growth of the world-system and the accompanying variations in economic structure.

Geographical divisions of labor are national, regional, and locally based economic specializations in primary, secondary, tertiary, or quaternary activities. The relationship between changing regional economic specialization and changing levels of prosperity has prompted the interpretation of economic development in distinctive stages. In reality, however, various pathways exist to development, as well as various processes and outcomes of development. A new international division of labor was initiated in the 1970s as a result of a major wave of corporate globalization. This new international division of labor has resulted in three main changes: the decline of the United States as an industrial producer, the decentralization of manufacturing production from the world's core regions to some semiperipheral and peripheral countries, and the emergence of new specializations in high-tech manufacturing and producer services within the core regions of the world-system.

■ Interpret how regional cores of economic development are created, following some initial advantage, through the operation of several basic principles of spatial organization.

Any significant initial local economic advantage—existing labor markets, consumer markets, frameworks of fixed social capital, and so on—tends to be reinforced through a process of cumulative causation, a spiral buildup of advantages that occurs in specific geographic settings as a result of the development of external economies, agglomeration effects, and localization economies. The agglomeration effects that are associated with various kinds of economic linkages and interdependencies—the cost advantages that accrue to individual firms because of their location among functionally related activities—are particularly important in driving cumulative causation. Spirals of local growth tend to attract people and investment funds from other areas. According to the basic principles of spatial interaction, these flows tend to be strongest from nearby regions and those with the lowest wages, fewest job opportunities, or least attractive investment opportunities.

■ Explain how spirals of economic development can be arrested in various ways, including the onset of disinvestment and deindustrialization.

Core-periphery patterns and relationships can be modified by changes that can slow or modify the spiral of cumulative causation. The main factor is the development of agglomeration diseconomies, the negative economic effects of urbanization, and the local concentration of industry. Spirals of cumulative causation can also be undermined by changes in the relative costs of the factors of land, labor, or capital; by the obsolescence of infrastructure and technology; or by the process of import substitution, whereby goods and services previously imported from core regions come to be replaced by locally made goods and locally provided services. The capital made available from disinvestment in core regions becomes available for investment by entrepreneurs in new ventures based on innovative products and innovative production technologies. This process is often referred to as creative destruction.

■ Demonstrate how globalization has resulted in patterns and processes of local and regional economic development that are open to external influences.

The globalization of the world economy involves a new international division of labor in association with the internationalization of finance, the deployment of a new technology system, and the homogenization of consumer markets. This new framework for economic geography has meant that the lives of people in different parts of the world have become increasingly intertwined. Transnational corporations now control a large fraction of the world's productive assets, and the largest of them are more powerful, in economic terms, than most sovereign nations. As these corporations have restructured their activities and redeployed their resources among different countries, regions, and places, they have created many new linkages and interdependencies among places and regions around the world. Even small and medium-sized companies are involved in the myriad global assembly lines and supply chains that characterize the contemporary world economy, linking the fortunes of diverse and often distant local economies.

KEY TERMS

agglomeration diseconomies (p. *278*)
agglomeration effects (p. *278*)
ancillary industries (p. *278*)
autarky (p. *269*)
backward linkages (p. *278*)
backwash effects (p. *278*)
carrying capacity (p. *266*)
conglomerate corporations (p. *283*)
creative destruction (p. *280*)
cumulative causation (p. *278*)
deindustrialization (p. *280*)
dependency (p. *273*)
ecological footprint (p. *266*)
elasticity of demand (p. *273*)
export-processing zones (EPZs) (p. *286*)
external economies (p. *277*)

flexible production systems (p. *285*)
Fordism (p. *285*)
foreign direct investment (p. *268*)
forward linkages (p. *278*)
geographical path dependence (p. *277*)
gross domestic product (GDP) (p. *261*)
gross national income (GNI) (p. *261*)
growth poles (p. *281*)
import substitution (p. *274*)
initial advantage (p. *277*)
international division of labor (p. *268*)
just-in-time production (p. *285*)
localization economies (p. *277*)
neo-Fordism (p. *285*)
neoliberal policies (p. *273*)
newly industrializing countries (p. *268*)

primary activities (p. *268*)
producer services (p. *282*)
purchasing power parity (p. *261*)
quaternary activities (p. *268*)
secondary activities (p. *268*)
spread effects (p. *278*)
strategic alliances (p. *286*)
sustainable development (p. *266*)
terms of trade (p. *274*)
tertiary activities (p. *268*)
trading blocs (p. *269*)
transnational corporations (p. *269*)
vertical disintegration (p. *286*)

REVIEW & DISCUSSION

1. Could your group measure the ecological footprint of your specific place? What data would you need to make an informed statement? Is your city considered a "sustainable" or "green" city? Why or why not? List three things in your community that might contribute to sustainable or unsustainable development.

2. Research the businesses in the area connected to your campus. Create a list of the stores. Determine whether or not the stores are locally owned. What is the ratio of locally owned to big-box retail or chain stores? As a comparative study, consider other neighborhoods that are close to yours. Create a list of the locally owned stores and big-box retail or chain stores. How many similarities are there, in terms of businesses? How many differences? What does this say about the economic structures of your campus? Do most people in your community work for chain or big box stores, or are there many entrepreneurs and small business owners?

3. After rereading the section on Fair Trade in the chapter, walk with your group to your on-campus food store and create a list of fair trade products. What is the percentage of fair trade products versus not fair trade products in the store? What does this say about the economic structure of your campus food store? Does it make a difference to you as a consumer whether a product is fair trade or not?

UNPLUGGED

1. While India's per capita income is well below that of the United States (Figure 8.1), India has more people who earn the equivalent of $70,000 or above a year than the United States does. How can you explain this, and what might be some of the consequences from the point of view of economic geography?

2. Write a short essay (500 words, or two double-spaced typed pages) on any local specialized manufacturing region or office district with which you are familiar. Describe the different kinds of firms that are found there, and suggest the kinds of linkages between them that might be considered examples of agglomeration effects. (*Hint:* You might consider the manufacturing of cars. How many parts are needed to make an entire car? What other firms are needed in the production? Do they come from the same place or are they manufactured in different locations? Are the parts manufactured in one place, region, or country? How would this be an example of agglomeration effects?)

3. Consider the economic structure of the city or town you live in. List two primary, secondary, tertiary, and quaternary activities in your region. How might these activities lead to one part of your city or town being more economically developed than others? Finally, consider where you fit into these economic structures. If you work in a community business, what is your labor used to produce?

DATA ANALYSIS

Working for the Few
http://goo.gl/um28xo

In this chapter, we have looked at global and domestic economic energies, to answer among other questions: Where in the world is money made and retained? In January 2014, the World Economic Forum met in Davos, Switzerland. At the same time, Oxfam International released a report called, "Working for the Few" in which several startling statistics were reported, including that the 85 richest people in the world have as much wealth as the 3.5 billion poorest. What do you think are some of the causes of increasing disparity, reflecting on what you learned in this chapter? Go to Oxfam's report summary at http://www.oxfam.org/sites/www.oxfam.org/files/bp-working-for-few-political-capture-economic-inequality-200114-summ-en.pdf and answer these questions:

1. What percentage of the world's population own half the world's wealth?

2. In the United States, what percentage of the population has experienced financial growth since 2008? What percentage has gotten poorer?

3. Where does the wealthiest percentage of the world's population live?

4. Where do the poorest live?

Now that you have the global forecast, consider the results of a global economy on domestic workforces and personal incomes. How do most Americans make a living? Let's look closer at the U.S. economy and create a projection based on current data. Start at the U.S. Bureau of Labor Statistics

Web site (http://www.bls.gov/home.htm) and look up "Economic Indicators" to answer these questions:

5. What are the indicators economists have chosen to determine the health of the economy?

6. Compare the categories for "Employment Situation" to "Consumer Price Index" and "Producer Price Index." Why do these categories comprise major pieces of the economy?

7. Next look at the unemployment rates for the past year. How is unemployment dispersed throughout the United States, that is, are some areas more unemployed then others?

Compare this data to the Bureau for Labor Statistics' data graphs for NPR called "What America Does for Work" at http://www.npr.org/blogs/money/2012/03/20/149015363/what-america-does-for-work.

What America Does for Work
http://goo.gl/FRlYaf

8. What sectors employ the majority of Americans?

9. Look at these numbers and compare the first graph to the second graph of where most Americans worked in 1972. What has changed?

10. In your opinion, do you think there is such a thing as a domestic economy in a globalized world? Why or why not?

11. Finally analyze how you think growing disparity will affect the health and harmony of the planet. Why are policymakers and scholars concerned about the "growing tide of inequality"?

Mastering Geography™

Looking for additional review and test prep materials? Visit the Study Area in MasteringGeography™ to enhance your geographic literacy, spatial reasoning skills, and understanding of this chapter's content by accessing a variety of resources, including **MapMaster** interactive maps, Videos, *In the News* RSS feeds, flashcards, web links, self-study quizzes, and an eText version of *Human Geography*.

LEARNING OUTCOMES

- *Compare* and contrast traditional agriculture practices across the globe.

- *Describe* the three revolutionary phases of agricultural development, from the domestication of plants and animals to the latest developments in biotechnology and industrial innovation.

- *Analyze* the ways the forces, institutions, and organizational forms of globalization have transformed agriculture.

- *Explain* the organization of the agro-commodity system from the farm to the retail outlet, including the different economic sectors and corporate forms.

- *Understand* the ways agriculture has transformed the environment, including soil erosion, desertification, deforestation, soil and water pollution, and plant and animal species degradation.

- *Probe* the current issues and opportunities food-policy experts, national governments, consumers, and agriculturalists face with respect to the availability and quality of food.

▲ Roadside vendors sell food to Muslims to break fast after sundown on the first day of the holy month of Ramadan in Old Dhaka.

9

GEOGRAPHIES OF FOOD AND AGRICULTURE

Visit most any city in the periphery and you'll find vendors along the street selling locally prepared food—often called street food—from temporary stands and carts. In the core, large cities sometimes host their own set of such food vendors, though the choice of food is usually far more limited—pretzels, hot dogs, roasted chestnuts in winter, ice cream in summer—than in the periphery. In Mexico City, for instance, food vendors—known as *ambulantes*—sell fresh cut fruit (in plastic bags that look like transparent ice cream cones) sprinkled with lime juice and cayenne pepper, as well as tacos, *tortas* (sandwiches), *tlacoyos* (masa patties filled with cheese or beans), *camotes* (sweet potatoes), *chicharones* (pork rinds), roasted ears of corn, fresh fruit juices, and a wide assortment of other treats. The food is frequently delicious, very portable, and quite inexpensive, largely due to the low overhead costs of selling it from a non-permanent site. This food and these vendors are highly important to the local economy providing inexpensive choices to well-off and poor locals and tourists who write endless blogs about where the best street food is around the globe.

The street vendors of Mexico City provide a window into the many ways and types of places food is available for consumption from carts, to outdoor markets, to supermarkets (including grocery chains and superstores Walmart and Target), pickup truck beds, farm stands, corner grocery stores, online providers, vending machines, and buyers' clubs like Costco and Sam's Club in the United States.

In the core, an increasingly popular form of edibles vending is the food truck, a hipster variation on a previous trend that began in the late nineteenth century. Contemporary food trucks—vehicles that both transport and are the site for selling food—are also exploding in number of cities in parts of Asia and in cities across Australia, and in Canada, Belgium, France, the United Kingdom and Mexico. A few offer frozen or pre-packaged food, like the traditional ice cream truck; most are more like restaurant kitchens on wheels. In the United States, the current trend in food trucks emerged in Los Angeles with the company, Kogi Korean BBQ. A high-end culinary experience, Kogi is a fusion of Korean and Mexican food. And once Kogi, which generated a clientele through Twitter, hit the streets, other creative culinary food trucks soon followed suit so that they have become a staple of the urban landscape in large cities in the United States from coast to coast.

The first food truck in the United States is reported to have been put into service when, in 1872, a vendor in Rhode Island cut windows into a small horse-drawn freight wagon and started selling comestibles to late-night newspaper workers at the *Providence Journal.* These vehicles came to be known as lunch wagons and were soon being custom-produced to include sinks, refrigerators, and cooking stoves. Before long, the lunch wagons lost their wheels and became stationary dining cars and ultimately became the classic American diner. These locally owned and operated restaurants serving homemade food have largely disappeared from the landscape and been replaced by imitation diners that are part of national chains, such as Denny's, which calls itself "America's diner."

Like the diner, which persists across the U.S. landscape despite the small total number, the descendants of early food trucks also thrived on the edges of urban growth and change, though none of them as upscale or adventurous as the current ones. For instance, "roach coaches," aluminum-sided vending trucks still operate providing breakfast and lunch staples—coffee and pastries, packaged sandwiches, candy bars, chips and soda—to distant construction sites or places where plain food is is not readily available. In fact, these shiny trucks are a model for some of the retro food trucks of today that offer more elaborate culinary delights.

How we get our food, and where we eat it is as much a part of food and agricultural geographies as the other areas of interest we cover in this chapter. From agricultural production, processing, distribution, and financing to issues of food consumption, such as access and quality, geographers are interested in understanding this most essential element of human life.

TRADITIONAL AGRICULTURAL GEOGRAPHY

In this chapter, we examine the geography of food and agriculture from the global to the household and individual level. We begin by exploring traditional agricultural practices and proceed through the three major revolutions of agricultural change. Much of the chapter is devoted to understanding the ways geographers have investigated the dramatic transformations in

agriculture over 50 plus years as it has become increasingly industrialized through technological, political, social, and economic forces. We also treat the effects globalization has had on producing, marketing, delivering, and consuming food.

The study of agriculture has a long tradition in geography. Because of geographers' interest in the relationships between people and land, it is hardly surprising that agriculture has been a primary concern. Geography is committed to viewing the physical and human systems as interactively linked. This approach combines an understanding of spatial differentiation and the importance of place as well as the recognition that agriculture practices are affected by local, regional, national, and globally extensive structures and processes.

One of the most widely recognized and appreciated contributions that geography has made to the study of agriculture is the mapping of the factors that shape it. Geographers map soil, temperature, and terrain, as well as the areal distribution of different types of agriculture and the relationships among and between agriculture and other practices or variables. They also map a host of other aspects of food and agriculture from the distribution of processing plants and grocery stores to the availability of fresh produce and the number of school and community gardens in poor urban neighborhoods.

Major changes in agriculture worldwide have occurred in the last five decades. Of these, the possibility that increases in food production will not keep pace with increases in population and that the planet is at the limit of arable land and fresh water is perhaps the most worrisome. Add to that very difficult challenge, the decline in the number of people employed in farming in both the core and the periphery and the impacts of climate change on agriculture and the result is a global agricultural system, and the physical environment on which it depends, that is gravely endangered (**Figure 9.1**).

Before we begin to understand the present state of the agricultural system, it's valuable to learn some of the basic terminology for talking about food and agriculture as well as to review our history as food producers more generally. That review will provide a foundation for appreciating and recognizing the value of some of the alternative agricultural practices emerging in response to a whole host of food production, access, and consumption issues that we also deal with later in this chapter.

Agricultural Practices

Agriculture is a science, an art, and a business directed at the cultivation of crops and the raising of livestock for sustenance and profit. The unique and ingenious methods humans have learned to transform the land through agriculture are an important reflection of the two-way relationship between people and their environments (**Figure 9.2a and 9.2b**). Just as geography shapes our choices and behaviors, so we are able to shape the physical landscape to meet our needs.

In examining agricultural practices, geographers have sought to understand the myriad ways humans have learned to modify the natural world around them to nourish themselves, their families, and ultimately the global community. In addition to understanding agricultural systems, geographers

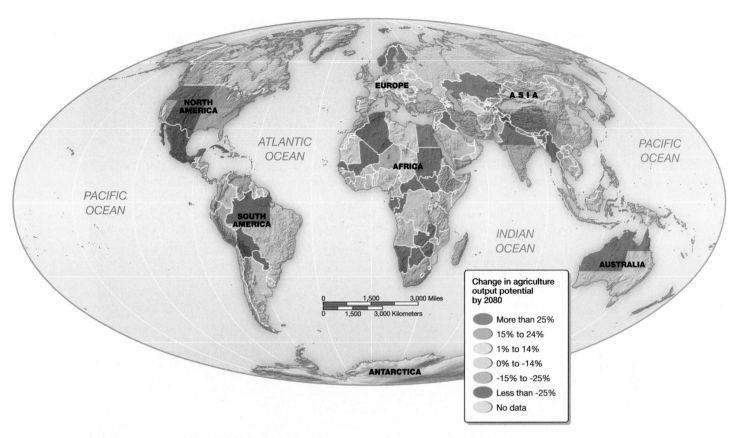

▲ **Figure 9.1 Effect of climate change on global farming** Crop forecasts show that some countries further from the equator will see increased production but most others, as the map shows, will be far worse off by 2080.

are also interested in investigating the lifestyles and cultures of different agricultural communities. They and other social scientists often use the adjective **agrarian** to describe the way of life that is deeply embedded in the demands of agricultural production. *Agrarian* not only defines the culture of distinctive agricultural communities but also refers to the type of tenure (or landholding) system that determines who has access to land and what kind of cultivation practices are employed there.

Origins and Change in Agriculture

While there is no definitive answer as to where agriculture originated, we know that before humans discovered the advantages of agriculture, they procured their food through hunting (including fishing) and gathering. **Hunting and gathering** characterize activities through which people feed themselves by killing wild animals and marine species and gathering fruits, roots, nuts, and other edible plants. Subsistence agriculture replaced hunting and gathering activities in many parts of the globe when people came to understand that the domestication of animals and plants could enable them to settle in one place rather than having to go off routinely in search of edible animals and plants (**Figure 9.3**). **Subsistence agriculture** is a system in which agriculturalists consume most of what they produce.

During the twentieth century, the dominant agricultural system in the core countries became **commercial agriculture**, a system in which farmers produce crops and animals primarily for sale rather than for direct consumption by themselves and their families (**Figure 9.4**). Worldwide, subsistence agriculture is diminishing as increasing numbers of places are incorporated into a globalized economy with a substantial commercial agricultural sector. Still widely practiced in the periphery, however, subsistence activities usually follow one of three dominant forms: shifting cultivation, intensive subsistence agriculture, and pastoralism.

Shifting Cultivation

In **shifting cultivation**, farmers aim to maintain soil fertility by rotating the fields they cultivate. Shifting cultivation contrasts with another method of maintaining soil fertility, **crop rotation**, in which the fields under cultivation remain the same, but the crops planted are changed to balance the types of nutrients withdrawn from and delivered to the soil.

Shifting cultivation is globally distributed in the tropics—especially in the rain forests of Central and West Africa; the Amazon in South America; and much of Southeast Asia, including Thailand, Burma, Malaysia, and Indonesia—where climate, rainfall, and vegetation combine to produce soils lacking nutrients. The practices involved in shifting cultivation have changed very little over thousands of years (**Figure 9.5**). Shifting cultivation requires only human energy as compared with modern forms of farming, though, not surprisingly it can successfully support only low population densities since the total amount of crop produced is not large.

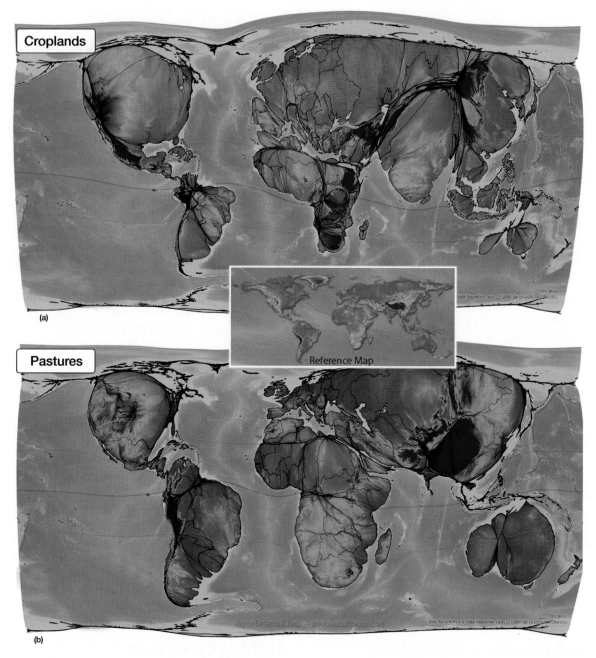

▲ **Figure 9.2 Cropland and pastures cartograms** (a) This cartogram shows the amount of cropland available by continent. As a topographic map, the areas of elevation are also shown in dark brown color, providing some insight into land that is unlikely to be suitable for food production. (b)This cartogram shows the amount of land available for raising animals and thus where high-protein food can be grown.

The typical agrarian system that supports shifting cultivation is one in which small groups of villagers hold land in common tenure. Through collective agreement or a ruling council, sites are distributed among village families and then cleared for planting by family members. As villages grow, tillable sites must be located farther and farther away. When population growth reaches a critical stage, several families within the village normally split off to establish another village in one of the more remote sites.

The typical method for preparing a new site is through **slash-and-burn** agriculture. Existing plants are cropped close to the ground, left to dry for a period, and then ignited. The burning process adds valuable nutrients to the soil, such as potash, which is about the only fertilizer that is readily available and free of charge. Once the land is cleared and ready for cultivation, it is known as **swidden** (see Chapter 2). While generally an agricultural practice that is workable when undertaken by small populations on limited portions of land, slash-and-burn is also seen to be ecologically destructive when large numbers of farmers participate, especially in areas with vulnerable and endangered species, such as rain forests.

From region to region the kinds of crops grown and their arrangement in the swidden vary depending upon local taste and plant domestication histories. In the warm, humid tropics, tubers—sweet potatoes and yams—predominate, while grains such as corn or rice are more widely planted in the subtropics.

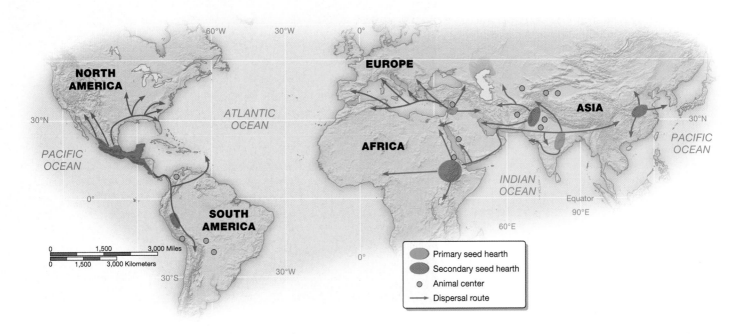

▲ **Figure 9.3 Areas of plant and animal domestication** The origins of plant and animal domestication, however, are not definitively known, and much of what is represented on this map is speculative. Primary seed hearths are those places where domestication is believed to have first begun. Secondary seed hearths followed soon after.

(*Source:* Adapted from J. M. Rubenstein, *The Cultural Landscape: An Introduction to Human Geography,* 7th ed., Prentice Hall © 2003, p. 319.)

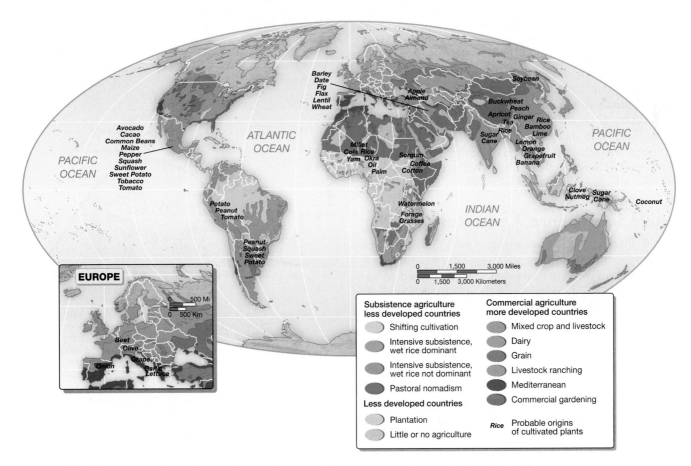

▲ **Figure 9.4 Global distribution of agriculture** This map shows agriculture practices as they are distributed across the globe. Notice the difference between core and periphery with respect to the distribution of commercial versus subsistence agriculture.

► **Figure 9.5** **Shifting cultivation** One way to maintain soil fertility is by rotating the fields within which cultivation occurs. Land rotation—also called shifting cultivation—is pictured here as agriculturalists prepare a new field so that the old fields can "rest" and be regenerated. It is also possible to plant certain crops to restore soil fertility. That is, a crop that leaches the soil of one kind of nutrient is followed during the next growing season by a dissimilar crop that returns that nutrient to the soil.

The practice of mixing different seeds and seedlings in the same swidden is called **intertillage** (**Figure 9.6**). Not only are different plants cultivated, but their planting is usually staggered so that harvesting can continue throughout the year. Staggered planting and harvesting reduces the risk of disasters from crop failure and increases the nutritional balance of the diet. Shifting cultivation also frequently involves a gender division of labor that may vary from region to region (**Figure 9.7**). For the most part, men are largely responsible for the initial tasks of clearing away vegetation, cutting down trees, and burning the stumps. Women are typically involved with sowing seeds and harvesting crops.

Although sometimes heralded as an ingenious, well-balanced response to the environmental constraints of the tropics and subtropics, shifting cultivation is not without limitations. Its most obvious limitation is that it can be effective only with small populations. Population growth and the greater need for increased outputs per acre have led to its replacement by more intensive forms of agriculture.

Intensive Subsistence Agriculture

The second dominant form of subsistence activity is **intensive subsistence agriculture**, a practice involving the effective and efficient use of a small parcel of land to maximize crop yield; a considerable expenditure of human labor and application of fertilizer are also usually involved. Unlike shifting cultivation, intensive subsistence cultivation can often support larger populations. Shifting cultivation is more characteristic of low agricultural densities, whereas intensive subsistence normally reflects high agricultural density. Consequently, intensive subsistence usually occurs in Asia, especially in India, China, and Southeast Asia.

Recall that shifting cultivation involves the application of a relatively limited amount of labor and other resources to cultivation. Conversely, intensive subsistence agriculture involves fairly constant human labor to achieve high productivity from a small amount of land. In the face of fierce population pressure and limited arable land, intensive subsistence agriculture reflects the inventive ways humans confront environmental constraints and

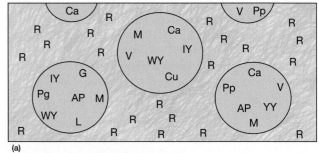

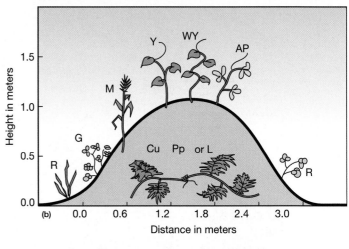

R	= Rice	IY	= White Yam
Ca	= Cassava	WY	= Water Yam
Cu	= Melon	AP	= Air Potato
G	= Groundnut	YY	= Yellow Yam
L	= Gourds	Pp	= Pumpkin
M	= Maize	Pg	= Pigeon pea

▲ **Figure 9.6** **Intertillage** Planting different crops together in the same field has many benefits, including the spreading out of food production over the farming season, reduction of disease and pest loss, greater protection from loss of soil moisture, and control of soil erosion. This diagram illustrates what an intertilled site might contain and how the planting is arranged. Hill-planted seeds have tall stalks and a deeper root system; those planted on flat earth tend to be spreading plants that produce large leaves for shading. Part A is meant to show how the plants would look as you gaze down on them. Part B shows a side view.

▲ Figure 9.7 Gender division of labor in rice processing in Tamil Nadu, India Rice production is a major source of livelihood for both low-income men and women in India. Women thresh the rice while men look on from the rice bales they have delivered. Women also process, clean, select, and store the seeds for next year's crop, and are primarily responsible for preparing and cooking rice for household consumption and sometimes for market sale.

reshape the landscape in the process. In fact, the landscape of intensive subsistence agriculture is often a distinctive one, including raised fields and terraced hillside plots (**Figure 9.8**).

Intensive subsistence agriculture is able to support large rural populations. In contrast to shifting cultivation, fields are planted year after year as fertilizers and other soil enhancers are applied to maintain soil nutrients. For the most part, the limitations on the size of plots have more to do with the size of the population than geographical conditions. In Bangladesh and southern China, for example, where a significant proportion of the population is engaged in intensive subsistence agriculture, land is passed down from generation to generation—usually from fathers to sons—so that each successive generation, if there are multiple male offspring, receives a smaller and smaller share of the family holdings. Yet even with shrinking plot size each family must produce enough to sustain itself.

Generally speaking, the crops that dominate intensive subsistence agriculture are rice and other grains. Rice production predominates in those areas of Asia—South China, Southeast Asia, Bangladesh, and parts of India—where summer rainfall is abundant. In drier climates and places where the winters are too cold for rice production, other sorts of grains—among them wheat, barley, millet, sorghum, corn, and oats—are grown for subsistence. In both situations, the land is intensively used. In fact, it is not uncommon in milder climates for fields to be planted and harvested more than once a year, a practice known as **double cropping**.

▶ Figure 9.8 Intensive subsistence agriculture Where usable agricultural land is at a premium, agriculturalists have developed ingenious methods for taking advantage of every square inch of usable terrain. Landscapes like this one—of a terraced rice field in Bali, Indonesia—can be extremely productive when carefully tended and can feed relatively large rural populations.

Pastoralism

Although not obviously a form of agricultural production, pastoralism is a form of subsistence activity associated with a traditional way of life and agricultural practice. **Pastoralism** involves the breeding and herding of animals to satisfy the human needs for food, shelter, and clothing. Usually practiced in the cold and/or dry climates of savannas (grasslands), deserts, and steppes (lightly wooded, grassy plains), where subsistence agriculture is impracticable, pastoralism can be either sedentary (pastoralists live in settlements and herd animals in nearby pastures) or nomadic (pastoralists travel with their herds over long distances, never settling in any one place for very long). Although forms of commercial pastoralism exist—the regularized herding of animals for profitable meat production, as among Basque Americans in the Great Basin and Range regions of Utah and Nevada and among the gauchos of the Argentinean grasslands—we are concerned here with pastoralism as a subsistence activity.

Pastoralism is largely confined to parts of North Africa and the savannas of central and southern Africa, the Middle East, and central Asia. Pastoralists generally graze cattle, sheep, goats, and camels, although reindeer are herded in parts of Eurasia. The type of animal herded is related to the culture of the pastoralists, as well as the animals' adaptability to the regional topography and foraging conditions (**Figure 9.9**).

Nomadism is a form of pastoralism that involves the systematic and continuous movement of groups of herders, their families, and the herds in search of forage. Most pastoralists practice **transhumance**, the movement of herds according to seasonal rhythms: warmer, lowland areas in the winter, and cooler, highland areas in the summer. In addition, women and children in pastoralist groups may be involved with cultivation. They usually split off from the larger group and plant crops at fixed locations in the spring. However, the distinguishing characteristic of pastoralists is that they depend on animals, not crops, for their livelihood.

Pastoralism as a subsistence activity is on the decline as more and more pastoralists have become integrated into a global economy that requires more efficient and regularized forms of production. Pastoralists have also been forced off the land by competition from other land uses and the state's need to track citizens for taxation and military reasons.

Culture and Society in Agriculture

It is important to appreciate that these basic forms of food production are predicated on cultural and social norms that are key to their success. In shifting cultivation, the division of tasks between men and women (and sometimes children) results from traditional practices and understandings of men's and women's roles in relationship to food, family, and resource use. Further, access to land is socially negotiated through well-established customs and traditions.

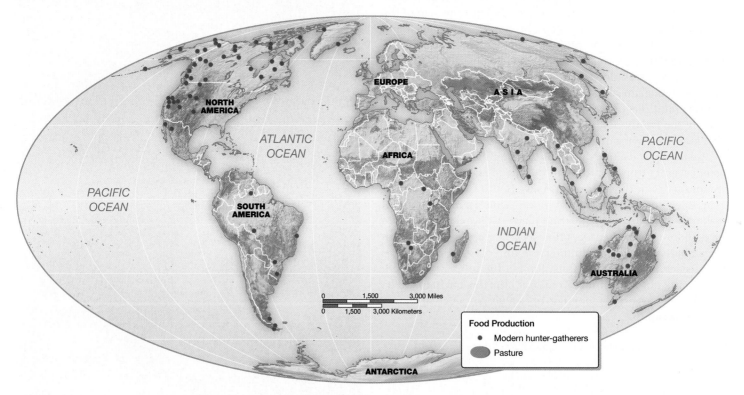

▲ **Figure 9.9 Global distribution of pastoralists and hunter-gatherers** This map makes clear that pastoralism and hunting-gathering continue to be livelihoods pursued throughout the world. Often these activities are undertaken by indigenous people.

Cultural practices are also key to access to nutrition because the general appropriateness of certain foodstuffs, food taboos, and food distribution are often determined by cultural norms related to age and gender as is land ownership and the rules of inheritance. Also, a community's organization of access to fisheries and the use of forests are shaped by cultural perceptions and traditions. Finally, indigenous cultural knowledge of species and conservation techniques are important for the successful preservation of the complex of species that are needed to maintain the health and sustainability of the spaces of food resource use and production.

Pastoralism, shifting cultivation, and intensive subsistence cultivation, are not simply subsistence activities but part of a social system as well. Pastoralist family groups are governed by a leader or chieftain. Groups are divided into units that follow different routes with the herds. The routes are well known; group members are intimately conversant with the landscape, watering places, and opportunities for contact with sedentary groups. The success of pastoralist activities to sustain the nutritional needs of the group depends not only on sophisticated cultural knowledge of the landscape, but also on a system of resource use that will sustain that environment for future needs.

Although declining throughout the world, the success of these ancient agricultural systems has been predicated on a complex of cultural practices that have enabled small groups of producers to harvest adequate amounts of food for survival. While these types of agriculture are certainly not sufficient to "feed the world," they do provide instruction and inspiration for a global agricultural system that is itself confronting profound obstacles to its ability to provide adequate nutrition to the world's population.

APPLY YOUR KNOWLEDGE

1. How is pastoralism both an agricultural practice and a sociopolitical system?

2. Take a closer look at the diversity of transhumance around the globe with this photo-essay map, Nomad's Life, at http://www.nomadslife.nl/en/. Click on the map and read about each group's lifestyle. Where are nomadic groups located? What do each of these groups share? How do they relate to the sedentary communities in the nation-states they inhabit? Do you think nomadic life is viable today? Why or why not?

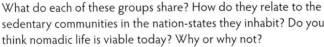

Global Exchange

http://goo.gl/Mo4e36

AGRICULTURAL REVOLUTION AND INDUSTRIALIZATION

Geographers and others scholars view world agricultural practices as having proceeded through "revolutionary" phases, just as manufacturing did. And as in manufacturing, agricultural practices have not been transformed everywhere at the same time; consequently, some parts of the world are still largely unaffected by certain aspects of change. By seeing agriculture in this light, we can recognize that, as in manufacturing, geography and society have changed as the global community has moved from predominantly subsistence to predominantly capital-intensive, commercial practices. This history has proceeded in alternating cycles: long periods of very gradual change punctuated by short, explosive periods of radical change, resulting in three distinct revolutionary periods.

The First Agricultural Revolution

The first agricultural revolution is commonly recognized as having been founded on the development of seed agriculture and the use of the plow and draft animals (**Figure 9.10**).

Aspects of this transformation were discussed in Chapter 2. Seed agriculture, which emerged through the domestication of crops, such as wheat and rice, and animals, such as sheep and goats, replaced hunting and gathering as a way of living and sustaining life. Seed agriculture arose during roughly the same period in several regions around the world. The result was a broad belt of cultivated lands across southwest Asia, from Greece in the west into present-day Turkey and part of Iran in the east, as well as in parts of Central and South America, northern China, northeast India, and East Africa.

The domestication of plants and animals allowed for the rise of settled ways of life. Villages were built, creating types of social, cultural, economic, and political relationships that differed from those that dominated hunter-gatherer societies. On the floodplains along the Tigris, Euphrates, and Nile rivers, important complex civilizations were built upon the fruits of the first agricultural revolution (**Figure 9.11**). Over time, the knowledge and skill underlying seed agriculture and the domestication of plants diffused outward from these original areas, having a revolutionary impact throughout the globe.

▼ **Figure 9.10** **Plowing with yoked oxen and camels, India** In many parts of the world, agriculturalists rely on draft animals to prepare land for cultivation. Animals were an important element in the first agricultural revolution. By expanding the amount of energy applied to production, draft animals enabled humans to increase food supplies. Many contemporary traditional farmers view draft animals as their most valuable possessions.

▲ Figure 9.11 Agricultural fields along the Nile River, Luxor, Egypt The lush fertile farmland in the foreground is enabled by the silt deposited by the Nile River. At a distance from the river valley, however, dry land not suitable for farming is clearly visible.

The Second Agricultural Revolution

A great deal of debate exists as to the timing and location of the second agricultural revolution. Though most historians agree that it did not occur everywhere at the same time, they disagree over which elements were essential to the fundamental transformation of subsistence agriculture. Important elements included

- dramatic improvements in outputs, such as crop and livestock yields;

- innovations such as the improved yoke for oxen and the replacement of the ox with the horse; and

- new inputs to agricultural production, such as fertilizers and field drainage systems.

The apex of the second agricultural revolution coincided historically and geographically with the Industrial Revolution in England and Western Europe. Although many important changes in agriculture preceded the Industrial Revolution, none had more of an impact than the rise of an industrialized manufacturing sector.

On the eve of the Industrial Revolution—in the middle of the eighteenth century—in Western Europe and England, subsistence peasant agriculture dominated, though partial integration into a market economy was underway. Many peasants were utilizing a crop-rotation system that, in addition to the application of natural and semiprocessed fertilizers, improved soil productivity and led to increased crop and livestock yields. Additionally, the feudal landholding system—a social and economic system based on peasant service to a lord in exchange for access to land—was breaking down and yielding to a new agrarian system based on an emerging system of private-property relations. Communal lands were being replaced by enclosed, individually owned land or land worked independently by tenants or renters.

Such a situation was a logical response to the demands for food production that emerged from the dramatic social and economic changes accompanying the Industrial Revolution. Perhaps most important of all these changes was the emergence—through the creation of an urban industrial workforce—of a commercial market for food. Many innovations of the Industrial Revolution, such as improvements in transportation technology, had substantial impacts on agriculture. Innovations applied directly to agricultural practices, such as the new types of horse-drawn farm machinery, improved control over—as well as the quantity of—yields.

APPLY YOUR KNOWLEDGE

1. Why was the Industrial Revolution so important to the second agricultural revolution? How did manufacturing technologies change agricultural technologies?

2. Research a technology that debuted in the Industrial Revolution that affected both manufacturing and agricultural practices. What kinds of changes did it produce in agricultural production?

The Third Agricultural Revolution

The third agricultural revolution is fairly recent; it began in the late nineteenth century and gained momentum throughout the twentieth century. Each of the third agricultural revolution's important developmental phases originated in North America. They include mechanization, chemical farming with synthetic fertilizers, and globally widespread food manufacturing. **Mechanization** is the replacement of human farm labor with machines. Tractors, combines, reapers, pickers, and other forms of motorized machines have, since the 1880s and 1890s, progressively replaced human and animal labor in the United States. In Europe, mechanization did not become widespread until after World War II. **Figure 9.12** shows the global distribution of tractors as a measure of the mechanization of worldwide agriculture.

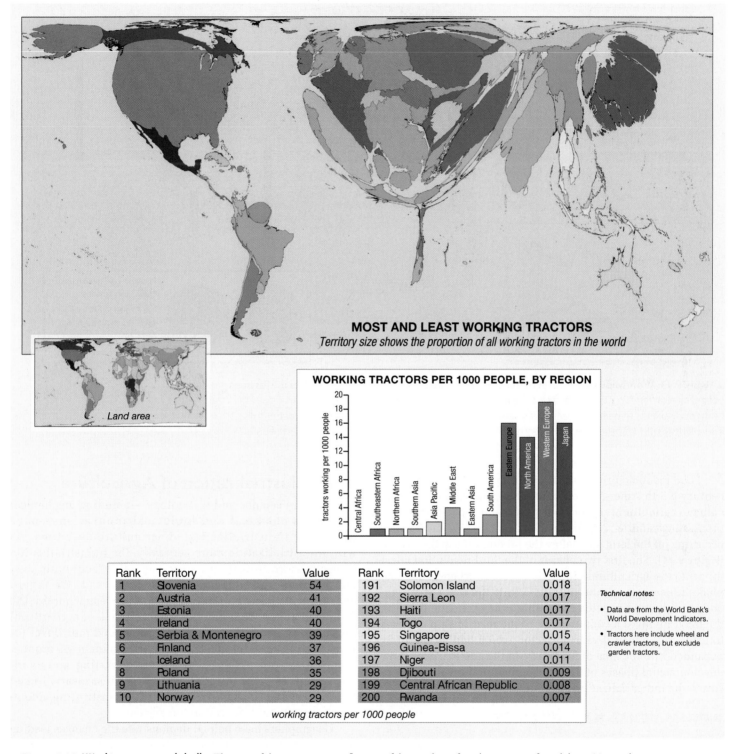

MOST AND LEAST WORKING TRACTORS
Territory size shows the proportion of all working tractors in the world

Land area

WORKING TRACTORS PER 1000 PEOPLE, BY REGION

Rank	Territory	Value		Rank	Territory	Value
1	Slovenia	54		191	Solomon Island	0.018
2	Austria	41		192	Sierra Leon	0.017
3	Estonia	40		193	Haiti	0.017
4	Ireland	40		194	Togo	0.017
5	Serbia & Montenegro	39		195	Singapore	0.015
6	Finland	37		196	Guinea-Bissa	0.014
7	Iceland	36		197	Niger	0.011
8	Poland	35		198	Djibouti	0.009
9	Lithuania	29		199	Central African Republic	0.008
10	Norway	29		200	Rwanda	0.007

working tractors per 1000 people

Technical notes:

• Data are from the World Bank's World Development Indicators.

• Tractors here include wheel and crawler tractors, but exclude garden tractors.

▲ **Figure 9.12 Working tractors globally**   The size of the territory is a reflection of the number of working tractors found there. Notice the uneven distribution between the top countries possessing tractors (all in Europe) and the bottom ten (all in Africa) It is important to note that the tractor count is based on working tractors and not simply tractors since tractors that don't work are of no use to farmers.

Chemical farming is the application of synthetic fertilizers to the soil and herbicides, fungicides, and pesticides to crops to enhance yields (**Figure 9.13**). Becoming widespread in the 1950s in the United States, chemical farming diffused to Europe in the 1960s and to peripheral regions of the world in the 1970s. The widespread application of synthetic fertilizers and their impact on the environment is what Rachel Carson wrote about in her highly influential book *Silent Spring*.

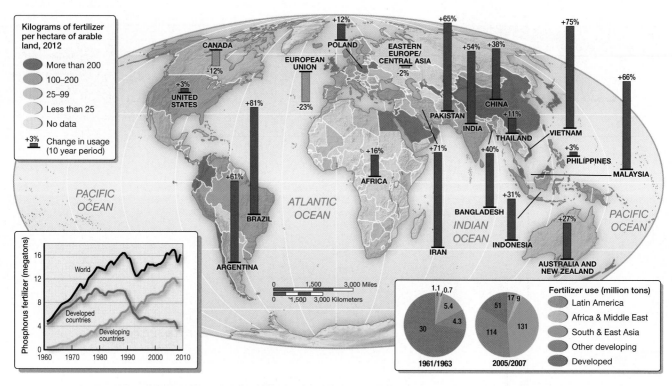

▲ **Figure 9.13 Worldwide Growth in Fertilizer Use, 2005–2007** As the map indicates, fertilizer use has been growing in peripheral countries faster than in the core, though core countries are still the largest users of fertilizer. One of the biggest problems, and one that is only expected to grow, is increased runoff from fertilizer and resultant dead zones along ocean shores.

(*Source:* Adapted from *New York Times*, http://www.nytimes.com/imagepages/2008/04/30/business/20080430_fertilizer_graphic_html accessed September 4, 2014.)

Food manufacturing also had its origins in late-nineteenth-century North America. **Food manufacturing** adds economic value to agricultural products through a range of treatments—processing, canning, refining, packing, packaging, and so on—occurring off the farm and before the products reach the market (**Figure 9.14**). The first two phases of the third revolution affect inputs to the agricultural production process, whereas the final phase affects agricultural outputs. While the first two are related to the modernization of farming as an economic practice, the third involves a complication of the relationship of farms to firms in the manufacturing sector, which had increasingly expanded into the area of food early in the 1960s. These three developmental phases of the third agricultural revolution constitute the industrialization of agriculture.

APPLY YOUR KNOWLEDGE

1. What were the major developments in agriculture after World War II?

2. Take a trip to your local grocery store or open up a kitchen cupboard and read the labels on at least four different products. Identify the various processes involved in making each of the products ready for the market. What processing has occurred? Has refining or cooking been involved? What additional ingredients have been added to the product and why (to improve flavor, to extend shelf-life, or for some other reason)? How does the packaging (container, images of the product, colors of the labeling, etc.) enhance the attractiveness of the product?

The Industrialization of Agriculture

Advances in science and technology—including mechanical as well as chemical and biological innovations—have driven the industrialization of agriculture over time. As with industrialization more generally, the industrialization of agriculture unfolded as the capitalist economic system became more advanced and widespread. We regard **agricultural industrialization** as the process whereby the farm has moved from being the centerpiece of agricultural production to being one part of an integrated multilevel (or vertically organized) industrial process that includes production, storage, processing, distribution, marketing, and retailing. Experts in the study of agriculture see it as clearly linked to industry and the service sector, thus constituting a complex agro-commodity production system.

Geographers have helped demonstrate the changes leading to the transformation of an agricultural product into an industrial food product. This transformation has been accomplished not only through the indirect and/or direct altering of agricultural outputs (such as tomatoes or wheat) but also through changes in rural economic activities. Agricultural industrialization involves three important developments:

■ Changes in rural labor activities as machines replace and/or enhance human labor

■ Introduction of innovative inputs—fertilizers and other agrochemicals, hybrid seeds, and biotechnologies—to supplement, alter, or replace biological outputs

◄ Figure 9.14 Food manufacturing Salmon processing usually occurs around several product forms from fillets to salmon roe. Most salmon processed in the U.S. comes from the coastal waters off Alaska.

The Green Revolution and the Blue Revolution

Recall that the industrialization of agriculture has not occurred simultaneously throughout the globe. Changes in the global economic system affect different places in different ways as different states and social groups respond to and shape these changes. For example, the use of fertilizers and high-yielding seeds occurred much earlier in core-region agriculture than in the periphery, where many people still farm without them. In the late 1960s, however, core countries began exporting fertilizers and high-yielding seeds to regions of the periphery (largely in Asia and Mexico) in an attempt to boost agricultural production. In a development known as the **Green Revolution**, they also sent new machines and institutions, all designed to increase global agricultural productivity, as described in Box 9.1, "Window on the World: The Green Revolution, Then and Now."

The Green Revolution was accompanied later on by a **Blue Revolution**, which affected world fisheries by introducing larger and more sophisticated vessels into wild fisheries and expanded the capacities of aquaculture. **Aquaculture** is the growing of aquatic creatures in ponds on shore or in pens suspended in water. The Blue Revolution has shifted primary-sector activities toward a greater dependence on capitalized inputs—credit, machinery, fuel, feeds, fertilizers, and pesticides—instead of human labor and nature productivity. One the one hand, the Blue and Green revolutions have increased food production in many places, but on the other they have engendered conflict over how the new practices redistribute power over and access to food resources.

Economic crises, a decrease in government programs, and reduced trade barriers have slowed the progress of the Green and Blue revolutions in many countries. For example, fertilizer use in countries such as Brazil and Mexico has declined in the face of high prices, fewer subsidies, and increased competition from imported corn and wheat, especially from the United States.

Nontraditional agricultural exports

Many governments have shifted from giving top priority to self-sufficiency in basic grains to encouraging crops that are more competitive in international trade with a higher profit margin, such as fruit, vegetables, and flowers (**Figure 9.15**).

▼ Figure 9.15 Commercial flower production Shown here is a large greenhouse, part of U.S. corporation with eight branches, that produces flowers for a U.S. market with its largest market being along the eastern seaboard.

311

9.1 Window on the World

The Green Revolution Then and Now

The origin

In 1943, the Rockefeller Foundation provided funds to a group of U.S. agricultural scientists to set up a research project in Mexico aimed at increasing that country's wheat production through expansion of irrigation infrastructure, modernization of management techniques, distribution of hybridized seeds, synthetic fertilizers, and pesticides. Within 7 years, scientists were able to distribute the first modified wheat seeds to Indian farmers. Known as the Green Revolution, the project was eventually expanded to include research on maize as well. By 1967, Green Revolution scientists were exporting their work to other parts of the world and had added rice to their research agenda. Norman Borlaug, one of the founders of the Green Revolution, went on to win the Nobel Peace Prize in 1970 for promoting world peace through the elimination of hunger (**Figure 9.A**).

While the initial focus of the Green Revolution was on the development of seed varieties that would produce higher yields than those traditionally used in the target areas, eventually it came to constitute a package of inputs: new "miracle seeds," water, fertilizers, and pesticides. Farmers who use all of the inputs—and use them properly—can achieve the yields that scientists produced in their experimental plots, which are two to five times larger than those of traditional crops. In some countries, the resulting yields are high enough to enable export trade, generating important sources of foreign exchange. In addition, the creation of varieties that produce faster-maturing crops has allowed some farmers to plant two or more crops per year on the same land, increasing their individual production—and wealth—considerably.

The effects

Thanks to Green Revolution innovations, rice production in Asia grew 66 percent between 1965 and 1985. India, for example, became largely self-sufficient in rice and wheat. Worldwide, Green Revolution seeds and agricultural techniques accounted for almost 90 percent of the increase in world grain output in the 1960s and about 70 percent in the 1970s. In the late 1980s and 1990s, at least 80 percent of the additional production of grains could be attributed to Green Revolution techniques. Thus, although hunger and famine persist, many argue that they would be much worse if the Green Revolution had never occurred (**Figures 9.B.1 and 9.B.2**).

The Green Revolution has not been an unqualified success, however. One important reason is that wheat, rice, and maize are unsuitable crops in many global regions, and research on more suitable crops, such as sorghum and millet, has lagged behind. In Africa, poor soils and lack of water make progress more difficult to achieve. Another important factor is the vulnerability of the new seed strains to pest and disease infestation, often after only a couple of years of planting. Traditional varieties sometimes have a built-in resistance to the pests and diseases characteristic of an area but genetically engineered varieties often lack such resistance.

A social effect of the Green Revolution technology has been a decreased need for human labor. In southeastern Brazil, machines replaced farm workers, creating significant unemployment. Green Revolution technology and training have also tended to exclude women, who play important roles in traditional food production. In addition, the new agricultural chemicals, especially pesticides, have contributed to ecosystem pollution and worker poisonings, and the more intensive use of irrigation has created salt buildup in soils (*salinization*) and water scarcity (**Figure 9.C**).

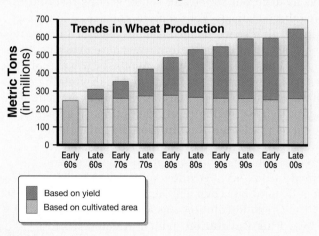

▲ **Figure 9.B1 Wheat production** This table shows the increases in yield achieved on either constant or decreasing amounts of land.

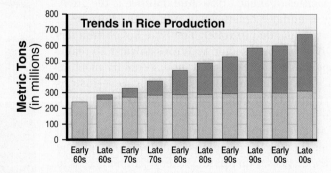

▲ **Figure 9.B2 Rice production** As with the companion table on wheat, this table shows a trend of increasing rice yield, though rice yields absorb more land than wheat.

▲ **Figure 9.A Norman Borlaug** Known as the father of the Green Revolution, Borlaug was called "the man who saved a billion lives."

	Biochemical	Mechanical	Social
Innovations	Hybrid seed selection, use of fertilizers, herbicides and pesticides	Diesel and electric pump powered irrigation, mechanization, transport improvements	Land reforms, loans, changes in distributive system
Consequences	Increased yields, weeds and pests controlled, increased costs for farmer, possible environmental degradation	Water supply controlled, less labor needed, increase in arable area, increase in access to markets	Farm consolidation, better seeds and other inputs available to poor farmers

▲ **Figure 9.C Three elements of the Green Revolution** Notice that the Green Revolution was not just a change in agricultural practices but a social change as well affecting not just the larger social system but individual farmers and farming households.

Criticism of the model

Critics have argued that the Green Revolution has magnified social inequities by allowing wealth and power to accrue to a small number of agriculturalists while causing greater poverty and landlessness among poorer segments of the population. In Mexico, a black market developed in Green Revolution seeds, fertilizers, and pesticides. Poorer farmers, coerced into using them, accrued high debts that they could not even begin to repay. Many lost their lands and became migrant laborers or moved to the cities and joined the urban poor.

The new seed varieties sometimes produce grains that are less nutritious, less palatable, or less flavorful. The chemical fertilizers and pesticides that must be used are derived from fossil fuels—mainly oil—and are thus subject to the vagaries of world oil prices. Furthermore, the use of these chemicals, as well as monocropping practices, has produced environmental contamination and soil erosion. Water developments have benefited some regions, but less well-endowed areas have experienced an exaggeration of already existing regional inequities. Worse, pressures to build water projects and to acquire foreign exchange to pay for importation of Green Revolution inputs have increased pressure on countries to grow even more crops for export, often at the expense of production for local consumption.

There are two final criticisms that have raised concern about the overall benefits of the Green Revolution. The first is that it has decreased the production of biomass fuels—wood, crop residues, and dung—traditionally used in many peripheral areas of the world. The second is that the Green Revolution has contributed to a worldwide loss of genetic diversity by replacing a wide range of local crops and varieties with a narrow range of high-yielding varieties of a small number of crops. Planting single varieties over large areas (monocultures) has made agriculture more vulnerable to disease and pests (**Figure 9.D**).

In response to these criticisms scientists are developing plants that will increase production of biomass in the form of animal fodder and fuel residues, as well as of food, and that will give optimal yields when intertilled—a very common practice in Africa. In the Sahel, scientists are working on crops that mature more quickly to compensate for the serious drop in the average length of the rainy season the region has recently experienced.

Despite criticisms, it is clear the global agricultural system has grown spectacularly. And yet, the rapid growth in agricultural output that was the hallmark of the twentieth century has declined in the twenty-first century to the point where demand is vastly outstripping supply, driven not just by population growth but also by changing food preference among developing countries like China. With consumption outstripping supply, stockpiles of wheat, rice, soybeans, and grain are also being diminished, creating serious concern among policymakers. As shown in Chapter 4, climate change has been identified as one of the most harmful and least easily remediable factors behind lessening food supplies. Drought in California, Ukraine, and Australia, linked to climate change, has cut agricultural production in these places and raised the price of food. Thus, while the Green Revolution has come under much justified criticism over the years, its main objective of finding innovative new ways to feed the world's peoples must be respected. In the process, the world's agricultural system has been expanded into hitherto very remote regions, and important knowledge has been gained about how to conduct science and how to understand the role that science plays in improving agriculture.

The **Borlaug hypothesis** (named after Norman Borlaug), states that because global food demand is on the rise, restricting crop usage to traditional low-yield methods (such as organic farming) requires either the world population to decrease or the further conversion of forest land into cropland. While there are signs that world population is beginning to decline, the Borlaug hypothesis—which is controversial—proposes that high-yield biotechnological techniques aimed at saving forest ecosystems from destruction are essential for saving the planet from ecological crisis. We discuss biotechnology and agriculture later in this chapter as well as the relative merits of sustainable organic agriculture and biotechnological commercial agriculture in the Future Geography section.

1. What is the Green Revolution? How is it different from the Blue Revolution?

2. What are three aspects of the Green Revolution that have made it controversial?

▲ **Figure 9.D Rice paddy, India** The introduction of high-yielding, semi-dwarf types of rice, starting in 1962 with the Green Revolution, emphasized the intensive use of fertilizers and pesticides. Rice production increased substantially. However, this achievement was made at a cost to the environment where semiaquatic organisms, including wild fish, frogs, shrimps, clams, and snails, which have always been part of these ecosystems, have disappeared. Moreover, to keep soil salinity low, a large quantity of additional water is needed but is seldom available resulting in soil degradation as well as species loss.

Nontraditional agricultural exports (NTAEs) such as these, contrast with traditional exports, such as sugar and coffee. NTAEs have become increasingly important in areas of Mexico, Central America, Colombia, and Chile, replacing grain production and traditional exports, such as coffee and cotton. These new crops obtain high prices but also require heavy applications of pesticides and water to meet export-quality standards. They also require fast, refrigerated transport to market and are vulnerable to climatic variation and to the vagaries of the international market, including changing tastes for foods and health scares about pesticide or biological contamination.

APPLY YOUR KNOWLEDGE

1. What are some of benefits of the Green Revolution? What are some of the criticisms? Which arguments do you find most persuasive?

2. In 2014, the 100th birthday of Norman Borlaug was celebrated and his granddaughter Julie Borlaug gave an interview. Go to this profile http://www.desmoinesregister.com/story/money/agriculture/2014/03/22/borlaug-gmo-pr-falls-short/6727245/?from=global&sessionKey=&autologin= and read her arguments, and then watch the short video about Borlaug's legacy. What does Julie Bourlaug say about the "anti-" movement? What do you think about her argument?

Contract Farming

One increasingly significant aspect of industrial agricultural transformation is contract farming. **Contract farming** is an agreement between farmers and processing and/or marketing firms for the production, supply, and purchase of agricultural products—from beef, cotton, and flowers to milk, poultry, and vegetables. The legal arrangement requires the firm to provide specified support through, for example, the supply of fertilizer or seeds and the provision of technical advice. The farmer is in turn obliged to produce a specific commodity in quantities and at quality standards determined by the firm. A great deal of agricultural production in the contemporary global system proceeds according to contracts.

Contract farming has been seen by some as a strategy for maintaining the livelihoods of small farmers marginalized by the growth of more corporate forms of farming. For farmers, contractual arrangements can provide access to services and credit as well as new technology. And the price offered for the product can help reduce risk and uncertainty. On the other hand, there are risks associated with the contract. The most worrisome is the indebtedness that often occurs when farmers are unable to meet the conditions of the contract or when firms fail to honor their agreements.

Biotechnology and Agriculture

In addition to the Green Revolution, agriculture has also undergone a Biorevolution. The **Biorevolution** involves the genetic engineering of plants and animals and has the potential to outstrip the productivity increases of the Green Revolution. Ever since the nineteenth century, when Austrian botanist Gregor Mendel identified hereditary traits in plants and French chemist Louis Pasteur explained fermentation, the manipulation and management of biological organisms has been key to the development of agriculture.

Biotechnology

The central feature of the Biorevolution is **biotechnology**, which is any technique that uses living organisms (or parts of organisms) to improve, make, or modify plants and animals or to develop microorganisms for specific uses. Recombinant DNA techniques, tissue culture, cell fusion, enzyme and fermentation technology, and embryo transfer are some of the most talked-about aspects of biotechnology in agriculture.

A common argument for applying biotechnology to agriculture is the belief that these techniques can help reduce agricultural production costs and serve as a kind of resource management (where certain natural resources are replaced by manufactured ones) practice. Biotechnology has been hailed as a way to address growing concern over the rising costs of cash-crop production, surpluses and spoilage, environmental degradation from chemical fertilizers and overuse, soil depletion, and related challenges now obstructing profitable agricultural production.

Genetically Modified Organisms

A **genetically modified organism,** or **GMO** as it is more commonly known, is any organism that has had its DNA modified in a laboratory rather than through cross-pollination or other forms of evolution. Examples of GMOs include a bell pepper with DNA from a fish added to make it more drought-tolerant, a potato that releases its own pesticide, and a soybean that has been genetically engineered to resist fungus.

Genetic modification has both critics and supporters. Proponents argue that it allows great advances in agriculture (e.g., plants that are more resistant to certain diseases or water shortages), as well as other beneficial creations, such as the petroleum-eating bacteria that can help clean up oil spills. But opponents worry that genetically modified organisms may have unexpected and irreversible effects on human health and the environment and result in maturation problems in children or in mutant plant and animal species.

In the United States, genetic modification is permitted on the principle that there is no evidence yet that it is dangerous. GMO foods are fairly common in the United States and estimates of their market saturation vary widely. It is not easy to identify GMOs in the grocery store, as there are no labeling requirements (**Figure 9.16**). While the U.S. food-safety establishment maintains that GMOs are safe until proven otherwise, countries in Europe have taken the opposite position: genetic modification has not been proven safe, so they will not accept genetically modified food from the United States or any other country. The World Trade Organization has determined that

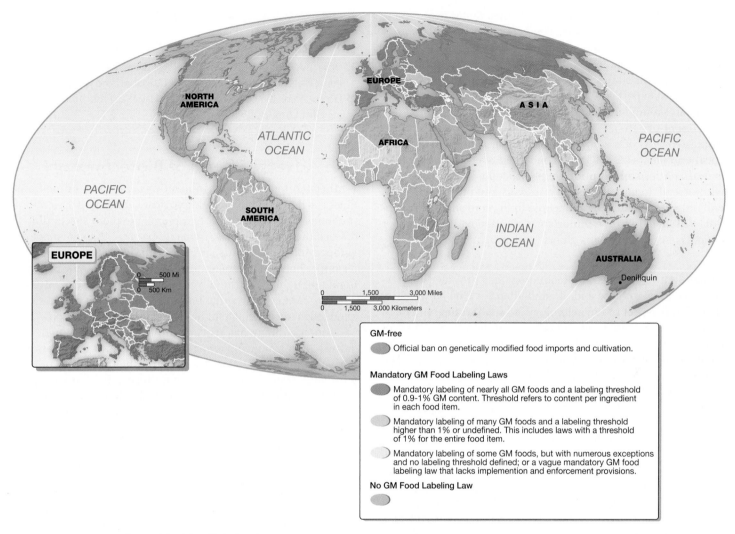

▲ **Figure 9.16 Genetically modified food labeling laws** While some countries such as the United States do not require GMO products to be labeled, there are only two countries that have official bans on genetically modified (or genetically enginereed, GE) food. It's noteworthy that many countries have followed the US's lead in not having any bans whatsoever on GMO food.

not allowing GMO food into a country creates an unnecessary obstacle to international trade.

GMOs represent perhaps the most highly technological way to date in which nature and society come together in the global food system. At present, so little is known about the impacts of GMOs on human health, the environment, or even the wider global economic system that it is difficult to sort the costs of their increasing incorporation into global food production from the benefits. What is clear is that genetic modification is not just a passing fad and the debate cannot be reduced to a simple question of "good" versus "bad." GMOs are neither entirely good nor entirely evil; certain applications may be widely beneficial, while others may not. Regulatory structures are crucial to protecting human health and the environment, as well as spreading whatever benefits may accrue from GMOs beyond the core and into the periphery. But regulatory structures are not easily implemented.

Opposition to GMOs

Protests for the introduction of GMO regulatory structures have been most effective in Europe and parts of Asia, Africa, and Latin America as well as in Canada and Mexico. In these regions, national governments are devising regulations to control or publicize the entry of GMO commodities into the food system and are requiring more research to better understand the long-term effects of GMO consumption on humans as well as the impact of GMOs on the food chain. Widespread protests against GMO foods have also been organized in the United States, but the government has not responded with support for these movements. Congress has passed domestic legislation making the labeling of GMOs voluntary, and at the international level the U.S. government has threatened action against countries who are arguing for the mandatory labeling of GMO foods traded in the global marketplace. It is important to be aware that the position taken by the United States is pro-trade, rather than strictly pro-GMO. The U.S. government is mostly concerned with the economic impact GMO labeling is likely to have and is reluctant to put the trade of U.S. agricultural products at risk. This strategy is followed because the United States is the largest exporter of agricultural goods in the world as well as the largest producer of GMO foods.

Global debate and activism—both popular and governmental—over GMOs is still in its early stages. Some of the most vocal opponents of GMOs are concerned that engineered food is destined for consumption by poor people in the periphery, while "real" food, produced by artisans and organic growers, will only be available to rich people in the core and little attention will be paid to the importance of reducing world hunger or access to a safe diet as a global human right.

Biopharming Indeed, biotechnology has provided impressive responses to these and other challenges. One particularly spectacular aspect of biotechnology is **biopharming**, in which genes from other life forms (plant, animal, fungal, bacterial, or human) are inserted into host plants. In essence, this is bioengineering plants so that they produce pharmaceuticals. The resulting "pharma crops" are effectively genetically modified organisms that contain pharmaceuticals, antibodies, and industrial enzymes. The possibilities include pharma plants for treating diarrhea, heart disease, cancer, AIDS, and other illnesses, and vaccines that would be administered by consuming modified grain or fruit such as an apple that contains a flu vaccine. Pharma crops are still in the experimental stage and the research is highly confidential. It is estimated that there are over 400 pharma crops lowered currently at the experimental stage, with many of them being anonymously field tested. One of the expected benefits is lowered costs of medication, though it is too soon to tell how this decrease might work in practice.

Among the most controversial aspects of biopharming is how little knowledge there is about the effects of growing and ingesting pharma crops. Because pharma crops involve open-air field-testing there is great concern about contamination. Formal responses to field-testing have emerged from scientific, agricultural, consumer protection, environmental, and food manufacturing groups. The concerns are related to:

- the unavoidability of food contamination as drugs and industrial chemicals are transported into non-biopharm fields.

- the health consequences to humans of biopharm crops.

- the inevitability of wider environmental contamination for mammals, birds, and insects that feed on biopharm crops.

- the financial liability to farmers and food manufacturers whose crops are contaminated without their knowledge.

One way to allay the concerns of scientists and other relevant groups is to ban outdoor field trials. Other groups wish to ban biopharming all together. At the moment, there are a set of expanding protocols in place under the jurisdiction of the U.S. Department of Agriculture (USDA) that are minimal, requiring a wide border around the field as well as inspections.

APPLY YOUR KNOWLEDGE

1. What is the difference between a GMO and biopharm crop?

2. What counties require GMO labeling? Do an Internet search for maps that show which countries require labels and which do not. Do you think the United States should label any foodstuff containing GMO products? Why or why not?

The Adverse Effects of Biotechnology

While such technological innovations can seem miraculous, there is a downside to biotechnological solutions to agricultural problems. For example, cloned plants are more susceptible to disease than natural ones, probably because they have not developed tolerances. This susceptibility leads to an increasing need for chemical treatment. And while industry may reap economic benefits from the development and wide use of tissue cultures, farmers may suffer because they lack the capital or the knowledge to participate in biotechnological applications.

Biotechnology has truly revolutionized traditional agriculture. Its proponents argue that it provides a new pathway to sustainable production. By streamlining the growth process with such innovations as tissue cultures, disease- and pest-resistant and fertilizer-independent plants, optimists believe that the Biorevolution can maximize global agricultural production to keep up with requirements of population and demand. Moreover, the intensification of agriculture, which the Biorevolution (along with the Green and Blue revolutions) enables, has reduced the risk that increased forest resources will be converted into agricultural land.

Just as with the Green Revolution, however, biotechnology may have ill effects on peripheral countries (and on poor laborers and small farmers in core countries) (**Figure 9.17**). **Table 9.1** compares the impacts of the Biorevolution and the Green Revolution on various aspects of global agricultural production. For example, biotechnology has enabled the development of plants that can be grown outside of their natural or currently most suitable environment. Yet location-specific cash crops are critical to economic stability for many peripheral nations—such as cotton in India, bananas in Central America and the Caribbean, sugar in Cuba, and coffee in Kenya, Colombia, and Ethiopia. These and other export crops are threatened by the development of alternative sites of production or by multinational agricultural corporations entering foreign markets. Transformations in agriculture have ripple effects throughout the world system.

A tragic story of the impact of foreign multinationals on domestic cash-crop producers' centers around the cotton farmers in India. In 2011, the New York University School of Law Center for Human Rights and Global Justice released a report called *Every Thirty Minutes: Farmer Suicides, Human Rights, and the Agrarian Crisis in India*. The report states:

Economic reforms and the opening of Indian agriculture to the global market over the past two decades have increased costs, while reducing yields and profits for many farmers, to the point of great financial and emotional distress. As a result, smallholder farmers are often trapped in a cycle of

TABLE 9.1   Biorevolution compared with Green Revolution

Characteristics	Green Revolution	Biorevolution
Crops affected	Wheat, rice, maize	Potentially all crops, including vegetables, fruits, agro-export crops, and specialty crops
Other sectors affected	Pesticides, fertilizer, energy, seeds, and irrigation	Pesticides, animal products, pharmaceuticals, processed food products, energy, mining, warfare
Territories affected	Developing countries	All areas, all nations, all locations, including marginal lands
Development and dissemination of technology	Largely public or quasi-public sector, international agricultural research centers (IARCs); millions of dollars for R&D	Largely private sector, especially corporations; billions of dollars for R&D
Proprietary considerations	Plant breeders' rights and patents generally not relevant	Genes, cells, plants, and animals as well as the techniques used to produce them are patentable
Capital costs of research	Relatively low	Relatively high for some techniques; relatively low for others
Access to information	Relatively easy, due to the public funding of research and development	Restricted due to privatization and proprietary considerations
Research skills required	Conventional plant breeding and parallel agricultural sciences	Molecular and cell biology expertise as well as conventional plant-breeding skills
Crop vulnerability	High-yielding varieties relatively uniform; high vulnerability	Tissue culture crop propagation produces exact genetic copies; even more vulnerable
Side effects	Increased monoculture and use of farm chemicals; marginalization of small farmer; ecological degradation. Increased foreign debt due to decrease in biomass fuels and the increasing reliance on costly, usually imported, petroleum	Crop substitution replacing periphery exports; herbicide tolerance; increasing use of chemicals; engineered organisms might affect environment; further marginalization of small farmer

Source: Adapted from M. Kenney and F. Buttel, "Biotechnology: Prospects and Dilemmas for Third-World Development," *Development and Change 16*, 1995, p. 70; and H. Hobbelink, *Biotechnology and the Future of World Agriculture: The Fourth Resource.* London: Red Books, 1991.

▲ **Figure 9.17** Mothers and children protest GMOs in Quezon City, Philippines. "Green Moms" shown here are advocating organic foods and breastfeeding practices in opposition to a genetically modified rice variety known as "Golden Rice".

debt. During a bad year, money from the sale of the cotton crop might not cover even the initial cost of the inputs, let alone suffice to pay the usurious interest on loans or provide adequate food or necessities for the family. The only way out might be to take on more loans and buy more inputs, which in turn can lead to even greater debt.

In fact, what many of these indebted farmers do out of sheer desperation is commit suicide. Over the last 20 years, a quarter of a million Indian farmers have committed suicide, making this the largest wave of recorded suicides in human history. The bottom line is that global agricultural transformations are far from benign and it will require an enormous amount of political cooperation to begin to address their profound effect.

APPLY YOUR KNOWLEDGE

1. What are some of the positive effects of biotechnology on global agriculture?

2. The industrialization of agriculture has had negative effects on traditional farmers across the globe, including the United States. Referring to the U.S. Census of Agriculture, compare the number of family farms, the mechanization of farming, the application of chemicals to farmland, and the amount of agricultural output for every decade between 1910 and 2010. What observations can you make about these changes and how they have shaped the U.S. food production and distribution system?

GLOBAL CHANGE IN FOOD PRODUCTION AND CONSUMPTION

When geographers talk about the globalization of agriculture, they are referring to the incorporation of agriculture into the world economic system of capitalism. A useful way to think about the term **globalized agriculture** is to recognize that, as both an economic sector and a geographically distributed activity, modern agriculture is increasingly dependent on an economy and set of regulatory practices that are global in scope and organization.

Forces of Change

Several forces, institutions, and organizational forms play a role in the globalization of agriculture. Technology, economics, and politics have played a central role in propelling national and regional agricultural systems into becoming global in scope. One important way these forces of change have been harnessed is through new global institutions, especially trade and financial organizations. The result is an integrated, globally organized, agro-production system.

The globalization of agriculture has dramatically changed relationships among and within different agricultural production systems. Important outcomes of these changed relationships have been the elimination of some forms of agriculture and the erosion or alteration of some agricultural systems as they become integrated into the global economy. Two examples

include the current decline of traditional agricultural practices, such as shifting cultivation, and the erosion of a national agricultural system based on family farms (**Figure 9.18**).

Agriculture is one part of a complex and interrelated worldwide economic system. Important changes in the wider economy—whether technological, social, political, or otherwise—affect all sectors, including agriculture. National problems in agriculture, such as production surpluses, soil erosion, and food price stability, affect other economic sectors globally, nationally, and locally in different ways. The same is true of global factors, such as the price and availability of oil and other petroleum products critical to commercialized agriculture, the stability of the dollar in the world currency market, and recession or inflation.

Because of the systemic impact of many problems, integration and coordination of the global economy is needed to anticipate or respond to them. In the last several decades, global and international coordination efforts among states have occurred. These include policies advanced by the World Trade Organization (WTO), as well as the formation of supranational economic organizations such as the European Union (EU) and the Association of Southeast Asian Nations (ASEAN).

National and International Agricultural Policy

For all countries, the production of food is thus critically important for sustaining its population and enabling the production of food and other aspects of the economy to function and grow. Because food is also traded as a global commodity, international treaties, policies, agreements, and markets have coalesced regulate it. There are three key questions critical to understanding national and international agricultural policy: Why are there agricultural policies? What do these policies look like? What effects do they have?

Why Have Domestic and International Agricultural Policies?
At the domestic level governments implement agricultural policies not only to assure adequate foodstuffs for their national population but also to support the smooth operation of agricultural product markets. This smooth operation most frequently involves a supply adequate to needs and demand, price stability, product quality and selection and issues of labor and land use. These quite legitimate concerns require governments to become involved in a range of policies including:

- education of a labor pool and research to support improvements in agricultural practices,

- adequate access to resources such as land and water,

- technological support for everything from business planning to risk management,

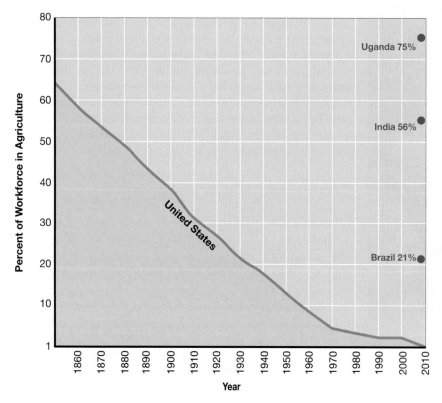

▼ Figure 9.18 **Family farms in decline** This graph shows the precipitous decline in the percentage of the workforce involved in agriculture in the United States as industrial agriculture was adopted. In most poor countries, farming is still a significant source of livelihood.

- infrastructure to move agricultural goods including roads, ports, irrigation, and communication,

- biosecurity to combat threatening pests and diseases, and

- a trading environment that allows for national competitiveness in a global market.

What Are Some Key Agricultural Policies?

In the United States, agricultural policy, as in other countries and regional government organizations like the EU, is complex reflecting the change in agricultural practices over time as well as the growth of a global market for agricultural goods. Early agricultural policy tended to focus on developing the agricultural sector and supporting the stability and growth of farming as national economic activity. In the United States, this early period was followed by a dramatic economic crash—the Great Depression in 1929–1945—that required more direct support for farmers in the form of income and price guarantees. These measured were meant to ensure that farmers remained in the agricultural sector and certain staple food were produced. Over time, other kinds of domestic policies become necessary including market and trading regulation.

It is at the international level that the growth in agricultural policies has been the greatest, especially over the last 25 years. At this level, multilateral trading systems are in place meant to govern import and export restrictions, promote food security and innovation, disseminate new technologies and growing and breeding techniques, and just generally increase agricultural productivity and profit.

What Are Some Important Impacts of National and International Policies?

The impacts of agricultural policy have not always been positive, despite their stated intent. Perhaps the most controversial domestic policy in the United States has been the subsidy system, which though meant to support struggling farmers in the 1930s, now routinely goes to corporate farms. Subsidies have also been shown to promote overproduction and can end up costing taxpayers both higher taxes and higher food costs. The point is not that all domestic agricultural regulation is problematic, but that it can have both positive and negative impacts.

At the international level, perhaps the most controversial regulatory body has been the World Trade Organization (WTO). A product of the rapid globalization of the 1990s, the WTO is a multilateral organization made up of 160 members (largely nation-states) and 24 observer governments that constitute 94% of global trade and almost 97% global GDP. The WTO deals with all global trade, with agriculture being one of its most significant areas. While most countries have laws meant to regulate trade outside of the national territory, the WTO's regulations supersede national ones for member nations and thereby make it one of the most powerful regulation and enforcement bodies on the planet.

With its aim to "ensure that trade flows as smoothly, predictably, and freely as possible," it has often generated significant opposition, not just from small farmers and less powerful groups, but governments as powerful as the United States, the European Union, and Australia, just to name a few. Organizations such as Global Exchange argue that the WTO is fundamentally undemocratic and that it is deaf to labor and human rights in its push for creating a smooth surface for trade. Further, they and others point out that the WTO is increasing world hunger because its leading principle, the Agreement on Agriculture, requires that market forces control agricultural policies instead of national policies that aim to guarantee domestic food security and the protection of farmers.

Opposition to Globalized Agriculture

These new forms of cooperation have other opponents as well. African activists have been involved in frequent and sustained protests against the involvement of European Union companies in biofuel farming there (which we describe in more detail in the material that follows). Similarly, Indian tribal women in the state of Orissa have been involved for several years in a campaign to prevent foreign and national companies from planting genetically modified (GM) crops. Exhibiting 500 indigenous rice varieties, these women argued before the state assembly in 2005 that the government would put at risk its rich array of rice species if genetically modified organisms entered the region. These and many other protests across the globe demonstrate that the global transformation of agriculture faces resistance.

Agricultural Subsidies

As mentioned earlier, one way that governments try to maintain the profitability of the agricultural sector while keeping food prices affordable is through direct and indirect subsidies to agricultural producers (**Figure 9.19**). For example, the U.S. government subsidizes agriculture in a number of ways. One is by paying farmers not to grow certain crops that are expected to be in excess supply. Another is by buying up surplus supplies and guaranteeing a fixed price for them.

Billions of dollars are paid out each year in agricultural subsidies, the effects of which are complex and global in impact. Government efforts, while perhaps stabilizing agricultural production in the short term, can lead to problems within the larger national and international agricultural system. For instance, guaranteeing a fixed price for surplus food can act as a disincentive for producers to lower their production, so the problem of overproduction continues. Once in possession of the surplus, governments must find ways to redistribute it. The U.S. government often sells or donates its surplus to foreign governments, where the "dumping" of cheap foodstuffs may undermine the local price structure for food and reduce economic incentives for local farmers to farm.

Many reasons exist for state intervention in agriculture. Governments routinely intervene in one economic sector or

▲ **Figure 9.19 Direct subsidies for animal products and feed** This graphic provides data on animal and feed subsidies only and covers only core countries.

another, for instance, to correct wider problems of inflation or depression. For years, bread subsidies were the norm across the Middle East, used as a way of placating the population. In Tunisia in 2011, civil unrest was predicated in part on the high cost of food because national grain subsidies had not kept pace with soaring prices. It is important to keep in mind that the revolution in Tunisia is more than just about bread. Basic human rights are the most pressing demands across the Middle East from Tunisia to Yemen. When a government fails to provide those rights for the majority of its citizens, instead using handouts or subsidies as a substitute for democratic or economic reforms, bread becomes a powerful symbol of all these citizens lack.

APPLY YOUR KNOWLEDGE

1. What are some reasons different groups would protest national agricultural policies? Think about how policies are set and the role of the WTO, and how GMOs might threaten indigenous seed varieties.

2. Research agricultural subsidies in the United States. Which growers receive the most subsidies? Once you have established this fact, compare the U.S. data with two other countries—one from the periphery and one from the core. How much in government subsidies do farmers in other countries receive and how does it compare with subsidies in the United States. What impacts might this have on the agriculture industry of periphery nations?

The Organization of the Agro-Food System

Although the changes that have occurred in agriculture worldwide are multifaceted, certain elements that serve as important indicators of change can help us understand them. Geographers and other scholars interested in contemporary agriculture have noted three prominent and interconnected forces that signal a dramatic departure from previous forms of agricultural practice: agribusiness, food chains, and integration of agriculture with the manufacturing, service, finance, and trade sectors.

Agribusiness

The concept of agribusiness has received a good deal of attention in the last three decades, and in the popular mind it has come to be associated with large corporations, such as ConAgra or Del Monte. Our definition of agribusiness departs from this conceptualization. Although multi- and transnational corporations (TNCs) are certainly involved in agribusiness, **agribusiness** is a system rather than a kind of corporate entity. It is a set of economic and political relationships that organizes food production from the development of seeds to the retailing and consumption of the agricultural product. Defining agribusiness as a system does not mean that corporations are not critically important to the food production process. In the core economies, the transnational corporation is the dominant player, operating at numerous strategically important stages of the food production process. TNCs have become dominant for a number of reasons, but mostly because of their ability to negotiate the complexities of production and distribution in many geographical locations. That capability requires special knowledge of national, regional, and local regulations and pricing factors.

Food Supply Chain

The concept of a food supply chain (a special type of commodity chain; see Chapter 2) is a way to understand the organizational structure of agribusiness as a complex political and economic system of inputs, processing and manufacturing, and outputs. A **food supply chain** is composed of five central and connected sectors (inputs, production, processing, distribution, and consumption) with four contextual elements acting as external mediating forces (the state, international trade, the physical environment, and credit and finance). **Figure 9.20** illustrates these linkages and relationships, including how state farm policies shape inputs, prices, farm structure, and even the physical environment.

The food supply chain concept illustrates the network of connections among producers and consumers and regions and places. Consider for example, the linkages that connect cattle production in the Amazon and Mexico, the processing of canned beef along the United States–Mexico border, the

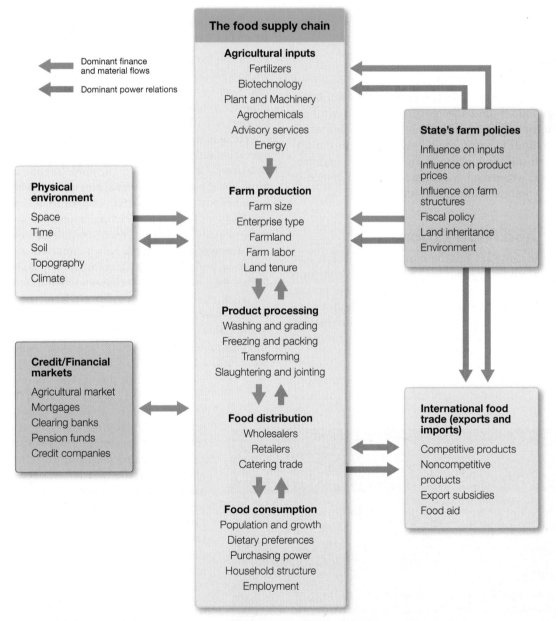

▲ **Figure 9.20 The food supply chain** The production of food has been transformed by industrialization into a complex system that comprises distinctly separate and hierarchically organized sectors. Mediating forces (the state, the structure and processes of international trade, credit and finance arrangements, and the physical environment) influence how the system operates at all scales of social and geographical resolution.
(*Source: Bowler, I (ed.), The Geography of Agriculture in Developed Market Economies. New York: J. Wiley & Sons, 1992, p. 12.*)

availability of frozen hamburger patties in core grocery stores, and the construction of McDonald's restaurants in Moscow (**Figure 9.21**). Because of complex food chains such as this, it is now common to find that traditional agricultural practices in peripheral regions have been displaced by expensive, capital-intensive practices.

That agriculture is not an independent or unique economic activity is not a particularly new realization. Beginning with the second agricultural revolution, agriculture began slowly but inexorably to be transformed by industrial practices. What is different about the current state of the food system is how farming has become just one stage of a complex and multidimensional economic process. This process is as much about distribution and marketing—key elements of the service sector—as it is about growing and processing agricultural products.

Food Regimes and Alternative Food Movements

A **food regime** is a specific set of links that exist between food production and consumption. Like the agricultural revolutions already described, food regimes emerge during key historical periods, when different cultural, political, and economic forces are in operation. In contrast to a food chain, which describes the network within which specific food items are produced, manufactured, and marketed, a food regime indicates the ways a particular type of food item is dominant during a specific time period.

Colonialism and Food Exports

Although hundreds of food chains may be in operation at any one time, agricultural researchers believe that only one food regime dominates each particular period. During the decades surrounding the turn of the nineteenth century, an independent system of nation-states emerged and colonization expanded (see Chapters 2 and 10). At the same time, the

industrialization of agriculture began. These two forces of political and economic change were critical to the fostering of the first food regime, in which colonies became important sources of exportable foodstuffs by supplying the industrializing European states with cheap food in the form of wheat and meat. The expansion of the colonial agriculture sectors, however, created a crisis in production. The crisis stemmed from the higher cost efficiency of colonial food production, which undercut the prices of domestically produced food, put domestic agricultural workers out of work, and forced members of the agricultural sector in Europe to improve cost efficiency. Industrialized agriculture helped to drive down operating costs and restabilized the sector (reducing even more the need for farm workers). This movement toward the integration of agriculture and industry is also referred to as agro-industrialization.

Fresh Fruits and Vegetables

From the turn of the nineteenth century until the 1960s, the United States consolidated a wheat and livestock food regime that was epitomized in the meat-heavy diets of the typical American family meal. This food orientation can be explored through a wide range of sources from agricultural censuses to cookbooks, household expenditures on food, to restaurant and special event menus. Food researchers believe that presently, however, a fresh fruit and vegetable regime is dominant. The more perishable agro-commodities of fresh fruits and vegetables have become central to diets of people in the core. Integrated networks of food chains, using integrated networks of refrigeration systems, deliver fresh fruits and vegetables from all over the world to Europe, North America, and parts of South America, and several Asian countries (**Figure 9.22**). Echoing the former food networks that characterized nineteenth-century imperialism, peripheral production systems supply core consumers with fresh, often exotic and off-season, produce. Consumers in the core regions have come to expect the full range of fruits and vegetables to be available year-round, and unusual and exotic produce has become increasingly popular.

Alongside the emergence of a core-oriented food regime of fresh fruits and vegetables, it is important to note an additional aspect of food production practices that has been taking hold in core regions and accelerating especially over the last ten years: sustainable agriculture. One goal of sustainability in agriculture has been increased commitment to organic crops—both proteins (beef, eggs, soy) and produce (fruits and vegetables). **Organic farming** describes farming or animal husbandry that occurs without the use of commercial fertilizers, synthetic pesticides, or growth hormones. It is important to point out that organic food production is not the primary mode of practice, but rather that it has become a growing force alongside the dominant **conventional farming** (an approach that uses chemicals in the form of plant protectants and fertilizers and intensive, hormone-based practices to breed and raise animals) (**Table 9.2**). We discuss issues of food safety in more detail later in

▼ **Figure 9.21 World's largest cattle feedlot** This feedlot near Greeley, Colorado, houses over 120,000 head of cattle and is a subsidiary of the food giant ConAgra.

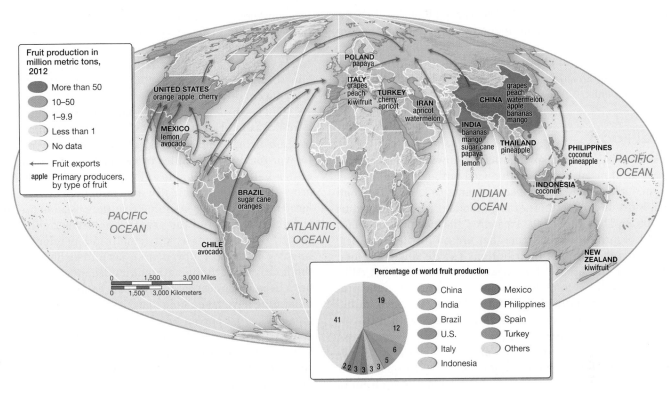

▲ **Figure 9.22 Global flows of fresh fruits**   Shown here are both the flows of fresh fruit as well as the world's largest producers of fruit.

TABLE 9.2   Conventional versus alternative food

Conventional	Alternative
Cheap food to most of the Western world, where the amount of money spent on food in the household budget has steadily decreased.	Logic of sustainable development, not mass production.
Quantity over quality where agriculture is part of consumption; food is a commodity like any other.	The idea of (real or idealized) "quality" predominates—expressed in such parameters as farmers' markets, CSAs, free-range, organic, GMO-free, integrated pest-management schemes, authenticity, traceability.
Fast, convenient foods: for nutritionists, the year 2000 was significant since it was the first time the number of overweight people matched the number of under-nourished people—at 1.1 billion.	Slow, pleasurable food and consumers with an ethical consciousness; a new moral economy surfaces where issues of provenance take a central role and consumers demand knowledge of where their food comes from, whether it contains additives, and whether pesticides were used in production.
"Hard power"—retailer-led supply chains, emasculation of primary producers as supermarkets drive down costs to below that of even production.	"Soft power"—food sells itself through ethical persuasion and the new "moral economy."
Corporate capital, combined with science and technology enable large-scale food processing and a reliance on large retailers—at the expense of small, independent farms.	Social, as opposed to corporate capital. Direct farmer-to-consumer interaction, relational, trust-based, sense of "community," and attempts to restore small farm enterprises.
Food is "liberated" from nature; technological fixes predominate.	Food is relocated in specific ecological niches and the externalized costs of the conventional system—such as soil degradation, water pollution, animal welfare issues, and health care costs—are addressed.
Logic of deterritorialization—where geographies are a result of the intensification and stretching of the links and networks between production and consumption.	Logic of reterritorialization—where geographies are shaped by the resurgence of local and regional practices as central forces in new articulations of production and consumption.
Geographic specialization, monoculture, and spatial homogenization in North America and Europe produce a large quantity of standardized products.	The "local" as a set of social networks and cultural contexts that are harnessed in place—which provide the foundation for development, innovation, and economic relations.
Placeless landscape of food production (United Kingdom, United States) versus place-based landscapes (France, Italy)—where region-specific foods are much more prevalent and provenance is understood as a selling point.	Transparent food chain, traceable, where producers attempt to regain power against the conventional price-driven "race to the bottom" and consumers' knowledge of provenance helps to produce local networks and promotion of social capital.

the chapter, but safety issues as well as overall health concerns and an evolving commitment to food as a gastronomic experience among mostly white, middle-class households is at the center of a number of alternative movements that are driving organically produced food as well as the local food and slow food movements.

Local Food

Local food is usually also organically grown and its designation as local means that it is produced within a fairly limited distance from where it is consumed, an area known as a **food shed**. Most understandings of local food set a 100-mile radius as a border. Thus, if one is a "locavore," the food one consumes should be produced on farms that are no further than 100 miles from the point of distribution. The distance that food travels from the farm to the consumer is known as **food miles** (**Figure 9.23**). Local food movements have resulted in the proliferation of communities of individuals who have joined together to support the growth of new farms within the local food shed (the area within the 100-mile radius). These individuals have been behind a movement known as Community Supported Agriculture, or CSA. The CSA operates by collecting subscriptions and then paying a farmer up front to grow food locally for CSA members for a season or a specified time frame. The farmer then is able to purchase seeds, hire workers, cultivate produce and livestock, and deliver the harvest without relying on interest-bearing loans. The CSA members, in return for their investment in the farm, receive weekly shares of produce (and sometimes meats, eggs and cheese, flowers, and milk) that reflect the season and local growing conditions.

Thus, while a CSA in Tucson, Arizona, might receive verdolagas (a green that is native to the region) in addition to squash, snow peas, beets, and carrots, they are unlikely to ever receive pineapples or celery because neither can be grown naturally in the desert. CSA farmers are expected to grow their food organically and members accept what the farmers grow at the same time that they assume the financial risks that are part of the vagaries of any agricultural enterprise. CSA farms are usually small, independent, and labor-intensive. The CSA movement is seen to be helping to restore the family farm to the national landscapes from which they have been disappearing for 30 years as more corporate forms of farming have taken hold. The CSA movement originated in Japan but it has spread to other core regions, including North America and Europe. In addition to the emergence of CSAs, it is also important to note that urban agriculture is also on the rise. **Urban agriculture** is the establishment or performance of agricultural practices in or near an urban or city-like setting (see Box 9.2 "Visualizing Geography: The Growth of Urban Agriculture").

Although local food and CSAs, and organic and ecologically sustainable agricultural practices are movements that signal a shift in food production and consumption, they in no way are challenging the dominance of more conventionally produced, distributed, marketed, and consumed food. Moreover, as a number of critics of the movements have pointed out, these alternative practices are largely organized and promoted by white, middle-class members of core regions and exclude, often simply through cost and associated accessibility, low-income people of color. The latter are denied these purportedly healthier eating opportunities because they lack the information, income, and proximity to access them. In core countries, the result is that low-income people turn to cheap, easily accessible food, also known as fast food.

Related to food security is the concept of **food sovereignty**, the right of peoples, communities, and countries to define their own agricultural, labor, fishing, food, and land policies that are ecologically, socially, economically, and culturally appropriate to their unique circumstances.

Today in the United States, where food is abundant and overeating is a national problem, 10 percent of the population is undernourished or experiences food-security problems at one time or another each year. In the periphery, where food availability is more limited than in core countries like the United States, undernutrition is far more pervasive. Furthermore, in

▲ **Figure 9.23 Food shed** This combination photo and graphic shows how a food shed reflects the way that a local diet can be based on food raised and/or processed from within a 100 mile radius.

some parts of the world, among some social classes, there are higher levels of undernutrition among women and girls than among men and boys. This is largely because different cultural and social norms favor men and boys, who eat first and leave the leftovers for females, or who eat certain high-status foods, such as proteins like meat or fish, that women are not allowed to eat.

Food sovereignty is a movement that recognizes that minority populations in the core and periphery have had the least amount of control over the food they purchase and consume. Local movements to achieve food sovereignty are usually organized by people of color, peasants, and low-income people who point out that access to healthy food is shaped by both class and by race. There are six basic principles of food sovereignty:

- The importance of sufficient, healthy, and culturally appropriate food for all

- The significance of all food providers from farmers and pastoralists to cooks

- The necessity of a close link between food providers and consumers

- The requirement to control land as well as all the other inputs to food production

- The centrality of local knowledge and skill to food production

- The need for food production to support ecosystem balance and resilience

Related to food sovereignty is the concept of **food justice**, which is about enabling communities to enact the principles just mentioned, which is simply to be able to grow, eat, and sell healthy food and care for the well-being of the local ecosystem in culturally appropriate ways. It is also about ensuring that people who work in the food industry are treated fairly and have access to affordable and healthy food as well (**Figure 9.24**).

Organic Farming

http://goo.gl/OsezQB

APPLY YOUR KNOWLEDGE

1. What are some of the alternative food movements today? Think about where and how different food is produced, and by and for whom it is produced.

2. Look closer at a group that resists globalized agriculture by visiting the La Via Campesina: International Peasant's Movement Web site at http://viacampesina.org/en/. Select one of the recent stories. What is the issue and where in the world is the story based? What is "international peasant solidarity"? How does this movement relate to global agriculture? How does food justice relate to the environment, human-society relations, and health?

Fast Food

Fast food was born in the United States as a product of the post–World War II economic boom and the social, political, and cultural transformations that occurred in its wake. The concept of **fast food**—edibles that can be prepared and served very quickly in packaged form in a restaurant—actually preceded the name for it. What made fast food *fast,* was the adoption of industrial organizational principles applied to food preparation in the form of the Speedee Service System. The system, invented by Richard and Maurice McDonald in their San Bernardino, California, McDonald Brothers Burger Bar Drive-In in 1948, revolutionized the restaurant business. To preclude the labor and material costs of standard restaurant food preparation, each Burger Bar worker was assigned one task in an assembly-line operation—cooking the burger, placing it on the bun, or packaging it for take-away, for instance—and the product was standardized so that the same condiments—ketchup, onions, mustard, and two pickles—were added to each patty.

As Eric Schlosser wrote in his *New York Times* best-selling book *Fast Food Nation,* once Raymond Kroc came along in 1954 and convinced the McDonald brothers to agree to franchise—which involved the licensing of trademarks such as the golden arches and the adherence to established methods of doing business—the fast-food concept began to expand all

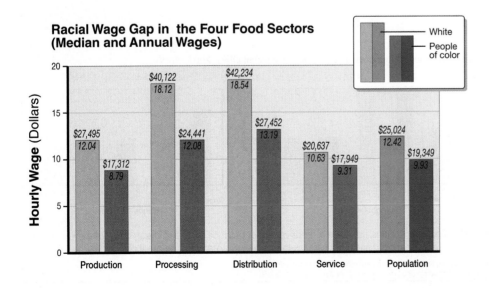

◀ **Figure 9.24 Racial wage gap in four food sectors** This graph shows median and annual wages for different sectors of the food retail industry differentiating by race. In no sector are people of color doing as well as white people.

Racial Wage Gap in the Four Food Sectors (Median and Annual Wages)

9.2 Visualizing Geography

Urban Agriculture

We usually think of agriculture as a rural activity, but urban agriculture made possible the emergence of the world's first cities. Until recently, urban agriculture was largely ignored in the development of urban economic policies; it's produce was seen as part of the informal sector of the local economy and not significant in terms of income-generating potential.

9.1 Urban Agriculture Around the World

Most definitions of urban agriculture focus on the establishment or performance of agricultural practices in cities. Since the Industrial Revolution, many countries have discouraged urban agriculture as arable land has been used for real estate development or seriously degraded through industrial processes.

Urban residents across the globe are increasing their participation in growing crops and raising livestock, for reasons ranging from food security to income production to taste and health concerns. A recent publication by the Food and Agricultural Organization of the United Nations has estimated that 800 million people are actively engaged in urban agriculture around the world.

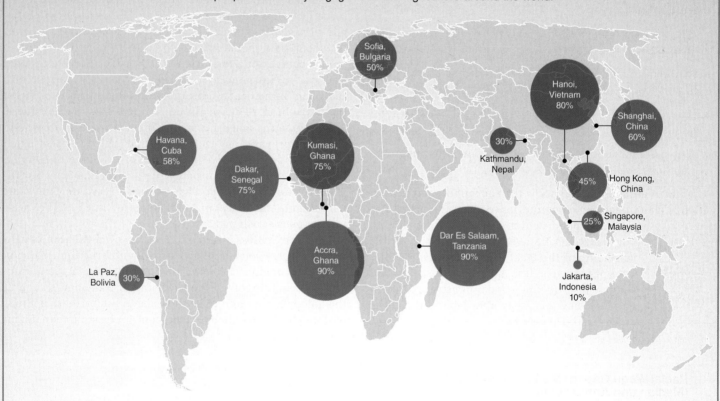

Figure 9.1.1 Percentage of fresh vegetables from urban farming in selected cities, 2007

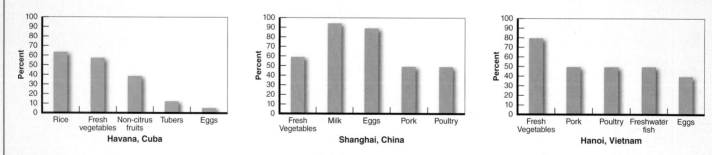

Figure 9.1.2 Percentage of various foods produced by urban farming. Some cities generate a great variety of produce including meat, dairy, and even fish in addition to fresh vegetables. Havana, Hanoi, and Shanghai are three cities with diverse urban agriculture practices.

Sources: United Nations Food and Agricultural Organization and Department of Economic and Social Affairs, Five Borough Farm, USDA

9.2 Global food consumption

According to the latest UN projections, world population will rise from over 7 billion today to 9.1 billion in 2050. Nearly all of the population growth will occur in developing regions like sub-Saharan Africa and East and South East Asia.

The demand for food is expected to continue to grow as a result both of population growth and rising incomes. Demand for cereals (for food and animal feed) is projected to reach some 3 billion tons by 2050. Annual cereal production will have to grow by almost a billion tons, and meat production by about 200 million tons to reach a total of 470 million tons in 2050

2005/2007
2050
Kg/person/year
250
225
200
175
150
125
100
75
50
25
0
Cereals
Roots/tubers
Sugar
Veg. oils
Meat
Milk
World
Developed
Developing

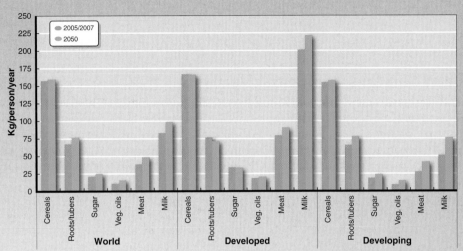

Figure 9.2.1 **Projected food consumption per capita**

9.3 Benefits of urban agriculture

Although urban agriculture is a growing movement in core countries, it is in the periphery where it more often is the sole means of economic and personal survival. As wage cuts, inflation, job loss, civil strife, and natural disasters become more frequent, urban agriculture in the periphery has become a way to address greater food insecurity.

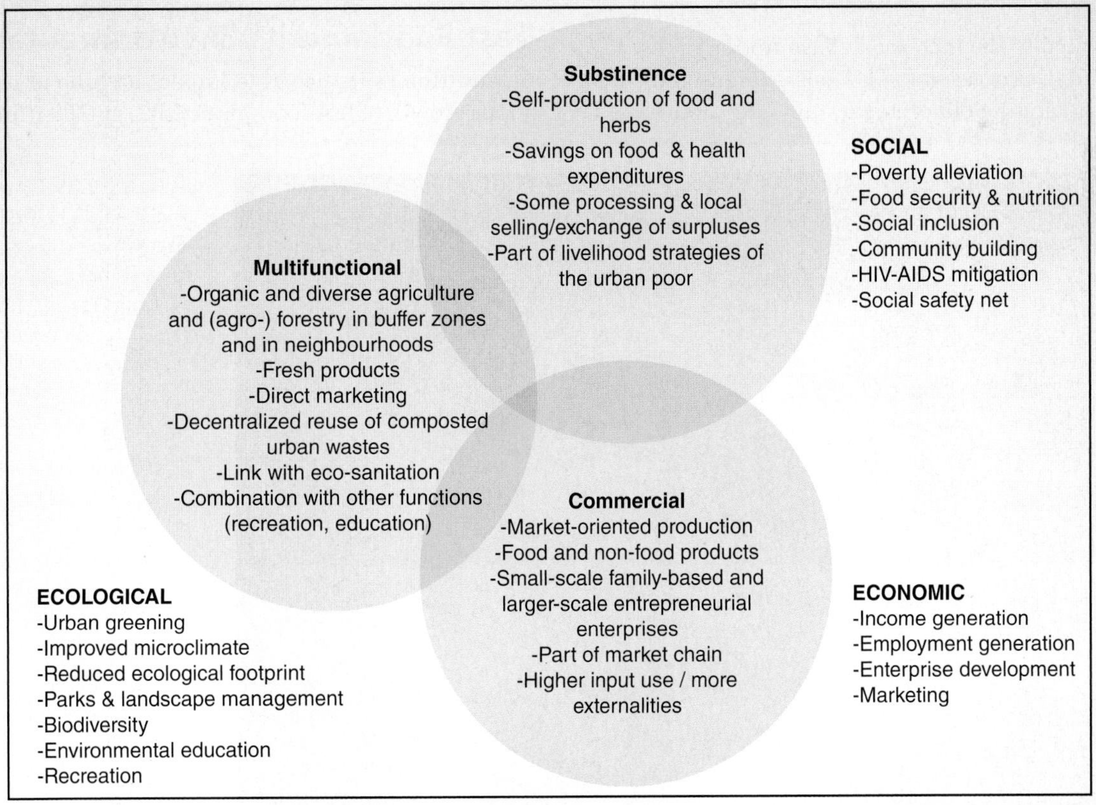

Substinence
-Self-production of food and herbs
-Savings on food & health expenditures
-Some processing & local selling/exchange of surpluses
-Part of livelihood strategies of the urban poor
SOCIAL
-Poverty alleviation
-Food security & nutrition
-Social inclusion
-Community building
-HIV-AIDS mitigation
-Social safety net
Multifunctional
-Organic and diverse agriculture and (agro-) forestry in buffer zones and in neighbourhoods
-Fresh products
-Direct marketing
-Decentralized reuse of composted urban wastes
-Link with eco-sanitation
-Combination with other functions (recreation, education)
Commercial
-Market-oriented production
-Food and non-food products
-Small-scale family-based and larger-scale entrepreneurial enterprises
-Part of market chain
-Higher input use / more externalities
ECOLOGICAL
-Urban greening
-Improved microclimate
-Reduced ecological footprint
-Parks & landscape management
-Biodiversity
-Environmental education
-Recreation
ECONOMIC
-Income generation
-Employment generation
-Enterprise development
-Marketing

International Network for Urban Agriculture

http://goo.gl/pOjbe2

Figure 9.3.1 **Main types of urban farming and their benefits**

1. Why did urban agriculture decline in core countries? 2. Why is urban agriculture important for overall food security?

over the United States, so that by 1951, *Merriam-Webster's Dictionary* included it as a new word; and by 1964, there were 657 McDonald franchises. In time, other enterprising individuals seized the fast-food idea and opened hundreds of In-N-Out Burger, Burger King, Chik-fil-A, and Subway restaurants (**Figure 9.25**).

Fast Food and its Health Effects

Health practitioners, alternative food activists, journalists, and others have turned a critical eye on fast food to expose some of its dietary, labor-related, and ecological shortcomings. The concerns about the effects of fast food on health have risen as our consumption of (or, as some health experts would argue, addiction to) fast food increases and our levels of routine exercise decrease (**Figure 9.26**). Worries focus on the increased incidence of obesity (an amount of fatty body mass that is a danger to health) and diabetes (a disorder that affects the human body's use of food for energy), particularly among children.

Since the late 1970s, the rate of obesity and diabetes among U.S. children has doubled. Recent data shows that more than 34 percent of children in North America, 38 percent in Europe, 22 percent in the western Pacific, and 22 percent in southeast Asia are overweight or obese. These numbers, though certainly alarming, are not surprising given how heavily the fast-food industry advertises to children, a large percentage of whom can recognize the golden arches of McDonald's before they know their own name.

Diseases related to the increased consumption of fast food and many prepared, processed foods—energy-dense, nutrient-poor foods with high levels of sugar and saturated fats—as well as reduced physical activity, have also led to adult obesity rates that have risen threefold or more since 1980 across North America, the United Kingdom, Eastern Europe, the Middle East, the Pacific Islands, Australasia, and China. Moreover, the obesity epidemic is often increasing faster in rapidly developing peripheral countries than in the core. The explanation for this epidemic reflects the impacts of economic growth, modernization, and globalization on daily nutritional practices.

The World Health Organization's assessment of the obesity and diabetes problem indicates that growing prosperity in the developing world is not entirely positive. Evidence suggests that as rural people move to cities and their incomes improve, their diets, previously high in complex carbohydrates, come to contain higher proportion of fats, saturated fats, and sugars. At the same time, urban life ordinarily means a move from highly demanding physical labor toward less physically demanding work and the increasing use of automated transport and technology in the home. These shifts have a significant effect on overall body fat as well.

Obesity is of concern to the international health community because it poses a major risk for serious diet-related chronic diseases, including type 2 diabetes, cardiovascular problems, hypertension and stroke, and certain forms of cancer. The consequences range from increased risk of premature death to grave chronic conditions that reduce the overall quality of life.

Overweight and Obesity Visualization

http://goo.gl/KTsmHU

Fast Food and its Environmental Impacts

In addition to its negative impact on human health, the rapacious growth of fast-food production and distribution processes

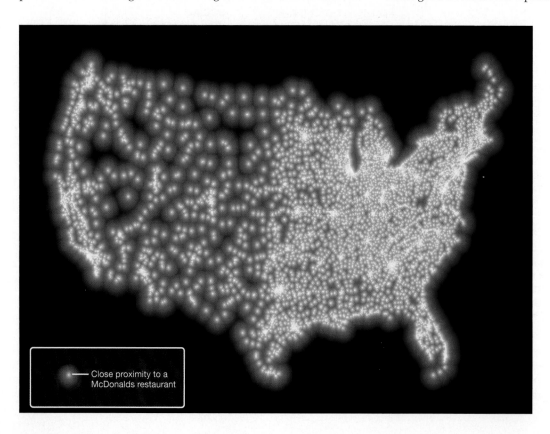

— Close proximity to a McDonalds restaurant

◄ **Figure 9.25** Distance to McDonald's in the United States This visualization shows the United States virtually covered with McDonald's restaurants. The spot in the United States furthest from the golden arches is 115 miles away.

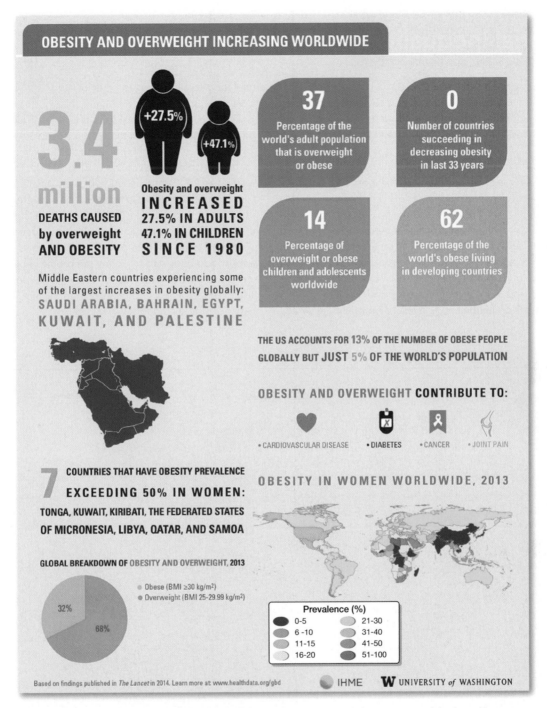

▲ Figure 9.26 Obesity on the increase globally Notice the variation in obesity rates worldwide and how some countries possess a large percentage of obese people while having a relatively smaller percentage of the global population.

is dramatically affecting Earth's resources, especially forests and farmland. It also has significant effects on hunger in the periphery (**Figure 9.27**).

The impacts of fast-food production and consumption on people and places are complex, from the convenience for busy consumers and the provision of play spaces for children to the elimination of forests and the debilitating health effects of overconsumption. How we manage these effects and address the causes have wide-ranging implications for people and the planet.

APPLY YOUR KNOWLEDGE

1. What are some of the impacts of fast-food culture on human health and the environment?

2. Determine how many fast-food restaurants are in your town. Pick three of these and research their menu and the nutritional value of several popular items. This information is usually available on the Internet. Note the prices for these items. What can you conclude about the nutritional value of each item and its price?

9.3 Spatial Inequality   Food Deserts

A **food desert** is a geographic area where access to affordable and nutritious food is highly limited, especially for individuals without automobiles. The term describes neighborhoods that have little or no retail food outlets, especially grocery stores that offer fresh produce and meats at reasonable cost.

Research on nutrition and geography has shown that food deserts are most likely to occur in low-income urban neighborhoods and rural areas where population density is much lower than in cities. It is generally agreed that food deserts are a result of land-use policies that have facilitated the tremendous growth of wealthy, white suburbs where large food retailers compete to serve this population. The same retailers are often reluctant to open up outlets in low-income neighborhoods because of a concern about lack of demand. In inner cities, low-income people have less to spend on nutritious food. In rural areas poverty will be complicated by low numbers of consumers. As a result, typical inner-city and rural food retailers tend to be associated with gas stations and convenience stores.

Distance from nutritional food retailers—as measured from a central point in a neighborhood to the nearest grocery store—is the main criterion for determining if a neighborhood qualifies as a food desert (**Figure 9.E**). For rural people, the calculation is 10 to 20 miles to the nearest supermarket (**Figure 9.F**). For urban residents, the standard appropriate distance to a supermarket is one mile (**Figure 9.G**). A highly regarded study found that people of low socioeconomic status—whether in urban or rural areas—ultimately spend up to 37 percent more on their food purchases, due to smaller weekly food budgets and poorly stocked neighborhood retail food outlets. A diet that relies on these types of retailers consists largely of processed foods high in calories, sugars, salt, fat, and artificial ingredients.

USDA Food Access Research Atlas

http://goo.gl/cgekQh

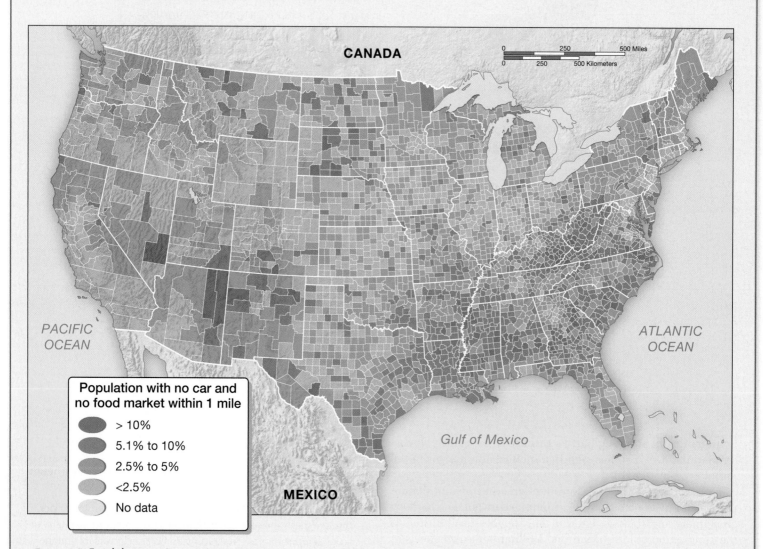

Population with no car and no food market within 1 mile

- > 10%
- 5.1% to 10%
- 2.5% to 5%
- <2.5%
- No data

▲ **Figure 9.E Food deserts** This map shows the distribution of food deserts across the United States measured by the absence of a household vehicle and the residence having no supermarket within a mile's distance.

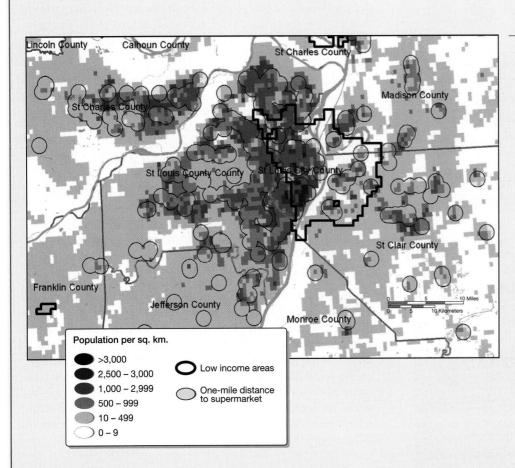

Population per sq. km.

- >3,000
- 2,500 – 3,000
- 1,000 – 2,999
- 500 – 999
- 10 – 499
- 0 – 9

- ⬭ Low income areas
- ◯ One-mile distance to supermarket

1. What is a food desert and where are they most likely to occur?

2. How do rural food deserts differ from urban food deserts? What are the differences and similarities in the populations how reside in them?

◀ **Figure 9.F Urban food desert, St Louis, Missouri**
This map provides a look at a food desert at the urban level and shows both density of the population—the darker, the more dense—and light circles where there are populations that are within one mile of a supermarket. Those areas outside the circle are also outside that one-mile radius and thus are further from a reasonably priced source of affordable, fresh food.

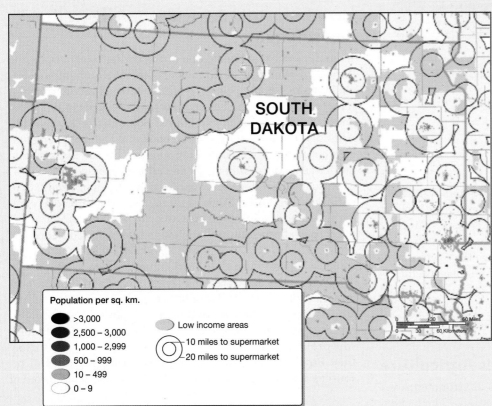

SOUTH DAKOTA

Population per sq. km.

- >3,000
- 2,500 – 3,000
- 1,000 – 2,999
- 500 – 999
- 10 – 499
- 0 – 9

- ⬭ Low income areas
- ◯ 10 miles to supermarket
- ◯ 20 miles to supermarket

◀ **Figure 9.G Rural food desert, South Dakota** In contrast to urban food deserts, rural food deserts are defined by the presence of a supermarket within 10 to 20 miles. Again, those areas outside the circles have no access to reasonably priced fresh food without having to travel up to 20 miles for it.

1/4 lb. of hamburger beef imported from Latin America requires the clearing of **6 sq. yd.** of rain forest and the destruction of **165 lbs.** of living matter, including **20 to 30** different plant species, **100** insect species, and **dozens** of bird, mammal, and reptile species.

Cattle degrade the land by stripping vegetation and compacting the earth. Each **1 lb.** of feedlot steak costs about **35 lbs.** of eroded topsoil.

Nearly **half** of the **total** amount of water used annually in the United States goes to grow feed and provide drinking water for cattle and other livestock. Cattle also contribute to a **significant percentage** of pollutants to U.S. rivers, lakes, streams and aquifers.

Grain-fed cattle are a **significant factor** in the generation of carbon dioxide, methane and nitrous oxide, the **three** major gasses responsible for global warming.

U.S. cattle production has caused a **significant** loss of biodiversity on both public and private lands. **More** plant species in the United States have been eliminated or threatened by livestock grazing than by any other cause.

▲ **Figure 9.27 Fast-food production's impact** This infographic shows the impact of beef production and consumption on the environment.

THE ENVIRONMENT AND AGRICULTURAL INDUSTRIALIZATION

Agriculture always involves the interaction of biophysical and human systems. This relationship makes agriculture distinct from forms of economic activity that do not depend so directly on the environment. This relationship also necessitates that communities determine how best to manage the environment to facilitate the continued production of food. Because the relationships between the human system of agriculture and the biophysical system of the environment are highly interactive, it is important to look at the ways they shape each other.

How the Environment Shapes Agriculture

Farmers have increasingly managed the environment over the course of the three agricultural revolutions. The current widespread use of fertilizers, irrigation systems, pesticides, herbicides, and industrial greenhouses even suggests that

agriculture has become an economic practice that can ignore the limitations of the physical environment. Because agriculture is an economic activity, its management of the environment in which it occurs becomes critical. As geographer Martin Parry writes:

> Soil, terrain, water, weather, and pests can be modified and many of the activities through the farming year, such as tillage and spraying, are directed toward this. But these activities must be cost-effective; the benefits of growing a particular crop, or increasing its yield by fertilizing, must exceed the costs of doing so. Often such practices are simply not economic, with the result that factors such as soil quality, terrain, and climate continue to affect agriculture by limiting the range of crops and animals that can profitably be farmed. In this way the physical environment still effectively limits the range of agricultural activities open to the farmer at each location.[1]

The effect of the environmenta on industrialized agricultural practices may not at first seem obvious, but it is important to appreciate that there are many. Examples include the availability of water, the quality of soil, air temperature, and length of the growing season, among other factors. Of course, on the other hand, there are also many readily observable contemporary and historical examples of the ways that agriculture destroys, depletes, or degrades the environmental resources on which its existence and profitability depend, as we discuss in the material that follows.

How Industrial Agriculture Affects the Environment

As mentioned previously, one of the earliest treatises on the impact of chemical pesticides on the environment was Rachel Carson's *Silent Spring*, which identified the detrimental impacts of synthetic chemical pesticides—especially DDT—on the health of human and animal populations (**Figure 9.28**). Although the publication of the book and the environmental awareness that it generated led to a ban on the use of many pesticides in most industrialized nations, chemical companies continued to produce and market these types of products in peripheral countries. Some of these pesticides, aimed at combating malaria and other insect-borne diseases, are applied to crops that are sold in the markets of developed countries. Thus, a kind of "circle of poison" has been set into motion, encompassing the entire global agricultural system.

Among the most pressing issues facing agricultural producers today are soil degradation and denudation, which are occurring at rates more than a thousand times the natural rates. Although we in the United States tend to dismiss soil problems, such as erosion, as an artifact of the 1930s' Dust Bowl, the effects of agriculture on worldwide soil resources are dramatic, as **Table 9.3** illustrates. Most forms of

[1] M. Parry, "Agriculture as a Resource System," in I. Bowler (ed.), *The Geography of Agriculture in Development Market Economics.* Harlow, England: Longman Scientific and Technical, 1992, p. 208.

The quantity and quality of soil worldwide are important determining factors for the quantity and quality of food that can be produced (**Figure 9.29**). The loss of topsoil worldwide is a critical problem because topsoil is a fixed resource that cannot be readily replaced. It takes, on average, 100 to 500 years to generate 10 millimeters (1/2 inch) of topsoil, and it is estimated that nearly 50,000 million metric tons (55,000 million tons) of topsoil are lost each year to erosion.

The nature–society relationship previously discussed in Chapter 4 is very much at the heart of agricultural practices. Yet as agriculture has industrialized, its impacts on the environment have multiplied and in some parts of the globe are at crisis stage. In some regions the agricultural system leads to overproduction of foodstuffs, but in others the quantity and quality of water and soil severely limit the ability of a region's people to feed themselves.

Union of Concerned Scientists: Industrial Agriculture

http://goo.gl/ayDezu

▲ Figure 9.28 **Impact of pesticides on pollinators** Pollination is the transfer of pollen from one flower to another and is critical to fruit and seed production. Pollination is frequently provided by insects seeking nectar, pollen, or other floral rewards. Some scientists believe that both wild and managed pollinators are disappearing at alarming rates as a result pesticide poisoning, as well as habitat loss, diseases, and pests.

agriculture tend to increase soil degradation. Although severe problems of soil degradation persist in the United States—which has a federal agency devoted exclusively to managing soil conservation—more severe problems are occurring in peripheral countries.

The problem of soil degradation and loss is particularly critical in the humid tropical areas of the globe—especially in South America and Asia. Arguably the most critical hotspot with respect to soil and related environmental impacts is the moist Brazilian *cerrado*, or grassland, which is being converted to soybean production to feed the growing biofuels industry (which we discuss in more detail in the next section). In addition to soil erosion and degradation, the conversion of the *cerrado* is also threatening the species richness of birds, fishes, reptiles, amphibians, and insects. Some local animal and plant species face extinction.

APPLY YOUR KNOWLEDGE

1. What is soil degradation and why is it such a pressing issue for agriculture?

2. Choose three common commercial crops (e.g., coffee, cotton, citrus, lettuce, potatoes, or peanuts) that are grown in the United States. Identify the areas in the country where these three crops are grown and consult soil, temperature, and rainfall maps to learn what the local environmental conditions are that enable these crops to be grown in those areas. Next, identify what effect crop cultivation has on that same environment. Which of the crops has the most negative impact on the local environment? Which has the least?

EMERGING CHALLENGES AND OPPORTUNITIES IN THE GLOBAL FOOD SYSTEM

In this final section, we examine two problematic issues in the world food system. These cases certainly do not illuminate the

TABLE 9.3 Global Soil Degradation

Region	Degrading area (km²)	mi²	% Territory	% Global degrading area	% Total population	Affected people
Asia	9,128,498	3,524,533	25.39	0.890	23.58	1,070,737,071
Africa	6,596,641	2,546,977	26.65	0.441	29.31	230,604,253
Europe	656,007	253,286	12.88	0.089	9.75	48,457,913
South America	4,719,162	1,822,079	24.86	0.457	22.90	149,905,245
Oceania	2,364,959	913,116	34.68	1.224	26.29	5,494,554
North America	3,968,971	1,532,428	20.25	9.755	14.24	36,654,152
World Total (land, excluding inland water body)	35,058,104	1,35,360,010	23.54	100.000	23.89	1,537,679,148

Source: Adapted from Z. G. Bai, D. L. Dent, L. Olsson, and M. E. Schaepman, *Global Assessment of Land Degradation and Improvement 1: Identification by Remote Sensing.* Report 2008/01, FAO/ISRIC, Rome and Wageningen, 2008, p. 24, Table 1.

▲ **Figure 9.29** **Desertification, Gansu, China** A woman walks along the leading edge of the Kumtag Desert as it threatens to engulf her onion farm. The sand is advancing at the rate of up to 4 meters (13 feet) a year.

myriad challenges and possibilities facing food producers and policymakers today, but they do provide a sense of the broad range of issues.

Food and Hunger

We have spent most of this chapter describing the ways food is cultivated, processed, engineered, marketed, financed, and consumed throughout the world. What we have yet to do is talk about access to this most essential of resources. Although there is more than enough food to feed all the people who inhabit Earth, access to food is uneven, and many millions of individuals in both the core and the periphery have had their lives shortened or harmed because war, poverty, or natural disaster has prevented them from securing adequate nutrition. In fact, hunger is very likely the most pressing problem facing the world today though the good news is that it has been dropping fairly steadily since the early 2000s. These positive signs of hunger abatement mean the global community must continue their efforts toward a hunger-free world. (**Figure 9.30**).

Nutrition Issues

Hunger can be chronic or acute. Chronic hunger is nutritional deprivation that occurs over a sustained period of time: months or even years. Acute hunger is short-term and is often related to catastrophic events—personal or systemic. Chronic hunger, also known as **undernutrition**, is the inadequate intake of one or more nutrients and/or of calories. Undernutrition can occur in individuals of all ages, but its effects on children are dramatic, leading to stunted growth, inadequate brain development, and a host of other serious physical ailments.

Globally, 24,000 people a day die from the complications brought about by undernutrition. Children are the most frequent victims. Children who are poorly nourished suffer up to 160 days of illness each year, and poor nutrition plays a role in at least half of the 10.9 million child deaths each year. Moreover, undernutrition magnifies the effect of every disease, including measles and malaria. Malnutrition can also be caused by diseases, such as the diseases that cause diarrhea, by reducing the body's ability to convert food into usable nutrients. **Malnutrition** is the condition that develops when the body does not get the right amount of the vitamins, minerals, and other nutrients it needs to maintain healthy tissues and organ function. A person with malnutrition can be either under- or overnourished.

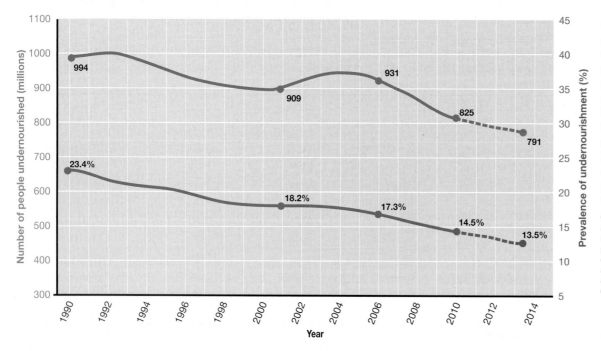

◀ **Figure 9.30** **Number of hungry people worldwide** Almost one in seven people worldwide is chronically undernourished. After steadily declining, global hunger began to rise again in the 1990s. The number of hungry people peaked in 2009 at 1 billion and then dropped in 2010 and is expected to continue to drop in the period 2012-2014.

(*Source:* Adapted from *Growing a Better Future: Food Justice in a Resource-Constrained World*, 2011, with the permission of Oxfam GB, www.oxfam.org.uk, FAO, http://www.fao.org/hunger, and Financial Times, http://cachef.ft.com/cms/s/0/68b31de6-329e-11e)-97ca-00144feabdc),s01=2.html.)

Famine, Food Security and Food Safety

Perhaps the most widely publicized examples of acute hunger are famines, especially those that have occurred in parts of the periphery over the last few decades. **Famine** is acute starvation associated with a sharp increase in mortality. The most widely publicized famines of the late twentieth century occurred in Bangladesh in 1974 and Ethiopia in 1984–1985. The causes of these two famines (and the other twentieth-century famines that preceded them) were complex. The crisis of starving people so often publicized by the news media is usually just the final stage of a process that has been unfolding for a far longer period, sometimes years or even decades. Experts who study famine argue that there are at least two critical factors behind long-standing vulnerability to famine. The first has to do with a population's command over food resources in terms of their livelihood. The second has to do with a trigger mechanism, which may be a natural phenomenon, like drought, or a human-made situation, such as civil war.

People who study famine and other forms of hunger have come to conceptualize nutritional vulnerability in terms of the notion of food security. **Food security** for a person, a household, or even a country is assured access to enough food at all times to ensure active and healthy lives. And while famine is a dramatic reminder of the precarious nature of food security, it is important to appreciate that chronic hunger resulting from food insecurity is a far more widespread and devastating problem than famine, which tends to be shorter in duration and more contained geographically.

At the same time, hunger is a serious problem for peripheral populations worldwide (see Box 9.4 Geography Matters: Hunger and Poverty in the United States). For both core and peripheral populations, issues of food safety have also grown in significance. Over the last several years, "food scares" caused by tainted foodstuffs have been routinely in the headlines: California spinach tainted with *E. coli* bacteria, *salmonella* in canned tomatoes from Mexico and in hummus in the United Kingdom, toxic chemicals routinely added to pickled vegetables in China, *E. coli* tainted cucumbers in Germany, and potentially cancer-causing dioxins (which entered grazing pastures through illegally dumped toxic waste) in buffalo milk used in the production of Italian mozzarella.

APPLY YOUR KNOWLEDGE

1. According to figures from *Food First* there is enough wheat, rice, and other grains produced to provide every human being on the planet with 3,500 calories a day. Given this, why are there more than 900 million hungry people on the planet?

2. Why are African American, Hispanic, and older households most likely to experience food insecurity in the United States?

The impact of food scares is widespread, affecting not only those who have become ill or died but also farmers and grocers as well as the credibility of government agencies meant to protect consumers from tainted and toxic foods. Mostly, the recent apparent escalation in food scares reminds consumers of the vulnerability of the global food system.

Land Grabs

One of the factors thought to be adding to global hunger and food security issues is the increasing amount of cropland around the globe now being redirected to raising **biofuels**, renewable fuels derived from biological materials that can be regenerated. The practice is known as a land grab—large-scale land acquisition in developing countries by domestic and foreign companies, governments, and individuals. Land grabs occur to enable both biofuel as well as food production.

Land grabs for biofuels are happening across Asia, Latin America, and Africa and often involve violence. For example, over the last several years in Tanzania, the government has handed over 9,000 hectares of land on which over 11,000 people depend for their livelihoods to the U.K. firm Sun Biofuels PLC for a jatropha (a shrubby plant that produces seeds that are 40 percent oil) plantation (**Figure 9.31**). A recent report by the World Bank states that biofuel production has forced global food prices up by 75 percent. It is our view that the price of food is a complex variable affected by causes that include the emergence and growth of biofuel production but is also related to rising demand among increasingly more affluent populations, commodities speculation, rapidly escalating energy prices, and poor harvests.

But it's not just for biofuels that land is being grabbed by investors from some of the world's poorest countries. Since 2000, at least 31 million hectares (77 million acres) of land has been acquired by foreign investors who wish to improve their own food security back home. One way that researchers have been attempting to characterize the magnitude of the problem is to calculate how much food is being lost to the local population when crops are grown on land that may

▼ **Figure 9.31 Jatropha plantation** Workers are shown here in a field of jatropha plants, a source of bio-diesel, at the International Crops Research Institute for the Semi-Arid Tropics, in Hyderabad, India. Bio-diesel contains no petroleum, but it can be blended with petroleum diesel to create a bio-diesel blend or can be used in its pure form for fuel.

9.4 Geography Matters

Hunger and Poverty in the United States

By Nik Heynen, University of Georgia

Perhaps you grew up hearing "clear your plate—there are children in the world who do not have enough to eat!" Now that you are older, do you ever think about this while eating dinner? Have there been times over the last year when you did not have enough money to feed yourself? According to the United States Department of Agriculture (USDA), one in six residents in the United States (48.9 million people) battled against hunger in 2012.

As geographers, we can think about hunger in many different ways.

From a demographic perspective Chapter 3, we can see that nearly 100 million people living in the United States existed on less than 185 percent of the U.S. federal poverty level in 2012. Because of their low income, these individuals qualified for almost all federal programs that provide food and nutrition to the hungry. Children in the United States often suffer the most, as a percentage of the population from hunger—and the United States has the most powerful economy on the planet. In 2012, approximately 16.3 million children in the United States (nearly a quarter of all U.S. children) were living in poverty and nearly 31 million U.S. children (42%) lived in households far enough below the poverty level to qualify for federal food programs.

As geographers we also try to understand the way these numbers map across the United States. **Figure 9.H** shows responses to the question "Have there been times in the past twelve months when you did not have enough money to buy food that you or your family needed?" As this map illustrates, the state of Mississippi has the highest rate of hunger across the United States with almost a quarter of all households struggling against food hardship. It also shows that the state with the lowest rate, North Dakota, still had one in ten households reporting food hardship.

Geographers also research how groups have organized against hunger and poverty within the United States. One example is how the Black Panther Party (BPP) contributed to feeding hungry children across the United States through their Free Breakfast Program for children (**Figure 9.I**). The BPP's first Free Breakfast Program for children was initiated at St. Augustine's Church in Oakland, California in September 1968, and by the end of 1969 the BPP set up kitchens in many other U.S. cities. At the breakfast program's peak, BPP members and other volunteers were feeding approximately 250,000 children a day before they went to school across the country. The BPP's program was also an impetus for the currently federally funded school breakfast programs in existence within the United States today. California's school breakfast programs resulted from the pressure the BPP's breakfast program put on then-Governor Ronald Reagan. His administration eventually spearheaded the implementation of the State's development of a school breakfast program.

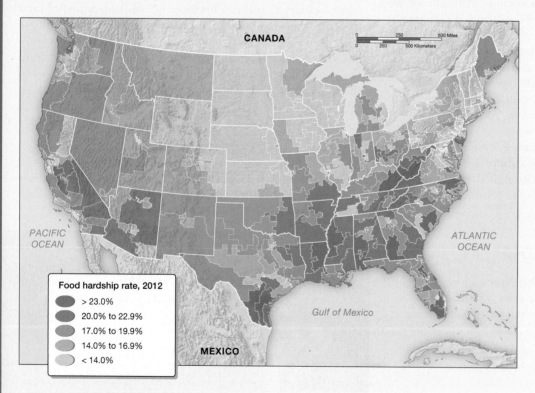

Food hardship rate, 2012
- > 23.0%
- 20.0% to 22.9%
- 17.0% to 19.9%
- 14.0% to 16.9%
- < 14.0%

▲ Figure 9.H Food hardship This map shows responses to the question "Have there been times in the past twelve months when you did not have enough money to buy food that you or your family needed?"

As the history of the BPP demonstrates, there are many ways that organized citizens can contribute to combating poverty and hunger. Within this context, it is important to understand the larger structural issues that produce poverty and allow hunger to persist. If we really want to bring an end to hunger in the United States, we have to tackle the uneven development of economic opportunity and social power relations that allow inequality to persist. Foremost in this effort would be to think about the distribution of wealth and the unequal access to employment opportunities. Poverty and hunger could certainly be erased if more citizens worked toward these goals.

1. According to figures from Food First enough wheat, rice, and other grains are produced to provide every human being on the planet with 3,500 calories a day. Given these statistics, why are there more than 900 million hungry people on the planet (http://foodfirst.org/)?

2. Using Figure 9.H, describe five spatial relationships related to food hardship within the U.S.

See: Food Research and Action Center's web site: http://frac.org/

▲ **Figure 9.I Black Panthers' Free Breakfast Program for children** Black Panther Bill Whitfield provides free breakfast to children in Kansas City in 1969.

have been taken without their consent and is then exported. The study added up every known land grab larger than 200 hectares between 2000 and 2013 and found that grabbed land could support 300–550 million people through industrial production or 190–370 million people otherwise. The most chilling aspect of the land grab phenomenon is that land is being taken in countries that already have extremely high levels of malnutrition. Couple this with the fact that there is often no national policy in place to prevent investors from exporting their crops and the situation is especially alarming and could become disastrous (**Figure 9.32**).

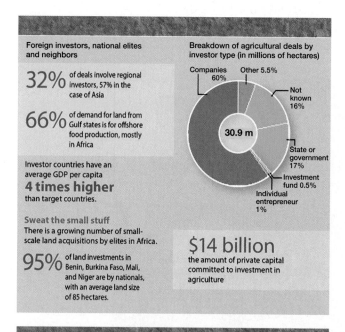

Foreign investors, national elites and neighbors

32% of deals involve regional investors, 57% in the case of Asia

66% of demand for land from Gulf states is for offshore food production, mostly in Africa

Investor countries have an average GDP per capita
4 times higher than target countries.

Sweat the small stuff
There is a growing number of small-scale land acquisitions by elites in Africa.

95% of land investments in Benin, Burkina Faso, Mali, and Niger are by nationals, with an average land size of 85 hectares.

Breakdown of agricultural deals by investor type (in millions of hectares)

Companies 60%   Other 5.5%
Not known 16%
30.9 m
State or government 17%
Investment fund 0.5%
Individual entrepreneur 1%

$14 billion
the amount of private capital committed to investment in agriculture

Big land deals jeopardize food security.

2 billion people depend on samllholder farms for their food and livelihood

Two thirds
of investments in agricultural land by foreign investors (2000-2010) were in countries with serious hunger problems

of foreign investors' crops are destined for export

75% proportion of Liberia's land allocated or promised to large investors in 2012

24% proportion of Liberian children who are malnourished

Empty land myth

The World Bank says Africa is home to more than half the world's available yet 'unused' land. But investors often seek occupied, fertile land, close to infrastructure.

Of agricultural deals since 2000:

50%+ are for land less than a day trip away from the nearest city

20% are in densely populated areas

46% are on land formerly used for smallholder agriculture

Land targeted for deals, by surface area

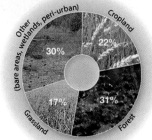

Other (bare areas, wetlands, peri-urban) 30%
Cropland 22%
Forest 31%
Grassland 17%

Addressing Hunger

While it is important to understand the compound causes and repercussions of hunger and undernutrition, perhaps the most relevant point to take away from this discussion is that it is a problem that can be solved. Neither short-term nor long-term hunger is due to inadequate supply, so the solution must lie in improving access to and distribution of supplies. This could occur, under a radical scenario, through a massive redistribution that would give all the people access to the same amount of food resources. Barring such a dramatic restructuring of the current economic system, the solution to the problem of hunger lies in improving access to livelihoods that pay well enough that adequate nutrition becomes a human right and is not dependent upon the vagaries of economic or natural systems.

APPLY YOUR KNOWLEDGE

1. What is "land grabbing"? What is the difference between land grabbing and imperialism, a concept you have studied throughout the text?

2. Search the Internet for news stories on recent protests against GMO food. What are the protesters' concerns and how do they go about pressing their demands? Write a paragraph stating whether or not you believe their concerns are justified. Provide scientific evidence to support your argument.

FUTURE GEOGRAPHIES

There are many factors that will affect the future of global agriculture, including climate change, adequate and safe water, and how to achieve a more just distribution of nutritious food. But an increasing chorus of scientists, policymakers, governments and farmers are moving to imagine a different future for farming that recognizes the strengths of both industrialized agriculture and more sustainable, organic methods. Such a future would provide greater support than is currently available to organic farmers to improve their yields. As the study contends, in the future knowledge will be key to producing adequate and healthy food supplies:

Conventional farming requires knowledge of how to manage what farmers know as inputs—synthetic fertilizer, chemical pesticides, and the like—as well as fields laid out precisely via global positioning systems. Organic farmers, on the other hand, must learn to manage an entire ecosystem geared to producing food—controlling pests through biological means, using the waste from animals to fertilize fields, and even growing one crop amidst another.

Better science and technological support to organic farmers will enable healthier soils, which will be critical to food production in the face of climate change. But a global population that

◀ **Figure 9.32 Land Grabbing and the effect on hunger** This graphic shows the strong relationship between the sale or lease of land in poor parts of the globe and rich countries who are acquiring the land. While it seems as if poor countries have plenty of land to spare, most of the land sold or leased is highly arable or close to infrastructure, such as roads or rail connections.

will continue to increase through 2050 will require more food and conventional methods aided by Green Revolution techniques, biotechnology, and genetically modified crops (that are able to endure drought conditions) may also be critical to producing an adequate yield. The point is that a "one size fits all solution" is not a likely one. Instead a hybrid model will allow adaptation to different local conditions and build on the strengths—and hopefully mitigate the weaknesses—of all of them.

The future of the global food system is being shaped at this very moment in food science laboratories, in corporate boardrooms, on the street in organized protests, and in settlements throughout the world. It seems poised to continue to produce vast quantities of grain, enough to feed the world. But feeding the world will also require the recognition that not all people have the resources to afford the increasing cost of food on the world market. Issues of redistribution of food also need to be confronted. Food revolutions such as biotechnology are likely to continue to expand, but the results of new approaches are

not likely to be equally affordable across the socioeconomic spectrum. Therefore, social revolutions in how food is made accessible to all will also need to be considered.

APPLY YOUR KNOWLEDGE:

1. What are the new approaches to food production suggested by scientists, politicians, and farmers, working together?

2. What do you think the major differences are between conventional farming and organic farming and how do you think they affect food supply? Think about the processes and technologies on the growing side, as well as the final products. Do an Internet search on food production near you—are there more organic or conventional farms? Are there any hybrid models? What do you think is the most effective way to feed our planet?

 # CONCLUSION

Agriculture has become a highly complex, globally integrated system. While traditional forms of agricultural practices, such as subsistence farming, continue to exist, they are overshadowed by the global industrialization of agriculture. This industrialization includes not only mechanization and chemical applications but also the linking of the agricultural sector to the manufacturing, service, and finance sectors of the economy. In addition, states have become important players in the regulation and support of agriculture at all levels, from the local to the global.

The dramatic changes that have occurred in agriculture affect different places and different social groups. Households in both the core and the periphery have strained to adjust to these changes, often disrupting existing patterns of authority and access to resources. Just as people have been affected by

the transformations in global agriculture, so have the land, air, and water.

The geography of agriculture today is a far cry from what it was 100 or even 50 years ago. As the globalization of the economy has accelerated in the last few decades, so has the globalization of agriculture. The changes in global agriculture do not necessarily mean increased prosperity in the core, nor are the implications of these changes simple. For example, the production of oranges in Florida is directly influenced by the newer Brazilian orange industry. Both industries affect the prices of oranges in the marketplaces of Europe and Asia. Additionally, other forces, such as social reactions to genetically engineered foods, agricultural research, trade concerns, and a host of other factors, have repercussions throughout the world food system.

LEARNING OUTCOMES REVISITED

- **Compare and contrast traditional agriculture practices across the globe.**

  Although traditional agricultural practices, including subsistence farming, shifting cultivation, and pastoralism, no longer dominate agricultural practices on a global scale, they are still engaged in throughout the world, in some cases alongside more mechanized forms. Globalization, in addition to restructuring entire national farming systems, has also transformed farming households in the core, periphery, and semiperiphery. Thus, traditional forms of agriculture are waning as more and more places are irresistibly drawn into a globalized economy supported by a strong commercial agricultural sector.

- **Describe the three revolutionary phases of agricultural development, from the domestication of plants and animals to the latest developments in biotechnology and industrial innovation.**

The three revolutionary phases have not occurred simultaneously throughout the globe, but have been adopted and adapted to differing degrees, based on levels of development, culture, and physical geography. The first phase involved the domestication of seeds and animals. The second revolved around the improvement of outputs and innovations for making farming more efficient, such as an improved yoke for oxen. The third was based on inputs to production, such as fertilizers and field drainage systems.

- **Analyze the ways that the forces, institutions, and organizational forms of globalization have transformed agriculture.**

  Two of the most important forces behind agricultural transformation are multinational and transnational corporations and states. Institutions like the World Trade Organization as well as regional associations like the European Union have also been important influences. And the organization of agriculture itself has experienced significant changes as it has moved from a family-oriented business model to a corporate undertaking that stretches across national boundaries.

■ **Examine the organization of the agro-commodity system from the farm to the retail outlet, including the different economic sectors and corporate forms.**

The farm is no longer the central piece in the chain of agricultural organization, but one of several important components that include seed and fertilizer manufacturers, food processors, food distributors, and consumers. The organizational structure of agriculture is composed of five central and connected sectors (inputs, production, processing, distribution, and consumption) with four contextual elements acting as external mediating forces (the state, international trade, the physical environment, and credit and finance).

■ **Understand the ways that agriculture has transformed the environment, including soil erosion, desertification, deforestation, soil and water pollution, and plant and animal species degradation.**

While most of the core countries have instituted legislation to address some of the environmental problems associated with agriculture, these problems exist throughout the global system to greater and lesser degrees. As agriculture has industrialized, its impacts on the environment have multiplied and spread so that some parts of the globe are at crisis stage. In some regions, the agricultural system has led to overproduction of foodstuffs, but in others the quantity and quality of food production is severely limited by physical constraints and environmental degradation. The challenge of the twenty-first century is to work toward a more sustainable relationship between humans and the environment, especially with respect to food production.

■ **Probe the current issues food-policy experts, national governments, consumers, and agriculturalists face with respect to the availability and quality of food as well as the alternative practices that are emerging to address some of these issues in a world where access to safe, healthy, and nutritious foodstuffs is unevenly distributed.**

Genetic modification is one way of improving productivity, though it does not address issues of access to food. Opportunities for the world's poor—who are increasingly residing in urban settings—to grow their own food is another way. Still another way to improve food availability and quality is to recognize access to food as a human right and work toward more even distribution. Other responses include promoting more sustainable farming practices and supporting small farmers' efforts to produce efficiently and effectively.

KEY TERMS

agrarian (p. *301*)

agribusiness (p. *320*)

agricultural industrialization (p. *310*)

agriculture (p. *300*)

aquaculture (p. *311*)

biofuels (p. *335*)

biopharming (p. *316*)

Biorevolution (p. *314*)

biotechnology (p. *314*)

Borlaug hypothesis (p. *313*)

chemical farming (p. *309*)

commercial agriculture (p. *301*)

contract farming (p. *314*)

conventional farming (p. *322*)

crop rotation (p. *301*)

double cropping (p. *305*)

famine (p. *335*)

fast food (p. *325*)

food desert (p. *330*)

food justice (p. *325*)

food manufacturing (p. *310*)

food miles (p. *324*)

food regime (p. *322*)

food security (p. *335*)

food shed (p. *324*)

food sovereignty (p. *324*)

food supply chain (p. *321*)

globalized agriculture (p. *318*)

GMO (genetically modified organism) (p. *314*)

Green Revolution (p. *311*)

hunting and gathering (p. *301*)

intensive subsistence agriculture (p. *304*)

intertillage (p. *304*)

local food (p. *324*)

malnutrition (p. *334*)

mechanization (p. *308*)

nontraditional agricultural exports (NTAEs) (p. *314*)

organic farming (p. *322*)

pastoralism (p. *306*)

shifting cultivation (p. *301*)

slash-and-burn (p. *302*)

subsistence agriculture (p. *301*)

swidden (p. *302*)

transhumance (p. *306*)

undernutrition (p. *334*)

urban agriculture (p. *324*)

REVIEW & DISCUSSION

1. Search the term "Green Revolution" on the Internet. From what you find, create a time line of the Green Revolution identifying its most significant milestones as well as the countries it has most greatly affected. Make a list of the pros and cons of the Green Revolution and decide where you stand with respect to the controversy and why.

2. Visit your local supermarket and identify seven different types of produce. Determine which company produced them and where and how they were grown. You may need to ask the produce manager where they come from but the tiny labels (with the PLU codes) will tell you whether they are commercially grown, organic, or GMO crops. Once you have determined the source of the produce, calculate how far it traveled (food miles) to reach your town.

3. Explore further the impact of the environment on agriculture by researching the projections that are being made

about climate change and agricultural change. How will climate change affect the traditional areas of the globe where staples (e.g., rice, wheat, potatoes) are grown? How will it affect the areas where more popular fruits (e.g., oranges, apples, and wine grapes) are grown? Speculate on what you think the political and economic effects of the changing geography of global agriculture will be.

UNPLUGGED

1. The Food and Agriculture Organization (FAO) has published a range of yearbooks containing statistical data on many aspects of global food production since the mid-1950s. Using these *State of Food and Agricultural Production* yearbooks, compare the changes that have occurred in agricultural production between the core and the periphery since the middle of the last century. You can use just two yearbooks for this exercise, or you may want to use several to get a better sense of when and where the most significant changes have occurred. Once you have identified where the changes have been most significant, craft an explanation as to why these changes may have occurred.

2. The USDA also provides statistics on food and agricultural production, though, of course, data is limited to the United States. Contained in volumes called *Agricultural Statistics* are a range of important variables, from what is being grown and where, to who is working on farms, and what kinds of subsidies the government is providing. Drawing on the USDA's *Agricultural Statistics,* examine the changing patterns of federal subsidies. Using a map of the United States, show which states have received subsidies from the 1940s (immediately following the Great Depression) to the present day. Have subsidies increased for some parts of the country and not others? Have subsidies increased or decreased overall for the entire country? Which farm sectors and, therefore, which regions have most heavily benefited from federal agricultural subsidies? Why? Summarize and explain your findings.

3. Odds are your usual breakfast is the result of the activities of a whole chain of producers, processors, distributors, and retailers whose interactions provide insights into both the globalization of food production and the industrialization of agriculture. Consider the various foods you consume in your typical breakfast and describe not only where (and by whom) they were produced (grown and processed) and how they were transported (by whom) from the processing site but also where and by whom they were retailed. Summarize how the various components of your meal illustrate both the globalization and the industrialization of agriculture.

DATA ANALYSIS

In this chapter we have looked at a central component of human-environmental interactions: the geography of food and agriculture, from the global to the household and individual level. In looking at this basic aspect of life—producing and consuming food—the issue of space, economy, and politics play a huge role as seen in the debates over the Green Revolution, the Biorevolution, food sovereignty, anti-GMO resistance movements and the concept of "food deserts." To look closer at how and where we produce food, watch the story of Ron Finley, a "guerilla gardener" in South Central Los Angeles and answer these questions.]

LA Guerilla Gardeners

http://goo.gl/RgNYY0

1. What does Ron Finley say about fast food *versus* drive-by-shootings in his communities?

2. Why is "food the problem, and food is the solution"?

3. Where in the city does Finley plant his gardens?

4. What is Los Angeles Green Grounds and how do they work?

5. How is gardening like art? How does Finley talk about soil?

6. How does guerilla gardening change a community? How are children a vital component of this process?

7. What does Finley say about flipping the script and making gardening "gangster"?

8. Do an Internet search on "guerrilla gardeners." What other cities have guerrilla gardener groups? Does your city? Would you consider starting a guerilla garden?

MasteringGeography™

LEARNING OUTCOMES

- *Describe* the geopolitical model of the state and explain how it links geography and state practices with respect to the key issues of power and territory.

- *Compare* and contrast the ways that different contemporary theorists—from Deleuze to Althusser—approach the state as a political and geographical entity.

- *Interpret* how imperialism, colonialism, heartland theory, domino theory, the end of the Cold War, and the emergence of the new world order are key examples of ways geography has influenced politics and politics has influenced geography.

- *Demonstrate* how the growth and proliferation of international and supranational organizations created the foundation for the emergence of global forms of governance.

- *Recognize* how events of international political significance are often the result of East/West and North/South divisions, whereas national and local political issues emerge out of tensions related to colonialism, regionalism, and sectionalism.

- *Describe* the difference between the politics of geography and the geography of politics.

342

▲ Palestinians celebrating their bid for United Nations statehood recognition. A photo of Palestinian Authority President Mahmoud Abbas is prominently displayed.

POLITICAL GEOGRAPHIES

10

Imagine what it would feel like if you had no nationality, no attachment to a place through citizenship. How would your world change? At first, it would seem like a minor inconvenience. You could still buy your morning coffee, surf the Internet, hang out with friends. Before long, however, you'd come to realize that having citizenship status is your access to a whole set of rights. Without citizenship you'd live in an identity limbo unable to register for a new semester, get a job, drive legally, travel by air or open a checking account, just to name the most obvious limits.

Nearly 12 million people around the world are stateless, lacking a legal attachment to the country in which they dwell or to any other nation. For example, more than 100,000 Bedoon in Kuwait live without a nationality despite the fact that they constitute ten percent of the country's population. In Kuwait, the term *Bedoon* means "without a state." Some Bedoon were originally nomadic people who lived and prospered in the deserts for centuries but have since been settled into villages and cities. The government is concerned that such people don't legally belong in Kuwait and should be returned to Iran or Saudi Arabia or elsewhere. Whatever the case, Kuwait's Bedoon live there without any legal rights despite the fact that they made up over 80 percent of the Kuwaiti Army during the 1990s, the period in which Kuwait was at war with Iraq.

The 1954 UN Convention defines statelessness as "a person who is not considered a national by any state under the operation of its laws." Unfortunately, few signatories to the convention ever formulated policies to enable stateless people within their territories to establish legal nationality. This lack of legislative action—mostly deliberate—leaves millions without rights.

Other stateless populations include the four million Palestinians living in Israel and the Occupied Territories (discussed in more detail later in this chapter) who technically have no citizenship. There are 90,000 Saharawis, inhabitants of Western Sahara, who live as exiles in Algeria, where they have no legal status. Kurds are another widely known as a population that lacks their own state, though their case is an encouraging one as it seems that Kurdistan, "the land of the Kurds" a geo-cultural region where Kurdish people form a prominent minority, will be able to establish an independent nation state in the near future. The Roma are a European stateless population of between roughly 70,000–80,000 who live or have lived in Czechia, Yugoslavia, Bosnia and Herzegovina, Montenegro, Serbia, Slovenia, Macedonia, and Italy. They originated in India and were brought by Muslims as slaves between the 6th and 11th centuries. As people without citizenship, the Roma (also known pejoratively as Gypsies) are discriminated against with the average Roma far behind a European counterpart on social indicators like level of education and quality of employment and housing. The life expectancy of the Roma people is ten years lower than European Union citizens.

One thing that countries can do to improve the lives of the millions of stateless people globally is establish laws that provide a pathway to citizenship or recognize the appeal for independence by those people who occupy their own territory such as the Kurds and the Palestinians. In 2013, the United Kingdom passed a law that would allow stateless people within its borders to become citizens. Yet in Kuwait, the fate of the Bedoon still remains uncertain, despite years of agitation and a parliament that, though it passed an act in 2012 that would allow the naturalization of 4,000 stateless people, has still not enforced it.

THE DEVELOPMENT OF POLITICAL GEOGRAPHY

This chapter explores how new maps and geopolitical arrangements continue to emerge around the globe at the same time that established boundaries persist. Exploration, imperialism, colonization, decolonization, and the Cold War between East and West are powerful forces that have created and transformed national boundaries. Much of the political strife that currently grips the globe involves local or regional responses to the impacts of globalization of the economy. Political geography is not just about global or international relationships. It is also about the many other geographic and political divisions that stretch from the globe to the neighborhood and to the individual body.

Political geography is a long-established subfield of geography. The ancient Greek philosopher Aristotle is often considered the first political geographer. His model of the state is based upon factors such as climate, terrain, and the relationship between population and territory. Other important political geographers have, over the years, promoted theories of the state that incorporate elements of the landscape and the physical environment as well as the population characteristics of regions. They believed that states consolidated and fragmented based on complex relationships among and between factors such as population size and composition, agricultural productivity, land area, and the role of the city.

The factors were deemed important to state growth and change. Why these factors were identified as central

undoubtedly had much to do with the widespread influence of Charles Darwin on intellectual and social life. His theory of competition inspired political geographers to conceptualize the state as a kind of biological organism that grew and contracted in response to external factors and forces. It was also during the late nineteenth century that foreign policy as a focus of state activity began to be studied. This new field came to be called *geopolitics*.

The Geopolitical Model of the State

Geopolitics is the state's power to control space or territory and shape international political relations. Within the discipline of geography, geopolitical theory originated with Friedrich Ratzel (1844–1904). Ratzel was greatly influenced by the theories embodied in *social Darwinism* that emerged in the mid-nineteenth century. His model portrays the state as behaving like a biological organism, with its growth and change seen as natural and inevitable. Ratzel's views continue to influence theorizing about the state today through the conviction that geopolitics stems from the interactions of power and territory.

Although it has evolved since Ratzel first introduced the concept, geopolitical theory has become one of the cornerstones of contemporary political geography and state foreign policy more generally. And, although the organic view of the state has been abandoned, the twin features of power and territory still lie at the heart of political geography.

Figure 10.1 illustrates Ratzel's conceptualization of the interaction of power and territory through the changing face of Europe from the end of World War I to the present. The transformations in the maps reflect the unstable relationship between power and territory, especially some states' failure to achieve stability. The most recent map of Europe portrays the precarious nation-state boundaries in the post–Cold War period. Estonia, Latvia, and Lithuania have regained their sovereign status. Czechoslovakia has dissolved into Czechia and Slovakia. Yugoslavia has dissolved into seven states, but not without much civil strife and loss of life. The former Soviet Union is now the Commonwealth of Independent States, with Russia the largest and most powerful. And still the map is not fully settled.

The tensions within the former Soviet Union, for instance, are especially high as new republics such as Ukraine voted to move politically closer to Europe and the United States and further away from Russia. But in February of 2014, a series of violent events consolidated into the "Ukrainian Revolution" and resulted in Ukraine's then president fleeing the country for Russia. Soon after the revolution, Russia invaded and seized control of the Crimean Peninsula and has declared it part of the Russian territory (**Figure 10.2**). Thus the map of Europe continues to shift even today.

Boundaries and Frontiers

Boundaries enable territoriality to be defined and enforced and allow conflict and competition to be managed and channeled. The creation of boundaries is, therefore, an important element

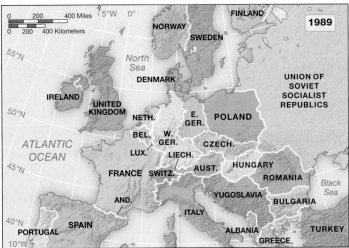

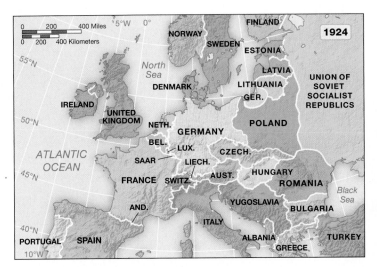

East Germany to 1990		Independent, 2008	
Independent, 1991		Former Soviet satellite states	
Independent, 1992		U.S.S.R. to 1991	
Independent, 1993		Yugoslavia to 1991	
Independent, 2006		Czechoslovakia to 1991	

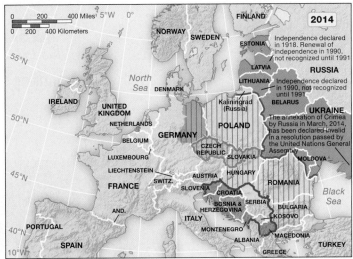

▲ **Figure 10.1 The changing map of Europe: 1924, 1989, 2014** The boundaries of the European states have undergone dramatic changes since World War I. The changing map of Europe illustrates the instability of international politics and the dynamism in the geography of political boundaries.

in making political territories. Boundaries are normally *inclusionary*. That is, they are constructed to regulate and control specific sets of people and resources within them. Encompassed within a clearly defined territory, all sorts of activity can be controlled and regulated—from birth to death. The delimited area—which may include both land and water, and even airspace—over which a state exercises control and which is recognized by other states is **territory**. The space over which a state rules should not be confused with a non-sovereign geographic unit that possesses rights and protections through its relationship with a state. This entity is identified by the word *territory* as well. For instance, Guam, Puerto Rico and Samoa, are territories of the United States. The rules by which territories are able to participate in the government of their respective states differ from one state to another.

Boundaries can also be *exclusionary*. They are designed to control people and resources outside them. National boundaries, for example, can control the flow of immigrants or imported goods. Municipal boundaries may separate different tax structures or even access to alcohol; land-use

zoning boundaries can regulate access to upscale neighborhoods; field boundaries can regulate access to pasture, and so on (**Figure 10.3**).

One aspect of boundaries that is of increasing interest to geographers is how they are organized and policed. Since 9/11, for example, the relevance of policing national borders with respect to international travel has grown dramatically. States across the world, led by front-runners like Israel, the United Kingdom, and the United States, have been instituting a wide range of practices to secure borders by electronic screening. By way of passports that contain RFID (radio frequency identification) chips, as well as retinal scanning and related biometrics (electronic technologies for recognizing an individual based on one or more physical or behavioral elements), state agencies are increasingly able to know more and more personal information about travelers. These practices have the effect of creating virtual (technologically generated) boundaries that fortify the territorial ones.

Once established, boundaries tend to reinforce spatial exclusion as well as differentiation. The outcomes result

▶ **Figure 10.2 Ukraine and the Crimean Peninsula** This map shows the Crimean Peninsula seized by Russia in Spring of 2014. The peninsula contains a majority population of ethnic Russians. The United Nations and the United States have refused to recognize Russia's claim to Crimea.

The annexation of Crimea by Russia in March, 2014, has been declared invalid in a resolution passed by the United Nations General Assembly

▶ **Figure 10.3 Boundary between rural and urban places** Some boundaries signal differences in settlement activities that may actually be governed by land-use regulations. The division between agricultural activities and suburban living is clearly shown in this image.

partly from different sets of rules, both formal and informal, that apply within different territories, and partly because boundaries often restrict contact between people and foster the development of stereotypes about "others." This restricted contact, in turn, reinforces the role of boundaries in regulating and controlling conflict and competition between territorial groups.

Boundaries can be established in many ways and with differing degrees of permeability. At one extreme are informal, implied boundaries that are set by markers and symbols but never delineated on maps or in legal documents. Good examples are the "turf" of a city gang, the "territory" of an organized crime "family," or the range of a pastoral tribe. At the other extreme are formal boundaries

permeability. The boundaries between the states of the European Union, for example, have become quite permeable, and people and goods from member states can now move freely between them with no customs or passport controls.

APPLY YOUR KNOWLEDGE

1. Why are boundaries an important concept in geopolitics? What are some issues that develop from the inclusionary aspect of borders? What are some issues that can result from the exclusionary aspect of borders?

2. Using Google Earth or aerial photographs, compare and contrast the international boundaries between the United States and Canada and between the United States and Mexico. How does the U.S.-Mexican boundary differ from the U.S.-Canadian boundary? How are they similar? Speculate as to why these differences and similarities exist.

Frontier Regions
Frontier regions occur where boundaries are very weakly developed. They involve zones of underdeveloped territoriality, areas that are distinctive for their marginality rather than for their belonging. In the nineteenth century, vast frontier regions still existed—major geographic realms that had not yet been conquered, explored, and settled (such as Australia, the American West, the Canadian North, and sub-Saharan Africa). All of these are now largely settled, with boundaries set at a range of jurisdictional levels from individual land ownership to local and national governmental borders (**Figure 10.5**). Only Antarctica, virtually absent of permanent human settlement, exists today as a frontier region in this strict sense of the term (see Box 10.1 **Geography Matters: Politics at the Poles**).

▲ Figure 10.4 United States boundaries (a) The U.S.–Mexico border heavily patrolled and lined with barbed-wire chain-link fences along the highly urbanized parts. Aerial surveillance is also extensive along the border. This photo shows the U.S.–Mexico border along the Tijuana River estuary. (b) The U.S-Canada border crosses vast forests and is not as heavily securitized, even around large population centers, as the southern border with Mexico.

established in international law, delimited on maps, demarcated on the ground, fortified, and aggressively defended against the movement not only of people but also of goods, money, and even ideas. An example of this sort of boundary is the ones between the United States and Mexico and the United States and Canada (**Figure 10.4a & 10.4b**). There are also formal boundaries that have some degree of

There remain, nevertheless, many regions that are still somewhat marginal in that they have not been fully settled or do not have a recognized economic potential, even though their national political boundaries and sovereignty are clearcut. Such regions—the Sahara Desert, for example—often span national boundaries simply because they are inhospitable, inaccessible, and (at the moment) economically unimportant. Political boundaries are drawn through them because they represent the line of least resistance for the nations involvd. Of course, these frontier are most contentions at the local level. An example is the boundary between wealthy and poorer sections of a city or town where a train track or a river may divide them.

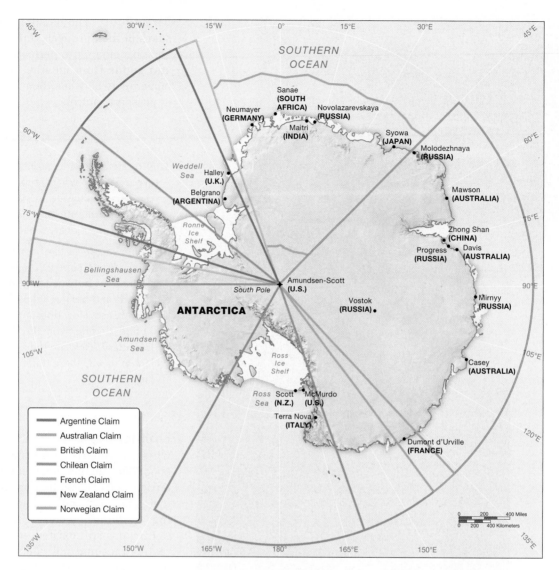

▲ **Figure 10.5 The Last Frontier, Antarctica** Even the uninhabitable terrain of Antarctica has become a site for competition among states. The radial lines delineating the various claims bear no relationship to the physical geography of Antarctica; rather, they are cartographic devices designed to formalize and legitimate colonial designs on the region. The overlay—the radiating lines from the center of the landmass—reflect claims by those states to have exclusive access to the mineral wealth the mineral wealth of the continent and the drive behind much of the territorial claims.

APPLY YOUR KNOWLEDGE

1. What makes a specific region a frontier?

2. Examine a map of the Gobi Desert. Identify which countries are part of it. How might this frontier region have strategic political significance for those countries? How does the desert terrain contribute to that strategic significance?

Boundary Formation Generally speaking, formal boundaries tend first to follow natural barriers, such as rivers, mountain ranges, and oceans. Good examples of countries with important mountain-range boundaries include France and Spain (the Pyrenees); Italy and France, Switzerland, and Austria (the Alps); and India and Nepal and China (the Himalayas). Chile,

though, provides the ultimate example: a cartographic freak, restricted by the Andes to a very long and relatively thin strip along the Pacific coast. Examples of countries with boundaries formed by rivers include China and North Korea (the Yalu and Tumen), Laos and Thailand (the Mekong), and Zambia and Zimbabwe (the Zambezi). Similarly, major lakes divide Canada and the eastern United States (along the Great Lakes), France and Switzerland (Lake Geneva), and Kenya and Uganda (Lake Victoria).

Where no natural features occur, formal boundaries tend to be fixed along the easiest and most practical cartographic device: a straight line. Examples include the boundaries among Egypt, Sudan, and Libya, between Syria and Iraq, and between the western United States and Canada. Straight-line boundaries are also characteristic of formal boundaries established through colonization, which is the outcome of a particular form of territoriality. The reason, once again, is practicality. Straight lines are easy to survey and even easier to delimit on maps of territory that remain to be fully charted, claimed, and settled. Straight-line boundaries were established, for example, in many parts of Africa during European colonization in the nineteenth century.

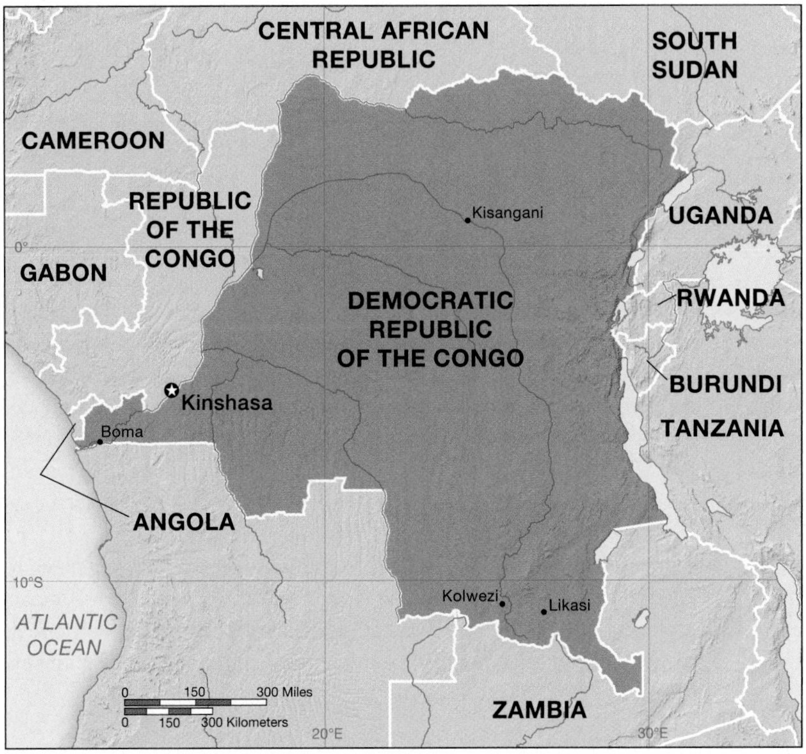

In detail, however, formal boundaries often detour from straight lines and natural barriers to accommodate special needs and claims. Colombia's border, for instance, was established to contain the source of the Orinoco River; the Democratic Republic of Congo's border was established to provide a corridor of access to the Atlantic Ocean (**Figure 10.6**).

Territories delimited by formal boundaries—national states, states, counties, municipalities, special districts, and so on—are known as *de jure* spaces or regions. *De jure* means "legally recognized." Historically, the word referred to a loose patchwork of territories (with few formally defined or delimited boundaries). More recently it has evolved to describe nested hierarchies (**Figure 10.7**) and overlapping systems of legally recognized territories.

▲ Figure 10.6 **Democratic Republic of Congo's access to the sea** Ocean access is a key need of states as it allows them to send and receive goods by way of contemporary container technologies, among other things.

▼ Figure 10.7 **Nested hierarchy of *de jure* territories** *De jure* territories are constructed at various spatial scales. Administrative and governmental territories are often "nested," with one set of territories fitting within the larger framework of another, as in this example of region, department, arrondissement, canton, and commune in France.

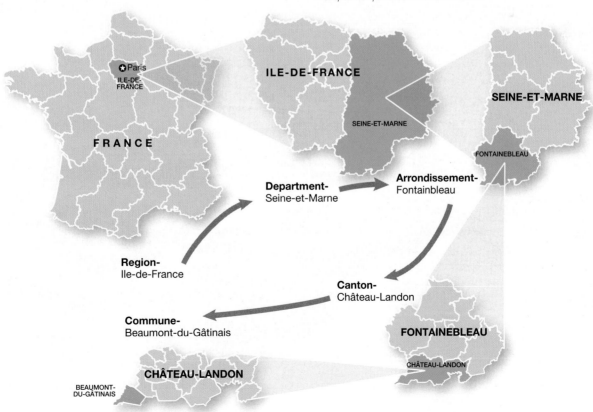

10.1 Geography Matters

Politics at the Poles

By Philip E. Steinberg, Durham University

The Land-Ocean Divide

Look at any world map and what do you see? A series of state territories defined by boundary lines. If you take a closer look, though, you'll notice that these territories cover only 30% of the world's surface. The other 70%—the portion covered by salt water—is beyond the territory of any individual state. According to the political geography that underpins international law, land can be developed and policed, and therefore may be claimed by individual states. Water is different. It can't be adequately developed or policed by any single state; it is designated as a global commons where ships of any state can navigate freely.

What happens, though, when water is frozen, and takes on some of the properties of land? What happens when land is frozen and becomes somewhat more like the ocean?

Antarctica: Politics on Frozen Land

Antarctica is a continent with no indigenous inhabitants. It is almost impossible to reach for long portions of the year due to weather conditions; agriculture is out of the question; no military could police its vast distances; and if resource extraction were ever attempted it would prove exceptionally challenging. Antarctica is probably more remote and hostile to human habitation than most of the ocean. Nonetheless, when national governments began to recognize that there was a mass of land beneath the Antarctic ice cap, they sent scientists and explorers to claim it (**Figure 10.A**).

By the middle of the twentieth century, Antarctica was beginning to emerge as a site of geopolitical competition. Recognizing that the impact of an all out war for Antarctica could be catastrophic while the benefits of "winning" the war would be few, treaty negotiations commenced.

▲ **Figure 10.A McMurdo Station** A U.S. Antarctic research center located on the southern tip of Ross Island in Antarctica, operated through the United States Antarctic Program, a branch of the National Science Foundation.

Negotiators quickly came to terms with the fact that although Antarctica was land and thus hypothetically parts of it could be claimed by individual states, the continent as a whole was well suited for a collective governance system. That collective governance put a limit on individual claims so that all states could continue nonexclusive use, much like the governance system that prevails in the ocean. The resulting Antarctic Treaty of 1961 established Antarctica as a scientific preserve where military activities are banned and the rights of scientists from all nations are guaranteed.

The Arctic: Politics on Frozen Ocean

The situation in the Arctic is the opposite of Antarctica. Whereas Antarctica is a frozen, uninhabited continent surrounded by water, the Arctic is a frozen ocean surrounded by inhabited land (**Figure 10.B**). If this situation existed anywhere else in the world, the central ocean would be governed just like any other ocean, under the rules of the UN Convention on the Law of the Sea, which are designed to preserve access by the community of states. Meanwhile, the inhabited areas of land around the ocean's edges would fall within the sovereign territory of each bordering state.

Things are not so simple, however. Although the Arctic is an ocean surrounded by land, that ocean and land are frozen for much of the year. And for the indigenous inhabitants of the area as well as other people living in the region, this makes a world of difference.

Indigenous peoples in the Arctic make their homes on both land and water. Their "home" is not the territory of one of the five states with Arctic land (Canada, Denmark, Norway, Russia, and the United States). Rather, it is a complex, continually freezing and thawing mosaic of land, water, and ice that cannot be represented by a typical Western map that defines land as territory and water as beyond territory's limits.

The largest political grouping of northern indigenous people, the Inuit Circumpolar Council (ICC), recognizes this disjuncture between the geography of the world map and the geography of their own lives.

The ICC acknowledges the sovereignty of the five states that control pieces of Arctic land. However, it maintains that these states' sovereignty should legally recognize that the home-space of the Inuit includes ice, water, and land that exist across and outside—as well as within—the borders of individual states. For the ICC, the Arctic presents a new political geography that requires a new kind of map.

---

1. The modern state system emerged in temperate Europe, where there is a clear distinction between land (where everyday activities occur) and water (which is understood as hostile to human habitation). What might the world map have looked like if instead it had emerged in the Arctic where frozen conditions complicate this distinction between land and water?

2. The Inuit Circumpolar Council calls for a new notion of sovereignty that recognizes the ways in which livelihoods cross between land and water. How might this notion be taken up by low-lying island states whose land territory is threatened by sea-level rise?

3. What other spaces could be peacefully governed by a treaty modelled after the Antarctic Treaty?

Web sites:

Antarctic Treaty Secretariat: http://www.ats.aq/seleccion.htm.
Inuit Circumpolar Council: http://www.inuitcircumpolar.com.
Especially see: ICC Declaration on Sovereignty: http://www.inuitcircumpolar.com/sovereignty-in-the-arctic.html
ICC, "The Sea Ice is Our Highway": http://www.inuitcircumpolar.com/uploads/3/0/5/4/30542564/20080423_iccamsa_finalpdfprint.pdf

◀ **Figure 10.B** **Studying the frozen sea** A scientist drags a sled full of scientific instruments across the frozen Arctic Ocean outside Barrow, Alaska, after drilling ice samples to measure how light and heat penetrate through different layers of ice.

These *de jure* territories are often used as the basic units of analysis in human geography, largely because they are both convenient and significant units of analysis. They are often, in fact, the only areal units for which reliable data are available. They are also important in their own right because of their status as units of governance or administration. A lot of regional analysis and nearly all attempts at regionalization, therefore, are based on a framework of *de jure* spaces.

GEOPOLITICS AND THE WORLD ORDER

There is, arguably, no other concept to which political geographers devote more attention than the state. The state is one of the most powerful forces—if not the most powerful—implicated in the process of globalization. The state effectively regulates, supports, and legitimates the globalization of the economy.

States, Nations and Citizenship

The **state** is an independent political unit with recognized boundaries, although some of these boundaries may be in dispute. In contrast to a state, a **nation** is a group of people sharing certain elements of culture, such as religion, language, history, or political identity. Members of a nation recognize a common identity, but they need not reside within a common geographical area. For example, the Kurdish nation includes members of Kurdish culture throughout the world, regardless of their places of origin and their lack of their own state. The term **nation-state** refers to an ideal form consisting of a homogeneous group of people governed by their own state. In a true nation-state, no significant group exists that is not part of the nation. and no significant portion of the nation is left outside of its territorial boundaries. **Sovereignty** is the exercise of state power over people and territory; that power is recognized by other states and codified by international law. **Government** is the body or group of persons who run the administration of a country. There are various forms of government that differ according to where the decision-making power is held, who elects the decision-making power, and how power is distributed (**Table 10.1**). Different from government, **governance** refers to the norms, rules and laws that are invoked to regulate a people or a state. Governance is also thought by scholars to exceed explicit regulations and include strategies, tactics, and processes for controlling the population of a nation or some other organization of individuals. Governance includes a wide range of public and private entities from states to NGOs.

What is a citizen?

Citizenship is a category of belonging to a state that includes civil, political, and social rights. Following the overthrow or decline of monarchies in Europe in the late eighteenth to mid-nineteenth centuries, a number of new republics were created. Republican government, as distinct from monarchy, requires the democratic participation and support of its population. Monarchical political power is derived from force and subjugation; republican political power derives from the support

TABLE 10.1 Forms of Government

Authoritarian	This form is characterized by absolute obedience to a formal authority, opposed to individual freedom and related to an expectation of unquestioning obedience.
Autocracy	A system of government in which supreme power is concentrated in the hands of one person, whose decisions are not subject to external legal restraints nor regularized mechanisms of popular control.
Dictatorship	This type of government is based on political authority that is monopolized by a single person or a political party, and exercised through various oppressive mechanisms.
Democracy	A form of government in which all eligible citizens participate equally—either directly or indirectly through elected representatives—in the proposal, development, and creation of laws. The term comes from the Greek and means 'rule of the people.'
Monarchy	A system of government where authority is embodied in one individual. In governments where the monarch has no or few legal restraints, it is an absolute monarchy and is a form of autocracy. Monarchs usually have hereditary rights to the throne.
Oligarchy	A form of government where power is concentrated in the hands of a few people (and this can be based on wealth, royalty, family, education, religious, corporate or military control).
Regime	This is a system with a set of rules, cultural or social norms that regulate the operation of the government with society.
Republic	A system in which power resides in the people who elect leaders to run the government.
Theocracy	A form of government in which a god or deity is recognized as the civil ruler and religious leaders administer policy and interpret laws based on a divine commission.
Totalitarian	In this system, the state holds total authority over the society and seeks to control all aspects of public and private life wherever possible. Totalitarianism differs from dictatorships in that totalitarianism concerns the scope of the governing power, whereas dictatorships are defined by the source of the governing power.

▲ **Figure 10.8** English Defense League (EDL), 2011 Pictured are members of the right-wing EDL in a march in the city of Luton. The group opposes the spread of Islamism in England.

of the governed. By creating a sense of nationhood, the newly emerging states of Europe were attempting to homogenize their multiple and sometimes conflicting constituencies so that they could govern with the population's active cooperation according to a sense of common purpose.

Modern citizenship as a political category was a product of the popular revolutions—from the English Civil War to the American War of Independence—that transformed monarchies into republics. In the process, these revolutions produced the need to reimagine the socially and culturally diverse population occupying the territory of the state. Where once they were considered subjects with no need for a unified identity, nationhood required of these same people a sense of

▼ **Figure 10.9** Independence for Tibet Exiled Tibetans and supporters of a free Tibet protest in a demonstration in Vancouver, Canada.

an "imagined community," one that rose above divisions of class, culture, and ethnicity. This new identity was called *citizen,* and citizenship came to be based on a framework of civil, political, and social rights and responsibilities. As our chapter opener makes clear, citizenship is bound up with a complex legal apparatus that determines who can and who cannot join the nation.

In the United States, for example, foreign-born individuals who wish to migrate temporarily are able to enter the country through more than 75 different kinds of visas. The expectation is that temporary migrants will return to their countries of origin after the period of their approved visit is complete. For immigrants who wish to reside in the United States permanently, however, they must secure what is informally known as a green card, or Lawful Permanent Residence. Obtaining full citizenship in the United States depends upon a number of factors but is an option for most green card holders. Significantly, it is estimated that about 700,000 individuals arrive in the United States without the proper documents. It is not known how many undocumented migrants currently live in the United States. This situation is a controversial one with no widely acceptable response being offered at the federal government level.

Multinational States

Given that nations were created out of very diverse populations, it is not surprising that almost no entirely pure nation-states exist today. Rather, multinational states—states composed of more than one regional or ethnic group—are the norm. Spain is such a multinational state (composed of Catalans, Basques, Gallegos, and Castilians), as are France, Kenya, the United States, and Bolivia. Indeed, a very limited number of nation-states are uni-national; Iceland is one of the few that are. Since World War I, it has become increasingly common for groups of people sharing an identity different from the majority, yet living within the same political unit with the majority, to agitate to form their own state. This has been the case with the Québecois in Canada and the Basques in Spain. It is out of this desire for autonomy that the term *nationalism* emerges. **Nationalism** is the feeling of belonging to a nation, as well as the belief that a nation has a natural right to determine its own affairs. Nationalism can accommodate itself to very different social and cultural movements, from the white supremacy movements in the United States and Europe (**Figure 10.8**) to the movements for independence in China such as Tibet (**Figure 10.9**) The impact of minority nationalism on the world map was especially pronounced during the twentieth century.

APPLY YOUR KNOWLEDGE

1. Define citizenship. What is the difference between being a subject and a citizen? How does the "imagined community" relate to being a citizen?

2. Nationalism as a political movement is on the rise around the globe. Identify two national movements in places other than the United States—one that is left-wing (politically progressive) and one that is right-wing (politically conservative)—and trace their origins and current missions. How do these two movements differ? In what ways are they the same?

Russia's State and National Transformation

The history and the present status of the former Soviet Union illustrates the tensions among and between states, nations, and nationalism. Both enduring nationalism and the desire for sovereignty are evident in the history of the Russian Empire. Russia's strategies to bind the 100-plus nationalities (non-Russian ethnic peoples) into a unified Russian state were often punitive and seldom successful. Non-Russian nations were simply expected to conform to Russian cultural norms. Those that did not were more or less persecuted. The result was opposition and, among many if not most of the nationalities, sometimes rebellion and refusal to bow to Russian cultural dominance.

Such was the legacy that Vladimir Lenin and the Bolsheviks inherited from the Russian Empire following the overthrow of the aristocracy in 1917. The solution to the "national problem" orchestrated by Lenin was *recognition* of the many nationalities through the newly formed Union of Soviet Socialist Republics (USSR). Lenin believed that a *federal system,* with *federal units* delimited according to the geographic extent of ethnonational communities, would ensure political equality among at least the major nations in the new state. This political arrangement recognized the different nationalities and provided them a measure of independence. Federation was also a way of bringing reluctant areas of the former Russian Empire into the Soviet fold. A **federal state** allocates power to units of local government within the country. The United States is a federal state with its system of state, county, and city/town government. A federal state can be contrasted with a **unitary state**, in which power is concentrated in the central government. The Russian state under the czar was a unitary state but a federated state as the USSR.

By 1991, the relatively peaceful breakup of the Soviet Union was underway, and new states emerged to claim their independence. **Figure 10.10** is a recent map of the former USSR, including all the newly independent states now in the Commonwealth of Independent States (CIS) and the Baltic states of Estonia, Latvia, and Lithuania, which chose not to be members of the CIS.

The CIS is a **confederation**, a group of states united for a common purpose. The newly independent states that chose confederation did so mostly for economic (and to a lesser extent, for military) purposes. A similar case can be found in the Confederate States of America, the 11 southern states that seceded from the United States between 1860 and 1861 to achieve economic and political solidarity. This secession led ultimately to the Civil War, a bloody conflict that caused a massive loss of American lives for both the Union and the Confederacy.

With the fall of communism in Eastern Europe in 1991, regions such as the Balkans (the mountainous isthmus of land between the Danube River and the plains of northern Greece that includes Albania, Bulgaria, continental Greece, southeast Romania, European Turkey, and most of the territories formerly organized as Yugoslavia) also experienced national movements resulting in the redrawing of political boundaries. And while the breakup of the former USSR had been primarily peaceful (with significant exceptions such as Georgia, Ukraine and

▼ **Figure 10.10 Independent states of the former Soviet Union** For the most part, the administrative structure of the USSR has remained in place. Until recently, the autonomous regions and republics had more than nominal local control, and the former federal republics had become independent states (pictured here in various colors).

(*Source:* Reprinted with permission from Prentice Hall, from J. M. Rubenstein, *The Cultural Landscape: An Introduction to Human Geography,* 5th ed. © 1996, p. 318.)

◀ **Figure 10.11 Celebrating Kosovo's independence** This sculpture, erected in Pristina, the capital of Kosovo, is repainted frequently. Here citizens are invited to comment. It originally was painted with flags to honor all those countries that supported Kosovo's independence.

Chechnya), the redrawing of national boundaries in the Balkans resulted in bitter and widespread ethnic conflict. The region is situated at a geopolitical crossroads where East meets West, Islam meets Christendom, the Ottoman Empire met the Austro-Hungarian Empire, and communism once confronted capitalist democracy.

The most extreme conflict in the region occurred during the late 1990s in Kosovo, a region within Serbia that lost its autonomy in 1989 when Serbian nationalist leader Slobodan Milosevic placed it under military occupation. A year later Kosovo's parliament was abolished and its political leaders fled. Milosevic orchestrated numerous attempts to rid the region of ethnic Albanians, and in 1995 Serb refugees from Croatia began flooding into Kosovo. Fueled by the racist rhetoric and military support of Milosevic, a civil war between Serbs and ethnic Albanians eventually erupted. By late 1999, 800,000 ethnic Albanians had fled Kosovo and tens of thousands had been massacred. An 11-week air war by NATO helped to bring the conflict to a halt, although atrocities continued as ethnic Albanians retaliated against Serbs. Kosovo became an independent state in 2008 (**Figure 10.11**).

State Theory

The definition of the state provided in the previous section is a static one. The state, through its institutions—such as the military or the educational system—can act to protect national territory and harmonize the interests of its people. Therefore, the state stands behind a *set of institutions* for the protection and maintenance of society. A state is not only a place, a bounded territory, it is also an active entity that operates through the rules and regulations of its various institutions, from social-service agencies to governing bodies to the courts, to shape a country's populations.

Althusser

An especially influential state theorist whose work has inspired a large number of political geographers is Louis Althusser who developed an approach that identifies the operations of the state in two ways:

- Recognizing the state as an ideological force operating through the institutions of the schools, the media, the family, and religion to produce citizens who conform to state expectations

- Viewing the state as repressive through different institutions like the courts, army, and the police, to compel citizens to comply with its rules

These two related aspects of a model for understanding the state have been employed by political geographers in a number of ways, but perhaps the most common is to explore how the spaces of these various institutions, such as the school or the family, are produced and operate to shape citizens.

Foucault

A theorist who has also been a significant influence on political geography is Michel Foucault who pushed Althusser's ideas further by exploring just *how* the various institutions of the state operate to do the work of shaping citizens. Foucault examined how power, knowledge, and discourse operate in concert to produce particular kinds of subjects: citizens, women, soldiers, terrorists.

For Foucault, **discourse** is made through the rules, identities, practices, exclusions, and a range of other elements that form a way of thinking about something. For example, an army turns civilians into soldiers through the combination of power (in the form of both force and cooperation) and knowledge (accumulated insights about training, discipline, and warfare). The result is an institutional way of thinking that make large numbers of individuals function as an effective group.

Deleuze

Another state theorist important to geographers is Giles Deleuze. Different from Althusser and Foucault but building on their work, he recognizes that

- the state is not just a set of institutions—the courts, the legislature, the military, and so on—but a force; and

- the force is greater than formal institutions at the same time that it works through them.

For Deleuze, the state is not a thing but a principle that works by way of authority. Deleuze rejects the commonly held view that the state was "created" during a particular period of human history and then expanded its power over time. He contends that the state has always existed in different forms, even before the emergence of the institutions by which we most clearly recognize it now. Deleuze believes that the state is best thought of as a *machine,* with its purpose being to regulate and dominate. This machine of the state operates through mundane practices that produce a population willing to submit to authority whether a police officer, a forest ranger, a teacher, or a president.

Biopolitics

Biopolitics is the extension of state power over the physical and political bodies of a population. Most often attributed to Foucault, biopolitics or biopower is also associated with political theorists who preceded him as well as those who wrote simultaneously with and after him. The key objective of biopolitics is to regulate the national population—the social body—by shaping the individuals that constitute it.

Biopolitics operates through monitoring, recording, categorizing and policing to optimize the vitality of the population and ensure a healthy workforce. As a result, the population is sorted according to differences such as age, gender, ethnicity, race, health status, income, household size, and so on. The census, as discussed in Chapter 3, is a massive undertaking meant to produce statistical data on these differences. A contemporary extension of demographic monitoring might be seen in the increasing ability to manipulate life at the molecular level through recombinant genetics or the ability to select for particular traits through assisted reproductive technology.

Geographers have incorporated these and other theories of the state to explain the unfolding impact of different kinds of political spaces, from the organization and influence of the classroom to the development and effects of international laws. For political geographers, theories and models of geopolitics have been their most prominent contributions to understanding the role and behavior of the state. The revolutionary fervor that spread across North Africa and the Middle East in 2011 is an illustration of geopolitics and the relationship of a national population to its state (see Box 10.2, "Window on the World: The Arab Spring and its Aftermath").

APPLY YOUR KNOWLEDGE

1. What is the state according to Althusser, Foucault, and Deleuze? Compare and contrast each theorist's contribution by reflecting on what they say about power and the role of institutions.

2. Based on what you have leaned about Foucault's theory of discourse and Deleuze's conception of the state, write a paragraph describing the Arab Spring from each of these theorist's perspectives.

Imperialism and Colonialism

Geopolitics may involve the extension of power by one group over another. Two ways this may occur are through the related processes of imperialism and colonialism. Recall from Chapter 2, Imperialism is the extension of state authority over the political and economic life of other territories. Over the last 500 years, imperialism has resulted in the political or economic domination of strong core states over the weaker states of the periphery. Imperialism does not necessarily imply formal governmental control over the dominated area; it may also involve a process by which some countries pressure the independent governments of other countries to behave in certain ways. This pressure can take many forms, such as military threat, economic sanctions, or cultural domination. Imperialism always involves some form of *authoritative control* of one state by another.

Imperialism As discussed in Chapter 2, the process of imperialism begins with exploration (**Figure 10.12**), often prompted by the state's perception that there is a scarcity or lack of a critical natural resource. It culminates in development via colonization or the exploitation of indigenous people and resources, or both. In the first phases of imperialism, the core exploits the periphery for raw materials. As the periphery becomes developed, colonization may occur and economies based on money transactions—or "cash economies"—may be introduced where none previously existed. The periphery may also become a market for the manufactured goods of the core. Eventually, though not always, the periphery—because of the availability of cheap labor, land, and other inputs to production—can become a new arena for large-scale capital investment. Some peripheral countries improve their status and become semiperipheral or even core countries. **Figure 10.13** is a map of the colonies created by European imperialism in Africa.

Colonialism is a form of imperialism. It involves the formal establishment and maintenance of rule by a sovereign power over a foreign population through the establishment of settlements. The colony does not have any independent standing within the world system and instead is considered an adjunct of the colonizing power. From the fifteenth to the early twentieth centuries, colonialism constituted an important component of core expansion. Between 1500 and 1900, the primary colonizing states were Britain, Portugal, Spain, the Netherlands, and France. **Figure 10.14** illustrates the imperialism and the post-colonization situation in South America.

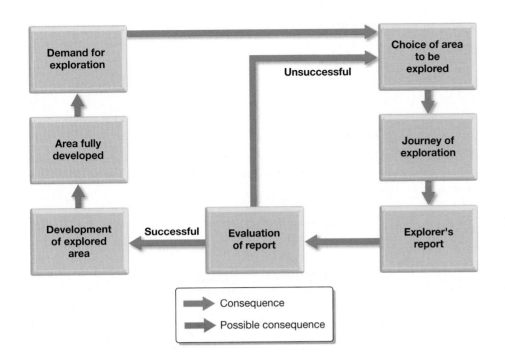

◀ Figure 10.12 Principal steps in the process of exploration This diagram illustrates the main elements in the process of exploration, beginning with a need in the home country. Geographers figured prominently in the process of exploration by identifying areas to be explored as well as traveling to those places and cataloging resources and people. Exploration is a component of the process of imperialism.

(Adapted from J. D. Overton, "A Theory of Exploration," *Journal of Historical Geography, 7,* 1981, p. 57.)

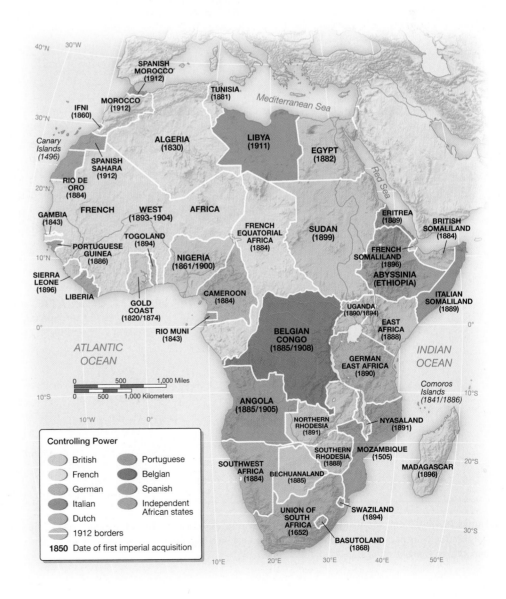

▶ Figure 10.13 European colonies in Africa, 1496–1912 Lying within easy reach of Europe, Africa was the most likely continent for early European expansion. The Belgian, Italian, French, German, and Portuguese states all laid claim to various parts of Africa and in some cases went to war to protect those claims. The partitioning of the African continent by the imperial powers created a crazy quilt that cross-cut preexisting affiliations and alliances among the African peoples.

(Adapted from *Harper Atlas of World History.* New York: Harper-Collins, 1992, p. 139.)

▲ Figure 10.14 Colonization and Independence in South America and the Caribbean, 1800–1850 The Spanish and Portuguese dominated the colonization and settlement of South America. The Dutch, French, and English were a minor and largely tentative presence. South American colonization yielded rich commodity and mineral returns.

(Adapted from *Rand McNally Atlas of World History*. Skokie, IL: Rand McNally.)

▲ **Figure 10.15 Curry restaurants, Brick Lane, London** The British presence in India affected culture, politics, the economy, and the layout of cities, as well as numerous other aspects of everyday life. Importantly, Indian culture has come to influence British culture even more directly through immigrants who've migrated to the Great Britain.

Important states more recently involved in colonization and imperialist wars include the United States and Japan. Although colonial penetration often results in political dominance by the colonizer, such is not always the case. For example, Britain may have succeeded in setting up British colonial communities in China, but it never succeeded in imposing British administrative or legal structures in any widespread way. And at the end of the colonial era, a few former colonies, such as the United States, Canada, and Australia, became core states themselves. Others, such as Rwanda, Bolivia, and Cambodia, remain firmly within the periphery. Some former colonies, such as Mexico, India and Brazil, have come close to the core but have not fully attained core status and are categorized as being within the semiperiphery.

The Effects of Colonialism An example of colonialism is the extension of British rule in India, which began with the establishment of the East India Trading Company in the mid-eighteenth century. The British government gave the company the power to establish forts and settlements, as well as to maintain an army. The company soon established settlements—including factories—in Mumbai (formerly Bombay), Chennai (formerly Madras), and Calcutta. What began as a small trading and manufacturing operation burgeoned over time into a major military, administrative, and economic presence by the British government and did not end until Indian independence in 1947. During that 200-year period, Indians were brutalized and killed and their society transformed by British influence. That influence permeated all aspects of daily life and institutions from language and judicial procedure to cultural identity.

Migration from South Asia over the last 50 years has been dramatic with Indians, Bangladeshis, and Pakistanis living all

over the world with an especially high number of migrants in the United Kingdom. There they have had their own significant impact on all aspects of British life with the most popular dish now being curry (**Figure 10.15**).

APPLY YOUR KNOWLEDGE

1. What were some of the goals of the Arab Spring uprisings? How do they compare to the realities after the uprisings? Why do you think this might be?

2. Select one of the countries that experienced the Arab Spring and do an Internet search on the current political climate. What do you find? Have there been any changes that support the goals of the uprisings? In what part of government and/or society do you see progress?

Orientalism and Post-colonialism

The reasons Britain was able to be so callous in its colonial practices are complex. Theorist Edward Said has proposed the concept of Orientalism to explain them, at least in part. For Said, **Orientalism** is a discourse that positions the West as culturally superior to the East seeing Eastern societies as unchanging and undeveloped and therefore inferior. This mischaracterization allows the West to possess as well as deploy patronizing perceptions of the Middle East, Asian, and North African societies (the East or Orient). The colonized are seen as inferior and in need of disciplining to the superior and enlightened colonizer. The colonizer has a moral obligation to colonize, according to this logic. Said was a humanities scholar and his criticism was leveled at literature but his ideas have been embraced by social scientists as well. Importantly, Said's book, *Orientalism* (1978) was a major contributor to the then emerging field of **postcolonial studies**, an academic discipline that focuses on the cultural legacies of colonialism and imperialism.

The postcolonial history of the Indian subcontinent has included partition and repartition, as well as the eruption of regional and ethnic conflicts. In 1947, Pakistan split off from India and became a separate Muslim state. In 1971, Bangladesh, previously part of Pakistan, declared its independence. Regional conflicts include radical Sikh and Islamic movements for independence in the states of Punjab and Kashmir, respectively. Ethnic conflicts include decades of physical violence between Muslims and Hindus over religious beliefs and the privileging of Hindus over Muslims in the national culture and economy. It would be misleading, however, to attribute all of India's current strife to colonialism. The Hindu caste system, which distinguishes social classes based on heredity and plays a significant role in political conflict, preceded British colonization and persists to this day.

Since the turn of the nineteenth century, the effects of colonialism have continued to be felt as peoples all over the globe struggle for political and economic independence. The 1994 civil war in Rwanda is a sobering example of the ill effects of colonialism. As in India, where an estimated 1 million Hindus and Muslims died in a civil war when the British pulled out, the exit of Belgium from Rwanda left colonially created

10.2 Window on the World

The Arab Spring and Its Aftermath

Despite the fact that Iran is not Arab, and its "Green Revolution" occurred in 2009, the protests that gathered momentum in Tehran and swept across the country, can be seen as a powerful inspiration for what became known as the Arab Spring—massive popular protests that occurred across North Africa and parts of the Middle East in Spring 2011 (**Figure 10.C**).

Tunisia's was the first uprising to occur in the region and there is no doubt that its "Jasmine Revolution" served to embolden other national protests. Sustained public anger there, which forced the president out of office after nearly a quarter century, was ignited by the suicide of a young, unemployed man who set fire to himself after officials prevented him from selling fruits and vegetables on the street.

In Egypt, the world watched in awe as President Hosni Mubarak, in power for three decades, resigned in response to the demands of millions of protestors as well as pressure from the international community. Instrumental in sparking the protests was the video blogging of a 26-year-old woman who urged the Egyptian people to join her in Tahrir Square to bring down Mubarak's regime. Her posts went viral and the event attracted 80,000 attendees.

In Morocco—different from its neighbors in that the country has a successful economy, an elected parliament, and a reformist monarchy—the protesters were pushing for reform that would "restore dignity and end graft." Behind the facade of prosperity is a country of impoverished youth with little chance of employment and a government elite living obscenely lavish lives. In Mauritania, peaceful protests called attention to similar issues as well as an end to human rights abuses.

Algeria is also a country that on the surface seems generally prosperous, based on its sizeable oil and gas reserves. Protesters there demanded that the Algerian constitution limit presidential terms.

Saudi Arabia experienced no sizeable protests, probably because opposition movements are banned there; the country's vast oil reserves make it

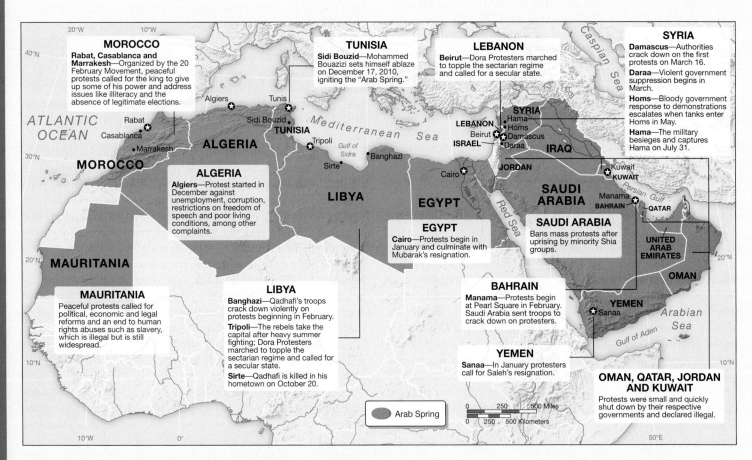

▲ **Figure 10.C Arab Spring 2010–2011** Shown in this map are the countries involved in the Arab Spring movement as well as some of the actions that surrounded them and the reactions to them.

one of the wealthiest of the Arab states; and the country is deeply conservative with a population that supports King Abdullah.

In Bahrain, people died in protests demanding that King Hamad take action to increase political freedoms and eliminate job discrimination that favors the governing Sunni Muslim minority. As small groups of demonstrators gathered in the center of the capital city Manama, King Hamad swiftly imposed a state of emergency and cleared the protesters' camps in a brutal show of force aided by Saudi and United Arab Emirates soldiers.

In both Jordan and Oman, protests were limited but the demands of the protestors were familiar: job creation, controls on food prices, an end to government corruption, and more democratic government. The same demands issued from quickly shut down protests in Qatar and Kuwait. Young people across the Middle East and North Africa were the key elements of the protests.

In Libya, Yemen, and Syria, revolutions turned into armed insurrection. Uncharacteristic of the larger movement in the region, where nonviolent protest was the explicit objective, protests in these three countries were directed at the intransigence of the leaders and the deep frustration and righteous anger of the protesters.

Libyans rose up against their ruler, Colonel Muammar Gaddafi, one of the most autocratic and longest-ruling dictators in the region. Citizens demanded a return to a multiparty democracy, a reduction in unemployment and poverty, freedom from domestic surveillance, and the right to gather peacefully to express dissent. Gaddafi ordered the police to fire on the peaceful protesters.

The response was an armed uprising that spread across the country, pitting government forces against civilians. In October 2011 Colonel Gaddafi was killed by Libyan rebel forces.

The situations in Syria and Yemen also turned violent. Yemen is the Arab world's poorest nation where about half of its 23 million people live on $2 a day or less and one-third suffer from chronic hunger. Yemeni protest led to the ouster of their president. The violence against peaceful protesters there was brutal.

The Syrian government, too, met peaceful protest with an armed response. Government forces deployed tanks, shelled residential areas, and used snipers on rooftops to kill protesters. A bloody civil war has ensued and well over 100,000 Syrians have died and over 2 million refugees have fled the country. As described in a later section, the situation in Syria has become complicated by the emergence of an extremist movement that has claimed the establishment of an Islamic state in parts of Iraq and Syria.

After the Arab Spring

Upheaval across the Arab World in 2011 has not led to many positive changes with some countries engaged in brutal civil wars. Many observers believe that the region's long-standing political order, consisting largely of military governments or rule by postcolonial dynasties, are unlikely to let go of the monopolies on economic resources they control. It seems clear that movement toward democratic rule—which would include accountability, transparency, and equality—will be very slow:

- Egypt's military ousted a democratically elected president and replaced him with a general who was recently elected to hold that seat.

- On the Arabian Peninsula (Saudi Arabia, Yemen, Oman, and Kuwait) as well as in Jordan, the ruling monarchs have managed to maintain the status quo. In the Gulf States, generous handouts such as state jobs and discounted housing were distributed in exchange for political passivity. Bahrain is the exception where protest by the majority Shiites against the Sunni-ruled kingdom continue.

- The presence of al-Qaeda in Yemen as well as separatist challenges to government authority have made that country an increasingly unstable place. Many observers believe Yemen is in danger of becoming a failed state.

- In Morocco and Algeria, little if anything has changed.

- The one bright spot is Tunisia, which adopted a new constitution, one that for the first time includes rights for women. Compared to the coup in Egypt, the horrific civil war in Syria, and suppression in Bahrain and instability in Yemen, Tunisia's revolution stands out among "Arab Spring" countries as having survived.

---

1. What is the Arab Spring? What makes is it a geopolitical phenomenon?

2. Why are there different types of protests and different responses by the the various government to them among the many countries that participated in the Arab Spring.

◀ **Figure 10.16 Refugees returning to Rwanda** Fleeing civil unrest in their own country, Rwandans from the Hutu tribe increasingly sought refuge in the Democratic Republic of Congo (formerly Zaire) when the Tutsi-led government assumed power in 1994. Two-and-a-half years later, over half a million Rwandan refugees in the Democratic Republic of Congo occupied some of the largest refugee camps in the world. In late 1996, they began streaming back into Rwanda when the Tutsi-led government urged them to come help rebuild the country.

APPLY YOUR KNOWLEDGE

1. Define the concept of Orientalism. Why do you think this mentality created political problems that continue to exist in former colonies?

2. On the Internet, find three different recent news articles from reputable sources about a political issue in the Middle East. Consider the articles from the perspective of Orientalism. How do the articles depict the West and the Middle East? Is the West depicted as being culturally superior to the Middle East? If so, show how that is accomplished. If not, explain how the journalists avoid adopting an Orientalist explanation.

tribal rivalries unresolved and seething. Although the Germans were the first to colonize Rwanda, the Belgians, who arrived after World War I, established political dominance among the Tutsi by allowing them special access to education and the bureaucracy.

Previously, a complementary relationship had existed between the Tutsi, who were cattle herders, and the more numerous Hutu, who were agriculturists. In effect, colonialism introduced difference into an existing political and social structure that had operated more or less peacefully for centuries. In 1959, the Hutus rebelled and the Belgians abandoned their Tutsi favorites to side with the Hutus. In 1962, the Belgians ceded independence to Rwanda, leaving behind a volatile political situation that has erupted periodically ever since, most tragically in 1994's civil war. After a year of violence in which over half a million Tutsis were killed, the Hutus were driven across the border to the Democratic Republic of Congo (DRC) and a new Tutsi-led Rwandan government was formed. The Hutu refugees gathered in UN camps that gradually came to be controlled by armed extremists, who transformed them into virtual military bases and used them to attack the Tutsis in Rwanda (**Figure 10.16**). When Rwanda's Tutsi-led military, with Uganda's support, invaded the DRC to break up the camps, over a million refugees were released. Refugees have since dispersed across Central Africa.

To help address the culture of impunity that angered so many Rwandans, the UN established a truth commission, known as "Gacaca Court System," based on Rwandan traditional justice. Although its work has had some important moral impacts, the commission's aim—enabling Tutsis and Hutus to work together to rebuild their society and economy—has not been realized. Hundreds of thousands of children have become orphans as their parents either died in the atrocities or their mothers were infected with HIV/AIDS through rape and have since died. The country spends more on debt repayments to international banks than it does on education and health. Seventy percent of all households in Rwanda live below the national poverty line.

The North/South Divide
The colonization of Africa, South America, parts of the Pacific, Asia, and smaller territories scattered throughout the Southern Hemisphere resulted in a political geographic division of the world into North and South, known as the **North/South divide**. In the North are the imperialist states of Europe, the United States, Russia, and Japan. In the South are colonized states. Though the equator is often used as a dividing line, some so-called Southern territories, such as Australia and New Zealand, actually are part of the North in an economic sense.

The crucial point is that a relation of dependence was set up between countries in the South (the periphery) and those in the North (the core) that began with colonialism and persists even today. Only a few peripheral countries have become prosperous and economically competitive since achieving political autonomy, though that is beginning to change, as this book addresses in Chapter 8. Political independence is markedly different from economic independence, and the South remains very much oriented to the economic demands of the North.

Decolonization
The reacquisition by colonized peoples of control over their own territory is known as **decolonization**. In many cases, sovereign statehood has been achievable only through armed conflict. From the Revolutionary War in the United States to the twentieth-century decolonization of Africa, the world map created by the colonizing powers has repeatedly been redrawn. Many former colonies achieved independence after World War I. Deeply desirous of averting wars like the one that had just ended, the victors (excluding the United States, which entered a period of isolationism following the war) established the League of Nations. One of the first international organizations ever formed, the League of Nations had a goal of international peace and security. An **international organization** is one that includes two or more

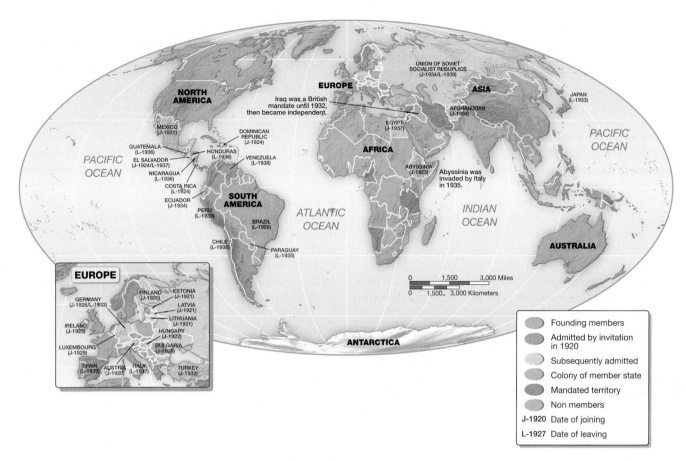

▲ **Figure 10.17 Countries participating in the League of Nations** Although the United States never joined the League of Nations, it was central in its creation, the first international organization of the twentieth century. Britain and France played important roles in the League but were never able to secure arms limitations and security agreements among the membership. The League was dissolved in 1946.

states seeking political and/or economic cooperation with each other. **Figure 10.17** shows the member countries of the League.

Within the League, a system was designed to assess the possibilities for independence of colonies and to ensure that the process occurred in an orderly fashion. Known as the *colonial mandate system,* it had some success in overseeing the dismantling of numerous colonial administrations. **Figures 10.18** illustrates decolonization during the twentieth century in Asia and the South Pacific. Although the League of Nations proved effective in settling minor international disputes, it was unable to prevent aggression by major powers and was dissolved in 1946. It did, however, serve as the model for the United Nations.

Neocolonialism Decolonization does not necessarily constitute an end to domination within the world system, however. Even though a former colony

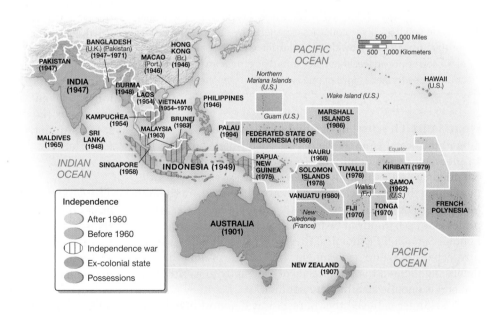

▲ **Figure 10.18 Independence in Asia and the South Pacific, before and after 1960** Decolonization and independence are not uniform phenomena. The form of colonial domination that was imposed is as much a factor as the composition and level of political organization that existed in an area before colonization occurred. Mostly, decolonization and political independence forced societies into a nation-state mold for which they had little, if any, preparation.

(Adapted from *The Harper Atlas of World History, Revised Edition,* Librairie Hachette, p. 283. Copyright © 1992 by HarperCollins Publishers, Inc.)

may exhibit all the manifestations of independence, including its own national flag, governmental structure, currency, educational system, and so on, its economy and social structures may continue to be dramatically shaped in a variety of ways by core states. Participation in foreign aid, trade, and investments from core countries subjects the periphery to relations that are little different from those they experienced as colonial subjects. This process is known as **neocolonialism**, the use of economic, political, and cultural arrangements to influence a country, in the absence of more direct control. Ghanaian president Kwame Nkrumah coined the term to describe how promoting a capitalist culture in a formerly colonized country helps assimilate the local people into the norms of the neocolonizer, which enables the opening of the country to foreign corporate entry.

In former British colony Kenya, for example, core countries' provision of foreign aid monies, development expertise, and educational opportunities to selected individuals has created a class of native civil servants that is in many ways more strongly connected to core processes and networks than those operating within Kenya. This relatively small group of men and women, often foreign educated, emerged as the first capitalist middle class in Kenyan history. Their cultural ties become complicated and their loyalties divided as they form an elite class that transmits global corporate culture even more deeply into their country. One aspect of elite governing classes in some formerly colonized countries has been extreme corruption. While corruption occurs in core and peripheral regions, its effects are felt most significantly by the poor (See Box 10.4 Spatial Inequality: Global Corruption).

APPLY YOUR KNOWLEDGE

1. What are the three types of corruption?

2. Why do you think some countries experience higher levels of corruption than others? How do you think this data compares with levels of corruption in wealthier, developed governments? Do some countries experience more of one type of corruption than others?

Heartland Theory   By the end of the nineteenth century, numerous formal empires were well established and imperialist ideologies were dominant. To justify the strategic value of colonialism and explain the dynamic processes and possibilities behind the new world map created by imperialism, Halford Mackinder (1861–1947) developed a theory. Mackinder was a professor of geography at Oxford University and director of the London School of Economics. Given his background in geography, economics, and government, it is not surprising that his theory highlighted the importance of geography to world political and economic stability and conflict.

Mackinder believed that Eurasia was the most likely base from which a successful campaign for world conquest could be launched. He considered its closed heartland to be the "geographical pivot," the location central to establishing global control. Mackinder premised his model on the conviction that the age of maritime exploration, beginning with Columbus, was drawing to a close. He theorized that land transportation technology,

especially railways, would reinstate land-based power, and sea prowess would no longer be as essential to political dominance. Eurasia, which had been politically powerful in earlier centuries, would rise again because it was adjacent to the borders of so many important countries and it was strategically buttressed by an inner and outer crescent of landmasses (**Figure 10.19**).

When Mackinder presented his geostrategic theory in 1904, Russia controlled a large portion of the Eurasian landmass protected from British sea power. In an address to the British Royal Geographical Society, Mackinder suggested that the "empire of the world" would be in sight if one power, or combination of powers, came to control the heartland. He believed that Germany allied with Russia and China organized by Japan were alliances to be feared. Mackinder's theory was a product of the age of imperialism. To understand why Britain adopted this theory, it is important to remember that antagonism was increasing among the core European states, leading to World War I a decade later.

Mackinder's geopolitical theory has certain resonance today. An example is the aggressive nationalism of Vladimir Putin, the current prime minister of Russia. As mentioned earlier in this chapter, the breakup of the Soviet Union created extensive regional realignments in the former confederation. Putin's response to the Western leanings of Estonia and Georgia and Russia's invasion of Chechnya (a region that is part of Russia) and more recently Ukraine have some observers asserting that a new round of Russian empire-building is underway. Although Putin is not likely to attempt expansion into Central Europe, he could undermine those countries by his capture of the Crimean Peninsula, which he accomplished in Spring 2014. As their next-door neighbor, a re-Russified Ukraine, is a worrisome prospect. Expanding into Ukraine has given Russia access to the Mediterranean, the transit route between the Caspian Sea and the Black Sea. Further, it has given Russia complete access to known fuel reserves in the maritime zone around the peninsula worth trillions of dollars. Many political geographers today believe that Mackinder's theories are outdated, but Prime Minister Putin's recent aggressions and animosity toward what he regards as the "traitorous" former republics and breakaway regions underscore the current relevance of his geostrategic formulations.

The East/West Divide   In addition to a North/South divide based on imperialism and colonialism, the world order of states can also be viewed as dividing along an East/West split. The **East/West divide** refers to the gulf between communist and noncommunist countries, respectively. The East/West divide has played a significant role in global politics since at least the end of World War II in 1945 and, perhaps more accurately, since the Russian Revolution in 1917. By the second decade of the twentieth century, the major world powers were backing away from colonization. Still, many were reluctant to accelerate decolonization for fear that independent countries in Africa and elsewhere would choose communist political and economic systems instead of some form of Western-style capitalism.

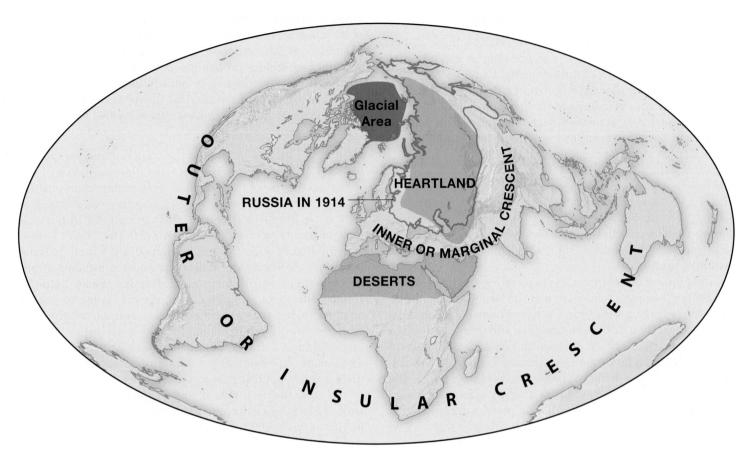

▲ **Figure 10.19 The heartland** A quintessential geographical conceptualization of world politics, Mackinder's heartland theory has formed the basis for important geopolitical strategies throughout the decades since the theory's inception. While the pivot area (the heartland) of Eurasia is wholly continental, the outer crescent is mostly oceanic and the inner crescent part continental and part oceanic.

Cuba provides an interesting illustration of an East/West tension that persists despite the official end of the **Cold War**: the state of heightened military and political tension as well as economic competition between the former Soviet Union and its satellite states and the United States and its allies, lasting from 1946 to 1991. Although Cuba did not become independent from Spain until 1902, U.S. interest in the island dates back to the establishment of trade relations in the late eighteenth century. In the first half of the twentieth century, U.S. economic imperialism replaced Spanish colonialism and wrought major changes in the island's global linkages. Cuba, at that time, was experiencing a series of reform and revolutionary movements that culminated in the rise of Fidel Castro, a committed communist, in the 1950s. This period coincided with a moment in U.S. history when anticommunist sentiment was at its peak. Castro came to power by 1959 and the United States initially responded with a militarily aggressive stance toward Cuba. Since then U.S. relations with Cuba have been premised on economic sanctions. Since the turn of the new century, economic sanctions were softened through the efforts of President George W. Bush. President Barack Obama has reduced travel restrictions but failed to lift the economic embargo (**Figure 10.20**).

▼ **Figure 10.20 U.S. tourists in Cuba** Tourists visit the Bodeguita del Medio, a bar in Havana, Cuba frequented by U.S. novelist Ernest Hemingway.

China and the East/West Divide

The West's anxieties about communism also included China, though after World War II it was not seen as important to Western interests nor did Europe or the United States have the resources to challenge the growing communist movement there. Even as the communist revolution took control of China, no Western military action was taken to challenge it. Instead, as concern over the power of the USSR was rising, the West made efforts to limit Soviet/Chinese collaboration by maintaining diplomatic ties with China. President Richard Nixon was the first to make significant steps toward improving relations with China. He was followed by Presidents Carter, Reagan, and George H.W. Bush. Each also helped to expand economic and cultural relations with China so that by the late 1980s a rapprochement (a diplomatic term for reconciliation) with China was fully established.

It is important to appreciate that during the half century (including the present moment) that the West has been diplomatically engaging with China, it has remained a communist country. Its embrace of liberal capitalist and centrally-directed economic practices should not be mistaken for anything like an equal enthusiasm for democracy. China is a one-party communist state known as the People's Republic of China (PRC). The Chinese Communist Party (CCP) is a tightly organized, hierarchical bureaucracy with authoritarian power concentrated at the top. The CCP controls all levels of society and there are no elections (**Figure 10.21**). While fostering the miraculous growth of its economy, it has kept the majority of wealth in the hands of an elite. At the same time, the antidemocratic CCP exercises outright censorship through its control of the media and has been particularly effective in mollifying the Chinese public by distracting them from the real inequities in society. The model of government in China is best described at authoritarian capitalism.

Domino Theory Although the United States was unable to stop Castro's rise to power in Cuba, the end of World War II marked the rise of the United States to a dominant position among countries of the core. Following the war, the tension that arose between East and West translated into a U.S. foreign policy pitched in opposition to the Soviet Union. Domino theory underlay that foreign policy, which included economic, political, and military objectives directed at preventing Soviet world domination. **Domino theory** held that if one country in a region chose, or was forced to accept, a communist political and economic system, neighboring countries would fall to communism as well, just as one falling domino in a line of dominos causes all the others to fall. The means of preventing the domino-like spread of communism was often military aggression (**Figure 10.22**).

Domino theory first took root in 1947, when the postwar United States feared communism would spread from Greece to Turkey to Western Europe. It culminated in U.S. wars in

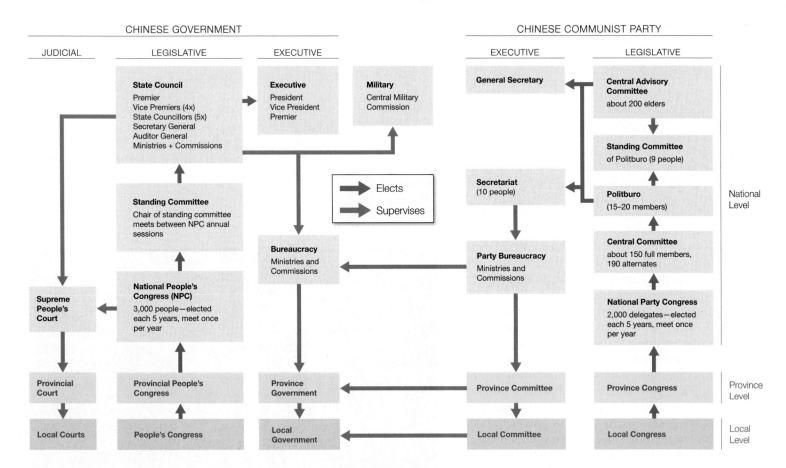

▲ **Figure 10.21 Organizational chart of the Chinese Communist Party** As the ruling party of an authoritarian government, the Chinese Community Party is highly bureaucratized with units of organization all the way to the neighborhood or village level.

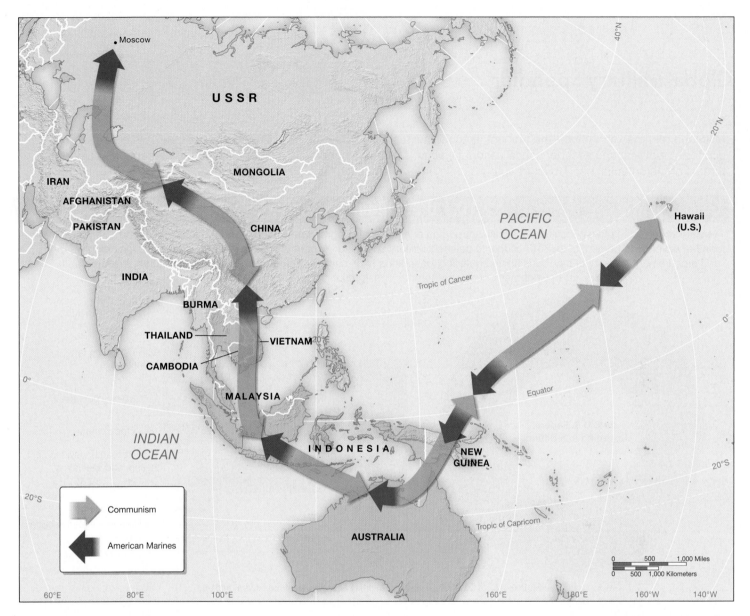

▲ **Figure 10.22 Domino Theory** As this graphic shows, the fear in the capitalist world was that if one country fell to communism its neighbors would as well. Domino theory is still invoked today though not necessarily in terms of communism. (*Source:* Bunge's Atlas of Nuclear way; http://indiemaps.com/blog/2010/03/wild-billbunge/)

Korea, Vietnam, Laos, Nicaragua, and El Salvador and geno-cide in Cambodia, when Khmer Rouge communist revolu-tionaries overcame the U.S.-backed military government in 1975. However, preventing the domino effect was based not just on military aggression. Cooperation was also emphasized. For example, the international military alliance NATO (North Atlantic Treaty Organization) was established in 1949 with the stated purpose of *safeguarding* the West against Soviet aggres-sion. After World War II, the core countries set up a variety of foreign aid, trade, and banking organizations to open for-eign markets and bring peripheral countries into the global capitalist economic system. The strategy not only improved productivity in the core countries but was also seen as a way of strengthening the position of the West in its confrontation with the East.

The New World Order

With the fall of the Berlin Wall in 1989, the Cold War came to an end. Socialist and communist countries, such as China and the USSR had already been warming to Western-style capitalist economic development. With the collapse of the latter in December 1991, President George H.W. Bush made a speech referring to a "new world order." The **new world order** assumes that with the triumph of capitalism over communism, the United States is the world's only superpower and therefore its policing force. With the political, economic, and cultural dominance of the United States comes the worldwide promo-tion of liberal democracy and of a global economy predicated on transnational corporate growth through organizations like the World Bank and the World Trade Organization. Military prowess is also an important aspect of being the world's police

10.3 Visualizing Geography

Global Military Spending

Military or defense spending—the amount of economic resources devoted to military purposes—reflects a range of economic, political and security concerns for any state. Defense spending is often a reflection of perceptions of external threats or the potential need to exert aggression in order to establish legitimacy or superiority.

10.1 U.S. Military Spending

Military expenditure often constitutes a significant portion of a national budget. The United States, which is the world's largest military spender, devoted 18 percent of its 2013 budget to defense. Military spending includes a wide range of budget items including weapons and procurement, personnel pay, housing and outfitting, operations (war, security maneuvers, etc.), administration, research and development, nuclear programs, construction and other things.

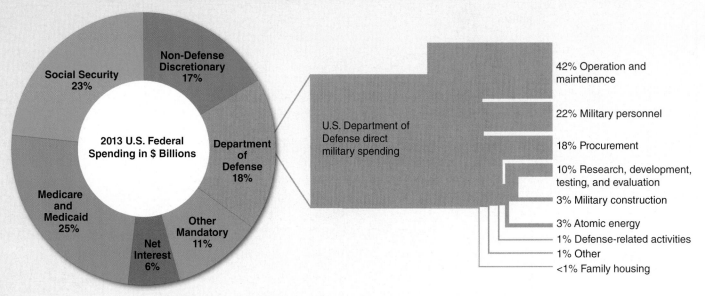

Figure 10.1.1 2013 U.S. military spending as a portion of the Federal budget.

While the U.S. population—Democratic, Republican and independent voters—would all like to see a reduction in military spending, Congress rarely responds positively to the American peoples' preferences on defense spending issues.

The U.S. spends more on defense than the next nine biggest spenders combined.

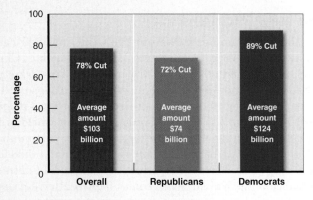

Figure 10.1.2 U.S. defense spending cuts made by survey respondents.

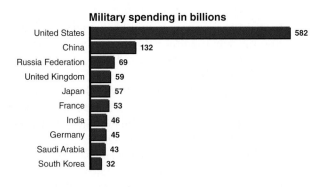

Figure 10.1.3 U.S. military spending compared to other countries.

10.2 Global Military Spending

In 2013, the latest year for which data are available, global military spending was $1.747 trillion. Military spending continues to increase in peripheral countries, likely because of a combination of increasing economic growth and a related concern for security but also because of the preponderance of autocratic regimes in the region and a possibly emerging regional arms race.

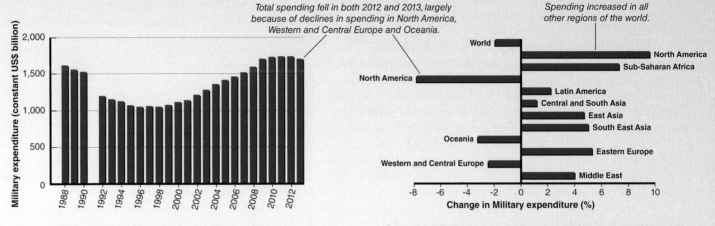

Total spending fell in both 2012 and 2013, largely because of declines in spending in North America, Western and Central Europe and Oceania.

Spending increased in all other regions of the world.

Figure 10.2.1 **World military expenditure, 1988-2013**
Information from the Stockholm International Peace Research Institute (SIPRI), http://www.sipri.org/

Figure 10.2.2 **Changes in military expenditure by region, 2012-2013**
Information from the Stockholm International Peace Research Institute (SIPRI), http://www.sipri.org/

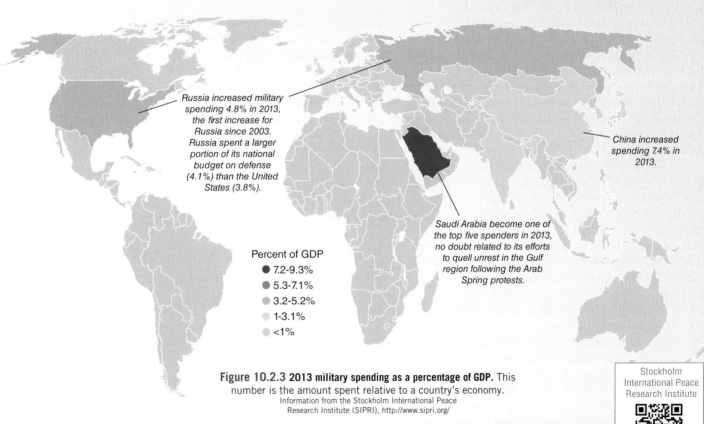

Russia increased military spending 4.8% in 2013, the first increase for Russia since 2003. Russia spent a larger portion of its national budget on defense (4.1%) than the United States (3.8%).

China increased spending 7.4% in 2013.

Saudi Arabia become one of the top five spenders in 2013, no doubt related to its efforts to quell unrest in the Gulf region following the Arab Spring protests.

Percent of GDP
- 7.2-9.3%
- 5.3-7.1%
- 3.2-5.2%
- 1-3.1%
- <1%

Figure 10.2.3 **2013 military spending as a percentage of GDP.** This number is the amount spent relative to a country's economy.
Information from the Stockholm International Peace Research Institute (SIPRI), http://www.sipri.org/

Stockholm International Peace Research Institute

http://goo.gl/d7wcr7

Sources: *Consulting the American People on National Defense Spending*, by the Program for Public Consultation, in collaboration with the Stimson Center and the Center for Public Integrity (2012), http://www.publicintegrity.org/2012/05/10/8856/public-overwhelmingly-supports-large-defense-spending-cuts, International Institute for Strategic Studies, Stockholm International Peace Research Institute (SIPRI), Congressional Budget Office, U.S. Department of Defense

1. What is the correlation between the top military spending nations in the world and their economic status? What do you think the link is between security, wealth and military spending, or is there one?

2. Why do you think there is unilateral support to reduce military spending in the U.S., but a reluctance to do so? Why is education budgeted so low in comparison? What about a nation's budget reflects its priorities?

force as **Box 10.3** Visualizing Geography: Global Military Spending, makes clear. It also suggests that there are countries vying for dominance.

Terrorism   However, the move toward liberal, Western-style democracies and capitalist consumption practices necessary to the success of the new global economy has created instability in many parts of the world. This instability is especially problematic where the Cold War struggle between the United States and the Soviet Union was once waged, in countries that once appeared ripe for succumbing to communism.

With the emergence of a new world order, radical forms of warfare and political practices replaced more conventional ones. The attacks on the World Trade Center and the Pentagon on September 11, 2001, and the resulting war on terrorism are prime examples of the political and economic restructuring impulses of the new world order. The terrorist attacks and the subsequent response by the United States and other governments make clear that terrorism is the pivotal factor in current global geopolitics.

Terrorism is a complicated concept whose definition very much depends on the social and historical context. A very simple definition is that **terrorism** is the threat or use of force to bring about political change. It is most commonly understood as actions by individuals or groups of individuals against civilian populations to undermine state practices or institutions. But the state can also be an agent of terrorism. Terrorism involves violent acts directed against society—whether by antigovernment actors, governments themselves, angry mobs or militants, or even psychotic individuals—and it will always mean different things to different people.

The term *terrorism* was first used during the French Revolution (1789–1795) to describe the new revolutionary government's repression of its people during the "Reign of Terror." By the mid-twentieth century, some used the term to describe many left-wing groups, as well as subnationalists (minority groups within the nation-state), or radical ethnic groups. In the 1980s, the violent activity of hate groups—anti-gay, racist, neo-Nazi organizations—in the United States was described by some as terrorist. At the same time, terrorism internationally was identified as a brand of ethnic or subnational warfare sponsored by rogue regimes.

Ethnic and subnational terrorism affects many countries today, including Russia, Uzbekistan, India, China, Colombia, the Philippines, Israel, and Rwanda. In Sri Lanka, the Tamil minority rebelled in the early 1980s against the oppressive Sinhalese majority. Originally a political movement, eventually the Liberation Tigers of Tamil Eelam launched a brutal campaign of terror that left over 65,000 people dead. Another ethnic or subnational violent struggle involves the Russian region of Chechnya, where the population has demanded independence. For the Muslim Chechens, Russia is acting as a terrorist state by employing military force to keep them from gaining independence. Chechens believe they have little in common politically, historically, or culturally with Russia.

Faith-based Terrorism   While subnational resistance organizations using terrorist tactics continue to operate throughout the world, the most widely recognized terrorism of the new century has religious roots. The September 11 attacks have helped to bring the realities of religious terrorism sharply into public focus. The connection between religion and terrorism is nothing new as terrorism has been perpetrated by religious fanatics for more than 2,000 years. Indeed, words like *zealot, assassin,* and *thug* all stem from fundamentalist religious movements of previous eras. And while the links between Muslims and terrorism in the world today are especially strong and geographically widespread, it is critical to understand that Muslim terrorism is not the only form of terrorism and that domestic terrorism sparked by fundamentalist Christian organizations has taken the lives of hundreds of innocent people and continues to be a threat in the United States and elsewhere.

A chilling example of domestic terrorism in the United States is the bombing of the Alfred P. Murrah Federal Office Building in Oklahoma City in April 1995, when 168 people were killed (**Figure 10.23**). In June 1997, Timothy McVeigh, a U.S. Army veteran and Christian white supremacist, was convicted in federal court of perpetrating the attack. McVeigh was executed by lethal injection in June 2001, and his accomplice, Terry L. Nichols, was sentenced to life in prison. Both McVeigh and Nichols were connected to the Christian Identity movement through a group that came to be known as the Michigan Militia Corps, a paramilitary survivalist organization. The Christian Identity movement is just one of several religious extremist groups in the United States, including other forms of white supremacy movements, apocalyptic cults, and Black Hebrew Israelism. The Christian Identity movement is based on a belief in the superiority of whiteness as ordained by God.

The War on Terror in Iraq   The United States responded to the terrorist attacks of September 11, 2001, by declaring a global war against terrorism and identifying first Afghanistan and then Iraq as the greatest threats to U.S. security (**Figure 10.24**). Although the evidence of involvement in the 9/11 attacks by Iraq and its leader, Saddam Hussein, was highly questionable, on March 19, 2003, after amassing over 200,000 U.S. troops in the Persian Gulf region, President George W. Bush ordered the bombing of the city of Baghdad, Iraq. The declaration of war

◀ Figure 10.23 The Oklahoma City Bombing Memorial Until the attacks of September 11, 2001, the Oklahoma City bombing was the worst terrorist incident—domestic or foreign—to be perpetrated on U.S. soil. The blast was caused by a deadly combination of fuel oil and fertilizer placed in a van parked outside the building and detonated remotely.

▶ Figure 10.24 The assault on the World Trade Center on September 11, 2001 Shown here are the burning towers of the World Trade Center just moments after highjacked jet liners were deliberately flown into them.

and invasion occurred without the explicit authorization of the UN Security Council, and some legal authorities take the view that the action violated the UN Charter. Some of the staunchest U.S. allies (Germany, France, and Canada) as well as Russia opposed the attack and hundreds of thousands of antiwar protestors repeatedly took to the streets throughout the world for the weeks and months preceding and following the onset of war launched by the coalition forces of the United Kingdom and the United States.

Though the West withdrew all of its troops from Iraq by the end of 2011, the legacy of the war persists today. For the United States, the war resulted in 4,459 deaths. Another distressing and long-term aspect of the Iraq war, as well as other U.S. wars, is its impact on returning veterans' mental health; the Department of Veteran Affairs director has stated that there are about 1,000 suicide attempts per month among the nation's 25 million veterans (from all U.S. wars). Besides the mental health impacts, there are, of course, also very significant physical ones, including lost limbs and serious brain and organ injuries. The U.S. Treasury estimates the direct cost of the war at $845 billion. Nobel-prize winning economist Joseph Stiglitz and his coauthor Linda Bilmes argue that the true cost of the war is closer to $3 trillion when the indirect costs are also

10.4 Spatial Inequality   Global Corruption

Corruption can be defined in a number of ways but a simple definition provides the most insight into the variety of its expression as well as its effects. **Corruption** is any abuse of a position of trust (in either the public or private sector) in order to gain an unfair advantage. Corrupt actions include a wide range of activities from accepting bribes or gifts to manipulating elections, laundering money, or defrauding investors. These different forms are frequently categorized as different degrees of corruption.

- Petty corruption includes small acts of abuse such as small gifts or using personal connections to obtain favors. This form of corruption often happens where poorly paid officials take bribes to supplement their incomes.

- Grand corruption occurs at the highest levels of government often in those countries ruled by dictators or where the policing apparatus is weak. Rich countries are not immune from grand corruption, however.

- Systemic corruption is such that corruption has become the rule rather than the exception to it. This form of corruption happens when there is a lack of transparency in governing or where there is a culture of impunity.

The spatial unevenness of corruption reflects the fact that corruption is most likely to occur in countries besieged by conflict and poverty. While the correspondence is not perfect, data suggests the poorer the country the higher the level of corruption, or at least, according to the corruption watch organization Transparency International, perceptions of corruption are highest in countries with the greatest percentage of very poor people or people in highly insecure situations (**Figure 10.D**).

Transparency International

http://goo.gl/mBpDxv

1. What are the different sorts of corruption, especially as it occurs in the public sector?

2. Why should the global community be worried about corruption; isn't it a victimless crime?

▼ **Figure 10.D Global Corruption Perception** This graph links perceptions of public sector corruption to unemployment and inflation. Black lettering along the horizontal axis shows a country's GDP. The scale goes from 0 (highly corrupt) to 100 (very clean).

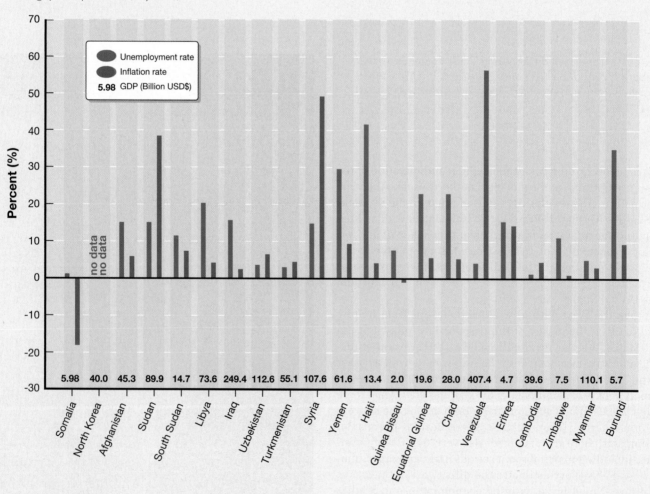

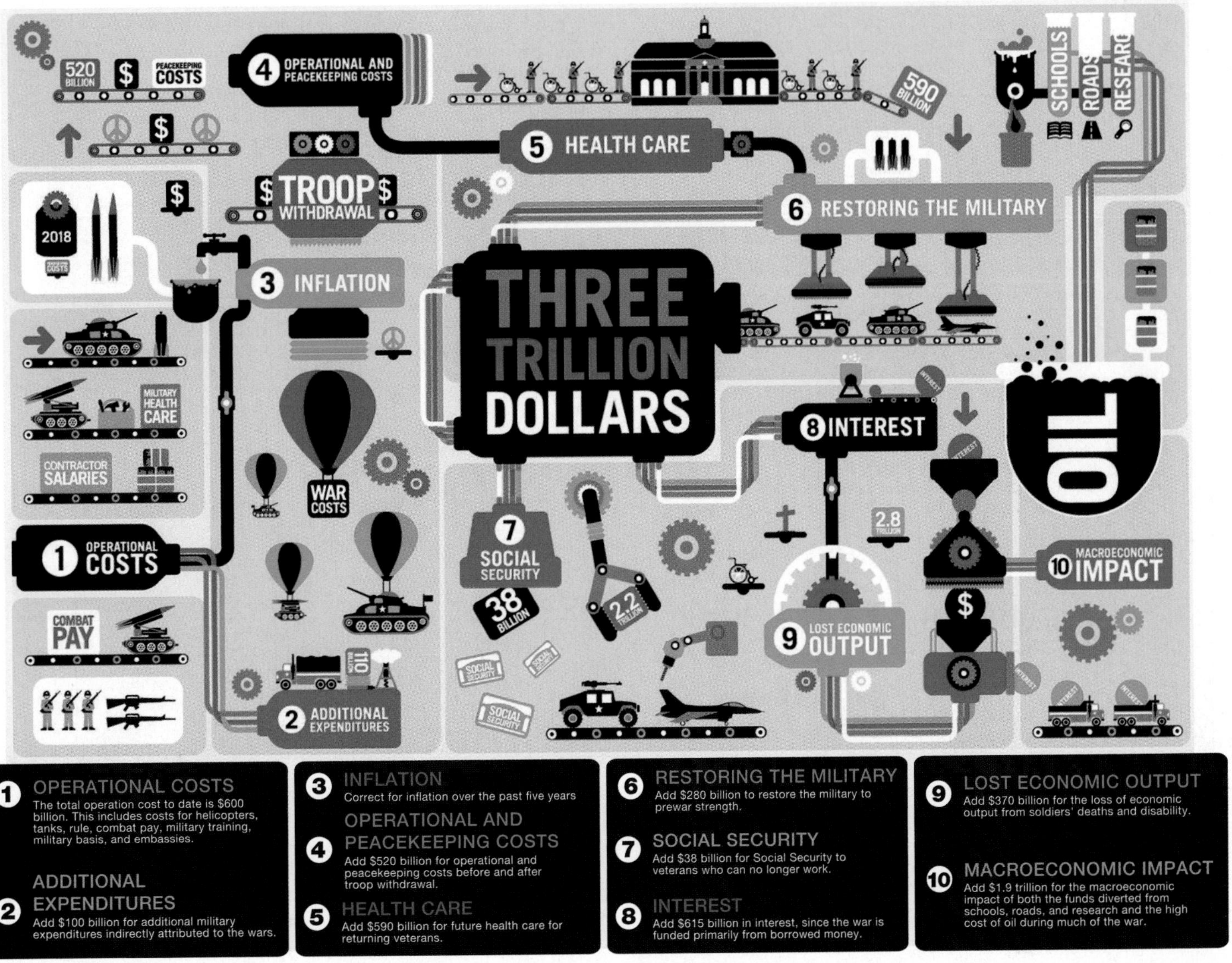

▲ Figure 10.25 Cost of the Iraq and Afghanistan Wars The cost of war is not only in human lives and expenditures on military materiel. Related costs include loss of economic output, increasing health care costs and social security expenditures for some veterans. (*Source:* Based on infographic: "The Three-Trillion Dollar War: Its Cost in Ten Steps" by Jason Bishop and Matt Owens, Good Magazine, November 6, 2008. http://magazine.good.is/infographics/the-three-trillion-dollar-war-its-cost-in-ten-steps)

accounted for, including interest on the debt raised to fund the war, the rising cost of oil, health care costs for returning veterans, and the cost of replacing destroyed military hardware and degraded operational capacity (**Figure 10.25**).[1]

After the Iraq War

For Iraq, somewhere between 100,000 and 110,000 combined Iraqi civilian and military deaths occurred due to the war. Despite the fact that "major combat" has ended, peace has not returned to Iraq. In fact, the country is experiencing a dramatic resurgence of conflict as the Islamic State of Iraq and Syria (ISIS) has moved through the northern part of the country capturing cities and surrounding territory. Discrimination against Sunnis since the fall of Saddam Hussein enabled this Sunni jihadist militia group to consolidate and gain support. And though al-Qaeda has distanced itself from ISIS, the group does adhere to al-Qaeda's global jihadist principles including anti-Westernism, strict interpretation of Islam, and the promotion of brutal religious violence against Shia Muslims and Christians and videotaped beheadings of Westerners.

The aim of ISIS is to return to an earlier, and they would say more pure, period of Islam by establishing its own rule through conquered territory. On June 24, 2014, after securing a significant stretch of territory, a caliphate—an Islamic state led by a Muslim religious and political leader—was declared with Abu Bakr al-Baghdi as its caliph (**Figure 10.26**). The keys to the success of the Islamic state are said by experts to be its military acumen, its production of effective propaganda and control of social media, particularly Twitter, its ability to raise a significant war chest

[1]J. Stiglitz and L. Blimes, *The Three Trillion Dollar War.* New York: WW Norton, 2008.

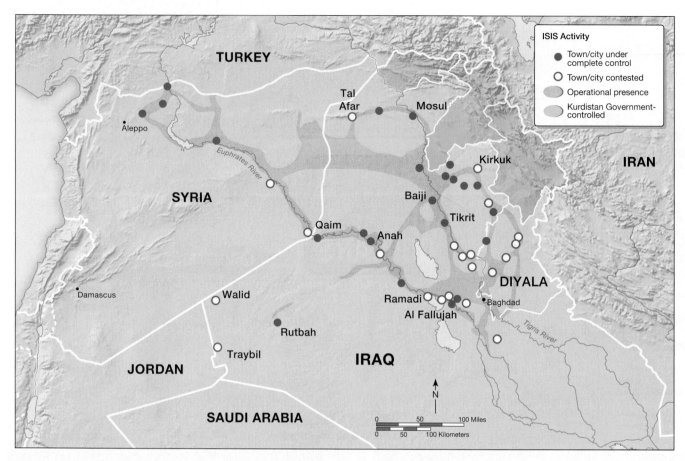

▲ **Figure 10.26 The Caliphate of the Islamic State** This map shows the extent of the territory of ISIS as well as the cities that have been taken under complete control by the ISIS army as of late 2014. (*Source:* Data from Institute for the Study of War)

(estimated to be around $2 billion) and its capture of weapons and vehicles provided by the United States to the Iraqi Army.

APPLY YOUR KNOWLEDGE

1. What is the new world order and what is terrorism? Why is terrorism considered to be a complex concept? What is the difference between terrorist militias and state military forces?

2. Conduct an Internet search on the Iraq war and how it has affected your city and state. How many soldiers from your area have been killed? In economic terms, how much has the war cost your city and state? (To determine the economic impact of the war, consult the National Priorities Project database.)

INTERNATIONAL AND SUPRANATIONAL ORGANIZATIONS AND NEW REGIMES OF GLOBAL GOVERNANCE

Just as states are key players in political geography, international and supranational organizations have become important participants in the world system in the last century. These organizations have become increasingly important means of

achieving goals that could otherwise be blocked by international boundaries. These goals include the freer flow of goods and information and more cooperative management of shared resources, such as water.

Transnational Political Integration

Perhaps the best-known international organization operating today is the United Nations (**Figure 10.27**). The period since World War II has seen the rise and growth not only of large international organizations but also of new regional arrangements. These arrangements vary from local ones, such as the Swiss-French cooperative management of Basel-Mulhouse airport, to more extensive ones, such as the North American Free Trade Agreement (NAFTA), which joins Canada, the United States, and Mexico into a single trade region.

Regional organizations and arrangements now address a wide array of issues, including the management of international watersheds and river basins, such as the Great Lakes of North America and the Danube and Rhine rivers in Europe. They also oversee the maintenance of health and sanitation standards and coordinate regional planning and tourism. Such regional arrangements seek to overcome the barriers to the rational solution of shared problems posed by international boundaries. They also provide larger arenas for the pursuit of political, economic, social, and cultural objectives.

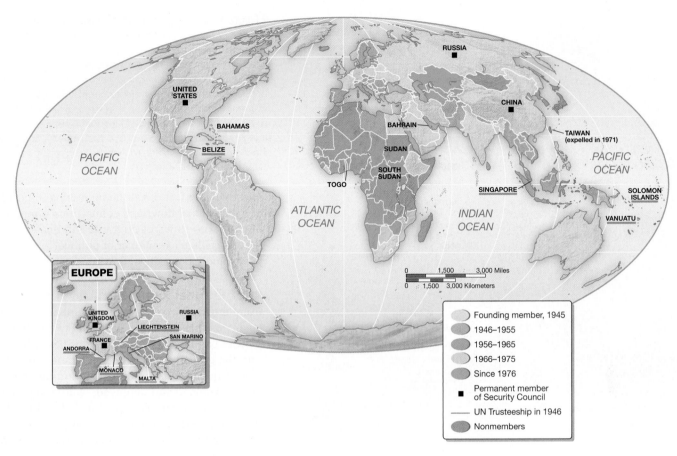

▲ **Figure 10.27 The United Nations** The U.N. Charter was approved by the U.S. Senate in July 1945, raising hopes for a more long-lived organization than the ineffective League of Nations. Located in New York City, the United Nations is composed of a Security Council, which includes permanent members, the United States, Britain, China, France, and Russia, and a General Assembly, which includes all those countries identified on the map.

Globalization, Transnational Governance, and the State

As we have already noted, globalization has been as much about restructuring geoeconomics as it has been about reshaping geopolitics. In fact, some globalization scholars believe that the impact of globalization on politics has been so profound that globalization is leading to the diminution of the powers of the modern state, if not its ultimate disappearance. These scholars believe that because the modern state is organized around a bounded territory and because globalization is creating a new economic space that is transnational, the state is increasingly incapable of responding to the needs of the new transnational economy. Although we do not subscribe to this position, we do recognize that the state is undergoing dramatic changes that are restructuring its role with respect to both local, domestic concerns as well as global, transnational ones.

Supranational organizations

A collection of individual states with a common goal that may be economic and/or political in known as a **supranational organization**. By organizing and regulating designated operations of the individual member states, these organizations diminish, to some extent, individual state sovereignty in favor of the collective interests of the entire membership. For example, at the end of World War II, European leaders realized that Europe's fragmented state system was insufficient to the demands and levels of competition coalescing within the world political and economic system. They responded by creating an entity that would preserve important features of state sovereignty and identity. They also intended to create a more efficient intra-European marketing system and an entity more competitive in global transactions.

Figure 10.28 shows the original member countries of the European Economic Community, which evolved into the EU in 1992, the existing members of the EU, and those countries that are currently candidates for admission. The EU holds elections, has its own parliament and court system, and decides whether and when to allow new members to join. Generally speaking, the EU aims to create a common geographical space within Europe in which goods, services, people, and information move freely and in which a single monetary currency prevails. Whether an EU foreign policy will ever be accomplished remains to be seen, but a common European currency, the euro, has been in circulation since 2002.

The Decline of the Superpower?

In the twentieth century, from the end of World War II until 1989, when the Berlin Wall was dismantled, world politics was organized around two

◄ **Figure 10.28** **Membership in the European Union** The goal of the European Union is to increase economic integration and cooperation among member states. The treaty established European citizenship for citizens of each member state, enhanced customs and immigration agreements, and allowed for the establishment of a common currency, the euro, which is currently in circulation among all of the original 12 members except Denmark and the United Kingdom. Newer members, including Finland, Austria, Slovakia, Latvia, Lithuania, and Malta, also use the euro in addition to their own currencies.

(*Source:* Data from Institute for the Study of War)

effectively. In fact, what the increasing importance of transnational flows and connections—from flows of capital to flows of migrants—indicates is that the state is less a container of political or economic power and more a site of flows and connections.

International Regimes As flows and connections multiply, contemporary globalization has made possible a steadily shrinking world. In short, politics can move beyond the confines of the state into the global political arena, where rapid communications enable complex supporting networks to be developed and deployed, facilitating interaction and decision making. One indication of the increasingly global nature of politics outside of formal political institutions is the rise of environmental organizations whose purview and membership are global, as discussed in Chapter 4.

What is interesting about the institutionalization of global politics is that it has been less involved with the traditional preoccupations of relations between states and military security issues and more involved with issues of economic, ecological, and social security. The massive growth in flows of trade, foreign direct investment, financial commodities, tourism, migration, crime, drugs, cultural products, and ideas has been accompanied by the emergence and expansion

superpowers. The capitalist West rallied around the United States, and the communist East around the Soviet Union. But with the fall of the Berlin Wall signaling the "end of communism," the bipolar world order came to an end. The concept of the new world order, organized around global capitalism, emerged and has increasingly solidified around a new set of political powers and institutions, which have recast the role of the state.

The increasing importance of trade-facilitating organizations, such as the EU, NAFTA, the Association of Southeast Asian Nations (ASEAN), the Organization of Petroleum Exporting Countries (OPEC), and the World Trade Organization (WTO), is the most telling indicator that the world, besides being transformed into one global economic space, is also experiencing global geopolitical transformations (**Figure 10.29**). These organizations are unique in modern history, as they aim to treat the world and different regional clusters as seamless trading areas free of the rules that ordinarily regulate national economies.

The state must now contend with a whole new set of processes and other important political actors on the international stage as well as within its own territory. For instance, the transnational financial network that was established in the 1980s is far beyond the control of any one state, even a very powerful state like the United States, to regulate

▼ **Figure 10.29** **Members of the World Trade Organization, 2014** A global economy requires that countries provide a relatively barrier-free space for the efficient movement of goods, ideas and capital.

(*Source:* Data from World Trade Organization http://www.wto.org/english/thewto_e/countries_e/org6_map_e.htm)

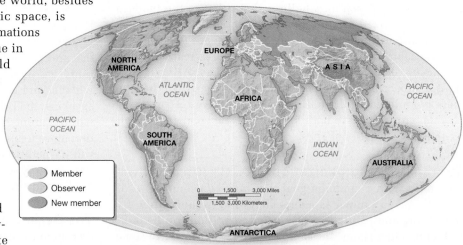

of global and regional institutions to manage and regulate these flows. The modern state has been drawn increasingly into this complex of global, regional, and multilateral systems of governance. And as the state has been drawn into these new activities, it has shed or deemphasized some of its previous responsibilities, such as maintenance of social welfare.

The involvement of the state in new global activities, the growth of supranational and regional institutions and organizations, the critical significance of transnational corporations to global capital, and the proliferation of transnational social movements and professional organizations are captured by the term **international regime**. The term reflects the fact that the arena of contemporary politics is now international, so much so that even city governments and local interests groups—from sister-city organizations to car clubs—are making connections and conducting their activities both within and beyond the boundaries of their own states. An example of this is the human rights movement that has gained ascendancy over the last four to five decades.

Human Rights   **Human rights**, including the rights to justice, freedom, and equality, are considered by most societies to belong automatically to all people. Until World War II, safeguarding human rights was the provenance of states whose rules and regulations legislated the proper treatment of its citizens, from schoolchildren to prisoners. Since the late 1940s and 1950s, nearly all states have come to accept the importance of a comprehensive political and legal framework that focuses on human rights and that allows an international organization to intervene in the operations of a sovereign state that is in violation of the International Bill of Human Rights adopted by the United Nations in 1948 (**Figure 10.30**).

In 1998, the United Nations realized another step in the protection of human rights by adopting a treaty to establish a permanent International Criminal Court (ICC). In establishing the court, Kofi Anan, then secretary-general of the United Nations, stated: "Our hope is that, by punishing the guilty, the ICC will bring some comfort to the surviving victims and to the communities that have been targeted. More important, we hope it will deter future war criminals, and bring nearer the day when no ruler, no State, no junta and no army anywhere will be able to abuse human rights with impunity." By forming the treaty, the United Nations aimed to create a permanent mechanism to bring to justice the perpetrators of such crimes as genocide, ethnic cleansing, and sexual slavery and put an end to the impunity so often enjoyed by those in positions of power.

The court has the mandate to try individuals, not states, for crimes committed in the present (after July 2002). What makes the ICC unique is the principle of complementarity, which means that the court can only exercise its jurisdiction when a national court is unable or unwilling to genuinely do so itself. Thus, the importance of the international system has increased over the last 50 years of the twentieth century and appears likely to continue to do so into the twenty-first century. It is important to note, however, that not all states support the ICC. Seven UN members voted

against the treaty to establish the ICC. These were the United States, China, Iraq, Israel, Libya, Qatar, and Yemen.

An example of global human rights legislation is the UN Declaration on Rights of Indigenous People, approved in 2007 when 143 countries voted for it; 11 countries abstained; and four—Australia, Canada, New Zealand, and the United States,

▼ **Figure 10.30 International Declaration of Human Rights** While the International Bill of Human Rights is over five and a half decades old, it is still being amended and extended.

all countries with very significant numbers of indigenous people—voted against it. The declaration emphasizes the rights of indigenous peoples to maintain and strengthen their own institutions, cultures, and traditions and development in keeping with their own needs and aspirations. It also prohibits discrimination against them and promotes their full and effective participation in all matters that concern them. The four "no" voting countries expressed concern about the vague language of the declaration and the impact this might have on existing, already settled treaties.

Children's Rights

An additional and often overlooked aspect of human rights is children's rights. In 1989, the United Nations adopted the Convention on the Rights of the Child, and despite the convention's nearly universal ratification (only the United States and Somalia have not ratified it), many of the most basic rights of children are still not being protected. The convention promises to support **children's rights** around the world—the fundamental rights to life, liberty, education, and health care. Among the fundamental safeguards the convention provides are protection of children in armed conflict; protection from discrimination; protection from torture or cruel, inhuman, or degrading treatment or punishment; protection within the justice system; and protection from economic exploitation.

And yet, despite the existence of the convention, around the world street children are killed or tortured by police, recruited or kidnapped to serve as soldiers in military forces or labor under extremely difficult conditions, forced into prostitution, and imprisoned. Refugee children, often separated from their families, are vulnerable to exploitation, sexual abuse, and domestic violence. Keeping the promises made in the Convention on the Rights of the Child is one of the biggest challenges of the twenty-first century.

The emergence of human rights as a globally relevant issue has occurred as groups and organizations, both governmental and nongovernmental, have been able to debate and discuss issues that concern all people everywhere and can do so at the international level through conferences, e-mail, listservs, and direct action at international events. The phenomenon of different people and groups across the world in common cause is known as global civil society. **Global civil society** is composed of the broad range of institutions that operate between the private market and the state.

APPLY YOUR KNOWLEDGE

1. What are some examples of how global geopolitics is changing the flow of political and economic power? How has the role of the state changed over time? Think about the role of the international regime, supranational organizations, and the issue of human rights.

2. Research arguments made by the United States about why it has not ratified the UN Convention on the Rights of the Child or joined the International Criminal Court. Do you find the arguments compelling? Explain in a paragraph your reaction to the arguments and whether or not you support them and why.

The Two-Way Street of Politics and Geography

Political geography can be viewed according to two contrasting orientations. The first orientation sees it as the *politics of geography*. This perspective emphasizes that *geography*—the areal distribution/differentiation of people and objects in space—has a very real and measurable impact on politics. Regionalism, sectionalism, and irredentism illustrate how geography shapes politics. This politics-of-geography orientation is also a reminder that politics occurs at all levels of the human experience, from the international order down to the neighborhood, household, and individual body.

The second orientation sees political geography as the *geography of politics*. This approach analyzes how *politics*—the tactics or operations of the state—shapes geography. Mackinder's heartland theory and the domino theory, discussed earlier in this chapter, attempt to explain how the geography of politics works at the international level. In the heartland theory, the state expands into new territory to relieve population pressures. In the domino theory, as communism seeks new members, it expands geographically to incorporate new territories. As an example of the geography-of-politics orientation, consider Palestine and Israel. An examination of a series of maps of Palestine/Israel since 1923 reveals how the changing geography of this area is a response to changing international, national, regional, and local politics.

The Politics of Geography

Territory is often regarded as a space to which a particular group attaches its identity. Related to this concept of territory is the notion of **self-determination**, which refers to the right of a group with a distinctive politico-territorial identity to determine its own destiny, at least in part, through the control of its own territory. The Palestinians in Israel are a prime example of the urge for self determination.

Self-Determination and the Israel-Palestine Conflict

The Israel-Palestine conflict is a modern territorial dispute with its roots in post–World War I colonial politics. Under the auspices of the League of Nations, the British government assumed control of Palestine and its population in 1923, marking the beginning of the British Mandate of Palestine, which included the areas of Trans-Jordan and Palestine. Within the borders of the Palestine mandate, the conflict began to take shape. **Figure 10.31** maps out how subsequent conflicts have created territorial changes over time.

Subsequent to the mandate period, three particular events stand out as defining moments for the geopolitical shape of the former territory of Palestine and contemporary Israel: the 1947 U.N. Partition Plan, the Palestine War of 1948, and the Six-Day War in 1967. For more details on these events, refer to the time line of the conflict in **Figure 10.32**. At its very core, the conflict is between the Israelis and the Palestinians. On the one side, efforts by the Jewish community to establish an exclusively Jewish state in Palestine have

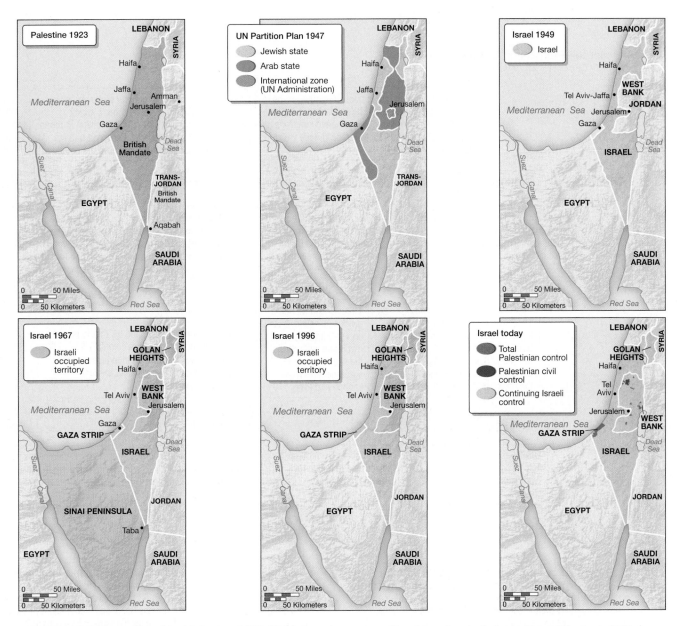

▲ **Figure 10.31 Maps of Israel and Palestine, 1923–2014** Since the creation of Israel out of much of what had been Palestine in 1947, the geography of the region has undergone significant transformation. A series of wars between Israelis and Arabs and a number of political decisions by Israel have produced the changing geographies we see here. Importantly, at a smaller scale, of Israel today would not show continuous regions of Palestinian territory, but highly fragmented ones

(*Source:* Adapted with permission from Prentice Hall, from J. M. Rubenstein, *The Cultural Landscape: An Introduction to Human Geography*, 5th ed., 1996, p. 233.)

largely succeeded, as is clear by the creation of the modern state of Israel. On the other side, the Palestinian community continues to struggle against the effects of displacement and colonization and remains committed to the founding of an independent Palestinian state.

The conflict is essentially about land. Currently, Israel controls over 75 percent of former Palestine; this includes a military and settler occupation of the West Bank. Israel also controls land, sea, and flight access to the Gaza Strip. Despite this external control, Palestinians remain firmly engaged in a national project of establishing their own independent state in the remaining 25 percent of the land. This struggle is

inseparable from Palestinian efforts to resist the Israeli military occupation of the West Bank and Gaza Strip.

The Palestinians' passionate desire for self-determination coupled with Israel's security needs make political negotiations at the governmental level complex, protracted, and emotionally loaded. The upwelling of violence in 2014 in Israel and Palestine is evidence of this.

The kidnapping and subsequent killing of three male teenagers living in an illegal (according to international law) Jewish settlement in the Palestinian West Bank is assumed to have been conducted by Hamas, a militant organization founded to liberate Palestine from the Israeli

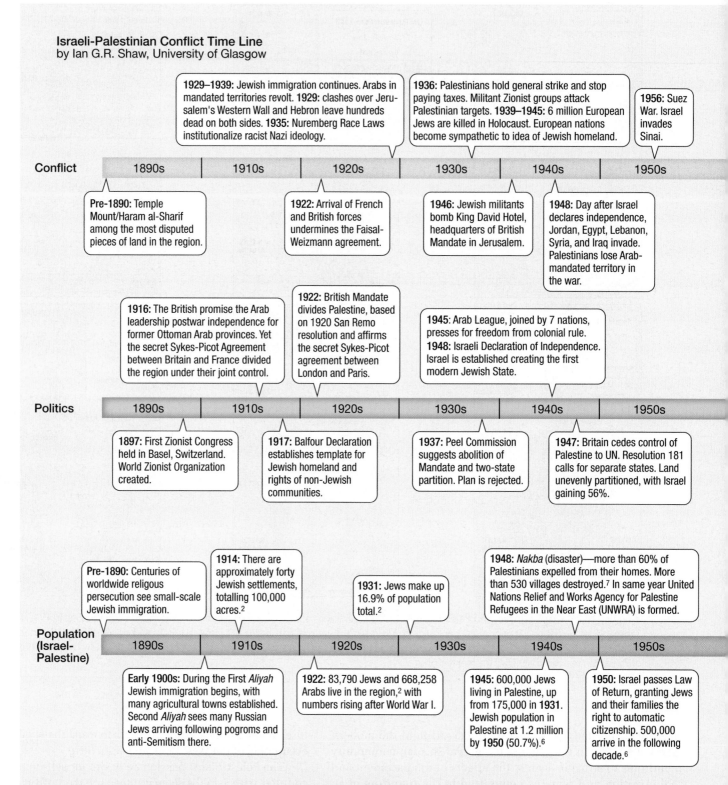

Israeli-Palestinian Conflict Time Line
by Ian G.R. Shaw, University of Glasgow

1929–1939: Jewish immigration continues. Arabs in mandated territories revolt. **1929:** clashes over Jerusalem's Western Wall and Hebron leave hundreds dead on both sides. **1935:** Nuremberg Race Laws institutionalize racist Nazi ideology.

1936: Palestinians hold general strike and stop paying taxes. Militant Zionist groups attack Palestinian targets. **1939–1945:** 6 million European Jews are killed in Holocaust. European nations become sympathetic to idea of Jewish homeland.

1956: Suez War. Israel invades Sinai.

Conflict | 1890s | 1910s | 1920s | 1930s | 1940s | 1950s

Pre-1890: Temple Mount/Haram al-Sharif among the most disputed pieces of land in the region.

1922: Arrival of French and British forces undermines the Faisal-Weizmann agreement.

1946: Jewish militants bomb King David Hotel, headquarters of British Mandate in Jerusalem.

1948: Day after Israel declares independence, Jordan, Egypt, Lebanon, Syria, and Iraq invade. Palestinians lose Arab-mandated territory in the war.

1916: The British promise the Arab leadership postwar independence for former Ottoman Arab provinces. Yet the secret Sykes-Picot Agreement between Britain and France divided the region under their joint control.

1922: British Mandate divides Palestine, based on 1920 San Remo resolution and affirms the secret Sykes-Picot agreement between London and Paris.

1945: Arab League, joined by 7 nations, presses for freedom from colonial rule. **1948:** Israeli Declaration of Independence. Israel is established creating the first modern Jewish State.

Politics | 1890s | 1910s | 1920s | 1930s | 1940s | 1950s

1897: First Zionist Congress held in Basel, Switzerland. World Zionist Organization created.

1917: Balfour Declaration establishes template for Jewish homeland and rights of non-Jewish communities.

1937: Peel Commission suggests abolition of Mandate and two-state partition. Plan is rejected.

1947: Britain cedes control of Palestine to UN. Resolution 181 calls for separate states. Land unevenly partitioned, with Israel gaining 56%.

Pre-1890: Centuries of worldwide religous persecution see small-scale Jewish immigration.

1914: There are approximately forty Jewish settlements, totalling 100,000 acres.[2]

1931: Jews make up 16.9% of population total.[2]

1948: *Nakba* (disaster)—more than 60% of Palestinians expelled from their homes. More than 530 villages destroyed.[7] In same year United Nations Relief and Works Agency for Palestine Refugees in the Near East (UNWRA) is formed.

Population (Israel-Palestine) | 1890s | 1910s | 1920s | 1930s | 1940s | 1950s

Early 1900s: During the First *Aliyah* Jewish immigration begins, with many agricultural towns established. Second *Aliyah* sees many Russian Jews arriving following pogroms and anti-Semitism there.

1922: 83,790 Jews and 668,258 Arabs live in the region,[2] with numbers rising after World War I.

1945: 600,000 Jews living in Palestine, up from 175,000 in **1931.** Jewish population in Palestine at 1.2 million by **1950** (50.7%).[6]

1950: Israel passes Law of Return, granting Jews and their families the right to automatic citizenship. 500,000 arrive in the following decade.[6]

Sources: [1] ProCon.org 2011 http://israelipalestinian.procon.org/view.resource.php?resourceID=000635#israel2005 ProCon.org is a nonprofit public charity with no government affiliation. It contains an amalgam of population and statistical data on deaths for both Israelis and Palestinians, including multiple sources (e.g., UN, Israeli Ministry of Foreign Affairs, and so on). [2] ProCon.org http://israeli palestinian.procon.org/viewresource.asp?resourceID=000636 [3] BBC News http://news.bbc.co.uk/1/shared/spl/hi/middle_east/03/v3_ip_timeline/html/1967.stm [4] Palestinian Centre for Human Rights 2009 http://www.pchrgaza.org/files/PressR/English/2008/36-2009.html [5] BBC News 2009 http://news.bbc.co.uk/1/hi/world/middle_east/7838618.stm [6] Council on Foreign Relations 2009 http://www.cfr.org/publication/15268/ [7] The Electronic Intifada 2007 http://electronicintifada.net/bytopic/197.shtml [8] UNWRA 2008 http://www.un.org/unrwa/publications/pdf/rr_countryandarea.pdf [9] Congressional Research Service 2008 http://www.un.org/unrwa/publications/pdf/rr_countryandarea.pdf. All accessed June 29, 2009.

▲ **Figure 10.32** Time line of Israeli/Palestinian conflict and change since 1890 The history of modern Israel/Palestine is one of enduring conflict as Israel increasingly occupies Palestinian territory and as Palestinians continue to fight to keep their homelands. This time line shows the different conflicts as well as the larger political and population issues around which these conflicts have been played out.

(Source: Ian shaw.)

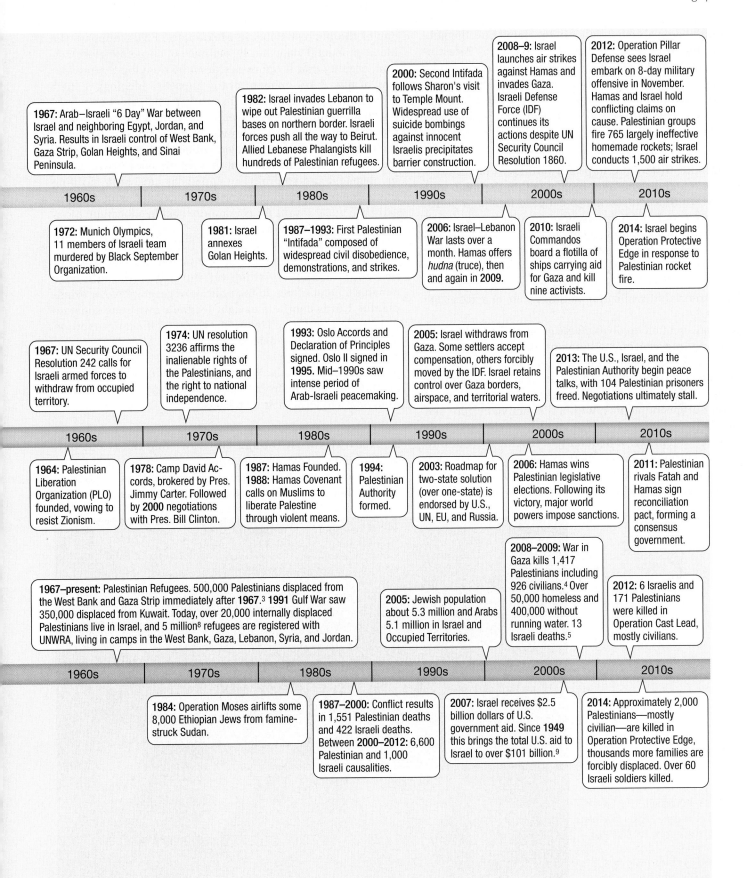

2008–9: Israel launches air strikes against Hamas and invades Gaza. Israeli Defense Force (IDF) continues its actions despite UN Security Council Resolution 1860.

2012: Operation Pillar Defense sees Israel embark on 8-day military offensive in November. Hamas and Israel hold conflicting claims on cause. Palestinian groups fire 765 largely ineffective homemade rockets; Israel conducts 1,500 air strikes.

2000: Second Intifada follows Sharon's visit to Temple Mount. Widespread use of suicide bombings against innocent Israelis precipitates barrier construction.

1982: Israel invades Lebanon to wipe out Palestinian guerrilla bases on northern border. Israeli forces push all the way to Beirut. Allied Lebanese Phalangists kill hundreds of Palestinian refugees.

1967: Arab–Israeli "6 Day" War between Israel and neighboring Egypt, Jordan, and Syria. Results in Israeli control of West Bank, Gaza Strip, Golan Heights, and Sinai Peninsula.

1960s   1970s   1980s   1990s   2000s   2010s

1972: Munich Olympics, 11 members of Israeli team murdered by Black September Organization.

1981: Israel annexes Golan Heights.

1987–1993: First Palestinian "Intifada" composed of widespread civil disobedience, demonstrations, and strikes.

2006: Israel–Lebanon War lasts over a month. Hamas offers *hudna* (truce), then and again in **2009**.

2010: Israeli Commandos board a flotilla of ships carrying aid for Gaza and kill nine activists.

2014: Israel begins Operation Protective Edge in response to Palestinian rocket fire.

1967: UN Security Council Resolution 242 calls for Israeli armed forces to withdraw from occupied territory.

1974: UN resolution 3236 affirms the inalienable rights of the Palestinians, and the right to national independence.

1993: Oslo Accords and Declaration of Principles signed. Oslo II signed in **1995.** Mid–1990s saw intense period of Arab-Israeli peacemaking.

2005: Israel withdraws from Gaza. Some settlers accept compensation, others forcibly moved by the IDF. Israel retains control over Gaza borders, airspace, and territorial waters.

2013: The U.S., Israel, and the Palestinian Authority begin peace talks, with 104 Palestinian prisoners freed. Negotiations ultimately stall.

1960s   1970s   1980s   1990s   2000s   2010s

1964: Palestinian Liberation Organization (PLO) founded, vowing to resist Zionism.

1978: Camp David Accords, brokered by Pres. Jimmy Carter. Followed by 2000 negotiations with Pres. Bill Clinton.

1987: Hamas Founded. **1988:** Hamas Covenant calls on Muslims to liberate Palestine through violent means.

1994: Palestinian Authority formed.

2003: Roadmap for two-state solution (over one-state) is endorsed by U.S., UN, EU, and Russia.

2006: Hamas wins Palestinian legislative elections. Following its victory, major world powers impose sanctions.

2011: Palestinian rivals Fatah and Hamas sign reconciliation pact, forming a consensus government.

2008–2009: War in Gaza kills 1,417 Palestinians including 926 civilians.[4] Over 50,000 homeless and 400,000 without running water. 13 Israeli deaths.[5]

2012: 6 Israelis and 171 Palestinians were killed in Operation Cast Lead, mostly civilians.

1967–present: Palestinian Refugees. 500,000 Palestinians displaced from the West Bank and Gaza Strip immediately after **1967**.[3] **1991** Gulf War saw 350,000 displaced from Kuwait. Today, over 20,000 internally displaced Palestinians live in Israel, and 5 million[8] refugees are registered with UNWRA, living in camps in the West Bank, Gaza, Lebanon, Syria, and Jordan.

2005: Jewish population about 5.3 million and Arabs 5.1 million in Israel and Occupied Territories.

1960s   1970s   1980s   1990s   2000s   2010s

1984: Operation Moses airlifts some 8,000 Ethiopian Jews from famine-struck Sudan.

1987–2000: Conflict results in 1,551 Palestinian deaths and 422 Israeli deaths. Between **2000–2012:** 6,600 Palestinian and 1,000 Israeli causalities.

2007: Israel receives $2.5 billion dollars of U.S. government aid. Since **1949** this brings the total U.S. aid to Israel to over $101 billion.[9]

2014: Approximately 2,000 Palestinians—mostly civilian—are killed in Operation Protective Edge, thousands more families are forcibly displaced. Over 60 Israeli soldiers killed.

occupation and establish a Palestinian state. And the subsequent kidnapping and murder of a 16-year old Palestinian youth has been labeled a revenge attack undertaken by three Jewish settlers, two of them teenagers. These events confirm the extreme state of anger, violence, and fear that surrounds the Palestinian bid for self-determination and their own state.

Regionalism Different groups with different identities—religious or ethnic—sometimes coexist within the same state boundaries. At times, discordance between legal and political boundaries and the distribution of populations with distinct identities leads to movements to claim or reclaim particular territories. These movements, whether conflictual or peaceful, are known as *regional movements*. **Regionalism** is a feeling of collective identity based on a population's politico-territorial identification within a state or across state boundaries.

Regionalism often involves ethnic groups who seek autonomy from an interventionist state and the development of political power. The Basque provinces of northern Spain and southern France have sought autonomy from those states for most of the twentieth century. The Basque people are one of the oldest European peoples, with a distinctive culture and language.

Another separatist movement is the EZLN (Ejército Zapatista de Liberación Nacional), the Zapatista Army of Liberation, which practices a different type of revolutionary autonomy. Composed largely of indigenous peasants from Chiapas,

one of the poorest states in Mexico, the EZLN opposes the Mexican federal state and its embrace of corporate globalization, arguing that it oppresses them by denying the peasant way of life. A key element of the Zapatista ideology is their aspiration to practice politics in a truly participatory way, from the "bottom-up," by guaranteeing the right of indigenous peoples to form and govern their own municipalities traditionally. In doing so, the Zapatistas reject official authorities and elect their own at the same time that they refuse federal government involvement and control. The Zapatistas have been building this autonomous form of governing since 1994 with dramatically beneficial results for community health, education, housing, and general welfare.

We need only look at the long list of territorially based conflicts that have emerged in the post–Cold War world to realize the extent to which territorially based ethnicity remains a potent force in the politics of geography. For example, the Kurds continue to fight for their own state separate from Turkey and Iraq (**Figure 10.33**). A significant proportion of Québec's French-speaking population, already accorded substantial autonomy, continues to advocate complete independence from Canada. Regionalism also underlies efforts by Scotland to separate from the United Kingdom and Catalonia from Spain to form their own states.

Sectionalism Although the two sometimes coexist, **sectionalism**, an extreme devotion to local interests and customs, should not be confused with regionalism. Sectionalism has been identified as an overarching explanation for

▶ **Figure 10.33 Kurdistan** The "Land of the Kurds" is a geocultural region that refers to parts of eastern Turkey, northern Iraq, northwestern Iran and northeastern Syria, mainly inhabited by Kurds. Many Kurds would like to see this area become the territorial basis for an independent nation-state where Kurds would constitute a majority.

(*Source:* Data from Worldtribune.com http://www.worldtribune. com/wpcontent/uploads/2013/10/web_kurdistanmap.jpg)

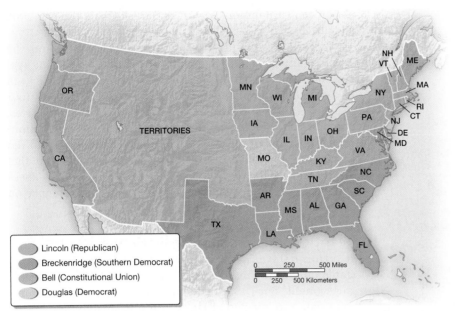

▲ Figure 10.34 The 1860 presidential election The U.S. presidential election of 1860 graphically illustrates the role sectionalism plays in determining who gets votes from which geographical regions. In a four-way race, Abraham Lincoln failed to win the support of any of the slave states. (Adapted from *Presidential Elections Since 1789*, 4th ed. Washington, DC: Congressional Quarterly Inc., 1987.)

the U.S. Civil War. It was an attachment to the institution of slavery and to the political and economic way of life slavery made possible that prompted the southern states to secede from the Union. Although the Civil War was waged around the real issue of permitting or prohibiting slavery, it also involved disagreement over the power of the states. The Union went to war to ensure that sectional interests would not take priority over the unity of the whole and states' rights would not undermine the power of the federal government. As **Figure 10.34** illustrates, the election of Abraham Lincoln to the presidency in 1860 reflected the sectionalism that dominated the country: He received no support from slave states.

The politics of geography, in terms of sectionalism, also finds strong focus today in rural versus urban politics. In France, for example, attitudes about birth control (and birth rates themselves) are significantly different between the urbanized north of the country and the more rural south. In England and Wales, a rural-versus-urban conflict over foxhunting persists. In the countryside, foxhunting is largely seen as a ritual of upper class rural life. In the early 2000s, a largely urban constituency of animal welfare proponents opposed foxhunting and pressured the government to act to protect the foxes. The result was the Hunting Act of 2004 that made it illegal to hunt a mammal using a dog. The act pitted a group called the Countryside Alliance, which continues to seek repeal of the act, against animal rights activists, mostly headquartered in large British cities, and the London-based Parliament.

Irredentism Any move or urge by a state to capture territories administered by another state on the grounds of

commonality, such as ethnicity, or on prior historical possession is known as **irredentism**. There are many examples of irredentism around the globe, including Bolivia's claim to land that was ceded to Chile during the war of the Pacific (1883) (which left Bolivia landlocked), as well as Macedonia, which promotes the view that all ethnic Slavs must be joined together through a United Macedonia claiming parts of Greece, Bulgaria, Albania, and Serbia.

The most recent example of irredentism that was acted on successfully was Russia's claim to the Crimean Peninsula, until February 2014. Russia occupied and later declared Crimea to be part of the Russian Federation and is currently administering the area from Moscow. The newly claimed Russian state of Crimea has not yet been recognized by the UN General Assembly or by many countries including the United States. The situation in Ukraine and Russia's aggression there is being widely recognized by political experts as the emergence of a new Cold War.

APPLY YOUR KNOWLEDGE

1. What is self-determination? How does this relate to ideas of national identity and territory?

2. Use the Internet to identify a self-determination movement somewhere in the world. Collect information on who is involved in the movement, what its aims are, how long the movement has been active, and any other information you feel is pertinent. Describe the role territory and power play in the aims of the group you have identified.

The Geography of Politics and Geographical Systems of Representation

An obvious way to show how politics shapes geography is to demonstrate how systems of political representation are geographically anchored. For instance, the United States has a political system in which democratic rule and territorial organization are linked by the concept of territorial representation.

Electoral divisions **Democratic rule** describes a system in which public policies and officials are directly chosen by popular vote. **Territorial organization** is a system of government formally structured by area, not by social groups. Thus, voters vote for officials and policies that will represent them and affect them *where they live*. The territorial bases of the U.S. system of representation are illustrated in **Figure 10.35**. The United States is a federation of 50 states, which are subdivided into over 3,000 counties or parishes. Counties and parishes are further broken down

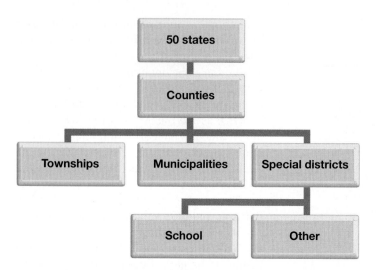

▲ **Figure 10.35 U.S. geographical basis of representation** This diagram is a breakdown of the types of voter districts at each level of political representation. Each type of district is territorially defined, creating a complicated and overlapping pattern of political units.

into municipalities, townships, and special districts, which include school districts, water districts, library districts, and others.

The electoral divisions established for choosing elected officials in the United States range from precincts and wards to congressional districts and states. State power is applied within geographical units, and state representatives are chosen from geographical units. The bottom line is that in the United States—as in many other representative democracies—politics is geography. People and their interests gain representation in government through the location of their interests in particular places and through their relative ability to capture political control of *geographically based* political units.

For example, election of the president involves a popular vote carried out at the precinct level but totaled at the state level. Thus, even though particular precincts, cities, or counties may vote for one candidate over another, if the majority of votes at the state level supports the opposing candidate, then that person is declared the winner in that state. The president is not elected by the popular vote, but by the electoral college.

Electoral college
The United States possesses a unique political-geographic body known as the **electoral college** composed of a specified number of delegates allocated to each state based on that state's population as of the most recent official census. In short, it is the state-level voting tally that drives the process. A candidate may win the countrywide popular vote but lose in the electoral college if that candidate fails to win enough states to secure the required majority of the electoral votes (**Figure 10.36**). The geographical implications of the U.S. presidential voting arrangement are crucial to candidates' campaign strategies. To be a winner requires concentrating time and energy on capturing a majority of votes in some of the

nation's most populous states, as Barack Obama did in 2008 and 2012.

Other systems of representation exist throughout the world. Many electoral systems are based on representing special constituencies in the legislative branch of government. In Pakistan, for example, there are four seats for Christians, four seats for Hindus and people belonging to the scheduled castes, and one seat each for the Sikh, Buddhist, and Parsi communities. Systems of representation are very much tied to the history of a country, with some very sensitive to the way that history and geography (who lives where) come together. These systems are both a product of and an important influence on the political culture of a country.

Reapportionment and Redistricting The U.S. Constitution determines the allocation of legislators among states, guaranteeing each state a representative system of government. For U.S. presidential and senatorial elections, candidates are elected at large within each state, not on the basis of electoral districts. U.S. representatives, however, are elected from congressional districts of roughly equal population size. This is also the case for state senators, representatives, and, often, other elected officials, from city council members to school board members. It is the responsibility of each state's legislature to create the districts that will elect most federal and state representatives. Other levels of government—from counties to special districts—also establish their own electoral districts. The result is that representatives are elected at any number of levels of government in a collection of districts that is complicated, extensive, and by no means systematic.

Problems of the proper "fit" between political representation and territory emerge when population changes. Because most forms of representation are based on population, it often becomes necessary to change electoral district boundaries to distribute the total population more evenly among districts. **Reapportionment** is the process of allocating electoral seats to geographical areas. **Redistricting** is the defining and redefining of territorial district boundaries. For example, the number of congressional representatives in the United States as a whole is fixed at 435. These 435 seats must be reapportioned in accordance with population change every 10 years. (Recall from Chapter 3 that the federal government is required to count the U.S. population every 10 years. One of the chief reasons for this is to maintain the proper match between population and representation.) Both reapportionment and redistricting are political, geographical, and statistical exercises. As geographer Richard Morrill writes:

> The process is *political* in that the design and approval of systems of districts is usually done by bodies of elected representatives; the balance of power between groups and areas is often involved; identification of citizens with a traditional electoral territory is altered; and the incumbency of individuals is usually at stake... . The process is *geographic* in that

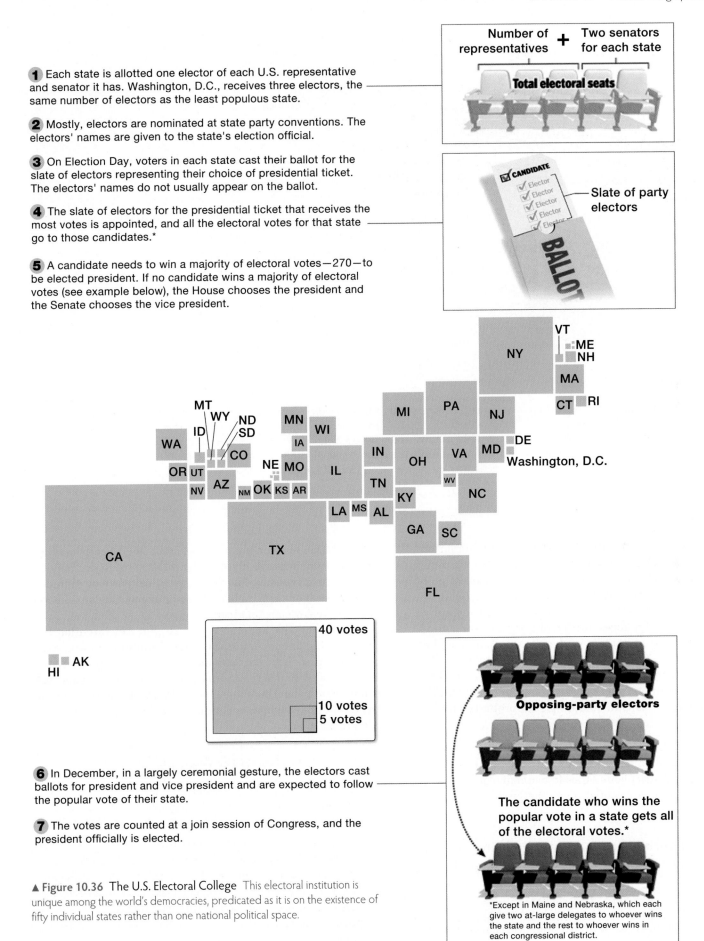

1 Each state is allotted one elector of each U.S. representative and senator it has. Washington, D.C., receives three electors, the same number of electors as the least populous state.

2 Mostly, electors are nominated at state party conventions. The electors' names are given to the state's election official.

3 On Election Day, voters in each state cast their ballot for the slate of electors representing their choice of presidential ticket. The electors' names do not usually appear on the ballot.

4 The slate of electors for the presidential ticket that receives the most votes is appointed, and all the electoral votes for that state go to those candidates.*

5 A candidate needs to win a majority of electoral votes—270—to be elected president. If no candidate wins a majority of electoral votes (see example below), the House chooses the president and the Senate chooses the vice president.

Number of representatives **+** Two senators for each state

Total electoral seats

Slate of party electors

CANDIDATE
Elector
Elector
Elector
Elector
Elector

BALLOT

40 votes

10 votes
5 votes

AK
HI

6 In December, in a largely ceremonial gesture, the electors cast ballots for president and vice president and are expected to follow the popular vote of their state.

7 The votes are counted at a join session of Congress, and the president officially is elected.

Opposing-party electors

The candidate who wins the popular vote in a state gets all of the electoral votes.*

▲ **Figure 10.36 The U.S. Electoral College** This electoral institution is unique among the world's democracies, predicated as it is on the existence of fifty individual states rather than one national political space.

*Except in Maine and Nebraska, which each give two at-large delegates to whoever wins the state and the rest to whoever wins in each congressional district.

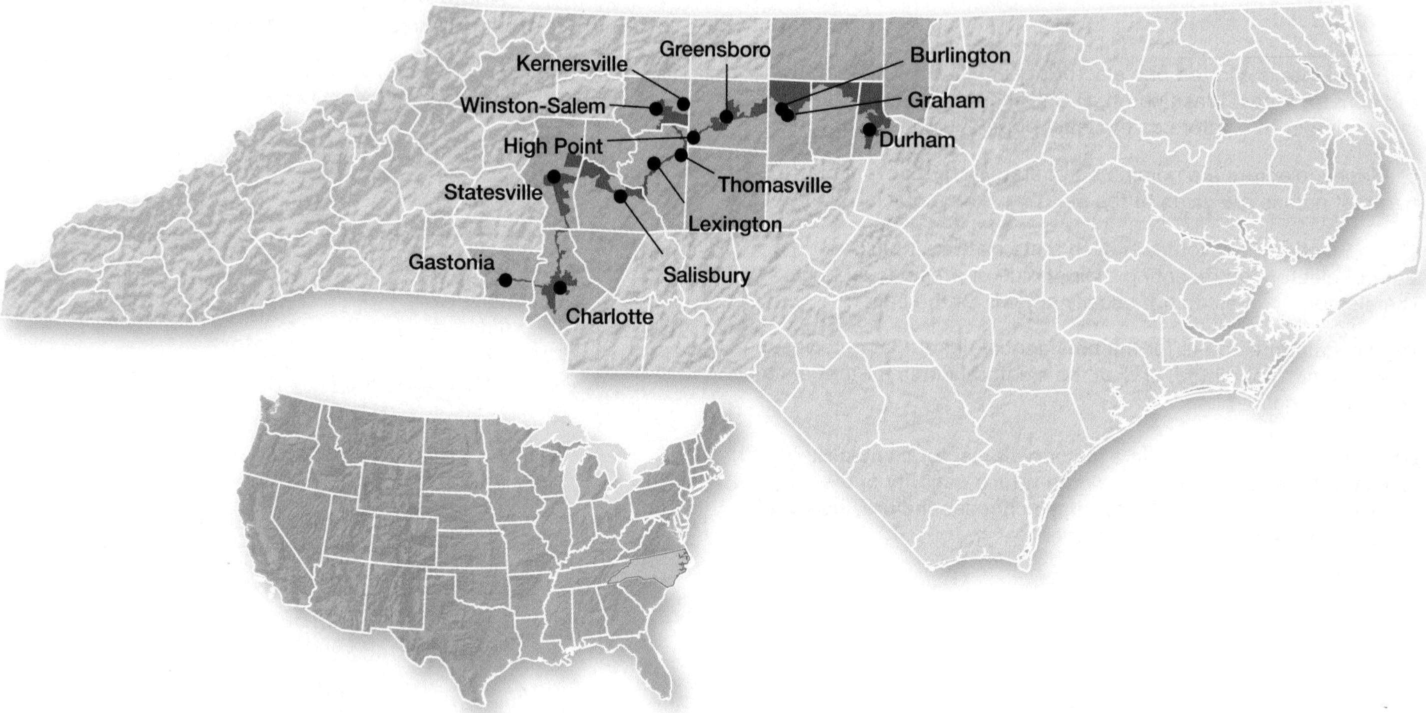

▲ **Figure 10.37 North Carolina's Twelfth Congressional District** The drawing of electoral district boundaries remains a politically volatile exercise. A 1996 case before the Supreme Court concerned the redrawn Twelfth Congressional District of North Carolina, whose shape (in red on the map) was as contorted as a salamander. Accusations of gerrymandering circulated around the drawing of the district, which was intended to consolidate African American voting strength. Although this redistricting was immediately challenged legally, the Supreme Court upheld its constitutionality in a 2001 decision.

areas must be allocated to districts and boundaries drawn (or territories partitioned into districts); communities of interest, which may have arisen in part from pre-existing systems of districts, may well be affected; restructuring of basic electoral geography is altered; and accessibility of voters to their representatives or centers of decision-making may be changed... . Redistricting is also *statistical* or mathematical in that there is a requirement for reasonably current and accurate data on population and its characteristics and, sometimes, of property and its valuation.[2]

The U.S. Census is a driver of redistricting activities as population growth and geographical shifts will result in the redrawing of district boundaries to maintain equal representation on the basis of population so that each person's vote is equally weighted.

Gerrymandering The purpose of redistricting is to ensure the equal probability of representation among all groups.

[2]R. Morrill, *Political Redistricting and Geographic Theory.* Washington, DC: Association of American Geographers, 1981, p. 1.

Redistricting for partisan purposes is known as **gerrymandering**. Gerrymandering involves boundaries of districts being redrawn to advantage a particular political party or candidate or to prevent or ensure a loss of power of a particular subpopulation (like African Americans).

Although the Federal Voting Rights Act intended that redistricting enhance minority representation in Congress, a fine line exists between "enhancement" and creating a district solely to ensure that a minority person is elected. **Figure 10.37** shows North Carolina's reengineered Twelfth District, the constitutionality of which was upheld by a Supreme Court decision in 2001.

APPLY YOUR KNOWLEDGE

1. Define both reapportionment and redistricting. What is the difference? How are both exercises "geographical, political, and statistical"?

2. Use the Internet to forge a basic understanding of how the U.S. electoral college works. Identify what you feel are the advantages and disadvantages of the system. Establish a position on whether the United States should retain the current electoral college or change it and indicate why you have taken that position, making sure that you address the *geographical issues* involved.

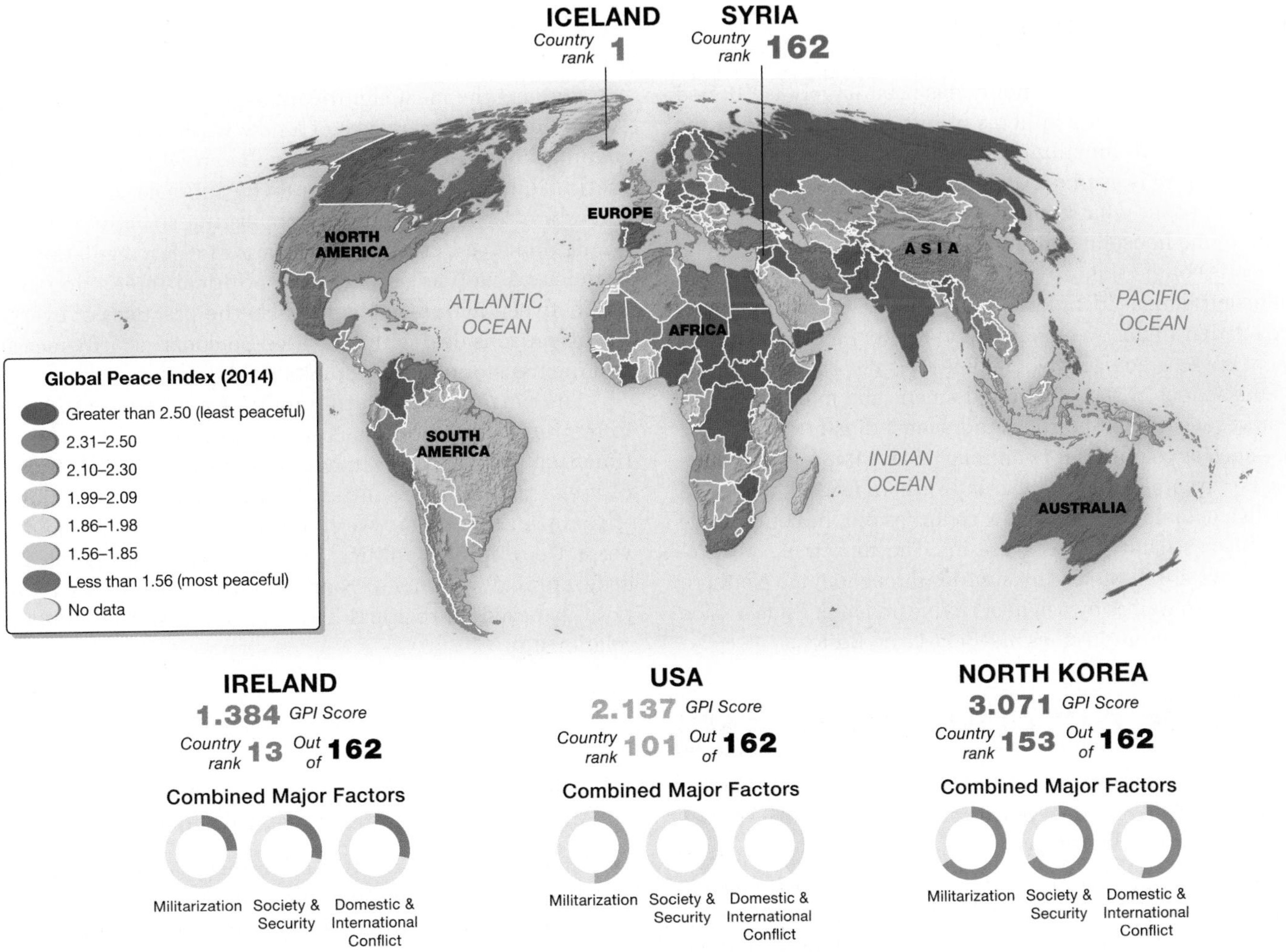

ICELAND
Country rank **1**

SYRIA
Country rank **162**

Global Peace Index (2014)

- Greater than 2.50 (least peaceful)
- 2.31–2.50
- 2.10–2.30
- 1.99–2.09
- 1.86–1.98
- 1.56–1.85
- Less than 1.56 (most peaceful)
- No data

IRELAND
1.384 GPI Score
Country rank **13**   Out of **162**

Combined Major Factors

Militarization | Society & Security | Domestic & International Conflict

USA
2.137 GPI Score
Country rank **101**   Out of **162**

Combined Major Factors

Militarization | Society & Security | Domestic & International Conflict

NORTH KOREA
3.071 GPI Score
Country rank **153**   Out of **162**

Combined Major Factors

Militarization | Society & Security | Domestic & International Conflict

▲ **Figure 10.38 Global Peace Index** This map indicates peace rankings for individual countries based on the Global Peace Index score. Dark red countries are the least peaceful while lighter yellow to green to purple countries are the most peaceful, based on variables such as militarization and conflict.

FUTURE GEOGRAPHIES

The end of the first Cold War and the possible emergence of another one, the widespread availability of telecommunications, the transnational reorganization of industry and finance, and the liberalization of trade all suggest that a new global political and economic order is accelerating in strength and extent. With the rapid growth of India and China and the prominence of the European Union, it would seem that a multipolar world is emerging, one that includes the United States but may not necessarily be so wholly dominated by the United States. This new, multipolar world might come close to what President George H. W. Bush heralded as a "new world order." And while the prospects for political stability in core countries are likely to remain strong, for peripheral ones the future looks far less bright.

The prospect for increasing ethnic rivalry and conflict is high. For instance, despite the recent popular vote for the creation of South Sudan, conflict there between Muslim majority and Christian and animist minorities is still ongoing. Internal conflicts such as this risk escalation into regional conflicts and possibly even the failure of states. Weak states can become hosts to violent drug cartels (as in Mexico) and terrorist groups (as in Afghanistan). While **bioterrorism**, the deliberate use of microorganisms or toxins from living organisms to induce death or disease is certainly a concern, the stability of the global collection of states especially those in the Middle East and Africa is also likely to remain an issue with respect to global capitalist stability. Peace seems scarce in the second decade of the twenty-first century. The Global Peace Index for 2014 shows a continuation of a seven-year trend in declining levels of peace globally (**Figure 10.38**). The world faces a range of challenges from climate change to political instability and without peace, it is difficult to foster cooperation and establish the social equity required to meet these challenges.

Global Peace Index
http://goo.gl/SM04Rg

CONCLUSION

The globalization of the economy has been largely facilitated by states extending their spheres of influence and paving the way for the smooth functioning of markets and industries. Political geography is as much about what happens at the global level as it is about what happens from the region to the neighborhood to the household and the individual.

Theories of the state have been one of geography's most important contributions to understanding politics. Ratzel's emphasis on the relationship between power and territory and Mackinder's model of the geographical pivot remind us that space and territory shape the actions of states in both dramatic and mundane ways. Time and space shape politics, and events distant in time and space—such as colonialism—continue to have impacts long afterwards. For example, the civil war in Northern Ireland, instigated by English colonial practices now centuries old, has only in the last decade or so shown signs of ceasing. The impacts of core colonization have also been felt in countries throughout the Northern Hemisphere, as well as by neighbors living unhappily side by side for several generations in cities like Belfast and Boston.

Perhaps the most significant aspect of contemporary globalization is the emergence of a new world order and a resulting focus on terrorism as well as the growth of transnational institutions of governance and the increase in war. These forces are reshaping not only governing structures and economic processes (creating new layers of rules, regulations, and policies, as well as new ways of political interaction among and between nation-states) but also the practices of everyday life (as citizens deal with increased personal security measures and the transformation of human rights).

The pairing of the terms *politics* and *geography* serves to remind us that politics is clearly geographical at the same time that geography is unavoidably political. The divisions of area into states, counties, cities, towns and special districts mean that where we live shapes our politics, and vice versa. Geography is politics, just as politics is geography. And geographical systems of representation, as well as identity politics based on regional histories, confirm this interactive relationship.

LEARNING OUTCOMES REVISITED

- **Express the geopolitical model of the state and explain how it links geography and state practices with respect to the key issues of power and territory.**

The ancient Greek philosopher Aristotle is often considered the first political geographer because his model of the state is based upon factors such as climate, terrain, and the relationship between population and territory. Other important political geographers have promoted theories of the state that incorporated elements of the landscape and the physical environment, as well as population characteristics of regions. Later scholars theorized that the state operated cyclically and organically. Twentieth-century theorists such as Foucault, Althusser, and Deleuze have shifted focus away from viewing the state as a set of institutions; they are more concerned with how state power is assembled and deployed.

- **Compare and contrast the ways that different contemporary theorists—from Deleuze to Althusser—approach the state as a political and geographical entity.**

The state is also a set of institutions for the protection and maintenance of society. A state is not only a place, a bounded territory, it is also an active entity that operates through the rules and regulations of its various institutions. State theorist Louis Althusser views the state as both ideological and repressive. Michel Foucault, another state theorist, has explored the ways that power, knowledge, and discourse operate to produce particular kinds of state subjects. Giles Deleuze sees the state as a force that is greater than the formal institutions that constitute it.

Deleuze believes that the state is best thought of as a *machine* whose purpose is to regulate and dominate.

- **Interpret how imperialism, colonialism, heartland theory, domino theory, the end of the Cold War, and the emergence of the new world order are key examples of ways geography has influenced politics and politics has influenced geography.**

Geopolitics may involve the extension of power by one group over another. There are many different manifestations of this phenomenon. Imperialism and colonialism involve occupation and control by one state over another. Heartland theory recognizes that a central location is pivotal to political and geographical control, whereas domino theory reflects the significance of proximity in the extension of power and control. During the Cold War, blocks of the global political system—capitalist versus communist—were in direct and indirect conflict. The current new world order is a manifestation of the decline of those old conflicts and the emergence of new ones.

- **Demonstrate how the growth and proliferation of international and supranational organizations created the foundation for the emergence of global forms of governance.**

Just as states are key players in political geography, so too have international and supranational organizations become important participants in the world system in the last century. These organizations have become increasingly important means of achieving goals such as the freer flow of goods and information and more cooperative management of shared resources, such as water.

- Recognize how events of international political significance are usually the result of East/West and North/South divisions, whereas national and local political issues emerge out of tensions related to regionalism and sectionalism and to some extent, previous experiences with colonialism.

  Capitalist colonialism and imperialism were key factors in producing global state divisions around capitalism versus communism (East/West) and rich versus poor (North/South). More recently, local divisions have emerged that reflect differences in ethnicity, political orientations, and economic commitments, among others. These are expressed through regionalism and sectionalism.

- Describe the difference between the politics of geography and the geography of politics as manifestations of the two-way relationship between politics and geography.

  Political geography can be viewed through two contrasting lenses. The first orientation sees it as the *politics of geography*. This perspective emphasizes that *geography*—the areal distribution/differentiation of people and objects in space—has a very real and measurable impact on politics. The second orientation sees political geography as the *geography of politics*. This approach analyzes how *politics*—the tactics or operations of the state—shapes geography.

KEY TERMS

biopolitics *(p. 356)*
bioterrorism *(p. 387)*
children's rights *(p. 378)*
citizenship *(p. 352)*
Cold War *(p. 365)*
confederation *(p. 354)*
corruption *(p. 372)*
decolonization *(p. 362)*
discourse *(p. 355)*
democratic rule *(p. 383)*
domino theory *(p. 366)*
East/West divide *(p. 364)*

electoral college *(p. 384)*
federal state *(p. 354)*
geopolitics *(p. 344)*
gerrymandering *(p. 386)*
global civil society *(p. 378)*
governance *(p. 352)*
government *(p. 352)*
human rights *(p. 377)*
international organization *(p. 362)*
international regime *(p. 377)*
irredentism *(p. 383)*

nation *(p. 352)*
nation-state *(p. 352)*
nationalism *(p. 353)*
neocolonialism *(p. 364)*
new world order *(p. 367)*
North/South divide *(p. 362)*
Orientalism *(p. 359)*
post-colonial studies *(p. 359)*
reapportionment *(p. 384)*
redistricting *(p. 384)*
regionalism *(p. 382)*
sectionalism *(p. 382)*

self-determination *(p. 378)*
sovereignty *(p. 352)*
state *(p. 352)*
supranational organization *(p. 375)*
territorial organization *(p. 383)*
territory *(p. 345)*
terrorism *(p. 370)*
unitary state *(p. 354)*

REVIEW & DISCUSSION

1. Research organizations that assist refugees in your city or state. Are refugees who come to your area typically from a specific location? If so, once you have identified the state or region a majority of the refugees come from, list four geopolitical factors that might have led to the displacement of these people. To do so, research the specific history of a region and map out the refugees' journey.

2. As a group, choose an international declaration or organization and research its efforts. For example, you might research the International Bill of Human Rights, the UN Declaration on the Rights of Indigenous People, or the International Criminal Court. Outline the history of the international regime and the impact it has had on international politics. Once you have done this, compare and contrast these international regimes with your own local government and laws. List three similarities and three differences.

3. Conduct an Internet search to determine if redistricting has occurred in your city or state. Create a time line demonstrating the frequency of redistricting where you live. Also determine the actors primarily responsible for redistricting and the arguments they employ. Finally, determine the political consequences of redistricting. Explain in what ways redistricting has (or has not) changed your local political landscape.

UNPLUGGED

1. International boundaries are a prominent feature of the political geography of the contemporary world. In this exercise, you will explore the impact of a boundary on nationalist attitudes and behaviors. You will need to use *The New York Times Index* to complete this assignment. Using the United States–Mexico border as your key word, describe the range of issues that derive from this juxtaposition of two very different nations. Concentrate on a five-year period, indicating which issues grew in importance, which issues declined, and which issues continued to have a consistent news profile throughout the period.

2. National elections usually tell a story about the ways regional ideas and attitudes shape the political agenda. In the 2000 and 2004 U.S. presidential elections, for example, pollsters considered religion an important issue. Using national election results data available through the Government Documents Division of your college or university library, describe the political geography of fundamentalist or evangelical Christians. Who did Christians vote for in all regions of the country in 2012? If not, George Romney, which ones did not, and what might explain the regional distribution of this powerful voting bloc?

3. U.S. presidential elections provide a snapshot of the changing political geography of the country. Compare the 1992 results with those for the year 2012 using maps that show the voting results. What are the most significant differences between the two maps? What are some reasons for these differences? If you are able to get maps that are disaggregated by race, ethnicity, or gender, what further explanations can you offer for the differences in the maps based on these additional variables?

4. Compare two maps of Europe (not including Russia and the former Soviet Union), one from 1930 and one from the present. How do issues of ethnicity, religion, and political-economic system (communist, capitalist) help to explain the changes in boundaries that have occurred? Identify any areas on the map that you feel may be the sites of future border changes and explain why.

DATA ANALYSIS

In this chapter we have looked at different forms of human organization around the world and how geography influences politics and vice versa. We have also looked at what the "state" is and how this geopolitical model relates to power and territory. But what is it like to operate outside the state? What does it mean to have no passport or national identity and try to maneuver the globe?

To look closer at the realities for stateless people, go to the U.N. High Commissioner for Refugees (UNHCR) website and explore the pages on "stateless persons" to answer these questions.

1. Read this page and define what it means to be "stateless."

2. What are the causes of statelessness? *Hint*: Look at the different pages connected to statelessness, like "The Causes of Statelessness."

3. In 2014, the UNHCR celebrated the 60th Anniversary of the 1954 Convention relating to the Status of Stateless Persons. Download the handbook relating to the 1954 Convention at http://www.unhcr.org/53b698ab9.html and read the introduction.

4. According to the handbook, do most countries have ways of determining whether a person is stateless? What is the UN's criteria?

5. Watch the story of Railya Abulkhanova, "I Am Stateless".

6. What is Railya's place of origin? How did she become stateless?

7. What is Railya's occupation? Does her level of education help her stateless status? Why or why not?

8. How does Railya describe being stateless?

9. What are some of the challenges Railya faces as a stateless person?

10. How would your life change if you suddenly became stateless? How linked is your citizenship to your identity? To your sense of security? To your freedom of movement in your country of residence and around the world?

MasteringGeography™

Looking for additional review and test prep materials? Visit the Study Area in MasteringGeography™ to enhance your geographic literacy, spatial reasoning skills, and understanding of this chapter's content by accessing a variety of resources, including **MapMaster** interactive maps, Videos, *In the News* RSS feeds, flashcards, web links, self-study quizzes, and an eText version of *Human Geography*.

LEARNING OUTCOMES

- *Explain* how the urban areas of the world are the linchpins of human geographies at the local, regional, and global scales.

- *Describe* how the earliest towns and cities developed independently in the various hearth areas of the first agricultural revolution.

- *Explain* how the expansion of trade around the world, associated with colonialism and imperialism, established numerous gateway cities.

- *Assess* why and how the Industrial Revolution generated new kinds of cities—and many more of them.

- *Interpret* how a small number of "world cities," most of them located

within the core regions of the world-system, have come to occupy key roles in the organization of global economics and culture.

- *Compare* and contrast the differences in trends and projections between the world's core regions and peripheral regions.

▲ Downtown Singapore

11

URBANIZATION AND THE GLOBAL URBAN SYSTEM

Singapore is one of the success stories of the global urban system. From a nineteenth-century British trading post of just a thousand or so people, mostly indigenous Malays employed by British traders, it has grown to a city-state of 5.3 million. People of Malayan origin are now in the minority: The descendants of Chinese rubber plantation workers are the majority. There is also a minority population of Indian descent, whose presence, like the Chinese, was originally connected to the Malayan rubber plantations and Singapore's role as a gateway city, dominating trade into and out of the Malaysian Peninsula.

For a couple of years after independence from the British in 1963, Singapore was part of the new state of Malaysia. Ethnic and cultural differences with the Malay majority led to Singapore's expulsion, and it became a modern city-state of just 274 square miles: just over one-tenth the size of Delaware. Politically isolated, but situated at the crossroads of trade routes in Southeast Asia, Singapore set about creating a favorable business climate.

Singapore's initial growth was export-led, taking advantage of cheap labor and its extensive port facilities. Even today the costs of hourly compensation for production workers in manufacturing industries in the city are only about 20 percent of those for U.S. workers. The globalization of economic activity in the 1970s brought significant flows of capital into the city along with the financial and business services that were beginning to reshape world economic geography and, with it, the global urban system. The city met the challenges of limited space, few natural resources, and rapid growth by establishing a highly centralized planning system. The country's leadership realized that a favorable business climate would require a highly efficient infrastructure and an exceptional quality of life as well as an advantageous labor and financial environment.

Today Singapore is a city of global importance. Its affluence and amenities have attracted talent from around the world, and its highly regulated economy and society have made it one of the safest and best integrated cities of its size. Because of the limited space and natural resources of the country, urban planning has played a significant role, and Singapore is now considered to be one of the best-planned metropolises in the world. It has been framed around high-density residential and commercial developments centered on multiple transport nodes. There are seamless connections between bus and rail services and a strict vehicle quota system that controls the amount of traffic in the city. There are cutting-edge water recycling plants and waste-to-energy facilities, and a citywide recycling program that recycles 60 percent of waste. Environmental policies and regulations have made it one of the leading "green cities" in the world. Despite the rapid urbanization over the past 50 years, greenery still covers more than half of the surface area of the city, including more than 450 public parks and gardens and four large nature reserves designed to preserve its rich biodiversity.

URBAN GEOGRAPHY AND URBANIZATION

Urbanization is one of the most important geographic phenomena in today's world. The proportion of the world's population living in urban settlements is growing at a rapid rate, and the world's economic, social, cultural, and political processes are increasingly being played out within and between the world's systems of towns and cities. The trend is now irreversible because of the global shift to technological-, industrial-, and service-based economies. In this chapter, we describe the extent and pattern of urbanization across the world, explaining its causes and the resultant changes wrought in people and places.

Studying Urbanization

Urban geography is concerned with the development of towns and cities around the world, with particular reference to the similarities and differences both *among* and *within* urban places. For urban geographers, some of the most important questions include:

- What attributes make towns and cities distinctive?

- How did these distinctive identities evolve?

- What are the relationships and interdependencies between particular sets of towns and cities?

- What are the relationships between cities and their surrounding territories?

- Do significant regularities exist in the spatial organization of land use within cities, in the patterning of neighborhood populations, or in the layout and landscapes of particular kinds of cities?

- What kinds of problems result from different patterns and processes of urbanization?

- How do environmental factors influence the character of urbanization, and how might climate change affect cities?

Urban geographers also want to know about the causes of the patterns and regularities they find. How, for example, do specialized urban subdistricts evolve? Why does urban growth occur in a particular region at a particular time? And why does urban growth exhibit a distinctive physical form during a certain period? In pursuing such questions, urban geographers have learned that the answers are ultimately to be found in the wider context of economic, social, cultural, and political life. Towns and cities must be viewed as part of the economies and societies that maintain them.

Urbanization, therefore, is not simply the population growth of towns and cities. It also involves many other changes, both quantitative and qualitative. Urbanization increases the proportion of a country or a region's population living in cities (as opposed to villages and rural settlements). It implies an increase in the size of many (but not all) cities, and it implies changes in economic structure and ways of life. Geographers conceptualize these changes in several different ways. One of the most important of these is examining the attributes and dynamics of urban systems. An **urban system**, or city system, is any interdependent set of urban settlements within a given region. For example, we can speak of the Spanish urban system (**Figure 11.1**), the African urban system, or even the global urban system (**Figure 11.2**). As urbanization takes place, urban systems reflect the increasing numbers of people living in ever-larger towns and cities. They also reflect other important changes, such as changes in the relative size of cities, changes in their functional relationships with one another, and changes in their employment base and population composition. Every town and city is part of the interlocking urban systems that link regional-, national-, and global-scale human geographies in a complex web of interdependence. These urban systems organize space through hierarchies of cities of different sizes and functions. Many of these hierarchical urban systems exhibit distinctive attributes and features, particularly in the relative size and spacing of individual towns and cities and the roles that individual cities play. The changing dynamics of urban systems have both reflected and shaped the world's geographies, as we shall see in this chapter.

Other important aspects of change associated with urbanization processes concern urban form, and we examine these in more detail in Chapter 11. **Urban form** refers to the physical structure and organization of cities in their land use, layout, and built environment. As urbanization takes place, not only do towns and cities grow bigger physically, extending upward and outward, but they also become reorganized, redeveloped, and redesigned in response to changing circumstances.

These changes, in turn, are closely related to another aspect of change: transformations in patterns of urban ecology. **Urban ecology** is the social and demographic composition of city districts and neighborhoods. Urbanization not only brings more people to cities, it also brings a greater variety of people. As different social, economic, demographic, and racial

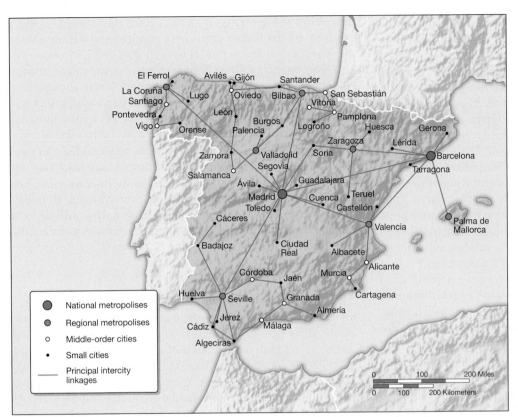

◀ **Figure 11.1 The Spanish urban system**
Note how the smaller cities tend to be linked to middle-order cities, which in turn are linked to regional metropolises, which are linked to the national metropolises, Madrid and Barcelona.

(*Source:* Adapted from L. Bourne, R. Sinclair, M. Ferrer, and A. d'Entremont (eds.), *The Changing Geography of Urban Systems*. Navarra, Spain: Department of Human Geography, Universidad de Navarra, 1989, fig. 2, p. 46.)

subgroups become sorted into different territories, distinctive urban ecologies emerge. As new subgroups arrive or old ones leave, these ecologies change.

Yet another aspect of change associated with urbanization concerns people's attitudes and behavior. New patterns of social interaction and new ways of life in cities have given rise to the concept of urbanism, which refers to the distinctive nature of social and cultural organization in cities. **Urbanism** describes the way of life fostered by urban settings, in which the number, physical density, and variety of people often result in distinctive attitudes, values, and patterns of behavior. Geographers are interested in urbanism because of the ways in which it varies both within and among cities of different types.

Urbanization and Changing Human Geographies

From small market towns and fishing ports to megacities of millions of people, the urban areas of the world are the linchpins of human geographies. They have always been a crucial element in spatial organization and the evolution of societies, but today they are more important than ever.

▲ **Figure 11.2 Earth at night, Asia**

How Urban? It is difficult to say just how urbanized the world has become. In many areas, urban growth is taking place at such a pace and under such chaotic conditions that experts can provide only informed estimates. Different countries use very different definitions of what constitutes an "urban" place. Some countries (e.g., Australia and Canada) count any settlement of 1,000 people or more as urban; others (including Italy and Jordan) use 10,000 as the minimum for an urban settlement, and Japan uses 50,000 as the cutoff. This, by the way, tells us something about the nature of urbanization itself: It is a relative phenomenon. In countries like Peru, where the population is thinly distributed and scattered, a settlement of 2,000 represents a significant center. In countries like Japan, however, with more residents, higher population densities, and a tradition of centralized agricultural settlement, a much larger concentration of people is required to count as "urban."

To get around these national differences in definitions of what counts as "urban," the World Bank has developed a uniform definition of what constitutes an urban area based on an "agglomeration index" that identifies an area of 1 square kilometer as urban if its population density exceeds 150 and it has access to a settlement of more than 50,000 inhabitants within 60 minutes by road. According to this measure, the overall level of urbanization in the world in 2010 was 51 percent. North America is the most urbanized continent in the world, with 82 percent of its population living in urban areas. In contrast, Africa is less than 40 percent urban. To put these figures in perspective, only 30 percent of the world's population was urbanized in 1950. In that year, there were only 83 metropolitan areas of a million or more, and only 8 of 5 million or more existed; in 2010, there were 468 metropolitan areas of a million or more people, and 51 with over 5 million.

Currently, some 200,000 people are added to the world's urban population every day. It is estimated that by 2030, six out of every ten people worldwide will live in a city, and by 2050 this proportion will increase to seven out of ten people. Almost all this urban population growth will occur in cities of developing countries. Many of the largest cities in the periphery are growing at annual rates of between 4 and 7 percent. At the higher rate, their populations will double in 10 years; at the lower rate, they will double in 17 years. The *doubling time* of a city's population is the time needed for it to double in size, at current growth rates. To put the situation in numerical terms, metropolitan areas like Mexico City and São Paulo are adding half a million persons to their population each year, nearly 10,000 every week, even taking into account losses from deaths and out-migration. It took London 190 years to grow from half a million to 10 million. It took New York 140 years. By contrast, Mexico City, São Paulo, Buenos Aires, Kolkata (Calcutta), Rio de Janeiro, Seoul, and Mumbai (Bombay) all took less than 75 years to grow from half a million to 10 million inhabitants. Urbanization on this scale is a remarkable geographical phenomenon—one of the most important processes shaping the world's landscapes.

The Functions of Urban Settlements Towns and cities are centers of cultural innovation, social transformation, and political change. They are also engines of economic development. An estimated 80 percent of global gross domestic product (GDP) is generated in cities, and the GDP of large cities like London, Los Angeles, Mexico City, and Paris is roughly equivalent to that of entire countries like Australia and Sweden. Although they often pose social and environmental problems, towns and cities are essential elements in human economic and social organization. Experts on urbanization point to four fundamental aspects of the role of towns and cities in human economic and social organization:

- *The mobilizing function of urban settlement.* Urban settings, with their physical infrastructure and their large and diverse populations, are places where entrepreneurs can get things done. Cities provide efficient and effective environments for organizing labor, capital, and raw materials and for distributing finished products. It is no accident that corporations large and small, banks and investment houses, universities and research laboratories, and specialized manufacturing all congregate in cities. The proximity of roadways, rails, ports, and airports in urban areas reduce time, travel, and energy costs in a way that rural areas simply cannot match. In developing countries, urban areas produce as much as 60 percent of total GDP with just one-third of the population.

- *The decision-making capacity of urban settlement.* Corporate offices, political assemblies, and government agencies all typically locate in cities. This, in turn, makes cities great magnets for the best thinkers and innovators, and the most qualified managers and administrators. As a result, the decision-making machinery of public and private institutions and organizations are brought together, making cities the nerve centers of national economies.

- *The generative functions of urban settlement.* The concentration of people in urban settings makes for much greater interaction and competition, which facilitates the generation of innovation, knowledge, and information. Cities are important arenas of cultural production, seedbeds of cultural innovation, centers of fashion, and the creation of taste. As Jane

▼ Figure 11.3 **Urban subculture** Goths in Bolkow, Poland

Jacobs pointed out long ago in her famous book, *The Death and Life of Great American Cities*,[1] the density and diversity of urban populations generate serendipity, unexpected encounters and "new combinations" that lead to innovation. Personal communications devices and social networking sites have enhanced the generative capacity of contemporary cities, but density, diversity, and face-to-face interactions are still important, especially in facilitating the "creative buzz" associated with the most vibrant and creative city districts.

■ *The transformative capacity of urban settlement.* The size, density, and variety of urban populations tend to have a liberating effect on people, allowing them to escape the rigidities of traditional, rural society and to participate in a variety of lifestyles and behaviors (**Figure 11.3**). The demographic heterogeneity and cultural hybridity of urban populations contributes to an atmosphere of freedom and possibility.

APPLY YOUR KNOWLEDGE

1. Provide two examples of how the transformative capacity of urban settlements can be liberating for some people.

2. Think about traditional identities and social mores around the world, and share examples of how the rural to urban move can change gender roles, career choices, arts, invention, social expectations, and politics.

In certain circumstances, combinations of these qualities result in distinctive "golden ages," for example: Athens 500–400 B.C.E., Rome 50 B.C.E. to 100 C.E., Florence 1400–1500, Vienna 1780–1910, London 1825–1900, Paris 1870–1910, and Berlin 1918–1933. Geographer Peter Hall[2] has written about these and other cities, noting that all of them led their respective states or empires, which made them magnets for the immigration of talent. As they drew talent from the far corners of the empires they controlled, it made them cosmopolitan. They were all also wealthy trading cities. Out of trade came new ways of economic organization, and out of those came new forms of production. Perhaps the most important factor, though, was that all were in the process of rapid economic and social transformation. As a result, they were in a state of uneasy tension between conservative forces and values—aristocratic, hierarchical, religious, conformist—and radical values that were the exact opposite: open, rational, skeptical. These radical values were articulated more often than not by creative people who felt themselves outsiders because they were young or provincial or even foreign, or because they did not belong to the established order of power and prestige. In such circumstances, the ferment of new movements in the arts, new philosophies, new political ideals, and new cultural practices lead to "golden ages," with wealthy individuals and well-funded institutions providing patronage for the avant-garde.

[1]Jacobs, J., *The Death and Life of Great American Cities*. New York: Random House, 1961.

[2]Hall, P., *Cities in Civilization*. London: Weidenfeld and Nicolson, 1998.

Of course, these are the extreme cases, and elsewhere it is not all positive. Urbanization has enormous environmental consequences, both global and local. With about half the world's population, cities account for two-thirds of global energy demand and a similar proportion of the world's greenhouse gas emissions. Sprawling urban development consumes arable land and vital green spaces. The efficiency of cities as places of production, administration, and service provision can be compromised by congestion and the cost of living. Political ideologies, misguided policies, or structural economic change can make cities become places of deprivation, inequality, and exclusion instead of being places of opportunity and prosperity. Demographic heterogeneity and cultural differences can lead to tension and conflict instead of positive change and so on.

Nevertheless, the mobilizing and generative functions of cities, together with their decision-making and transformative capacities mean that they are, on balance, places of prosperity. In this context, the idea of prosperity goes beyond income and material possessions (see Box 11.1: "Spatial Inequality: The Prosperity of Cities").

FOUNDATIONS OF THE GLOBAL URBAN SYSTEM

It is important to put the geographic study of towns and cities in historical context. After all, many of the world's cities are the product of long periods of development. We can only understand a city, old or young, if we know something about the reasons behind its growth, its rate of growth, and the processes that have contributed to this growth.

In broad terms, the earliest urbanization developed independently in the various hearth areas of the first agricultural revolution (see Chapter 2). The very first region of independent urbanism was in the Middle East, in the valleys of the Tigris and Euphrates (in Mesopotamia), and in the Nile Valley from around 3500 B.C.E. (see Chapter 4). Together, these intensively cultivated river valleys formed the so-called Fertile Crescent. In Mesopotamia, the growth in size of some of the agricultural villages located on the rich alluvial soils of the river floodplains formed the basis for the large rival city-states of the Sumerian empire. They included Ur (in present-day Iraq), the capital from about 2300 to 2180 B.C.E., as well as Eridu, Uruk, and Erbil. These fortified city-states contained tens of thousands of inhabitants; social stratification, with religious, political, and military classes; innovative technologies, including massive irrigation projects; and extensive trade connections. By 1885 B.C.E., the Sumerian city-states had been taken over by the Babylonians and then the Neo-Babylonians, who governed the region from their capital city, Babylon.

In Egypt, which became a unified state as early as 3100 B.C.E., large irrigation projects controlled the Nile's waters for agricultural and other uses, supporting a series of capital cities that included Thebes, Akhetaten (Tell el-Amarna), and Tanis. Internal peace in Egypt meant that there was no need for massive investments in these cities' defensive fortifications.

By 2500 B.C.E., cities had appeared in the Indus Valley, and by 1800 B.C.E., they were established in northern China. Other areas of independent urbanism include Mesoamerica

11.1 Spatial Inequality   The Prosperity of Cities

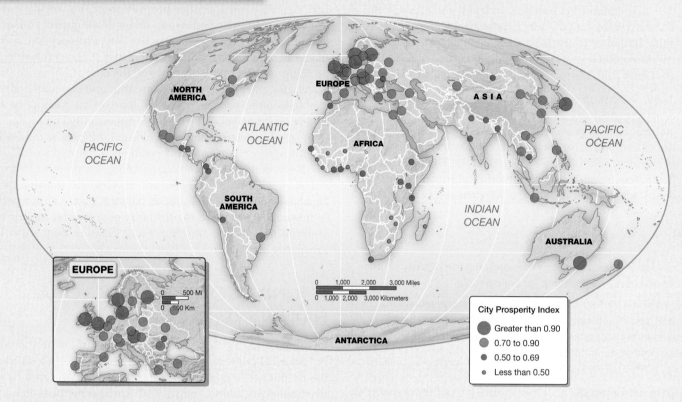

▲ Figure 11.A Prosperity index

The United Nations Human Settlements Programme has emphasized that prosperity means having a set of conditions that enable urban residents to flourish, feel happy and healthy, and in which business can thrive, institutions develop, and physical spaces become more integrated and diverse. The UN's "City Prosperity Index" is calculated to reflect:

- infrastructure and amenities—adequate water, sanitation, power supply, road network, information and communications technology;

- social services—education, health, recreation, safety, and security;

- environmental quality—energy efficiency, air pollution;

[3]United Nations Human Settlements Programme (UN-Habitat), *State of the World's Cities 2012/2013*. New York: Routledge, 2013.

- equity and social inclusion—the extent of poverty and inequality, and civic participation in the social, political and cultural spheres;

- income and employment that afford adequate living standards.[3]

The results are shown in **Figure 11.A**. The cities in the world's core economies are solidly prosperous; at the other extreme, a majority of African cities have very weak scores. Most of the Asian and Latin American cities included in the UN analysis fall in the middle of the index.

---

1. Identify two cities that do not seem to meet these conditions. What changes would be required for them to meet those conditions?

2. Do you agree that the conditions outlined by the UN are the most important ones for happiness and prosperity?

(from around 100 B.C.E.) and Andean America (from around 800 C.E.). Meanwhile, the original Middle Eastern urban hearth continued to produce successive generations of urbanized world-empires, including those of Greece, Rome, and Byzantium.

Explanations of these first transitions from subsistence societies to city-based economies differ. Most experts agree that changes in social organization were an important precondition for urbanization. Specifically, urbanization required the emergence of groups who were able to exact tributes, impose taxes, and control labor power, usually through some form of religious persuasion or military coercion. Once established, this elite group provided the stimulus for urban development by using its wealth to build palaces, arenas, and monuments to display its power and status. This activity not only created the basis for the physical core of ancient cities but also required an increased degree of specialization in nonagricultural activities—construction, crafts, administration, the priesthood, soldiery, and so on—which could be organized effectively only in an urban setting.

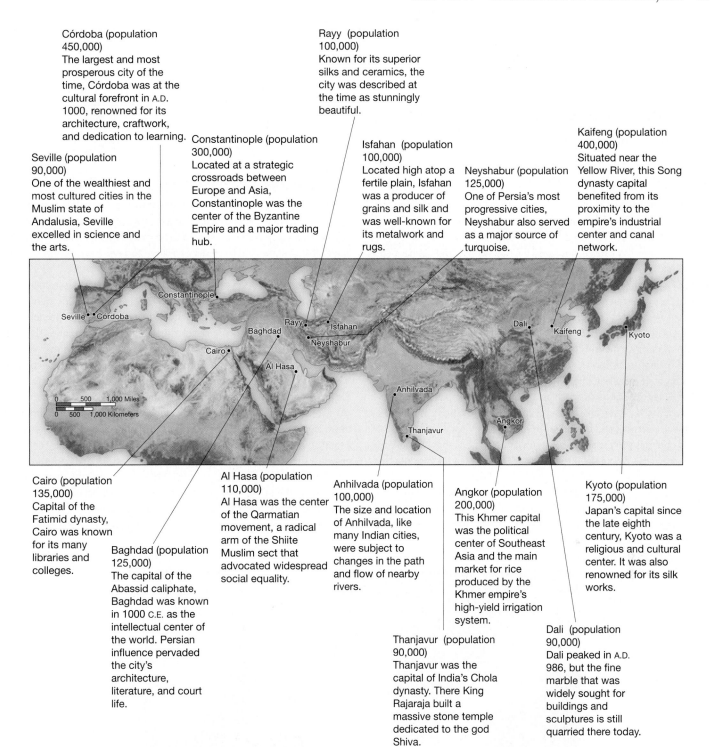

Córdoba (population 450,000)
The largest and most prosperous city of the time, Córdoba was at the cultural forefront in A.D. 1000, renowned for its architecture, craftwork, and dedication to learning.

Seville (population 90,000)
One of the wealthiest and most cultured cities in the Muslim state of Andalusia, Seville excelled in science and the arts.

Constantinople (population 300,000)
Located at a strategic crossroads between Europe and Asia, Constantinople was the center of the Byzantine Empire and a major trading hub.

Rayy (population 100,000)
Known for its superior silks and ceramics, the city was described at the time as stunningly beautiful.

Isfahan (population 100,000)
Located high atop a fertile plain, Isfahan was a producer of grains and silk and was well-known for its metalwork and rugs.

Neyshabur (population 125,000)
One of Persia's most progressive cities, Neyshabur also served as a major source of turquoise.

Kaifeng (population 400,000)
Situated near the Yellow River, this Song dynasty capital benefited from its proximity to the empire's industrial center and canal network.

Cairo (population 135,000)
Capital of the Fatimid dynasty, Cairo was known for its many libraries and colleges.

Baghdad (population 125,000)
The capital of the Abassid caliphate, Baghdad was known in 1000 C.E. as the intellectual center of the world. Persian influence pervaded the city's architecture, literature, and court life.

Al Hasa (population 110,000)
Al Hasa was the center of the Qarmatian movement, a radical arm of the Shiite Muslim sect that advocated widespread social equality.

Anhilvada (population 100,000)
The size and location of Anhilvada, like many Indian cities, were subject to changes in the path and flow of nearby rivers.

Angkor (population 200,000)
This Khmer capital was the political center of Southeast Asia and the main market for rice produced by the Khmer empire's high-yield irrigation system.

Thanjavur (population 90,000)
Thanjavur was the capital of India's Chola dynasty. There King Rajaraja built a massive stone temple dedicated to the god Shiva.

Kyoto (population 175,000)
Japan's capital since the late eighth century, Kyoto was a religious and cultural center. It was also renowned for its silk works.

Dali (population 90,000)
Dali peaked in A.D. 986, but the fine marble that was widely sought for buildings and sculptures is still quarried there today.

▲ **Figure 11.4 The most important cities in 1000 C.E.** The most Important cities in 1000 C.E. were the seats of world-empires—the Islamic caliphates, the Byzantine empire, the Chinese empire, and Indian kingdoms. These cities had developed well-established civilizations with urban systems based on regional trade and protected by strong military rule.

(*Source*: Data from T. Chandler, *Four Thousand Years of Urban Growth: A Historical Census*. Washington, D.C.: Worldwatch Institute, 1987; "The Year 1000," *U.S. News & World Report*, August 16, 1999, pp. 66–70.)

By 1000 C.E., city-based societies had emerged in Europe, the Middle East, and China, including a dozen major cities with populations of 100,000 or more (**Figure 11.4**).

These early urbanized economies were a precarious phenomenon, however, and many of them lapsed into ruralism before being revived or recolonized. In a number of cases, the decline of urban systems was a result of demographic setbacks associated with wars or epidemics. Such disasters left too few people to maintain the social and economic infrastructure necessary for urbanization.

11.2 Geography Matters

Cities and Economic Development

By Peter Taylor, Northumbria University

Two cities that saved rock n'roll

Rock n'roll dead? By the early 1960s the excitement generated by the first rock n'roll musicians was gone and the popular music industry in the United States was desperately searching for their replacements. That's when the 'British Invasion' led by The Beatles saved the day. But how did this happen? One city had an immense influence on this rescue: Liverpool. In the 1950s Liverpool had regular passenger liner services to and from New York, the centre of the American music business (**Figure 11.B**). English stewards on the ships came home to Liverpool with records from the US. This transfusion of musical material enabled a British city to rejuvenate a very American musical genre by inspiring the early Beatles. But this was just part of the story. Detroit was making a similarly valuable contribution to the music industry. This great automobile manufacturing city attracted streams of migrants, especially from the American South, establishing a creative cosmopolitan mix of cultures out of which the remarkably successful Motown Records was born. Between them, Liverpool and Detroit saved rock n'roll.

At its core this story is about two innovative music scenes arising from two different cities. Each music scene illustrates two key features of cities that enable innovation: inter-city relations (New York-Liverpool) through which products flow to be developed into new products; and migration flows (the American South to Detroit) that produce vibrant urban places where new products can be developed. Cities are central to economic development, as this example illustrates.

How economic development occurs

We often think of economic development as something that happens to countries. But what if countries are *not* the geographical units through which economic development is generated? The great urbanist thinker Jane Jacobs argues that countries are the result of military and political practices that define and maintain borders—and that economic development is a very different process that occurs in and through cities. In other words, she states that countries are political entities geographically defined by their *borders*, while cities are economic entities geographically defined by their *connections*.

Let's look at how economic development occurs through cities. Although economic development is generally equated with economic growth, it is actually a *specific* form of growth that increases the complexity of an economy. We can see complexity expressed through an economy's division of labor—the range and types of jobs in that economy. For example, a large successful city like New York has an immensely complex division of labor – think of the myriad of different jobs people do in that city. Simpler settlements, like a rural community, have a small range of jobs. Therefore simple economic growth such as doubling the output of a single factory does not qualify as development because it just increases old existing work: the division of labor stays the same. However, starting production of a different commodity creates new work, which is added to the existing division of labor making it more complex—this is economic development, and it occurs primarily in cities.

▲ Figure 11.B Liverpool docks in the 1950s

▲ **Figure 11.C** Morning commuters in a Tokyo subway.

locations, leaving some cities in economic decline. This may especially happen in cities that have just one or two dominant economic sectors, making a city's division of labor less complex. The simpler a city's division of labor, the more vulnerable it is to economic change: it can even experience economic 'un-development'. For example, Detroit originally had a wide range of industries but came to be dominated by automotive production. So the buzz that made 'Motown' music was affected by changes in the automotive industry when foreign producers like Honda and Toyota became successful. These declining industrial cities in the US and other countries are commonly called 'shrinking cities'.

But some large cities associated with a particular type of work do not decline because they maintain a diverse range of other vital jobs. New York, Los Angeles, and Washington, DC are famous for finance, cinema, and politics respectively—but each city retains its buzz beyond what it is famous for. This tells us that economic development is a dynamic process, and cities can only succeed through the continual renewal of their divisions of labor.

The 'buzz' of cities

Economic development occurs in cities because new work is more than likely to occur there due to the clustering of large numbers of people in a small area. This means there are innumerable contacts daily that provide the basic condition for generating new ideas that can be translated into innovations that generate new work for the city. This process is made especially dynamic through new contacts from people visiting from other cities (as in Liverpool) and from migrants arriving from other cultural regions (as in Detroit). Both provide the fertile conditions for borrowing ideas to develop and improvise, generating more new work for the city. The end result is the city's division of labor is enhanced— and economic development occurs.

Successful economic development in a city is sometimes referred as 'buzz' — like hives of activity, successful cities are perennially busy with rush hour commuters on area roads and railways streaming into areas of city work (**Figure 11.C**). There were once predictions that electronic communications would destroy these hives of activity since economic links can be made anywhere via the web (**Figure 11.D**). But the opposite has occurred: economic globalization is premised on a new age of great cities that are growing across the world because dense face-to-face communications are as important as ever for economic development.

Losing—or maintaining— buzz

But not all cities can continuously generate economic development. Different types of economic production come and go over time and may move to different

1. When you graduate, what difference do you think it would make to your prospects between starting a new job in a big city or in a much smaller place?

2. Check out the range of destinations you can get to from airports of two cities of different sizes. How do you think the two ranges might affect a small business trying to grow its market geographically?

3. Why are the main offices of the largest banks in the world concentrated in major cities such as New York, London, and Tokyo?

▲ **Figure 11.D Connectivity** This roofscape in Dubai, United Arab Emirates, reflects the importance of telecommunications in contemporary cities.

European Urban Expansion

In Europe the urban system introduced by the Greeks and re-established by the Romans almost collapsed during the Dark Ages of the early medieval period (476–1000 C.E.). During this time, feudalism gave rise to a fragmented landscape of inflexible and inward-looking world-empires. **Feudalism** was a rigid, rurally oriented form of economic and social organization based on the communal chiefdoms of Germanic tribes that had invaded the disintegrating Roman empire. From this unlikely beginning, an elaborate urban system developed, its largest centers eventually growing into what would become the nodal centers of a global world-system.

From the eleventh century onward, the European feudal system faltered and disintegrated in the face of successive demographic, economic, and political crises. These crises arose from steady population growth in conjunction with only modest technological improvements and limited amounts of cultivable land. To bolster their incomes and raise armies against one another, the feudal nobility began to levy increasingly higher taxes. Peasants were consequently obliged to sell more of their produce for cash on the market. As a result, a more extensive money economy developed, along with the beginnings of a pattern of trade in basic agricultural produce and craft manufactures. Some long-distance trade even began in luxury goods, such as spices, furs, silks, fruit, and wine. Towns began to increase in size and vitality on the basis of this trade.

The regional specializations and trading patterns that emerged provided the foundations for a new phase of urbanization based on merchant capitalism (**Figure 11.5**). Beginning with networks established by the merchants of Venice, Pisa, Genoa, and Florence (in northern Italy) and the trading partners of the Hanseatic League (a federation of city-states around the North Sea and Baltic coasts), a trading system of immense complexity soon spanned Europe from Bergen to Athens and from Lisbon to Vienna. By 1400, long-distance trading was well established, based not on the luxury goods of the pioneer merchants but on bulky staples such as grains, wine, salt, wool, cloth, and metals. Milan, Genoa, Venice, and Bruges had all grown to populations of 100,000 or more. Paris was the dominant European city, with a population of about 275,000. Europe stood poised to extend its grasp on a global scale.

Between the fifteenth and seventeenth centuries, a series of changes occurred that transformed not only the cities and city systems of Europe but the entire world economy. Merchant capitalism increased in scale and sophistication; economic and social reorganization was stimulated by the Protestant Reformation and the scientific revolution. Meanwhile, aggressive overseas colonization made Europeans the leaders, persuaders, and shapers of the rest of the world's economies and societies.

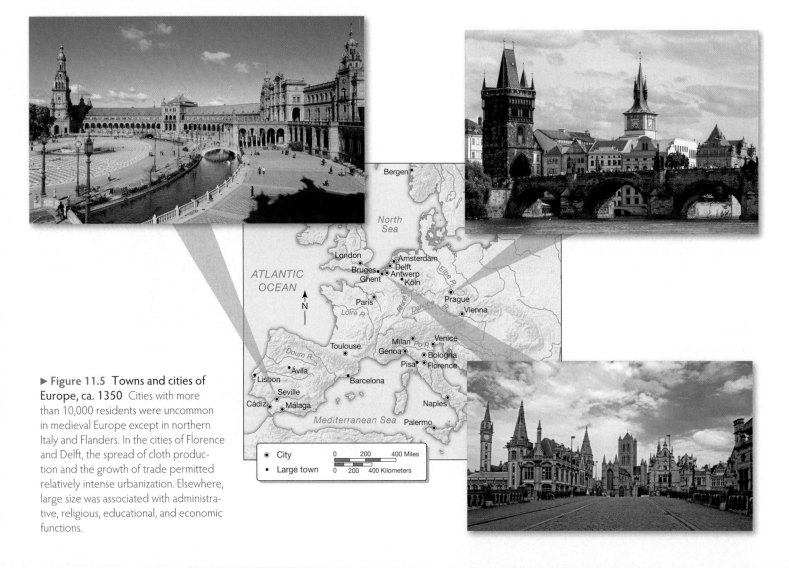

▶ Figure 11.5 Towns and cities of Europe, ca. 1350 Cities with more than 10,000 residents were uncommon in medieval Europe except in northern Italy and Flanders. In the cities of Florence and Delft, the spread of cloth production and the growth of trade permitted relatively intense urbanization. Elsewhere, large size was associated with administrative, religious, educational, and economic functions.

Colonialism and Urbanization Spanish and Portuguese colonists were the first to extend the European urban system into the world's peripheral regions. They established the basis of a Latin American urban system in just 60 years, between 1520 and 1580. Spanish colonists founded their cities on the sites of Indian cities (in Oaxaca and Mexico City, Mexico; Cajamarca and Cuzco, Peru; and Quito, Ecuador) or in regions of dense indigenous populations (in Puebla and Guadalajara, Mexico and Arequipa and Lima, Peru). These colonial towns were established mainly as administrative and military centers from which the Spanish Crown could occupy and exploit the New World. Portuguese colonists, in contrast, situated their cities—Recife, Salvador, São Paulo, and Rio de Janeiro—with commercial rather than administrative considerations in mind. They, too, were motivated by exploitation, but their strategy was to establish colonial towns in locations best suited to organizing the collection and export of the products of their mines and plantations.

The most important aspect of urbanization during this period, however, was the establishment of gateway cities around the rest of the world (**Figure 11.6**). **Gateway cities** serve as a link between one country or region and others because of their physical situation. These control centers command entrance to, and exit from, their particular country or region. European powers founded or developed thousands of towns in other parts of the world as they extended their trading networks and established their colonies. The great majority of these urban enclaves were ports. Protected by fortifications and European naval power, they began as trading posts and colonial administrative centers. Before long, they developed manufacturing of their own to supply the pioneers' needs, along with more extensive commercial and financial services.

As colonies were developed and trading networks expanded, some of these ports grew rapidly, acting as gateways for colonial expansion into continental interiors. Into their harbors came waves of European settlers; through their docks were funneled the produce of continental interiors. Rio de Janeiro (Brazil) grew on the basis of gold mining; Accra (Ghana) on cocoa; Buenos Aires (Argentina) on mutton, wool, and cereals; Kolkata (India, formerly Calcutta) on jute, cotton, and textiles; São Paulo (Brazil) on coffee; and so on. As these cities grew into major population centers, they became important markets for imported European goods, adding even more to their functions as gateways for international transport and trade.

APPLY YOUR KNOWLEDGE

1. From an atlas map, create a list of probable gateway cities along the Atlantic seaboard of North America.

2. Use the Internet to check on the early histories of these cities, noting their principal imports and exports. Compare your knowledge of the big cities along the East Coast against the peaks and falls of certain types of transportation, agricultural products, and industries like cotton, steel, and shipmaking.

In Europe itself, reorganization during the Renaissance saw the centralization of political power and the formation of national states, the beginnings of industrialization, and the funneling of plunder and produce from distant colonies. In this new context, the port cities of the North Sea and Atlantic coasts enjoyed a decisive locational advantage. By 1700, London had grown to 500,000 people, while Lisbon and Amsterdam had each grown to about 175,000. The cities of continental and Mediterranean Europe expanded at a more modest rate. By 1700, Venice had added only 30,000 to its 1400 population of 110,000, and Milan's population did not grow at all between 1400 and 1700.

Industrialization and Urbanization

It was not until the late eighteenth century, however, that urbanization began to become an important dimension of the world-system in its own right. In 1800, less than 5 percent of the world's 980 million people lived in towns and cities. But by 1950, however, 16 percent of the world's population was urban, and more than 900 cities of 100,000 or more existed around the world. The Industrial Revolution and European imperialism had created unprecedented concentrations of humanity that were intimately linked in networks and hierarchies of interdependence.

Cities became synonymous with industrialization. Industrial economies could be organized only through the large pools of labor; the transportation networks; the physical infrastructure of factories, warehouses, stores, and offices; and the consumer markets provided by cities. As industrialization spread throughout Europe in the first half of the nineteenth century and then to other parts of the world, urbanization increased at a faster pace. The higher wages and greater variety of opportunities in urban labor markets attracted migrants from surrounding areas. The countryside began to empty. In Europe the *demographic transition* caused a rapid growth in population as death rates dropped dramatically (see Chapter 3). This growth in population provided a massive increase in the labor supply throughout the nineteenth century, further boosting the rate of urbanization, not only within Europe but also in Australia, Canada, New Zealand, South Africa, and the United States as emigration spread industrialization and urbanization to the frontiers of the world-system.

During the Industrial Revolution and for much of the twentieth century, a close and positive relationship existed between rural and urban development in the core regions of the world (see Box 11.2, "Visualizing Geography: The Urbanization Process"). The appropriation of new land for agriculture, together with mechanization and the innovative techniques that urbanization allowed, resulted in increased agricultural productivity. This extra productivity released rural labor to work in the growing manufacturing sector in towns and cities. At the same time, it provided the additional produce needed to feed growing urban populations. The whole process was further reinforced by the capacity of urban labor forces to produce agricultural tools, machinery, fertilizer, and other products that made for still greater increases in agricultural productivity. This kind of urbanization is a special case of *cumulative causation* (see Chapter 7), in which a spiral buildup of advantages is enjoyed by particular places as a result of the development of external economies, agglomeration effects, and localization economies.

New York, at first a modest Dutch fur-trading port, became the gateway for millions of European immigrants and for a large volume of U.S. agricultural and manufacturing exports.

Boston first flourished as the principal colony of the Massachusetts Bay Company, exporting furs and fish and importing slaves from West Africa, hardwoods from central America, molasses from the Caribbean, manufactured goods from Europe, and tea (via Europe) from South Asia.

Salvador, Brazil, was the landfall of the Portuguese in 1500. They established plantations that were worked by slave labor from West Africa. Salvador became the gateway for most of the 3.5 million slaves who were shipped to Brazil between 1526 and 1870.

Guangzhou was the first Chinese port to be in regular contact with European traders—first Portuguese in the sixteenth century and then British in the seventeenth century.

Nagasaki was the only port that feudal Japanese leaders allowed open to European traders, and for more than 200 years Dutch merchants held a monopoly of the import-export business through the city.

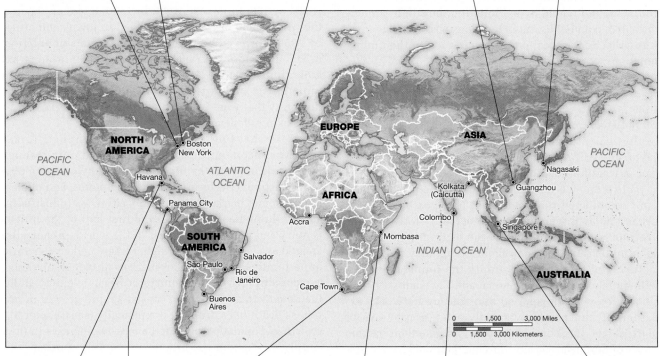

Havana was founded and developed by the Spanish in 1515 because of its excellent harbor. It was used as the assembly point for annual convoys returning to Spain.

Panama City, founded by the Spanish in 1519, became the gateway for gold and silver on its way by galleon to Spain.

Cape Town was founded in 1652 as a provisioning station for ships of the Dutch East India Company. Later, under British rule, it developed into an import-export gateway for South Africa.

Mombasa (in present-day Kenya) was already a significant Arab trading port when Vasco da Gama visited it in 1498 on his first voyage to India. The Portuguese used it as a trading station until it was recaptured by the Arabs in 1698. It did not become an important gateway port until it fell under British Imperial rule in the nineteenth century, when railroad development opened up the interior of Kenya, along with Rwanda, Uganda, and northern Tanzania.

Colombo's strategic situation on trade routes saw it occupied successively by the Portuguese, the Dutch, and the British. It became an important gateway after the British constructed an artificial harbor to handle the exports from tea plantations in Ceylon (now Sri Lanka).

Singapore, The British East India Company established Singapore as a trading post in 1819, and later became the chief port for exports from Malaysian rubber plantations. It was also the main British naval base in Southeast Asia.

▲ **Figure 11.6 Gateway cities in the evolving world-system periphery** Many of the world's most important cities grew to prominence as gateway cities because they commanded routeways into and out of developing colonies.

Manchester: Shock City   The outcome was the creation of the modern industrial city, and nowhere was this more striking than in the emergence of Manchester, England, as the "shock city" of nineteenth-century Europe. (A **shock city** is seen as the embodiment of surprising and disturbing changes in contemporary economic, social, and cultural life. Chicago was the shock city of the early twentieth century, and Los Angeles was the shock city of the late twentieth century.) Manchester grew from a small town of 15,000 in 1750 to a city of 70,000 in 1801, a metropolis of 500,000 in 1861, and an internationally renowned city of 2.3 million by 1911. Manchester's phenomenal growth was based on new textile manufacturing technologies, imported cotton from Britain's colonies and former colonies, and the migration of labor from Ireland, Italy, and the rural regions of northern England (**Figure 11.7**). The city's first cotton mill was built in the early 1780s, and by 1830 there were 99 cotton-spinning mills. The opening of the Suez Canal in 1869 halved the travel time between Britain and India. It ruined the Indian domestic cotton textile industry, but it allowed India to export its raw cotton to Manchester. Around the same time, British colonialists established cotton plantations in Egypt and Uganda, providing another source of supply.

Manchester was the archetypal form of an entirely new kind of city—the *industrial city*—whose fundamental reason for existence was not, as it was for earlier generations of cities, to fulfill military, political, ecclesiastical, or trading functions. Rather, it existed simply to assemble raw materials and to fabricate, assemble, and distribute manufactured goods. Manchester had to cope with unprecedented rates of growth and the unprecedented economic, social, and political problems that were a consequence of its growth.

Colonial Cities

As the example of Manchester shows, the industrialization of the core economies was highly dependent on the exploitation of peripheral regions. Inevitably, the new international division of labor that resulted from this relationship also had a significant impact on patterns and processes of urbanization in the periphery. European trade and imperialism led to the creation of new gateway cities in peripheral countries and, as Europeans raced to establish economic and political control over continental interiors, these colonial cities were established as centers of administration, political control, and commerce.

Colonial cities are those that were deliberately established or developed as administrative or commercial centers by colonial or imperial powers. Geographers often distinguish between two types of colonial city. The pure colonial city was usually established, or "planted," by colonial administrations in a location where no significant urban settlement had previously existed. Such cities were laid out expressly to fulfill colonial functions, with ceremonial spaces, offices, and depots for colonial traders, plantation representatives, and government officials; barracks for soldiers; and housing for colonists. Subsequently, as these cities grew, they added housing and commercial space for local peoples drawn by the opportunity to obtain jobs as servants, clerks, or porters. Examples of pure colonial cities are the original settlements of Mumbai (Bombay), Kolkata (Calcutta), Ho Chi Minh City (Saigon), Hong Kong, Jakarta, Manila, and Nairobi.

In the other type of colonial city, colonial functions were grafted onto an existing settlement, taking advantage of a good site and a ready supply of labor. Examples include Delhi, Mexico City, Shanghai, and Tunis. In these cities, the colonial imprint is most visible in and around the city center in the formal squares and public spaces, the layout of avenues, and the presence of colonial architecture and monuments.

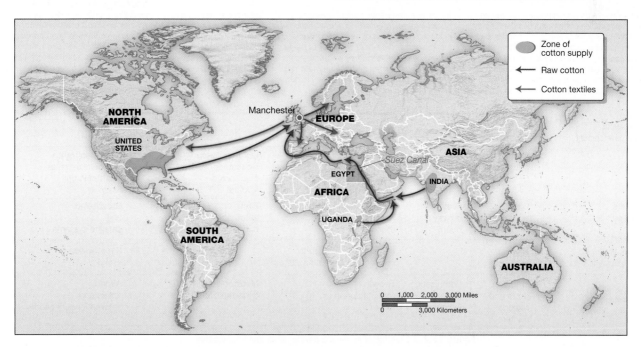

▲ **Figure 11.7 Manchester in the nineteenth-century world economy**   The opening of the Suez Canal in 1869 halved the traveling time between Britain and India. The canal ruined the Indian domestic cotton textile industry, but allowed India to export its raw cotton to Manchester.

11.3 Visualizing Geography

The Urbanization Process

11.1 Economic Development and Urbanization

There have been several broad phases in the nature of economic development in Europe and North America, and each new phase called for new kinds of cities, while existing cities had to be modified. At the same time, cities themselves played important roles in the transformation national economies. As centers of innovation, cities have traditionally functioned as engines of economic growth.

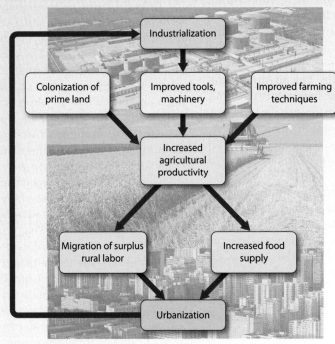

Figure 11.1.1 The urbanization process in Western Europe and North America. The system is centered around agricultural productivity.

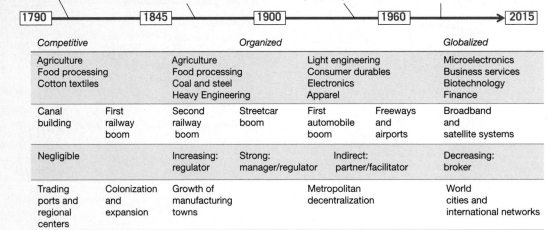

The earliest phase of industrial development was a phase of competitive enterprise, characterized by competition among small family businesses and with few constraints imposed by governments or public authorities. The dynamism of the entire system rested on the profitability of agriculture and industrial production based on mechanization, and intensively-used skilled labor.

By the late nineteenth century, business became more organized as corporations set out to serve regional or national consumer markets rather than local ones. Labor markets became more organized as wage norms spread, and governance became more organized as the need for regulation in public affairs became increasingly apparent.

After World War II, there was a shift away from industrial production and toward services, particularly sophisticated business and financial services, as the basis for profitability.

Increasing globalization of the economy since the 1970s allowed huge transnational corporations to move production and assembly operations to lower-cost, less developed parts of the world. There was a rapid decline of the old manufacturing base and the onset of a "new economy" based on digital technologies and business services.

	1790	1845	1900	1960	2015
Major phases of Economic development	*Competitive*		*Organized*		*Globalized*
Key Industries	Agriculture Food processing Cotton textiles		Agriculture Food processing Coal and steel Heavy Engineering	Light engineering Consumer durables Electronics Apparel	Microelectronics Business services Biotechnology Finance
Infrastructure development	Canal building	First railway boom	Second railway boom	Streetcar boom	First automobile boom	Freeways and airports	Broadband and satellite systems
Role of central government in urban development	Negligible		Increasing: regulator	Strong: manager/regulator	Indirect: partner/facilitator	Decreasing: broker
Epochs of urban system development	Trading ports and regional centers	Colonization and expansion	Growth of manufacturing towns		Metropolitan decentralization	World cities and international networks

Figure 11.1.2 Timeline of economic development and urbanization

11.2 Productivity Fuels Urbanization

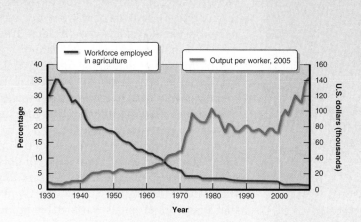

Figure 11.2.1 U.S. employment and productivity in agriculture 1930-2008. Improvements in farming methods have increased agricultural productivity and decreased the number of employees.

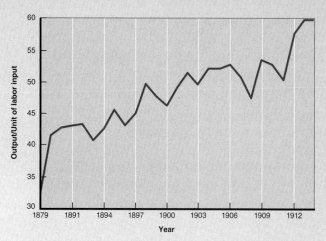

Figure 11.2.2 U.S. productivity in manufacturing, 1879–1914.

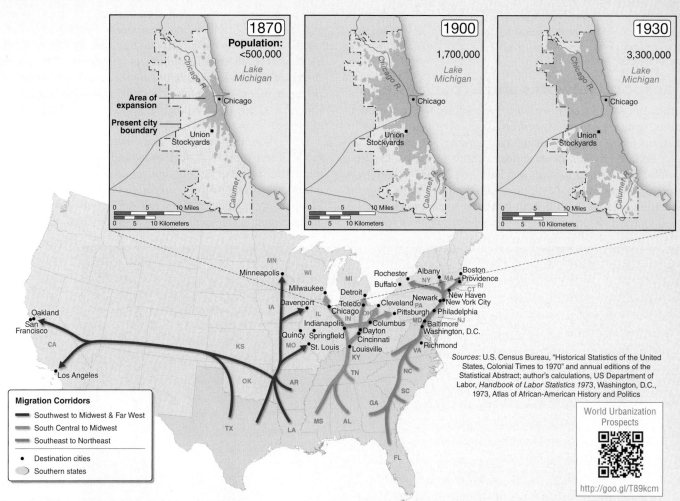

Sources: U.S. Census Bureau, "Historical Statistics of the United States, Colonial Times to 1970" and annual editions of the Statistical Abstract; author's calculations, US Department of Labor, *Handbook of Labor Statistics 1973*, Washington, D.C., 1973, Atlas of African-American History and Politics

World Urbanization Prospects

http://goo.gl/T89kcm

Figure 11.2.3 Migration of African-Americans 1916-1930. Thousands of African-Americans moved from rural southern areas to urban centers in the north and west. Chicago's expansion during this time was due to a number of factors, including an influx of African-Americans. The city experienced explosive growth between 1870 and 1930.

1. Why do you think racial groups are more mixed in some cities, compared to others? Compare historical factors (civil rights movement, WWII, American Indian relocation) and immigration polices with what you have learned.

2. Analyze the demographic concept of "minority-majority" geographies by doing an Internet search on "minority-majority cities/counties/states." Where are these cities/counties/states located and what do you observe about the political affiliations, immigration policies, social programs and economy compared to white majority areas?

▲ **Figure 11.8 Canal systems** The canal systems that opened up the interiors of Europe and North America in the eighteenth century were initially dependent on horsepower. This example shows part of the Grand Union canal, in England.

Transport Networks and Urban Systems

Within the world's core regions, meanwhile, the transformation of urban systems hinged on successive innovations in transport technology that opened up agrarian interiors and intensified intercity and interregional trading networks. The first phase of this transformation was based on an old technology: the canal (**Figure 11.8**). Merchant trade and the beginnings of industrialization in both Britain and France were underpinned by extensive navigation systems that joined one river system to another. The Industrial Revolution provided both the need and the capital for a spate of additional canal building that began to integrate emerging industrial towns. Market towns and hill towns that were not connected to canal systems were isolated from commerce and immediately fell behind.

In the United States, the opening of the Erie Canal in 1825 enabled New York, a colonial gateway port, to reorient itself toward the nation's growing interior. The Erie Canal was so profitable that it set off a "canal fever" that resulted in the construction of some 2,000 kilometers (1,240 miles) of navigable waterways in the next 25 years. But the scale of the United States was so great that a network of canals was a viable proposition only in more densely settled regions. The effective colonization of the interior and the growth of American cities did not take place until the development of steam-powered transportation—first riverboats and then railroads. The first steamboats, developed in the early 1800s, offered the possibility of opening up the vast interior by way of the Mississippi and its tributaries. The heyday of the river steamboat was between 1830 and 1850. During this period, river ports such as New Orleans (**Figure 11.9**), St. Louis, Cincinnati, and Louisville grew rapidly, extending the frontier of industrialization and modernization.

By 1860, the railroads had taken over as the dominant mode of transportation, further extending the frontier of settlement and industrialization and intensifying the growth of existing cities. The railroad originated in Britain, where George Stephenson engineered the world's first commercial railroad, a 20-kilometer (12.4-mile) line between Stockton and Darlington

◀ **Figure 11.9 New Orleans riverboats** This 'bird's-eye' view, drawn in 1885, shows the city of New Orleans and shipping on the Mississippi River.

◀ **Figure 11.10** Railroad yards, Chicago
The city's phenomenal growth in the late nineteenth and early twentieth century could not have taken place without its railroad network.

that was opened in 1825. The *Rocket*, the first locomotive for commercial passenger trains, was designed mainly by Stephenson's son Robert for the Liverpool and Manchester line, which opened 4 years later. The economic success of this line sparked two railroad-building booms that eventually created a highly integrated urban system and allowed Britain's manufacturing industry to flourish.

In other core countries, where sufficient capital existed to license (or copy) the locomotive technology and install the track, railroad systems led to the first full stage of urban system integration. Yet, while the railroads integrated the economies of entire countries and allowed vast territories to be colonized, they also brought some important regional and local restructuring and differentiation. In the United States, for example, the railroads led to the mushrooming of Chicago as the focal point for railroads (**Figure 11.10**). This extended the Manufacturing Belt's dominance over the West and South. The reorientation of the nation's transportation system effectively ended the role of the South's cotton regions as outliers of the British trading system. Instead, they became outliers of the U.S. Manufacturing Belt, supplying factories in New England and the Mid-Atlantic Piedmont. The phenomenal growth of New Orleans, which had thrived on cotton exports, came to an abrupt end.

In the twentieth century, the internal combustion engine powered further rounds of urban system development and integration. The development of trucks in the 1910s and 1920s suddenly released factories from the need to locate near railroads, canals, and waterfronts. Trucking allowed goods to be moved farther, faster, and cheaper than before. As a result, trucking made it feasible to locate factories on inexpensive land on city fringes and in smaller towns and peripheral regions where labor was cheaper. As we saw in Chapter 2, however, the single most important innovation with regard to the

international division of labor—and therefore also the global urban system—was the development of metal-hulled, ocean-going steamships that had vastly improved carrying capacity, speed, range, and reliability. The Suez Canal (opened in 1869) and the Panama Canal (opened in 1914) were also critical, providing shorter and less hazardous routes between core countries and colonial ports of call (Figure 2.10).

APPLY YOUR KNOWLEDGE

1. How have changing transportation technologies affected the history of the town or city in which you live?

2. What current transportation changes are operating that will continue to change your town, for example, bike lanes, public transit, connections to other metropolitan areas?

Primacy and Centrality in the Global Urban System

Towns and cities function as market centers and this results in a hierarchical system of central places. A **central place** is a settlement where certain types of products and services are available to consumers. **Central place theory** seeks to explain the tendency for central places to be organized in hierarchical systems, analyzing the relative size and geographic spacing of towns and cities as a function of consumer behavior. A fundamental tenet of central place theory is that the smallest settlements in an urban system provide only those goods and services that meet everyday needs, (e.g. bakery and dairy products and groceries) and that these small settlements are situated relatively close to one another because consumers, assumed to be spread throughout the countryside, are not prepared to travel far for such items. On the other hand, people are willing to travel farther for more expensive, less frequently purchased items. This means that the larger the settlement, with a broader variety of more specialized goods and services, the farther it will be from others of a similar size. The urban systems of most regions do exhibit a clear hierarchical structure, with many smaller towns, fewer mid-sized towns and cities, and a small number of large cities.

Urban systems also exhibit clear *functional* differences within such hierarchies. This is yet another reflection of the interdependence of places. The geographical division of labor resulting from such processes of economic development (Chapter 7) means that many medium- and larger-size cities perform specialized economic functions and so acquire distinctive characters. Thus the industrial era produced steel towns (e.g., Pittsburgh, Pennsylvania; Sheffield, England), textile towns (e.g., Lowell, Massachusetts; Manchester, England), and auto-manufacturing towns (e.g., Detroit, Michigan; Turin, Italy; Toyota City, Japan).

The Rank-Size Rule In many countries, the functional interdependency between places within urban systems tends to result in a distinctive relationship between the population size of cities and their rank within the overall hierarchy. This relationship is known as the **rank-size rule**, which describes a certain statistical regularity in the city-size distributions of countries and regions. The relationship is such that the *n*th largest city in a country or region is 1/*n* the size of the largest city in that country or region. Thus, if the largest city in a particular system has a population of 1 million, the fifth-largest city should have a population one-fifth as big (i.e., 200,000); the hundredth-ranked city should have a population one-hundredth as big (i.e., 10,000), and so on. Plotting this relationship on a graph with a logarithmic scale for population sizes would produce a perfectly straight line. The actual rank-size relationships for urban systems at all levels of economic development come close to this (**Figure 11.11**). Over time, the slope moves to the right on the graphs, reflecting the growth of towns and cities at every level in the urban hierarchy.

In some urban systems, the rank-size distribution is distorted as a result of the disproportionate size of the largest (and sometimes also the second-largest) city. According to the rank-size rule, the largest city should be just twice the size of the second-largest city. In the United Kingdom, London is more than 9 times the size of Birmingham, the second-largest city. In France, Paris is more than 8 times the size of Marseilles, France's second-largest city. In Mexico, the capital, Mexico City is more than 5 times the size of Guadalajara, the second-largest city; while Bangkok, in Thailand, is 15 times larger than Nanthaburi, the country's next-largest city. Geographers call this condition **primacy**, occurring when the population of the largest city in an urban system is disproportionately large in relation to the second- and third-largest cities in that system. Cities like London and Buenos Aires are termed *primate* cities.

When a city's economic, political, and cultural function is disproportionate to its population, the condition is known as **centrality**. Centrality refers to the functional dominance of cities within an urban system. Cities that account for a disproportionately high share of economic, political, and cultural activity have a high degree of centrality within their urban system. Very often primate cities exhibit this characteristic, but cities do not necessarily have to be primate to be functionally dominant within their urban system. **Figure 11.12** shows some examples of centrality, revealing the overwhelming dominance of some cities within the world-system periphery. Bangkok, for instance, with around 10 percent of the Thai population, accounts for approximately 38 percent of the country's overall GDP; over 85 percent of the country's GDP in banking, insurance, and real estate; and 75 percent of its manufacturing.

At the global scale, the most highly centralized cities are world cities (sometimes referred to as *global cities*) that play key roles in organizing space beyond their own national boundaries. Recall from Chapter 7 that such cities have existed ever since the evolution of a world-system in the sixteenth century. In the first stages of world-system growth, these key roles involved the organization of trade and the execution of colonial, imperial, and geopolitical strategies.

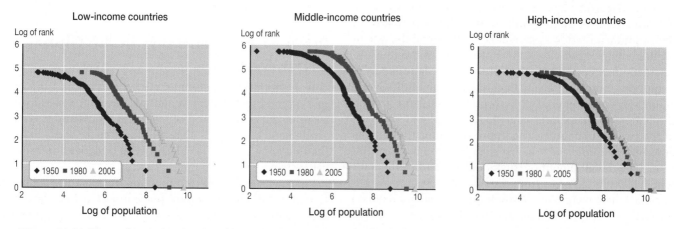

▲ **Figure 11.11 The rank-size distribution of cities, 1950–2005** These graphs indicate that urban systems in countries at all levels of development tend to conform fairly consistently to the rank-size rule. As urbanization brought increased populations to cities at every level in the urban hierarchy, the rank-size graphs move to the right.

(*Source: The World Bank, Reshaping economic Geography: World Development Report 2009. Washington, D.C., 2009, p. 51.*)

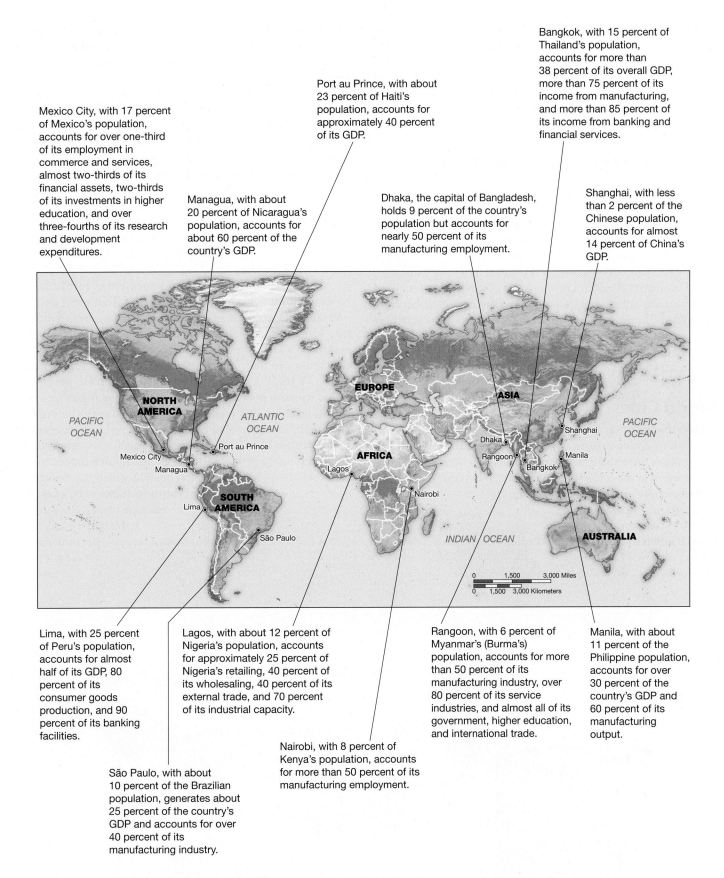

Mexico City, with 17 percent of Mexico's population, accounts for over one-third of its employment in commerce and services, almost two-thirds of its financial assets, two-thirds of its investments in higher education, and over three-fourths of its research and development expenditures.

Managua, with about 20 percent of Nicaragua's population, accounts for about 60 percent of the country's GDP.

Port au Prince, with about 23 percent of Haiti's population, accounts for approximately 40 percent of its GDP.

Dhaka, the capital of Bangladesh, holds 9 percent of the country's population but accounts for nearly 50 percent of its manufacturing employment.

Bangkok, with 15 percent of Thailand's population, accounts for more than 38 percent of its overall GDP, more than 75 percent of its income from manufacturing, and more than 85 percent of its income from banking and financial services.

Shanghai, with less than 2 percent of the Chinese population, accounts for almost 14 percent of China's GDP.

Lima, with 25 percent of Peru's population, accounts for almost half of its GDP, 80 percent of its consumer goods production, and 90 percent of its banking facilities.

Lagos, with about 12 percent of Nigeria's population, accounts for approximately 25 percent of Nigeria's retailing, 40 percent of its wholesaling, 40 percent of its external trade, and 70 percent of its industrial capacity.

Rangoon, with 6 percent of Myanmar's (Burma's) population, accounts for more than 50 percent of its manufacturing industry, over 80 percent of its service industries, and almost all of its government, higher education, and international trade.

Manila, with about 11 percent of the Philippine population, accounts for over 30 percent of the country's GDP and 60 percent of its manufacturing output.

São Paulo, with about 10 percent of the Brazilian population, generates about 25 percent of the country's GDP and accounts for over 40 percent of its manufacturing industry.

Nairobi, with 8 percent of Kenya's population, accounts for more than 50 percent of its manufacturing employment.

▲ **Figure 11.12 Examples of urban centrality** The economic, political, and cultural importance of some cities is disproportionate to their population size.

The world cities of the seventeenth century were London, Amsterdam, Antwerp, Genoa, Lisbon, and Venice. In the eighteenth century, Paris, Rome, and Vienna also became world cities, while Antwerp and Genoa became less influential. In the nineteenth century, Berlin, Chicago, Manchester, New York, and St. Petersburg became world cities, while Venice became less influential.

WORLD URBANIZATION TODAY

Much of the developed world, though, has become almost completely urbanized (**Figure 11.13**), and in many peripheral and semiperipheral regions the current rate of urbanization is without precedent (**Figure 11.14**). The dominant cities in the global urban system are the world cities, but they are not necessarily the largest. The largest cities in the global urban system, with populations of 10 million or more, are megacities; a few have become meta-cities, with populations in excess of 20 million. Meanwhile, for every world city, megacity, and meta-city, there are thousands of large cities and tens of thousands of towns and smaller cities, all linked in complex networks of trade and communications.

World Cities

Today, the globalization of the economy has resulted in the creation of a global urban system in which the key roles of world cities are concerned less with the deployment of imperial power and the orchestration of trade and more with transnational corporate organization, international banking and finance, supranational government, and the work

of international agencies. World cities have become the control centers for the flows of information, cultural products, and finance that collectively sustain the economic and cultural globalization of the world.

A great deal of synergy exists among the various functional dimensions of world cities. A city like London, for example (**Figure 11.15**), attracts transnational corporations because it is a center of culture and communications. It attracts specialized business services because it is a center of corporate headquarters and of global markets, and so on. These interdependencies represent a special case of the geographical *agglomeration effects* that we discussed in Chapter 7. In the case of New York City, corporate headquarters and specialized legal, financial, and business services cluster together because of the mutual cost savings and advantages of being close to one another.

At the same time, different world cities fulfill different roles within the world-system, making for different emphases and combinations (i.e., differences in the nature of their world-city functions) as well as for differences in the absolute and relative localization of particular world city functions (i.e., differences in their degree of importance as world cities; see **Figure 11.16**). For example, Brussels is relatively unimportant as a corporate headquarters location but qualifies as a world city because it is the administrative center of the European Union and has attracted a large number of nongovernmental organizations (NGOs) and advanced business services that are transnational in scope. Milan is relatively dependent in terms of corporate control and advanced business services but has global status in terms of cultural influence (especially fashion and design) and is an important regional financial center.

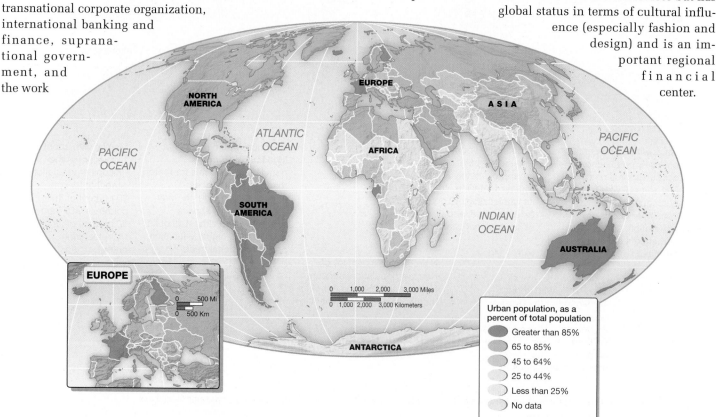

▲ Figure 11.13 Percentage of population living in urban settlements, 2009

(*Source:* Data from the World Bank, http://data.worldbank.org/indicator/SP.URB.TOTL.IN.ZS.)

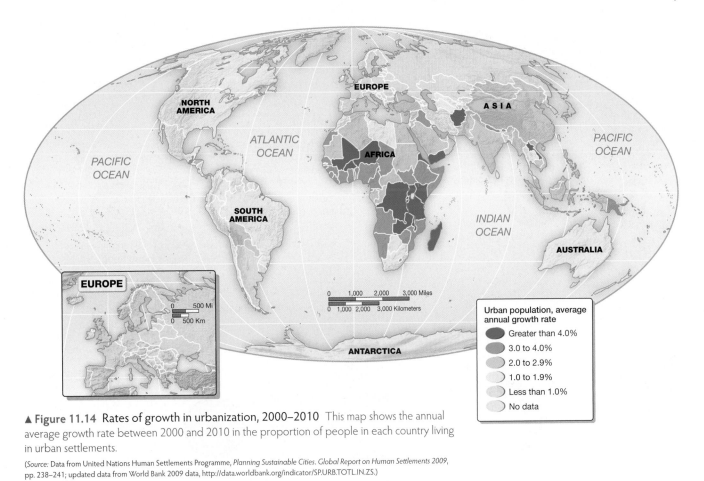

▲ **Figure 11.14 Rates of growth in urbanization, 2000–2010** This map shows the annual average growth rate between 2000 and 2010 in the proportion of people in each country living in urban settlements.

(*Source:* Data from United Nations Human Settlements Programme, *Planning Sustainable Cities. Global Report on Human Settlements 2009*, pp. 238–241; updated data from World Bank 2009 data, http://data.worldbank.org/indicator/SP.URB.TOTL.IN.ZS.)

▼ **Figure 11.15 London's pivotal role in global finance has changed the city's skyline.** The "square mile" of the City in the center of London contains around 500 banks, many of them specializing in areas such as foreign exchange markets, Eurobonds, and energy futures. The district also accounts for a quarter of the world market for marine insurance and over a third of the market in aviation risks.

Top 25 cities in the global cities index 2010

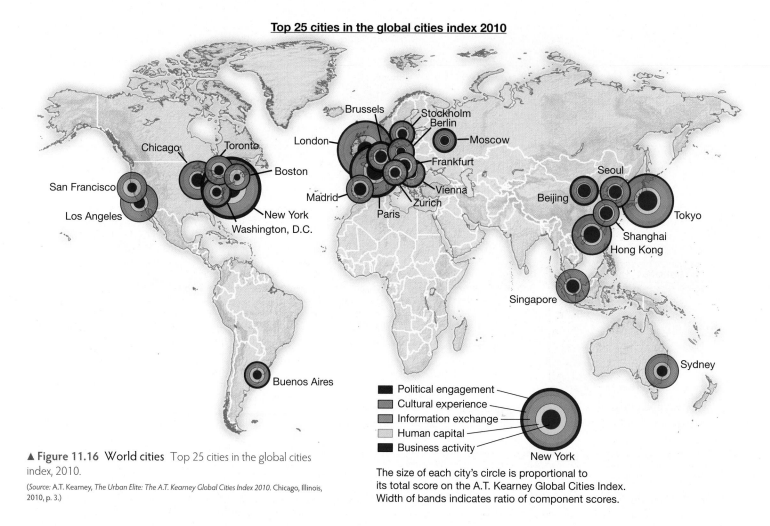

▲ **Figure 11.16** World cities Top 25 cities in the global cities index, 2010.

(*Source:* A.T. Kearney, *The Urban Elite: The A.T. Kearney Global Cities Index 2010.* Chicago, Illinois, 2010, p. 3.)

The size of each city's circle is proportional to its total score on the A.T. Kearney Global Cities Index. Width of bands indicates ratio of component scores.

Thus, there is a geographical complexity to world cities' roles that cannot be reduced to a simple "hierarchy" or ranking. World cities are not simply mini-Londons and little New Yorks. World cities exist in networks of flows among firms and institutions, networks that are complex and multilayered. As a result world cities are connected in different ways and integrated to different degrees in the global urban system.

World cities also provide an interface between the global and the local. They contain the economic, cultural, and institutional apparatus that channels national and provincial resources into the global economy and that transmits the impulses of globalization back to national and provincial centers. As such, world cities possess several functional characteristics. They are the sites of the following:

■ Most of the leading global markets for commodities, commodity futures, investment capital, foreign exchange, equities, and bonds

■ Clusters of specialized, advanced business services, especially those that are international in scope and that are attached to finance, accounting, insurance, advertising, property development, and law

■ Concentrations of corporate headquarters—not just of transnational corporations but also of major national firms and large foreign firms

■ Concentrations of national and international headquarters of trade and professional associations

■ Most of the leading NGOs and intergovernmental organizations (IGOs) that are international in scope (e.g., the World Health Organization; United Nations Educational, Scientific, and Cultural Organization (UNESCO); the International Labor Organization, and the International Federation of Agricultural Producers)

■ The most powerful and internationally influential media organizations (including newspapers, magazines, book publishing, and satellite television); news and information services (including news wires and online information services); and culture industries (including art and design, fashion, film, and television)

■ Many terrorist acts because of their importance and visibility

Megacities

In contrast to the world's core regions, where urbanization has largely resulted from economic growth, the urbanization of peripheral regions has been a consequence of demographic growth that preceded economic development. Although demographic transition is a fairly recent phenomenon in the peripheral regions of the world (see Chapter 3), it has generated large increases in population well in advance of any significant levels of industrialization or rural economic development. In the most massive population movement the world has ever seen, millions of people continue to pour out of rural villages each year, hoping for fresh economic opportunity in the world's cities.

The result, for the mainly rural populations of peripheral countries, has been a very different set of processes and outcomes in terms of urbanization (**Figure 11.17**). Problems with agricultural development (see Chapter 8) mean an apparently hopeless future of drudgery and poverty for fast-growing rural populations. Emigration provided a safety valve in the past, but as the frontiers of the world-system closed, the more affluent core countries have put up barriers to immigration. The only option for the growing numbers of impoverished rural residents in peripheral countries is to move to larger towns and cities, where at least there is the hope of employment and the prospect of access to schools, health clinics, piped water, and the kinds of public facilities and services that are often unavailable in rural regions. Cities also have the lure of modernization and the appeal of consumer goods—attractions that rural areas are now directly exposed to via satellite TV. Overall, the metropolises of the periphery have absorbed four out of five of the 1.5 billion city dwellers added to the world's population since 1970.

Rural migrants have poured into cities out of desperation and hope rather than being drawn by jobs and opportunities. Because these migration streams have been composed disproportionately of teenagers and young adults, an important additional component of urban growth has followed—exceptionally high rates of natural population increase. In most peripheral countries, the rate of natural increase of the population in cities exceeds that of net in-migration. On average, about 60 percent of urban population growth in peripheral countries is attributable to natural increase.

The consequence of all this urban population growth is **overurbanization**, which occurs when cities grow more rapidly than they can sustain jobs and housing. In such circumstances, urban growth produces instant slums—shacks set on unpaved streets, often with open sewers and no basic utilities (see Chapter 11). In general, it is the rate, not simply the level, of urbanization that produces slums; higher levels of urbanization tend to be associated with relatively fewer slums. The most extensive slums, according to UN statistics, are in the cities of sub-Saharan Africa, where over 60 percent of the population lives in unfit accommodations. The United Nations International Children's Fund (UNICEF) has blamed "uncontrollable urbanization" in less developed countries for the widespread creation of "danger zones" in which increasing numbers of children become beggars, prostitutes, and laborers before reaching their teens.[4] Pointing out that urban populations are growing at twice the general population rate, UNICEF has concluded that too many people are being squeezed into cities that do not have enough jobs, shelter, or schools to accommodate them.

Overurbanization, together with centrality, has been a major contributor to the development of megacities. **Megacities** are very large cities characterized by both primacy and a high degree of centrality within their national economy. Their most important common denominator is their sheer size—most of them number 10 million or more in population. This, together with their functional centrality, means that in many ways they have more in common with one another than with the smaller metropolitan areas and cities within their own countries. Megacities include Bangkok, Dhaka, Jakarta (**Figure 11.18**), Lagos, Manila, Mumbai, New Delhi, São Paulo (**Figure 11.19**), and Teheran. Each has more inhabitants than 100 of the member countries of the United Nations. Although most of them do not function as world cities, they do serve important intermediate roles between the upper tiers of the system of world cities and the provincial towns and villages of large regions of the world. They not only link local and provincial economies with the global economy but also provide a point of contact between the traditional and the modern, and between formal and informal economic sectors (see Box 11.3, "Window on the World: The Pearl River Delta: An Extended Metropolis").

APPLY YOUR KNOWLEDGE

1. Conduct an Internet search to find two megacities. List two ways a megacity differs from a world city. Is either (or both) of your megacities a former colonial city?

2. How might that history have contributed to the current status of being a megacity?

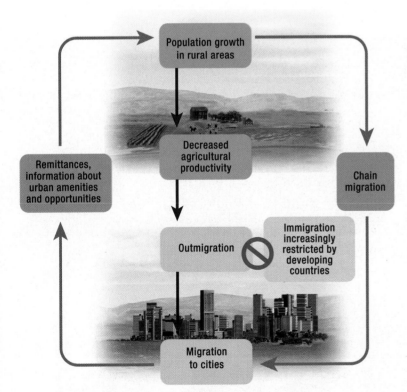

▲ **Figure 11.17** The urbanization process in the world's peripheral regions

[4]*The State of the World's Children 2011.* New York: United Nations International Children's Fund (UNICEF), 2011.

Mature Metropolises

The high levels of urbanization and relatively slow rates of urban growth within the world's core regions are reflected in relatively stable urban systems. There is constant change, nevertheless, in patterns and processes of urbanization as the metropolises, cities, and towns adjust to the opportunities of new technologies and new industries and to the constraints of obsolescent urban infrastructure and land-use conflicts. New rounds of urbanization are initiated in the places most suited to new technologies and new industries. Those places least suited are likely to suffer spirals of deindustrialization and urban decline.

▼ **Figure 11.19 Sao Paulo** The largest metropolitan area in Brazil, with a population of more than 11 million.

Deindustrialization and Agglomeration Diseconomies

Deindustrialization involves a decline in industrial employment in core regions as firms scale back their activities in response to lower levels of profitability (see Chapter 7). Such adversity has particularly affected cities like Detroit, Pittsburgh, and Cleveland (United States), Sheffield and Liverpool (United Kingdom), Lille (France), and Liège (Belgium)—places where heavy manufacturing constituted a key economic sector. Cities like these have suffered substantial reductions in employment since the 1970s and 1980s when better and more flexible transport and communications networks allowed many industries to choose from a broader range of potential locations.

In many instances, deindustrialization has been intensified by the dampening effects of *agglomeration diseconomies* (Chapter 7) on the growth of larger metropolitan areas. Agglomeration diseconomies, the negative effects of urban size and density, include noise, air pollution, increased crime, high commuting costs, inflated land and housing prices, traffic congestion, and crowded port and railroad facilities. They also include higher taxes levied to rebuild decaying infrastructure and to support services and amenities previously considered unnecessary—traffic police, city planners, and homeless shelters, for example.

The result of deindustrialization has been a *decentralization* of jobs and people from larger to smaller cities within the urban systems of core countries, and from metropolitan cores to suburban and ex-urban fringes. In some cases, routine production activities relocated to smaller metropolitan areas or to rural areas with lower labor costs and more hospitable business climates. In other cases, these activities moved overseas—as part of the new international division of labor (see Chapter 2)—or were eliminated entirely.

Counterurbanization and Reurbanization

The combination of deindustrialization in core manufacturing regions, agglomeration diseconomies in major metropolitan areas, and the improved accessibility of smaller towns and rural areas can give rise to the phenomenon of counterurbanization. **Counterurbanization** occurs when cities experience a net loss of population to smaller towns and rural areas. This process results in the deconcentration of population within an urban system. This is what happened in the United States, Britain, Japan, and many other developed countries in the 1970s and 1980s. Metropolitan growth slowed dramatically, while the growth rates of small- and medium-size towns and of some rural areas increased. In these countries, counties that for decades had recorded stable populations grew by 15 or 20 percent. Some of the strongest gains were registered in counties that were within commuting range of metropolitan areas, but some remote counties also registered big population increases.

Counterurbanization was a major reversal of long-standing trends, but it seems to have been a temporary adjustment rather than a permanent change. The globalization of the economy and the growth of postindustrial activities in revamped and expanded metropolitan settings have restored the trend toward the concentration of population within urban systems. Most of the cities that were declining fast in the 1970s and 1980s

are now either recovering (New York, London) or bottoming out (Paris, Chicago), while most of those that were growing only slowly (Tokyo, Barcelona) are now expanding more quickly. This trend of **reurbanization** involves the growth of population in metropolitan central cores following a period of absolute or relative decline in population.

Migration Drives Reurbanization Two principal kinds of migration are driving reurbanization in Europe and North America. One consists of less skilled migrants and immigrants from peripheral regions and countries, who move into cheaper central districts of major metropolitan areas because of proximity to the availability service jobs. A second, very different migration stream consists of "baby boomers" (see Chapter 3) electing to pursue urban rather than suburban lifestyles. They tend to move into older, central neighborhoods with character, "gentrifying" them in the process (see Chapter 11), or into newly-built condominiums or converted industrial spaces.

Reurbanization has also been fostered by city policies and real estate investments aimed at the regeneration of derelict and declining city districts. **Regeneration** involves the physical redevelopment of land where the existing buildings are no longer useful or profitable. It often takes place on **brownfield sites**—former industrial or commercial land where future use is affected by real or perceived environmental contamination. Because of the significant costs and risks involved, urban regeneration strategies are usually undertaken in the form of **public–private partnerships** between city governments and real estate developers. Regeneration strategies take various forms. Some cities have focused on "high" culture and the provision of museums, galleries, theatres, and concert halls. Bilbao, Spain, has been very successful in this way. Some cities emphasize tradition, history, and monuments in their regeneration strategies, while others focus on their package of nightlife—theaters, concert halls, nightclubs, cafés, and restaurants. Sports venues are also key to many regeneration strategies. In London, for example, the Olympic Park for the 2012 Olympics (**Figure 11.20**) was

▼ Figure 11.20 London Olympic Park

11.3 Window on the World

The Pearl River Delta: An Extended Metropolis

The Pearl River Delta (**Figure 11.E**) is one of the fastest-growing urban regions in the world. Anchored by the major metropolitan centers of Guangzhou, Hong Kong, Macau, Shenzhen, and Zhuhai, it is an extended metropolitan region of nearly 50 million people. It is one of three extended metropolitan regions—Beijing-Tianjin and Shanghai are the others—that have been fostered by the Chinese government to be engines of capitalist growth since liberal economic reforms were introduced in the late 1970s.

Hong Kong (**Figure 11.F**) was a British colony until 1997. It is now a metropolis of 7.4 million with a thriving industrial and commercial base that is recognized as a capitalist economic dynamo by the Chinese government, which has created a Special Administrative District for the metropolis. As a result, Hong Kong's citizens have retained their British-based legal system and its

▲ **Figure 11.F The City of Hong Kong** The city is home to thousands of companies located there simply for the purpose of doing business with China. As a result, Hong Kong remains a major world city—a major financial hub with a thriving commercial sector and a population of 7.4 million.

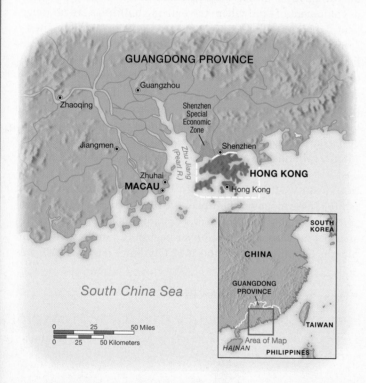

▲ **Figure 11.E Pearl River Delta** One of the fastest-growing regions of the world, the Pearl River Delta is an extended metropolitan region of more than 50 million people.

guaranteed rights of property ownership and democracy. Hong Kong is the world's largest container port, the third-largest center for foreign exchange trade, the seventh-largest stock market, and the tenth-largest trading economy.

Hong Kong's financial success encouraged the Chinese government to establish two of its first Special Economic Zones (SEZs) in nearby Shenzhen and Zhuhai. Designed to attract foreign capital, technology, and management practices, these SEZs were established as export-processing zones that offered cheap labor and land, along with tax breaks, to transnational corporations. Investors from Hong Kong and Taiwan responded quickly and enthusiastically. By 1993, more than 15,000 manufacturers from Hong Kong alone had set up businesses in Guangdong Province, and a similar number had established subcontracting relationships, contracting out assembly-line work to Chinese companies in the Pearl River Delta. Meanwhile, the Chinese government designated the entire delta region an Open Economic Region, where local governments, individual enterprises, and farm households enjoy a high degree of autonomy in economic decision making.

The relaxation of state control over the regional economy allowed the region's dense and growing rural population to migrate to urban areas in search of assembly-line jobs or to stay in rural areas and diversify agricultural production from paddy-rice cultivation to more profitable activities such as

constructed on brownfield sites in Stratford, an area of east London that had been previously rundown. The site is now used to accommodate low-cost housing as well as leisure activities, and its existence has attracted private investment to sites nearby.

APPLY YOUR KNOWLEDGE

1. What is the difference between counterurbanization and reurbanization?

2. Identify an American city that has experienced either counterurbanization or reurbanization, and the factors that contributed to this change.

▲ **Figure 11.G Infrastructure investment** Heavy holiday traffic on the new Yuegan Highway in south China's Guangdong Province.

economic growth for much of the past two decades. The region's GDP grew from just over US$8 billion in 1980 to nearly US$270 billion in 2010. During that period, the average real rate of GDP growth in the Pearl River Delta Economic Zone exceeded 16 percent, well above the People's Republic of China national figure of 9.8 percent. By 2010 and with only 3.5 percent of the country's population, the region was contributing 10 percent of the country's GDP and 29 percent of its total trade.

Guangzhou is a megacity with a 2010 population of around 10 million (**Figure 11.H**). Shenzhen has grown from a population of just 19,000 in 1975 to 8.1 million in 2010, with an additional 2 million in the surrounding municipalities. The southern border of the Shenzhen Special Economic Zone adjoins Hong Kong, but the northern border is walled off from the rest of China by an electrified fence to prevent smuggling and to keep back the mass of people trying to migrate illegally into Shenzhen and Hong Kong.

1. How have economic policies created this booming region?

2. Can you identify any other megacities that have experienced similar economic success?

market-farming activities, livestock husbandry, and fishery. Economic freedom also facilitated rural industrialization—mostly low-tech, small-scale, labor-intensive, and widely scattered across the countryside. The area between Guangzhou, Hong Kong, and Macau has quickly emerged as an especially important zone because of its relatively cheap land and labor and because of significant levels of investment by regional and local governments in the transport and communications infrastructure (**Figure 11.G**).

The metropolitan cores of the region, aiming to increase their competitiveness and prominence in the globalizing world economy, have invested heavily in infrastructure improvements. The Guangzhou municipal government, for example, invested more than $10 billion between 1998 and 2004 in infrastructure construction—including a metro system and an elevated railway network to link the city's new international airport, railway stations, and port. Throughout the region, enormous investments have been made in showpiece infrastructure projects geared to the needs of local and international capital. These include major airports, high-speed tolled highways, satellite ground stations, port installations, metro and light-rail networks, and new water-management systems. In turn, these projects have attracted business and technology parks, financial centers, and resort complexes in a loose-knit sprawl of urban development.

Today, the Pearl River Delta provides a thriving export-processing platform that has driven double-digit annual

▲ **Figure 11.H Guangzhou, China** Guangzhou is an ancient Chinese city known as Canton by European traders. Guangzhou has grown rapidly in recent decades, its modern architecture almost completely replacing the old city.

Cities and Climate Change

Natural and human-made disasters have been on the rise worldwide since the 1950s, coinciding with the rise in world urban population (**Figure 11.21**). Cities can be fragile environments, their concentrations of population making them vulnerable to natural disasters such as earthquakes, hurricanes, tsunamis, fires, and floods as well as to technological disasters. Rapidly-growing cities in less-developed regions are especially vulnerable. As we have seen (Chapter 4), disasters such as landslides, floods, windstorms, and wildland urban interface (WUI) fires are clearly linked to climate change, and they will almost certainly occur with greater frequency and intensity. Many urban

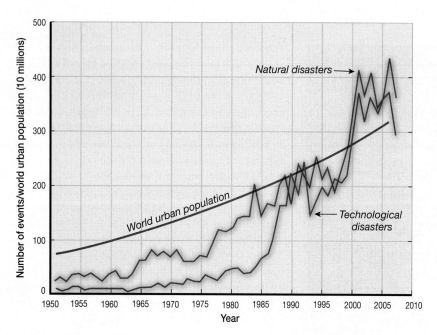

▲ **Figure 11.21** World urban population and recorded natural and technological disasters, 1950–2007

(*Source:* UN Human Settlements Programme, *Cities and Climate Change. Global Report on Human Settlements 2011.* London: Earthscan, 2011, Figure 4.2, p. 88.)

in volume of ocean water as it warms as a result of climate change is considered to be the principal cause of sea-level rise, but melting ice sheets are becoming increasingly important. The consequences of sea-level rise include increased storm flooding and damage, inundation, coastal erosion, increased salinity in estuaries and coastal aquifers, rising coastal water tables, obstructed drainage, and changes to ecosystems such as wetlands, mangrove swamps, and coral reefs that form natural protections for some coastal cities.

Cities, in turn, are a major contributor to climate change. The main sources of greenhouse gas emissions from urban areas are related to the consumption of fossil fuels for electricity supply, transportation and industry. In the United States, domestic energy use accounts for between one-quarter and one-third of cities' greenhouse gas emissions, with private automobiles accounting for a similar proportion. Worldwide, there is great disparity in the level of urban greenhouse gas emissions per capita, a direct reflection of core-periphery disparities in wealth and consumption (**Table 11.1**).

Yet while urbanization is a major contributor to climate change and its associated risks, the large populations, high densities, and collective wealth of cities also creates the potential for resilience to climate-related hazards, for example, through infrastructure developments

dwellers already live under significant risk from hurricanes, flooding, landslides, and drought (**Figure 11.22**). Coastal cities are at risk in addition from sea-level rise (**Figure 11.23**). The increase

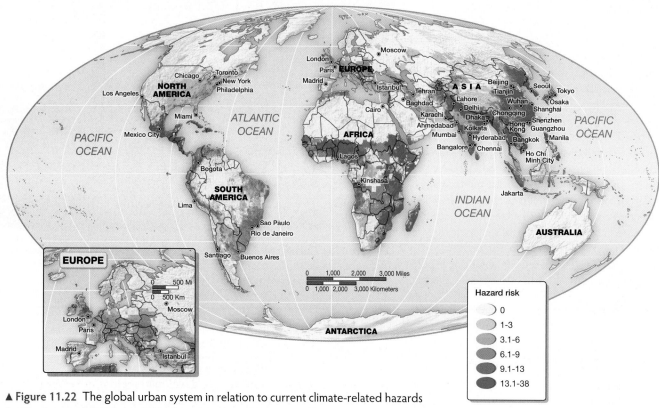

▲ **Figure 11.22** The global urban system in relation to current climate-related hazards

Note: The urban areas included in this figure have populations greater than 1 million. The hazard risk represents a cumulative score based on risk of cyclones, flooding, landslides and drought. A score of "0" denotes "low risk" and "10" denotes "high risk."

(*Source:* UN Human Settlements Programme, *Cities and Climate Change. Global Report on Human Settlements 2011.* London: Earthscan, 2011 (based on A. de Sherbinin et al., "The vulnerability of global cities to climate hazards," *Environment and Urbanization* 19(1): 39–64, (2007) Figure 1))

▲ **Figure 11.23 Sea-level rise and storm surge** Rising sea levels, combined with intense storms, put many coastal towns and cities at risk. This photograph shows the town of Dawlish, in England, where a 2014 storm washed away the railroad tracks and damaged homes.

TABLE 11.1 Greenhouse gas emissions per ton of CO_2 equivalent

Washington, D.C.	19.7
Los Angeles	13.0
Toronto	1.6
Glasgow	8.4
Toronto	8.2
Shanghai	8.1
Geneva	7.8
New York City	7.1
Beijing	6.9
London	6.2
Tokyo	4.8
Seoul	3.8
Barcelona	3.4
Rio de Janeiro	2.3
São Paulo	1.5

Source: UN Human Settlements Programme, *Cities and Climate Change. Global Report on Human Settlements 2011.* London: Earthscan, 2011.

that provide physical protection, or well-designed communications and early warning systems that can help to evacuate people swiftly when tropical storms approach. More important, cities—especially the high-emission cities of the core countries—have the potential to contribute significantly to the mitigation of pollution and climate change through "green" policies and practices.

APPLY YOUR KNOWLEDGE

1. What are some of the "green" strategies devised in different cities around the world to combat the effects of climate change?

2. Research on the Internet concepts like multipurpose architecture, urban agriculture, carbon offsetting, solar power, and water reuse. Look at what cities incorporate these and similar "green" ideas. How does your city compare?

Gender and Vulnerability The UN Human Settlements Programme's report on *Cities and Climate Change*[5] highlights significant differences between women and men in terms of their exposure to climate-related hazards, and their capacity to avoid or cope with or adapt to them. In general, women, especially poor women, are more likely than men to suffer injuries or death when a natural disaster occurs. Around 70 percent of the deaths resulting from severe cyclones that have battered the coastline of Bangladesh in recent years, for example, have

[5] UN Human Settlements Programme, *Cities and Climate Change. Global Report on Human Settlements 2011.* London: Earthscan, 2011.

421

been females. There are many factors that lead to this kind of gender imbalance in vulnerability:

- Women tend to experience unequal access to resources, credit, insurance, services, and information.

- Women's cultural roles and caregiving responsibilities often prevent them from migrating and seeking shelter before and after disaster events.

- When homes are destroyed or damaged, it often affects women's incomes more than men's as they tend to engage in income-generating activities from home and therefore lose income when homes are destroyed.

- In the aftermath of disasters, women have traditionally had difficulty receiving relief aid because the need to care for children at home makes it difficult for them to wait in long lines at recovery centers.

- The priority in disaster relief programs is often to reintegrate men into the workforce, and if men leave their families, as frequently occurs following a natural disaster, women are rendered ineligible for public assistance or may go unrecognized by systems geared to male heads of household.

Green Cities The pace and level of contemporary urbanization puts enormous pressure on energy and water resources, waste management, sewer systems, and transport networks. There is an increasing realization of the importance of "green" policies and practices that conserve resources, reduce energy consumption, and mitigate the human contribution to climate change. Because national and international protocols and discussions have had very limited success, the tasks of tackling climate change, avoiding lasting damage to vital ecosystems and improving the health and well-being of billions of people must be sought at the municipal level.

Not surprisingly, some of the most sophisticated approaches are to be found in the more affluent cities of core countries. Copenhagen, Denmark (**Figure 11.24**), is widely regarded as the most progressive large city in the world in this context, with aggressive goals and innovative policies. In 2009 the city set a target to become CO_2 neutral by 2025, which if met would make it the first large carbon-neutral city in the world. Almost all residents of Copenhagen live within 350 yards of a public transport stop, and the city aims to achieve a significant part of its CO_2 reductions through construction and renovation projects, with plans to upgrade all municipal buildings to the highest standards for energy efficiency.

Other cities that scored well on an overall "green city" index devised by the Economist Intelligence Unit[6] and covering cities' CO_2 emissions, energy consumption, building standards, land-use policies, public transport, water and sanitation, waste management, air quality, and environmental governance include Amsterdam, Berlin, Helsinki, New York City, Oslo, San Francisco, Seattle, Stockholm, Vancouver, Vienna, and Zurich. Meanwhile, there are many cities in the world's affluent core countries that score poorly on the index—Cleveland, Detroit, Miami, and Phoenix, for example—while there are some very progressive green cities in peripheral and semiperipheral countries. Curitiba, Brazil, for example (**Figure 11.25**), is the birthplace of an innovative bus rapid transit (BRT) system and the city established the country's first major pedestrian-only street. Its environmental policies and their oversight are consistently strong. Since 2009, for example, the city's environmental authority has been conducting an ongoing study on the CO_2 absorption rate in Curitiba's green spaces, as well as evaluating total CO_2 emissions in the city.

APPLY YOUR KNOWLEDGE

1. Look at one peripheral or semiperipheral country that is rapidly urbanizing. What industries have prompted some of the rural to urban migration? Are they local businesses, multinational corporations or a combination?

2. How do you think the global economic factors will continue to impact local communities?

FUTURE GEOGRAPHIES

The United Nations Human Settlements Program (UN-Habitat) estimates that by 2030, more than 65 percent of the world's population will be living in urban areas, and there will be around 575 cities with a population of a million or more, including about 50 cities of 5 million or more. The number of megacities—those with a population of 10 million or more—will increase, and the populations of most of them will swell significantly (**Figure 11.26**). The single most important aspect of future patterns of world urbanization is the striking difference in trends and projections between the core regions and the semiperipheral and peripheral regions. In 1950, two-thirds of the world's urban population was concentrated in the more developed countries of the core economies. By 2030, around 80 percent of all city dwellers will be in peripheral and

▲ Figure 11.24 Copenhagen, Denmark Copenhagen has become famous as a bicycle-friendly city.

[6]Economist Intelligence Unit, *The Green City Index*. Munich: Siemens AG, 2012.

▲ **Figure 11.25 Curitiba, Brazil** The city pioneered the world's first rapid-transit bus system in the 1970s and has become well-known for its progressive environmental policies.

semiperipheral countries. In 1950, 21 of the world's largest 30 metropolitan areas were located in core countries—11 of them in Europe and 6 in North America. By 1980, the situation was completely reversed, with 19 of the largest 30 located in peripheral and semiperipheral regions. By 2030, all but 2 or 3 of the 30 largest metropolitan areas are expected to be located in peripheral and semiperipheral regions.

Asia provides some of the most dramatic examples of this trend. From a region of villages, Asia is fast becoming a region of cities and towns. Between 1950 and 2005, for example, Asia's urban population rose more than tenfold, to over 1.5 billion people. Half of the world's urban population now lives in Asia. This region has accounted for about 65 percent of the demographic expansion of all urban areas across the world since the beginning of the twenty-first century. By 2020, about two-thirds of Asia's population will be living in urban areas. Large population concentrations in megacities will remain a prominent feature in urban Asia: By 2020, another three Asian cities—Beijing, Dhaka, and Mumbai—will join Delhi and Shanghai as "meta-cities," massive urban concentrations of more than 20 million people.

Nowhere is the trend toward rapid urbanization more pronounced than in China, where for decades the communist government imposed strict controls on where people were allowed to live, fearing the transformative and liberating effects of cities. By tying people's jobs, school admission, and even the right to buy food to the places where people were registered to live, the government made it almost impossible for rural residents to migrate to towns or cities. As a result, more than 70 percent of China's 1 billion people still lived in the countryside in 1985. Now, however, China is rapidly making up for lost time. The Chinese government, having decided that towns and cities can be engines of economic growth within a communist system, has not only relaxed residency laws but also drawn up plans to establish over 430 new cities.

Whatever the current level of urbanization in peripheral countries, almost all are forecast to experience high rates of urbanization, with growth forecasts of unprecedented speed and unmatched size. Karachi, Pakistan, a metropolis of 1.03 million in 1950, reached 8.5 million in 1995 and is expected to reach 16.2 million by 2015. Likewise, Cairo, Egypt, grew from 2.44 million to 9.7 million between 1950 and 1995 and is expected to reach 13.1 million by 2015. Mumbai (India; formerly Bombay), Delhi (India), Mexico City (Mexico), Dhaka (Bangladesh), Jakarta (Indonesia), Lagos (Nigeria), São Paulo (Brazil), and Shanghai (China) are all projected to have populations in excess of 17 million by 2015. For the most part, this growth will be a consequence of the onset of the demographic transition (see Chapter 3), which has produced fast-growing rural populations in regions that face increasing problems with agricultural development (see Chapter 8). As a response, many people in these regions will continue to migrate to urban areas seeking a better life.

What is the most significant policy change that has contributed to the growth of cities in China? Can you foresee any reasons why growth in peripheral and semiperipheral countries might slow in the future?

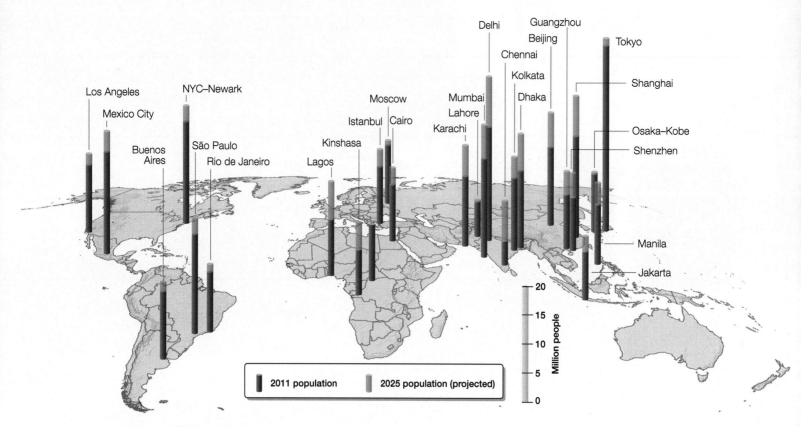

▲ **Figure 11.26 Megacity growth**   Megacities in South Asia are experiencing the most rapid increases in population.

■ CONCLUSION

Urbanization is one of the most important geographic phenomena. Cities can be seedbeds of economic development and cultural innovation. Cities and groups of cities also organize space—not just the territory immediately around them, but in some cases national and even international space. The causes and consequences of urbanization, however, are very different in different parts of the world. The urban experience of the world's peripheral regions stands in sharp contrast to that of the developed core regions, for example. This contrast is a reflection of some of the demographic, economic, and political factors that we have explored in previous chapters.

Much of the developed world has become almost completely urbanized, with highly organized systems of cities. Today, levels of urbanization are high throughout the world's core countries, while rates of urbanization are relatively low. At the top of the urban hierarchies of the world's core regions are world cities such as London, New York, Tokyo, Paris, and Zürich, which have become control centers for the flows of information, cultural products, and finance that collectively sustain the economic and cultural globalization of the world. In doing so, they help to consolidate the hegemony of the world's core regions.

Few of the metropolises of the periphery, on the other hand, are world cities occupying key roles in the organization of global economics and culture. Rather, they operate as connecting links between provincial towns and villages and the world economy. They have innumerable economic, social, and cultural linkages to their provinces on one side and to major world cities on the other. Almost all peripheral countries, meanwhile, are experiencing high rates of urbanization, with forecasted growth of unprecedented speed and unmatched size. In many peripheral and semiperipheral regions, current rates of urbanization have given rise to unintended metropolises and fears of "uncontrollable urbanization," with urban "danger zones" where "work" means anything that contributes to survival. The result, as we shall see in Chapter 11, is that these unintended metropolises are quite different from the cities of the core as places in which to live and work.

LEARNING OUTCOMES REVISITED

■ *Describe* how the earliest towns and cities developed independently in the various hearth areas of the first agricultural revolution.

The very first region of independent urbanism, in the Middle East, produced successive generations of urbanized world-empires, including those of Greece, Rome, and Byzantium. By 2500 B.C.E., cities had appeared in the Indus Valley, and by 1800 B.C.E., urban areas were established in northern China. Other areas of independent urbanism include Mesoamerica (from around 100 B.C.E.) and Andean America (from around 800 C.E.). The classical archaeological interpretation emphasizes the availability of an agricultural surplus large enough to allow the emergence of specialized, nonagricultural workers. Some urbanization, however, may have resulted from the pressure of population growth.

■ *Explain* how the expansion of trade around the world, associated with colonialism and imperialism, established numerous gateway cities.

European powers founded or developed literally thousands of towns as they extended their trading networks and established their colonies. The great majority of the towns were ports that served as control centers commanding entrance to, and exit from, their particular country or region. The great majority of them were ports. Protected by fortifications and European naval power, they began as trading posts and colonial administrative centers. Before long, they developed manufacturing of their own to supply the pioneers' needs, along with more extensive commercial and financial services. As colonies developed and trading networks expanded, some of these ports grew rapidly, acting as gateways for colonial expansion into continental interiors.

■ *Assess* why and how the Industrial Revolution generated new kinds of cities—and many more of them.

Industrial economies required the large pools of labor; the transportation networks; the physical infrastructure of factories, warehouses, stores, and offices; and the consumer markets provided by cities. As industrialization spread throughout Europe in the first half of the nineteenth century and then to other parts of the world, urbanization increased at a faster pace. The higher wages and greater variety of opportunities in urban labor markets attracted migrants from surrounding areas. The countryside began to empty. In Europe, the *demographic*

transition caused a rapid growth in population as death rates dropped dramatically. This growth in population provided a massive increase in the labor supply, further boosting the rate of urbanization, not only within Europe itself but also in Australia, Canada, New Zealand, South Africa, and the United States as emigration spread industrialization and urbanization to the frontiers of the world-system.

■ *Interpret* how a small number of "world cities," most of them located within the core regions of the world-system, have come to occupy key roles in the organization of global economics and culture.

At the top of a global urban system, these cities experience growth largely as a result of their role as key nodes in the world economy. World cities are the control centers for the flows of information, cultural products, and finance that collectively sustain the economic and cultural globalization of the world. The globalization of the economy has resulted in a global urban system in which the key roles of world cities are concerned with transnational corporate organization, international banking and finance, supranational government, and the work of international agencies. World cities also provide an interface between the global and the local. They contain the economic, cultural, and institutional apparatus that channels national and provincial resources into the global economy and transmits the impulses of globalization back to national and provincial centers.

■ *Compare* and contrast the differences in trends and projections between the world's core regions and peripheral regions.

In 1950, two-thirds of the world's urban population was concentrated in the more developed countries of the core economies. Since then, the world's urban population has increased threefold, the bulk of the growth having taken place in the less developed countries of the periphery. The world's core regions are highly urbanized, with slow rates of urban growth. Peripheral regions, although less highly urbanized, have been experiencing exceptionally high rates of urban growth, partly due to rural–urban migration and partly due to natural population increase. Much of the resulting urbanization has taken the form of "megacities" of 10 million people or more. Overurbanization has occurred where cities have grown more rapidly than they have been able to generate jobs or housing.

KEY TERMS

brownfield sites *(p. 417)*

central place *(p. 409)*

central place theory *(p. 409)*

centrality *(p. 410)*

colonial city *(p. 405)*

counterurbanization *(p. 417)*

feudalism *(p. 402)*

gateway city *(p. 403)*

megacity *(p. 415)*

overurbanization *(p. 415)*

primacy *(p. 410)*

public-private partnerships *(p. 417)*

rank-size rule *(p. 410)*

regeneration *(p. 417)*

reurbanization *(p. 417)*

shock city *(p. 405)*

urban ecology *(p. 394)*

urban form *(p. 394)*

urbanization *(p. 394)*

urban system *(p. 394)*

urbanism *(p. 395)*

REVIEW & DISCUSSION

1. Go to the library or conduct an Internet search for pictures of what your campus looked like 5, 10, and 30 years ago. How have the urban forms changed? Note five examples of how the campus has changed between then and now. How have the physical structures changed? How is the land used differently? Is the university organized differently? If so, why do you think this is the case? If not, explain why things are the same.

2. Consider the term *world cities*. Choose two cities that you consider to be world cities. Now read over the functional characteristics of world cities that are bulleted on page 402. Pick two bullet points and describe the characteristics in terms of the cities you have chosen. For example, the second bullet point notes how world cites have "clusters of specialized, advanced business services." Do an Internet search and identify the clusters of specialized business services in the cities you have chosen.

3. Counterurbanization refers to the process of cities losing population to smaller towns or rural areas. In light of the recent economic recession, conduct a library search to determine if any American cities or towns have experienced recent counterurbanization. List three reasons for the counterurbanization. Consider if some cities are experiencing reurbanization. List three reasons for this type of population movement.

UNPLUGGED

1. Figures 11.1 and 11.2 show that the United States, like most core countries, is already highly urbanized and has a relatively low rate of urbanization Nevertheless, some U.S. cities have been growing much faster than others. Which have been the fastest-growing U.S cities in recent times, and what reasons can you suggest for their relatively rapid growth? (*Hint:* The U.S. Bureau of the Census publishes data on population change by urban area, as does the Population Division of the UN Department of Economic and Social Affairs, in a volume entitled *World Urbanization Prospects.*)

2. Go to the U.S. Bureau of the Census Web page and find the population of the town or city you know best. Do the same for every census year, going from 2010 back to 2000, 1990, 1980, and so on, all the way back to 1860. Then plot the populations on a simple graph. List four explanations for the pattern that the graph reveals. Now draw a larger version of the same graph, annotating it to show the landmark events that might have influenced the city's growth (or decline).

3. The following cities all have populations in excess of 2 million. How many of them could you locate on a world map? Their size reflects a certain degree of importance, at least within their regional economy. What can you find out about each? Compile for each a 50-word description that explains its chief industries and a little of its history.

Poona	Ibadan	Recife
Bangalore	Turin	Ankara

4. Figure 11.6 features two colonial gateway cities on the eastern seaboard of the United States—Boston and New York. Two other colonial gateway cities were Charleston, South Carolina, and Savannah, Georgia. What can you find out about the commodities and manufactures that Charleston and Savannah imported and exported in pre-Revolutionary times? Where did their exports go? List three reasons why they went to those places. On the import side, where did imports come from? List three reasons why Charleston and Savannah needed to import these particular goods. Which geographic concepts do you consider to be useful in explaining these facts?

DATA ANALYSIS

State of the
World's Cities

http://goo.gl/16WvXC

As the globe becomes increasingly urbanized, cities are growing with disparate realities for their people. Look at the UN Habitat (unhabitat.org) Web site to discover more about cities in peripheral regions of the world. Download the publication *State of the World's Cities 2010/2011: Bridging the Urban Divide*. Read and answer these questions:

1. What is the urban divide?

2. How are slums the product of both physical space and social inequalities?

3. What is an inclusive city and why is that important?

4. Choose one city in each of the three study areas of Latin America, Asia, and Africa to compare statistics from the report's tables and graphs. Find out which of your cities is

   (1) the fastest in annual growth rate;

   (2) the most "unequal" or "equal" in income;

   (3) the most politically inclusive; and

   (4) the most culturally inclusive.

Using the UN Habitat's data e-sources, go to the "Urban Indicators" to compare the countries of each of your cities for the annual slum population growth rate versus the annual population growth rate.

5. What do the data you found on both city and country levels tell you about urbanization processes?

MasteringGeography™

Looking for additional review and test prep materials? Visit the Study Area in MasteringGeography™ to enhance your geographic literacy, spatial reasoning skills, and understanding of this chapter's content by accessing a variety of resources, including **MapMaster** interactive maps, Videos, *In the News* RSS feeds, flashcards, web links, self-study quizzes, and an eText version of *Human Geography*.

- *Assess* how the internal structure of cities is shaped by competition for territory and location.

- *State* the ways in which social patterns in cities are influenced by human territoriality.

- *Describe* the spatial structure of a typical North American city.

- *Compare* and contrast urban structures in different regions of the world.

- *Explain* the nature and causes of the problems associated with urbanization in various world regions.

▲ A street scene in the Spitalfields district of London, England

12

CITY SPACES: URBAN STRUCTURE

There is a plain building on a corner of Brick Lane, in London's Spitalfields district. It was built as La Neuve Eglise (the "New Church") in 1743 by French-speaking Protestant Huguenot refugees who had been driven out of France by the Catholic king Louis XIV. After the French immigrants began to move away from Spitalfields, the chapel closed and was taken over by a Methodist congregation in 1819. It became an orthodox Jewish synagogue in 1898, serving Yiddish-speaking Russian and East European Jews who had come to London to escape the pogroms that followed the assassination of the Tsar of Russia in 1881. Today it is the community mosque for the Bangladeshi population that has come to constitute three-quarters of the district's population.

There is little remaining evidence of Brick Lane's previous function as the high street for Spitalfields' Jewish community: just a couple of bagel shops selling salt beef sandwiches. Jewish households began to move out of Brick Lane in the 1950s, leaving behind inexpensive rental accommodation and a garment industry dependent on cheap, hard-working labor. Their place was taken by immigrants from East Pakistan (now Bangladesh). Chain migration, with families from the Sylhet region of Bangladesh moving into Spitalfields through word-of-mouth contact with friends and relatives already settled there, gradually displaced the Jewish population. Today, between two-thirds and three-quarters of the population of Spitalfields are of Bangladeshi origin, and Brick Lane has become a tourist stop and nighttime entertainment strip as well as a vibrant ethnic community. It is often cited as an example of London's cosmopolitanism. While locals shop in sari fabric stores and Bengali minimarkets, outsiders frequent the curry houses, Internet cafés, and indie clothes and record stores. Branded as Banglatown, Brick Lane has Chinatown-style ornamental gateway arches along with new signage and new, brighter street lamps, custom-designed to incorporate "Asian" motifs.

The Bangladeshi community itself has been able to secure public funding for mosques, *madrassas* (Islamic schools), and Islamic community organizations by leveraging its sheer weight of numbers in local elections. In this way, Bangladeshis and other local Muslims have been able to assert their own control over local space. The Brick Lane community now has a Bengali New Year Festival (*Baishakhi Mela*) and a two-week-long Brick Lane Curry Festival. Ties with Bangladesh, and Sylhet in particular, remain strong. Sylhet's weekly paper, the *Sylheter Dak*, a London branch office. The Sylhet region has meanwhile been changed by its association with the Brick Lane community. Sylhet itself is now one of the most prosperous towns in Bangladesh, with new shopping centers and other building projects funded by "Londoni" money.

Each city has its own set of internal geographies, and these are expressed in a mosaic of distinctive districts. We can recognize rich districts, poor districts, ethnic districts (like the Brick Lane area of Spitalfields), residential districts, commercial districts, industrial districts, transitional districts, and many other kinds. Each has its own distinct character and story, the product of successive phases of development and of demographic, social, cultural, and political change. Each chapter in a city's history leaves its mark, for better or worse, in the layout of its streets, the fabric of its buildings, the nature of its institutions, and the cultural legacies of its residents. It is possible to make some generalizations about the nature and spatial arrangement of city districts, though these differ in important ways according to the history and geography of different world regions. We begin by considering the characteristics of cities in North America.

SPATIAL PATTERNS AND PROCESSES IN NORTH AMERICAN CITIES

At the turn of the twenty-first century, the Society for American City and Regional Planning History asked a large sample of urban historians, social scientists, planning faculty, and working planners and architects to nominate the most important influences on the American metropolis over the past 50 years (**Table 12.1**). The list reflects the overwhelming impact of the federal government on the American metropolis, especially through policies that have promoted suburbanization and **sprawl,** sometimes intentionally and sometimes unintentionally. The two items at the top of the list have been of overwhelming importance in shaping the sprawling, polycentric metropolitan areas that are characteristic of contemporary North American urbanization. The 1956 Interstate Highway Act created a 41,000-mile Interstate Highway System that

TABLE 12.1 The Top 10 Influences on the American Metropolis 1950–2000

1. The 1956 Interstate Highway Act and the dominance of the automobile.
2. Federal Housing Administration mortgage financing and subdivision regulation.
3. Deindustrialization of central cities.
4. Urban renewal: downtown redevelopment and public housing projects (1949 Housing Act).
5. Levittown (the mass-produced suburban tract house).
6. Racial segregation and job discrimination in cities and suburbs.
7. Enclosed shopping malls.
8. Sunbelt-style sprawl.
9. Air conditioning.
10. Urban riots of the 1960s.

Source: Robert Fishman (2000) "The American metropolis at century's end: Past and future influences," *Housing Policy Debate, 11*:(1), p. 200.

transformed the American metropolis in ways its planners never anticipated. The system was supposed to support downtown districts by rescuing them from traffic congestion while facilitating high-speed long-distance travel from city to city. But their construction devastated many urban neighborhoods and the new peripheral beltways, originally intended to enable long-distance travelers to bypass crowded city centers, became shortcuts to sprawl.

Second on the list is the Federal Housing Administration's mortgage financing and regulation of subdivisions. The Federal Housing Administration (FHA) was created in the 1930s in the midst of the Depression and one of its key tasks was to stimulate the labor-intensive construction industry by underwriting fixed-rate mortgages with low down payments. In order to prevent the kind of shoddy construction that had been all too common in the early growth of American cities, the FHA also established and enforced minimum standards for the housing financed by its guaranteed loans. But the FHA refused to insure mortgages on older houses in older urban neighborhoods and its 1939 *Underwriting Manual* openly recommended that subdivision developers use restrictive covenants to prevent the sale of homes to minorities. It was not until 1949 that discriminatory covenants were declared unconstitutional. By then the FHA had firmly established the framework for future urban growth: one with a pronounced bias toward detached single-family owner-occupied housing in the suburbs, for white households.

The Multi-Nodal City

The suburban sprawl that is so characteristic of North American metropolitan areas is typically arranged around one or more "central cities". **Central cities** are the original, core jurisdictions of metropolitan areas. They are organized around a traditional downtown, known as the **central business district,** or **CBD**: the nucleus of commercial land uses. It traditionally contains the densest concentration of shops, offices, and warehouses and the tallest nonresidential buildings (**Figure 12.1**).

◄ **Figure 12.1 The central city**
This photograph of Chicago shows the concentration of high-rise office buildings on the skyline that is typical of central business districts (CBDs) in North American metropolitan areas.

It originally developed at the nodal point of transportation routes so that it also contains bus stations, railway terminals, and hotels. The CBD usually is surrounded by a zone of mixed land uses: warehouses, small factories and workshops, specialized stores, apartment buildings, public housing projects, and older residential neighborhoods. This zone is often referred to as the **zone in transition** because of its mixture of growth, change, and decline. Beyond this zone are residential neighborhoods, often highly segregated in terms of race and ethnicity (see **Box 12.1**: "Visualizing Geography: Racial Segregation"). The advent of trucks and automobiles allowed both industry and households to become mobile, and the Interstate Highway Act and FHA policies allowed both industry and households to decentralize. New, automobile-based suburban nodes of commercial and industrial activity began to emerge. They were not arranged in any predictable fashion except in relation to surrounding land uses. They might develop around a government center, a university, a transit stop, or a highway intersection. If the nodes were office and retailing centers, they tended to attract middle-income residential development, whereas if they were industrial centers, they would attract working-class residential development. In this way, American cities began to develop an irregular-shaped patchwork of land uses and sprawl around multiple nodal centers.

Accessibility versus Living Space It is in many ways counterintuitive that affluent suburbanites have to travel long distances to work in nodal centers from distant suburbs while disadvantaged households live close by the jobs and urban amenities that cluster in nodal centers. According to American Community Survey data, over 220 million adults average an hour and a half a day in their cars. Economists would expect people to seek to maximize the *utility* they derive from a particular location. The utility of a specific place or location refers to its usefulness to a particular person or group.

The price people are prepared to pay for different locations—what economists call the bid-rent—is a direct reflection of this utility. In general, the utility of a location is a function of its *accessibility*. Commercial land users want to be accessible to one another, to markets, and to workers; private residents want to be accessible to jobs, amenities, and friends; and public institutions want to be accessible to clients. In an idealized, theoretical city built on an isotropic surface, the point of maximum accessibility is the city center. An **isotropic surface** is a hypothetical, uniform plane: flat, and with no variations in its physical attributes. Under these conditions, accessibility decreases steadily with distance from a nodal center. Likewise, utility decreases, *but at different rates for different land users*. This idealized pattern of decreasing accessibility and utility is reflected in concentric zones of different mixes of land use (**Figure 12.2**).

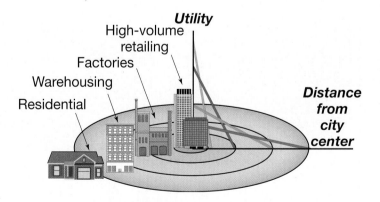

▲ **Figure 12.2 Accessibility, bid-rent, and urban structure** Competition for accessible sites near the city center is an important determinant of land-use patterns. Different users are prepared to pay different amounts—the bid-rents—for locations at various distances from the city center. The result is often a concentric pattern of land uses.

(*Source:* Reprinted with permission of Prentice Hall, from P. L. Knox, *Urbanization* © 1994, p. 99.)

431

12.1 Visualizing Geography

Spatial Segregation

The racial and ethnic composition of the United States has steadily become more diverse – a result of immigration and differential birth rates. Studies of racial segregation within U.S. metropolitan areas have found that, overall, the degree of segregation peaked between 1960 and 1970. Between 1980 and 2010 racial segregation declined, but at a very slow pace. In 2010 the average white person in metropolitan America lived in a neighborhood that was 75 percent white, while the average African American lived in a neighborhood that was only 35 percent white and as much as 45 percent black. Hispanics and Asians are considerably less segregated than African Americans, and their segregation levels have remained steady for several decades. In addition, since both these groups are growing, there is a tendency for their neighborhoods to become more homogeneous. As a result, these groups live in more isolated settings now than they did in the 1980s.

12.1.1 Diversity Experienced in Each Group's Typical Neighborhood

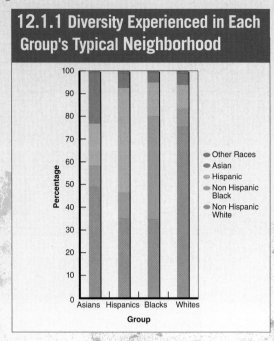

12.1.3 Sacramento

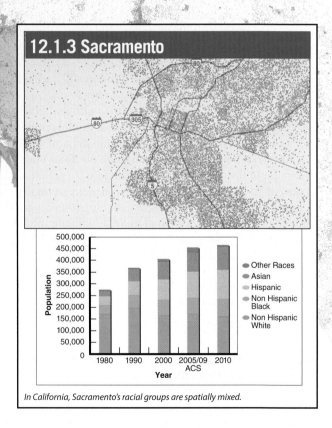

In California, Sacramento's racial groups are spatially mixed.

12.1.2 Los Angeles

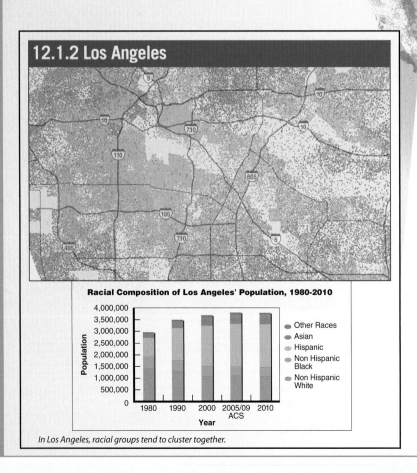

In Los Angeles, racial groups tend to cluster together.

Source: John R. Logan and Brian Stults. 2011. "The Persistence of Segregation in the Metropolis: New Findings from the 2010 Census" Census Brief prepared for Project US2010.

Source: 2010 Census, 2010-SF1a, QT-P6 Hispanic or Latino Origin by Race, United States Census Bureau.

12.1.4 Chicago

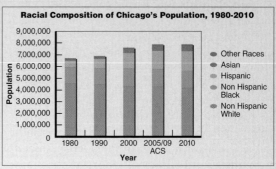

Racial Composition of Chicago's Population, 1980-2010

- Other Races
- Asian
- Hispanic
- Non Hispanic Black
- Non Hispanic White

In Midwestern cities, the racial divide tends to be quite sharp. In Chicago, bands of different racial groups, blacks, and Latinos radiate out from the city center, echoing the models of urban social geography developed in the 1930s.

12.1.5 Atlanta

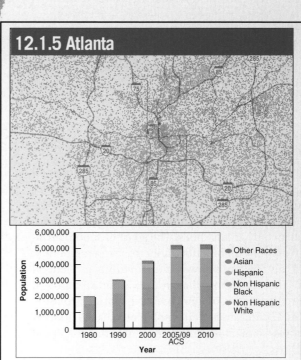

- Other Races
- Asian
- Hispanic
- Non Hispanic Black
- Non Hispanic White

In Atlanta, there is a simple north-south division, with whites (and a few enclaves of Asians) in the northern half of the metro area, and African Americans in the southern half.

One dot = 50 people
- White
- Black
- Asian
- Hispanic
- Other/Native American/Multi-racial

One dot = 100 people on full U.S. map

America's Racial Segregation

http://goo.gl/q771LK

1. Why do you think racial groups are more mixed in some cities, compared to others? Compare historical factors (civil rights movement, WWII, American Indian relocation) and immigration polices with what you have learned.

2. Analyze the demographic concept of "minority-majority" geographies by doing an Internet search on "minority-majority cities/counties/states." Where are these cities/counties/states located and what do you observe about the political affiliations, immigration policies, social programs and economy compared to white majority areas?

The implication of this model is that the poorest households will end up occupying the periphery of the city. Although this is true in some parts of the world, we know that in North America it is not the case. In fact, the farthest suburbs are generally the territory of wealthier households, while the poor usually occupy more accessible locations nearer city centers. To reflect reality, some of the assumptions of the economist's model must be modified. In this case, we must assume that wealthier households trade off the convenience of accessibility for the greater utility of being able to consume larger amounts of (relatively cheap) suburban space. Poorer households, unable to afford the recurrent costs of transportation, trade off living space for accessibility to jobs and end up in high-density areas, at expensive locations, near their jobs. Because of the presumed trade-off between accessibility and living space, this modified urban land-use model is often referred to as a **trade-off model**.

APPLY YOUR KNOWLEDGE

1. Identify three different trade-offs that you have made in terms of your own living space. List three ways these trade-offs have affected your access to school, transportation, and job.

2. Think about how these trade-offs have impacted your access to school, transportation and job; and share what motivated you to select one trade-off over another.

The Polycentric Metropolis

With continued urban growth, the irregular-shaped, multiple-nuclei patchwork of land uses of the mid-twentieth century city has scaled up into polycentric metropolitan regions. During the middle decades of the twentieth century, American cities were reshaped by automobiles. The spurt of suburbanization resulting from the combination of increased automobility, massive federal outlays on highway construction, and federal mortgage insurance programs that underpinned the growth of home ownership produced a polycentric spatial structure. Suburban nodes of office and retail employment each had their own **urban realms**, semiautonomous subregions that displaced

the simple core-periphery relationship between city centers and their suburbs. Initially, the shift to an expanded polycentric metropolis was most pronounced in the northeastern United States. Geographer Jean Gottmann captured the moment in 1961 with his conceptualization of "megalopolis"—his term for the highly urbanized region between Boston and Washington, D.C.

In the polycentric new metropolis (**Figure 12.3**), the system of nodes and realms is bound together with ever-expanding four-, six-, and eight-lane highways. It is interspersed with smaller clusters of decentralized employment, studded with micropolitan centers and filled out with booming stand-alone suburbs. Traditional nodal anchors—downtown commercial centers—remain very important, especially as settings for the advanced business services—advertising, banking, insurance, investment management, and logistics services. But in addition there are other nodes. These vary in character and include:

- **edge cities**, decentralized clusters of retailing and office development, often located on an axis with a major airport, sometimes adjacent to a high-speed rail station, and always linked to an urban freeway system;

- newer business centers, often developing in a prestigious residential quarter and serving as a setting for newer services such as corporate headquarters, the media, advertising, public relations, and design;

- outermost complexes of back-office and R&D operations, typically near major transport hubs 20 to 30 miles from the main core; and

- specialized subcenters, usually for education, entertainment, and sporting complexes and exhibition and convention centers.[1]

Some urbanized regions have been extended and reshaped into polycentric networks of half a dozen or more urban realms and as many as 50 nodal centers of different types and sizes (**Figure 12.4**). These centers are physically separate but functionally networked and they draw enormous economic strength

[1]P. Hall, "Global City-Regions in the Twenty-first Century." In A. J. Scott (ed.), *Global City-Regions: Trends, Theory Policy*. New York: Oxford University Press, 2001, pp. 59–77.

▶ **Figure 12.3 The twentieth-century metropolis** Consists of a central city, a ring of suburbs, and a series of far-flung urban realms, studded with edge cities (nodal concentrations of office and retail space) and "edgeless cities" of suburban and ex-urban office parks and shopping malls.

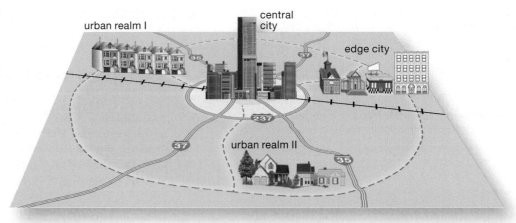

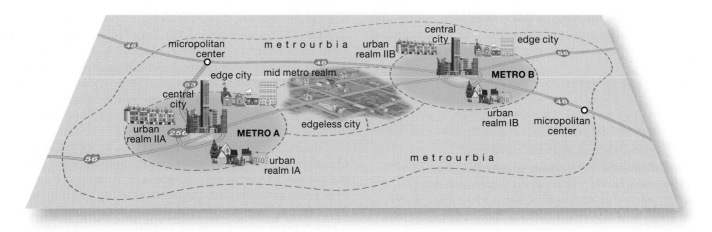

▲ **Figure 12.4 The new metropolis** The largest metropolitan regions are now "megapolitan," with coalescing metropolitan areas merging into disjointed and decentralized urban landscapes with varying-sized urban centers, subcenters, and satellites and unexpected juxtapositions of form and function. The residential settings in suburban and ex-urban areas are thoroughly interspersed with office employment and high-end retailing, creating "metroburban" landscapes.

from a new functional division of labor. Bound together through urban freeways, arterial highways, beltways, and inter-states, new metropolises coalesce into "megapolitan" regions that dominate national economies. The term **metroburbia** has emerged to capture the way that residential settings in suburban and ex-urban areas are thoroughly interspersed with office employment and high-end retailing.

Meanwhile, some older inner-urban districts have been redeveloped around mixed-use projects—complexes of shops, offices, and apartments. Some older, centrally located, working-class neighborhoods have also been invaded by higher-income households seeking the character and convenience of centrally located and (for them) less expensive residences—a

▼ **Figure 12.5 Gentrification** Neighborhoods of row houses in central locations, like this one in Philadelphia, have long been attractive to upwardly-mobile households.

process known as gentrification (**Figure 12.5**). **Gentrification** involves the renovation of housing in older, centrally located lower-income neighborhoods through an influx of more afflu-ent households seeking the character and convenience of less-expensive but well-located residences. Typically, the colonizing households are dominated by young professionals involved in the "new economy," together with teachers, lawyers, designers, artists, architects, writers, and creative staff in advertising firms. The increased pool of professional, administrative, managerial, and technical workers in the new economy has generated an expanding group of potential gentrifiers. They are attracted to (mildly) edgy districts with some history, human scale, and ethnic and architectural diversity, and they often use "sweat equity"—their own, do-it-yourself labor—for renovations and improvements to older properties. Their arrival in inner-city districts tends to displace poorer households as the new fash-ionability and buzz of the district pushes up rents and house prices, increases property taxes, and prompts the closure of stores specializing in inexpensive goods and produce. Incoming gentrifiers, meanwhile, contribute to the physical renovation or rehabilitation of the older and usually rather deteriorated housing stock while supporting new businesses such as upscale restaurants, coffee shops, delicatessens, wine bars, galleries, cloth-ing boutiques, and bookstores.

Because it brings about improve-ments to the built environment, en-courages new retail activity, and results in the expansion of the local tax base without necessarily drawing heavily on public funds, gentrification can be seen as a good thing, and has been widely encouraged by city governments. It has also been eagerly exploited by realtors and landlords seeking to turn a profit on older properties. But because it drives up house prices and results in the displacement of vulnerable and disadvantaged households, it is not always so straightforward an issue.

APPLY YOUR KNOWLEDGE

1. Create a map like the ones in Figures 12.3 and 12.4 for your metropolitan area and label urban realms, major highways, edge cities, secondary business centers, and neighborhoods.

2. Analyze which areas have experienced gentrification by comparing neighborhood renovations, property value, income averages, age and ethnicity of inhabitants, and use media sources to observe how the culture of the community may have changed.

Problems of North American Cities

For all their relative prosperity, North American cities have their share of problems. Suburban sprawl is generally regarded as a distinctively North American problem, though it has its apologists. The most acute problems, though, are localized in central city areas and are interrelated: fiscal problems, infrastructure problems, and localized cycles of poverty and spirals of neighborhood decay.

Sprawl The unplanned, *ad hoc* nature of most suburban development destroys millions of acres of wildlife habitat and agricultural land every year. Rationalized, standardized, and tightly zoned off-ramp subdivisions are placeless neighborhoods that lack visual, demographic, and social diversity (**Figure 12.6**). The politics and economics of private subdivision—the "sprawl machine"—lead to a lack of public open space, urban infrastructure, and civic amenities. Increased traffic, punishing commutes, and a chronic dependence on automobiles are all the result of low-density, single-family suburban developments. The environmental

costs of automobile dependency include air pollution—and in particular the generation of millions of tons of greenhouse gases from suburban commuters. Run-off from the roads and parking lots pollutes suburban watersheds. The automobile-dependent lifestyles associated with sprawl, meanwhile, lead to increases in rates of asthma, lung cancer, and heart problems. Stress resulting from commuting has adverse effects on marriages and family life.

The counterargument is that sprawl is a logical consequence of economic growth and the democratization of society, providing millions of people with the kinds of mobility, privacy, and choice that were once the prerogatives of the rich and powerful. Sprawl reflects market forces and represents the desires and preferences of the broad mass of people for living space. Joel Kotkin, a blogger and popular speaker on the business circuit, has persistently argued that most people seem to like living in the suburbs: so why all the fuss? Libertarian think tanks like the Reason Foundation and the Heritage Foundation have made efforts to counter antisprawl arguments on the ideological principle of freedom of markets and individual choice. What they overlook, from a geographical perspective, is that the benefits of sprawl—for example, more housing for less cost with higher eventual appreciation—still tend to accrue to Americans individually, while sprawl's cost in infrastructure building, energy generation, and pollution mitigation tends to be borne by society overall.

Fiscal Problems The term *fiscal* refers to government revenue or taxes. Economic restructuring and metropolitan sprawl have left central cities with a chronic "fiscal squeeze." A **fiscal squeeze** occurs when increasing limitations on tax revenues combine with increasing demands for expenditures on urban infrastructure and city services. The revenue-generating potential of most central cities has steadily fallen as metropolitan areas have lost both residential and commercial taxpayers to suburban jurisdictions. Growth industries, white-collar jobs, retailing, and more affluent households have moved out to suburban and ex-urban jurisdictions, taking their local tax dollars with them.

At the same time, growth in property-tax revenues from older, decaying neighborhoods has slowed as the growth of property values has slowed. At the same time, these older, decaying neighborhoods cost more to maintain and service. The residual populations of these neighborhoods, with high proportions of elderly and indigent households, are increasingly in need of municipal welfare services. Large numbers of low-income migrants and immigrants also bring increased demands for municipal services. Added to all this, central city governments are still responsible for services and amenities used by the entire metropolitan population: municipal galleries and museums, sports facilities, parks, traffic police, and public transport, for example. The net result is that many central city jurisdictions are in a precarious financial position. Stockton, California,

▼ **Figure 12.6 Urban sprawl** Albuquerque, New Mexico.

▲ **Figure 12.7 Infrastructure problems** Flooded roads around a broken water main at the University Village mall, Seattle in 2013.

filed for bankruptcy in 2012, followed by Detroit in 2013. In a constant drive to develop revenue-generating projects, cities compete fiercely with one another to finance and attract tourist developments, museums, sports franchises, and business and conference centers.

Infrastructure Problems

As fiscal problems have intensified, public spending on the urban infrastructure of roads, bridges, parking spaces, transit systems, communications systems, power lines, gas lines, street lighting, water mains, sewers, and drains has declined. Meanwhile, much of the original infrastructure, put in place 75 or 100 years ago with a design life of 50 or 75 years, is obsolete, worn out, and in some cases perilously near the point of collapse. Major disasters, such as the failure of New Orleans' levee system during Hurricane Katrina in 2005, the collapse of the I-35W bridge in Minneapolis in 2007, and the collapse of the Interstate 5 bridge over the Skagit River at Mount Vernon, Washington State in 2013, have highlighted the problem.

Infrastructure problems are easily overlooked because they build up slowly. Only when a bridge collapses or a water main bursts are infrastructure problems newsworthy. In Boston, three-quarters of the sewer system was built a century or more ago and has now deteriorated to the point where about 20 percent of the system's overall flow is lost to leaks. In Cleveland, the District of Columbia, and Philadelphia, the losses approach 25 percent. Hundreds of kilometers of water mains in New York City have been identified as being in need of replacement, at an estimated cost of over $5 billion. Overall, only 50 percent of all wastewater treatment systems in America are operating at 80 percent or more of capacity; one in nine of the nation's 607,380 bridges are rated as structurally deficient; 42 percent of major urban highways remain congested; and there are almost a quarter of a million water main breaks annually (**Figure 12.7**). The American Society of Civil Engineers

2013 Report Card on America's Infrastructure estimated that $3.6 trillion of investment is needed by 2020—about twice the current rate. [2]

Freshwater supplies are also at risk: Old water systems are unable to cope with the leaching of pollutants into city water. Common pollutants include chlorides, oil, phosphates, and nondegradable toxic chemicals from industrial wastewater; dissolved salts and chemicals from highway de-icing; nitrates and ammonia from fertilizers and sewage; and coliform bacteria from septic tanks and sewage. Many cities still use water-cleaning technology dating to World War I. About one-third of all towns and cities in the United States have contaminated water supplies, and about eight million people use water that is potentially dangerous.

Poverty and Neighborhood Decay

Inner-city poverty and neighborhood decay have become increasingly pronounced in the past several decades as manufacturing, warehousing, and retailing jobs have moved out to suburban and edge-city locations and as many of the more prosperous households have moved out to be near these jobs. The spiral of neighborhood decay typically begins with substandard housing occupied by low-income households that can afford to rent only a minimal amount of space. The consequent overcrowding not only causes greater wear and tear on the housing itself but also puts pressure on the neighborhood infrastructure of streets, parks, schools, and playgrounds. The need for maintenance and repair increases quickly and is rarely met. Individual households cannot afford it, and landlords have no incentive, because they have a captive market. Public authorities face a fiscal squeeze and are in any case often indifferent to the needs of such neighborhoods because of their relative lack of political power.

Shops and privately run services such as restaurants and hair salons are afflicted with the same syndrome of decay. With a low-income clientele, profit margins must be kept low, leaving little to spare for upkeep or improvement. Many small businesses fail, or relocate to more favorable settings, leaving commercial property vacant for long periods. In extreme cases, property is abandoned when the owners are unable to find renters or buyers. Residential buildings may also be left derelict.

Meanwhile, a dismal cycle of poverty intersects with these localized spirals of decay. The **cycle of poverty** involves the transmission of poverty and deprivation from one generation to another through a combination of domestic circumstances and local conditions. This cycle begins with an absence of employment opportunities and, therefore, a concentration of low incomes, poor housing, and overcrowded conditions. Such conditions are unhealthy. Overcrowding makes people vulnerable to poor health, which is compounded by poor diets. This contributes to absenteeism from work, which results in decreased income. Similarly, absenteeism from school because of illness contributes to the cycle of poverty by constraining educational achievement, limiting occupational skills, and leading to low wages. Crowding also produces psychological

[2]*Renewing America's Infrastructure.* American Society of Civil Engineers, 2013, http://www.infrastructurereportcard.org/

12.2 Geography Matters

Detroit's Open Geography: Problems and Potential

By Mitch Rose, University of Aberystwyth

In many big cities, a premiere tourist experience is to take the elevator to the top of an iconic tall building. By getting above the city, over the dense array of structures that block your eye, you can get a sense of the city's shape and form, its neighbourhoods and street systems, its morphology and feel: New York with its grids, Chicago and the lake, London and Paris with their mixture of old and new.

Detroit offers a similar opportunity. The tallest building in Detroit is the 73-story Renaissance Center, which rises along Detroit's riverfront. When you take the glass elevator up, you see a very distinctive urban landscape. Fields of low-rise housing heading out towards an endless horizon, wide functional boulevards crossing from east to west, and the Detroit river with its regular flow of tankers, railway lines and highways connecting cross-town neighbourhoods and factories now long abandoned. This is a city that was created to build cars. While Detroit had a wide array of manufacturing industries in the 1940s and 50s, fully 40% of them served the production of automobiles. Such an economy gave rise to two distinctive urban dynamics.

First, the city is organised almost like a series of linked factory towns, with small plants serving as the organising principal for the neighbourhoods and transport links that surround it. Unlike the more flexible industries of today, the factories that defined Detroit made single products with little variation and relied upon a steady stream of raw materials It is not uncommon to see railway lines and major thoroughfares running straight through the city. Second, these factories paid its workers relatively well. Given the lack of geographical impediments, single plot housing was the norm. So even though Detroit at its height had a population of over 1.8 million, there was none of the density found in cities like Chicago and New York. As the attached map shows, Detroit's urban footprint can accommodate Manhattan, Boston, and San Francisco combined (**Figure 12.A**). Even at its height, Detroit's urban density was around 13,000 people per square mile. The current density of Manhattan per square mile is around 70,000 people.

The Empty City

Detroit's unique geography has given rise to some of its biggest problems. While many cities have experienced economic and population decline over the last forty years, Detroit's particular form of urban development has meant that population loss has resulted in large areas of the city becoming abandoned and depopulated. Today almost forty square miles in Detroit are vacant, almost a third of the city's total land. Many of these empty lots are abandoned homes, but a number are former factories, warehouses and other commercial structures, some of which have been demolished and other still stand, slowly decaying. The checkerboard nature of vacancy means that neighbourhoods that are relatively well-populated feel empty, since they are often surrounded by, and interspersed with, abandoned homes and vacant land. This has created a number of dilemmas for the city. It is difficult to provide services, such as fire, police and rubbish collection to the small pockets of populations spread throughout the city's wide area; it is difficult to attract developers since land parcels are small and scattered and large lots difficult to assemble; it is difficult to attract shops and grocery stores since there are few sites where they can be assured of capturing a local market; and it is difficult to raise funds because tax revenues are low, land taxes are not being paid, and eminent domain laws are weak. Detroit's geography was well-suited to a particular economic project, but the imprint it left upon the landscape has been far more durable than the economy that created it. And while many of Detroit's problems are similar to those in other cities, the geographic dimension creates a number of unique challenges.

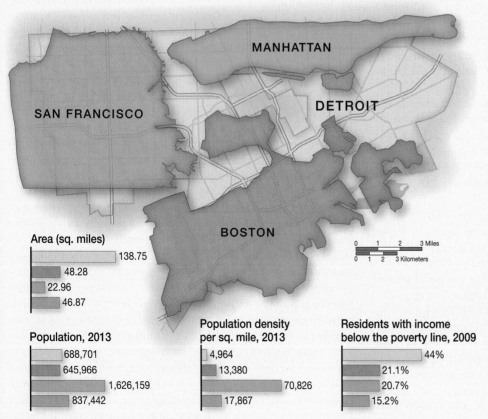

Area (sq. miles)
138.75
48.28
22.96
46.87

Population, 2013
688,701
645,966
1,626,159
837,442

Population density per sq. mile, 2013
4,964
13,380
70,826
17,867

Residents with income below the poverty line, 2009
44%
21.1%
20.7%
15.2%

▲ Figure 12.A Detroit's urban footprint

▲ **Figure 12.B** **Detroit urban farms.** Volunteers work at urban farm in Detroit to grow produce for Capuchin Soup Kitchen.

▲ **Figure 12.C** **Open-air urban markets.** Flower Day at Eastern Market in Detroit.

City Farms

Land is usually a precious commodity, especially in cities. Why is land the problem in Detroit? This is one of the paradoxes of capitalism: if something is not deemed to have market value it is not deemed to be valuable. It is not surprising, therefore, that the focus of Detroit's economic strategy has been to increase the value of its land. However, in the midst of these market driven strategies, a number of community lead initiatives are re-claiming the value of land by putting it to use in innovative ways. These include furniture making from local 'weed-woods' like *ailanthus altissima* (Tree-of-heaven), neighbourhood-scale art projects like the Heidelberg Project, and outdoor training labs that promote biologically appropriate landscape development. However, the initiative that has become most widespread in the city is urban farming.

Currently there are over 2,000 small to medium scale farms in Detroit, most of which are on vacant and abandoned land, run by local church groups, community organisations and state schools (**Figure 12.B**).

Farming in Detroit was begun primarily by local residents to accomplish two things: (1) make the empty lots between their homes productive and attractive, and (2) have access to fresh produce, as fruits and vegetables are scarce in a city with few grocery stores.

Building a New Economy

Given the popularity of urban farming in the city, one could argue that Detroit is not only developing a different valuation of land, it is developing an alternative mode of economic development. For example, Detroit's Eastern Market, an open-air wholesale and retail market, has been instrumental in providing local farmers an outlet for their produce (**Figure 12.C**). The market also encourages the proliferation of farms throughout the city. By providing food production services (such as freezing and jam making) and technical assistance (like soil sampling and equipment training), it hopes to have 20% of Detroit's food produced locally by 2020. The rationale for this push is to create a more robust approach to food security. Given Detroit's past, many in the city believe that its population's basic needs cannot be provided by the market or the state. Thus, Eastern Market is attempting to foster an economy whose primary aim is not to generate profits or improve the city's GDP, but to secure the city's basic needs. What is interesting about Eastern Market is not simply its ambition to create food security but its recognition that the economy is not 'one thing' – it is not simply a set of wage-labour relations designed to produce surplus capital. The idea than an economy can be *built* to produce food security opens our imagination to other outputs that an economy could potentially produce: security, biodiversity, public health, and community cohesion. Detroit's geography has created some unique challenges, it has also produced some innovative responses – and perhaps some thought-provoking questions about 'what an economy is' and 'what an economy can do'.

1. What other one-industry cities in the U.S. are struggling to survive? Are these cities unique in terms of land mass?

2. What other kinds of economies can you imagine in a city reinventing itself as Detroit is doing?

stress, which contributes to social disorganization and a variety of pathological behaviors, including crime and vandalism. Such conditions not only affect people's educational achievement and employment opportunities but also lead to *labeling* of the neighborhood, whereby all residents may find their employment opportunities affected by the poor image of their community.

One of the most important elements in the cycle of neighborhood poverty is the educational setting. Schools, obsolete and physically deteriorated like their surroundings, are unattractive to teachers—partly because of the physical environment and partly because of the social and disciplinary environment. Fiscal squeeze leaves schools resource-poor, with relatively small budgets for staff, equipment, and materials. Over the long term, poor educational resources translate into poor education, however positive the values of students and their parents. Poor education limits occupational choice and, ultimately, results in lower incomes. Students, faced with evidence all around them of unemployment or low-wage jobs at the end of school careers, find it difficult to be positive about school. The result becomes a self-fulfilling prophecy of failure, and people become trapped in areas of concentrated poverty (**Figure 12.8**).

Many areas of concentrated poverty are also racial **ghettos**, although not all ghettos are poverty areas. Recall that ethnic and racial **congregation** can mitigate the effects of poverty. Nevertheless, discrimination is usually the main cause of ghettoization. In the United States, discrimination in housing markets is illegal, but it nevertheless takes place in a variety of ways. One example of housing-market discrimination by banks and other lending institutions is the practice of redlining. **Redlining** involves marking off bad-risk neighborhoods on a city map and then using the map to determine lending policy. This practice results in a bias against minorities, female-headed households, and other vulnerable groups who tend to be localized in low-income neighborhoods. Redlining tends to become another self-fulfilling prophecy, since neighborhoods starved of property loans become progressively more run-down and therefore increasingly unattractive to lenders. Discrimination affects education and labor markets as well as housing markets. In the case of ghetto poverty, all three types of discrimination come together, reinforcing the cycle of poverty and intensifying the disadvantages of the minority poor.

APPLY YOUR KNOWLEDGE

1. Identify two fiscal, industrial, and/or neighborhood problems in your city.

2. Brainstorm some strategies to remedy these problems and compare them to existing projects in your city; for example, sales tax increase for schools; neighborhood-based garden programs; and funds supporting female and minority owned businesses. What do you think is an effective way to end a cycle of poverty?

Urban Design and Planning

Urban planning in North America is relatively weak, reflecting the importance of property rights in local and national politics, the unpopularity of government intervention of any kind, and the fragmentation of planning responsibilities among multiple urban and suburban jurisdictions in most metropolitan regions. Much of suburbia is, in practice, planned by developers and governed by private homeowner associations with their own sets of rules. In central cities, planning has morphed into public-private cooperation, with local governments increasingly behaving like businesses in their attempts to attract economic development and balance the books.

▼ **Figure 12.8 Poverty areas** Concentrations of poverty are found not only in decaying inner-city areas but also in newer public housing projects and in older suburbs that have filtered down the housing scale, as in this example in Washington, DC.

New Urbanism All this is not to say that there are no attempts at planning better cities. In particular, the problems of sprawl have given rise to a distinctively North American planning movement, the so-called New Urbanism. It is based on the observation that walkable, diverse, urban places like Nantucket, Massachusetts, Alexandria, Virginia, Georgetown, Washington, D.C., historic Charleston, South Carolina, and Savannah, Georgia, are enduringly popular. New urban spaces, it is argued, should be designed to be like these by specifying design codes that replicate the best features of such places, with special emphasis on traditional vocabulary boulevards, plazas, monuments, and the pedestrian scale of streets and public spaces. Well-designed front porches, narrow streets, and smaller lot sizes will actively produce community by bringing neighbors together more closely and more frequently; and therefore be inclined to greet and chat with each other more

often, leading to a feelings of community, civility, and sense of place. The whole approach is based on a great deal of accumulated wisdom about urban planning but has been given an updated angle because of its emphasis on design codes. These have been specified in detail in a **SmartCode** approved by a nonprofit organization, the Congress for the New Urbanism.

New Urbanism has certainly enlivened interest in planning and urban design, and brought fresh ideas to what had become routinized and bureaucratized issues of land use and zoning. It has also reinforced sense of place, livability, sustainability, and quality of life as important policy issues, and helped to resurrect the idea of a definable public interest. New Urbanism appeals to developers because its closely written codes alleviate many of the problems of customer choice associated with new suburban development and because it has a strong brand identity with sensibilities—a vision of community and a sense of certainty, respectability, and predictability—that appeal to the key market segment of upper-middle-class households. Nevertheless, New Urbanism has come in for a great deal of criticism as being backward-looking and superficial, with no real engagement with the physical infrastructure and political economy of the city as a whole. The principal underlying weakness of New Urbanism, though, is that it falls into the trap of environmental determinism, privileging spatial form over social process: design codes become behavior codes, ignoring the messy complexity and creativity of real urban life, its politics, stresses, and contradictions.

EUROPEAN CITIES

European cities, like North American cities, reflect the operation of competitive land markets and social congregation along ethnic lines. They also suffer from similar problems of urban management, infrastructure maintenance, and poverty. What makes most European cities distinctive in comparison with North American cities is that they are the product of several major epochs of urban development.

Features of European Cities

As we saw in Chapter 2, many of today's most important European cities were founded in the Roman period, and it is not uncommon for the outlines of Roman and medieval urban development to be preserved in their street plans. Many distinctive features of European cities derive from their long history. In the historic cores of some older cities, the layout of streets reflects ancient patterns of rural settlement and field boundaries. Beyond these historic cores, narrow, complex streets are the product of the long, slow growth of European cities in the pre-automobile era, when hand-pushed and horse-drawn carts were the principal means of transportation and urban development was piecemeal and small-scale.

Plazas and squares are another important historical legacy in many European cities. Greek, Roman, and medieval cities were all characterized by plazas, central squares, and marketplaces, and those elements are still important nodes of urban activity (**Figure 12.9**). European history also means that its cities bear the accessories and scars of war. The legacy of defensive hilltop and cliff-top sites and city walls has limited and shaped the growth of modern European cities, while in more recent times the bombings and shellings of World War II destroyed many city buildings (**Figure 12.10**).

The legacy of a long and varied history includes a rich variety of symbolism. Europeans are reminded of their past not only by large numbers of statues and memorials but also by cathedrals, churches, and monasteries; by guildhalls and city walls; by the palaces of royalty and the mansions of aristocracy; and by city halls and the libraries, museums, sports stadiums,

◀ **Figure 12.9 Vigevano, Italy** Widely considered to be one of the finest piazzas in Italy, the Piazza Ducale in Vigevano is a product of early Renaissance town planning, designed by Bramante for Ludovico Maria Sforza in 1492–93 as a noble forecourt to the Castello Sforzesco. Unified by the arcades that completely surround the square, the piazza provides an important social space for the citizens of the town.

▲ **Figure 12.10 The scars of war** The view from Dresden's town hall of the devastated Old Town after allied bombings in February 1945.

and galleries that are monuments to civic achievement. European cities are also typically compact in form, resulting in high densities of population. A long history of pre-automobile urban development and the constraints of peripheral defensive walls all made urban land expensive and encouraged a tradition of high-density living in tenements and apartment houses.

Other distinctive features of European cities include:

- **Low skylines**—Although the larger European cities have a fair number of high-rise apartment buildings and a sprinkling of office skyscrapers, they all offer a predominantly low skyline. This is partly because much of their growth came before the invention of the elevator and the development of steel-reinforced, concrete building techniques and partly because of master plans and building codes (some written as long ago as the sixteenth century) seeking to preserve the dominance of monumental buildings like palaces and cathedrals.

- **Lively downtowns**—The CBDs of European cities have retained their focal position in residents' shopping and social lives because of the relatively late arrival of the suburbanizing influence of the automobile and strong planning controls directed against urban sprawl.

- **Neighborhood stability**—Europeans change residence, on average, about half as often as Americans. In addition, the physical life cycle of city neighborhoods tends to be longer because of the past use of durable construction materials, such as brick and stone. As a result, European cities provide relatively stable socioeconomic environments.

- **Municipal socialism**—For decades, European welfare states have provided a broad range of municipal services and amenities, from clinics to public transit systems. Perhaps the most important to urban structure is social housing (public housing), which accounts for 20 to 40 percent of all housing in most larger English, French, and

German cities. In recent years, neoliberal policies have resulted in a reduction in public services, especially in France and Great Britain.

The richness of European history and the diversity of Europe's geography mean that there are important regional variations: The industrial cities of northern England, northeastern France, and the Ruhr district of Germany, for example, are quite different in character from the cities of Mediterranean Europe. One of the most interesting regional variations is in Eastern Europe, where the legacy of an interlude of 44 years of socialism (1945–1989) was grafted onto cities that had already developed mature patterns of land use and social differentiation. Major examples include Belgrade, Budapest, Katowice, Kraków, Leipzig, Prague, and Warsaw. State control of land and housing meant that huge public housing estates and industrial zones were created in outlying districts. The structure of the older cities was little altered, however, apart from the addition of socialist monuments and the renaming of streets.

European Urban Design and Planning

European city planning and design have a long history. Most Greek and Roman settlements were laid out on grid systems, within which the siting of key buildings and the relationship of neighborhoods to one another were carefully considered.

The roots of modern Western urban planning and design can be traced to the Renaissance and Baroque periods (between the fifteenth and seventeenth centuries) in Europe. Artists and intellectuals dreamed of ideal cities, and rich and powerful regimes used urban design to produce extravagant symbolizations of wealth, power, and destiny. Inspired by the classical art forms of ancient Greece and Rome, Renaissance urban design sought to recast cities to show off the power and the glory of the state and the Church. Spreading slowly from its origins in Italy at the beginning of the fifteenth century, Renaissance design had diffused to most of the larger cities of Europe by the end of the eighteenth century. Dramatic advances in military ordnance (cannon and artillery) brought a surge of planned redevelopment that featured impressive fortifications; geometric-shaped redoubts, or strongholds; and an extensive *glacis militaire*—a sloping, clear zone of fire. Inside new walls, cities were recast according to a new aesthetic of grand design (**Figure 12.11**)—fancy palaces and geometrical plans, streetscapes, and gardens that emphasized views of dramatic perspectives. These developments were often of such a scale that they effectively fixed the layout of cities well into the eighteenth and even into the nineteenth century, when walls and/or glacis eventually made way for urban redevelopment in the form of parks, railway lines, or beltways.

Order, Safety, and Efficiency
As societies and economies became more complex with the transition to industrial capitalism, national rulers and city leaders in Europe looked to urban design to impose order, safety, and efficiency, as well as to

▲ **Figure 12.11** Castello Sforza, Milan, Italy.

new, tree-lined avenues (**Figure 12.12**) and numerous public open spaces and monuments. In doing so, he made the city not only more efficient (wide boulevards meant better flows of traffic) and a better place to live (parks and gardens allowed more fresh air and sunlight into a crowded city and were held to be a "civilizing" influence) but also safer from revolutionary politics (wide boulevards were hard to barricade; monuments and statues instilled a sense of pride and civic identity).

The preferred architectural style for these new designs was the **Beaux Arts** style, which takes its name from L'École des Beaux Arts in Paris. In this school, architects were trained to draw on Classical, Renaissance, and Baroque styles to synthesize new designs for the Industrial Age. The idea was that the new buildings would blend artfully with the older palaces, cathedrals, and civic buildings that dominated European city centers. Haussmann's ideas were widely influential and extensively copied.

symbolize the new seats of power and authority. An important early precedent was set in Paris by Napoleon III, who presided over a comprehensive program of urban redevelopment and monumental urban design. The work was carried out by Baron Georges Haussmann between 1853 and 1870. Haussmann demolished large sections of old Paris to make way for broad,

Modern Urban Design
Early in the twentieth century there emerged a different intellectual and artistic reaction to the pressures of industrialization and urbanization. The **Modern movement** was based on the idea that buildings and cities should be designed and run like machines. Equally important to the Modernists was the idea that urban design should not simply reflect dominant social and cultural values, but, rather, help create a new moral and social order. The movement's

▼ **Figure 12.12** Boulevard des Italiens, Paris   Central Paris owes much of its character to the *grandes boulevards* that were key to the urban renewal schemes of Baron Georges-Eugène Haussmann.

best-known advocate was Le Corbusier, a Paris-based Swiss who provided the inspiration for technocratic urban design. Modernist buildings sought to dramatize technology, exploit industrial production techniques, and use modern materials and unembellished, functional design. Le Corbusier's ideal city (*La Ville Radieuse*) featured linear clusters of high-density, medium-rise apartment blocks, elevated on stilts and segregated from industrial districts; high-rise tower office blocks; and transportation routes—all separated by broad expanses of public open space.

After World War II, the International Style became pervasive in urban design. The box-like steel-frame buildings with concrete and glass facades were avant-garde yet respectable and, above all, comparatively inexpensive to build. This tradition of urban design, more than anything else, has imposed a measure of uniformity on cities around the world (**Figure 12.13**). Because of globalization International Style buildings have appeared in big cities in every part of the world. Furthermore, the International Style has often been the preferred style for large-scale urban design projects around the world. Modern urban design has had many critics, mainly on the grounds that it tends to take away the natural life and vitality of cities, replacing varied and human-scale environments with monotonous and austere settings. In response, historic preservation has become an important element of urban planning in every city that can afford it.

▲ Figure 12.13 **International Style** The Total office tower buildings at La Défense, Paris.

APPLY YOUR KNOWLEDGE

1. Search the Internet to find six images that best capture the typical features of European cities, such as narrow, complex streets, plazas and market squares, lively downtowns, neighborhood stability, and municipal socialism.

2. Compare the designs of three cities in Western Europe with three cities in Eastern Europe and analyze how those urban landscapes may have been impacted by political conflicts, social programs, and economic policies (socialism, capitalism).

ISLAMIC CITIES

Islamic cities provide good examples of how social and cultural values and people's responses to their environment are translated into spatial terms through the built environment. Because of similarities in cityscapes, layout, and design, geographers are able to talk about the Islamic city as a meaningful category. It is a category that includes thousands of towns and cities, not only in the Arabian Peninsula and the Middle East—the heart of the Islamic Empire under the prophet Muhammad (570–632 C.E.)—but also in regions into which Islam spread later: North Africa, coastal East Africa, South-Central Asia, and Indonesia. Most cities in North Africa and South-Central Asia are Islamic, and many elements of the classic Islamic city can be found in towns and cities as far away as Seville, Granada, and Córdoba in southern Spain (the western extent of Islam), Kano in northern Nigeria and Dar es Salaam in Tanzania (the

southern extent), and Davao in the Philippines (the eastern extent).

The fundamentals of the layout and design of the traditional Islamic city are so closely attached to Islamic cultural values that they are referenced in the Qur'an, the holy book of Islam. Although urban growth in the Islamic world does not have to conform to any overall master plan or layout, certain basic regulations and principles are intended to support Islam's emphasis on personal privacy and virtue, on communal well-being, and on the inner essence of things rather than on their outward appearance.

The most dominant feature of the traditional Islamic city is the *Jami*—the principal mosque (**Figure 12.14**). Located centrally, the mosque complex is a center not only of worship but also of education and the hub of a broad range of welfare functions. As cities grow, new, smaller mosques are built toward the edge of the city, each out of earshot of the call to prayer from the Jami and from one another. The traditional Islamic city was walled for defense, with several lookout towers and a *Kasbah*, or citadel (fortress), containing palace buildings, baths, barracks, and its own small mosque and shops.

Traditionally, gates controlled access to the city, allowing careful scrutiny of strangers and permitting the imposition of taxes on merchants. The major streets led from these gates to the main covered bazaars or street markets (*suqs*, **Figure 12.15**). The suqs nearest the Jami typically specialize in the cleanest and most prestigious goods, such as books, perfumes, prayer mats, and modern consumer goods. Those nearer the gates feature bulkier and less valuable goods such as basic foodstuffs,

◄ **Figure 12.14** Mosque The dominant feature of traditional Islamic cities is the *Jami*, or main mosque. This photograph shows the Celil Hayat Mosque in Arbil City, Kurdistan, Iraq.

building materials, textiles, leather goods, and pots and pans. Within the suqs, every profession and line of business had its own alley, and the residential districts around the suqs are organized into distinctive quarters, or *ahya'*, according to occupation (or sometimes ethnicity, tribal affiliation, or religious sect).

Privacy is central to the construction of the Islamic city. Above all, women must be protected, according to Islamic values, from the gaze of unrelated men. Traditionally, doors must not face each other across a minor street, and windows must be small, narrow, and above eye level. Cul-de-sacs (dead-end streets) are used where possible to restrict the number of persons approaching the home. Angled entrances prevent intrusive glances. Larger homes are built around courtyards, which provide an interior and private focus for domestic life.

The rights of others are also emphasized in Islamic urban design. The Qur'an specifies an obligation to neighborly cooperation and consideration—traditionally this consideration is interpreted as applying to a minimum radius of 40 houses. Roofs, in traditional designs, are surrounded by parapets to preclude views of neighbors' homes, and drainage channels are steered away from neighbors' houses. Refuse and wastewater are carefully recycled. Public thoroughfares were originally designed to be wide enough to allow two fully laden camels to pass each other and high enough to accommodate a camel and rider. The overall result is a compact, cellular urban structure within which it is possible to maintain a high degree of privacy **(Figure 12.16)**.

Because most Islamic cities are located in hot, dry climates, these basic principles of urban design have evolved in conjunction with certain practical solutions to intense heat and sunlight. Twisting streets, as narrow as permissible, help to maximize shade, as do latticework on windows and a cellular residential courtyard design. In some regions, local architectural styles include air ducts and roof funnels with adjustable shutters that can be used to create dust-free drafts **(Figure 12.17)**.

▼ **Figure 12.15** Bazaar-i Vakil suq in Shiraz, Iran The *suq*, a covered bazaar or open street market, is one of the most important distinguishing features of a traditional Islamic city. Typically, a suq consists of small stalls located in numerous passageways. Many important suqs are covered with vaults or domes.

▲ **Figure 12.16 Housing in Kalaa Sghira, Tunisia** Seen from above, the traditional Islamic city is a compact mass of residences with walled courtyards.

While all these features are still characteristic of Islamic cities, they are most evident in the old cores, or *medinas*. Like cities everywhere, however, Islamic cities also bear the imprint of globalization. Although Islamic culture is self-consciously opposed to many aspects of globalization, it has been unable to resist altogether the penetration of the world economy and the infusion of the Western-based culture of global metropolitanism. The result can be seen in international hotels, skyscrapers and office blocks, modern factories, highways, airports, and stores. Indeed, the leading cities of some oil-rich states have become the "shock cities" of the early twenty-first century, with phenomenal rates of growth characterized by breathtakingly ambitious architectural and urban design projects (see Box 12.3, "Window on the World: Doha, Qatar"). Meanwhile, Islamic culture and urban design principles have not always been able to cope with the pressures of contemporary rates of urbanization, so the larger Islamic cities in less-affluent states—cities such as Algiers, Cairo, Karachi, and Teheran—now share with other peripheral cities the common denominators of unmanageable size, shanty and squatter development, and low-income mass housing. In the next section of the chapter, we examine the problems of large and rapidly growing metropolises throughout the periphery.

APPLY YOUR KNOWLEDGE

1. Research an Islamic city that is approximate in size to your community and list two similarities and two differences between these cities.

2. Analyze the way culture impacts urban design by looking at market places, centers of worship, and differences in the value of privacy; for example, the use of front porches in American homes versus private courtyards in Islamic homes.

◀ **Figure 12.17 Islamic architecture** In Islamic societies elaborate precautions are taken through architecture and urban design to ensure the privacy of individuals, especially women. Entrances are L-shaped and staggered across the street from one another. Windows are often placed above pedestrian access. Architectural details also reflect climatic influences: window screens and narrow, twisting streets maximize shade, while air ducts and roof funnels create dust-free drafts, as in this photograph of traditional wind towers in the city of Yazd, Iran.

CITIES OF THE PERIPHERY: DUALISM

The cities of the world-system periphery, often still referred to as Third World cities, are numerous and varied. What they have in common is the experience of unprecedented rates of growth driven by rural "push"—overpopulation and the lack of employment opportunities in rural areas—rather than the "pull" of prospective jobs in towns and cities. Faced with poverty in overpopulated rural areas, many people regard moving to a city much like playing a lottery: You buy a ticket (in other words, go to the city) in the hope of hitting the jackpot (in other words, landing a good job). As with all lotteries, most people lose, and the net result is widespread underemployment. **Underemployment** occurs when people work less than full-time even though they would prefer to work more hours. Underemployment is difficult to measure with any degree of accuracy, but estimates commonly range from 30 to 50 percent of the employed workforce in many peripheral cities.

Because of their rapid growth and high underemployment, the peripheral metropolises of the world, the shock cities of the late twentieth century—Mexico City (Mexico), São Paulo (Brazil), Lagos (Nigeria), Mumbai (India; formerly Bombay, **Figure 12.18**), Dhaka (Bangladesh), Jakarta (Indonesia), Karachi (Pakistan), and Manila (the Philippines)—embodied the most remarkable and unprecedented changes in economic, social, and cultural life. Socioeconomic conditions in parts of these cities are still shocking, in the sense of being deplorably bad, but they are no longer shocking in the sense of being unprecedented. Rather, they have become all too familiar.

Recall from Chapter 10 that peripheral metropolises play a key role in international economic flows, linking provincial regions with the hierarchy of world cities and, thus, with the global economy. Within peripheral metropolises, this role results in a pronounced **dualism**, or juxtaposition in geographic space of the formal and informal sectors of the economy. This dualism is evidenced by the contrast between high-rise modern office and apartment towers and luxurious homes and the slums and shantytowns (**Figure 12.19**).

The Informal Economy

In many peripheral cities, more than one-third of the population is engaged in the informal sector of the urban economy, and in some—for example Chennai, India; Colombo, Sri Lanka; Delhi, India; Guayaquil, Ecuador; and Lahore, Pakistan—the figure is more than one-half. The **informal sector** of an economy involves a wide variety of economic activities whose common feature is that they take place beyond official record and are not subject to formalized systems of regulation or remuneration. People who cannot find regularly paid work must resort to various ways of gleaning a living. Some of these ways are imaginative, some desperate. Examples range from street vending (**Figure 12.20**), shoe shining, craft work, and street-corner repairs to scavenging in garbage dumps (**Figure 12.21**). Children represent a significant element of the informal economy. For hundreds of thousands of street kids in less developed countries, "work" means anything that contributes to survival: shining shoes, guiding cars into parking spaces, chasing other street kids away from patrons at an outdoor café, working as domestic help, making fireworks, or selling drugs. Occupations such as selling souvenirs, driving pedicabs, making home-brewed beer, writing letters for others, and dressmaking may seem marginal from the point of view of the global economy, but more than a billion people around the world must feed, clothe, and house themselves entirely from such occupations. Across Africa, the International Labor Office estimates, informal-sector employment is growing 10 times faster than formal-sector employment. The informal sector represents an important coping mechanism. For too many, however, coping means resorting to begging, crime, or prostitution.

◄ **Figure 12.18** Mumbai, India

◀ **Figure 12.19 Dualism** Slums adjacent to modern apartment buildings in Metro Manila, Philippines

◀ **Figure 12.20 Street market** Vegetable stalls and street vendors in Rawalpindi, Pakistan

In most peripheral countries, the informal labor force includes children. In environments of extreme poverty, every family member must contribute something, and so children are expected to do their share. Industries in the formal sector often take advantage of this situation. Many firms farm out their production under subcontracting schemes that are based not in factories but in home settings that use child workers. In these settings, labor standards are nearly impossible to enforce. In the Philippines, for example, batches of rural children are ferried by syndicates to work in garment-manufacturing sweatshops in urban areas.

▲ **Figure 12.21 Garbage picking** Scavengers picking out recyclable garbage at the Nong Khaem dump, Bangkok, Thailand.

The Informal Sector in Context Despite this side of the picture, the informal sector has positive aspects. Pedicabs, for example, provide an affordable, nonpolluting means of transportation in crowded metropolitan settings. Garbage picking provides an important means of recycling paper, steel, glass, and plastic products. One study of Mexico City estimated that as much as 25 percent of the municipal waste ends up being recycled by the 15,000 or so scavengers who work over the city's official dump sites. This positive contribution to the economy, though, scarcely justifies the lives of poverty and degradation experienced by the scavengers. Meanwhile, it should be noted that even in the formal sector wages can be extremely low and working conditions dangerous (see Box 12.4: "Spatial Inequality: Garment Workers in Dhaka").

APPLY YOUR KNOWLEDGE

1. What responsibility do corporations like H&M, Gap, and Apple have to workers in developing countries? How connected are the consumers of these products to the laborers who create them?

2. How does raising minimum wage impact workers' lives? Compare the minimum wage rates around the world and how workers live comparatively by looking at this article "Global Minimum Wage": http://www.reuters.com/subjects/global-minimum-wage

Urban geographers also recognize that the informal sector represents an important resource to the formal sector of peripheral economies. The informal sector provides a vast range of cheap goods and services that reduce the cost of living for employees in the formal sector, thus enabling employers to keep wages low. Although this network does not contribute to urban economic growth or help alleviate poverty, it does keep companies competitive within the context of the global economic system. For export-oriented companies, in particular, the informal sector provides a considerable indirect subsidy to production. This subsidy is often passed on to consumers in the core regions in the form of lower prices for goods and consumer products made in the periphery.

Consider, for example, the paper industry in Cali, Colombia. This industry is dominated by one company, Cartón de Colombia, which was established in 1944 with North American capital and subsequently acquired by the Mobil Oil Company. Most of the company's lower-quality paper products are made from recycled waste paper. Sixty percent of this waste paper is gathered by the 1,200 to 1,500 garbage pickers in Cali. Some work the city's municipal waste dump, some work the alleys and yards of shopping and industrial areas, and some work the routes of municipal garbage trucks, intercepting trash cans before the truck arrives. They are part of Cali's informal economy, for they are not employed by Cartón de Colombia nor do they have any sort of contract with the company or its representatives. They simply show up each day to sell their pickings. In this way the company avoids paying both wages and benefits and is able to dictate the price it will pay for various grades of waste paper. The company can operate profitably while keeping the price of its products down—the arrangement is a microcosm of core-periphery relationships.

APPLY YOUR KNOWLEDGE

1. Identify an informal economic function in your town or city and list three positive and three negative effects of the activities involved in this type of economy.

2. Measure the impact of these informal economies by looking at the services they provide and analyze how they connect to the formal industries.

Slums of Hope, Slums of Despair

The informal labor market is directly paralleled in informal shantytowns and squatter housing: Because there are insufficient jobs with regular wages in the cities of the periphery, many families cannot afford rent or house payments for sound housing. Unemployment, underemployment, and poverty mean overcrowding. In situations where urban growth has swamped the available stock of cheap housing and outstripped the capacity of builders to create affordable new housing, the inevitable outcome is makeshift shanty housing. Such housing has to be constructed on the cheapest and least desirable sites. Often this means building on bare rock,

12.3 Window on the World

Doha, Qatar

Doha, the capital of Qatar, owes its growth to the world's appetite for fossil fuels. For centuries it was a small port, relying on fishing and the trade routes of the Indian Ocean. (**Figure 12.D**) Qatar's oil and gas reserves, controlled by the country's ruling elite, began to transform the city in the mid-twentieth century but it was only in the 1990s that the Emir of Qatar, Hamad Bin Khalifa Al Thani, embarked on an extravagant strategy of economic and urban development based on a combination of oil and gas revenues and an aggressively pro-business regime based on lax labor laws and malleable financial regulations. Doha grew from around 500,000 inhabitants in the late 1990s to more than 1.5 million in 2013. Most of Doha's residents—about two-thirds of them—are foreigners: thousands

▲ **Figure 12.D** Doha waterfront skyline

▲ **Figure 12.E** Doha: Migrant workers' accommodations.

of professionals and tens of thousands of laborers who have been drawn to Qatar to build the city.

The real estate boom created a new skyline of office and residential towers in the space of less than 10 years. It also created a very unequal society, with super-affluent Qataris and well-paid foreign professionals but miserably paid migrant laborers. (**Figure 12.E**) Zones, partitions, walls, enclaves, and compounds are characteristic aspects of modern Doha. Government investment has created a series of planned districts, each with prescribed functions. These include the Pearl-Qatar, a vast urban development consisting of fifteen thousand upscale dwellings (**Figure 12.F**); a Diplomatic Quarter; Education City, a conglomeration of Euro-American universities and research institutions; an Aspire Zone, devoted to sports stadia and other athletic facilities (**Figure 12.G**); and an Industrial Area on the periphery of the city where tens of thousands of transnational laborers live in dormitory camps amidst a checkerboard of industry.

▲ **Figure 12.F** Doha: Pear-Qatar development

Meanwhile, Qatar's rulers have sought to elevate Doha's status from an oil- and gas-based welfare enclave to a world city with global recognition and influence. The strategy has been to establish the city as a global service hub, developing media, sports, and cultural facilities and events in order to brand the city as an international service center. One of the first developments, in the mid-1990s, was the establishment of the now-influential news group, Al Jazeera. Since then, the city has hosted academic, religious, political, and economic conferences and a broad range of high-profile sports events. Improbably, Qatar was selected to host the finals of the 2022 football World Cup after promising to splurge billions of dollars in building air-conditioned stadiums (with both the Qataris and the organizers conveniently neglecting the fact that much of what is enjoyable about a World Cup happens outside, in the streets, bars, and plazas of host cities).

The Qatari elite has also asserted its global influence through overseas investments. Barcelona's football club has been sponsored by the Qatar Foundation and Qatar Airways, while another prominent football club, Paris St-Germain, was bought in 2011 by the state-owned Qatar Sports Investments. The Qatar Investment Authority and its partners have meanwhile reached beyond Doha to acquire trophy real estate assets in major cities. In London, for example, these include the 87-story Shard building, the super-luxury One Hyde Park complex (where one buyer reputedly paid £140 million pounds—more than U.S. $225 million—for a penthouse apartment), Harrods department store, the entire 2012 Olympic Village, and the former U.S. embassy building.

Meanwhile, back in Doha, the attempt to buy international economic and cultural standing has been severely compromised by the country's questionable human rights record and the scandalous conditions endured by the low-paid migrants working on World Cup facilities and real estate developments. According to official figures, more than 430 Nepalese and 560 Indian workers died between January 2012 and mid-April 2014.

---

1. Looking at both Doha's wealth and labor practices, what are the benefits and negative effects of a pro-business regime built on an extractive economy and lax labor laws?
2. Do an Internet search to see how Qatar has responded to the international criticism of its labor and immigration laws (e.g., the kafala system) and analyze how you think reforms could impact the lifestyles of Qataris.

▲ **Figure 12.G Doha** The Aspire zone

over ravines, on derelict land, on swamps, or on steep slopes. Nearly always it means building without any basic infrastructure of streets or utilities. Shacks are constructed out of any material that comes to hand, such as planks, cardboard, tar paper, thatch, mud, and corrugated iron. Many of these instant slums are squatter settlements, built illegally by families who are desperate for shelter. **Squatter settlements** are residential developments on land that is neither owned nor rented by its occupants. Squatter settlements are often, but not always, slums. In Chile, squatter settlements are called *callampas*, "mushroom cities," while in Turkey, they are called *gecekondu*, meaning that they were built after dusk and before dawn. In India, they are called *bustees*; in Tunisia, *gourbevilles*; in Brazil, *favelas*; and in Argentina, simply *villas miserias.* These settlements typically account for well over one-third and sometimes up to three-quarters of the population of major cities (**Figure 12.22**).

Many of them are "slums of despair," where overcrowding, lack of adequate sanitation, and lack of maintenance lead to shockingly high levels of ill health and infant mortality and where social pathologies are at their worst. In many cities, more than half of the housing is substandard. The United Nations estimated in 2013 that more than a billion people worldwide live in inadequate housing in urban areas.

Dharavi
Nevertheless, there are many shanty and squatter neighborhoods where self-help and community organization emerges. Consider, for example, the squatter settlement of Dharavi, a "mega-slum" in Mumbai, India.[3] Dharavi is about 1 square mile in size and home to over a million people. This means that as many as 18,000 people crowd into each single acre of land. In a city where house rents are among the highest

in the world, Dharavi provides a cheap and affordable option to those who move to Mumbai to earn their living. Rents can be as low as 185 rupees ($4) per month. Recycling is one of the slum's biggest industries. In Dharavi nothing is considered garbage. Thousands of tons of scrap plastic, metals, paper, cotton, soap, and glass revolve through Dharavi each day. Ruined plastic toys are tossed into massive grinders, chopped into tiny pieces, and melted down into multicolored pellets, ready to be refashioned into knockoff Barbie dolls. Dharavi also houses about 15,000 hutment factories, each typically staffed with children as well as adults sewing cotton, melting plastic, hammering iron, molding clay, or producing embroidered garments, export-quality leather goods, pottery, and plastic. The aggregate annual turnover of these businesses is estimated to be more than $650 million a year.

Yet in spite of all these positives, conditions for Dharavi's residents are miserable. Most people have food intakes of less than the recommended minimum of 1,500 calories a day; 90 percent of all infants and children under 4 have less than the minimum calories needed for a healthy diet. More than half of the children and almost half the adults have intestinal worm infections. Infant and child mortality is high—though nobody knows just how high—with malaria, tetanus, diarrhea, dysentery, and cholera as the principal causes of death among under-fives.

Slum Clearance Programs
Faced with the growth of slums, the first response of many governments has been to eradicate them. Encouraged by Western development economists and housing experts, many cities sought to stamp out unintended urbanization through large-scale eviction and clearance programs. In Caracas (Venezuela), Lagos (Nigeria), Bangkok

▼ Figure 12.22 Hillside favela, Rio de Janeiro, Brazil.

[3]See *The Economist,* 2007, "A Flourishing Slum," http://www.economist.com/world/asia/displaystory.cfm?story_id=10311293, accessed August 5, 2008; M. Jacobson, "Mumbai Slum," *National Geographic, 2007,* http://ngm.nationalgeographic.com/ngm/0705/feature3/index.html, accessed August 5, 2008; C. W. Dugger, "Toilets Underused to Fight Disease, U.N. Study Finds," *New York Times,* 2006, http://www.nytimes.com/2006/11/10/world/10toilet.html?_r=1&ex=1189828800&en=905358c57769b677&ei=5070&oref=slogin, accessed February 4, 2010.

12.4 Spatial Inequality   Garment Workers in Dhaka

▲ **Figure 12.H Sweatshop production** Garment workers in Dhaka, Bangladesh.

Garment workers are the mainstay of the formal economy in Dhaka, the capital of Bangladesh. They stand at the beginning of a supply chain that ends with the customers of stores like H&M, Old Navy, and Banana Republic in Europe and North America. The circumstances of producers and consumers could hardly be more unequal. Dhaka's garment workers—mostly young women—work in sweatshop conditions, in dangerously overcrowded, rickety factories with nothing by way of health and safety protection (**Figure 12.H**). Their wages for an 8-hour day, 6-day week are less than $10 per week. Two or three hours of compulsory overtime just about every day can bring this figure up to $16–$22 per week. During big production runs and peak seasons, the women are required to work until 1 am or later and still expected to return the following morning at 8:00 am. They are not allowed to take sick days, and in some factories they are rarely paid on time.

Their terrible working conditions were widely publicized in the wake of the collapse of a factory building in April 2013 that killed 1,129 and injured a further 2,515. The disaster led to widespread discussions about corporate responsibility across global supply chains and demands for labor law reform. Thanks largely to local union organization, assisted by the AFL-CIO's Solidarity Center and the International Labor Organization, an increase of more than 75 percent in garment workers' minimum wage has been announced; though it has still to be enacted and even if it is, it will still be the lowest minimum wage in the world.

1. What kind of labor laws would help alleviate sweatshop conditions?

2. What might be the consequences of a significant increase in garment workers' minimum wage?

(Thailand), Kolkata (India; formerly Calcutta), Manila (the Philippines), and scores of other cities in the periphery, hundreds of thousands of shanty dwellers were ordered out on short notice and their homes bulldozed to make way for public works, land speculation, luxury housing, and urban renewal and, on occasion, to improve the appearance of cities for special visitors. Seoul, South Korea, has probably had the most forced evictions of any city in the world. Since 1966, as part of sustained government clean-up campaigns millions of people in Seoul have been forced out of accommodations that they owned or rented.

The Beijing Olympics in 2008 displaced 1.5 million people, according to the Geneva-based Centre on Housing Rights and Evictions (COHRE). Its report[4] blames the Chinese government for widespread forced evictions along with other human rights violations during preparations for the Olympics.

[4]Centre on Housing Rights and Evictions (COHRE), 2008, "One World, Whose Dream: Housing Rights Violations and the Beijing Olympic Games," http://www.cohre.org/store/attachments/One_World_Whose_Dream_July08.pdf, accessed February 4, 2010.

◄ **Figure 12.23 Self-help as a solution to housing problems** Self-help is often the only solution to housing problems because wages are so low and so scarce that builders cannot construct even the most inexpensive new housing and make a profit and because municipalities cannot afford to build sufficient quantities of subsidized housing. Municipal authorities can encourage self-help housing by creating the preconditions: clearing sites, putting in the footings for small dwellings, and installing a basic framework of water and sewage utilities. This "sites-and-services" approach has become the mainstay of urban housing policies in many peripheral countries. This photograph shows self-help housing in Ndola, Zambia.

Yet Seoul and Beijing, more than most other cities, could afford to build new low-income housing to replace the demolished neighborhoods. Most peripheral cities cannot do so, which means that displaced slum dwellers have no option but to create new squatter and shanty settlements elsewhere in the city. Most cities, in fact, cannot evict and demolish fast enough to keep pace with the growth of slums caused by in-migration. The futility of slum clearance has led to a widespread reevaluation of the wisdom of such policies. The thinking now is that informal-sector housing should be seen as a rational response to poverty. Shanty and squatter neighborhoods not only provide affordable shelter but also function as important reception areas for migrants to the city, with supportive communal organizations and informal employment opportunities that help them to adjust to city life. They can, in other words, be "slums of hope." City authorities, recognizing the positive functions of informal housing, are now increasingly disposed to be tolerant and even helpful to squatters rather than sending in police and municipal workers with bulldozers.

In fact, many informal settlements are the product of careful planning. In parts of Latin America, for example, it is common for community activists to draw up plans for invading unused land and then quickly build shanty housing before landowners can react. The activists' strategy is to organize a critical mass of people large enough to be able to negotiate with the authorities to resist eviction. It is also common for activists to plan their invasions for public holidays so that the risk of early detection is minimized. As the risk of eviction diminishes over time, some residents of informal housing are able to gradually improve their dwellings through self-help **(Figure 12.23)**.

The Challenges of Growth

Globalization has greatly stimulated the formal economic sector of many peripheral cities, bringing new levels of affluence and consumption that in turn have created significant challenges in terms of infrastructure and environmental quality.

Peripheral cities have always been congested, but in recent years the modernizing influence of growing formal-sector activities has turned the congestion into near gridlock. Sharp increases have occurred in the availability and use of automobiles. India, for example, has become one of the world's fastest-growing automobile markets. More than 2.7 million cars were sold there in 2013. Yet India's transportation infrastructure has not caught up with this sudden growth, nor have its drivers. The auto accident rate is the highest in the world. India has about 1 percent of the world's cars yet still manages to kill over 100,000 people in traffic accidents each year: about 10 percent of the entire world's traffic fatalities.

Not only are there more people and more traffic but the changing spatial organization of peripheral cities has increased the need for transportation. Traditional patterns of land use have been superseded by the agglomerating tendencies inherent in modern industry and the segregating tendencies inherent in modernizing societies. The greatest single change has been the separation of home from work, however, which has meant a significant increase in commuting.

Governments have invested in expensive new freeways and street-widening schemes, and some cities boast ultramodern transit systems **(Figure 12.24)**. But new freeways and transit systems still disgorge into a congested and chaotic mixture of motorized traffic, bicycles, animal-drawn vehicles, and hand-drawn carts. Some of the worst traffic tales come from Mexico City—where traffic backups total more than 90 kilometers (60 miles) each day, on average—and Bangkok, where the 24-kilometer (15-mile) trip into town from Don Muang Airport can take 3 hours. In São Paulo, Brazil, gridlock can span 160 kilometers (100 miles), rush-hour traffic jams average 85 kilometers (53 miles) in length, and 15-hour traffic jams are not unusual. The costs of these traffic backups are enormous. The annual costs of traffic delays in Singapore have been estimated at $305 million; in Bangkok, Thailand, they have been estimated at $272 million—the equivalent of around 1 percent of Thailand's gross national product.

▲ **Figure 12.24 Delhi Metro** A yellow line tube train on the Metro Rail system, Delhi, India.

Air Pollution Meanwhile, because of the speed of economic and demographic growth, environmental problems can escalate rapidly. Air pollution has reached very harmful levels in many cities. With the development of a modern industrial sector and the growth of automobile ownership, but without enforceable regulations on pollution and vehicle emissions, tons of lead, sulfur oxides, fluorides, carbon monoxide, nitrogen oxides, petrochemical oxidants, and other toxic chemicals are pumped into the atmosphere every day in large cities. The burning of charcoal, wood, and kerosene for fuel and cooking in informal-sector neighborhoods also contributes significantly to dirty air. In cities where sewerage systems are deficient, the problem is compounded by the presence of airborne dried fecal

matter. Worldwide, according to UN data, more than 1.1 billion people live in urban areas where air pollution exceeds healthful levels.

WHO studies demonstrate that it is unhealthy for human beings to breathe air with more than 100 to 120 parts per billion (ppb) of ozone contaminants for more than one day a year. Yet Mexico City residents breathe this level, or more, for over 300 days a year. In Bangkok, Thailand, where air pollution is almost as severe as in Mexico City, research has shown that lead-bearing air pollutants reduce children's IQ by an average of 3.5 points per year until they are 7 years old. It has also been estimated that Bangkok's pall of dust and smoke causes more than 1,400 deaths annually and $3.1 billion each year in lost productivity resulting from traffic- and pollution-linked illnesses. In Manila, the Philippines, the Asian Development Bank found levels of suspended particulate matter in the air to be 200–400 percent above guideline levels. The World Bank has cited China as having 16 of the 20 most air-polluted cities on Earth (**Figure 12.25**). China's spectacular economic growth has brought with it air pollution levels that are blamed for up to 1.2 million premature deaths a year. Until recently, China's domestic politics meant that public discussion of air pollution was suppressed. Chinese officials who had been collecting air pollution data had refused to release it until they came under pressure from the public who saw that the US Embassy in Beijing was measuring the levels hourly and posting the data in a Twitter feed, @BeijingAir.

APPLY YOUR KNOWLEDGE

1. Thinking about globalized economies and population patterns, where do you think megacities will be located in the next 10, 20 and 40 years—and why?

2. Research a specific megacity (as categorized by the United Nations) and examine it in terms of the presence of slum housing, environmental degradation, and infrastructure concerns. List and explain three causes of any one of these problems.

▼ **Figure 12.25 Air pollution** Hazardous levels of pollution in Beijing, China, mean that most bicycle and moped riders regularly wear masks.

FUTURE GEOGRAPHIES

The future of North American cities in the next 10–20 years or so seems relatively easy to predict. In spite of many ideas about urban planning and the prospect of new and improved transportation, the patterns and problems described in this chapter seem set to continue. The reasons are straightforward. Demographic trends are established and accessibility, agglomeration, territoriality, congregation, segregation, and sprawl will continue to shape people's behavior. In North American cities, continuing fiscal constraints dictate that infrastructure problems will continue and likely intensify. The political and financial interests behind the "sprawl machine" ensure a continuation of polycentric, metroburban development. The past 200 years of economic, social, and urban history provide no reason at all to expect any mitigation of urban poverty or neighborhood decay. This is borne out by the survey conducted by the Society for American City and Regional Planning History cited at the beginning of this chapter. Their respondents envisage a continuation and even intensification of the urban problems that have characterized the past 50 years, including growing disparities of wealth, a perpetual urban underclass, the physical deterioration of first-ring post-1945 suburbs, and continued sprawl into new peripheral edge cities (Table 12.2). Some respondents, however, predicted that smart growth policies would help preserve the environment and limit sprawl; this group also believed that cities would overcome racial and class divisions to become more diverse than they are now.

The basic templates of urban structure are equally well established in both European and Islamic cities. Change will likely be marginal, even as economic and demographic modifications take place and as the impact of new technologies—electric cars, perhaps—alter the locational behavior of people and businesses. We can apply a similar logic to the unintended metropolises of the periphery. Faced with continuing streams of migration as well as high rates of natural increase among their relatively youthful populations, the megacities of the periphery will continue to expand, virtually unchecked. Informal economic activities will continue to have an important role, and cityscapes will continue to be dominated by stark contrasts between the towers of international business and elite residences on the one hand, and slums and informal housing on the other. Sheer pressure of numbers will ensure continuing problems of congestion, water supply, sanitation, and environmental degradation. What is less predictable is how individual cities might change as a result of the introduction of progressive planning

TABLE 12.2 The 10 Most Likely Influences on the American Metropolis 2000–2050

1. Growing disparities of wealth.
2. Suburban political majority.
3. Aging of the baby boomers.
4. Perpetual "underclass" in central cities and inner-ring suburbs.
5. "Smart growth": environmental and planning initiatives to limit sprawl.
6. The Internet.
7. Deterioration of the "first-ring" post-1945 suburbs.
8. Shrinking household size.
9. Expanded superhighway system of "outer beltways" to serve new edge cities.
10. Racial integration as part of the increasing diversity in cities and suburbs.

Source: Robert Fishman (2000) The American metropolis at century's end: Past and future influences, *Housing Policy Debate, 11:*(1), p. 200.

policies or new transportation systems or—from a more pessimistic perspective—as a result of localized economic problems, political unrest, or environmental disasters.

APPLY YOUR KNOWLEDGE

1. Why do you think that the "past 200 years" have not seen a decline or mitigation of poverty in the United States, despite the considerable wealth of the country?

2. Thinking about technological innovations, economic resources, demographics, and openness to progressive social policies, name one city in each region, Asia, South America, Eastern Europe and Africa, that you predict could experience positive growth in the next 20 years, and why.

CONCLUSION

Patterns of land use and the functional organization of economic and social subareas in cities are partly a product of economic, political, and technological conditions at the time of the city's growth, partly a product of regional cultural values, and partly a product of processes of globalization. Geographers can draw on several perspectives in looking at patterns of land use within cities, including an economic perspective that emphasizes competition for space and a sociocultural perspective that emphasizes ethnic congregation and segregation. Nevertheless, urban structure varies considerably because of the influence of history, culture, and the different roles that different cities have played within the world-system.

The evolution of the unintended metropolis of the periphery has been very different from the evolution of metropolitan areas in the world's core regions. Similarly, the problems they have faced are very different. In the core regions, the consequences of an economic transformation to a postindustrial economy have dominated urban change. Traditional manufacturing and related activities have been moved out of central cities, leaving decaying neighborhoods and a residual population of elderly and marginalized people. New, postindustrial

activities have begun to cluster in redeveloped CBDs and in edge cities around metropolitan fringes. In a few cases, metropolitan growth has become so complex and extensive that 100-mile cities have begun to emerge, with half a dozen or more major commercial and industrial centers forming the nuclei of a series of interdependent urban realms.

In other parts of the world, traditional patterns of land use and the functional organization of economic and social subareas have been quite different, reflecting different historical legacies and different environmental and cultural influences. A basic trend affecting the cities of the world's periphery is demographic—the phenomenal rates of natural increase and in-migration. An ever-growing informal sector of the economy, in which people seek economic survival, is reflected in extensive areas of shanty housing. High rates of unemployment, underemployment, and poverty generate acute social problems, which are overwhelming for city governments that are understaffed and underfunded. If present trends continue, such problems are likely to characterize increasing numbers of the world's largest settlements. Meanwhile, globalization processes are recasting metropolitan structure and intensifying social and economic inequalities.

LEARNING OUTCOMES REVISITED

■ **Assess how the internal structure of cities is shaped by competition for territory and location.**

In general, all categories of land users—commercial and industrial, as well as residential—compete for the most convenient and accessible locations within the city. An important exception is that wealthier households tend to trade off the convenience of accessibility for the greater utility of being able to consume larger amounts of (relatively cheap) suburban space. Poorer households, unable to afford the recurrent costs of transportation, trade off living space for accessibility to jobs.

■ **State the ways in which social patterns in cities are influenced by human territoriality.**

Territoriality provides a means of establishing and preserving group membership and identity. Processes of congregation and discrimination often result in **segregation**, the spatial separation of specific subgroups within a wider population. Segregation varies a great deal in both intensity and form. Consider how social patterns are dynamic and how different communities can both occupy and transition from an area.

■ **Describe the spatial structure of the typical North American city.**

Most larger cities are structured around a central business district (CBD); a transitional zone; suburbs; secondary business districts and commercial strips; and industrial districts. In larger metropolitan areas, a polycentric structure is typical, with "edge cities," new business centers, and specialized subcenters. The internal organization of cities reflects the way that they function, both to bring certain people and activities together and to sort them out into neighborhoods and functional subareas. During the middle decades of the twentieth century, North American

cities were reshaped by the combination of increased automobility, federal outlays on highway construction, and federal mortgage insurance programs that promoted home ownership. The resulting spurt of city building produced a dispersed spatial structure and the emergence of a polycentric metropolitan structure.

■ **Compare and contrast urban structures in different regions of the world.**

Urban structure varies considerably because of the influence of history, culture, and the different roles that cities have played within the world-system. European cities have evolved under circumstances very different from those in American cities and consequently exhibit some distinctive characteristics that reflect their history. Islamic cities provide examples of how social and cultural values and people's responses to their environment are translated into spatial terms through urban form and the design of the built environment. The new cities of the world's peripheral regions are characterized and shaped by explosive growth.

■ **Explain the nature and causes of the problems associated with urbanization in various world regions.**

The most acute problems of the cities of the world's core regions are localized in the central city areas that have borne the brunt of restructuring from an industrial to a postindustrial economy, while the problems of the cities of the periphery stem from the way in which their demographic growth has outstripped their economic growth. The central city districts of cities in the world's core regions typically experience several interrelated problems: fiscal problems, infrastructure problems, and localized spirals of neighborhood decay and cycles of poverty. In peripheral cities high rates of long-term unemployment and underemployment, low and unreliable wages of informal-sector jobs, chronic poverty, and slum housing are common.

KEY TERMS

Beaux Arts (p. *443*)
central business district (CBD) (p. *430*)
central cities (p. *430*)
congregation (p. *440*)
cycle of poverty (p. *437*)
dualism (p. *447*)
edge cities (p. *434*)
fiscal squeeze (p. *436*)

gentrification (p. *435*)
ghettos (p. *440*)
informal sector (p. *447*)
isotropic surface (p. *431*)
metroburbia (p. *435*)
Modern movement (p. *443*)
redlining (p. *440*)
segregation (p. *432*)

sprawl (p. *430*)
squatter settlement (p. *452*)
trade-off model (p. *434*)
underemployment (p. *447*)
urban realms (p. *434*)
Zone in Transition (p. *431*)

REVIEW & DISCUSSION

1. Research the different ethnic, class, and gender makeup of your city or town. Examine the data through the lens of congregation and the territorial and residential clustering of specific groups or subgroups of people. Identify where these different groups congregated. Does a pattern of discrimination or segregation emerge? Historically, how has the clustering responded to different events like federal homestead policies, immigration, industry, natural disasters, and so on, and how have significant events changed the clustering in your city? Explain your observations in a paragraph.

2. Conduct research on the different ethnic groups in your city over the past 20 years. Map out and list three ways the congregation of these groups changed. Would you identify the change as "invasion and succession"? Has the infrastructure around these groups changed? Explain your observations in a paragraph.

3. Create a list of the various modes of transportation that are commonly available in your community. List three ways that the transportation shapes your city or town. Identify two major problems with the current transportation configuration of your city (*Hint:* Compare how long it takes a person to get from point A to B using various modes of transportation.) Think about whether certain types of transportation are facilitated more than others? In other words, are bus routes widespread and frequent? Are there separate streets for bicycles? Do you have high occupancy vehicle lanes for carpooling? Also consider the terrain itself: Do you have elevation or natural landscape considerations (mountains, water ways) to negotiate getting from point A to B? On a piece of paper map out how these problems could be corrected.

UNPLUGGED

1. Collect a week's worth of local newspapers and review the coverage of urban problems. What kinds of problems in what kinds of communities are covered? Compile a list of the different categories of problems, and then carefully analyze the content of the week's coverage, calculating the number of column inches devoted to each category. Compare how issues in different communities are reported and supported by local officials. Are some problems considered "fixable" and others just part of life in that community? How empowered are different communities in your city to fix these issues?

2. On a tracing-paper overlay of a street map of your town or city, plot the distribution of houses and apartments for sale or rent in different price brackets. (You can obtain the information from the real estate pages of your city's local newspaper; in smaller cities you may have to gather data from several issues of the paper—your local library will likely have back issues.) Explain the spatial distributions that you observe and monitor. Have any of these areas recently undergone gentrification? What do you notice about those areas?

3. Most cities consist of "ordinary" cityscapes that are strongly evocative because they are widely understood as being a particular kind of place. Write a brief essay (500 words, or two double-spaced, typed pages) describing an "ordinary" cityscape with which you are familiar. What are its principal features, and how might it be considered typical of a particular kind of place? What makes this cityscape "ordinary" and is there anything that makes the cityscape unique compared to other "ordinary" cityscapes? Why might we have a notion of an "ordinary" cityscape in the first place?

DATA ANALYSIS

World Cities Culture Report

http://goo.gl/OKP7Oy

In this chapter, we have looked at urban design and landscape spaces around the world. To look at the impact of different factors on cities Compare the designs of three different cities around the globe by choosing 1) one world city; 2) one megacity; and 3) one former colonial city. Start by using *Google Maps* and pull up each of your cities. How do the aerial and street views compare on each space?

Secondly, go to each of those city's websites and do an image search for each city you choose. As you pull up the data and images, answer these questions:

1. What are the main components of the urban design (*Hint:* this includes buildings, public space, streets, transport, and landscape)?

2. What are the transportation networks and flows of people in and out of each city to other cities and rural areas?

3. What are the religious traditions (Islamic, Buddhist, Christian, etc.); historical events (like wars, fires or natural disasters); and environmental space (mountains, beaches or deserts) of each city? How do those aspects influence urban design?

4. Are any of your cities connected to each other—economically, culturally, historically?

5. Which of your cities is incorporating green policies in its design either in infrastructure, transportation, architecture, urban gardening?

Finally, pull up the 2013 report, *World Cities Culture Report* (http://www.worldcitiescultureforum.com/sites/all/themes/wccr/assets/pdfs/WCCR2013_low.pdf)

6. What is the importance of culture to each of these 18 cities (Hint: Read the executive summary, pp. 8-11)?

7. Choose three of these 18 cities from different parts of the world. What makes them similar and what makes them different in their use of culture?

8. Based on your overall analysis, what do you think are the most significant influences (social, cultural, economic, religious, environmental) in the cities—and why?

MasteringGeography™

Looking for additional review and test prep materials? Visit the Study Area in MasteringGeography™ to enhance your geographic literacy, spatial reasoning skills, and understanding of this chapter's content by accessing a variety of resources, including **MapMaster** interactive maps, Videos, *In the News* RSS feeds, flashcards, web links, self-study quizzes, and an eText version of *Human Geography*.

Glossary

A

absolute space: a mathematical space described through points, lines, areas, planes, and configurations whose relationships can be fixed precisely through mathematical reasoning.

accessibility: the opportunity for contact or interaction from a given point or location in relation to other locations.

acid rain: the wet deposition of acids upon Earth created by the natural cleansing properties of the atmosphere.

actor-network theory: an orientation that views the world as composed of "heterogeneous things," including humans and nonhumans and objects.

affect: emotions that are embodied reactions to the social and physical environment.

age-sex pyramid: a representation of the population based on its composition according to age and sex.

agglomeration diseconomies: the negative economic effects of urbanization and the local concentration of industry.

agglomeration effects: interdependencies associated with various kinds of economic linkages, including the cost advantages that accrue to individual firms because of their location among functionally related activities.

agrarian: referring to the culture of agricultural communities and the type of tenure system that determines access to land and the kind of cultivation practices employed there.

agribusiness: a set of economic and political relationships that organizes agro-food production from the development of seeds to the retailing and consumption of the agricultural product.

agricultural density: ratio between the number of agriculturists per unit of arable land and a specific area.

agricultural industrialization: process whereby the farm has moved from being the centerpiece of agricultural production to becoming one part of an integrated string of vertically organized industrial processes including production, storage, processing, distribution, marketing, and retailing.

agriculture: a science, art, and business directed at the cultivation of crops and the raising of livestock for sustenance and profit.

amenity migration: a form a migration in which the migrant seeks not necessarily employment, but cultural, environmental, or social benefits in a new country or city.

ancillary industries: industries that manufacture parts and components to be used by larger industries.

anthropocene: the modern geological era during which humans have dramatically affected the environment.

aquaculture: the cultivation of fish and shellfish under controlled conditions, usually in coastal lagoons.

arithmetic density (crude density): total number of people divided by the total land area.

autarky: an economic policy or situation in which a nation is independent of international trade and not reliant upon imported goods.

azimuthal projection: a map projection on which compass directions are correct only from one central point.

B

baby boom: population of individuals born between the years 1946 and 1964.

backward linkages: develop as new firms arrive to provide the growing industry with components, supplies, specialized services, or facilities.

backwash effects: the negative impacts on a region (or regions) of the economic growth of some other region.

Beaux Arts: a style of urban design that sought to combine the best elements of all of the classic architectural styles.

biofuels: renewable fuels derived from biological materials that can be regenerated.

biomass: fuels made from biological material from living or recently living organisms that include wood, waste, gas, and alcohol fuels.

biometric census: a census in which the government photographs and fingerprints individuals to create a national database.

biopharming: an application of biotechnology in which genes from other life forms (plant, animal, fungal, bacterial, or human) are inserted into a host plant.

biopolitics: the extension of state power over the physical and political bodies of a population.

bioprospecting: the search for plant and animal species that may yield medicinal drugs and other commercially valuable compounds.

Biorevolution: the genetic engineering of plants and animals with the potential to exceed the productivity of the Green Revolution.

biotechnology: technique that uses living organisms (or parts of organisms) to make or modify products, to improve plants and animals, or to develop microorganisms for specific uses.

bioterrorism: deliberate use of microorganisms or toxins from living organisms to induce death or disease.

Blue Revolution: the introduction of new production techniques, processing technology, infrastructure, and larger, motorized boats as well as the application of transgenics into peripheral country fisheries.

Borlaug hypothesis: restricting crop usage to traditional low-yield methods (such as organic farming) in the face of rising global food demand would require either the world population to decrease or the further conversion of forest land into cropland.

brownfield sites: former industrial or commercial land where future use is affected by real or perceived environmental contamination.

C

capitalism: a form of economic and social organization characterized by the profit motive and the control of the means of production, distribution, and the exchange of goods by private ownership.

cargo cults: involve the belief that certain ritualistic acts will lead to a bestowing of material wealth cults.

carrying capacity: the maximum number of users that can be sustained, over the long term, by a given set of natural resources.

cartography: the body of practical and theoretical knowledge about making distinctive visual representations of Earth's surface in the form of maps.

census: a count of the number of people in a country, region, or city

central business district (CBD): the central nucleus of commercial land uses in a city.

central cities: the original, core jurisdictions of metropolitan areas.

central place: a settlement in which certain products and services are available to consumers.

central place theory: a theory that seeks to explain the relative size and spacing of towns and cities as a function of people's shopping behavior.

centrality: the functional dominance of cities within an urban system.

chemical farming: application of synthetic fertilizers to the soil—and herbicides, fungicides, and pesticides to crops—in order to enhance yields.

children's geographies: the spaces and places of the lives of youth and children disability

children's rights: the fundamental right of children to life, liberty, education, and health care codified in 1989 by the United Nations Convention on the Rights of the Child.

citizenship: a category of belonging to a nation-state that includes civil, political, and social rights.

climate change: any significant change in measures of climate (such as temperature, precipitation, or wind) lasting for an extended period (decades or longer).

cognitive distance: the distance that people perceive to exist in a given situation.

cognitive images (mental maps): psychological representations of locations that are made up from people's individual ideas and impressions of these locations.

cognitive space: space defined and measured in terms of the nature and degree of people's values, feelings, beliefs, and perceptions about locations, districts, and regions.

cohort: a group of individuals who share a common temporal demographic experience.

Cold War: the state of heightened military and political tension as well as economic competition between the former Soviet Union and its satellite states and the United States and its allies.

colonial city: a city that was deliberately established or developed as an administrative or commercial center by colonial or imperial powers.

colonialism: the establishment and maintenance of political and legal domination by a state over a separate and alien society.

colonization: the physical settlement of a new territory of people from a colonizing state.

Columbian Exchange: interaction between the Old World—originating with the voyages of Columbus—and the New World.

commercial agriculture: farming primarily for sale, not direct consumption.

commodity chain: network of labor and production processes beginning with the extraction or production of raw materials and ending with the delivery of a finished commodity.

comparative advantage: principle whereby places and regions specialize in activities for which they have the greatest advantage in productivity relative to other regions—or for which they have the least disadvantage.

confederation: a group of states united for a common purpose.

conformal projection: a map projection on which compass bearings are rendered accurately.

conglomerate corporations: companies that have diversified into various economic activities, usually through a process of mergers and acquisitions.

congregation: the territorial and residential clustering of specific groups or subgroups of people.

conservation: the view that natural resources should be used wisely and that society's effects on the natural world should represent stewardship and not exploitation.

contract farming: an agreement between farmers and processing and/or marketing firms for the production, supply, and purchase of agricultural products—from beef, cotton, and flowers to milk, poultry, and vegetables.

conventional farming: approach that uses chemicals in the form of plant protectants and fertilizers, or intensive, hormone-based practices in breeding and raising animals.

core regions: regions that dominate trade, control the most advanced technologies, and have high levels of productivity within diversified economies.

corruption: any abuse of a position of trust (in either the public or private sector) gain an unfair advantage.

cosmopolitanism: an intellectual and aesthetic openness toward divergent experiences, images, and products from different cultures.

cost/price squeeze: the simultaneous decrease in selling prices and rise in production costs that reduce a business's profit margin.

counterurbanization: the net loss of population from cities to smaller towns and rural areas.

creative destruction: the withdrawal of investments from activities (and regions) that yield low rates of profit in order to reinvest in new activities (and new places).

crop rotation: method of maintaining soil fertility in which the fields under cultivation remain the same but the crop being planted is changed.

crude birthrate (CBR): ratio of the number of live births in a single year for every thousand people in the population.

crude death rate (CDR): the number of deaths in a single year for every thousand people in the population.

crude density (arithmetic density): total number of people divided by the total land area.

cultural complex: combination of traits characteristic of a particular group.

cultural ecology: study of the relationship between a cultural group and its natural environment.

cultural geography: how space, place, and landscape shape culture at the same time that culture shapes space, place, and landscape.

cultural hearths: the geographic origins or sources of innovations, ideas, or ideologies.

cultural landscape: a characteristic and tangible outcome of the complex interactions between a human group and a natural environment.

cultural nationalism: an effort to protect regional and national cultures from the homogenizing impacts of globalization, especially from the penetrating influence of U.S. culture.

cultural region: the areas within which a particular cultural system prevails.

cultural space: the space of people with common ties, described through the places, territories, and settings whose attributes carry special meaning for particular groups map scale.

cultural system: a collection of interacting elements that taken together shape a group's collective identity.

cultural trait: a single aspect of the complex of routine practices that constitute a particular cultural group.

culture: a shared set of meanings that are lived through the material and symbolic practices of everyday life.

cumulative causation: a spiral buildup of advantages that occurs in specific geographic settings as a result of the development of external economies, agglomeration effects, and localization economies.

cuneiform: a writing system named for the wedge shape of its letters.

cycle of poverty: the transmission of poverty and deprivation from one generation to another through a combination of domestic circumstances and local, neighborhood conditions.

D

decolonization: the acquisition by colonized peoples of control over their own territory.

deep ecology: approach to nature revolving around two key components: self-realization and biospherical egalitarianism.

deforestation: the removal of trees from a forested area without adequate replanting.

deindustrialization: a relative decline in industrial employment in core regions.

democratic rule: a system in which public policies and officials are directly chosen by popular vote.

demographic collapse: phenomenon of near genocide of native populations.

demographic transition: replacement of high birth and death rates by low birth and death rates.

demography: the study of the characteristics of human populations.

dependency: high level of reliance by a country on foreign enterprises, investment, or technology.

dependency ratio: measure of the economic impact of the young and old on the more economically productive members of the population.

derelict landscapes: landscapes that have experienced abandonment, misuse, disinvestment, or vandalism.

desertification: the degradation of land cover and damage to the soil and water in grasslands and arid and semiarid lands.

dialects: regional variations in standard languages.

diaspora: spatial dispersion of a previously homogeneous group.

digital divide: inequality of access to telecommunications and information technology, particularly the Internet.

discourse: institutionalized ways of constituting knowledge.

distance-decay function: the rate at which a particular activity or process diminishes with increasing distance.

division of labor: the specialization of different people, regions, or countries in particular kinds of economic activities.

domino theory: the theory that if one country in a region chooses or is forced to accept a communist political and economic system, then neighboring countries would be irresistibly susceptible to communism.

double cropping: practice used in the milder climates whereby intensive subsistence fields are planted and harvested more than once a year.

doubling time: measure of how long it will take the population of an area to grow to twice its current size.

dualism: the juxtaposition in geographic space of the formal and informal sectors of the economy.

E

East/West divide: communist and noncommunist countries, respectively.

ecofeminism: view that patriarchal ideology is at the center of our present environmental malaise.

ecological footprint: measure of the human pressures on the natural environment from the consumption of renewable resources and the production of pollution indicating how much space a population needs compared to what is available.

ecological imperialism: introduction of exotic plants and animals into new ecosystems.

eco-migration: population movement caused by the degradation of land and essential natural resources.

economies of scale: cost advantages to manufacturers that accrue from high-volume production, since the average cost of production falls with increasing output.

ecosystem: community of different species interacting with each other and with the larger physical environment that surrounds it.

ecotheology: the view that it is necessary to address the current environmental crisis through belief systems that will overcome the inadequacies of humanly created institutions.

edge cities: nodal concentrations of shopping and office space situated on the outer fringes of metropolitan areas, typically near major highway intersections.

elasticity of demand: degree to which levels of demand for a product or service change in response to changes in price.

electoral college: a unique political-geographic body that the United States possesses, composed of a specified number of delegates allocated to each state based on that state's population as of the most recent official.

emigration: move from a particular location.

environmental determinism: doctrine holding that human activities are controlled by the environment.

environmental ethics: philosophical perspective on nature that prescribes moral principles as guidance for our treatment of it.

environmental justice: movement reflecting a growing political consciousness, largely among the world's poor, that their immediate environs are far more toxic than those in wealthier neighborhoods.

equal-area (equivalent) projection: map projection that portrays areas on Earth's surface in their true proportions.

equidistant projection: map projection that allows distance to be represented as accurately as possible.

established churches: recognized by law as the official church of the state.

ethnicity: socially created system of rules about who belongs and who does not belong to a particular group based upon actual or perceived commonality.

ethnocentrism: attitude that one's own race and culture are superior to others'.

ethology: scientific study of the formation and evolution of human customs and beliefs.

export-processing zones (EPZs): small areas within which especially favorable investment and trading conditions are created by governments in order to attract export-oriented industries.

external arena: regions of the world not yet absorbed into the modern world system.

external economies: cost savings that result from circumstances beyond a firm's own organization and methods of production.

F

famine: acute starvation associated with a sharp increase in mortality.

fast food: edibles that can be prepared and served very quickly, sold in a restaurant, and served to customers in packaged form.

federal state: form of government in which power is allocated to units of local government within the country.

feudalism: a rigid, rurally oriented form of economic and social organization based on the communal chiefdoms of Germanic tribes that had invaded the disintegrating Roman empire.

fiscal squeeze: increasing limitations on city revenues, combined with increasing demands for expenditure.

flexible production systems: ability of manufacturers to shift quickly and efficiently from one level of output to another, or from one product configuration to another.

folk culture: traditional practices of small groups, especially rural people with a simple lifestyle who are seen to be homogeneous in their belief systems and practices.

food chain: five central and connected sectors (inputs, production, product processing, distribution, and consumption) with four contextual elements acting as external mediating forces (the state, international trade, the physical environment, and credit and finance).

food desert: a geographic area where access to affordable and nutritious food is highly limited, especially for individuals without automobiles.

food justice: enabling communities to enact the principles just mentioned, which is simply to be able to grow, eat, and sell healthy food and care for the well-being of the local ecosystem in culturally appropriate ways.

food manufacturing: adding value to agricultural products through a range of treatments—such as processing, canning, refining, packing, and packaging—that occur off the farm and before products reach the market.

food miles: the distance that food travels from the farm to the consumer.

food regime: specific set of links that exists among food production and consumption and capital investment and accumulation opportunities.

food security: assured access by a person, household, or even a country to enough food at all times to ensure active and healthy lives.

food shed: Local food is usually also organically grown and its designation as local means that it is produced within a fairly limited distance from where it is consumed, an area known as a food shed.

food sovereignty: right of peoples, communities, and countries to define their own agricultural, labor, fishing, food, and land policies that are ecologically, socially, economically, and culturally appropriate to their unique circumstances.

food supply chain: a special type of commodity chain composed of five central and connected sectors with four contextual elements acting as external mediating forces.

forced migration: movement of an individual against his or her will.

Fordism: principles for mass production based on assembly-line techniques, scientific management, mass consumption based on higher wages, and sophisticated advertising techniques.

foreign direct investment: total of overseas business investments made by private companies.

forward linkages: develop as new firms arrive to take the finished products of the growing industry and use them in their own processing, assembly, finishing, packaging, or distribution operations.

friction of distance: deterrent or inhibiting effect of distance on human activity.

functional illiteracy: an individual's reading and writing skills are inadequate to manage daily living or hold down a job that requires reading skills beyond a basic level.

functional regions: regions with some variability in certain attributes but with an overall coherence to the structure and dynamics of economic, political, and social organization.

fusion language: a language that is influenced by so many other languages.

G

gateway city: serves as a link between one country or region and others because of its physical situation.

gender: social differences between men and women rather than the anatomical differences that are related to sex.

genetically modified organism (GMO): any organism that has had its DNA modified in a laboratory rather than through cross-pollination or other forms of evolution.

genre de vie: functionally organized way of life that is seen to be characteristic of a particular cultural group.

gentrification: invasion of older, centrally located, working-class neighborhoods by higher-income households seeking the character and convenience of less expensive and well-located residences.

geodemographic analysis: practice of assessing the location and composition of particular populations.

geodemographic research: study of census data and commercial data (such as sales data and property records) about the populations of small districts to create profiles of those populations for market research.

geographic information system (GIS): organized collection of computer hardware, software, and geographic data that is designed to capture, store, update, manipulate, and display geographically referenced information.

geographical imagination: capacity to understand changing patterns, changing processes, and changing relationships among people, places, and regions.

geographical path dependence: historical relationship between the present activities associated with a place and the past experiences of that place.

geopolitics: state's power to control space or territory and shape the foreign policy of individual states and international political relations.

gerrymandering: practice of redistricting for partisan purposes.

ghetto: an area of a city inhabited by a minority group, sometimes by choice but more often as a result of social, legal, or economic discrimination.

global change: combination of political, economic, social, historical, and environmental problems at the world scale.

global civil society: set of institutions, organizations, and behaviors situated between the state, business world, and family, including voluntary and nonprofit organizations, philanthropic institutions, and social and political movements.

Global Positioning System (GPS): system of satellites that orbit Earth on precisely predictable paths, broadcasting highly accurate time and locational information.

globalization: increasing interconnectedness of different parts of the world through common processes of economic, environmental, political, and cultural change.

globalized agriculture: system of food production increasingly dependent upon an economy and set of regulatory practices that are global in scope and organization.

governance: refers to the norms, rules and laws that are invoked to regulate a people or a state.

government: the body or group of persons who run the administration of a country.

graffiti: refers to the inscriptions—largely figure drawings—scratched on walls in ancient Rome.

Green Revolution: export of a technological package of fertilizers and high-yielding seeds from the core to the periphery to increase global agricultural productivity.

greenhouse gases (GHG): any gas that absorbs infrared radiation in the atmosphere, including, but not limited to, water vapor, carbon dioxide (CO_2), methane (CH_4), and nitrous oxide (N_2O).

greening: adding biomass, including grasses and trees, through rainfall to an area that was formerly a desert.

gross domestic product (GDP): estimate of the total value of all materials, foodstuffs, goods, and services produced by a country in a particular year.

gross migration: total number of migrants moving into and out of a place, region, or country.

gross national income (GNI): similar to GDP, but also includes the value of income from abroad.

growth poles: economic activities that are deliberately organized around one or more high-growth industries.

guest workers: individuals who migrate temporarily to take up jobs in other countries.

H

hajj: religious pilgrimage.

hearth areas: geographic settings where new practices have developed and from which they have subsequently spread.

health density: the ratio of the number of physicians to the total population.

health geography: starts with health as an initial condition and focuses on the dynamic relationship between health, people, and place.

hinterland: sphere of economic influence of a town or city.

hegemony: domination over the world economy exercised by one national state in a particular historical epoch through a combination of economic, military, financial, and cultural means.

historical geography: geography of the past.

human geography: study of the spatial organization of human activity and of people's relationships with their environments.

human rights: people's individual rights to justice, freedom, and equality, considered by most societies to belong automatically to all people.

humanistic approach: point of view that places the individual—especially individual values, meaning systems, intentions, and conscious acts—at the center of analysis.

hunting and gathering: activities whereby people feed themselves through killing wild animals and fish and gathering fruits, roots, nuts, and other edible plants to sustain themselves.

hybridity: a mixing of different types; in cultural geography, hybridity is most often associated with movements across a binary of, for instance, the racial categories of black and white such that identities are more multiple and ambivalent.

I

identity: sense that people make of themselves through their subjective feelings based on their everyday experiences and wider social relations.

immigration: move to another location.

imperialism: extension of the power of a nation through direct or indirect control of the economic and political life of other territories.

import substitution: process by which domestic producers provide goods or services that formerly were bought from foreign producers.

infant mortality rate: annual number of deaths of infants under 1 year of age compared to the total number of live births for that same year.

inflation: increased supply of printed currency that leads to higher prices and international financial differentials.

informal sector: economic activities that take place beyond official record, not subject to formalized systems of regulation or remuneration.

infrastructure (or fixed social capital): underlying framework of services and amenities needed to facilitate productive activity.

initial advantage: critical importance of an early start in economic development; a special case of external economies.

intensive subsistence agriculture: practice that involves the effective and efficient use—usually through a considerable expenditure of human labor and application of fertilizer—of a small parcel of land in order to maximize crop yield.

internal migration: move within a particular country or region.

internally displaced persons (IDPs): individuals who are uprooted within the boundaries of their own country because of conflict or human rights abuse.

international division of labor: specialization, by countries, in particular products for export.

international migration: move from one country to another.

international organization: group that includes two or more states seeking political and/or economic cooperation with each other.

international regime: orientation of contemporary politics around the international arena instead of the national.

intersectionality: a recognition of the ways that different forms or systems of oppression, domination, or discrimination overlap.

intersubjectivity: shared meanings among people, derived from their lived experience of everyday practice.

intertillage: practice of mixing different seeds and seedlings in the same swidden.

invasion and succession: process of neighborhood change whereby one social or ethnic group succeeds another.

irredentism: assertion by the government of a country that a minority living outside its formal borders belongs to it historically and culturally.

Islam: Arabic term that means submission to God's will.

Islamism: anticolonial, anti-imperial, and generally anticore political movement.

isotropic surface: hypothetical, uniform plain that is flat and has no variations in its physical attributes.

J

jihad: sacred struggle.

just-in-time production: manufacturing process in which daily or hourly delivery schedules of materials allow for minimal or zero inventories.

K

kinship: relationship based on blood, marriage, or adoption.

L

land reform: redistribution of land by the state with a goal of increasing productivity and reducing social unrest.

landscape as text: idea that landscapes can be read and written by groups and individuals.

language: communicating ideas or feelings by means of a conventionalized system of signs, gestures, marks, or articulate vocal sounds.

language branch: collection of languages that possess a definite common origin but have split into individual languages.

language family: collection of individual languages believed to be related in their prehistorical origin.

language group: collection of several individual languages that are part of a language branch, share a common origin, and have similar grammar and vocabulary.

language hearths: a subset of cultural hearths; they are the source areas of languages.

language tree: a representation of the relationships of languages to each other.

latitude: angular distance of a point on Earth's surface, measured north or south from the equator, which is 0°.

law of diminishing returns: tendency for productivity to decline, after a certain point, with the continued application of capital and/or labor to a given resource base.

leadership cycles: periods of international power established by individual states through economic, political, and military competition.

life expectancy: average number of years a newborn infant can expect to live.

lifeworld: taken-for-granted pattern and context for everyday living through which people conduct their lives.

linguistic drift: process of random change inherent to all languages.

linguistic weathering: sort of "wearing out" of words.

literacy: at a very basic level, the ability to read and write.

local food: food that is organically grown and produced within a fairly limited distance from where it is consumed.

localization economies: cost savings that accrue to particular industries as a result of clustering together at a specific location.

longitude: angular distance of a point on Earth's surface, measured east or west from the prime meridian (the line that passes through both poles and through Greenwich, England, and that has the value of 0°).

M

malnutrition: the condition that develops when the body does not get the right amount of the vitamins, minerals, and other nutrients it needs to maintain healthy tissues and organ function.

map projection: systematic rendering on a flat surface of the geographic coordinates of the features found on Earth's surface.

materialism: a theory which emphasizes that the material world—its objects and nonhuman entities—is at least partly separate from humans and possesses the power to affect humans; materialism attempts to understand the ways that specific properties of material things affect the interactions between humans and nonhuman entities.

mechanization: replacement of human farm labor with machines.

medical geography: subarea of the discipline that specializes in understanding the spatial aspects of health and illness.

megacity: very large city characterized by both primacy and high centrality within its national economy.

metroburbia: suburban and exurban areas where residential settings are thoroughly interspersed with office employment and high-end retailing.

middle cohort: members of the population 15 to 64 years of age who are considered economically active and productive.

migration: move beyond the same political jurisdiction, involving a change of residence—either as emigration or as immigration.

minisystem: society with a single cultural base and a reciprocal social economy.

minority groups: population subgroups that are seen—or that see themselves—as somehow different from the general population.

mobility: ability to move, either permanently or temporarily.

Modern movement: architectural movement based on the idea that buildings and cities should be designed and run like machines.

modernity: forward-looking view of the world that emphasizes reason, scientific rationality, creativity, novelty, and progress.

mother tongue: a language that a person has learned from birth or the first few years of life.

multiple-nucleii model: model of urbanization proposed by Chauncy Harris and Edward Ullman in which decentralized nodes of different categories of land use end up in many different configurations, depending on local conditions.

mutually intelligible: When speakers of different but related varieties of languages are able to understand each other, the languages are said to be mutually intelligible.

Muslim: member of the Islamic community of believers whose duty is obedience and submission to the will of God.

N

nation: group of people often sharing common elements of culture, such as religion or language or a history or political identity.

nationalism: feeling of belonging to a nation as well as the belief that a nation has a natural right to determine its own affairs.

nation-state: ideal form consisting of a homogeneous group of people governed by their own state.

natural decrease: difference between CDR and CBR, which is the deficit of births relative to deaths.

natural increase: difference between the CBR and CDR, which is the surplus of births relative to deaths.

nature: social creation as well as the physical universe that includes human beings.

neocolonialism: economic and political strategies by which powerful states in core economies indirectly maintain or extend their influence over other areas or people.

neo-Fordism: economic principles in which the logic of mass production coupled with mass consumption is modified by the addition of more flexible production, distribution, and marketing systems.

neoliberal policies: economic policies that are predicated on a minimalist role for the state, assuming the desirability of free markets as the ideal condition not only for economic organization but also for political and social life.

neoliberalism: reduction in the role and budget of government, including reduced subsidies and the privatization of formerly publicly owned and operated concerns, such as utilities.

net migration: gain or loss in the total population of a particular area as a result of migration.

new world order: triumph of capitalism over communism, wherein the United States becomes the world's only superpower and therefore its policing force.

newly industrializing countries (NICs): countries formerly peripheral within the world system that have acquired a significant industrial sector, usually through foreign direct investment.

non-representational theory: an approach to human (and non-human) practices that explores how they are performed and what are their effects such as how music produces in humans both remembering and forgetting.

nontraditional agricultural exports (NTAEs): new export crops that contrast with traditional exports.

North/South divide: differentiation made between the colonizing states of the Northern Hemisphere and the formerly colonized states of the Southern Hemisphere.

nutritional density: ratio between the total population and the amount of land under cultivation in a given unit of area.

O

official language: the language in which the government, including the courts, the legislature, and the administrative branch, conducts its business.

offshore financial centers: islands or microstates that have become a specialized node in the geography of worldwide financial flows.

old-age cohort: members of the population 65 years of age and older who are considered beyond their economically active and productive years.

ordinary landscapes (vernacular landscapes): everyday landscapes that people create in the course of their lives.

organic farming: farming or animal husbandry done without commercial fertilizers, synthetic pesticides, or growth hormones.

Orientalism: discourse that positions the West as culturally superior to the East.

organized religions: When belief systems and associated rituals are systematically arranged and formally established, they are referred to as **organized religions**.

overurbanization: condition in which cities grow more rapidly than the jobs and housing they can sustain.

P

pandemic: an epidemic that spreads rapidly around the world with high rates of illness and death.

pastoralism: subsistence activity that involves the breeding and herding of animals to satisfy the human needs of food, shelter, and clothing.

peripheral regions: regions with undeveloped or narrowly specialized economies with low levels of productivity.

physical geography: subarea of the discipline that studies Earth's natural processes and their outcomes.

place: specific geographic setting with distinctive physical, social, and cultural attributes.

place name: one way that language and geography come together.

plantation: large landholding that usually specializes in the production of one particular crop for market.

pictograms: pictures meant to represent words.

political ecology: approach to cultural geography that studies humans in their environment through the relationships of patterns of resource use to political and economic forces.

popular culture: practices and meaning systems produced by large groups of people whose norms and tastes are often heterogeneous and change frequently, often in response to commercial products.

population policy: official government policy designed to effect any or all of several objectives, including the size, composition, and distribution of population.

postmodernity: view of the world that emphasizes an openness to a range of perspectives in social inquiry, artistic expression, and political empowerment.

preservation: approach to nature advocating that certain habitats, species, and resources should remain off-limits to human use, regardless of whether the use maintains or depletes the resource in question.

primacy: condition in which the population of the largest city in an urban system is disproportionately large in relation to the second- and third-largest cities.

primary activities: economic activities that are concerned directly with natural resources of any kind.

producer services: services that enhance the productivity or efficiency of other firms' activities or that enable them to maintain specialized roles.

proto-languages: The form of the language tree is a node-link diagram that contains branch points, or nodes, from which the daughter languages—offspring of older languages—descend by different links. The nodes are **proto-languages**, also known as *common languages*.

proxemics: study of the social and cultural meanings that people give to personal space.

public–private partnerships: Because of the significant costs and risks involved, urban regeneration strategies are usually undertaken in the form of **public–private partnerships** between city governments and real estate developers.

pull factors: forces of attraction that influence migrants to move to a particular location.

purchasing power parity (PPP): measures how much of a common "market basket" of goods and services each currency can purchase locally, including goods and services that are not traded internationally.

push factors: events and conditions that impel an individual to move from a location.

Q

quaternary activities: economic activities that deal with the handling and processing of knowledge and information.

R

race: problematic classification of human beings based on skin color and other physical characteristics.

racialization: practice of categorizing people according to race or of imposing a racial character or context.

rank-size rule: statistical regularity in size distributions of cities and regions.

reapportionment: process of allocating electoral seats to geographical areas.

redistricting: defining and redefining of territorial district boundaries.

redlining: practice whereby lending institutions delimit "bad-risk" neighborhoods on a city map and then use the map as the basis for determining loans.

refugee: individual who crosses national boundaries to seek safety and asylum.

regeneration: involves the physical redevelopment of land where the existing buildings are no longer useful or profitable.

region: larger-sized territory that encompasses many places, all or most of which share similar attributes in comparison with the attributes of places elsewhere.

regional geography: study of the ways unique combinations of environmental and human factors produce territories with distinctive landscapes and cultural attributes.

regionalism: feeling of collective identity based on a population's politico-territorial identification within a state or across state boundaries.

regionalization: classification of individual places or areal units.

religion: belief system and set of practices that recognize the existence of a power higher than humans.

remote sensing: collection of information about parts of Earth's surface by means of aerial photography or satellite imagery designed to record data on visible, infrared, and microwave sensor systems.

resilience: the ability of people, organizations, or systems to prepare for, respond, recover from and thrive in the face of hazards.

reurbanization: growth of population in metropolitan central cores, following a period of absolute or relative decline in population.

risk society: contemporary societies in which politics is increasingly about avoiding hazards.

rites of passage: ceremonial acts, customs, practices, or procedures that recognize key transitions in human life, such as birth, menstruation, and other markers of adulthood such as marriage.

romanticism: philosophy that emphasizes interdependence and relatedness between humans and nature.

S

sacred space: area recognized by individuals or groups as worthy of special attention as a site of special religious experiences or events.

sacred spaces: physical settings recognized by individuals or groups as worthy of special attention because they are the sites of special religious experiences and events.

secondary activities: economic activities that process, transform, fabricate, or assemble the raw materials derived from primary activities or that reassemble, refinish, or package manufactured goods.

sectionalism: extreme devotion to local interests and customs.

segregation: spatial separation of specific population subgroups within a wider population.

self-determination: right of a group with a distinctive politico-territorial identity to determine its own destiny, at least in part, through the control of its own territory.

semiotics: practice of writing and reading signs.

semiperipheral regions: regions that are able to exploit peripheral regions but are themselves exploited and dominated by core regions.

sense of place: feelings evoked among people as a result of the experiences and memories that they associate with a place and the symbolism that they attach to it.

sex: the biological and physiological characteristics that differentiate males and females at birth, based on bodily characteristics such as anatomy, chromosomes, and hormones

sexuality: set of practices and identities that a given culture considers related to each other and to those things it considers sexual acts and desires.

shifting cultivation: system in which farmers aim to maintain soil fertility by rotating the fields within which cultivation occurs.

shock city: city that is seen as the embodiment of surprising and disturbing changes in economic, social, and cultural life.

site: physical attributes of a location—its terrain, its soil, vegetation, and water sources, for example.

situation: location of a place relative to other places and human activities.

slang: language that consists of nonstandard words and phrases and is a common occurrence among most languages.

slash-and-burn (swidden): system of cultivation in which plants are cropped close to the ground, left to dry for a period, and then ignited.

slow food: attempt to resist fast food by preserving the cultural cuisine and the associated food and farming of an ecoregion.

society: sum of the inventions, institutions, and relationships created and reproduced by human beings across particular places and times.

socioeconomic space: the space that can be described in terms of sites and situations, routes, regions, and distribution patterns. In these terms, spatial relationships are fixed through measures of time, cost, profit, and production, as well as through physical distance.

sovereignty: exercise of state power over people and territory, recognized by other states and codified by international law.

spatial analysis: study of geographic phenomena in terms of their arrangement as points, lines, areas, or surfaces on a map.

spatial diffusion: way that things spread through space and over time.

spatial interaction: movement and flows involving human activity.

spatial justice: fairness of the distribution of society's burdens and benefits, taking into account spatial variations in people's needs and in their contribution to the production of wealth and social well-being.

sprawl: the unplanned, *ad hoc* nature of most suburban development destroys millions of acres of wildlife habitat and agricultural land every year.

spread effects: positive impacts on a region (or regions) of the economic growth of some other region.

squatter settlements: residential developments that occur on land that is neither owned nor rented by its occupants.

standard language: *see* official language.

state: an independent political units with territorial boundaries that are internationally recognized by other states.

strategic alliances: commercial agreements between transnational corporations, usually involving shared technologies, marketing networks, market research, or product development.

subsistence agriculture: farming for direct consumption by the producers; not for sale.

suburbanization: growth of population along the fringes of large metropolitan areas.

supranational organizations: collections of individual states with a common goal that may be economic and/or political in nature.

sustainability: the interdependence of the economy, the environment, and social well-being.

sustainable development: vision of development that seeks a balance among economic growth, environmental impacts, and social equity.

swidden: land that is cleared using the slash-and-burn process and is ready for cultivation.

symbolic landscapes: representations of particular values or aspirations that the builders and financiers of those landscapes want to impart to a larger public.

T

technology: physical objects or artifacts, activities or processes, and knowledge or know-how.

technology systems: clusters of interrelated energy, transportation, and production technologies that dominate economic activity for several decades at a time.

terms of trade: ratio of prices at which exports and imports are exchanged.

territorial organization: system of government formally structured by area, not by social groups.

territoriality: specific attachment of individuals or peoples to a specific location or territory.

territory: delimited area over which a state exercises control and which is recognized by other states.

terrorism: threat or use of force to bring about political change.

tertiary activities: economic activities involving the sale and exchange of goods and services.

time-space convergence: rate at which places move closer together in travel or communication time or costs.

topological space: connections between, or connectivity of, particular points in space.

topophilia: emotions and meanings associated with particular places that have become significant to individuals.

total fertility rate (TFR): average number of children a woman will have throughout the years that demographers have identified as her childbearing years, approximately ages 15 through 49.

trade-off model: a modified urban land-use model that describes how poorer households, unable to afford the recurrent costs of transportation, trade off living space for accessibility to jobs and end up in high-density areas, at expensive locations, near their low-wage jobs.

trading blocs: groups of countries with formalized systems of trading agreements.

transcendentalism: philosophy in which a person attempts to rise above nature and the limitations of the body to the point where the spirit dominates the flesh.

transgender: the term that refers to a person whose self-identity does not conform to conventional notions of the male or female gender.

transhumance: movement of herds according to seasonal rhythms: warmer, lowland areas in the winter; cooler, highland areas in the summer.

transnational corporations: companies with investments and activities that span international boundaries and with subsidiary companies, factories, offices, or facilities in several countries.

transnational migrant: migrants who set up homes and/or work in more than one nation-state.

tribe: form of social identity created by groups who share a set of ideas about collective loyalty and political action.

U

underemployment: situation in which people work less than full-time even though they would prefer to work more hours.

undernutrition: inadequate intake of one or more nutrients and/or calories.

undocumented workers: those individuals who arrive in the United States without official entry visas and are considered by the government to be in the country illegally.

unitary state: form of government in which power is concentrated in the central government.

urban agriculture (peri-urban agriculture): establishment or performance of agricultural practices in or near an urban or citylike setting.

urban ecology: social and demographic composition of city districts and neighborhoods.

urban form: physical structure and organization of cities.

urban system: interdependent set of urban settlements within a specified region.

urbanism: way of life, attitudes, values, and patterns of behavior fostered by urban settings.

urbanization: increasing concentration of population into growing metropolitan areas.

urban realm: semiautonomous subregions that displaced the simple core-periphery relationship between city centers and their suburbs.

utility: usefulness of a specific place or location to a particular person or group.

V

vertical disintegration: evolution from large, functionally integrated firms within a given industry toward networks of specialized firms, subcontractors, and suppliers.

virgin soil epidemics: conditions in which the population at risk has no natural immunity or previous exposure to a disease within the lifetime of the oldest member of the group.

virtual water: water embedded in the production of the food and other things we consume.

vital records: information about births, deaths, marriages, divorces, and the incidence of certain infectious diseases.

vocal fry: the practice of speaking in the lowest voice register to produce a popping or creaky sound at a very low frequency.

voluntary migration: movement by an individual based on choice.

W

white privilege: advantages that accrue to white people beyond what is commonly experienced by people of color.

white supremacy: the belief that white people are superior to other races—sit at the heart of the racist practices described here.

world city: city in which a disproportionate part of the world's most important business is conducted.

world-empire: minisystems that have been absorbed into a common political system while retaining their fundamental cultural differences.

world music: musical genre defined largely in response to the sudden increase of non-English-language recordings released in the United Kingdom and the United States in the 1980s.

world-system: interdependent system of countries linked by economic and political competition.

Y

youth cohort: members of the population who are less than 15 years of age and generally considered to be too young to be fully active in the labor force.

Z

zone in transition: area of mixed commercial and residential land uses surrounding the CBD.

Photo Credits

Chapter 1 Pages 2–3: Bhaskar Krishnamurthy/Robert Harding World Imagery. Page 5 (t): Pidjoe/Getty Images. Page 5 (b): Heeb Christian/Prisma Bildagentur AG/Alamy. Page 6: Thomas Hartwell/AP Images. Page 7: Paul J. Fusco/Science Source. Page 14 (l): BasPhoto/Fotolia. Page 14 (r): Tony Watson/Alamy. Page 15: Stephen Finn/Fotolia. Page 22: LianeM/Shutterstock. Page 23 (t): Matthi/123RF. Page 23 (b): Mubus7/Shutterstock. Page 24 (t): Paul Knox. Page 24 (b): Fotomatador/Alamy. Page 25: Paul Knox. Page 26: 28/Ocean/Corbis.

Chapter 2 Pages 30–31: Yang Liu/Corbis. Page 34 (l): Vovez/Fotolia. Page 34 (r): Gabriele Maltinti/Fotolia. Page 35: Cristian Partenie Borda/123RF. Page 38: Dylan Martinez/Reuters. Page 40: Departure from Lisbon for Brazil, the East Indies and America. Illustration from "Americae Tertia Pars," Theodore de Bry (1592). Engraving. Bridgeman Art Library/Getty Images. Page 45: Pictorial Press, Ltd./Alamy. Page 53: Apple, Inc. Page 54: Katvic/Fotolia. Page 56: Lucas Jackson/Reuters. Page 57 (t): Artcphotos/Shutterstock. Page 57 (b): Jesse Allen/Suomi NPP VIIRS/Cooperative Institute for Meteorological Satellite Studies/Earth Observatory (CIMSS)/NASA. Page 59: Beawiharta/Landov.

Chapter 3 Pages 64–65: Imaginechina/AP Images. Page 66: Adam Davis/Epa/Newscom. Page 67: Manpreet Romana/Newscom. Page 70: Otto Stadler/Getty Images. Page 84: Stacy Walsh Rosenstock/Alamy. Page 92 (t): DBimages/Alamy. Page 92 (b): Jim West/Alamy. Page 93: Sunsinger/Fotolia. Page 95: Caro/Alamy. Page 97: Northwest Arctic Borough via The Anchorage Daily News/AP Images.

Chapter 4 Pages 106–107: Andrew Burton/Getty Images. Page 109: Reuters. Page 110: SFM Press Reporter/Alamy. Page 116: Pearson Education, Inc. Page 117: Lucaar/Fotolia. Page 118: Reinhard Krause/Reuters. Page 122: View of Tenochtitlan in the Lake of Mexico (Vista de Tenochtitlán en el Lago de México) (1964), Luis Covarrubias. Mural. Museo Nacional de Antropologia, Mexico City/Schalkwijk/Art Resource, New York. Page 125: Antonio Scorza/AFP/Getty Images. Page 127: Richard Shephard/Abacausa.com/Newscom. Page 130: Kim Jae-Hwan/AFP/Getty Images/Newscom. Page 133 (t): NASA. Page 133 (b): Xinhua Press/Corbis. Page 139: Oleg Znamenskiy/Fotolia. Page 140: Jan Wlodarczyk/Alamy. Page 141: Toshifumi Kitamura/AFP/Newscom. Page 143: Godard Space Flight Center/Ozone Hole Watch/NASA. Page 145: James Forte/National Geographic/Corbis. Page 147: Andy Wong/AP Images. Page 148: Expedition 35 Crew/ISS Crew Earth Observations/Image Science & Analysis Laboratory, Johnson Space Center/NASA.

Chapter 5 Pages 152–523: Stephen Shaver/ZUMA Press/Newscom. Page 154: Arne Dedert/dpa/Newscom. Page 155: Jeremy Sutton-Hibbert/Alamy. Page 157 (t): Courtesy of Mithun Photography. Page 157 (b): Gavriel Jecan/Terra/Corbis. Page 158 (tl): Pierre-Jean Durieu/123RF. 158 (tc): Anthony Asael/Danita Delimont/Alamy. Page 158 (tr): Omdim/Fotolia. Page 158 (bl): Denboma/Fotolia. Page 158 (bc): Santiago Llanquin/AP Images. Page 158 (br): Paul Liu/Fotolia. Page 159 (t): Eddie Gerald/Alamy. Page 159 (b): Olga Lipatova/Shutterstock. Page 161 (t): Walter Bibikow/Getty Images. Page 161 (b): Paul Harris/John Warburton-Lee Photography/Alamy. Page 162: SCFotos/Stuart Crump Visuals/Alamy. Page 163 (t): Randy Duchaine/Alamy. Page 163 (b): Dydia DeLyser. Page 164 (t): Martyn Evans/Alamy. Page 164 (b): Shin Young-kyun/Yonhap/AP Images. Page 167: Hand-colored plate by F. Courbain from *Fashion in Paris* by M. Loyd/Heinemann, London (1898). Robana/British Library Board/Art Resource, New York. Page 168: Atlantide Phototravel/Corbis. Page 170 (t): Kyodo/Newscom. Page 170 (b): Peter Schickert/Imagebroker/Alamy. Page 171: Will Burgess/Reuters. Page 172: Sframe/Fotolia. Page 173: Lisa Sciascia/Alamy. Page 174: Eric Audras-Onoky/Photononstop/Alamy. Page 175 (t): Beth A. Keiser/AP Images. Page 175 (b): Eric Lafforgue/Alamy. Page 176: Justin Kase z12z/Alamy. Page 178: Seyllou/AFP/Getty Images/Newscom. Page 179: Rachel Youdelman/Pearson Education, Inc. Page 180: Press Association/AP Images. Page 182: National Park Service.

Chapter 6 Pages 186–187: Khanm/Reuters. Page 188: Sanjeev Verma/Hindustan Times/Newscom. Page 196: Tim Roberts Photography/Shutterstock. Page 198: Paul Adams. Page 199: Hemis/Alamy. Page 203: Ivy Close Images/Alamy. Page 210 (t): Toby Melville/Reuters/Corbis. Page 210 (b): A.F. Kersting/AKG Images/Newscom. Page 214: Christopher Nicholson/Alamy. Page 215 (t): Heiti Paves/Alamy. Page 215 (b): Michael Freeman/Corbis. Page 216: Paul Harris/BWP Media/Atticus Images/Newscom. Page 217 (l): Insadco Photography/Alamy. Page 217 (c): Mai Chen/Alamy. Page 217 (r): Bob Kreisel/Alamy. Page 220 (t): Paul Thompson Images/Alamy. Page 220 (b): Iain McGillivray/Shutterstock. Page 221: Huci/Fotolia. Page 223: Boris Diakovsky/Shutterstock.

Chapter 7 Pages 228–229: Frederic Soltan/Corbis. Page 230: Bloomberg/Getty Images. Page 232: Chuck Pefley/Alamy. Page 234 (tl): Ian Paterson/Alamy. Page 234 (tr): Darrin Jenkins/Alamy. Page 234 (bl): Ian Gavan/Getty Images. Page 234 (br): Black Hand. Page 235: Kikisora/Fotolia. Page 240 (tl): Scaliger/Fotolia. Page 240 (tr): Paul Liu/Fotolia. Page 240 (cl): Paul Liu/Fotolia. Page 240 (cr): QQ7/Shutterstock. Page 240 (bl): Visions Of America LLC/123RF. Page 240 (br): David R. Frazier/Danita Delimont/Newscom. Page 241: Paul Knox. Page 242 (l): Heeb Christian/Prisma Bildagentur AG/Alamy. Page 242 (c): Audreylinh/Fotolia. Page 242 (r): ReinhardT/Fotolia. Page 242 (b): Heeb Christian/Prisma Bildagentur AG/

Alamy. Page 243 (t): Franck Fotos/Alamy. Page 243 (b): Alan Copson/Jon Arnold Images, Ltd./Alamy. Page 244 (l): Jon Arnold Images Ltd/Alamy. Page 244 (tr): John Harper/Getty Images. Page 244 (br): Tim Graham/Alamy. Page 245: *The Cornfield* (1826), John Constable. Oil on canvas, 143 × 122 cm. Presented by subscribers, including Wordsworth, Faraday and Sir William Beechey, 1837. National Gallery, London. World History Archive/Alamy. Page 246 (tl): Erin Paul Donovan/Alamy. Page 246 (tr): Krista Rossow/Getty Images. Page 246 (bl): Len Holsborg/Alamy. Page 246 (br): Connie J. Spinardi/Getty Images. Page 247: Briquet Nicolas/ABACA/Newscom. Page 249 (l): Bluered/CuboImages srl/Alamy. Page 249 (r): Heike Mayer. Page 250: Elizabeth Currid-Halkett/University of Southern California, Sarah Williams/Massachusetts Institute of Technology, and Gilad Ravid/Ben-Gurion University of the Negev. Page 251: Elizabeth Currid-Halkett/University of Southern California, Sarah Williams/Massachusetts Institute of Technology, and Gilad Ravid/Ben-Gurion University of the Negev. Page 253 (t): Nicolas Randall/Expuesto/Alamy. Page 253 (c): Santi Rodriguez/Fotolia. Page 253 (b): Tyrone Siu/Reuters/Corbis. Page 254: Directphoto Collection/Alamy.

Chapter 8 Pages 259–260: Pacific Press/Alamy. Page 262: Keith Srakocic/AP Images. Page 263: Richard Carey/Fotolia. Page 264: Pearson Education, Inc. Page 265 (r): Bloomua/Shutterstock. Page 265 (l): Paul Solski/Fotolia. Page 266: Konstantin Kokoshkin/Corbis. Page 267: Paul Knox. Page 269: SeongJoon Cho/Bloomberg/Getty Images. Page 270: Seanpavonephoto/Fotolia. Page 271 (t): Karl Johaentges/Alamy. Page 271 (b): Feng Li/Getty Images. Page 273: Erik De Castro/Reuters/Corbis. Page 274: Marco Vacca/Alamy. Page 275: Simon Rawles/Alamy. Page 281 (l): Rebecca Cook/Reuters. Page 281(r): Marco Secchi/Alamy. Page 282: Donvictorio/Shutterstock. Page 284: Richard Vogel/AP Images. Page 285: Benetton USA Corp. Page 286: Keith Dannemiller/Alamy. Page 288 (t): Tsering Topgyal/AP Images. Page 288 (b): Gautam Singh/AP Images. Page 289: Pcruciatti/Shutterstock. Page 292: Namas Bhojani/Bloomberg/Getty Images. Page 293: Sjon Heijenga/Buiten-Beeld/Alamy.

Chapter 9 Pages 298–299: Andrew Biraj/Reuters. Page 304: Rob Huibers/HollandseHoogte/Redux. Page 305 (t): V. Muthuraman/SuperStock/Alamy. Page 305 (b): Elvira Oomens/iStockphoto/Getty Images. Page 307: James P. Blair/National Geographic/Getty Images. Page 308: Rieger Bertrand/Hemis/Alamy. Page 311 (t): Imagebroker/Alamy. Page 311 (b): Hansenn/Fotolia. Page 312: Bill Meeks/AP Images. Page 313: Hypocritus/Fotolia. Page 317: Bullit Marquez/AP Images. Page 322: Collection10/GlowImages/Alamy. Page 324: The Lexicon of Sustainability. Page 333: ArtMediaPix/Alamy. Page 334: Peter Parks/AFP/Getty Images. Page 335: Mahesh Kumar/AP Images. Page 337: William P. Straeter/AP Images.

Chapter 10 Pages 342–343: Majdi Mohammed/AP Images. Page 346: Image Source/Alamy. Page 347 (t): Ted Wood/Aurora Photos/Alamy. Page 347 (b): Jennifer DeMonte/AP Images. Page 350: Steve Ringman/Seattle Times/KRT/Newscom. Page 351: David Hosking/Alamy. Page 353 (t): Leon Neal/AFP/Getty Images. Page 353 (b): GMPhoto/Alamy. Page 355: Christian Kober/Robert Harding World Imagery/Alamy. Page 359: Andy Lane/Alamy. Page 362: Jenny Matthews/Alamy. Page 365: Adalberto Roque/AFP/Getty Images. Page 371 (t): J. Bicking/Shutterstock. Page 371 (b): Laperruque/Alamy.

Chapter 11 Pages 392–393: Then Chih Wey/Xinhua News Agency/Newscom. Page 395: Earth Observatory/NASA. Page 396: Witold Skrypczak/Alamy. Page 400: Mary Evans Picture Library/The Image Works. Page 401 (t): Chad Ehlers/Alamy. Page 401 (b): James D. Morgan/Rex/AP Images. Page 402 (tl): Adrian Nunez/Shutterstock. Page 402 (tr): Alexandr/Fotolia. Page 402 (b): Mino21/Fotolia. Page 406: Huyangshu/Shutterstock. Page 408: James David Photo/Fotolia. Page 408 (b): Niday Picture Library/Alamy. Page 409: Frank Tozier/Alamy. Page 413: Marc Zakian/Alamy. Page 416 (t): Marcaux/Getty Images. Page 416 (b): Florian Kopp/Image Broker/Glow Images. Page 417: Amer Ghazzal/Demotix/Corbis. Page 418: Xaume Olleros/Alamy. Page 419 (t): Huang Guobao/Xinhua/Photoshot/Newscom. Page 419 (b): Pavel Losevsky/Fotolia. Page 421: Ben Birchall/AP Images. Page 422: Niels Quist/Alamy. Page 423: Dircinha Welter/Flickr Vision/Getty Images.

Chapter 12 Pages 428–429: Steve Vidler/Photoshot. Page 431: Spirit of America/Shutterstock. Page 435: Monica Almeida/The New York Times/Redux. Page 436: Trekandphoto/Fotolia. Page 437: Zuma Press, Inc./Alamy. Page 439 (l) Jim West/Alamy. Page 439 (r) Paul Warner/Getty Images. Page 440: Christopher Pillitz/Getty Images. Page 441: Paul Knox. Page 442: Walter Hahn/AFP/Getty Images. Page 443 (t): Taras Vyshnya/Fotolia. Page 443 (b): Postcard depicting the Boulevard des Italiens, ca.1910 (color lithograph). Archives Charmet/Bridgeman Images. Page 444: Carolyn Clarke/Alamy. Page 445 (t) Sadik Gulec/Shutterstock. Page 445 (b) Robert Preston Photography/Alamy. Page 446 (t): Ian M. Butterfield/Alamy. Page 446 (b): Terry Bruce/Alamy. Page 447: Frederic Soltan/Sygma/Corbis. Page 448 (t): Ivan Nesterov/Alamy. Page 448 (b): Travelib Pakistan/Alamy. Page 449: Caroline Penn/Alamy. Page 450 (t): Philipus/Alamy. Page 450 (b): Amos Chapple/Getty Images. Page 451 (t): Sophie James/Alamy. Page 451 (b): Nadine Rupp/Getty Images. Page 452: PeskyMonkey/E+/Getty Images. Page 453: Mohammad Asad/Pacific Press/Alamy. Page 454: The World Bank. Page 455 (t): Image/Alamy. Page 455 (b): Lou Linwei/Alamy.

Index

469

World States

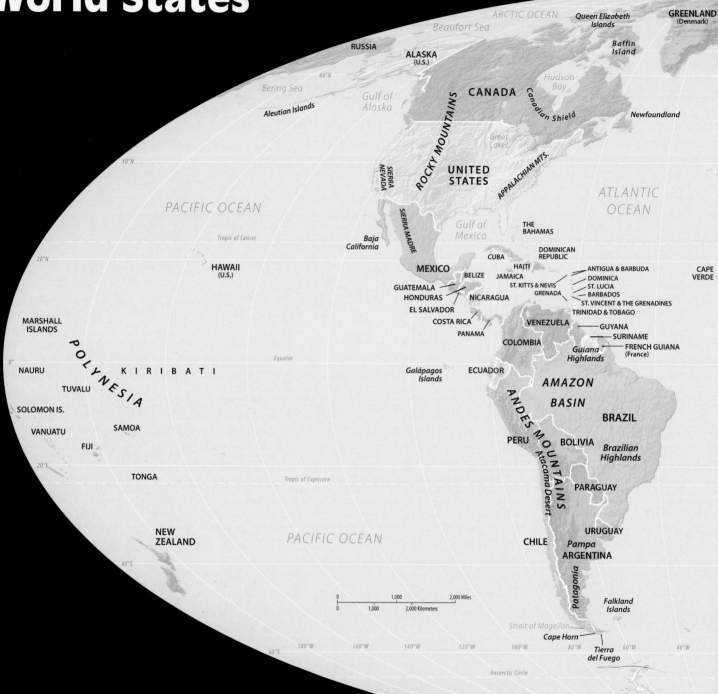

ARCTIC OCEAN
Beaufort Sea
Queen Elizabeth Islands
GREENLAND (Denmark)
RUSSIA
ALASKA (U.S.)
60°N
Baffin Island
Bering Sea
Gulf of Alaska
Aleutian Islands
CANADA
Hudson Bay
Canadian Shield
Newfoundland
ROCKY MOUNTAINS
Great Lakes
40°N
SIERRA NEVADA
UNITED STATES
APPALACHIAN MTS.
ATLANTIC OCEAN
PACIFIC OCEAN
SIERRA MADRE
Gulf of Mexico
THE BAHAMAS
Tropic of Cancer
Baja California
20°N
HAWAII (U.S.)
MEXICO
CUBA
HAITI
DOMINICAN REPUBLIC
ANTIGUA & BARBUDA
CAPE VERDE
BELIZE
JAMAICA
DOMINICA
GUATEMALA
ST. KITTS & NEVIS
ST. LUCIA
MARSHALL ISLANDS
HONDURAS
NICARAGUA
GRENADA
BARBADOS
ST. VINCENT & THE GRENADINES
EL SALVADOR
TRINIDAD & TOBAGO
COSTA RICA
VENEZUELA
GUYANA
POLYNESIA
PANAMA
COLOMBIA
SURINAME
FRENCH GUIANA (France)
0°
Guiana Highlands
Equator
NAURU
KIRIBATI
Galápagos Islands
ECUADOR
AMAZON BASIN
TUVALU
BRAZIL
SOLOMON IS.
ANDES MOUNTAINS
PERU
BOLIVIA
Brazilian Highlands
VANUATU
SAMOA
FIJI
20°S
TONGA
Tropic of Capricorn
Atacama Desert
PARAGUAY
URUGUAY
NEW ZEALAND
PACIFIC OCEAN
CHILE
Pampa
ARGENTINA
40°S
Patagonia
Falkland Islands
0   1,000   2,000 Miles
0   1,000   2,000 Kilometers
Strait of Magellan
Cape Horn
Tierra del Fuego
60°S
180°W   160°W   140°W   120°W   100°W   80°W   60°W   40°W
Antarctic Circle

ARCTIC OCEAN

Barents Sea

Arctic Circle

SIBERIA

RUSSIA

URAL MTS.

See Europe inset map below

KAZAKHSTAN

MONGOLIA

GOBI (DESERT)

GEORGIA

ARMENIA

TURKEY

AZER.

UZBEKISTAN

TIAN SHAN

KYRGYZSTAN

TURKMENISTAN

TAJIKISTAN

NORTH KOREA

SOUTH KOREA

JAPAN

40°N

PACIFIC OCEAN

LEBANON

SYRIA

ISRAEL

JORDAN

IRAQ

IRAN

AFGHANISTAN

HIMALAYAS

CHINA

BHUTAN

ATLAS MTS.

TUNISIA

MOROCCO

ALGERIA

LIBYA

SAHARA

WESTERN SAHARA (Morocco)

EGYPT

SAHEL

KUWAIT

BAHRAIN

QATAR

Arabian Peninsula

UNITED ARAB EMIRATES

PAKISTAN

NEPAL

Thar Desert

INDIA

Deccan Plateau

BANGLADESH

BURMA (MYANMAR)

LAOS

TAIWAN

Tropic of Cancer

20°N

East China Sea

South China Sea

MAURITANIA

MALI

NIGER

CHAD

SUDAN

SAUDI ARABIA

OMAN

ERITREA

YEMEN

DJIBOUTI

Arabian Sea

THAILAND

VIETNAM

CAMBODIA

PHILIPPINES

SENEGAL

THE GAMBIA

GUINEA

BURKINA FASO

TOGO

BENIN

NIGERIA

CENTRAL AFRICAN REPUBLIC

SOUTH SUDAN

Ethiopian Highlands

ETHIOPIA

SOMALIA

Bay of Bengal

SRI LANKA

MALDIVES

BRUNEI

MALAYSIA

PALAU

MICRONESIA

COTE D'IVOIRE

GHANA

EQ. GUINEA

CAMEROON

UGANDA

KENYA

SINGAPORE

INDONESIA

LIBERIA

SAO TOME & PRINCIPE

GABON

REPUBLIC OF THE CONGO

RWANDA

BURUNDI

SIERRA LEONE

GUINEA-BISSAU

DEMOCRATIC REPUBLIC OF THE CONGO

TANZANIA

SEYCHELLES

COMOROS

INDIAN OCEAN

PAPUA NEW GUINEA

SOLOMON ISLANDS

Equator

0°

ANGOLA

MALAWI

ZAMBIA

MADAGASCAR

MAURITIUS

TIMOR LESTE

20°S

NAMIBIA

ZIMBABWE

BOTSWANA

Kalahari Desert

MOZAMBIQUE

SWAZILAND

ATLANTIC OCEAN

Tropic of Capricorn

AUSTRALIA

Great Victoria Desert

SOUTH AFRICA

LESOTHO

0°

20°E

40°E

60°E

80°E

140°E

40°S

ANTARCTICA

20°W

0°

20°E

40°E

60°E

ICELAND

Arctic Circle

SWEDEN

FINLAND

RUSSIA

0   250   500 Miles

0   250   500 Kilometers

NORWAY

60°N

UNITED KINGDOM

DEN.

ESTONIA

LATVIA

RUSS.

LITH.

IRELAND

NETH.

BELG.

LUX.

GERMANY

CZECH REP.

LIECH.

SWITZ.

AUS.

SLOVE.

POLAND

BELARUS

UKRAINE

MOLDOVA

ATLANTIC OCEAN

FRANCE

ITALY

SLVK.

HUNGARY

CRO.

BOS. & HERZ.

SERB.

MONT.

KOS.

MAC.

ALB.

ROMANIA

BULGARIA

Black Sea

40°N

PORTUGAL

SPAIN

ANDORRA

GREECE

TURKEY

MOROCCO

ALGERIA

TUNISIA

MALTA

CYPRUS

EUROPE